Subcellular Biochemistry

Volume 113

The book series SUBCELLULAR BIOCHEMISTRY is a renowned and well recognized forum for disseminating advances of emerging topics in Cell Biology and related subjects. All volumes are edited by established scientists and the individual chapters are written by experts on the relevant topic. The individual chapters of each volume are fully citable and indexed in Medline/Pubmed to ensure maximum visibility of the work.

David A. D. Parry • Pradeep K. Luther
Editors

Fibrous Proteins: Structures and Mechanisms

Second Edition

Springer

Editors
David A. D. Parry
School of Natural Sciences
Massey University
Palmerston North, New Zealand

Pradeep K. Luther
National Heart and Lung Institute
Imperial College London
London, UK

ISSN 0306-0225 ISSN 2542-8810 (electronic)
Subcellular Biochemistry
ISBN 978-3-032-05272-8 ISBN 978-3-032-05273-5 (eBook)
https://doi.org/10.1007/978-3-032-05273-5

This Springer imprint is published by the registered company Springer Nature Switzerland AG
The registered company address is: Gewerbestrasse 11, 6330 Cham, Switzerland

Preface

Fibrous and globular proteins, though built from the same amino acid building blocks, differ in a number of important ways. For example, the amino acid sequences of the fibrous proteins generally display much greater regularity than do the globular proteins, and this feature frequently enables the general form of the structure to be specified with some accuracy from sequence data alone. Indeed, many computationally based techniques have been developed over recent years to ascertain structure and function, and artificial intelligence is now beginning to be employed to the same end. The fibrous proteins are also the main constituents of our bodies (skin, tendon, bone, muscle), and thereby define our structural form. They also provide us with an ability to move, defend ourselves, and forage for food. Furthermore, the fibrous proteins exist and function in both an intracellular (actin and myosin filaments and the intermediate filaments) and an extracellular environment (collagen fibrils). In recent years new methodologies, such as high-resolution cryo-electron microscopy (cryo-EM), cryo-electron tomography (cryo-ET), and associated computational analysis techniques, have been central in revealing the structure at atomic resolution of a number of the fibrous proteins presented in this book. Protein engineering has also led to the development of exciting new biomaterials based on fibrous biopolymers. It is, of course, not only the fibrous proteins that are crucial in providing living creatures with their ability to survive in a hostile climate. The globular proteins have an equally important role in vivo but this collection of review chapters will concentrate only on the fibrous ones.

This volume seeks to build on our earlier volume in this series—*Fibrous Proteins: Structures and Mechanisms*—that was published by Springer in 2017. As noted above much has happened in the past eight years, and we considered it opportune to update the exciting progress that has been made in the fibrous protein field. Considerable care has gone into selecting the leading researchers in their fields to write the 16 chapters for this second edition. The authors are world experts in their respective areas of research, and each has provided the reader with a "state-of-the-art" summary of current thinking. As such, the book is clearly written and liberally illustrated. Those fibrous proteins described in detail include the constituents of muscle, connective tissue, keratin, silk, fibrin, and biomaterials. The opening chapter by Wood raises a very important topic—that of artificial intelligence and the increasing role that it will play in future years in protein research. The next three chapters (Luther, Lehman, and Maeda) cover muscle and concentrate on thick and thin filament structure and function. These highlight the recent advances made in technology and image processing that have enabled new insights to be gained into the structure and working of the sarcomere and the effects that drugs designed to tackle disease have on this important motile element. In the following three chapters on the keratins, Parry considers the recent developments in the structures of the corneous alpha and beta proteins from the sauropsids (birds and reptiles), Medalia reveals the incredibly detailed structures of the intermediate filaments that can now be deduced from high-resolution cryo-electron microscopy, and Köster explains the structure-mechanics relationship in intermediate filaments. The next five chapters are devoted to important but diverse aspects of connective tissue. Bella reviews the marine collagens—a topic of increasing interest and relevance, Norris reveals that the glycosylation of collagen plays an important role in fibril assembly, Ruberti discusses how collagen provides a tissue with its optimum mechanical attributes, and Fertala reviews collagen diseases and fibrosis. Birch and Laird complete this section with a fascinating chapter on collagen aging. The last four chapters cover a range of diverse topics: Weisel describes the important developments that have been made on the structure and function of fibrin; Sutherland completes the exciting 60-year-old story of the coiled-coil structure of the silks of the bees, ants, and wasps and the development of new biomaterials based on this structure; and the final two chapters are devoted to the designability of α-helical protein filaments (Conticello) and functional plasticity through amyloid polymorphism (Landau).

This second edition of what proved to be a highly influential book when it was first published in 2017 will be a valuable resource for senior scientists and postgraduate students in academic institutions as well as those working in the medical and industrial fields. It will also establish the basis for future activities in this area of research.

Finally, we wish to dedicate this volume to our friend and colleague John Squire, who tragically passed away as a result of the Covid pandemic in January 2021. John was co-editor of the first edition of this volume and of several other books with one of us (DADP) as well as being the PhD supervisor of the other of us (PKL). We both owe him a great deal. His research contributions to muscle research were also of the highest possible quality. He is much missed.

Palmerston North, New Zealand David A. D. Parry
London, UK Pradeep K. Luther

Contents

Chapter 1
Sequence Prediction, Structural Modelling and the Challenges of Plasticity in Coiled Coils

Christopher W. Wood, Guto G. Rhys, and Evangelia Notari

Abstract The coiled coil is one of the most widespread and versatile protein folding motifs and is involved in a vast array of biological functions. The study and design of these assemblies have been shaped by the development of specialised computational tools. This review charts the landscape of these in silico methods, beginning with the foundational sequence-based algorithms for coiled-coil prediction and classification, before moving to the powerful parametric methods that enable the generation of idealised, atomic-resolution structural models. We argue that while indispensable, these tools can promote a view of coiled coils as static, singular structures. Challenging this, we highlight the structural plasticity of coiled coils, the remarkable and increasingly evident ability of these assemblies to adopt multiple distinct conformations and dynamically switch between states. To understand the energetic and mechanistic principles governing this complex behaviour, we explore the crucial role of molecular dynamics (MD) simulations in providing atomistic insights that are inaccessible to static modelling. Finally, we look forward, considering how the next generation of coiled-coil bioinformatics must evolve to address these challenges to take advantage of the opportunities presented by the current explosion in genomic sequence data and the proliferation of AI-predicted protein structures.

Keywords Coiled coil · Structural plasticity · Protein design · Bioinformatics · Structural modelling · Molecular dynamics · Protein switches · Deep learning

---

C. W. Wood (✉)
School of Biological Sciences, University of Edinburgh, Edinburgh, UK
e-mail: chris.wood@ed.ac.uk

G. G. Rhys
School of Chemistry, Cardiff University, Cardiff, UK

E. Notari
School of Chemistry, University of Edinburgh, Edinburgh, UK

D. A. D. Parry, P. K. Luther (eds.), *Fibrous Proteins: Structures and Mechanisms*, Subcellular Biochemistry 113, https://doi.org/10.1007/978-3-032-05273-5_1

Abbreviations

AI	Artificial Intelligence
CCTM	Coiled Coil Transmembrane
CNN	Convolutional Neural Network
CPU	Central Processing Unit
DNN	Deep Neural Network
EM	Electron Microscopy
GPU	Graphics Processing Unit
HIV	Human Immunodeficiency Virus
HMM	Hidden Markov Model
KIH	Knobs-into-Holes
LSTM	Long Short-Term Memory
MD	Molecular Dynamics
MSA	Multiple Sequence Alignment
NMR	Nuclear Magnetic Resonance
PLM	Protein Language Model
PSSM	Position-Specific Scoring Matrix
RMSD	Root Mean Square Deviation
RMSF	Root Mean Square Fluctuation
SAH	Single Alpha Helix
SASA	Surface Accessible Solvent Area

Introduction

The coiled coil is a foundational element in the history of structural biology. Pioneering X-ray diffraction studies of fibrous proteins like keratin [72] and tropomyosin [5] provided early glimpses into the fibrous structures of coiled coils, laying the groundwork for Francis Crick's seminal postulates on the mode, driving forces, and sequence biases underlying coiled-coil formation [26, 27]. Characterised by two or more α-helices wrapping around each other into a supercoiled structure, coiled coils are defined by a distinctive repeating pattern of typically seven amino acid residues, the heptad repeat (*abcdefg*). The specific arrangement of residues at the different positions dictates relative orientation of the helices, oligomerisation state, and overall stability.

Coiled coils represent one of the most widespread and versatile protein folding motifs, foundational to protein structure and critical for mediating a vast array of protein-protein interactions. Their functional diversity is intrinsically linked to their unique architecture and, as a result, predicting the assembly characteristics of coiled coils is critical. In natural systems, these features can dictate the stoichiometry of protein complexes, the relative positioning of interacting domains and thus modulate signalling and, ultimately, provide profound functional insights into the natural system. In the realm of *de novo* protein design, accurate prediction of assembly is paramount for both positive design (ensuring the formation of the desired target structure) and negative design (preventing the formation of undesired alternative assemblies or off-target interactions).

This review will navigate the landscape of computational strategies employed to understand coiled-coil assembly, from sequence analysis to dynamic structural modelling. We will first chart the history and recent progress of sequence-based algorithms, which provide the crucial first step of identifying potential coiled-coil domains from genomic data. From there, we will explore the diverse methods for structural modelling of coiled coils, from early parametric equations to sophisticated design suites, which allow for the generation of idealised atomic models. However, these powerful predictive and modelling tools often produce a single, static picture, reinforcing a historical view of coiled coils as rigid structures. We will challenge this misconception by discussing the significant and growing evidence for structural plasticity in coiled coils, detailing the spectrum of ways these assemblies can dynamically rearrange in response to their environment. To rationalise this complex behaviour, which static methods cannot capture, we will then examine the application of molecular dynamics simulations as an indispensable tool for probing the energetic landscapes and atomistic mechanisms that underpin this plasticity. Finally, we will turn our attention to the future, discussing the unmet needs and opportunities that lie ahead and arguing for the development of new computational frameworks capable of meeting the challenges posed by the unprecedented scale of both sequence and structural data in the post-AlphaFold era.

Sequence-Based Prediction of Coiled-Coil Domains

The identification of coiled-coil domains directly from amino acid sequences has been a long-standing goal in structural bioinformatics. These sequences have a characteristic heptad repeat pattern (*abcdefg*), where the *a* and *d* positions are usually hydrophobic residues that form the interface between helices, with the other positions usually being polar residues that are exposed to solvent. These amino acid preferences were identified as early as 1948 [8] and explained by Crick's initial model for coiled-coil assembly [27]. It is these sequence biases that form the basis for predictive methods for identifying coiled-coil sequences. Over the decades, a variety of computational tools have been developed, evolving from prediction of coiled coil domains in primary sequences to incorporating register (phase of the heptad repeat) information and oligomerisation state prediction (Fig. 1.1).

The development of sequence-based coiled-coil prediction algorithms has seen significant progress. Early methods focused on the statistical propensities of amino acids to occupy specific positions within the heptad repeat, while later approaches incorporated more complex models, including pairwise residue correlations, Hidden Markov Models (HMMs), and, more recently, deep learning techniques. Table 1.1 provides a chronological overview of key sequence-based algorithms, highlighting their methodologies and prediction capabilities.

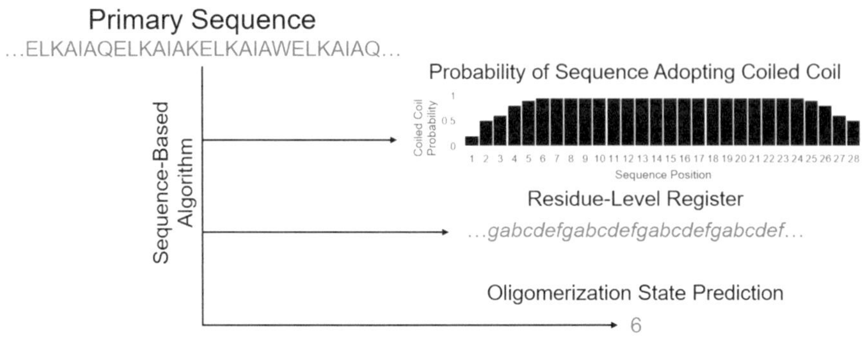

Fig. 1.1 Input and output of sequence-based prediction of coiled coils

Early Algorithms

Early coiled-coil prediction algorithms leveraged the distinct sequence periodicity inherent in these structures. A pioneering tool in this domain was COILS [70]. This algorithm operates by comparing an input amino acid sequence against a database of known coiled-coil sequences. The method analysed the frequency of each amino acid at each position within the heptad repeat of known coiled coils. By scoring a sliding window of the query sequence against the reference frequencies, COILS calculates the probability that a given segment will adopt a coiled-coil conformation. This approach provided a foundational probabilistic assessment of coiled-coil propensity and paved the way for more sophisticated methods.

Recognising that interactions between residues, not just individual residue propensities, significantly influence coiled-coil stability, Berger and colleagues developed PairCoil in 1995 [12]. This method marked an important advancement by explicitly considering pairwise residue correlations. Instead of analysing residues in isolation, PairCoil evaluated how specific pairs of amino acids at defined separations (e.g. positions i and i + 1, i and i + 3, i and i + 4, corresponding to interactions within and between helices) contribute to or detract from coiled-coil formation. This consideration of inter-residue relationships within the heptad repeat offered a more nuanced and accurate prediction. This approach was further developed in PairCoil2 [79]. This updated version incorporated further algorithmic refinements, an expanded and updated database of known coiled coils, and improved statistical models, leading to enhanced accuracy and reliability in identifying coiled-coil regions.

The application of Hidden Markov Models (HMMs) represented another significant step forward in capturing the statistical properties of coiled-coil sequences. Delorenzi and Speed introduced Marcoil in 2002 [30], which employs HMMs to identify coiled-coil domains. HMMs are well-suited for biological sequence analysis as they can represent the probability of transitions between different states

Table 1.1 Sequence-based coiled-coil prediction algorithms

Algorithm	Year	Method/principle	Key prediction capabilities
COILS [70]	1991	Compares sequence to a database of known coiled coils	Probability of coiled-coil formation
PairCoil [12]	1995	Considers pairwise residue correlations	Improved accuracy over single-residue analysis
SCORER [134]	1995	Focused on amino acid selection at *a* and *d* positions to influence oligomer-state selection	Predicting oligomerisation states (dimer, trimer, tetramer)
MultiCoil [128]	1997	Distinguishes between two- and three-stranded coiled coils	Predicts dimeric vs. trimeric states
Marcoil [30]	2002	Hidden Markov Models (HMMs)	Improved accuracy by modelling dependencies between neighbouring residues
PairCoil2 [79]	2006	Updated PairCoil with refinements	Improved accuracy and reliability for coiled-coil identification
CCHMM_PROF [9]	2009	HMM-based, uses profile (multiple sequence alignment) information	Predicts coiled-coil regions using evolutionary information
SPIRICOIL [95]	2010	HMMs on sequences represented by protein domains with validated coiled coils	Prediction based on structurally validated coiled-coil containing domains
MultiCoil2 [121]	2011	Updated MultiCoil for twilight zone sequences	Predicts oligomerisation state even with limited similarity to known coiled coils
SCORER 2.0 [4]	2011	Knowledge-based potential, improved log-odds-based scoring system, uses PSSMs and MultiCoil	Scoring the likelihood of a sequence to form a parallel two-stranded coiled coil; guiding de novo design; distinguishing parallel dimeric and trimeric coiled-coil sequences
LOGICOIL [122]	2013	Probabilistic model	Predicts oligomeric state (dimer, trimer, higher-order)
RFCoil (C. [64])	2015	Random Forest classifier	Predicts dimer vs. trimer coiled coils
Waggawagga [107]	2015	Comparative visualisation, SAH domain detection	Visualises predictions from multiple tools, identifies stable single α-helices (SAH)
DeepCoil/ DeepCoil2 [69]	2019	Deep learning (Convolutional Neural Networks (CNNs))	Prediction of coiled-coil domains and assigns register
CoCoPred [33]	2022	Deep learning (Bi-directional LSTM and CNNs)	Predicts coiled-coil regions, residue-level register, and oligomerisation state
CoCoNat [74]	2023	Protein language models and deep learning	Predicts coiled-coil regions, residue-level register, and oligomerisation state

(e.g. coiled-coil state vs. non-coiled-coil state) and the probability of observing particular amino acids in each state. By modelling the sequence as a Markov process, Marcoil effectively captures dependencies between neighbouring residues and the characteristic patterns of the heptad repeat, leading to more accurate predictions. Further leveraging the power of HMMs, CCHMM_PROF incorporated evolutionary information by using profiles derived from multiple sequence alignments [9]. This allows the model to consider conservation patterns across homologous sequences, thereby improving its ability to distinguish true coiled-coil signals from sequence noise. Another HMM-based approach, SPIRICOIL [95], specifically trains its models on sequences derived from protein domains that contain structurally validated coiled coils, aiming to capture the subtle features present in well-characterised examples.

While many early tools focused on identifying coiled-coil regions, predicting their specific oligomeric state remained a challenge. LOGICOIL [122] addressed this by using a probabilistic model specifically designed to predict whether a coiled coil is likely to form a dimer, trimer, or even a higher-order oligomer. By assigning probabilities to these different oligomeric states, LOGICOIL provided valuable insights into the quaternary structural organisation and potential protein-protein interactions mediated by these domains.

Beyond the initial wave, several other tools have focused on specific aspects of coiled-coil prediction or utility. The MultiCoil [128] was specifically designed to distinguish between two- and three-stranded coiled coils, assessing the likelihood of a sequence forming either a dimeric or a trimeric structure. Its updated version, MultiCoil2 [121] was optimised to improve oligomerisation state prediction, particularly for sequences "in the twilight zone" with limited similarity to known coiled coils. SCORER 2.0 [4] is a knowledge-based potential used for scoring the likelihood of a sequence to form a parallel two-stranded and three-stranded coiled coil. RFCoil [64] utilises a Random Forest machine learning approach to predict whether a coiled-coil sequence is likely to form a dimer or a trimer. For visualisation and comparative analysis, Simm et al. introduced Waggawagga [107], a tool that not only allows for the comparison of predictions from various algorithms but also notably detects stable single alpha-helices (SAH domains), which can sometimes be misidentified as coiled coils.

Modern Deep Learning-Based Approaches

The advent of advanced machine learning, particularly deep learning, has brought a new level of sophistication to coiled-coil prediction. DeepCoil, presented by Ludwiczak et al. [69], leverages deep learning techniques, specifically convolutional neural networks (CNNs). By training these networks on extensive datasets, DeepCoil captures complex, high-dimensional sequence patterns indicative of coiled-coil structures, leading to significantly improved predictive performance in speed and accuracy. A couple of years after the release of DeepCoil, the algorithm was significantly refined and updated, and released as DeepCoil2, which improved performance (https://github.com/labstructbioinf/DeepCoil/releases/tag/v2.0). The application of CNNs to coiled coil prediction has been further augmented with Long Short-Term Memory architectures [47] to create CoCoPred [33], which can predict coiled-coil regions, residue-level register, and oligomerisation state.

Language models are widely used in many domains such as Natural Language Processing and Generative AI, but they are also proving useful for many applications in biology, especially through the application of Protein Language Models [23]. The CoCoNat (Coiled-Coil kNowledge ATtention network) algorithm [74] employs protein language models (PLMs) and deep learning with attention mechanisms to identify coiled-coil regions but also to provide accurate predictions of residue-level register annotation and the oligomerisation state, with state-of-the-art performance, according to their benchmarks.

Strengths and Limitations of Current Prediction Methods

Current sequence-based coiled-coil prediction methods have advanced considerably, offering invaluable tools for proteome-wide analyses and hypothesis generation. Their primary strength lies in the ability to identify potential coiled-coil regions from primary sequence data alone, which is abundant and rapidly growing. Machine learning-based methods, in particular, have demonstrated enhanced accuracy by learning intricate sequence-structure relationships from large datasets.

However, limitations persist, as highlighted by a previous critical evaluation of coiled-coil prediction algorithms [108]. Many methods struggle with non-canonical coiled coils, structures that deviate from the classical heptad repeat containing insertions or deletions—known as stutters and stammers [20]—or have different repeat lengths. Another challenge is the prediction of short coiled-coil regions, which can be difficult to distinguish from other α-helical structures or random occurrences of hydrophobic/charged residue patterns, potentially leading to false positives or false negatives. False positives can also arise from certain sequence features, like polyglutamine tracts, which some algorithms may erroneously predict as coiled coils. Furthermore, while some tools attempt oligomerisation state prediction (e.g. LOGICOIL, MultiCoil2, CoCoNat), accurately determining whether a coiled coil is a dimer, trimer, tetramer, or another oligomeric form remains a significant challenge, especially for less common or mixed oligomeric states. Finally, predicting precise structural details, including the exact geometry, inter-helical packing details (e.g. parallel vs. antiparallel, specific angles), is generally beyond the capability of current sequence-based methods.

Structural Modelling of Coiled Coils

Understanding the three-dimensional architecture of coiled coils is crucial for elucidating their diverse biological functions. While experimental methods like X-ray crystallography and NMR spectroscopy provide high-resolution structural information, computational modelling has become an indispensable tool for predicting, designing, and analysing coiled-coil structures. Computational modelling has been instrumental in advancing our understanding of coiled-coil architecture, enabling both the prediction of naturally occurring structures and the de novo design of novel coiled-coil assemblies. The modelling of coiled-coil structures relies on a combination of mathematical models, computational algorithms, and, increasingly, the integration of machine learning (Table 1.2).

A foundation of structural modelling of coiled coils was established in Crick's seminal 1953 papers, providing the fundamental mathematical framework for describing the idealised geometry of coiled coils [26, 27]. Crick derived a set of parametric equations to describe the paths of the constituent α-helices, using parameters such as their radius, pitch, and the interface angle between them, enabling the precise modelling of their supercoiled arrangement (Fig. 1.2). This parameterisation ensured the formation of the characteristic "knobs-into-holes" packing of side chains at the helical interface and has served as the basis for many subsequent computational methods and remains a cornerstone for understanding coiled-coil geometry.

While Crick's original equations described ideal, regular coiled coils, many natural coiled coils exhibit deviations, such as variations in superhelical radius or pitch, or non-canonical heptad repeat patterns. Crick's equations were subsequently generalised to provide a more flexible mathematical framework capable of describing these non-canonical coiled coils [86]. By allowing parameters to vary along the coiled-coil axis, these generalised equations enable more accurate modelling of a wider range of coiled-coil architectures, including those with stutters, stammers, or other irregularities.

The accurate identification and characterisation of coiled-coil regions from atomic coordinates are also crucial. Early tools like SOCKET [123] were developed to identify coiled-coil structural motifs in PDB files by recognising the characteristic knobs-into-holes packing. For more detailed geometric analysis, TWISTER [110] quantifies parameters such as superhelical radius and pitch. More recently the SOCKET software has been updated to produce both iSOCKET and SOCKET2 [46, 56], with updated and improved functionality, allowing graph-based analysis and analysis of higher-order coiled-coils, respectively.

Platforms for De Novo Design and Model Building

Building on these parametric foundations, several computational tools were developed for prediction and design. The CCCP method, from DeGrado and colleagues, is

Table 1.2 Selected structural modelling and analysis techniques/tools for coiled coils

Technique/tool	Primary focus	Key features/ capabilities	Typical application(s)
Crick Parameterisation [26, 27]	Idealised geometric description	Mathematical equations for helix radius, pitch, interface angle	Foundational understanding, basis for modelling tools
Generalised Crick Equations [86]	Modelling non-canonical coiled coils	Allows parameters to vary along the axis	Modelling irregular coiled coils (stutters, stammers)
CCCP [38]	Design and prediction	Energetic evaluation of packing arrangements	De novo design, prediction of natural structures
SOCKET/ iSOCKET/ Socket2 [46, 56, 123]	Identification from atomic structures of proteins	Recognises knobs-into-holes packing, quantifies geometry	Structural bioinformatics, database annotation
TWISTER [110]	Detailed geometric analysis	Quantifies superhelical radius, pitch, residue parameters	In-depth structural characterisation
CLASSY [39]	Design of protein-interaction specificity	Computational design of peptides with selective binding	Designing selective bZIP-binding peptides
CCFold [42]	Structure prediction from sequence	Combines sequence prediction with fragment assembly	Predicting 3D structure when only sequence is known
CCBuilder/ CCBuilder 2 [131, 132]	Parametric modelling and design	User-friendly interface for α-helical coiled coils and collagen	De novo design, hypothesis testing
ISAMBARD [133]	Biomolecular structure analysis, design, model building, evaluation	Open-source Python package, modular and extensible	Custom modelling workflows, design, structural analysis
Rosetta [61]	Macromolecular modelling, structure prediction, design	Diverse protocols for de novo design, symmetric docking, refinement	De novo coiled-coil design, oligomer assembly, model refinement
SamCC-Turbo [113]	Assessment and redesign of coiled-coil interactions	Structure-based assessment, focus on interaction specificity	Redesigning coiled coils for specific interactions, specificity engineering
AlphaFold2 [50]	General protein structure prediction (with coiled-coil applicability)	Deep learning-based, high accuracy for many proteins	Predicting 3D structures, oligomeric state (with caveats for coiled coils)

a notable computational approach used for both designing and predicting coiled-coil structures [38]. It facilitates the creation of novel coiled-coil proteins with desired properties and aids in predicting naturally occurring coiled-coil structures by evaluating the energetic favourability of different packing arrangements.

This approach was further developed to provide a user-friendly interface for model building and design. CCBuilder [132] and its successor CCBuilder 2.0 [131] provide a suite of tools for the parametric modelling and design of both

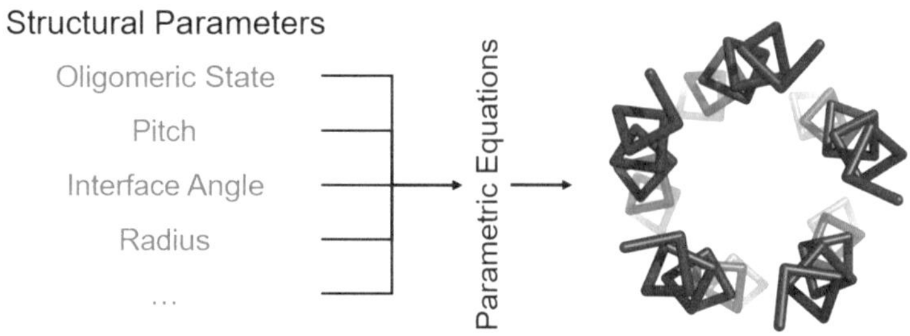

Fig. 1.2 Schematic of coiled coil model building. A mathematical description of the coiled coil is defined using a range of structural parameters that define the geometry of the backbone of the protein

α-helical coiled coils and collagen triple helices. These tools offer an intuitive interface for generating and manipulating models, making them accessible to a broad range of researchers. The modelling engine used in CCBuilder 2.0, ISAMBARD [133], is available as an open-source Python package and allows programmatic generation of models of coiled coils, exposing all parameters enabling the construction of more complex models at scale and allowing sophisticated custom workflows.

The Rosetta macromolecular modelling suite [61] also has significant functionality related to coiled-coil model building and design. Rosetta encompasses various protocols well-suited for coiled coils, including tools for the de novo design of helical bundles with specific geometries and stabilities. Its symmetric docking and assembly capabilities are particularly useful for building and refining oligomeric coiled-coil structures, and its refinement protocols can improve initial models generated by other methods.

Deep Learning Approaches and Modern Challenges

The advent of deep learning methods, particularly AlphaFold2 [50], has revolutionised protein structure prediction, and its applicability to coiled coils is an active area of research. While AlphaFold2 can predict the overall fold of many proteins with remarkable accuracy, coiled coils, with their repetitive nature and potential for multiple oligomeric states, present unique challenges. Research by Madaj et al. evaluated the performance of AlphaFold2 in modelling dimeric, trimeric, and tetrameric coiled-coil domains [73]. Their findings indicate that while AlphaFold2 can often accurately predict oligomeric states and model fine structural details, it exhibits a bias towards predicting parallel coiled-coil topologies, sometimes incorrectly modelling antiparallel structures. The study also explored the impact of multiple sequence alignments (MSAs), noting that while MSAs generally improve modelling, accurate models can sometimes be generated without them, particularly for simpler topologies.

To address some of AlphaFold's limitations with long or complex coiled coils, CCfrag was developed [78]. This computational tool improves coiled-coil modelling using AlphaFold through a fragment-based approach. Instead of processing the entire sequence at once, CCfrag breaks it into overlapping fragments, runs AlphaFold on each, and then integrates the results. This piecewise strategy aims to overcome issues like the generation of spuriously bent models for long fibrous coiled coils and difficulties in modelling non-canonical structures, providing a more detailed understanding of local features.

The prediction of specific interaction interfaces, oligomerisation states, and overall stability remains a key focus at the forefront of coiled-coil modelling. A recent comprehensive study, "An Atlas of Protein Homo-oligomerization Across Domains of Life," highlighted the prevalence of coiled coils in forming dimeric interfaces and the large cluster of natural sequences that adopt such structures [101].

Tools and Databases for Coiled-Coil Structure Analysis

A variety of specialised tools and databases aid researchers in the analysis of coiled-coil structures (Table 1.2). Beyond the prediction and modelling tools mentioned above (e.g. SOCKET, TWISTER, CCBuilder, ISAMBARD, Rosetta), databases like the CC+ Database [116] have provided curated sets of experimentally determined coiled-coil structures alongside assembly parameters, annotations for the heptad repeat and knobs-into-holes analysis. Szczepaniak et al. have taken a similar approach to create a database of coiled coils [113], utilising SamCC_Turbo to identify coiled-coil domains at scale.

Unmet Needs of Current Structural Modelling Methods

The parametric tools and computational workflows described above provide a computationally efficient means to generate high-fidelity, static models of canonical coiled-coil structures. They have proven invaluable for interpreting experimental data and for guiding protein design. However, a significant limitation of these methods is that they generally produce a single, idealised structural state. This approach inherently assumes that a given sequence will fold into one predetermined and rigid conformation. While true for many systems, this static view is an oversimplification that neglects the dynamic nature inherent to a large number of coiled coils. Consequently, a critical unmet need is the ability to account for and predict the structural heterogeneity that is increasingly observed in these systems. The following section will explore this issue, detailing the diverse manifestations of structural plasticity within the coiled-coil fold.

Structural Plasticity in Coiled Coils

Coiled-coil domains are often found in highly structured proteins, including fibrous proteins such as α-keratin [72], contributing to the rigidity of biological materials such as human hair and nails. However, much like other types of protein folds, coiled coils can be dynamic, with numerous possible arrangements of the component helices, which leads to structural plasticity. The interaction could be an exchange between a coiled-coil structure and one that is not, or between two different coiled-coil states. For the latter, different states can result from a change in the Crick parameters (i.e. a difference in pitch angle, superhelical radius, and interface angle), in addition to the relative orientation of helices (i.e. helical axial shift and parallel vs. antiparallel) and oligomeric state. Furthermore, changes in state can be a result of forming different heteromeric interactions. Therefore, while coiled coils have well-defined (knobs-into-holes) interactions, there are a myriad of ways that this can be achieved, which often leads to structural plasticity.

Early evidence for structural plasticity of coiled coils emerged from an engineered leucine zipper fragment of the GCN4 yeast transcription activator. In this work, Harbury et al. identified that specific variants of the leucine zipper peptide altered the oligomeric state of the homomers from the parallel dimers of the full construct, into parallel trimeric and tetrameric coiled coils [44]. However, one of the peptide variants containing valine and leucine at all *a* and *d* positions, respectively, showed heterogeneity, existing in two different states as a dimer and a trimer. This heterogeneity between dimers and trimers has been confirmed in many subsequent GCN4 variants and designed peptides [34, 87]. In our recent work, we obtained direct evidence that some designed coiled-coil peptides can coexist in two states [29]. Generally, peptide coiled coils can show structural plasticity because peptide-peptide interfaces are smaller, and they are sterically unencumbered making alternative states of interaction more accessible. In fact, their dynamic nature has even been demonstrated in some thermostable coiled-coil peptides, which undergo complete strand exchange within minutes [40, 41, 57].

The environment in which coiled coils form can have a profound impact on their protein-folding energy landscape, meaning that changing conditions can perturb the balance between coiled-coil state. This was demonstrated for the native GCN4 leucine zipper peptide that crystallises as either a parallel homodimer or a parallel homotrimer depending on crystallisation conditions [87]. Evidence for a shift in states has also been observed many times in higher oligomers designed by the Woolfson group where water-soluble and trans-membrane peptides show discrepancy between their solution-phase oligomeric state and their crystal structures [98, 102].

Dynamic coiled coils are not limited to peptide sequences (Fig. 1.3). In fact, there are many examples where nature has adopted conformational variability for function. Class-I viral fusion proteins are found in many viruses including hemagglutinin from influenza, spike proteins from coronaviruses, and glycoprotein 41 from HIV [125]. Upon binding to a host cell, these proteins irreversibly transition from

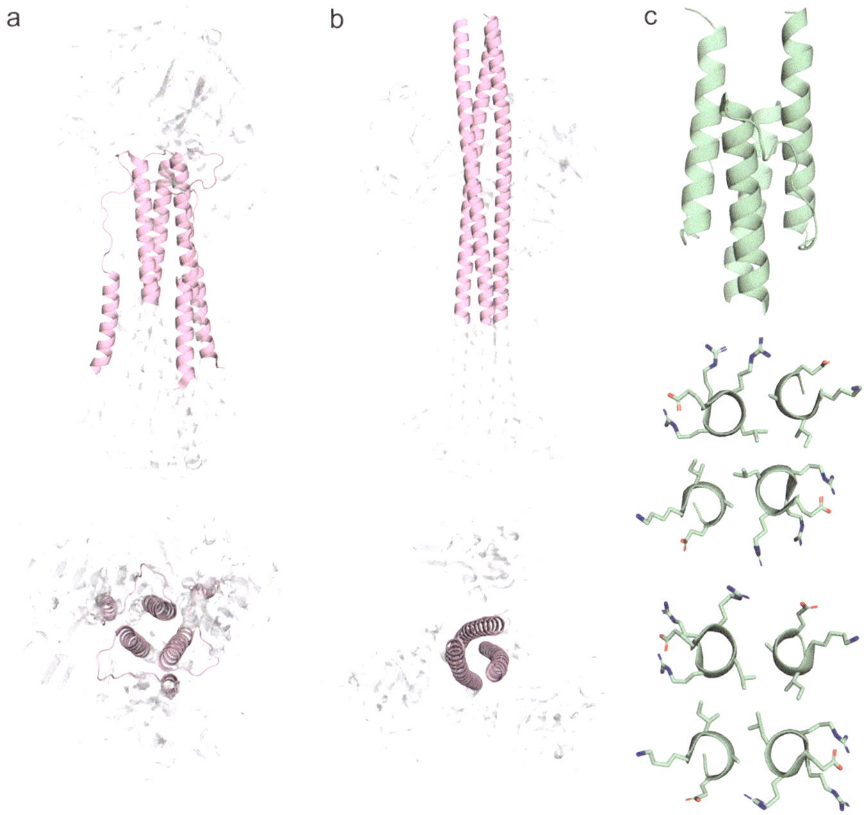

Fig. 1.3 Examples of dynamic natural coiled coils. The (**a**) pre- and (**b**) post-fusion states of influenza hemagglutinin (pdb: 6y5h & 6y5k). The region that forms the extended coiled coil in the post-fusion state is coloured pink. (**c**) A HAMP domain found in the chemotaxis Af1503 protein of Archaeoglobus fulgidus. (**c** top and middle) The wild-type sequence NMR structure shows complementary x-da packing of side chains (**c** bottom) whereas a crystal structure of the A291V mutant shows subtle rotation of helices to give canonical coiled-coil packing (pdb: 2t7h & 2y21)

compact prefusion helical-bundle states into elongated contiguous trimeric coiled coils, which drive fusion of viruses to the host cell (Fig. 1.3a). Another example is the HAMP domains found in bacterial chemoreceptor and sensor proteins (e.g. CheA, CheW) [90]. Signals such as chemicals or light are transduced from sensor domains to the peripheral intracellular HAMP domains. The HAMP domains propagate the signal by interconverting between two different parallel tetrameric helical states by simultaneous rotation of the helices (interface angle) [126]. The two states differ in their packing between forming complementary x-da and da-x layers and a classical coiled coil.

Therefore, both sequence and environment have a role to play in determining the dynamics of coiled coils. Natural proteins are often an inspiration for protein engineering and design, and there have been many attempts to embed dynamic elements into coiled coils to create functional proteins that respond to stimuli [2].

pH-Dependent Plasticity in Coiled Coils

Biological microenvironments can differ in their pH, and therefore, nature has evolved methods to sense and regulate pH. In some cases, coiled coils play a role in responding to pH changes (Fig. 1.4). For example, the influenza virus enters host cells via endosomes but must escape into the cytoplasm in order to replicate. This is achieved by the aforementioned class-1 viral protein hemagglutinin, which undergoes a structural rearrangement at the lower pH of endosomes [11]. Sensing occurs at a distal site to the coiled-coil-forming region, where a histidine residue is protonated, at low pH, that results in a hydrogen bond with a neighbouring asparagine [76]. In doing so, the histidine residue disrupts the hydrogen-bond network between monomeric units, weakening the interaction and initiating the fusion process. In the weakened dilated state, the constrained N-terminal helices are released, which rapidly fold into a contiguous trimeric coiled coil.

Another example of coiled coils that change state when exposed to endosomal conditions are class-A scavenger receptors, which are involved in binding various biomolecules and internalising them into endosomes [21]. These membrane-bound proteins contain coiled-coil domains that have been shown to dissociate and unfold

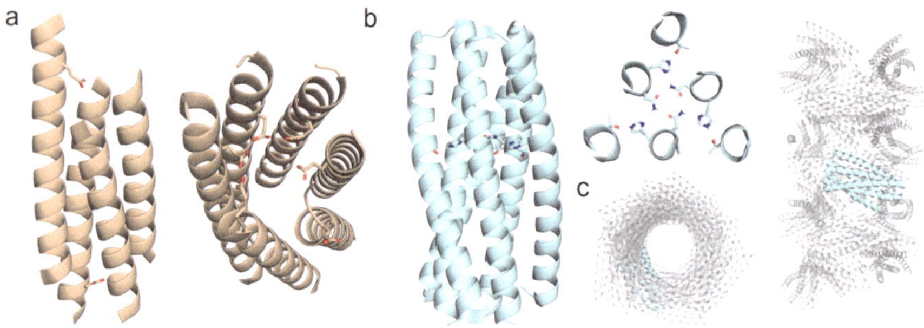

Fig. 1.4 Designed peptide and protein sequences that are responsive to pH. (**a**) Crystal structures of CCHex-L24E forming (left) an antiparallel tetramer under basic conditions and (**b**) forms a hexameric barrel under acidic conditions (pdb: 6q5i & 6q5j). In both structures there is some disorder of glutamate residues and subsequent C-termini due to charge repulsion of packing all glutamate residues. (**c**) A crystal structure of a designed trimeric bundle, pRO-2.3, containing an extensive hydrogen-bond network. The pH-responsive designs include buried histidine residues that can be reversibly protonated to disrupt the hydrogen-bond network, leading to random coil. (**c**) The structure of pRO-2.3 trimer was fused into a single chain, and the surface residues were redesigned to produce a range of designed proteins that self-associate into filaments. Here we see the cryo-EM structure of one of the designs, DpHF19, that undergoes reversible formation of fibres at pH 4.1 or lower (pdb: 8ubg)

at lower pH values [112]. In one example, the change from unfolded to a folded dimeric coiled coil could be attributed to buried glutamate residues that must be protonated in order to stabilise the homodimeric interface.

Taking inspiration from nature, both glutamate and histidine-based pH-responsive peptides and proteins have been developed. For glutamate-based switches, both unfolded to folded coil coils and switching between coiled-coil states have been demonstrated. Researchers in the Jerala and Koksch groups have developed two-helix coiled-coil peptides that contain peripheral or buried glutamate and histidine residues at the coiled-coil interface [6, 52]. The Jerala group's sequence (SwitCCh) was designed to bind Zn(II) ions via histidine residues at peripheral *e* positions to promote folding. In addition, the incorporation of several glutamate residues at *e* and *g* positions was a negative design feature, to destabilise coiled-coil formation in the absence of Zn(II) ions. At micromolar concentrations and pH 7.4 the peptide was a random coil, whereas at pH 4 the coulombic repulsion is diminished by the protonation of the glutamate residues such that both the SwitCCh and its designed precursor become highly alpha-helical dimers. Folding could also be partly induced at pH 7.4 by the addition of Zn(II) ions. The Koksch group designed sequences that optimally fold into an antiparallel monomeric helix-loop-helix at pH 5. In their most successful sequence (AP-LZ (EH8)), a buried glutamate at an *a* position forms a salt bridge with the opposing helix via a histidine at a *d* position. There are two unique features to this design. Firstly, the charged states of the residues induce folding rather than unfolding. Secondly, the synergistic use of both residues means that folding is induced over an optimal pH window rather than a binary transition between alkaline and acidic pH values.

In other examples, designed peptides can undergo transitions between two different coiled-coil states. In our work, we introduced a glutamate mutation into a designed hexameric alpha-helical barrel (CC-Hex) [99]. The glutamate mutation (CC-Hex-L24E) was introduced at an *a* position, which for alpha-helical barrels would lead to a ring of glutamates inside the channel. As expected, at low pH (4) the six glutamate residues are likely to be protonated forming a stable hydrogen-bond network, which resulted in the formation of a stable hexameric barrel. What was unexpected is that at higher pH (7.4) the peptide does not unfold but rather associates with weaker affinity into a different state, namely a tetramer. To confirm this, we were able to obtain crystal structures of CC-Hex-L24E under both alkaline and acidic conditions (Fig. 1.4a). These structures matched the solution-phase data and revealed that under acidic conditions the glutamates are tolerated, although two chains displayed fraying, presumably because not all six glutamate residues would fit in the cavity. The crystal structure under basic conditions forms a staggered antiparallel peptide. By staggering the interface between peptides, the anionic glutamate residues can avoid interactions with the hydrophobic interface, and being antiparallel instead of parallel reduces the number of glutamate residues from four to two at each end of the bundle. Analogous behaviour was observed by researchers in the André group who observed that a glutamate mutant of a sequence expected to form a pentameric barrel undergoes a transition to an alternate state at low pH, albeit a staggered trimer of dimers, rather than the antiparallel tetramer observed in our study [68].

Others have pursued designing buried histidine residues to achieve pH-responsive coiled coils. Researchers in the Conticello group adapted an engineered GCN4 zipper mutant, designed in the aforementioned work by Harbury et al., into a reversible fibril-forming peptide [137]. The discrete GCN4 zipper sequence was altered at the *e* and *g* positions to produce a new sequence that satisfies coulombic interaction when in a staggered configuration, which leaves exposed residues that can only be satisfied by the addition of multiple units. To make fibril assembly pH-responsive they introduced histidine residues at three *d* positions. Analysis of the peptide's α-helicity by circular dichroism revealed that the sequence was mostly a random coil at pH 5.6, showed an inflection point at pH 5.8, and became fully helical by pH 6.8. They reasoned that the internal structure of the fibril contained rings of three histidine residues that form stable hydrogen-bond networks at alkaline pH, whereas acidification causes protonation of the histidine residues ultimately leading to dissociation and unfolding of the fibril peptides.

Histidine residues have also been embedded into more complex hydrogen-bond networks of helical bundles with coiled-coil character. In research by Boyken et al., they applied the Crick parameters to generate ensembles of idealised four- or six-helix antiparallel bundles with consolidated cores [18]. Individual helices were then connected to form helix-loop-helix monomer units. Applying a search algorithm (HBNet) developed in previous research, Boyken et al. identified the small number of structures from the ensemble that could accommodate a hydrogen-bond network [17]. The sequences were optimised around the hydrogen-bond networks, and five designs were recombinantly expressed. Using size-exclusion chromatography and native mass spectrum, four of the designs showed a switch between monomer at low pH and the expected helix-loop-helix dimer or trimers at neutral pH. Impressively, taking the successful designs a step further, the repeating structure of the coiled coils allowed Boyken et al. to tune the pH set point of dissociation by increasing or decreasing the number of lateral layers that contain duplicated hydrogen-bond networks. By increasing the number of hydrogen-bond networks, the pH value of the set-point could be increased, leading to a range of designs with pH set points from 5 to 3.5. In more recent follow-up work, Shen et al. adapted one of the trimeric designs into a single-chain construct that could self-assemble into protein filaments [103]. Applying a docking algorithm that they had developed previously, they identified arrangements, and subsequently surface residue mutations, that should be conducive to fibril formation. Two of the 18 designs successfully formed filaments, and much like the earlier work the number and positioning of the buried hydrogen-bond network could alter the pH set point for filament formation [104].

Metal Binding

In natural proteins, coiled-coil domains are often found adjacent to metal-sensing domains, rather than being directly involved in metal binding. This is exemplified in the prokaryotic MerR transcriptional factors, which are homodimeric proteins that dimerise along an antiparallel dimer coiled coil [91]. At either side of the coiled-coil helix are the N-terminal DNA-binding domain and the C-terminal metal-binding domain. Upon metal ion or metal oxide ion binding the DNA-binding domains on either side of the coiled coils and the coiled coil itself undergo conformational changes that pucker the DNA leading to gene activation. The majority of the switch in conformation is observed in the DNA binding domain, with the coiled-coil domain showing a modest 3° shift in pitch angle. Another example found in natural proteins is troponin, which is a complex of three chains that control muscle contractions [77]. Two of the chains, troponin I and troponin T, form an "IT arm" that contains a parallel dimeric coiled coil. The IT arm is adjacent to four calcium-binding domains found in the third chain, troponin C, and is assumed to provide structural rigidity, rather than dynamics to the complex.

Unlike natural proteins, there are many examples of designed helical bundles, of which many are coiled coils, that bind to metal ions or clusters (Fig. 1.5). Here, we will provide some key examples that show switching behaviour and refer the reader to more comprehensive discussions elsewhere [28, 58, 63, 129]. For peptides, several research groups have developed sequences that switch between a random coil and dimeric, trimeric or tetrameric coiled coils upon the addition of various metals. The Pecararo group have extensively explored metal binding in their designed peptides named TRI and GRAND [54]. In particular, they have introduced single cysteine mutations in designs, resulting in a switch from random coil to trimeric coiled coils upon binding, in a trigonal-planar geometry, various soft metal ions such as Hg(II), Cd(II), As(III), Bi(III), and Pb(II) [100]. Similar results have also been achieved by researchers in the Tanaka group [65]. Going beyond trigonal-planar cysteine geometry, many have designed rubredoxin-like CXXC into peptides that switch from random coil to dimeric or trimeric helices that provide tetrahedral or higher metal coordination geometries [53, 71, 115].

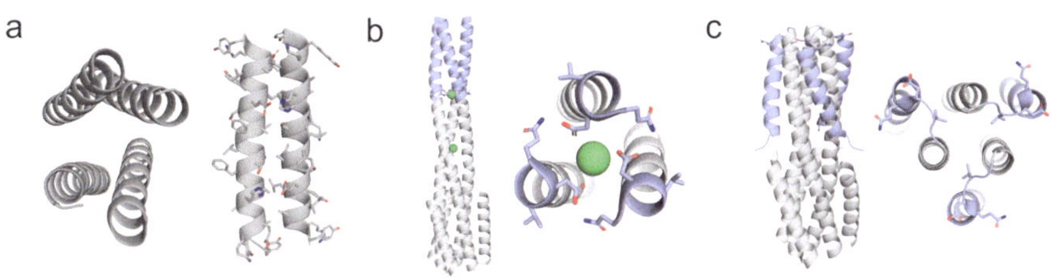

Fig. 1.5 Designed protein sequences that are responsive to metal ions. (**a** left) Tetrameric solid-state NMR structure (pdb: 2muz) and (**a** right) dimeric X-ray structure (pdb: 4p6j) of the designed Zn(II) ions and proton antiporter, ROCKER. The antiparallel arrangement brings the histidine and glutamate residues into close contact in order to bind Zn(II) ions, with central serine residues enabling transport through the core. (**b**) A crystal structure evidencing the extended state of the designed Ca(II)-responsive bundle XAA_GVDQ, with Ca(II) bound by aspartic acid from the hinge region GVDQ (pdb: 6o0c). (**c**) An NMR structure evidencing the contracted state of the designed Ca(II)-responsive bundle XAA_GVDQ, with the hinge region GVDQ motif highlighted (pdb: 6o0c)

Beyond soft metal ions, many have designed peptides that fold upon binding to intermediate softness metal ions. Researchers in the Tanaka group introduced two histidine residues in neighbouring core *a* and *d* positions in their designed trimeric coiled coil named IZ-3adH [111]. The addition of Co(II), Ni(II) and Zn(II) induced folding into well-defined trimeric coiled coils, whereas Cu(II) binding led to poorly defined species. They posited that the difference was due to Cu(II) preferentially binding four histidine residues, rather than the full six-coordinate octahedral geometry formed by the other metals. We have also designed trimeric coiled coils that fold upon addition of Ni(II) and Cu(II) by introducing two histidine mutations into a designed parallel trimer (HisAD) [19]. A crystal structure of the Cu(II)-bound peptide indeed revealed the binding of two histidine residues to each Cu(II) and also the preference for Cu(II) to form tetrahedral geometry. Curiously the tetrahedral geometry was achieved by splaying of the helices at the N-termini to accommodate a peripheral glutamate, in addition to a second glutamate from interdigitation with a second biological unit. This clearly demonstrates that metal-binding geometries can be in tension with side-chain conformations accessible to idealised coiled coils.

Finally, for peptides that switch from an unfolded to folded state, sequences have been designed to bind to hard metals. Researchers in the Peacock group have designed sequences that bind to a range of lanthanide ions (La(III), Ce(III), Pr(III), Nd(III), Sm(III), Eu(III), Gd(III), Tb(III), Dy(III), Er(III), Yb(III), Lu(III)) by introducing asparagine and aspartic acid residues at neighbouring core positions of trimeric coiled coils [15, 109]. The position of the pair of mutations has a significant impact on both the disorder of the peptide in the absence of metal and the alpha-helicity in their presence. Incorporating the polar mutations in the middle of the sequence is most disruptive to folding, presumably because of significant disruption to the, otherwise, contiguous hydrophobic core.

It is also worth briefly mentioning that sequences have been designed to bind metals via non-canonical amino acids. For lanthanide binding, Kohn et al. introduced two γ-carboxyglutamate (Gla) residues at peripheral *e* and *g* positions of their designed peptide Gla_2Nx [55]. Upon addition of La(III) or Yb(III), the sequence folds into a dimeric coiled coil by binding to the Gla residues.

Looking beyond sequences that unfold, peptides and proteins have been designed that switch between two folded states. Researchers in the Woolfson and Kuhlman groups published examples of designed peptides that switch upon binding Zn (II) from trimeric coiled coils into zinc-finger domains [3, 22]. The Kuhlman group approached the challenge by adapting the physics-based RosettaDesign algorithm. Here sequences were iteratively optimised by threading them onto fixed backbones of both a natural coiled coil and a zinc finger domain and then using Monte Carlo optimisation to identify sequences that gave the most favourable score for both structures. The Woolfson group took a different approach by using rational design. To achieve this, they analysed the sequences of a natural zinc-finger domain and identified an underlying pattern that could be overlaid to achieve a coiled-coil repeating pattern. Coincidentally both groups took inspiration from the structure of hemagglutinin for their designs. Researchers in the Kuhlman group used a fragment of hemagglutinin for their backbone design, whereas researchers from the Woolfson group realised that mapping a coiled-coil sequence onto the zinc-finger domain sequence would lead to non-canonical repeats, which they had observed in hemagglutinin. For the binding residues, the Kuhlman group used a canonical combination of two cysteine and two histidine residues, whereas the Woolfson group used a tetrahistidine sequence to avoid issues with cysteine oxidation. Using circular dichroism, both sequences demonstrated reversible formation of coiled coils in the absence of free zinc ions.

There are also two examples of designed sequences that switch between two coiled-coil states upon binding metal ions. The first was the design of a Zn(II) and proton transmembrane antiporter by researchers in the DeGrado group [49]. Here Joh et al. designed a peptide sequence that self-assembles into an antiparallel homotetramer in lipid membranes. Near the N-terminus they introduced one histidine and one glutamate residue in an EXXH motif in addition to a glutamate residue near the C-terminus. In an antiparallel tetramer arrangement, the polar mutations at each end of the tetramer would present binding sites for two Zn(II) ions comprising a histidine and three glutamates, provided that the glutamates can bridge between both metal ions. The sequence was designed to preferentially give an asymmetric structure that would only bind Zn(II) ions at one end of the tetramer at a time.

While the antiparallel and asymmetric nature of the design lends itself to two isoenergetic states, dynamic switching between the two is not an inevitable outcome, as the sequences could be thermodynamically trapped in a symmetric intermediate state. To overcome this potential issue, Joh et al. used negative design. Initially this included identifying sequences that gave the biggest difference in energy between sidechain relaxed molecular dynamics (MD) models of symmetrical and asymmetrical states. The top 1008 sequences were further refined by running an all atom-MD simulation of each sequence in both states, and then using their previously developed algorithm, VALOCIDY, to compare the absolute free energy of each state [37]. Finally manual inspection of the top-scoring sequences led to the selection of the ROCKER sequence. Using a combination of X-ray crystallography, and solution and solid-state NMR, Joh et al. confirmed the structure of ROCKER and ultimately could confirm its Zn(II) and proton antiporter properties using liposome-based flux assays.

The second more recent example was designed by researchers in the Baker group [124]. Wei et al. designed new sequences based on their previously designed six-helix trimer, 2L6HC3_13. To design two states, they cut the outer helix in half and reconnected the isolated helix via a loop, i.e. effectively a circular permutation. The hypothesis was that a break in the helix would enable the outer helices to dissociate and reorient as a continuous part of the inner helices. Analogous to hemagglutinin this would enable the sequence to flip between a compact shorter helical bundle and an elongated coiled-coil structure. Interestingly, in contrast to the Joh et al. approach to design a single multistate ROCKER sequence, Wei et al. attempted to design and confirm sequences that were selective for one state over the other. By doing so, they were able to iteratively confirm designs and gather key information of sequences that were close to having isoenergetic states. The sequences for each state were optimised using Rosetta fixed-backbone design and subsequently screened for optimal sequences using the "fold-and-dock" physical force-field-based approach. After three iterative rounds of testing sequences in the lab and further sequence designs, they arrived at a successful sequence that was confirmed in the long state by X-ray crystallography.

To tune the relative stability of the two states, Wei et al. used a similar design principle that was used to promote the Baker group's pH switching design. More specifically, by replacing hydrophobic layers of the coiled coil with buried hydrogen-bond networks, they could weaken the affinity of the outer helices to the inner helices, thereby decreasing the stability of the compact small state. By varying the position of layers of hydrogen-bond networks, Wei et al. observed by X-ray crystallography and NMR that each sequence would preferentially form one of two states, indicating that the design having as few as three amino acid differences would exist in alternate states. Finally, to develop an inducible switch with metal ions, Wei et al. introduced a GVDQ sequence into the hinge region. In the absence of calcium the sequence forms a loop, but in its presence the loop refolds into the contiguous helix forming a Ca(II) binding site. The apo state was confirmed by NMR; however, the holo-state had to be confirmed by X-ray crystallography as the sequence showed a propensity to aggregate under calcium conditions. This underscores the formidable challenges when trying to design sequences that are stable in multiple states.

Mechanoresponsive

In response to mechanical forces, coiled-coil proteins undergo significant structural changes, a phenomenon often referred to as mechanosensitive transformation. This is particularly evident in alpha-helical coiled-coil rich materials such as wool, hair, and hagfish slime threads, where stretching can induce a transition from an alpha-helix to a beta-sheet conformation [5, 10]. This "alpha-beta transition" (α-β transition) is a key deformation mechanism in these protein materials, leading to a notable increase in stiffness, strength, and energy dissipation capacity at large deformations. The transition has been observed across various conditions, including changes in pH, temperature, solvent composition, and mechanical deformation [80].

Research indicates that the occurrence of the α-β transition is controlled by the length of the constituent alpha-helical proteins [94]. Short proteins, with fewer than 26 amino acids (or 3.8 nm in length), primarily exhibit interprotein sliding when stretched. Conversely, longer proteins undergo the α-β transition. Molecular dynamics simulations support this, predicting that a coiled-coil to beta-strand transition occurs for coiled-coil proteins larger than a critical length of approximately 40 amino acids. This transition in longer coiled-coils involves the unravelling of turns and the formation of beta-sheet clusters, which act as "clamps" preventing further sliding and enforcing the unfolding of other alpha-helical turns along the chain. This is also known as a solid-solid phase transition and is driven by internal energy changes rather than entropic forces [80].

Transition Under Electric Field

Membrane-spanning coiled coils can be designed with defined assemblies through a combination of rational and computational de novo design, optimising the steric packing of coiled-coil helices. These designed peptides can spontaneously insert into membranes and form ion-conductive channels [82]. However, these coiled coils can exhibit significant structural rearrangement in response to the application of an electric field. Mahendran et al. designed a consensus peptide (cWza) that formed a monodisperse transmembrane α-helical barrel, exhibiting a low-conductance (L) state that transformed into a high-conductance (H) state at higher applied potentials, suggesting a switch in pitch and pore diameter, though not necessarily in oligomeric state [75]. Further research on de novo-designed coiled-coil transmembrane (CCTM) peptides revealed that at higher voltage potentials, there is a possible switch from a well-defined low-conductance, coiled-coil type-2 barrel to larger barrels with bigger pores, resulting in higher conductance [82].

Transition Due to PTMs

In the early exploration of coiled-coil (CC) dynamics, a foundational study by Szilák et al. notably demonstrated that phosphorylation could destabilise α-helices. Focusing on the leucine zipper dimerisation domain of Vitellogenin Binding Protein (VBP), a bZIP DNA binding protein, they introduced a PKA phosphorylation site within the leucine zipper. Their findings revealed that phosphorylation of a threonine residue in the *b* position, located on the outside of the coiled coil, away from the hydrophobic interface, destabilised the leucine zipper dimer, reducing DNA binding 100-fold. This was primarily attributed to the inherent change in the α-helix propensity of the phosphorylated threonine rather than a direct disruption of the protein-protein interface [114].

Building on these early insights, more recent work has further elucidated the mechanisms and expanded the scope of PTM-driven coiled-coil dynamics, which offers precise, switchable control over protein-protein interactions. De novo designed peptides that form well-characterised heterodimeric and heterotetrameric coiled coils, often starting from hyperthermostable scaffolds and then introducing specific modifications [45, 119, 127]. A key design strategy involves incorporating post-translational modification (PTM) sites, particularly phosphorylation sites (e.g. the PKA recognition motif R-R-X-S), to enable dynamic, reversible switching. These phosphorylation sites are strategically placed within the sequence to maximise the impact of the added negative charge and bulk of the phosphoryl group, thereby dictating assembly or disassembly. These engineered systems can be tuned to exhibit specific binding affinities (from nanomolar to micromolar ranges), and their switching kinetics can be modulated by varying the stability of the coiled-coil complex at the amino-acid level, or by adjusting temperature and concentration.

Beyond basic interaction control, these designed coiled coils have been fused with other modules, such as fluorescent proteins for FRET-based biosensors or membrane-targeting sequences, enabling applications like reversible membrane tethering or real-time kinase activity detection [45].

Challenges Posed by Coiled-Coil Dynamics and Structural Plasticity

The extensive evidence for coiled-coil plasticity, as detailed in the preceding section, presents a considerable challenge for both predictive bioinformatics and structural modelling. The sequence-based and parametric methods discussed earlier are largely predicated on identifying or building single, thermodynamically stable states. They often fall short when a single sequence can adopt multiple, isoenergetic conformations or switch between states in response to subtle environmental triggers. This inherent dynamism means that predicting the precise oligomeric state, orientation, or even the overall fold from sequence alone becomes exceptionally difficult, as the "correct" structure is context dependent. To bridge this gap between static models and dynamic reality, and to truly understand the principles governing conformational switching, computational tools that can probe the energetic landscape and simulate the explicit motions of these systems are required. Molecular dynamics (MD) simulations have, therefore, become an indispensable tool for exploring the assembly, stability, and plasticity of coiled coils at an atomistic level, providing insights that are often inaccessible to other methods.

Application of Molecular Dynamics to Understand Coiled-Coil Assembly

Molecular dynamics (MD) simulations have proven to be a powerful tool for the evaluation of designed proteins, as they provide energetic and dynamic insights at the atomistic level. Their ability to generate conformational ensembles can complement protein design efforts based on "static" experimental structures, e.g. from X-ray or cryo-EM experiments [24]. MD simulations are nowadays routinely used in protein design pipelines for elucidating the stability and dynamic properties of designed proteins, for understanding sequence-to-structure relationships, for optimising side-chain packing, and for identifying transition mechanisms between distinct conformational states [25, 88, 89]. Additionally, the development of tools making MD simulations more user-friendly has facilitated their widespread adoption for protein simulations [32, 48, 105]. This has consequently enabled their extensive use for coiled-coil modelling in an effort to complement and/or rationalise experimental data.

Early applications of MD simulations in coiled-coil design focused on the assessment of the structural stability of coiled-coil assemblies upon changes in topology, side-chain orientation, oligomerisation state, amino acid composition, and peptide length [62, 92, 136]. These early approaches simply involved running sets of MD simulations with different input structures representing the distinct conditions studied, and assessing stability via a range of metrics, such as root mean square deviation (RMSD) from the initial structures, root mean square fluctuation (RMSF) of individual residues, and solvent accessible surface area (SASA). Before accessible and accurate methods of parametric modelling [38, 131–133] and deep learning methods for sequence-to-structure prediction [1, 67, 78], MD simulations were also useful in generating 3D structures of coiled coils via simulated annealing protocols [84, 135]. Unbiased MD simulations are still used to this day to probe the conformational stability of coiled coils, to investigate the effect of mutations and the addition of ligands, to predict the rate of coiled-coil folding and unfolding, and to build and refine models of coiled-coil designs [7, 14, 51, 59, 118, 120].

On top of traditional MD simulation protocols, the increase in available computing power has translated into computationally expensive enhanced sampling techniques being more accessible for coiled-coil research. For instance, replica exchange simulations which require multiple CPUs and/or GPUs have been successfully used for modelling the dynamic behaviour of asparagine interface residues in coiled-coil dimers [117], for obtaining titration curves for ionisable residues [98], for probing the conformational and thermal stability of stapled coiled coils [93], and for benchmarking the structure and folding propensity of coarse-grained coiled-coil representations [96]. Steered MD simulations have been successfully used to study the mechanisms of coiled-coil shearing and dissociation and to evaluate the potential of coiled coils for use as building blocks [13, 35].

Other prominent enhanced sampling methods are metadynamics and umbrella sampling, which are commonly used to obtain free energy landscapes of processes of great importance to coiled-coil design, such as folding and binding. Metadynamics protocols have been employed to study the unfolding and dissociation mechanisms of coiled-coil trimers [43], to obtain the binding affinity of a coiled-coil homodimer in a membrane environment [16], and to predict the preferred topology and oligomerisation state of coiled-coil assemblies composed of two to six helices using a series of conformational restraints and sequential helix removal from each assembly, offering crucial insights into the energetics of coiled-coil dissociation [85]. Umbrella sampling has been successfully used to explain the selectivity of coiled coils from phage display screening [81], to probe the adsorption of coiled coils to lipid bilayers [130], to investigate rearrangement mechanisms [66], to model the energetics of sliding at homodimer interfaces [36], and to compute the energy required to unwind coiled-coil dimers [31].

Another MD-based method commonly used to explore free energy landscapes is MM-PBSA/MM-GBSA, which has been used for computing the binding affinities of coiled-coil dimers and for examining the effect of the force field and water model used on the free energies obtained, as well as for predicting the preferred oligomerisation state of dimeric to pentameric coiled coils [97, 138, 139]. However, the assumptions made in MM-PBSA/MM-GBSA calculations such as not properly accounting for entropy changes during binding, and the subsequent large statistical errors accompanying the obtained binding free energies, make this method less attractive than metadynamics or umbrella sampling protocols for free energy calculations [106].

MD simulations have also been used for modelling multi-state coiled coils, namely coiled coils that show structural plasticity and are able to access distinct conformational states under the same or different conditions. MD-based methods are advantageous for this task compared to other protein design frameworks, as they are inherently able to capture protein dynamics and can easily accommodate modelling of the appropriate environment that is necessary for the switching behaviour to be observed, e.g. different pH, solvent, temperature, redox potential and presence or absence of ligands. Applications of MD for coiled-coil multi-state modelling include describing bistable dimer interfaces [60], charting the free energy landscape of a coiled-coil hexamer with co-existing open and closed forms [29], successfully modelling the switching behaviour of the CC-Hex-L24E coiled-coil system at low and high pH conditions [85], and investigating the dynamics of the aforementioned CCTM peptides under an electric field, confirming that the structural dynamics of the peptide pores are significantly influenced by both the charges of the N-terminal tag sequences and the magnitude of the membrane potentials [83]. By demonstrating that a charge-neutral tag can stabilise a 10-mer barrel structure in the presence of an external electric field, whereas a positively charged tag can promote increasingly larger stoichiometries, it was shown that MD simulations with external electric fields enable more accurate screening of designed proteins that function under membrane potentials.

Considering the breadth of applications mentioned above, the broad applicability and versatility of MD simulations in coiled-coil research have made them a valuable tool in the protein designer's toolkit (Fig. 1.6). By providing atomistic insights into the conformational stability and dynamics of coiled coils, MD simulations can support the design of single- and multi-state de novo coiled coils for a variety of biotechnological applications. With the continuing advancement of computational power and more efficient algorithms, MD-based protein design protocols are bound to continue to inform the dynamic design and analysis of coiled-coil complexes.

Future Directions and Conclusion

The computational study of coiled coils is at a critical juncture, facing the challenges and opportunities presented by an unprecedented explosion in biological data. The sheer scale of information, from entire genomes to countless sequences designed by deep learning, is becoming problematic for traditional analysis and design pipelines. For the study of natural coiled coils, the future lies in moving beyond simple domain prediction towards high-throughput annotation that captures the full functional

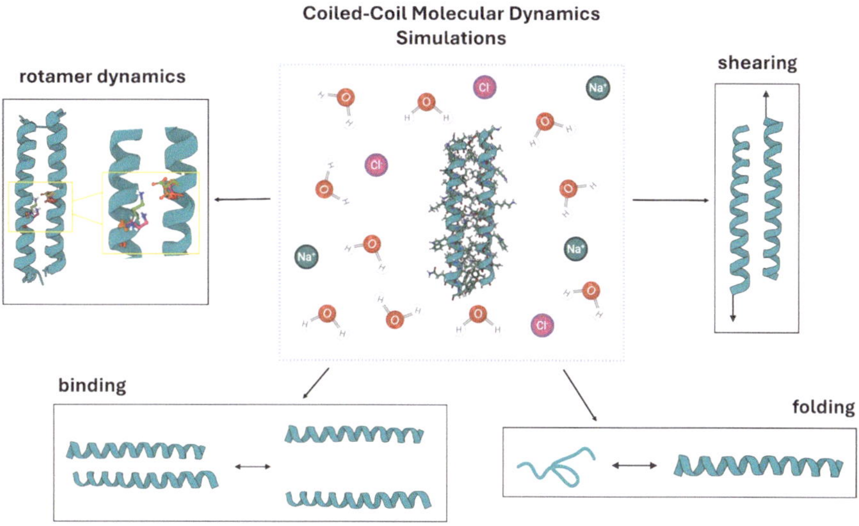

Fig. 1.6 Applications of molecular dynamics simulations in coiled-coil research. Middle: A coiled-coil dimer solvated in a water box with Na^{+}/Cl^{-} ions, as is typically done in a simulation setup. The blue outline represents the periodic boundary of the simulation box. Top left: MD simulations can provide insights into the dynamic behaviour of side chains, e.g. Glu and Lys which commonly form salt bridges in coiled coils. The different coloured side chains correspond to different simulation frames. Top right: Steered MD simulations are used to investigate coiled-coil shearing and interface sliding. Bottom: Enhanced sampling MD methods can be used to explore the free energy landscape of coiled-coil binding (left) and folding (right)

potential of a sequence. To truly understand the cellular role of coiled-coil-containing proteins, we need methods that can, at a genomic scale, infer information about oligomeric state, preferred orientation, and, crucially, the potential for dynamic behaviour or conformational switching. In parallel, the advent of deep learning-based protein design has enabled the generation of vast libraries of novel sequences. This creates a new bottleneck: the need for fast, computationally inexpensive methods to rapidly screen these putative coiled coils, filtering for candidates with desirable dynamic properties or, just as importantly, screening out those prone to unwanted aggregation or misfolding.

To meet these challenges, the tools themselves must evolve. The static outputs of current structural modelling tools, while useful, are insufficient for scrutinising the dynamic systems that are of increasing interest. A key future development will be the expansion of these modelling frameworks to capture, in a computationally efficient manner, some of the complex phenomena currently only accessible through expensive molecular dynamics simulations. Such next-generation models might predict ensembles of states, map rough energetic landscapes, or incorporate intrinsic flexibility, allowing for the rapid, fine-grained assessment of thousands of predicted natural structures or novel designs.

At the highest level of detail, molecular dynamics simulations will continue to provide unparalleled insight, but their computational cost remains a significant barrier. Future progress in this area may come from developments that increase computational efficiency without sacrificing accuracy. The application of machine learning potentials, for example, which are trained on quantum mechanical data or experimental results, could offer a path towards accurately simulating coiled-coil dynamics and assembly over longer timescales or in larger, more complex systems than are currently feasible.

In this review, we have charted the progression of computational tools for coiled-coil research, from the foundational analysis of sequence to the generation of idealised three-dimensional structures. We have highlighted that a critical limitation of this classical approach is its failure to account for the inherent dynamism of the motif and have positioned structural plasticity as a central theme for future investigation. As we have seen, molecular dynamics has been vital for probing these behaviours, but the challenges of scale and complexity necessitate a new generation of computational methods. Ultimately, developing a deeper, more dynamic, and predictive understanding of coiled-coil behaviour is a prerequisite for both deciphering their diverse roles inside the cell and unlocking their full potential as exquisitely programmable components for biotechnology and materials science.

References

1. Abramson J, Adler J, Dunger J, Evans R, Green T, Pritzel A, Ronneberger O et al (2024) Accurate structure prediction of biomolecular interactions with AlphaFold 3. Nature 630(8016):493–500. https://doi.org/10.1038/s41586-024-07487-w
2. Alberstein RG, Guo AB, Kortemme T (2022) Design principles of protein switches. Curr Opin Struct Biol 72:71–78. https://doi.org/10.1016/j.sbi.2021.08.004
3. Ambroggio XI, Kuhlman B (2006) Computational design of a single amino acid sequence that can switch between two distinct protein folds. J Am Chem Soc 128(4):1154–1161. https://doi.org/10.1021/ja054718w
4. Armstrong CT, Vincent TL, Green PJ, Woolfson DN (2011) SCORER 2.0: an algorithm for distinguishing parallel dimeric and trimeric coiled-coil sequences. Bioinformatics 27(14): 1908–1914. https://doi.org/10.1093/bioinformatics/btr299
5. Astbury WT, Woods HJ, Bragg WL (1948) X-ray studies of the structure of hair, wool, and related fibres. II.- the molecular structure and elastic properties of hair keratin. Philos Trans R Soc Lond A 232(707–720):333–394. https://doi.org/10.1098/rsta.1934.0010
6. Aupič J, Lapenta F, Jerala R (2018) SwitCCh: metal-site design for controlling the assembly of a coiled-coil homodimer. Chembiochem 19(23):2453–2457. https://doi.org/10.1002/cbic.201800578
7. Aupič J, Strmšek Ž, Lapenta F, Pahovnik D, Pisanski T, Drobnak I, Ljubetič A, Jerala R (2021) Designed folding pathway of modular coiled-coil-based proteins. Nat Commun 12(1): 940. https://doi.org/10.1038/s41467-021-21185-5
8. Bailey K (1948) Tropomyosin: a new asymmetric protein component of the muscle fibril. Biochem J 43(2):271–279. https://doi.org/10.1042/bj0430271
9. Bartoli L, Fariselli P, Krogh A, Casadio R (2009) CCHMM_PROF: a HMM-based coiled-coil predictor with evolutionary information. Bioinformatics 25(21):2757–2763. https://doi.org/10.1093/bioinformatics/btp539

10. Bendit EG (1957) The α–β transformation in keratin. Nature 179(4558):535–535. https://doi.org/10.1038/179535a0
11. Benton DJ, Gamblin SJ, Rosenthal PB, Skehel JJ (2020) Structural transitions in influenza haemagglutinin at membrane fusion pH. Nature 583(7814):150–153. https://doi.org/10.1038/s41586-020-2333-6
12. Berger B, Wilson DB, Wolf E, Tonchev T, Milla M, Kim PS (1995) Predicting coiled coils by use of pairwise residue correlations. Proc Natl Acad Sci 92(18):8259–8263. https://doi.org/10.1073/pnas.92.18.8259
13. Bergues-Pupo AE, Blank KG, Lipowsky R, Verde AV (2018) Trimeric coiled coils expand the range of strength, toughness and dynamics of coiled coil motifs under shear. Phys Chem Chem Phys 20(46):29105–29115. https://doi.org/10.1039/C8CP04896G
14. Berwick MR, Lewis DJ, Jones AW, Parslow RA, Dafforn TR, Cooper HJ, Wilkie J, Pikramenou Z, Britton MM, Peacock AFA (2014) De novo design of Ln(III) coiled coils for imaging applications. J Am Chem Soc 136(4):1166–1169. https://doi.org/10.1021/ja408741h
15. Berwick MR, Slope LN, Smith CF, King SM, Newton SL, Gillis RB, Adams GG et al (2016) Location dependent coordination chemistry and MRI relaxivity, in de novo designed lanthanide coiled coils. Chem Sci 7(3):2207–2216. https://doi.org/10.1039/C5SC04101E
16. Blazhynska M, Gumbart JC, Chen H, Tajkhorshid E, Roux B, Chipot C (2023) A rigorous framework for calculating protein–protein binding affinities in membranes. J Chem Theory Comput 19(24):9077–9092. https://doi.org/10.1021/acs.jctc.3c00941
17. Boyken SE, Chen Z, Groves B, Langan RA, Oberdorfer G, Ford A, Gilmore JM et al (2016) De novo design of protein homo-oligomers with modular hydrogen-bond network-mediated specificity. Science 352(6286):680–687. https://doi.org/10.1126/science.aad8865
18. Boyken SE, Benhaim MA, Busch F, Jia M, Bick MJ, Choi H, Klima JC et al (2019) De novo design of tunable, pH-driven conformational changes. Science 364(6441):658–664. https://doi.org/10.1126/science.aav7897
19. Boyle AL, Rabe M, Crone NSA, Rhys GG, Soler N, Voskamp P, Pannu NS, Kros A (2019) Selective coordination of three transition metal ions within a coiled-coil peptide scaffold. Chem Sci 10(31):7456–7465. https://doi.org/10.1039/C9SC01165J
20. Brown JH, Cohen C, Parry DA (1996) Heptad breaks in alpha-helical coiled coils: stutters and stammers. Proteins 26(2):134–145. https://doi.org/10.1002/(SICI)1097-0134(199610)26:2<134::AID-PROT3>3.0.CO;2-G
21. Canton J, Neculai D, Grinstein S (2013) Scavenger receptors in homeostasis and immunity. Nat Rev Immunol 13(9):621–634. https://doi.org/10.1038/nri3515
22. Cerasoli E, Sharpe BK, Woolfson DN (2005) ZiCo: a peptide designed to switch folded state upon binding zinc. J Am Chem Soc 127(43):15008–15009. https://doi.org/10.1021/ja0543604
23. Chen J-Y, Wang J-F, Hu Y, Li X-H, Qian Y-R, Song C-L (2025) Evaluating the advancements in protein language models for encoding strategies in protein function prediction: a comprehensive review. Front Bioeng Biotechnol 13:1506508. https://doi.org/10.3389/fbioe.2025.1506508
24. Childers MC, Daggett V (2017) Insights from molecular dynamics simulations for computational protein design. Mol Syst Des Eng 2(1):9–33. https://doi.org/10.1039/C6ME00083E
25. Crean RM, Gardner JM, Kamerlin SCL (2020) Harnessing conformational plasticity to generate designer enzymes. J Am Chem Soc 142(26):11324–11342. https://doi.org/10.1021/jacs.0c04924
26. Crick FHC (1953a) The Fourier transform of a coiled-coil. Acta Crystallogr 6(8–9):685–689. https://doi.org/10.1107/S0365110X53001952
27. Crick FHC (1953b) The packing of α-helices: simple coiled-coils. Acta Crystallogr 6(8–9):689–697. https://doi.org/10.1107/S0365110X53001964
28. Curtis RW, Chmielewski J (2021) A comparison of the collagen triple helix and coiled-coil peptide building blocks on metal ion-mediated supramolecular assembly. Pept Sci 113(2):e24190. https://doi.org/10.1002/pep2.24190

29. Dawson WM, Lang EJM, Rhys GG, Shelley KL, Christopher Williams R, Brady L, Crump MP, Mulholland AJ, Woolfson DN (2021) Structural resolution of switchable states of a de novo peptide assembly. Nat Commun 12(1):1530. https://doi.org/10.1038/s41467-021-21851-8
30. Delorenzi M, Speed T (2002) An HMM model for coiled-coil domains and a comparison with PSSM-based predictions. Bioinf (Oxf) 18(4):617–625. https://doi.org/10.1093/bioinformatics/18.4.617
31. Duan Z-W, Xie P, Li W, Wang P-Y (2012) Are coiled-coils of dimeric kinesins unwound during their walking on microtubule? PLoS One 7(4):e36071. https://doi.org/10.1371/journal.pone.0036071
32. Eastman P, Galvelis R, Peláez RP, Abreu CRA, Farr SE, Gallicchio E, Gorenko A et al (2024) OpenMM 8: molecular dynamics simulation with machine learning potentials. J Phys Chem B 128(1):109–116. https://doi.org/10.1021/acs.jpcb.3c06662
33. Feng S-H, Xia C-Q, Shen H-B (2022) CoCoPRED: coiled-coil protein structural feature prediction from amino acid sequence using deep neural networks. Bioinformatics 38(3):720–729. https://doi.org/10.1093/bioinformatics/btab744
34. Fletcher JM, Bartlett GJ, Boyle AL, Danon JJ, Rush LE, Lupas AN, Woolfson DN (2017) N@a and N@d: oligomer and partner specification by asparagine in coiled-coil interfaces. ACS Chem Biol 12(2):528–538. https://doi.org/10.1021/acschembio.6b00935
35. Goktas M, Luo C, Ruby MA, Sullan AE, Bergues-Pupo RL, Verde AV, Blank KG (2018) Molecular mechanics of coiled coils loaded in the shear geometry. Chem Sci 9(20):4610–4621. https://doi.org/10.1039/C8SC01037D
36. Gomez D, Gavrilov Y, Levy Y (2019) Sliding mechanism at a coiled-coil Interface. Biophys J 116(7):1228–1238. https://doi.org/10.1016/j.bpj.2019.02.026
37. Grigoryan G (2013) Absolute free energies of biomolecules from unperturbed ensembles. J Comput Chem 34(31):2726–2741. https://doi.org/10.1002/jcc.23448
38. Grigoryan G, DeGrado WF (2011) Probing designability via a generalized model of helical bundle geometry. J Mol Biol 405(4):1079–1100. https://doi.org/10.1016/j.jmb.2010.08.058
39. Grigoryan G, Reinke AW, Keating AE (2009) Design of protein-interaction specificity affords selective bZIP-binding peptides. Nature 458(7240):859–864. https://doi.org/10.1038/nature07885
40. Gröger K, Gavins G, Seitz O (2017) Strand displacement in coiled-coil structures: controlled induction and reversal of proximity. Angew Chem Int Ed 56(45):14217–14221. https://doi.org/10.1002/anie.201705339
41. Groth MC, Mathis Rink W, Meyer NF, Thomas F (2018) Kinetic studies on strand displacement in de novo designed parallel heterodimeric coiled coils. Chem Sci 9(18):4308–4316. https://doi.org/10.1039/C7SC05342H
42. Guzenko D, Strelkov SV (2018) CCFold: rapid and accurate prediction of coiled-coil structures and application to modelling intermediate filaments. Bioinformatics 34(2):215–222. https://doi.org/10.1093/bioinformatics/btx551
43. Hamed E, Keten S (2014) Hierarchical cascades of instability govern the mechanics of coiled coils: helix unfolding precedes coil unzipping. Biophys J 107(2):477–484. https://doi.org/10.1016/j.bpj.2014.06.009
44. Harbury PB, Zhang T, Kim PS, Alber T (1993) A switch between two-, three-, and four-stranded coiled coils in GCN4 leucine zipper mutants. Science 262(5138):1401–1407. https://doi.org/10.1126/science.8248779
45. Harrington L, Fletcher JM, Heermann T, Woolfson DN, Schwille P (2021) De novo design of a reversible phosphorylation-dependent switch for membrane targeting. Nat Commun 12(1):1472. https://doi.org/10.1038/s41467-021-21622-5
46. Heal JW, Bartlett GJ, Wood CW, Thomson AR, Woolfson DN (2018) Applying graph theory to protein structures: an atlas of coiled coils. Bioinformatics 34(19):3316–3323. https://doi.org/10.1093/bioinformatics/bty347

47. Hochreiter S, Schmidhuber J (1997) Long short-term memory. Neural Comput 9(8): 1735–1780. https://doi.org/10.1162/neco.1997.9.8.1735
48. Jo S, Kim T, Iyer VG, Im W (2008) CHARMM-GUI: a web-based graphical user Interface for CHARMM. J Comput Chem 29(11):1859–1865. https://doi.org/10.1002/jcc.20945
49. Joh NH, Wang T, Bhate MP, Acharya R, Yibing W, Grabe M, Hong M, Grigoryan G, DeGrado WF (2014) De novo design of a transmembrane Zn^{2+}-transporting four-helix bundle. Science 346(6216):1520–1524. https://doi.org/10.1126/science.1261172
50. Jumper J (2021) Highly accurate protein structure prediction with AlphaFold. Nature 596(7873):583–589. https://doi.org/10.1038/s41586-021-03819-2
51. Karami Y, Saighi P, Vanderhaegen R, Gerlier D, Longhi S, Laine E, Carbone A (2020) Predicting substitutions to modulate disorder and stability in coiled-coils. BMC Bioinformatics 21(19):573. https://doi.org/10.1186/s12859-020-03867-x
52. Kashiwada A, Tsuboi M, Takamura N, Brandenburg E, Matsuda K, Koksch B (2011) Design and characterization of endosomal-pH-responsive coiled coils for constructing an artificial membrane fusion system. Chem Eur J 17(22):6179–6186. https://doi.org/10.1002/chem.201003392
53. Kharenko OA, Ogawa MY (2004) Metal-induced folding of a designed metalloprotein. J Inorg Biochem 98(11):1971–1974. https://doi.org/10.1016/j.jinorgbio.2004.07.015
54. Koebke KJ, Pinter TBJ, Pitts WC, Pecoraro VL (2022) Catalysis and electron transfer in de novo designed Metalloproteins. Chem Rev 122(14):12046–12109. https://doi.org/10.1021/acs.chemrev.1c01025
55. Kohn WD, Kay CM, Sykes BD, Hodges RS (1998) Metal ion induced folding of a de novo designed coiled-coil peptide. J Am Chem Soc 120(6):1124–1132. https://doi.org/10.1021/ja973673z
56. Kumar P, Woolfson DN (2021) Socket2: a program for locating, visualizing and analyzing coiled-coil interfaces in protein structures. Bioinformatics 37(23):4575–4577. https://doi.org/10.1093/bioinformatics/btab631
57. Kurgan KW, Martin FJO, Dawson WM, Brunnock T, Orr-Ewing AJ, Woolfson DN (2025) Exchange, promiscuity, and orthogonality in de novo designed coiled-coil peptide assemblies. Chem Sci 16(4):1826–1836. https://doi.org/10.1039/D4SC06329E
58. Gatta L, Salvatore, and Vincent L. Pecoraro. (2025) Recent advances in de novo designed Metallopeptides as tailored enzyme mimics. Curr Opin Chem Biol 86:102586. https://doi.org/10.1016/j.cbpa.2025.102586
59. Lapenta F, Aupič J, Vezzoli M, Strmšek Ž, Da Vela S, Svergun DI, Carazo JM, Melero R, Jerala R (2021) Self-assembly and regulation of protein cages from pre-organised coiled-coil modules. Nat Commun 12(1):939. https://doi.org/10.1038/s41467-021-21184-6
60. Laughton CA, Luisi BF, Venkatesh Pratap J, Calladine CR (2008) A potential molecular switch in an α-helical coiled coil. Proteins Struct Funct Bioinf 70(1):25–30. https://doi.org/10.1002/prot.21596
61. Leaver-Fay A, Tyka M, Lewis SM, Lange OF, Thompson J, Jacak R, Kaufman K et al (2011) Rosetta3: an object-oriented software suite for the simulation and design of macromolecules. Methods Enzymol 487:545–574. https://doi.org/10.1016/B978-0-12-381270-4.00019-6
62. Lee H, R. G. and Larson. (2007) Prediction of the stability of coiled coils using molecular dynamics simulations. Mol Simul 33(6):463–473. https://doi.org/10.1080/08927020701370612
63. Leone L, De Fenza M, Esposito A, Maglio O, Nastri F, Lombardi A (2024) Peptides and metal ions: a successful marriage for developing artificial metalloproteins. J Pept Sci 30(10): e3606. https://doi.org/10.1002/psc.3606
64. Li C, Wang X-F, Chen Z, Zhang Z, Song J (2015) Computational characterization of parallel dimeric and trimeric coiled-coils using effective amino acid indices. Mol BioSyst 11(2): 354–360. https://doi.org/10.1039/c4mb00569d
65. Li X, Suzuki K, Kashiwada A, Hiroaki H, Kohda D, Tanaka T (2000) Soft metal ions, Cd (II) and Hg (II), induce triple-stranded α-helical assembly and folding of a de novo designed

peptide in their trigonal geometries. Protein Sci 9(7):1327–1333. https://doi.org/10.1110/ps.9.7.1327
66. Lin X, Noel JK, Wang Q, Ma J, Onuchic JN (2018) Atomistic simulations indicate the functional loop-to-coiled-coil transition in influenza hemagglutinin is not downhill. Proc Natl Acad Sci 115(34):E7905–E7913. https://doi.org/10.1073/pnas.1805442115
67. Lin Z (2023) Evolutionary-scale prediction of atomic-level protein structure with a language model. Science 379(6637):1123–1130. https://doi.org/10.1126/science.ade2574
68. Lizatović R, Aurelius O, Stenström O, Drakenberg T, Akke M, Logan DT, André I (2016) A de novo designed coiled-coil peptide with a reversible pH-induced oligomerization switch. Structure 24(6):946–955. https://doi.org/10.1016/j.str.2016.03.027
69. Ludwiczak J, Winski A, Szczepaniak K, Alva V, Dunin-Horkawicz S (2019) DeepCoil—a fast and accurate prediction of coiled-coil domains in protein sequences. Bioinformatics 35(16): 2790–2795. https://doi.org/10.1093/bioinformatics/bty1062
70. Lupas A, Van Dyke M, Stock J (1991) Predicting coiled coils from protein sequences. Science 252(5009):1162–1164. https://doi.org/10.1126/science.252.5009.1162
71. Luther P, Boyle AL (2024) Differences in heavy metal binding to cysteine-containing coiled-coil peptides. J Pept Sci 30(3):e3549. https://doi.org/10.1002/psc.3549
72. MacArthur I (1943) Structure of α-keratin. Nature 152(3845):38–41. https://doi.org/10.1038/152038a0
73. Madaj R, Martinez-Goikoetxea M, Kaminski K, Ludwiczak J, Dunin-Horkawicz S (2025) Applicability of AlphaFold2 in the modeling of dimeric, trimeric, and tetrameric coiled-coil domains. Protein Sci 34(1):e5244. https://doi.org/10.1002/pro.5244
74. Madeo G, Savojardo C, Manfredi M, Martelli PL, Casadio R (2023) CoCoNat: a novel method based on deep learning for coiled-coil prediction. Bioinformatics 39(8):btad495. https://doi.org/10.1093/bioinformatics/btad495
75. Mahendran KR, Niitsu A, Kong L, Thomson AR, Sessions RB, Woolfson DN, Bayley H (2017) A monodisperse transmembrane α-helical peptide barrel. Nat Chem 9(5):411–419. https://doi.org/10.1038/nchem.2647
76. Mair CM, Meyer T, Schneider K, Huang Q, Veit M, Herrmann A (2014) A histidine residue of the influenza virus hemagglutinin controls the pH dependence of the conformational change mediating membrane fusion. J Virol 88(22):13189–13200. https://doi.org/10.1128/jvi.01704-14
77. Marston S, Zamora JE (2020) Troponin structure and function: a view of recent progress. J Muscle Res Cell Motil 41(1):71–89. https://doi.org/10.1007/s10974-019-09513-1
78. Martinez-Goikoetxea M (2025) CCfrag: scanning folding potential of coiled-coil fragments with AlphaFold. Bioinf Adv 5(1):vbae195. https://doi.org/10.1093/bioadv/vbae195
79. McDonnell AV, Jiang T, Keating AE, Berger B (2006) Paircoil2: improved prediction of coiled coils from sequence. Bioinformatics 22(3):356–358. https://doi.org/10.1093/bioinformatics/bti797
80. Miserez A, Guerette PA (2013) Phase transition-induced elasticity of α-helical bioelastomeric fibres and networks. Chem Soc Rev 42(5):1973–1995. https://doi.org/10.1039/C2CS35294J
81. Mortier J, Nyakatura EK, Reimann O, Huhmann S, Daldrop JO, Baldauf C, Wolber G, Miettinen MS, Koksch B (2015) Coiled-coils in phage display screening: insight into exceptional selectivity provided by molecular dynamics. J Chem Inf Model 55(3):495–500. https://doi.org/10.1021/ci500689c
82. Niitsu A, Thomson AR, Scott AJ, Sengel JT, Jung J, Mahendran KR, Sodeoka M et al (2024) De novo design of α-helical peptide channels with designer stoichiometry. bioRxiv 318: 107380. https://doi.org/10.1101/2024.10.05.616771
83. Niitsu A, Jung J, Sugita Y (2025) Structural dynamics of a designed peptide pore under an external electric field. Biophys Chem 318:107380. https://doi.org/10.1016/j.bpc.2024.107380
84. Nilges M, Brünger AT (1991) Automated modeling of coiled coils: application to the GCN4 dimerization region. Protein Eng 4(6):649–659. https://doi.org/10.1093/protein/4.6.649

85. Notari E, Wood CW, Michel J (2025) Assessment of the topology and oligomerisation states of coiled coils using metadynamics with conformational restraints. J Chem Theory Comput 21(6):3260–3276. https://doi.org/10.1021/acs.jctc.4c01695
86. Offer G, Hicks MR, Woolfson DN (2002) Generalized Crick equations for modeling noncanonical coiled coils. J Struct Biol 137(1):41–53. https://doi.org/10.1006/jsbi.2002.4448
87. Oshaben KM, Salari R, McCaslin DR, Chong LT, Seth Horne W (2012) The native GCN4 leucine-zipper domain does not uniquely specify a dimeric oligomerization state. Biochemistry 51(47):9581–9591. https://doi.org/10.1021/bi301132k
88. Park H, DiMaio F, Baker D (2015) The origin of consistent protein structure refinement from structural averaging. Structure 23(6):1123–1128. https://doi.org/10.1016/j.str.2015.03.022
89. Park H, Lee GR, Kim DE, Anishchenko I, Cong Q, Baker D (2019) High-accuracy refinement using Rosetta in CASP13. Proteins Struct Funct Bioinf 87(12):1276–1282. https://doi.org/10.1002/prot.25784
90. Parkinson JS (2010) Signaling mechanisms of HAMP domains in chemoreceptors and sensor kinases. Ann Rev Microbiol 64:101–122. https://doi.org/10.1146/annurev.micro.112408.134215
91. Philips SJ, Canalizo-Hernandez M, Yildirim I, Schatz GC, Mondragón A, O'Halloran TV (2015) Allosteric transcriptional regulation via changes in the overall topology of the core promoter. Science 349(6250):877–881. https://doi.org/10.1126/science.aaa9809
92. Piñeiro Á, Villa A, Vagt T, Koksch B, Mark AE (2005) A molecular dynamics study of the formation, stability, and oligomerization state of two designed coiled coils: possibilities and limitations. Biophys J 89(6):3701–3713. https://doi.org/10.1529/biophysj.104.055590
93. Price JL (2025) Parallel temperature replica-exchange molecular dynamics simulations capture the observed impact of stapling on coiled-coil conformational stability. J Phys Chem B 129(3):866–875. https://doi.org/10.1021/acs.jpcb.4c06974
94. Qin Z, Buehler MJ (2010) Molecular dynamics simulation of the α-helix to β-sheet transition in coiled protein filaments: evidence for a critical filament length scale. Phys Rev Lett 104(19):198304. https://doi.org/10.1103/PhysRevLett.104.198304
95. Rackham OJL, Madera M, Armstrong CT, Vincent TL, Woolfson DN, Gough J (2010) The evolution and structure prediction of coiled coils across all genomes. J Mol Biol 403(3):480–493. https://doi.org/10.1016/j.jmb.2010.08.032
96. Ramirez DA, Hough LE, Shirts MR (2024) Coiled-coil domains are sufficient to drive liquid-liquid phase separation in protein models. Biophys J 123(6):703–717. https://doi.org/10.1016/j.bpj.2024.02.007
97. Rämisch S, Lizatović R, André I (2015) Exploring alternate states and Oligomerization preferences of coiled-coils by de novo structure modeling. Proteins Struct Funct Bioinf 83(2):235–247. https://doi.org/10.1002/prot.24729
98. Rhys GG, Wood CW, Lang EJM, Mulholland AJ, Leo Brady R, Thomson AR, Woolfson DN (2018) Maintaining and breaking symmetry in homomeric coiled-coil assemblies. Nat Commun 9(1):4132. https://doi.org/10.1038/s41467-018-06391-y
99. Rhys GG, Wood CW, Beesley JL, Zaccai NR, Burton AJ, Leo Brady R, Thomson AR, Woolfson DN (2019) Navigating the structural landscape of de novo α-helical bundles. J Am Chem Soc 141(22):8787–8797. https://doi.org/10.1021/jacs.8b13354
100. Ruckthong L, Deb A, Hemmingsen L, Penner-Hahn JE, Pecoraro VL (2018) Incorporation of second coordination sphere D-amino acids alters Cd(II) geometries in designed thiolate-rich proteins. JBIC J Biol Inorg Chem 23(1):123–135. https://doi.org/10.1007/s00775-017-1515-7
101. Schweke H, Pacesa M, Levin T, Goverde CA, Kumar P, Duhoo Y, Dornfeld LJ et al (2024) An atlas of protein homo-oligomerization across domains of life. Cell 187(4):999–1010.e15. https://doi.org/10.1016/j.cell.2024.01.022
102. Scott AJ, Niitsu A, Kratochvil HT, Lang EJM, Sengel JT, Dawson WM, Mahendran KR et al (2021) Constructing ion channels from water-soluble α-helical barrels. Nat Chem 13(7):643–650. https://doi.org/10.1038/s41557-021-00688-0

103. Shen H, Fallas JA, Lynch E, Sheffler W, Parry B, Jannetty N, Decarreau J et al (2018) De novo design of self-assembling helical protein filaments. Science 362(6415):705–709. https://doi.org/10.1126/science.aau3775
104. Shen H, Lynch EM, Akkineni S, Watson JL, Decarreau J, Bethel NP, Benna I et al (2024) De novo design of pH-responsive self-assembling helical protein filaments. Nat Nanotechnol 19(7):1016–1021. https://doi.org/10.1038/s41565-024-01641-1
105. Shrimpton-Phoenix E, Notari E, Kluonis T, Wood CW (2025) drMD: molecular dynamics for experimentalists. J Mol Biol 437(15):168918. https://doi.org/10.1016/j.jmb.2024.168918
106. Siebenmorgen T, Zacharias M (2019) Evaluation of predicted protein–protein complexes by binding free energy simulations. J Chem Theory Comput 15(3):2071–2086. https://doi.org/10.1021/acs.jctc.8b01022
107. Simm D, Hatje K, Kollmar M (2015) Waggawagga: comparative visualization of coiled-coil predictions and detection of stable single α-helices (SAH domains). Bioinformatics 31(5):767–769. https://doi.org/10.1093/bioinformatics/btu700
108. Simm D, Hatje K, Waack S, Kollmar M (2021) Critical assessment of coiled-coil predictions based on protein structure data. Sci Rep 11(1):12439. https://doi.org/10.1038/s41598-021-91886-w
109. Slope LN, Daubney OJ, Campbell H, White SA, Peacock AFA (2021) Location-dependent lanthanide selectivity engineered into structurally characterized designed coiled coils. Angew Chem Int Ed 60(46):24473–24477. https://doi.org/10.1002/anie.202110500
110. Strelkov SV, Burkhard P (2002) Analysis of alpha-helical coiled coils with the program TWISTER reveals a structural mechanism for stutter compensation. J Struct Biol 137(1–2):54–64. https://doi.org/10.1006/jsbi.2002.4454
111. Suzuki K, Hiroaki H, Kohda D, Nakamura H, Tanaka T (1998) Metal ion induced self-assembly of a designed peptide into a triple-stranded α-helical bundle: a novel metal binding site in the hydrophobic core. J Am Chem Soc 120(50):13008–13015. https://doi.org/10.1021/ja982768d
112. Suzuki K, Yamada T, Tanaka T (1999) Role of the buried glutamate in the α-helical coiled coil domain of the macrophage scavenger receptor. Biochemistry 38(6):1751–1756. https://doi.org/10.1021/bi9821014
113. Szczepaniak K, Bukala A, da Silva Neto AM, Ludwiczak J, Dunin-Horkawicz S (2021) A library of coiled-coil domains: from regular bundles to peculiar twists. Bioinformatics 36(22–23):5368–5376. https://doi.org/10.1093/bioinformatics/btaa1041
114. Szilák L, Moitra J, Krylov D, Vinson C (1997) Phosphorylation destabilizes α-helices. Nat Struct Biol 4(2):112–114. https://doi.org/10.1038/nsb0297-112
115. Tebo AG, Pinter TBJ, García-Serres R, Speelman AL, Tard C, Sénéque O, Blondin G et al (2018) Development of a rubredoxin-type center embedded in a de Dovo-designed three-helix bundle. Biochemistry 57(16):2308–2316. https://doi.org/10.1021/acs.biochem.8b00091
116. Testa OD, Moutevelis E, Woolfson DN (2009) CC+: a relational database of coiled-coil structures. Nucleic Acids Res 37(Database issue):D315–D322. https://doi.org/10.1093/nar/gkn675
117. Thomas F, Niitsu A, Oregioni A, Bartlett GJ, Woolfson DN (2017) Conformational dynamics of asparagine at coiled-coil interfaces. Biochemistry 56(50):6544–6554. https://doi.org/10.1021/acs.biochem.7b00848
118. Thomas F, Dawson WM, Lang EJM, Burton AJ, Bartlett GJ, Rhys GG, Mulholland AJ, Woolfson DN (2018) De novo-designed α-helical barrels as receptors for small molecules. ACS Synth Biol 7(7):1808–1816. https://doi.org/10.1021/acssynbio.8b00225
119. Thompson HF, Beesley JL, Langlands HD, Edgell CL, Savery NJ, Woolfson DN (2023) Rational design of phosphorylation-responsive coiled coil-peptide assemblies. ACS Synth Biol 12(4):1308–1319. https://doi.org/10.1021/acssynbio.3c00064
120. Torres-Sánchez A, Vanegas JM, Purohit PK, Arroyo M (2019) Combined molecular/continuum modeling reveals the role of friction during fast unfolding of coiled-coil proteins. Soft Matter 15(24):4961–4975. https://doi.org/10.1039/C9SM00117D

121. Trigg J, Gutwin K, Keating AE, Berger B (2011) Multicoil2: predicting coiled coils and their oligomerization states from sequence in the twilight zone. PLoS One 6(8):e23519. https://doi.org/10.1371/journal.pone.0023519
122. Vincent TL, Green PJ, Woolfson DN (2013) LOGICOIL—multi-state prediction of coiled-coil oligomeric state. Bioinformatics 29(1):69–76. https://doi.org/10.1093/bioinformatics/bts648
123. Walshaw J, Woolfson DN (2001) Socket: a program for identifying and analysing coiled-coil motifs within protein structures. J Mol Biol 307(5):1427–1450. https://doi.org/10.1006/jmbi.2001.4545
124. Wei KY, Moschidi D, Bick MJ, Nerli S, McShan AC, Carter LP, Huang P-S et al (2020) Computational design of closely related proteins that adopt two well-defined but structurally divergent folds. Proc Natl Acad Sci 117(13):7208–7215. https://doi.org/10.1073/pnas.1914808117
125. White JM, Ward AE, Odongo L, Tamm LK (2023) Viral membrane fusion: a dance between proteins and lipids. Annu Rev Virol 10:139–161. https://doi.org/10.1146/annurev-virology-111821-093413
126. Winski A, Ludwiczak J, Orlowska M, Madaj R, Kaminski K, Dunin-Horkawicz S (2024) AlphaFold2 captures the conformational landscape of the HAMP signaling domain. Protein Sci 33(1):e4846. https://doi.org/10.1002/pro.4846
127. Winter DL, Iranmanesh H, Clark DS, Glover DJ (2020) Design of tunable protein interfaces controlled by post-translational modifications. ACS Synth Biol 9(8):2132–2143. https://doi.org/10.1021/acssynbio.0c00208
128. Wolf E, Kim PS, Berger B (1997) MultiCoil: a program for predicting two- and three-stranded coiled coils. Protein Sci 6(6):1179–1189. https://doi.org/10.1002/pro.5560060606
129. Wolfe JA, Seth Horne W (2024) Application of artificial backbone connectivity in the development of metalloenzyme mimics. Curr Opin Chem Biol 81:102509. https://doi.org/10.1016/j.cbpa.2024.102509
130. Woo SY, Lee H (2016) All-atom simulations and free-energy calculations of coiled-coil peptides with lipid bilayers: binding strength, structural transition, and effect on lipid dynamics. Sci Rep 6(1):22299. https://doi.org/10.1038/srep22299
131. Wood CW, Woolfson DN (2018) CCBuilder 2.0: powerful and accessible coiled-coil modeling. Protein Sci 27(1):103–111. https://doi.org/10.1002/pro.3279
132. Wood CW, Bruning M, Ibarra AÁ, Bartlett GJ, Thomson AR, Sessions RB, Leo Brady R, Woolfson DN (2014) CCBuilder: an interactive web-based tool for building, designing and assessing coiled-coil protein assemblies. Bioinf (Oxf) 30(21):3029–3035. https://doi.org/10.1093/bioinformatics/btu502
133. Wood CW, Heal JW, Thomson AR, Bartlett GJ, Ibarra AÁ, Leo Brady R, Sessions RB, Woolfson DN (2017) ISAMBARD: an open-source computational environment for biomolecular analysis, modelling and design. Bioinf (Oxf) 33(19):3043–3050. https://doi.org/10.1093/bioinformatics/btx352
134. Woolfson DN, Alber T (1995) Predicting oligomerization states of coiled coils. Protein Sci 4(8):1596–1607. https://doi.org/10.1002/pro.5560040818
135. Yang P-K, Tzou W-S, Hwang M-J (1999) Restraint-driven formation of α-helical coiled coils in molecular dynamics simulations. Biopolymers 50(6):667–677. https://doi.org/10.1002/(SICI)1097-0282(199911)50:6<667::AID-BIP11>3.0.CO;2-2
136. Zhang L, Hermans J (1993) Molecular dynamics study of structure and stability of a model coiled coil. Proteins Struct Funct Bioinf 16(4):384–392. https://doi.org/10.1002/prot.340160407
137. Zimenkov Y, Dublin SN, Ni R, Tu RS, Breedveld V, Apkarian RP, Conticello VP (2006) Rational design of a reversible pH-responsive switch for peptide self-assembly. J Am Chem Soc 128(21):6770–6771. https://doi.org/10.1021/ja0605974

138. Zuo Z-L, Gandhi NS, Mancera RL (2010) Calculations of the free energy of interaction of the C-Fos–c-Jun coiled coil: effects of the solvation model and the inclusion of polarization effects. J Chem Inf Model 50(12):2201–2212. https://doi.org/10.1021/ci100321h
139. Zuo Z-L, Guo L, Mancera RL (2014) Free energy of binding of coiled-coil complexes with different electrostatic environments: the influence of force field polarisation and capping. Nat Prod Bioprospect 4(5):285–295. https://doi.org/10.1007/s13659-014-0036-0

Chapter 2
Structure of the Myosin Thick Filaments in Muscle

Pradeep K. Luther

Abstract Striated muscle is composed of overlapping arrays of thick myosin filaments and thin actin filaments. The thick filaments are composed of myosin molecules, which are hexamers of two heavy chains and two pairs of light chains. The heavy chain has an N-terminal head domain and a C-terminal helical rod domain. The latter dimerises to form a two-stranded coiled-coil rod. The distal two-thirds of these rods aggregate in parallel to form the filament backbone, while the heads lie on the surface to facilitate interactions with actin. The molecules aggregate in an antiparallel manner in the centre of the A-band to form the so-called bare zone. The proximal one-third of the rod can swivel and thereby allow the myosin heads to interact with actin. The atomic structure of the head, determined in the 1990s, was a major milestone in the muscle field. Over the next three decades, great strides were made in cryo-electron microscope technology and software. This led to the high-resolution structure of the insect flight muscle thick filament, showing the structure of the myosin tails at 6 Å resolution and the structure of the heads. There has been great excitement recently with the high-resolution structures of relaxed cardiac muscle thick filaments showing details of all the important players: three types of myosin crowns and the paths of their tails, the structure and interactions of cMyBP-C and the structure of two unique forms of titin and its role in filament assembly. Hypertrophic cardiomyopathy, which results from mutations in sarcomeric proteins, especially myosin and cMyBP-C, is a major health burden and insight gained from the new studies will help to devise new therapies.

Keywords Myosin structure · Cryo-electron microscopy · Cryo-EM · Structure and function of muscle · Myopathy · Hypertrophic cardiomyopathy · Mavacamten

---

P. K. Luther (✉)
Cardiac Function Section, National Heart and Lung Institute, Imperial College London, Hammersmith Campus, London, UK
e-mail: p.luther@imperial.ac.uk

D. A. D. Parry, P. K. Luther (eds.), *Fibrous Proteins: Structures and Mechanisms*, Subcellular Biochemistry 113, https://doi.org/10.1007/978-3-032-05273-5_2

Abbreviations

Cryo-EM	Cryogenic electron microscopy
EM	Electron microscopy
DQE	Detective quantum efficiency
ELC	Essential light chain
IFM	Insect flight muscle
IHM	Interacting heads motif
HMM	Heavy meromyosin (two myosin heads attached to S2)
LMM	Light meromyosin (distal 2/3 of myosin tail)
MyBP-C	Myosin-binding protein C
cMyBP-C	cardiac myosin-binding protein C
RLC	Regulatory light chain
LCD	Light chain domain
S1	Heavy meromyosin subfragment-1 (head)
S2	Heavy meromyosin subfragment-2 (tail region of myosin that links LMM to S1)
IQ motif	Calmodulin-binding motif (binds ELC and RLC to light chain domain)

Introduction

Here, we review the structure of the thick filament in muscle, a topic comprehensively addressed by John Squire and colleagues in the first edition of this book in 2017 [73]. Since then, however, there have been monumental discoveries in the structure of the thick filament, primarily through the use of cryo-electron microscopy (cryo-EM). This is the focus of the present review. Table 2.1 gives a summary of the timeline of structural milestones in the study of myosin thick filaments that are expounded in this review.

Striated muscle is a hierarchical collection of fibres comprising myofibrils, which are, in turn, composed of ~2 μm sarcomeres joined end-to-end. The striped appearance of striated muscle is one of the most iconic images in biology. It was first recorded by Leeuwenhoek soon after the invention of the microscope. Even without a microscope, however, the effect of the ~2 μm striations can sometimes be observed directly. For example, when light passes through the translucent body of a juvenile fish, a speckled pattern is observed arising from the diffraction of light on passing through the sarcomere striations.

The main components of the sarcomere are interdigitating arrays of thin actin-containing filaments and thick myosin-containing filaments (Fig. 2.1). The thin filaments emanate from either side of the Z-bands in a unipolar fashion, whereas the thick myosin filaments are bipolar and are tethered at the centre of the M-band. Myosin heads on the myosin filaments pull on actin filaments for the contraction of muscle powered by ATP hydrolysis. The structure of the thin actin-containing

Table 2.1 Key milestones in thick filament structure research

Year	Event	References
1954	Sliding filament model	[30, 31]
1957	Electron microscopy of muscle using plastic sections	[32]
1963	Structure of thick filament, bipolar model	[33, 41]
1982	Shutdown state of myosin	[9, 57, 78]
1993	Myosin S1 atomic structure	[63, 64]
2001, 2003	Identification of interacting heads motif in smooth muscle myosin on lipid layers	[48, 80, 81]
2005	Cryo-EM of tarantula thick filament, IHM	[82]
2008, 2013	IHM in cardiac thick filament, neg stain	[1, 89]
2016	6 Å structure of insect flight muscle thick filament backbone	[28]
2020, 2021	High-resolution structure of the shutdown state of myosin	[21, 67, 87]
2023	High-resolution structures of cardiac thick filament	[16, 75]

Key milestones in thick filament research. Important related events that must be noted are the atomic structure of g-actin determined in 1990 [39] and the cryo-EM structures of actomyosin interactions e.g. [58, 85] and the new time-resolved primed acto-myosin structure [43]

filaments is expertly covered in the chapters by Lehman (Chap. 3) and by Maeda and colleagues (Chap. 4) and will not be discussed further here.

Myosin Molecule and Assembly into Filaments

Myosin is a superfamily with more than 40 members. These are all characterised by actin binding, ATPase activity and force production [69]. The most studied of these is myosin II, which is a hexamer of two heavy chains and four light chains [7]. Myosin II is the only member of the myosin family in which the tails self-assemble into a filamentous structure [69]. Myosin molecules have a classic rod shape with two heads at one end (Fig. 2.2a). The molecule can be cleaved by trypsin into heavy meromyosin (HMM) and light meromyosin (LMM) (Fig. 2.2b). The latter comprises the distal two-third part of the rod portion of the myosin molecule. HMM comprises the two N-terminal heads (S1), each comprising a motor domain and light chain domain and the proximal 1/3 of the coiled-coil rod, known as subfragment 2 (S2). Huxley [33] showed that myosin molecules self-assemble with the rods packing parallel to one another to form the filament backbone with the heads lying on the surface of the structure thus formed. At the centre of the filament (the M-band), the myosin molecules pack in an antiparallel fashion, thereby producing a bipolar filament (Fig. 2.2c, d). In vertebrate striated muscle, myosin filaments are packed with three-fold symmetry in cross-section and in a pseudo-helical fashion axially with a subunit repeat of 143 Å and a helical repeat of 430 Å [70] (Fig. 2.2e).

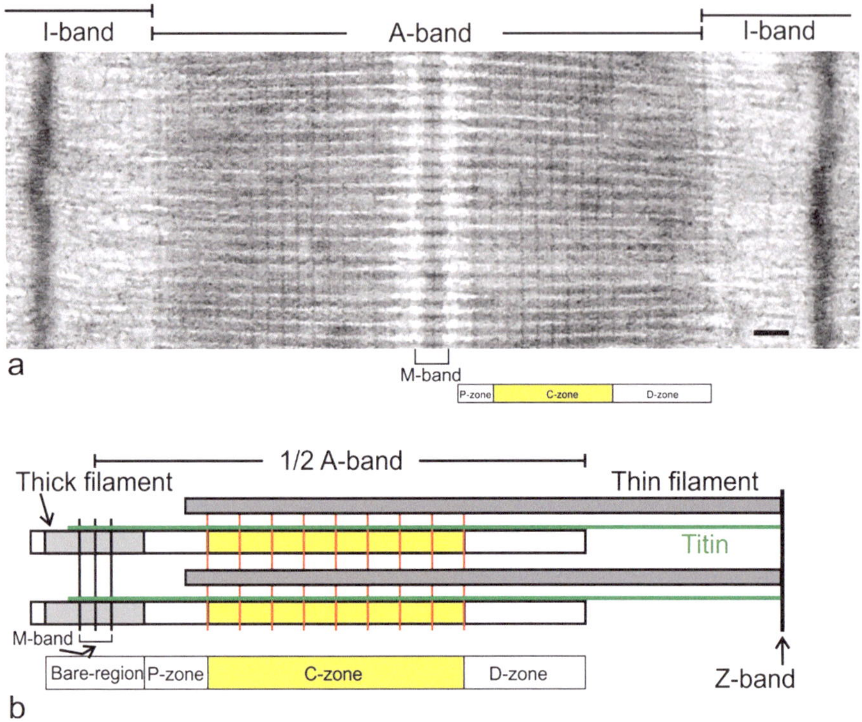

Fig. 2.1 Structure of the vertebrate striated muscle sarcomere. (**a**) An electron micrograph of a frog skeletal muscle sarcomere, (**b**) schematic view of a half sarcomere. Thick filaments centred at the M-band overlap with thin actin filaments, which emanate from the Z-bands. The A-band has three crossbridge regions: P-zone, C-zone and D-zone. The C-zone is the site of cMyBP-C (red bars) located over 9 stripes of separation 430 Å. Titin is a long giant molecule which contacts the thick filament and links the half A-band from the Z-band to the M-band. Scale bar = 100 nm. Adapted from Huang et al. [29]

The myosin head is tadpole-shaped with an N-terminal motor domain. The C-terminal of the head is a long α-helix (neck, lever) with two calmodulin binding motifs called IQ motifs in myosin II, around which are wrapped the essential light chain (ELC) proximal to the motor domain and the regulatory light chain (RLC) (Figs. 2.2b and 2.3a). Myosin motor domains have very similar structures in the different families [88].

Crystal Structure of Myosin S1

To understand the molecular basis of muscular contraction, we need to know the molecular structures of the proteins involved for each of their dynamic states. To determine these molecular structures, the method of choice has been

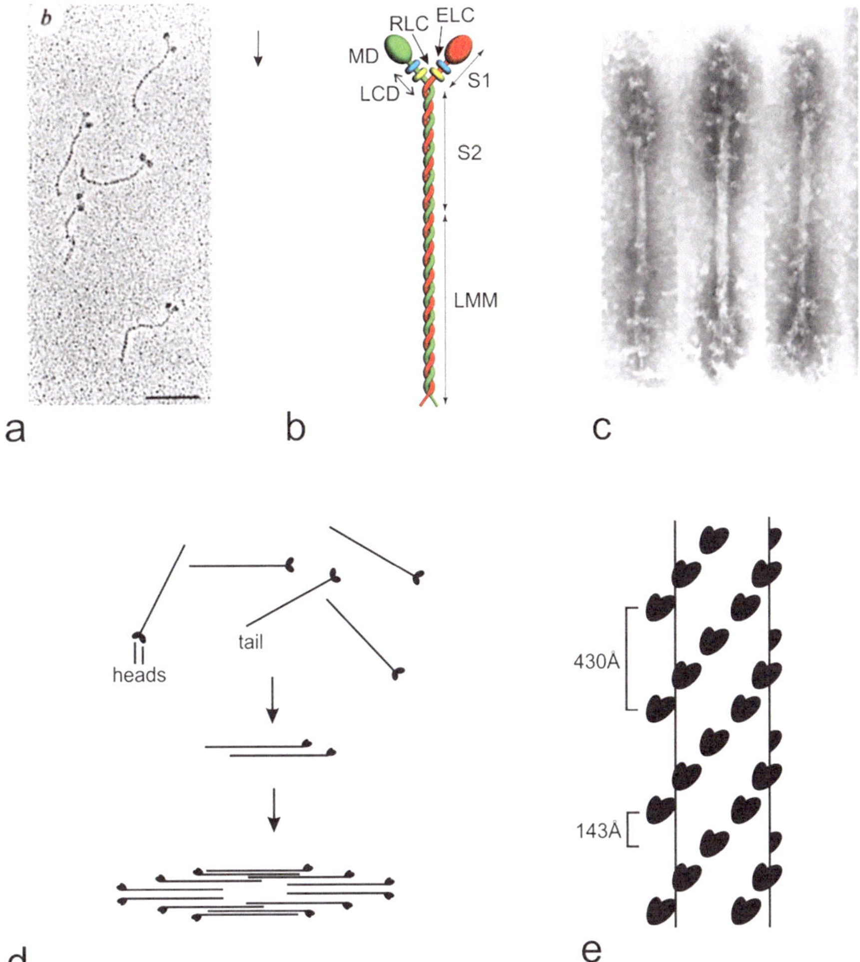

Fig. 2.2 Myosin molecule and assembly into filaments. (**a**) Metal shadowed electron micrograph of myosin molecules showing two bulbous heads at one end of the myosin tail. (**b**) Schematic drawing of a myosin molecule showing motor domain (MD) and light chain domain (LCD) forming heads (S1), the two light chains, RLC and ELC, α-helical coiled coil forming the tail, in which the proximal part is S2 and the distal part is LMM. (**c**) Synthetic myosin filaments showing central bare zone and projections on each side. (**d**) Model proposed by Huxley [33] to show the assembly of myosin molecules to form a thick filament with a central bare zone and crossbridges on the rest of the filament. (**e**) Diagram to show three-stranded myosin filament with subunit repeat 143 Å and helical repeat 430 Å. (Source of figures: (**a**) [50] with permission from Springer Nature, (**b**) [10] with permission from Elsevier, (**c**) [33] with permission from Elsevier)

X-crystallography. In the case of actin—the main co-partner of contraction—the crystal structure was first solved by Kabsch et al. [39]. A very significant breakthrough in muscle research was the determination of the crystal structure of myosin

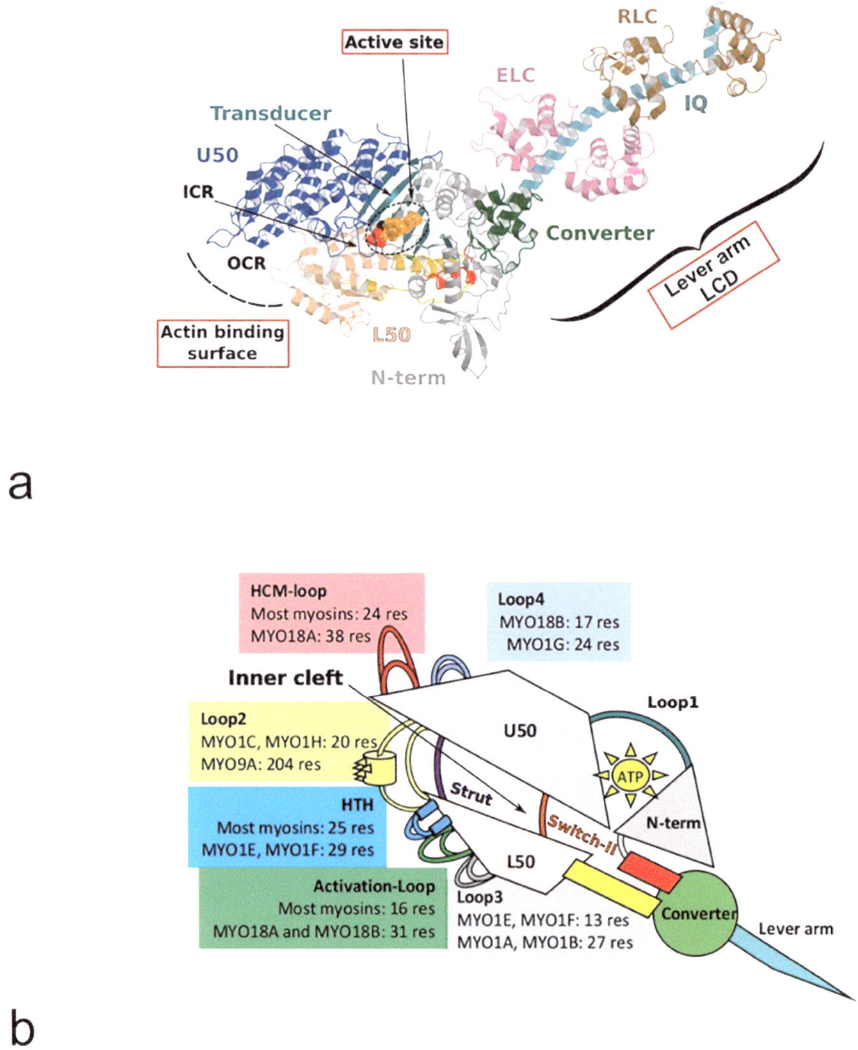

Fig. 2.3 Atomic structure and essential elements of myosin motor domain. (**a**) Ribbon drawing showing the main parts of the motor domain, (**b**) cartoon with labelled parts. The motor domain has four subdomains: upper 50 kDa (U50), lower 50 kDa (L50), N-terminal and converter. Actin binding is via conserved loops, loop 2, HCM loop, loop 3 and loop 4. ATP binding is in the core of the motor active site at switch 1 and the P loop. Force generated by ATP hydrolysis is transmitted from the transducer in the N-terminal subdomain to the relay and SH1 helix to the converter. The small angle change in the converter resulting from the release of hydrolysis products, ADP and Pi, is amplified by the connected lever arm. With permission from [66], (Copyright 2020 American Chemical Society)

S1 by Rayment and colleagues in 1993 [64]; a historical overview is covered by Rall [60]. The first sample studied was myosin from chicken fast skeletal muscle (pectoralis) lacking ATP, hence in the post-rigor state. Subsequent studies have been carried out on myosins from scallop, squid, vertebrate smooth muscle and *Dictyostelium discoideum* [14, 27, 62, 88].

The basic structure of S1 is shown as a ribbon drawing in Fig. 2.3a with a schematic view in 2.3b. The myosin head is composed of a motor domain and a lever arm. The motor domain has four subdomains: upper 50 kD (U50), lower 50 kD (L50), N-terminal and converter (Fig. 2.3a, b). The U50 and L50, parts of a 50 kD domain, are separated by a prominent cleft. Actin binding occurs on the outer face of both U50 and L50 with the cleft closed. This is via conserved loops emanating from U50 (cardiomyopathy (HCM) loop and loop 4) and from L50 (loop 3). The ATP-binding site is located in the core of the motor domain active site between two loops: switch 1 loop and the P-loop. The four domains are connected by highly conserved joints. Switch II connects U50 and L50 near the ATP site. The strut links U50 and L50 near the exit of the cleft, the relay joint links the convertor to the L50, and the SH1 helix links the convertor to the N-terminal domain. The convertor links directly to the lever arm; hence, it is in the right place to facilitate the tilting of the lever arm. The LCD (lever arm) is a passive extension of the converter domain. It amplifies the swinging movement of the converter that is driven by each cycle of ATP hydrolysis in the MD. The crystal structures of various myosins in different states and the cryo-EM structure of actomyosin show that the L50/U50 cleft closes upon actin binding [22]. ATP products are released with movement of the converter, which is amplified by the lever arm [26].

Contribution of Cryo-EM to the Study of Structure of Muscle

From the preceding sections, it is clear that X-crystallography has been a powerful tool to obtain the three-dimensional structure of macromolecules involved in muscular contraction at atomic resolution. The crystal structures of myosin in different states of the crossbridge cycle are a testament to its value [25]. However, obtaining suitable crystals of a sample is often a significant bottleneck of this technique. It requires relatively large quantities of protein of high purity and multiple trials to ascertain the regularity of the crystals obtained. Electron microscopy, on the other hand, does not have these same limitations. In the early days, heavy metal staining was used to improve the contrast, but these dried samples in a non-native environment yielded limited fine detail. Hence, cryogenic electron microscopy (cryo-EM) was developed to examine unstained frozen hydrated samples near liquid nitrogen temperature [15]. These samples were highly radiation-sensitive, and low-dose imaging was essential. The resulting noisy images then had to be averaged to improve the signal-to-noise ratio. Thus was born the technique of cryo-EM of isolated macromolecules with single particle analysis [19]. After a slow start at the turn of the century, cryo-EM has increasingly been the method of choice to determine the structure of important components of the sarcomere [5].

The term “resolution revolution” was coined in 2014 to mark the coming together of two important technical improvements in cryo-EM [4, 44]. The first of these was the development of direct electron detectors that displayed greatly improved detective quantum efficiency (DQE). These allowed image capture in "movie mode", which enabled examination of individual frames and the correction of electron beam-induced motion. The second significant technical achievement related to the development of new image processing software packages such as Relion [68], Frealign [20] and CryoSPARC [59], which incorporate a number of new algorithms that facilitate the recovery of high-resolution structure from large data sets. Three studies under the headings "post-cryo revolution" will be described in later sections, where near-atomic resolutions in key proteins in both invertebrate thick filaments and vertebrate cardiac thick filaments have been achieved.

Shutdown State of Myosin

An unexpected discovery in the 1980s produced some interest at the time [9, 57, 78], but it was to gain considerable momentum over the years. Indeed, very recently, it has provided considerable insight into the atomic structure and function of myosin. This is the story of the "shutdown" state of myosin.

Nature has deemed it imperative that myosins do not expend energy when relaxed, i.e. detached from actin and idle, and from the earliest time of evolution devised a shutdown state with the most minimal expenditure of energy. Electron microscopy by three groups, Craig et al., Onishi and Wakabayashi and Trybus et al. [9, 57, 78] showed that diverse myosin II molecules from smooth muscle and non-muscle (thymus), when phosphorylated, are active and have extended rods with two splayed heads at one end (Fig. 2.2a). However, when unphosphorylated in low salt and MgATP, they are depolymerised and fold into a compact form with the rod curling up in three parts (Fig. 2.4a, b). This is reminiscent of a pangolin rolling up to defend itself against predators. The sedimentation coefficient of the curled-up form of myosin is 10S. In contrast, the sedimentation coefficient of the rod form of myosin in high salt/phosphorylated is 6S. Cross et al. showed that the 10S folded form had ATP products bound (ADP and Pi) and that it is virtually totally inactive [12]. Upon phosphorylation, it straightens to become rod-like, the heads separate from one another, and myosin becomes active and ready to incorporate into filaments (Fig. 2.4c).

To understand the structure of shutdown myosin, Kathleen Trybus teamed up with Ken and Diane Taylor who used the novel method of lipid layer crystallisation to produce 2D crystals of HMM and myosin from smooth muscle in the unphosphorylated and thiophosphorylated states [80, 81]. They used cryo-EM to produce 3D structures of the 2D crystals and then used the Dominguez et al. [14] crystal structure of smooth muscle myosin head in the primed conformation and RLC from the Rayment crystal structure [64] to fit into their maps. They made the startling discovery that the two heads were in contact with each other in a highly asymmetrical fashion such that the actin binding site of one head (blocked head) was in contact with the converter domain in the other head (free head) preventing release

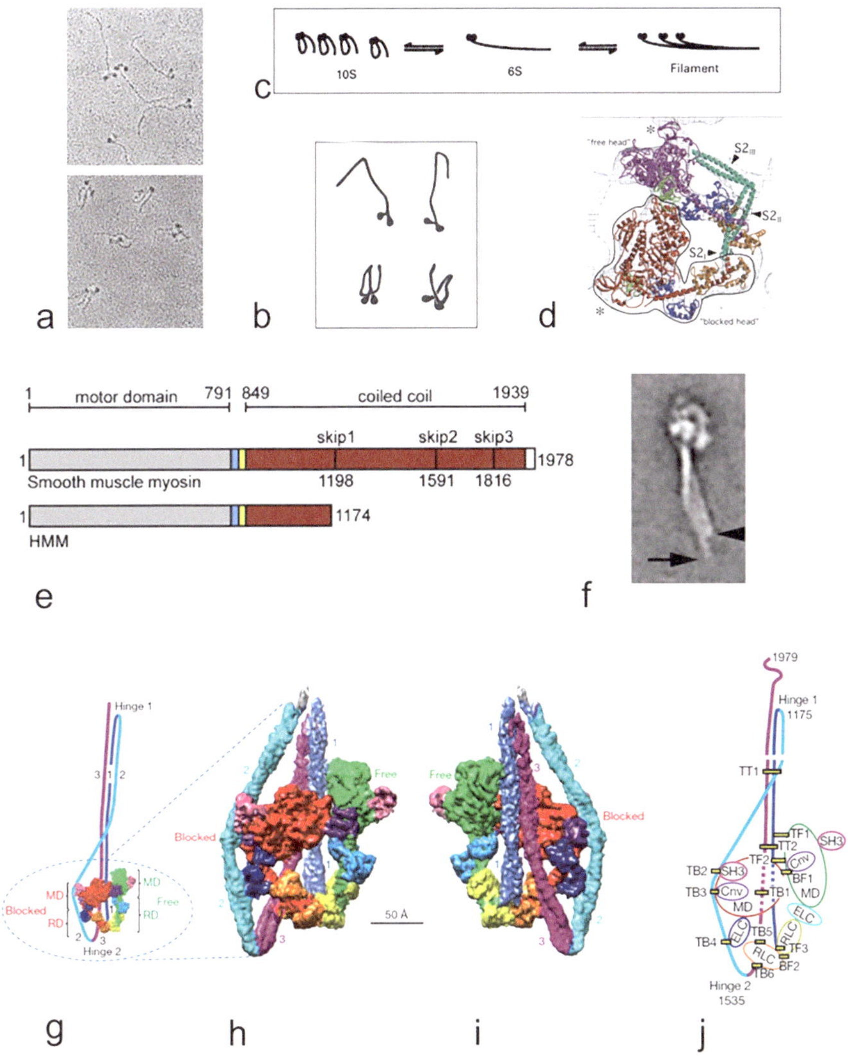

Fig. 2.4 The myosin molecule shutdown state: from pangolin form to atomic model. (**a**) Rotary-shadowed images of smooth muscle myosin: upper—extended form due to phosphorylated light chains, lower—non-phosphorylated rolled-up form. (**b**) Cartoon of the extended and rolled-up form. (**c**) Cycling of rolled-up 10S form to extended rod 6S form leading to filament formation. (**d**) Cryo-em reconstruction of the folded form of smooth muscle myosin prepared on lipid layers. This was the first demonstration of the asymmetric interaction of the two heads [81]. (**e**) Structural organisation of smooth muscle myosin heavy chain. The full molecule has three "skip" residues (inserted residues into the regular pattern of repeats). Only hinge 1 is at a skip residue (skip 1, residue 1175). (**f**) Negative stain electron micrograph of full-length smooth muscle myosin showing hinge 1 (arrowhead) and tip of the tail (arrow). (**g–j**) The atomic model of shutdown state of smooth muscle myosin from the study of Yang et al. [87], one of three papers achieving this breakthrough. (**g**) Organisation of a full-length molecule, ellipse shows the region of the molecule in the reconstruction. (**h, i**) MD of one head (blocked) has its actin binding interface in contact with the

of Pi and ADP, so effectively halting the ATPase cycle (Fig. 2.4d). The term "interacting heads motif", coined later, [2] aptly described the configuration.

It was still necessary, of course, to see the details of the shutdown molecule in a physiological state (i.e. non lipid-layer 2D crystal). Burgess et al. [6] examined vertebrate smooth muscle myosin and heavy meromyosin by electron microscopy of negatively stained samples, with ~2 nm resolution (Fig. 2.4f). Jung et al. [38] studied the scallop striated adductor muscle. Smooth muscle is activated by phosphorylation of S19 in RLC, and the scallop adductor muscle is activated by Ca^{2+} binding to the ELC [37]. Jung et al. [38] examined phosphorylation-regulated invertebrate tarantula and limulus striated muscles and expressed nonmuscle myosin IIA; they also examined vertebrate skeletal and cardiac muscle, which both lack the intrinsic myosin regulatory mechanism. Further negative stain study on smooth muscle myosin was undertaken by the Craig lab in 2019 [86]. These studies noted that the evolutionary separation between the myosins of 600 million years clearly shows that the shutdown state with minimum energy loss is a fundamental property of animal life. A comprehensive study of the evolutionary origin of the myosin IHM state was undertaken by the Craig lab who traced it to the earliest animal life [46].

The last part of the story of the shutdown state of myosin is the recent and very exciting work undertaken by three different groups. Each has achieved remarkable resolution in their structural studies of smooth muscle myosin using cryo-EM: Yang et al. 4–9 Å [87], Scarff et al. 4–6 Å [67], Heissler et al. 3.4 Å [21]. Their studies revealed myriad atomic interactions underlying the shutdown state. The higher resolution of Scarff et al. [67] and Heissler et al. [21] enabled them to visualise the ADP and inorganic phosphate. Here we illustrate the structure with the Yang et al. figures (Fig. 2.4g–j), which is colour coded for the different parts. The free head (FH, green) interacts with the blocked head (BH, red) and their respective light chains are shown. The tail bends into three parts. These are termed segments 1, 2 and 3, and they are connected by two hinges (hinge 1 and hinge 2). Hinge 2, between segments 2 and 3, is included in the map, but Hinge 1, between Segments 1 and 2, distant from the motor, is not resolved. The motor domains, tail segments and light chains all contribute to the shutdown state, and all the interactions between the tails and the heads are illustrated in Fig. 2.4j. The IHM is stabilised by actin-binding regions of BH sequestered by interactions with FH. The CM-loop-FH blocks the actin-binding interface of the FH. These head-head interactions account for the reduced actin binding and ATPase activity of the shutdown myosin. It is phosphorylation of S19 on both RLCs that unlocks the shutdown state.

Fig. 2.4 (continued) converter of the other head (free head). Additionally, the proximal part of S2 contacts the blocked head, adding further inhibition. (**j**) Details of the various interacting sites in the shutdown molecule [87]. (Source: (**a**, **b**) from [9] with permission from Springer Nature, (**c**) from [11] Springer Nature, (**d**) from [81], courtesy of Kenneth Taylor, (**e**, **f**) from [6] with permission from Springer Nature, (**g**–**j**) from [87] with permission from Springer Nature)

High-Resolution Structures of Relaxed Myosin Filaments

Challenge of Electron Microscopy of Helically Ordered Thick Filaments

In the 1960s and 1970s, it was well known from X-ray diffraction that myosin filaments had helically ordered crossbridges (see, for example, Huxley and Brown and Wray et al. [35, 84]), but EM researchers failed to image samples displaying such order. Here, we highlight the skill of Bob Kensler who managed to solve the helicity preservation problem [41]. He developed protocols for the gentle extraction of myosin filaments that preserved helical order amenable for viewing by negative stain electron microscopy. Optical diffraction patterns of the micrographs showed clear signs of helical structure comparable to X-ray patterns of live muscle. Kensler authored several influential papers on thick filament structures solved by helical averaging [1, 17, 40, 42].

Tarantula Leg Muscle Thick Filament (Cryo-EM)

Understanding the structure of isolated molecules and their mechanisms to achieve a relaxed state was important, but their structure in the actual thick filaments was also needed. John Wray had noticed that X-ray patterns of some invertebrate muscles showed much better crossbridge order than other muscles. On his recommendation, Roger Craig and Raul Padron, while post-docs at the MRC Centre in Cambridge, embarked on the EM analysis of tarantula leg muscle (Roger Craig, personal communication). They collaborated with Tony Crowther to study the thick filaments in this muscle by negative stain electron microscopy and found a four-stranded helical arrangement of crossbridges [13]. Craig and Padron continued to collaborate after they moved to Massachusetts and Venezuela, respectively. In 2005 they reported a phenomenally successful cryo-EM study of this muscle (Fig. 2.5a) [82]. At 25 Å resolution, this was the highest resolution structure of a thick filament at the time. The two groups recognised that the arrangement of the paired heads was reminiscent of the shutdown state that Wendt et al. had found in smooth muscle myosin on lipid layers [81].

Woodhead et al. described it as a J-motif (Fig. 2.5b) (later to be coined the interacting heads motif, IHM). Hence, the idea was born that in native relaxed muscle the myosin heads are "locked" against each other to minimise expenditure of energy. Woodhead et al. were able to fit unambiguously the model of Wendt et al. (PDB 1i84) into the map of the blocked and free heads and the α-helical neck with the light chains ELC and RLC (Fig. 2.5c). A critical discovery by Woodhead et al. was the finding that the heads were bent back onto S2, with interaction between S2 and the blocked head (Fig. 2.5d). S2 had not been visualised in the Wendt et al. structure.

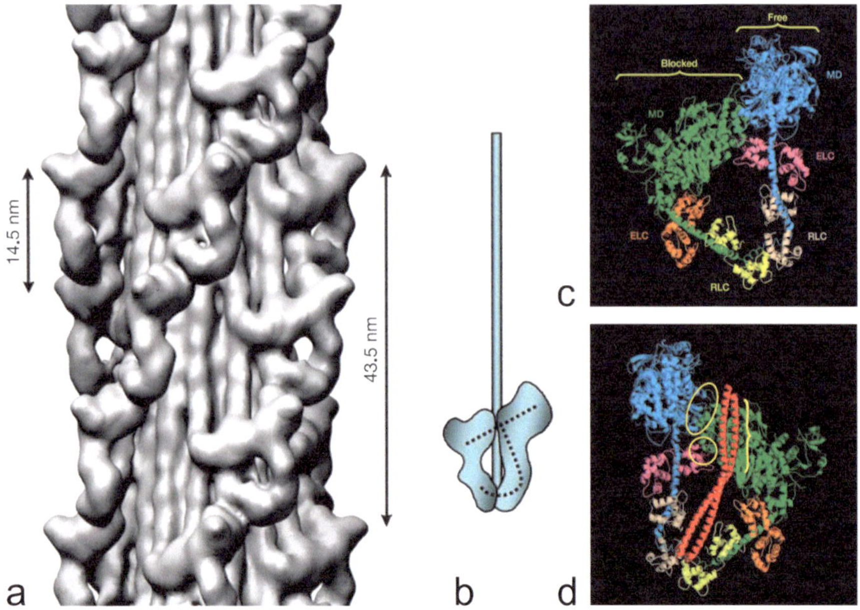

Fig. 2.5 Atomic model of tarantula leg muscle myosin filament in relaxed state. (**a**) The first high-resolution (~25 Å) reconstruction of a relaxed thick filament by cryo-EM of tarantula thick filament [82]. The reconstruction shows myosin heads arranged in IHM crowns with a separation of 145 Å and helical repeat of 435 Å. The structure of the tail and IHM resembles a J-motif (**b**). (**c**) Details of the IHM showing one head interacting with its actin-binding face, blocked head, to the convertor domain of the other head, free head, thereby inhibiting activity in both heads (**d**) Rotating (**c**) by 180°, shows the rear view with proximal part of S2 interacting with the blocked head. (From [82] with permission from Springer Nature)

Cardiac Muscle Thick Filament, Negative Stain Studies

The structure of thick filaments was investigated by Roger Craig and team from mouse heart muscle [89], and by AL-Khayat, Ed Morris and team from human heart muscle [1], both using single particle analysis of negatively stained samples. With resolutions of 40 Å and 28 Å, respectively, both studies identified myosin heads adopting the IHM state; in addition, detail was seen for the first time of titin and cardiac myosin-binding protein C (cMyBP-C). These studies are mentioned here for historical perspective as they have recently been surpassed by high-resolution cryo-EM studies (section "Post-cryo-revolution: Vertebrate (human and mouse) cardiac thick filament").

The tarantula and the two cardiac thick filament structures had a great impact in the muscle field at the time.

Post-cryo-revolution: Lethocerus *(Insect) Flight Muscle Thick Filament*

Ken Taylor and Mike Reedy had a long collaboration on the highly ordered flight muscle of the insect *Lethocerus indicus*, a giant water bug. Early in his career, Reedy recorded electron micrographs of this muscle in relaxed and rigor state, showing perpendicular and angled crossbridges between the myosin and actin filaments [65], and this result gave support for the tilting head hypothesis of muscular contraction [34]. Taylor and Reedy produced seminal papers on the structure of actin/myosin interaction in intact muscle [77]. In 2016, with the resolution revolution implemented, they undertook the cryo-EM investigation of the thick filament in *Lethocerus* flight muscle. A beautiful, detailed reconstruction was obtained at a resolution of 5.5 Å for the backbone (Fig. 2.7) and 10–20 Å for the more flexible heads (Fig. 2.6) [28]. The reconstruction showed, as expected, shelves of four-fold symmetric density at 145 Å intervals composed of the myosin heads—myosin crowns (Fig. 2.6a–d). The heads formed a clear IHM; however, this was not in the tilted angle found in tarantula and cardiac muscles but oriented perpendicularly away from the backbone (Fig. 2.6). The free head appears to make three contacts with the backbone through the motor domain, ELC and RLC. The blocked head appears to make no contact with the backbone, and this may account for its more disordered appearance. They identified the proximal part of S2 tethering myosin heads to the thick filament backbone.

The myosin tails could be followed in exquisite detail (Fig. 2.7a–c) and allowed assessment of the validity of the two tail packing models proposed in the 70 s. Wray [83] proposed that myosin rods pack into 40 Å subfilaments. John Squire, starting from the structure of vertebrate smooth muscle, proposed that within all myosin filaments, the tails pack side by side into layers with equivalent environments [71, 72]. Hu et al. [28] did not see any evidence of packing into subfilaments. Instead, they clearly found that the rods pack into curved layers (Fig. 2.7b, c), thereby fully vindicating the Squire model (see review by Taylor [76]).

Ken Taylor has continued his investigations of the high-resolution structures of thick filaments of insect families with asynchronous flight muscles and has found similar backbone structures with some differences in the head dispositions [47, 61].

Post-cryo-revolution: Vertebrate (Human and Mouse) Cardiac Thick Filament

A seminal pair of papers published back-to-back in *Nature* in 2023 described the near-atomic structure of relaxed cardiac thick filaments in humans and mice [16, 75]. These papers created considerable excitement as knowledge of cardiac muscle is important not only for basic science but also for therapy for cardiac disease due to the high prevalence of hypertrophic cardiomyopathy worldwide. Working

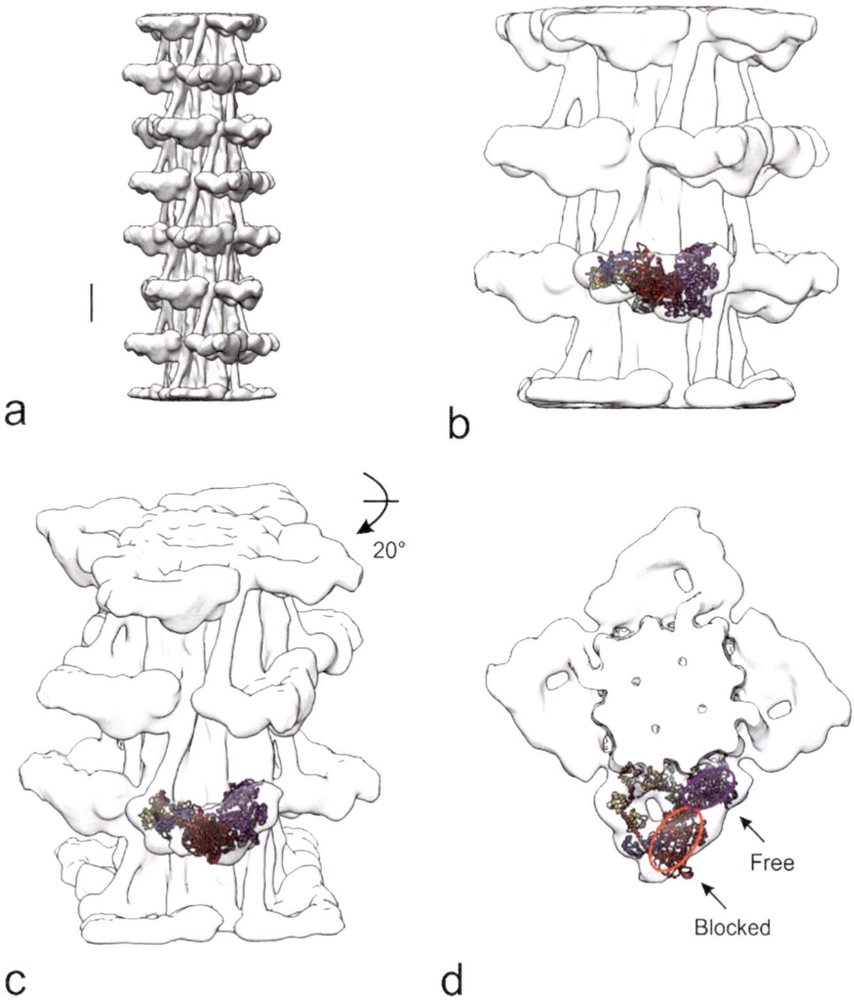

Fig. 2.6 Cryo-EM structure of IFM thick filament depicted here at 20 Å resolution from the study of [28]. (**a**) 3D reconstruction of relaxed thick filament showing myosin crowns at intervals of 145 Å. Scale bar 100 Å. (**b–d**) Illustrating the disposition of IHM, oriented perpendicular to the filament compared with angled in tarantula leg muscle and vertebrate cardiac muscle. One IHM is fitted with ribbon depiction of PDB 1i84 with blocked head shown in red and free head in purple (circled in (**d**)). The cross-section view in (**d**) illustrates the 4-stranded and four-fold symmetric composition of the IFM thick filaments

independently and using different cryo-EM techniques, both papers arrived at essentially the same conclusions. The two papers reveal the structure of all the important players: the myosin heads, the myosin tails, titin and cMyBP-C and how they interact with each other. Debabrata Dutta, in the group of Roger Craig and Raul

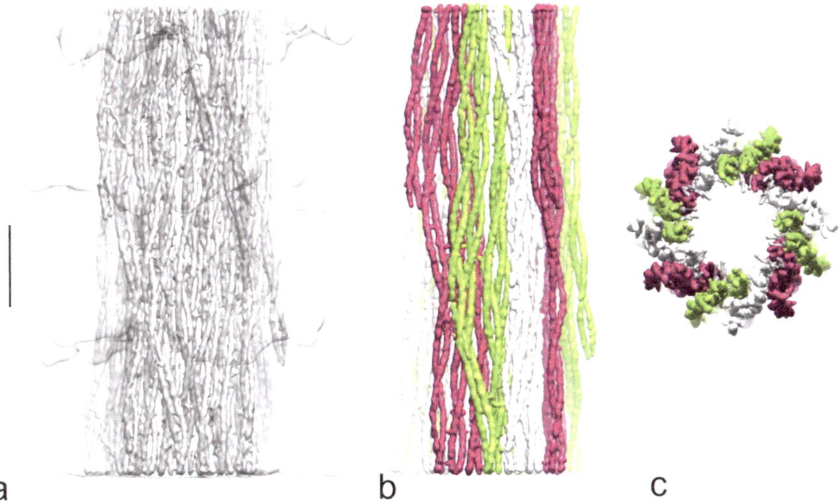

Fig. 2.7 Structure of IFM myosin filament backbone at 5.5 Å resolution from the study of Hu et al. [28]. At this resolution, the path of individual myosin tails can be followed in exquisite detail. (**a**, **b**) A 3 × 145 Å slab of the filament backbone showing the front half of the cylindrical backbone. (**a**) Grey-scale depiction of the 5.5 Å cryo-EM map showing the tails. To show the overall perspective of the map in relation to the heads, the 20 Å map is included in a faint transparent colour. (**b**) In this longitudinal view, the path of the three sets of tails is clearly revealed by segmenting in UCSF Chimera in pink, yellow and grey colours. Each tail is composed of coiled-coil α-helices. (**c**) Cross-section view showing the four-fold structure of the packing of the tails. The packing of the tails in ribbon-like sheets is immediately apparent from the cross-section view in (**c**) and from the two 90° apart views of the pink tails in (**b**): wide slab-view on the left and narrow edge-on view on the right. This vindicates Squire's model for packing of myosin tails in this muscle [72]. Scale bar (**a**–**c**) = 100 Å

Padron, used cryo-EM and single-particle analysis of isolated thick filaments of human heart muscle, enabling them to achieve a laudable resolution of 6 Å for the structure of a 430 Å repeat in the C-zone. In the second study, Davide Tamborrini, in the group of Stefan Raunser, studied the C-zone, P-zone and bare zone in mouse myofibrils by cryo-electron tomography. Although they achieved lower resolution for the thick filament part (~18 Å), the study is invaluable as it reveals the structure of the filament within the environment and influence of the sarcomere lattice.

For these studies, both groups used the drug, mavacamten, to maintain the thick filaments in the relaxed state. Mavacamten is a myosin inhibitor drug approved by the FDA to treat symptoms of HCM [3, 51, 55]. Both groups fitted the reconstructions with known crystal structures and with the aid of the artificial intelligence prediction tool AlphaFold [36].

Previous studies of cardiac thick filament structure in relaxed state (i.e. diastole) (discussed in section "Cardiac muscle thick filament, negative stain studies") had shown that the heads were arranged in IHM configuration [1, 89]. Roger Cooke introduced the concept of the "super-relaxed state" (SRX), where ATP turnover is greatly reduced and has been linked to inhibited IHM heads [23, 74]. Intermediate between active and SRX, another state has been found termed "disordered relaxed state" (DRX) with increased ATP use and likened to heads coming apart from IHM [8]. A significant part of the A-band is occupied by the distal D-zone (Fig. 2.1) in which titin has a repeating domain pattern, called super-repeats, of 7 Ig-like and Fn3-like domains. Some researchers believe that the heads in the C-zone are predominantly in SRX state and in the D-zone in DRX state [56]. The structure of the thick filament in the D-zone is not covered by the two studies. However, studies in the Author's laboratory [29] have shown that the underlying crossbridge repeat of 143 Å in the C-zone is also found in the D-zone.

The structure of the thick filament within the C-zone 430 Å repeat solved by Dutta et al. [16] and Tamborrini et al. [75] is highly complex. Within the volume of this repeat, there are nine IHM of myosin heads (each with blocked and free heads), six titins, three cMyBP-C, and 33 myosin tails. Nomenclature is essential to describe the different interactions between all of these. Because the two groups were working independently, they used different nomenclature and also different colour schemes in their figures. The structure of the 430 Å repeat in cardiac muscle myosin filament is shown in Fig. 2.8 (from Dutta et al. [16]). The myosin heads are arranged in IHM, and each of the three crowns is uniquely arranged. Dutta et al. labelled the three layers of crowns CrH, reflecting horizontal path, CrD, disordered i.e. noisy, weaker density appearance and CrT, tilted appearance. Tamborrini et al. [75] referred to these as crowns 1, 2 and 3 respectively, corresponding to the order of the crowns in the direction from the M-band to the Z-disc as in AL-Khayat et al. and Huang et al. [1, 29]. To account for their appearance and the ordering, it is perhaps appropriate to combine the nomenclatures and label them Crown 1H, Crown 2D and Crown 3T (or Cr1H, Cr2D and Cr3T). Unlike invertebrate muscle thick filaments as found in tarantula and insect flight muscle, which are fully helical with all the myosin tails packing in identical fashion, here the presence of titin and cMyBP-C interferes with symmetrical packing of the myosin tails hence the symmetry is quasi-helical. Figure 2.8b shows the cryo-EM reconstruction of five crowns of heads illustrating the three different crowns (one IHM is coloured in), myosin tails and densities that fit Ig domains for titin and cMyBP-C. The components are fully identified in Fig. 2.8c, d, including strands of titin, cMyBP-C and tails. Figure 2.8e shows a cross-section of the reconstruction at the level of the Crown 1H. Dutta et al. [16] point out that the heads are 40 Å further from the filament axis than the two previous studies of cardiac muscle that were based on negative stain samples, which suggests that those samples suffered shrinkage upon drying down [1, 89].

The myosin tails are visualised in exquisite detail (Fig. 2.8g–l). The tails have a unique arrangement that was not predicted before and show no sign of merging into subfilaments [83] or regular packing in planes [72] as found in invertebrate filaments [28]. Extending their map to 14 crowns, Dutta et al. [16] identified the complete paths of the tails originating from each of the crowns. The tail for Crown 2D originates at the surface of the filament and stays close to the surface along its whole length. Crown 1H tail moves to the interior for half of its length and then veers

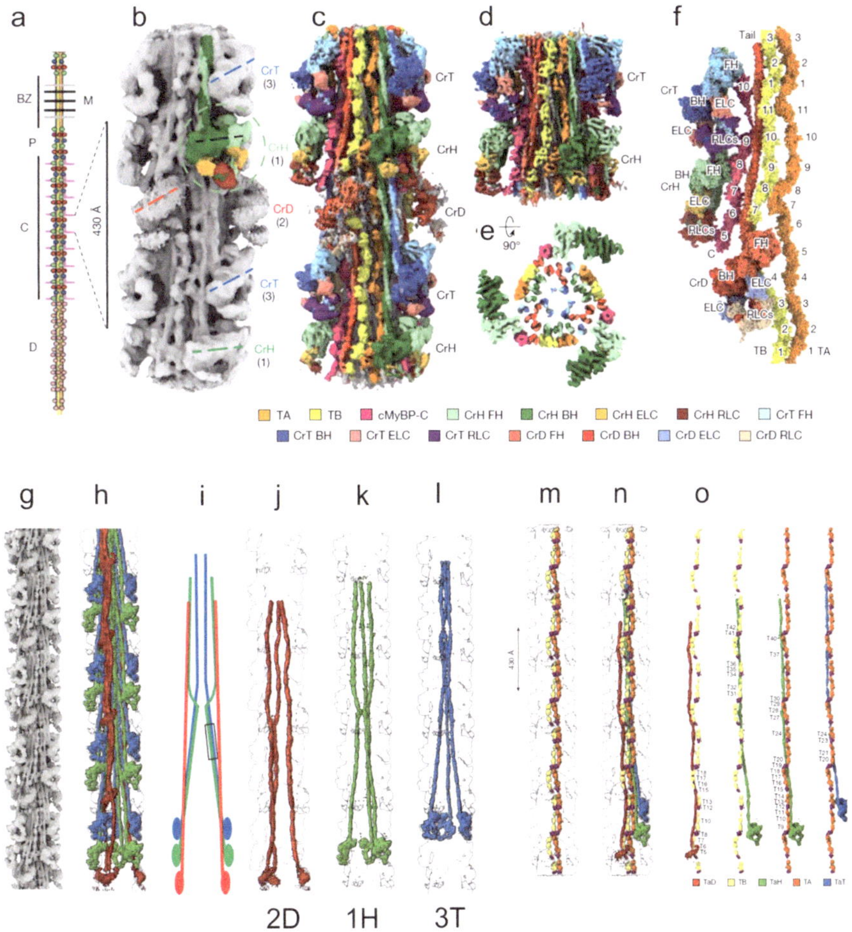

Fig. 2.8 3D reconstruction of cardiac thick filament C-zone illustrated here by the study of Dutta et al. [16]. (**a**) Diagram of myosin filament P, C and D zone with a 430 Å repeat highlighted. (**b**) Cryo-EM map showing the three different crowns, with one highlighted in colour. (**c**) All the components in the map are highlighted in the colour key below, showing the three different crowns, (Cr1H, Cr3T, and Cr2D), titin and cMyBP-C. (**d**) Close-up view of crowns Cr3T and Cr1H. (**e**) Edge-on view of (**d**) showing crown Cr1H free-head and blocked head, titin α and β. (**f**) Structure of one 120° three-fold segment showing crowns Cr3T and Cr1H, path of titin A and B and path of cMyBP-C from C-terminal domain 10 to 5. In these isolated thick filaments devoid of the sarcomere lattice, the N-terminal domains lack actin partners and are free to attach back to the thick filament backbone. In the tomographic reconstruction of Tamborrini et al. [75], only domains C10 to C7 are seen attached to the heads or tails. (**g–l**) Path of the tails associated with the 3 crowns. (**g**) cryo-EM map (**h**) the three crown heads and tails are coloured in. (**i**) Path of the tails of the three crowns. (**j**) Crown 2D tail is predominantly near the surface, (**k**) Crown 1H veers to the interior and then surfaces (**l**) Crown 3T is predominantly in the core of the filament. (**m–o**) Path of the strands of titin A and B in relation to myosin tails. (**m**) Titin Ig domains in both strands coloured purple show that the two strands run in register. (**n**) Path of crown tails relative to titin strands. (**o**) Both titin strands have extensive contacts with myosin tails. (From [16] with permission from Springer Nature)

off to the surface of the filament. Crown 3T tail veers into the interior and forms the core of the filament.

Titin has been an enigmatic component of the sarcomere since its discovery in the late 1970s [53, 54, 79]. Its presence was inferred from the elasticity of the sarcomeres and A-bands centralised in the sarcomere [24]. Liversage et al. proposed that six titins line each half of the thick filament [49]. The present studies reveal the role of titin in the A-band. Strings of ellipsoidal densities 40 Å long are observed with a repeat of 11 domains, the titin super repeat in the C-zone [45]. Six titins are indeed present in each half filament, as three pairs labelled titin α and titin β (labelled titin A and titin B by Dutta et al.). The authors found extensive links between the titins and myosin tails (Fig. 2.8m–o), demonstrating titin's role as a template for thick filament assembly by specific positioning of myosin molecules.

Surprisingly, no interaction was seen between cMyBP-C and titin, an interaction that had been expected based on in vitro studies [18]. cMyBP-C domains C10 to C7 interact instead with a sheet of tails formed by three consecutive Crown 2Ds (Fig. 2.9b, c). Both studies propose that this interaction targets cMyBP-C to the C-zone as this three-sheet grouping is only found in the C-zone. The studies also identified cMyBP-C binding directly to the heads: domains C10 and C8 of cMyBP-C bind to the free heads of crown 3T and 1H, respectively (Fig. 2.9b, c). Tamborrini et al. and Huang et al. [29] identified N-terminal domains of cMyBP-C binding to actin, which was first observed in a tomographic study of skeletal muscle by Luther et al. [52].

We are fortunate in the simultaneous publications of Dutta et al. [16] and Tamborrini et al. [75] studies. Both studies are a tour de force of cardiac muscle and cryo-EM research, giving satisfyingly similar results. They are synergistic as they contribute different essential details. For the single-particle approach of Dutta et al., they imaged isolated cardiac thick filaments and recorded ~56,000 movies from which they obtained three sets of particles ranging in number from ~100,000 to 150,000, where a particle is a 6-crown wide segment of myosin filament along the C-zone. Hence, the structure obtained was an average over the C-zone with 4.5–6 Å resolution. The tomography of myofibrils approach of Tamborrini et al. [75] is in theory superior to the single-particle analysis of isolated filaments as the former visualised the environment of the myosin filament and the important interaction between cMyBP-C and actin. Tamborrini et al. recorded 89 tomograms from which they extracted 5 separate subtomogram regions along the C-zone, P-zone and M-band, which were averaged separately. Their final particle (subtomogram) number was therefore much smaller ranging from 1200 to about 6000 for each of the 5 regions. However, this allowed them insight into the structure and variation along the filament. By careful denoising, they were able to visualise cMyBP-C forming links at various angles to actin (Fig. 2.9d). To achieve higher resolution in the tomography approach, many more tomograms would be needed.

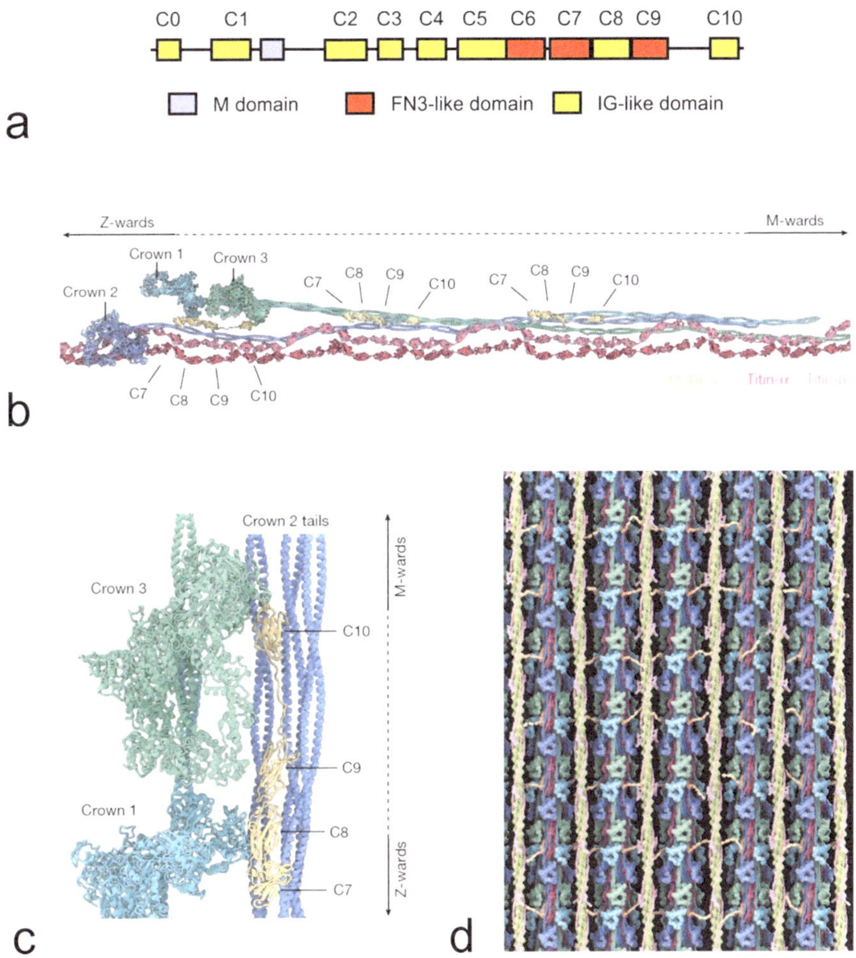

Fig. 2.9 cMyBP-C interactions and targeting. (**a**) Domain layout of cMyBP-C comprising Ig-like and Fn-like domains and M-domain, which has phosphorylation sites. (**b**, **c**) cMyBP-C binds to a sheet of three tails from three consecutive 2D crowns, a grouping found only in the C-zone, suggesting a targeting mechanism for cMyBP-C. Both studies discovered that cMyBP-C binds myosin heads: free-heads of crowns 3T and 1H bind to domains C10 and C8, respectively. Images from Tamborrini et al. [75]. (**d**) Simulation of a ~20 nm thick slice of cardiac C-zone showing a plane of myosin filaments (green and blue pastel colour IHMs) and at a different depth, a plane of actin filaments (yellow). cMyBP-C makes contact at variable angles with actin. (Image courtesy of Davide Tamborrini)

Summary

In the decade since the first edition of Fibrous Proteins with John Squire's comprehensive and authoritative chapter on thick filament structure, the cryo-EM revolution has been in full swing, and the application to the muscle thick filament structure has been a resounding success. What was previously modelled and speculated on has been laid bare and is crystal clear. The two cardiac muscle thick filament structures have revealed details beyond our wildest dreams. We now have insight into the structure of each of the three types of crowns, the complete paths of their tails, the complete path of the two types of titin strands and how they direct filament assembly and details of cMyB-C influence on contractility. From the new papers, many puzzles have been solved. For example, the controversy of cMyBP-C binding to both myosin and actin has been resolved: cMyBP-C binds with C-terminal domains to heads of two of the three crowns, and it binds to actin via its N-terminal domains. The two reconstructions have immense implications for therapeutic approaches for HCM. As noted, HCM is a massive economic burden and future cryo-EM studies, though highly complex, can contribute to the development of therapies and testing.

Acknowledgements I am immensely grateful to Peter Knight, Roger Craig and Ed Morris for insightful comments on the manuscript. Massive thanks go to Robert J. Edwards and Ken Taylor for sharing the Chimera session files for Figs. 2.6 and 2.7. This review is dedicated to the memory of my mentor and friend, John Squire, under whose guidance I started my research career with my PhD on the structure and symmetry of vertebrate myosin filaments. I am grateful to the British Heart Foundation for their stalwart support of my research.

References

1. Al-Khayat HA, Kensler RW, Squire JM, Marston SB, Morris EP (2013) Atomic model of the human cardiac muscle myosin filament. Proc Natl Acad Sci USA 110:318–323
2. Alamo L, Wriggers W, Pinto A, Bartoli F, Salazar L, Zhao FQ, Craig R, Padron R (2008) Three-dimensional reconstruction of tarantula myosin filaments suggests how phosphorylation may regulate myosin activity. J Mol Biol 384:780
3. Anderson RL, Trivedi DV, Sarkar SS, Henze M, Ma W, Gong H, Rogers CS, Gorham JM, Wong FL, Morck MM, Seidman JG, Ruppel KM, Irving TC, Cooke R, Green EM, Spudich JA (2018) Deciphering the super relaxed state of human beta-cardiac myosin and the mode of action of mavacamten from myosin molecules to muscle fibers. Proc Natl Acad Sci USA 115: E8143–E8152
4. Bai XC, McMullan G, Scheres SH (2015) How cryo-EM is revolutionizing structural biology. Trends Biochem Sci 40:49–57
5. Bradshaw M, Paul DM (2019) After the revolution: how is Cryo-EM contributing to muscle research? J Muscle Res Cell Motil 40:93–98
6. Burgess SA, Yu S, Walker ML, Hawkins RJ, Chalovich JM, Knight PJ (2007) Structures of smooth muscle myosin and heavy meromyosin in the folded, shutdown state. J Mol Biol 372: 1165–1178
7. Craig R, Padron R (2004) Molecular structure of the sarcomere. In: Engel AC, Franzini-Armstrong C (eds) Myology, 3rd edn. McGraw-Hill, New York, p 129
8. Craig R, Padron R (2022) Structural basis of the super- and hyper-relaxed states of myosin II. J Gen Physiol 154(1):e202113012
9. Craig R, Smith R, Kendrick-Jones J (1983) Light-chain phosphorylation controls the conformation of vertebrate non-muscle and smooth muscle myosin molecules. Nature 302:436–439
10. Craig R, Woodhead JL (2006) Structure and function of myosin filaments. Curr Opin Struct Biol 16:204–212
11. Cross RA (1988) What is 10S myosin for? J Muscle Res Cell Motil 9:108–110

12. Cross RA, Jackson AP, Citi S, Kendrickjones J, Bagshaw CR (1988) Active-site trapping of nucleotide by smooth and non-muscle myosins. J Mol Biol 203:173–181
13. Crowther RA, Padron R, Craig R (1985) Arrangement of the heads of myosin in relaxed thick filaments from tarantula muscle. J Mol Biol 184:429–439
14. Dominguez R, Freyzon Y, Trybus KM, Cohen C (1998) Crystal structure of a vertebrate smooth muscle myosin motor domain and its complex with the essential light chain: visualization of the pre-power stroke state. Cell 94:559–571
15. Dubochet J, Adrian M, Chang JJ, Homo JC, Lepault J, McDowall AW, Schultz P (1988) Cryo-electron microscopy of vitrified specimens. Q Rev Biophys 21:129–228
16. Dutta D, Nguyen V, Campbell KS, Padron R, Craig R (2023) Cryo-EM structure of the human cardiac myosin filament. Nature 623:853–862
17. Eakins F, Al-Khayat HA, Kensler RW, Morris EP, Squire JM (2002) 3D structure of fish muscle myosin filaments. J Struct Biol 137:154–163
18. Flashman E, Redwood C, Moolman-Smook J, Watkins H (2004) Cardiac myosin binding protein C: its role in physiology and disease. Circ Res 94:1279–1289
19. Frank J, van Heel M (1982) Correspondence analysis of aligned images of biological particles. J Mol Biol 161:134–137
20. Grigorieff N (2016) Frealign: an exploratory tool for single-particle Cryo-EM. Methods Enzymol 579:191–226
21. Heissler SM, Arora AS, Billington N, Sellers JR, Chinthalapudi K (2021) Cryo-EM structure of the autoinhibited state of myosin-2. Sci Adv 7:eabk3273
22. Holmes KC, Angert I, Kull FJ, Jahn W, Schröder RR (2003) Electron cryo-microscopy shows how strong binding of myosin to actin releases nucleotide. Nature 425:423–427
23. Hooijman P, Stewart MA, Cooke R (2011) A new state of cardiac myosin with very slow ATP turnover: a potential cardioprotective mechanism in the heart. Biophys J 100:1969–1976
24. Horowits R, Maruyama K, Podolsky RJ (1989) Elastic behavior of connectin filaments during thick filament movement in activated skeletal muscle. J Cell Biol 109:2169–2176
25. Houdusse A, Sweeney HL (2001) Myosin motors: missing structures and hidden springs. Curr Opin Struct Biol 11:182–194
26. Houdusse A, Sweeney HL (2016) How myosin generates force on actin filaments. Trends Biochem Sci 41:989–997
27. Houdusse A, Szent-Gyorgyi AG, Cohen C (1999) Scallop myosin has at least three conformational states. Acta Crystallogr Found Adv 55:21–21
28. Hu Z, Taylor DW, Reedy MK, Edwards RJ, Taylor KA (2016) Structure of myosin filaments from relaxed Lethocerus flight muscle by cryo-EM at 6 A resolution. Sci Adv 2:e1600058
29. Huang X, Torre I, Chiappi M, Yin Z, Vydyanath A, Cao S, Raschdorf O, Beeby M, Quigley B, de Tombe PP, Liu J, Morris EP, Luther PK (2023) Cryo-electron tomography of intact cardiac muscle reveals myosin binding protein-C linking myosin and actin filaments. J Muscle Res Cell Motil 44:165
30. Huxley AF, Niedergerke R (1954) Structural changes in muscle during contraction; interference microscopy of living muscle fibres. Nature 173:971–973
31. Huxley H, Hanson J (1954) Changes in the cross-striations of muscle during contraction and stretch and their structural interpretation. Nature 173:973–976
32. Huxley HE (1957) The double array of filaments in cross-striated muscle. J Biophys Biochem Cytol 3:631–648
33. Huxley HE (1963) Electron microscope studies on the structure of natural and synthetic protein filaments from striated muscle. J Mol Biol 7:281–308
34. Huxley HE (1969) The mechanism of muscular contraction. Science 164:1356–1365
35. Huxley HE, Brown W (1967) The low-angle x-ray diagram of vertebrate striated muscle and its behaviour during contraction and rigor. J Mol Biol 30:383–434
36. Jumper J, Evans R, Pritzel A, Green T, Figurnov M, Ronneberger O, Tunyasuvunakool K, Bates R, Zídek A, Potapenko A, Bridgland A, Meyer C, Kohl SAA, Ballard AJ, Cowie A, Romera-Paredes B, Nikolov S, Jain R, Adler J, Back T, Petersen S, Reiman D, Clancy E,

Zielinski M, Steinegger M, Pacholska M, Berghammer T, Bodenstein S, Silver D, Vinyals O, Senior AW, Kavukcuoglu K, Kohli P, Hassabis D (2021) Highly accurate protein structure prediction with AlphaFold. Nature 596:583

37. Jung HS, Burgess SA, Billington N, Colegrave M, Patel H, Chalovich JM, Chantler PD, Knight PJ (2008a) Conservation of the regulated structure of folded myosin 2 in species separated by at least 600 million years of independent evolution. Proc Natl Acad Sci USA 105:6022–6026
38. Jung HS, Komatsu S, Ikebe M, Craig R (2008b) Head-head and head-tail interaction: a general mechanism for switching off myosin II activity in cells. Mol Biol Cell 19:3234–3242
39. Kabsch W, Mannherz HG, Suck D, Pai EF, Holmes KC (1990) Atomic structure of the actin: DNase I complex. Nature 347:37–44
40. Kensler RW (2005) The mammalian cardiac muscle thick filament: crossbridge arrangement. J Struct Biol 149:303–312
41. Kensler RW, Levine RJC (1982) An electron-microscopic and optical diffraction analysis of the structure of Limulus telson muscle thick filaments. J Cell Biol 92:443–451
42. Kensler RW, Stewart M (1983) Frog skeletal muscle thick filaments are three-stranded. J Cell Biol 96:1797–1802
43. Klebl DP, McMillan SN, Risi C, Forgacs E, Virok B, Atherton JL, Harris SA, Stofella M, Winkelmann DA, Sobott F, Galkin VE, Knight PJ, Muench SP, Scarff CA, White HD (2025) Swinging lever mechanism of myosin directly shown by time-resolved cryo-EM. Nature 642: 519
44. Kuhlbrandt W (2014) Biochemistry. The resolution revolution. Science 343:1443–1444
45. Labeit S, Kolmerer B (1995) Titins: giant proteins in charge of muscle ultrastructure and elasticity. Science 270:293–296
46. Lee KH, Sulbarán G, Yang SX, Mun JY, Alamo L, Pinto A, Sato O, Ikebe M, Liu X, Korn ED, Sarsoza F, Bernstein SI, Padrón R, Craig R (2018) Interacting-heads motif has been conserved as a mechanism of myosin II inhibition since before the origin of animals. Proc Natl Acad Sci USA 115:E1991–E2000
47. Li JW, Rahmani H, Yeganeh FA, Rastegarpouyani H, Taylor DW, Wood NB, Previs MJ, Iwamoto H, Taylor KA (2023) Structure of the flight muscle thick filament from the bumble bee, at 6 Å resolution. Int J Mol Sci 24(1):377
48. Liu J, Wendt T, Taylor D, Taylor K (2003) Refined model of the 10S conformation of smooth muscle myosin by cryo-electron microscopy 3D image reconstruction. J Mol Biol 329:963–972
49. Liversage AD, Holmes D, Knight PJ, Tskhovrebova L, Trinick J (2001) Titin and the sarcomere symmetry paradox. J Mol Biol 305:401–409
50. Lowey S, Waller GS, Trybus KM (1993) Skeletal muscle myosin light chains are essential for physiological speeds of shortening. Nature 365:454–456
51. Luther PK, Marston SB (2024) Complex architecture of cardiac muscle thick filaments revealed. Trends Pharmacol Sci 45:191–192
52. Luther PK, Winkler H, Taylor K, Zoghbi ME, Craig R, Padron R, Squire JM, Liu J (2011) Direct visualization of myosin-binding protein C bridging myosin and actin filaments in intact muscle. Proc Natl Acad Sci USA 108:11423–11428
53. Maruyama K (1997) Connectin/titin, giant elastic protein of muscle. FASEB J 11:341–345
54. Maruyama K, Natori R, Nonomura Y (1976) New elastic protein from muscle. Nature 262:58–60
55. Nag S, Gollapudi SK, Del Rio CL, Spudich JA, McDowell R (2023) Mavacamten, a precision medicine for hypertrophic cardiomyopathy: from a motor protein to patients. Sci Adv 9: eabo7622
56. Nelson S, Beck-Previs S, Sadayappan S, Tong C, Warshaw DM (2023) Myosin-binding protein C stabilizes, but is not the sole determinant of SRX myosin in cardiac muscle. J Gen Physiol 155:e202213276
57. Onishi H, Wakabayashi T (1982) Electron-microscopic studies of myosin molecules from chicken gizzard muscle.1. The formation of the intramolecular loop in the myosin tail. J Biochem 92:871–879

58. Pospich S, Sweeney HL, Houdusse A, Raunser S (2021) High-resolution structures of the actomyosin-V complex in three nucleotide states provide insights into the force generation mechanism. elife 10:e73724
59. Punjani A, Rubinstein JL, Fleet DJ, Brubaker MA (2017) cryoSPARC: algorithms for rapid unsupervised cryo-EM structure determination. Nat Methods 14:290
60. Rall JA (2021) Investigation of the molecular motor of muscle: from generating life in a test tube to myosin structure over beers. Adv Physiol Educ 45:730–743
61. Rastegarpouyani H, Hojjatian A, Taylor KA (2024) Two forms of thick filament in the flight muscle of. Int J Mol Sci 25(20):11313
62. Rayment I (1996) The structural basis of the myosin ATPase activity. J Biol Chem 271:15850–15853
63. Rayment I, Holden HM, Whittaker M, Yohn CB, Lorenz M, Holmes KC, Milligan RA (1993a) Structure of the actin-myosin complex and its implications for muscle contraction. Science 261: 58–65
64. Rayment I, Rypniewski WR, Schmidt-Base K, Smith R, Tomchick DR, Benning MM, Winkelmann DA, Wesenberg G, Holden HM (1993b) Three-dimensional structure of myosin subfragment-1: a molecular motor. Science 261:50–58
65. Reedy MK, Holmes KC, Tregear RT (1965) Induced changes in orientation of the cross-bridges of glycerinated insect flight muscle. Nature 207:1276–1280
66. Robert-Paganin J, Pylypenko O, Kikuti C, Sweeney HL, Houdusse A (2020) Force generation by myosin motors: a structural perspective. Chem Rev 120:5–35
67. Scarff CA, Carrington G, Casas-Mao D, Chalovich JM, Knight PJ, Ranson NA, Peckham M (2020) Structure of the shutdown state of myosin-2. Nature 588:515–520
68. Scheres SH (2012) RELION: implementation of a Bayesian approach to cryo-EM structure determination. J Struct Biol 180:519–530
69. Sellers JR (2000) Myosins: a diverse superfamily. Biochim Biophys Acta 1496:3–22
70. Squire J (1981) The structural basis of muscular contraction. Plenum Press, New York
71. Squire JM (1971) General model for the structure of all myosin-containing filaments. Nature 233:457–462
72. Squire JM (1973) General model of myosin filament structure. 3. Molecular packing arrangements in myosin filaments. J Mol Biol 77:291–323
73. Squire JM, Paul DM, Morris EP (2017) Myosin and actin filaments in muscle: structures and interactions. In: Parry DA, Squire JM (eds) Fibrous proteins: structure and mechanism. Springer, Cham
74. Stewart MA, Franks-Skiba K, Chen S, Cooke R (2010) Myosin ATP turnover rate is a mechanism involved in thermogenesis in resting skeletal muscle fibers. Proc Natl Acad Sci USA 107:430–435
75. Tamborrini D, Wang Z, Wagner T, Tacke S, Stabrin M, Grange M, Kho AL, Rees M, Bennett P, Gautel M, Raunser S (2023) Structure of the native myosin filament in the relaxed cardiac sarcomere. Nature 623:863–871
76. Taylor KA (2023) John Squire and the myosin thick filament structure in muscle. J Muscle Res Cell Motil 44:143
77. Taylor KA, Schmitz H, Reedy MC, Goldman YE, Franzini-Armstrong C, Sasaki H, Tregear RT, Poole K, Lucaveche C, Edwards RJ, Chen LF, Winkler H, Reedy MK (1999) Tomographic 3D reconstruction of quick-frozen, Ca2+-activated contracting insect flight muscle. Cell 99: 421–431
78. Trybus KM, Huiatt TW, Lowey S (1982) A bent monomeric conformation of myosin from smooth muscle. Proc Natl Acad Sci USA 79:6151–6155
79. Wang K, McClure J, Tu A (1979) Titin: major myofibrillar components of striated muscle. Proc Natl Acad Sci USA 76:3698–3702
80. Wendt T, Taylor D, Messier T, Trybus KM, Taylor KA (1999) Visualization of head-head interactions in the inhibited state of smooth muscle myosin. J Cell Biol 147:1385–1390

81. Wendt T, Taylor D, Trybus KM, Taylor K (2001) Three-dimensional image reconstruction of dephosphorylated smooth muscle heavy meromyosin reveals asymmetry in the interaction between myosin heads and placement of subfragment 2. Proc Natl Acad Sci USA 98:4361–4366
82. Woodhead JL, Zhao FQ, Craig R, Egelman EH, Alamo L, Padron R (2005) Atomic model of a myosin filament in the relaxed state. Nature 436:1195–1199
83. Wray JS (1979) Structure of the backbone in myosin filaments of muscle. Nature 277:37–40
84. Wray JS, Vibert PJ, Cohen C (1975) Diversity of cross-bridge configurations in invertebrate muscles. Nature 257:561–564
85. Wulf SF, Ropars V, Fujita-Becker S, Oster M, Hofhaus G, Trabuco LG, Pylypenko O, Sweeney HL, Houdusse AM, Schröder RR (2016) Force-producing ADP state of myosin bound to actin. Proc Natl Acad Sci USA 113:E1844–E1852
86. Yang S, Lee KH, Woodhead JL, Sato O, Ikebe M, Craig R (2019) The central role of the tail in switching off 10S myosin II activity. J Gen Physiol 151:1081–1093
87. Yang S, Tiwari P, Lee KH, Sato O, Ikebe M, Padron R, Craig R (2020) Cryo-EM structure of the inhibited (10S) form of myosin II. Nature 588:521–525
88. Yang YT, Gourinath S, Kovács M, Nyitray L, Reutzel R, Himmel DM, O'Neall-Hennessey E, Reshetnikova L, Szent-Györgyi AG, Brown JH, Cohen C (2007) Rigor-like structures from muscle myosins reveal key mechanical elements in the transduction pathways of this allosteric motor. Structure 15:553–564
89. Zoghbi ME, Woodhead JL, Moss RL, Craig R (2008) Three-dimensional structure of vertebrate cardiac muscle myosin filaments. Proc Natl Acad Sci USA 105:2386–2390

Chapter 3
The Structural Role of Tropomyosin in Regulating Thin Filament Activation of Actin-Myosin Interaction

William Lehman

Abstract Myosin motor-heads projecting from muscle thick filaments interact cyclically with actin-based thin filament tracks, thereby driving inter-filament sliding that powers muscle contraction. Here, controlled recruitment of myosin heads from thick filaments leads pre-powerstroke myosin to bind weakly to actin. Myosin then isomerizes into strongly bound post-powerstroke conformations on actin, thus producing crossbridge motion in active muscles. In striated muscles, this process is regulated by a steric mechanism involving coiled-coil tropomyosin controlling access to myosin-binding sites on actin. Biochemical and structural studies suggest the regulatory mechanism involves tropomyosin occupying three average configurations on the actin thin filament, dependent on Ca^{2+}, troponin, and myosin binding. Once Ca^{2+}-activation of muscle occurs, tropomyosin pivots away from a troponin subunit-I (TnI) imposed B-state (myosin-blocking) position to a C-state position on actin, allowing initial weak myosin-binding to actin. The thin filament reconfiguration only partially relieves tropomyosin-troponin imposed steric inhibition of the myosin binding. However, the initial weak myosin-binding causes further tropomyosin translocation to an M-state position as myosin transitions from its pre-powerstroke to its post-powerstroke conformation, thereby fully activating the thin filament and resulting in contraction. This review summarizes the evolving structural evidence that has accumulated over many years, and which has shaped our current understanding of the troponin-tropomyosin steric regulatory mechanism that governs muscle contractility.

Keywords Actin · Cryo-EM · Molecular dynamics · Striated muscle · Myosin · Thin filaments · Tropomyosin · Troponin · X-ray fiber diffraction

W. Lehman (✉)
Department of Pharmacology, Physiology & Biophysics, Boston University Chobanian & Avedisian School of Medicine, Boston, MA, USA
e-mail: wlehman@bu.edu

D. A. D. Parry, P. K. Luther (eds.), *Fibrous Proteins: Structures and Mechanisms*, Subcellular Biochemistry 113, https://doi.org/10.1007/978-3-032-05273-5_3

Abbreviations

Cryo-EM	Cryogenic electron microscopy
cTermTnI	The C-terminal domain of troponin subunit I
TnC	Troponin subunit-C
TnI	Troponin subunit-I
TnT	Troponin subunit-T

Introduction

Striated muscle contraction is regulated by a Ca^{2+}-dependent reversible on-off switching mechanism [16, 48]. At low Ca^{2+}-concentrations in relaxed muscles, coiled-coil tropomyosin participates in steric-blocking myosin-binding sites on double-stranded helically arranged thin filament actin. Once muscles are activated at high Ca^{2+}, the tropomyosin constraint is relieved, and the myosin-binding sites on actin become open. The Ca^{2+}-dependent transition then is followed by ATP-driven myosin crossbridge cycling on actin, leading to force production and muscle contractility [16, 48]. This canonical tropomyosin-based "steric regulatory model" was first envisioned in the mid-1960s by Hanson and Lowy [19] as a control mechanism to govern muscle contractility. The steric model was then supported experimentally by Huxley and others, using X-ray fiber diffraction technology available at the time [20, 29, 54, 66]. Three decades later, their insights were largely substantiated by ~18–25 Å resolution EM-reconstructions generated by my laboratory working together with Peter Vibert and Roger Craig (Lehman et al. [39]; reviewed in Lehman [48]). While the steric model in its most elementary form is often regarded as the full story that depicts muscle regulation by tropomyosin and its partner troponin as participating in a straightforward binary on-off switching mechanism and formulated in textbook simplicity, this view detracts from a proper appreciation of the complexity of the process. Indeed, understanding troponin-tropomyosin-based thin filament regulation at a more nuanced level is of particular importance given current efforts by many researchers to assess and remedy effects of disease-related mutations borne by tropomyosin and troponin [38, 59, 82].

It is now well recognized that the tropomyosin-troponin thin filament-linked regulatory mechanism involves multiple steps, widely considered to be allosteric; the process is not just a simple two-step on-off switch [15, 47, 57]. A first step in the regulatory process involves Ca^{2+} binding to TnC of the troponin complex, thought to drag TnI away from its low-Ca^{2+} B-state "blocking" position on thin filament actin. This thin filament reorganization allows tropomyosin to transition on actin subunits along thin filaments to its C-state, i.e., Ca^{2+}-induced, position. However, the action of Ca^{2+} alone only partially exposes myosin-binding sites; thus, release of tropomyosin-troponin imposed steric inhibition is incomplete, though permitting weak transient myosin binding on actin. Full filament activation requires residual tropomyosin hindrance of actin-myosin binding to be relieved, thereby allowing myosin's weak to strong binding transition on actin to ensue ([48, 84]; cf. [15, 57]). Once initiated, myosin binding to actin causes further tropomyosin movement on neighboring actin subunits along thin filaments, facilitating additional myosin interaction and resulting in contraction. However, despite years of dedicated work and major advances in structural biology, a full atomic scale solution describing how troponin and its subunit TnI control the behavior of the tropomyosin coiled coil on actin and thus regulate contraction still remains incomplete. Moreover, the steps involved in tropomyosin movement from its C-state position on actin to its myosin-induced M-state have not been fully characterized.

The rapidly accelerating progress in structural biology, driven by advances in cryo-EM technology and by the power of computational chemistry, has led to what is coined the "Resolution Revolution" [10, 37]. Thus, the possibility of precisely defining dynamic protein-protein interactions of helically arranged thin filament components, specifically tropomyosin and troponin on actin, in different regulatory states and at near-atomic scale now has become possible, thus offering the prospect of a complete description of corresponding steric regulation. Most significantly, the 2020 publication of Yamada and colleagues' cryo-EM reconstructions of thin filaments captured at low and high Ca^{2+}-concentrations was a seminal moment in the long history of ever-increasing understanding of how the thin filament regulatory system governs the activity of striated muscle [83, 87]. In never before achieved detail, their ~8 Å resolution maps revealed the organization of troponin on actin that contributes to the Ca^{2+}-dependent repositioning of the tropomyosin coiled coil on the cardiac thin filament. The Yamada reconstructions supported and confirmed the long-held paradigm mentioned above that troponin-regulated tropomyosin movement on actin filaments controls myosin attachment and hence the myosin powerstroke on actin that drives muscle contraction. Computational work localizing troponin subunit-T (TnT) to actin and tropomyosin [67] quickly confirmed tenets of Yamada's work. In further detail, Rynkiewicz et al. [76] then modeled the entire tropomyosin-troponin assembly on actin based on the Yamada cryo-EM maps. Finally, complementary and even higher-resolution cryo-EM work by Risi et al. [73, 74] corroborated the Yamada study and provided an excellent foundation to begin to fully understand thin filament regulation.

This review summarizes our current perspective on the structural underpinnings of tropomyosin-troponin-based steric regulation and how we arrived at this understanding. Please note that Hitchcock-DeGregori and Barua [22] provided a detailed description of tropomyosin architecture in the previous edition of "Fibrous Proteins: Structures and Mechanisms" found in volume 82 of the Subcellular Biochemistry series.

In the Beginning There Was the Concept

The possibility of steric regulation of muscle contractility by coiled-coil tropomyosin was discussed by Hanson and Lowy at a meeting of the Royal Society in 1963. In their manuscript documenting the meeting's proceedings, Hanson and Lowy concluded that "If tropomyosin … forms some sort of complex with actin in the intact muscle, as seems very probable, then it might exert some control over the contractile function of actin, say by masking certain monomers in the filament and thereby selecting the sites for interaction with myosin" [19]. In other words, by forming an end-to-end linked cable extending continuously along actin-containing thin filaments, Hanson and Lowy proposed that coiled-coil tropomyosin is somehow forced, as we now know by Ca^{2+}-free troponin, to cover myosin-binding pockets along the filament. Then at elevated Ca^{2+} concentration, tropomyosin's grip over the myosin-binding site is weakened, thereby releasing its constraint on contractility. This possibility can be easily appreciated from the iconic model of the thin filament presented by Ebashi and colleagues (Fig. 3.1) [11, 12], who had discovered troponin and communicated their findings in 1963, closely tracking Hanson and Lowy's presentation.

As mentioned, Hugh Huxley and others [20, 27–30, 54, 66] provided the first experimental support for tropomyosin movement on thin filaments consistent with the steric blocking model. The brilliance of Huxley's initial observation and analysis cannot be emphasized enough. Studying low-angle X-ray diffraction patterns taken from muscle fibers, he noted trace differences in layer line intensities derived from muscle fibers that were monitored either in solutions containing activating Ca^{2+} or in EGTA-relaxing buffers [27–30]. An untrained eye let alone one unfamiliar with notions of thin filament regulation may have overlooked the faint Ca^{2+}-dependent differences in layer line intensities associated with muscle activation (Fig. 3.2). Instead, Huxley attributed them to helically distributed elongated tropomyosin, i.e., to densities distinct from actin but lying axially along the actin helix as illustrated in Fig. 3.1. Notably, Huxley realized that diffraction pattern differences

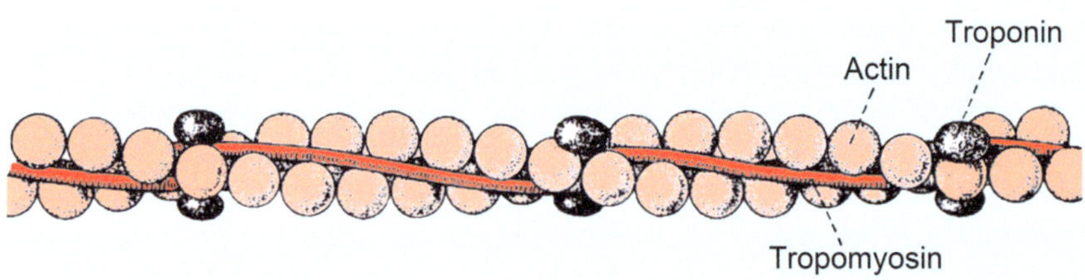

Fig. 3.1 Ebashi's model of troponin-tropomyosin on thin filaments. Note tropomyosin is shown as a strand lying continuously along successive actin subunits of the thin filament and positioned in a "groove" formed by adjacent spherically drawn actin subunits. Troponin is drawn as a nondescript but periodically localized mass on every seventh actin subunit. (Figure from Lehman [48] with permission)

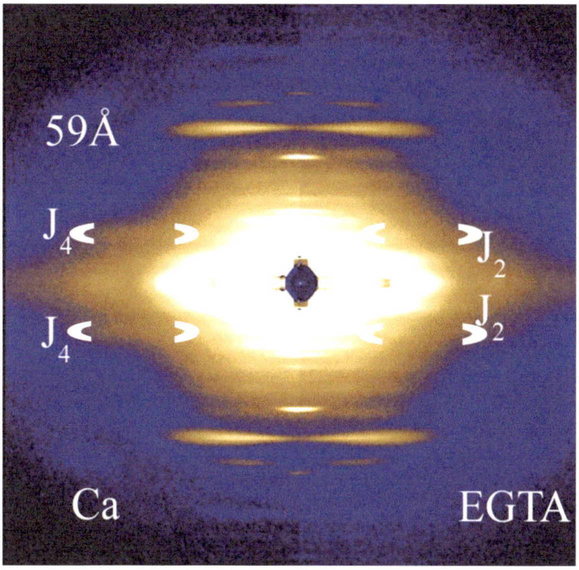

Fig. 3.2 Low-angle X-ray fiber diffraction of skinned rabbit psoas muscle fibers. The left panel quadrants show layer lines taken from Ca^{2+}-activated preparations, while the right quadrants show layer lines from relaxed preparations. Heavy central scattering dominates the pattern; the faint J4, second-order off-meridional layer line monitors changes in the azimuthal position of tropomyosin. Note the enhanced J4 intensity (between the curved white brackets) for fibers in calcium compared to corresponding intensities in low calcium determined by EGTA. The 59 Å layer line, derived from the F-actin genetic helix itself, is constant in all quadrants, as originally determined by Huxley [29]. (Figure from Poole et al. [70] with permission)

likely reflected Ca^{2+}-dependent changes in tropomyosin position along thin filaments. Parry and Squire [66] then modeled the data and, consistent with Huxley's view, concluded that Ca^{2+} activation resulted in the tropomyosin coiled-coil molecules moving from a peripheral position on actin subunits of thin filaments toward a notional "groove" found between adjacent subunits. The movement of tropomyosin chains into a "groove" would make the two-stranded actin-double helix now more obviously four-stranded, as shown in Fig. 3.3, and could account for the changes in diffraction patterns noted experimentally by Huxley.

For the following 25 years and again up to the present, the compelling steric model of muscle regulation has been repeatedly challenged. Some of the early misgivings were soon dismissed, but others have lingered only to be dispelled by the more recent cryo-EM work of Yamada and others. For example, it seemed odd at first that the radial distance between tropomyosin and the thin filament central axis remains virtually constant during Ca^{2+}-induced transitions, seemingly inconsistent with tropomyosin movement into a notional groove formed by double-stranded actin as seen in Fig. 3.3. This criticism fell away once the crystal structure of actin was solved [31], and then the structure of the assembled actin filament became evident [24]. Indeed, the surface of actin is rather flat (Fig. 3.4), and therefore, tropomyosin

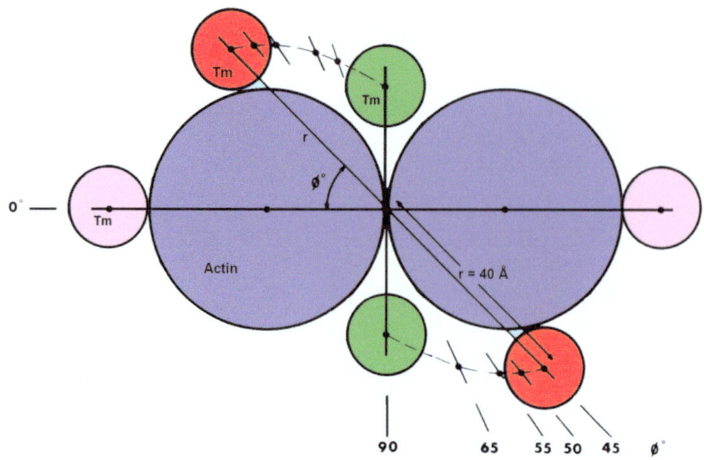

Fig. 3.3 Possible movement of tropomyosin (Tm, red, green, pink) around thin filament actin (Ac, purple), as depicted by Parry and Squire [66] to conform to the low-angle fiber diffraction data comparable to that described in Fig. 3.2. Illustrated well before the structure of the actin molecule was known. The filament is modeled with actin subunits represented as cross-sectioned spheres and tropomyosin slices through cylinders. In this representation, a fourfold filament symmetry is enhanced when tropomyosin is positioned at a 90° azimuth (green), i.e., in the groove between actin subunits, and is minimized as tropomyosin approaches a 0°, very peripheral, azimuth (pink). Red colored tropomyosin indicates its possible myosin-blocking location. The modeling predicted that the radial distance of tropomyosin (r) to the central axis of the filament increases incrementally as tropomyosin moves away from the filament groove toward the red steric inhibitory position. However, EM reconstruction has since shown that the tropomyosin radius is virtually constant during regulatory transitions, and that its regulatory movement occurs over a relatively flat surface of actin. Thus, while the fiber diffraction data in Fig. 3.2 indicates Ca^{2+}-dependent conformational shifting on the actin surface, this striking cartoon depicting possible tropomyosin movement on thin filaments is of significant historical importance since it helped to conceptualize steric regulation even though based on not yet definitive evidence. (Illustration adapted from Parry and Squire [66] with permission)

movement to block or expose the myosin-binding site actually occurs over a relatively planar interface, thus without any need to change radius during its regulatory transitions in response to Ca^{2+}-activation (Fig. 3.4). In fact, cartoons, such as those in Figs. 3.1 and 3.3, that depict actin molecules as nondescript spheres are misleading. Represented superficially as spheres, double helical actin produces a prominent groove in filament drawings, which bears little resemblance to features displayed by the now well-documented assembled filament [24]. Describing the regulatory transition of tropomyosin as "moving into the groove of actin" is outdated.

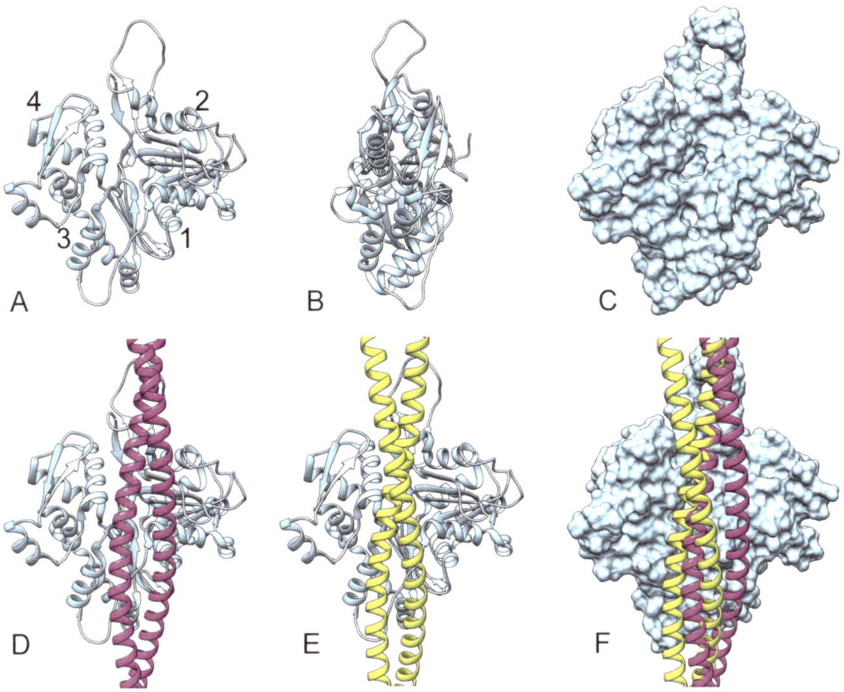

Fig. 3.4 Atomic structures of actin. (**a**) Face-on view of the broad surface of the actin molecule (ribbon representations, actin subdomains numbered). (**b**) Orthogonal view of (**a**), indicating the narrow depth of the non-spherical actin molecule. (**c**) Surface view of (**a**). (**d**) Position of B-state tropomyosin (magenta) over actin. (**e**) Position of C-state tropomyosin (yellow) over actin. (**f**) B- and C-state tropomyosin superposed over actin. Note that tropomyosin lies over the flat interface of actin. Structures based on coordinates taken from PDBs 7UTI and 7UTL

Thin Filaments Modeled as a Four-Stranded Helix

As is well known, actin molecules in actin filaments (F-actin) are assembled as a double helical array of subunits. However, as mentioned, the crystal structure of actin demonstrates that the molecule is rather flat, not spherical, and can be divided into four obvious major subdomains (Fig. 3.4). It follows that the helices in "double-helical" F-actin can be further divided into sub-strands. In essence, actin subdomains on neighboring actin subunits along F-actin stack vertically to form what can be regarded as individual helices, with actin subdomains 1 and 2 forming one helical "sub-strand" stack and actin subdomains 3 and 4 the second (Fig. 3.5a, b) [39, 69]. Hence, "doubled-helical" actin filaments can be considered four-stranded and thus F-actin a four-start helix, as readily noted in moderate resolution thin filament reconstructions [39, 40, 46, 69, 70]. This can be most easily recognized by viewing helical projections of actin filament EM densities onto a plane perpendicular to the filament axis (Fig. 3.5c, d) [40]. Consequently, Ca^{2+}-dependent

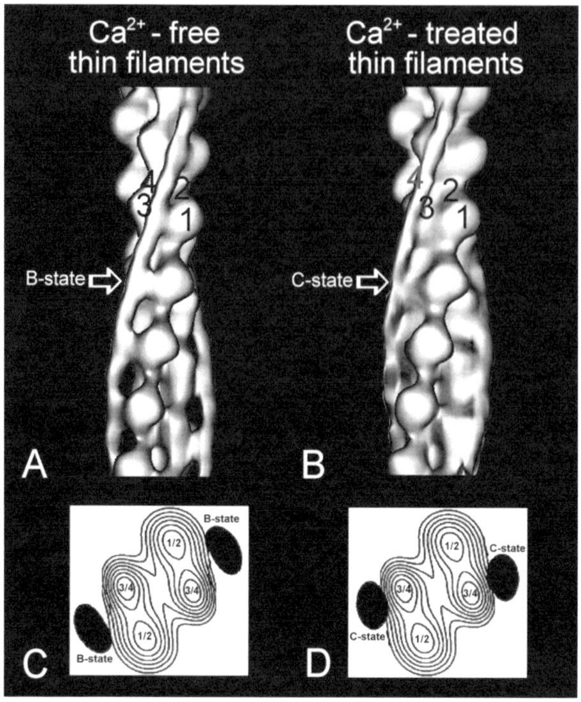

Fig. 3.5 Resolving tropomyosin strands on Ca^{2+}-free and Ca^{2+}-treated filaments. (**a**, **b**) 3D reconstructions of thin filament EM images carried out between 1994 and 2005 were the first to demonstrate Ca^{2+}-dependent tropomyosin strand movement for a variety of negatively stained thin filament types at ~20 Å resolution [39, 46, 69, 84]. Shown here are the first single particle reconstructions of cardiac thin filaments that define the double helical array of actin subunits as well as actin substructure, namely actin divided into subdomains (marked 1–4) and tropomyosin strands (arrows) moving from the Ca^{2+}-free, B-state in (**a**) to the Ca^{2+}-induced C-state positions in (**b**). (Note that the diameter of B-state tropomyosin strands is slightly wider than the diameter of C-state strands, as if extra material is present along the strand in Ca^{2+}-free conditions). (Figure from Pirani et al. [69] with permission). (**c**, **d**) Projecting thin filament densities as in (**a**) and (**b**) along their helical tracks onto a plane perpendicular to the filament axis. This process merges densities derived from actin subdomains 1 and 2 together and similarly those associated with subdomains 3 and 4 together as if originating from individual helical threads, thus locating the relative position of tropomyosin. Contour plots of such helical projections demonstrate the "four-strandedness" of double helical actin. The presence of the C-state tropomyosin density (black ellipse) close to the inner domain actin subdomains 3 and 4 contours in (**d**) reinforces the actin-based four-strandedness, easily seen in projections. In contrast, the presence of B-state density between the inner and outer domain subdomain contours in (**c**) reduces the apparent four-strandedness. (Adapted from Lehman [40] with permission)

tropomyosin repositioning on the actin surface, i.e., moving over the flat face of actin, simply alters actin's inherent four-stranded organization. In Ca^{2+}, the tropomyosin strand enhances the filament's four-strandedness by reinforcing actin's inner domain density (Fig. 3.5d), whereas in the absence of Ca^{2+}, it masks four-strandedness by moving between actin's inner and outer domain densities

(Fig. 3.5b). Thus, the thin filament EM data are compatible with Ca^{2+}-dependent changes observed in fiber diffraction patterns of intact muscle without needing to invoke groove-like filament features. Notably, conclusions arrived at by Huxley, his student Haselgrove, Vibert, and Parry and Squire [20, 29, 54, 66] invoking tropomyosin movement on the surface of the actin filament were spot on but based on unrealistic models of the actin molecule.

Layer lines derived from fiber diffraction of muscle, such as those attributed to tropomyosin in Fig. 3.2 [70], do not come labeled. Similarly, EM densities in moderate resolution reconstructions such as those in Fig. 3.5 [69] do not come named; rather their identification in each case was inferred largely from structural and biochemical data available at the time. For example, low-resolution structures attributed to tropomyosin actually might be a composite of tropomyosin and parts of the troponin complex. In fact, Squire and Morris [81] and Lehrer [88] argued some 25 years ago that segments of troponin might significantly contribute to what was considered to be tropomyosin strands on actin filaments (cf. Perry 2005; [68]). Thus, while generally discounted at the time, it remained possible, if not likely, that together tropomyosin and troponin, not just tropomyosin alone, contribute to strand density and cause steric interference of myosin binding sites on actin at low Ca^{2+}. Indeed, cryo-EM reconstructions extending resolution of thin filament tropomyosin to ~8 Å or better now indicate that much of TnI runs in tandem to the tropomyosin coiled coil on thin filaments and together may play a critical role in steric regulation (cf. Yamada et al. [87]).

Tropomyosin and Troponin Structure on the Thin Filament

Tropomyosin

Coiled-coil tropomyosin is a modular protein consisting of seven ~39 residue-long "pseudo-repeats" [4, 5, 22]. The composite shape and surface charge distribution associated with the molecule's string of pseudo-repeats are ideal for tropomyosin to form-fit to seven adjacent actin subunits along F-actin filaments [23, 49, 51]. In fact, each repeat contains specifically localized acidic residues to bind to Lys 326 and Lys 328 on successive actin subunits along the thin filament [4, 45, 53, 76]. In addition, the ends of the tropomyosin coiled coil splay to interact head-to-tail and form a 4-helix bundle, thereby polymerizing into cables that run continuously along the length of actin filament [18, 65]. Thus, not only is tropomyosin designed to form a fit to F-actin, but it is organized to build elongated cables that bind to successive actin subunits on F-actin. Despite this specificity, the electrostatic binding of tropomyosin to individual actin subunits is weak. Accordingly, even though the collective binding of the polymerized tropomyosin cable is strong, locally along the filament, tropomyosin interaction with individual actin subunits is weak. This dichotomy allows tropomyosin to change its local positioning around the actin filament in response to

Ca^{2+}, troponin, and myosin-head binding, while the elongated cable remains "tightly" bound to actin [23, 44, 62].

The periodic insertions of alanine clusters along the 40 nm-long tropomyosin coiled coil are largely responsible for tropomyosin's molecular curvature that is biased to the helical shape of the F-actin helix [4, 5, 22]. Once bound to F-actin, the high persistence length of tropomyosin (i.e., its flexural stiffness) [49–51] facilitates cooperative regulatory movement of the polymerized tropomyosin cable during its regulatory movement on thin filaments. Tropomyosin also displays high twisting stiffness, except at its C- and N-termini [41, 76]. Once the 4-helix bundle "overlapping domain" between successive coiled-coils forms and the tropomyosin cable assembles on actin, twisting across the overlapping domain is greatly dampened [41, 76]. The binding of TnT1 (the tail of TnT) to tropomyosin as well as post-translational phosphorylation of tropomyosin further enhances the stiffness of the tropomyosin overlapping-domain to approximate that of the rest of the coiled-coil [41, 42, 52, 80]. Thus, tropomyosin has properties of a mechanically continuous unjointed cable adapted to the F-actin surface [50, 51, 80].

While the preshaped design of the tropomyosin molecule epitomizes the phrase "form follows function," this cannot be said for the intrinsically disordered regulatory domains of TnI that modulate tropomyosin position on the thin filament.

Troponin I

TnI, like tropomyosin, has a segmented structure (Table 3.1), but unlike tropomyosin, the modules in TnI are not periodic; in fact, each component domain differs in structure, its interactions, and function [56]. Moreover, unlike tropomyosin, TnI is not simply an actin-binding protein, and its modules are not conserved across tissues, species, and phyla [55, 56, 78]. For example, the heart muscle TnI isoform contains a unique N-terminal domain that is partly responsible for cardiac lusitropy and is responsive to β-adrenergic stimulated PKA phosphorylation [6, 78, 79, 86]. The TnI midpiece contributes to the troponin core domain structure, where, together with TnT, it forms an "IT arm" scaffold to support TnC [83, 87]. The cardiac C-terminal regulatory domain of TnI (cTermTnI) (residues 137–210) of greatest interest here is

Table 3.1 Human cardiac TnI nomenclature

Residues	Region
2–32	N-terminal domain
42–136	Core domain IT arm
137–148	Inhibitory peptide
149–163	Switch peptide
164–185	Helix H_4
186–192	C-terminal linker
193–210	C-terminal tail

Adapted from Marston and Zamora [56]

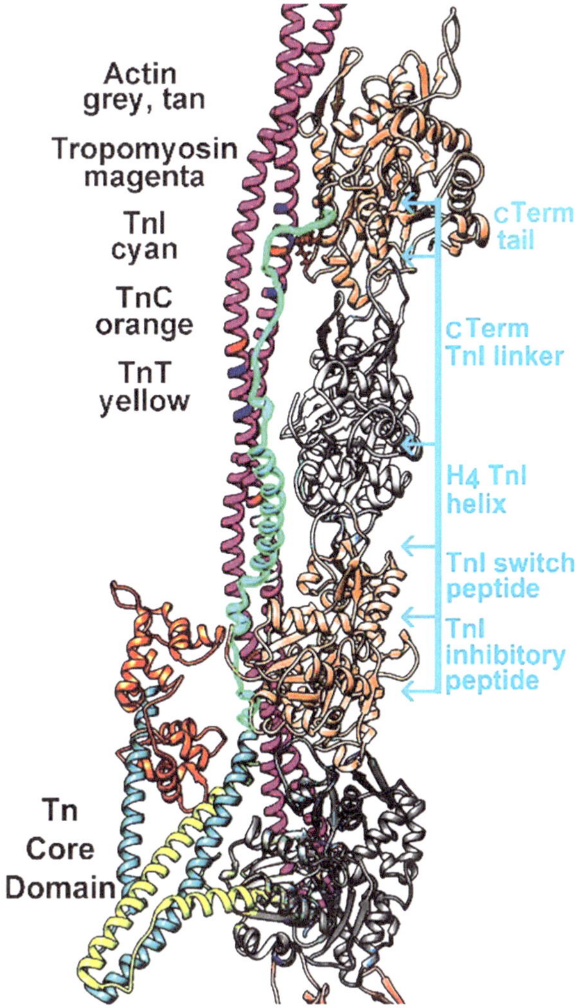

Fig. 3.6 Structure of TnI bound to thin filaments at low Ca^{2+}. Note contacts between the C-terminal domain of TnI (brackets) and actin and tropomyosin. (Model adapted from Lehman et al. [43] with permission)

unstructured in solution, but when associated with thin filaments, assumes metastable conformations [26, 60, 61, 73, 87]. At low Ca^{2+}, it runs as a single chain over a three-actin stretch of the thin filament and is located roughly parallel to tropomyosin (Fig. 3.6). At high Ca^{2+}, the "switch peptide" module of cTermTnI (residues 149–163) is drawn away actin-tropomyosin and associates with the N-lobe of TnC, while the rest of the cTermTnI dissociates from the thin filament [87]. Indeed, the molecular dynamics of cTermTnI built from the Yamada maps shows considerable oscillation of the cTermTnI peptide even in its low Ca^{2+} actin-tropomyosin bound form. Such variance occurs mostly in a linker region (residues 186–192) found between the "H_4 helix" (residues 164–185) and the very C-terminal tail of TnI (residues 193–210) [43]. TnI linker presumably facilitates binding to tropomyosin, which itself oscillates between B- and C-state positions even at low Ca^{2+} [69, 73]. Thus, the trajectory of cTermTnI detected by cryo-EM likely reflects an average of several binding modes, each loosely interacting with actin-tropomyosin. This view is consistent with the long-held notion that thin filament regulatory mechanisms involve multiple weak-binding interactions, permissive to low-energy barriers between states and thus ensuring that regulatory transitions occur at low-energy cost [23, 40]. This behavior makes cTermTnI an ideal vehicle to snag tropomyosin and bias the coiled coil to a low-Ca^{2+} position on actin.

It follows that any structure, such as cTermTnI, running roughly parallel to tropomyosin in resting muscle (as seen in Fig. 3.6), no doubt will affect layer line intensities observed in fiber diffraction patterns, which previously may have been interpreted as derived solely from helically arranged tropomyosin strands. By the same token, at high-Ca^{2+} with much of the cTermTnI domain released from actin and tropomyosin, the contribution of TnI to the tropomyosin layer line signal likely diminishes. Thus, modeling of fiber diffraction patterns deserves reevaluation.

Troponin-Induced Ca^{2+}-Dependent Tropomyosin Pivoting Over Fixed Sites on F-Actin

As mentioned, the cryo-EM of Yamada and colleagues describing the organization of troponin on actin-tropomyosin of cardiac thin filaments provided a significant advance in understanding thin filament regulation [87]. However, significant errors in the alignment of tropomyosin to actin and troponin in the Yamada EM reconstruction produced uninterpretable residue-to-residue contacts. Left uncorrected, i.e., without accurate modeling, this meant that interpreting the molecular interactions of troponin and tropomyosin with actin may have been inaccurate.

To correct this uncertainty, we systematically aligned the modular, pseudo-repeating tropomyosin to actin, TnT, and TnI domains in the Yamada cryo-EM map using molecular dynamics flexible fitting routines of properly vetted tropomyosin [76]. These corrections made possible near atomic analysis of thin filaments and more accurate structure/function relationships. Our atomic models of cardiac thin filaments in both low Ca^{2+} (B-state) and high Ca^{2+} (C-state) (PDBs 7UTL and 7UTI) fit to the Yamada and Risi cryo-EM maps [72–74, 87] with high precision [45, 76] (Fig. 3.7).

Following molecular fitting of tropomyosin to the Yamada reconstruction, our attention was drawn to the observation that intermolecular residue-to-residue salt-bridge formation between actin and tropomyosin is virtually the same for B- and C-state thin filament configurations [45]. How then is it possible for tropomyosin to reposition, let alone slide or roll over actin between B- and C-states, as variously proposed? Using molecular docking protocols and molecular dynamics simulations, it became obvious that tropomyosin pivots about relatively fixed points on actin (residues Lys 326 and Lys 328), and that this pivoting accounts for the B- to C-state

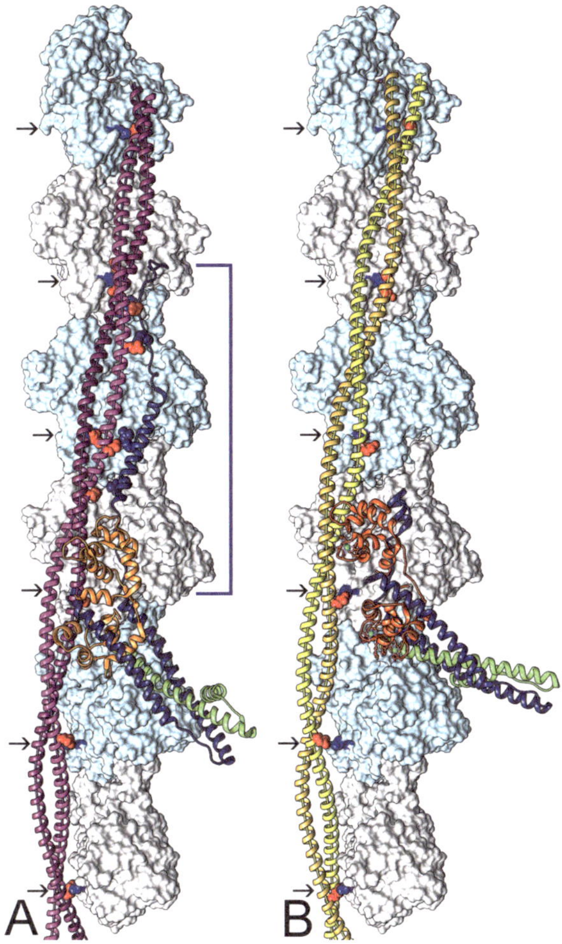

Fig. 3.7 Atomic models of thin filaments. (**a**) Low Ca^{2+} (from PDB: 7UTL); (**b**) High Ca^{2+} (from PDB: 7UTI). Note that the same tropomyosin pseudo-repeat residues in (**a**) and in (**b**) contact and pivot over actin residues 326 and 328 (arrows) at both low and high Ca^{2+} (red/blue spheres). Tpm—magenta/gold; TnT—green; TnC—orange; TnI dark blue; blue bracket indicates C-terminal TnI domain linked to actin-Tpm at low Ca^{2+}. (Adapted from Lehman et al. (2023) with permission)

change in the average center-of-mass of tropomyosin on actin. It followed that at low-Ca^{2+} concentration, C-terminal domains of TnI attract the tropomyosin coiled coil, which bends toward TnI to block myosin-binding (Fig. 3.8), while tropomyosin still maintains contact with its underlying actin-binding sites. Measurement of interaction energetics suggests that muscle activation and ensuing C-terminal TnI detachment from actin-tropomyosin involves tropomyosin pivoting back to an energetically favorable C-state orientation. Thus, Ca^{2+}-dependent pivoting of tropomyosin changes the average regulatory position of tropomyosin on the actin filament without any appreciable tropomyosin sliding or rolling on actin [45] as had been previously presumed [2, 25, 49–51, 58, 71, 85].

In principle, cTermTnI-induced toggling of tropomyosin between B- and C-state positions is straightforward. Indeed, molecular dynamics simulations showed that C-state tropomyosin spontaneously pivots toward cTermTnI (Fig. 3.8), likely driven by multiple weak TnI-tropomyosin interactions involving what may be the

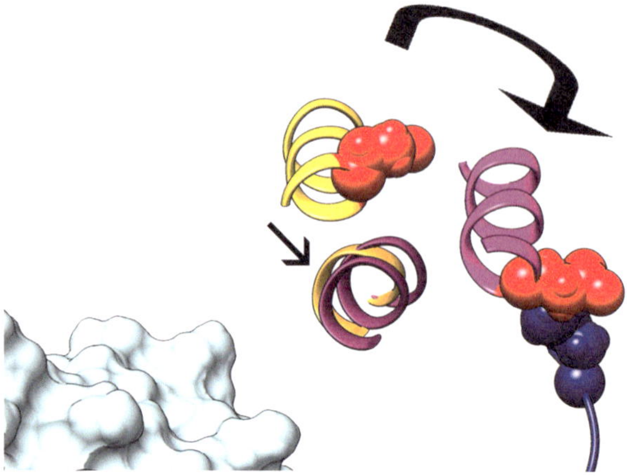

Fig. 3.8 Superposed cross-sections showing the two-chained tropomyosin coiled coil pivoting on actin (pale-blue) between C-state (yellow) and B-state (magenta) during MD. Note that the outer chains of coiled-coil tropomyosin pivot (large, curved arrow) about stationary inner chains (short arrow). Here, tropomyosin residue Glu 114 (red) is attracted to cTermTnI side chains of TnI Arg 192 (blue). (Adapted from Lehman et al. (2023) with permission)

composite effect of low-Ca^{2+} "Velcro-binding" by the cTerm H_4-helix, the C-terminal linker, and the C-terminal tail all linked to actin-tropomyosin ([43]; Lehman et al. 2023) (see Fig. 3.6). The reverse occurs at high-Ca^{2+}. Here, the TnI "switch-peptide" at the origin of cTermTnI domain binds to an opened Ca^{2+}-saturated TnC N-lobe, leading to an unzipping of cTermTnI from actin-tropomyosin [56], while tropomyosin pivots back to the C-state position [45, 87]. The TnI-dependent pivoting motion of tropomyosin over virtually fixed sites on actin represents movement of coiled-coil tropomyosin, although not the sliding or rolling motion on actin originally envisioned.

The concept of TnI-based Velcro-binding is consistent with the work of the Murphy laboratory on cardiac stunning (i.e., the cardiac response to ischemia followed by rapid perfusion) [63]. Here, the cleavage of the terminal 17 TnI residues during stunning does not prevent thin filament Ca^{2+} regulatory on-off switching in vitro or in vivo despite slowing relaxation kinetics. This observation suggests that if a part of "Velcro-bound" cTermTnI on actin-tropomyosin is detached or removed, then the structural route to relaxation can be compensated by secondary attachments [13, 64]. Indeed, tropomyosin still translates between thin filament B- and C-states despite the 17-residue deficit [14]. The premise that cTermTnI binding to actin-tropomyosin is multimodal and opportunistic is supported by recent work by Risi and Galkin, who sorted and categorized cryo-EM images of cardiac thin filaments, yielding several separate filament maps each showing well-defined distinct but related cTermTnI structures ([73]; cf. Sevrieva et al. [77]).

“Nothing Happens Until Something Moves” (Attributed to Albert Einstein)

The notion that tropomyosin moves across actin on Ca^{2+} activation is so engrained in our conceptual landscape that it is hard to shake. Indeed, tropomyosin does not move substantially over the surface of actin and away from actin residues 326 and 328 in response to Ca^{2+}-activation-linked cTermTnI release from actin [45]. Instead, as mentioned, the coiled coil pivots over a limited region of the thin filament. In contrast, tropomyosin only moves substantially in the regulatory process when myosin binds actin. Thus, muscle contraction is not regulated by a simple, reversible on-off switching mechanism as often depicted. In fact, the regulatory mechanism involves multiple steps [16, 17, 47, 48, 57, 84].

In summary, the first step in the regulatory process involves Ca^{2+} binding to TnC of the troponin complex, dragging TnI away from its low-Ca^{2+} blocked (B)-state position on actin. This thin filament reorganization allows tropomyosin to pivot on actin to its calcium (C)-state position, but it only partially exposes myosin-binding sites; thus, the release of TnI-tropomyosin-imposed steric inhibition is incomplete. Full filament activation requires neutralizing residual tropomyosin hindrance to actin-myosin binding. Here, myosin's weak-to-strong binding transition on actin causes further myosin-induced (M)-state tropomyosin movement, full filament activation, and then contraction. The semi-rigid property of tropomyosin ensures that local C→M state displacement of tropomyosin on actin will propagate along actin filaments by opening neighboring myosin binding sites.

Trust Movement (Attributed to Alfred Adler)

The structure of myosin motor-head (S1), once attached to the actin-tropomyosin filament, is revealed in great detail by cryo-EM reconstructions of myosin-decorated thin filaments. Captured in nucleotide-free (rigor) or ADP-bound myosin configurations [2, 7–9, 72, 74], these structures represent the last two steps of the myosin-powerstroke, with the myosin head remaining bound on thin filaments (Fig. 3.9, see PDBs: 8EFI-rigor; 8EFH-ADP) [9]. While the lever arm tail of the myosin as well as the myosin nucleotide-binding pocket show different conformations under these two conditions, in either case the contact interface between myosin and actin-tropomyosin is virtually indistinguishable at the 3.3–3.4 Å global resolution attained (Fig. 3.9).

The surface of the myosin-head displays several loop-like extensions involved in actin-binding. Notably, Loop-4 at the very tip of myosin 4 (white arrowhead in Fig. 3.9) approaches tropomyosin and is positioned to impact C- to M-state transitions [7, 8]. Here the charged side chain of myosin Arg 369 extends outward from the end of Loop-4 toward arginine residues on tropomyosin lying along the thin filament (Fig. 3.10). Thus, Arg 369 is likely to induce the movement of tropomyosin

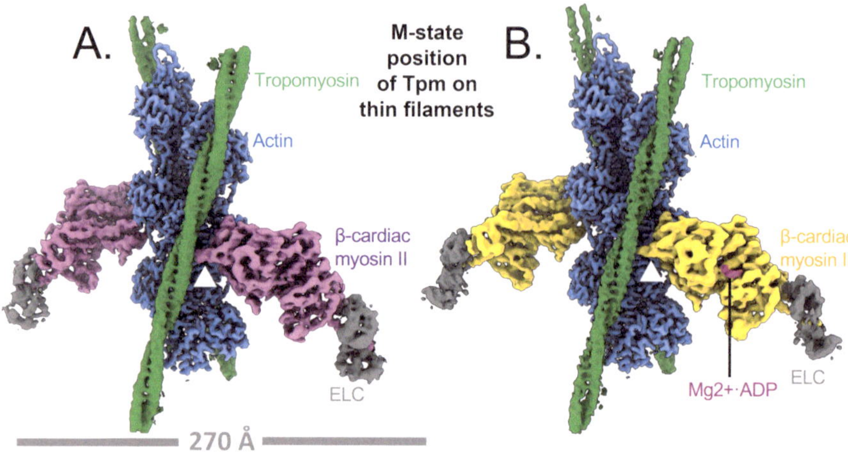

Fig. 3.9 Cryo-EM reconstructions of M-state tropomyosin on cardiac thin filaments following myosin-S1 binding. (**A**) S1—nucleotide-free ("rigor") post-powerstroke binding; (**B**) S1—ADP post-powerstroke binding. White arrowhead—Loop-4. (Figure from Doran et al. [9] with permission)

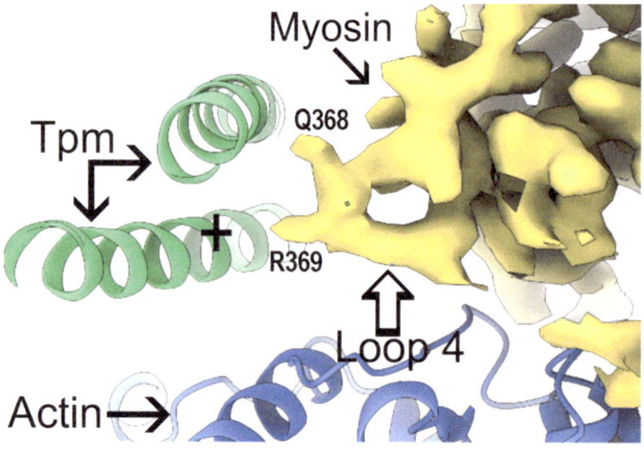

Fig. 3.10 Enlargement of M-state cryo-EM map showing myosin Loop-4 residue Arg 369 extending toward tropomyosin residue Arg 133 (+) $_{+}$on one tropomyosin chain, while the other chain is braced by Loop-4 Gln 368. (Figure adapted from Doran et al. [9] with permission)

across actin to the M-state by charge repulsion [7, 8, 75]. Similarly positioned Loop-4 Glu 370, lying near acidic residues on tropomyosin, may contribute to the tropomyosin transition toward the M-state. Not surprisingly, Arg 369 and Glu 370 are conserved among muscle myosins [7] and, when mutated, M-state myosin movement is incomplete [8].

Accounting for Stepwise Tropomyosin Translocation on Actin During Myosin-Binding

As mentioned, cryo-EM reconstructions of S1-decorated cardiac thin filaments suggest that Loop-4, at the tip of the myosin-S1 head, forces tropomyosin across actin filaments [7–9] while displacing tropomyosin from default binding interactions on actin residues Lys 326 and Lys 328 (Fig. 3.11). However instructive, cryo-EM only captures static structures of B- and C-state filaments [76, 87] (PDBs 7UTL, 7UTI), i.e., prior to myosin-binding, and at the end of the myosin-induced tropomyosin translocation, i.e., in the M-state [9] (PDBs 8EFI, 8EFH). By contrast, weak-binding transient intermediates between these states during the pre- to post-powerstroke myosin transition are not readily captured by cryo-EM [33, 75]. Thus, tropomyosin's movement during the C- to M-state transition is uncertain, and most inferences made about the myosin-induced C-state to the M-state tropomyosin translocation on actin are speculative [1, 21, 32]. Fortunately, computational chemistry has begun to fill the void left by current experimental work. Here, targeted MD has been used to steer cardiac myosin S1 in silico from pre- to post-powerstroke conformation on actin, while assessing the impact on tropomyosin position. Tropomyosin responds to the myosin transition by repositioning stepwise on thin filament actin toward its M-state location. Again, charge-repulsion between polar residues on myosin Loop-4 and on tropomyosin appears to facilitate the tropomyosin movement at each step, thereby completing the translocation of the coiled coil across the actin surface [75] (Fig. 3.12).

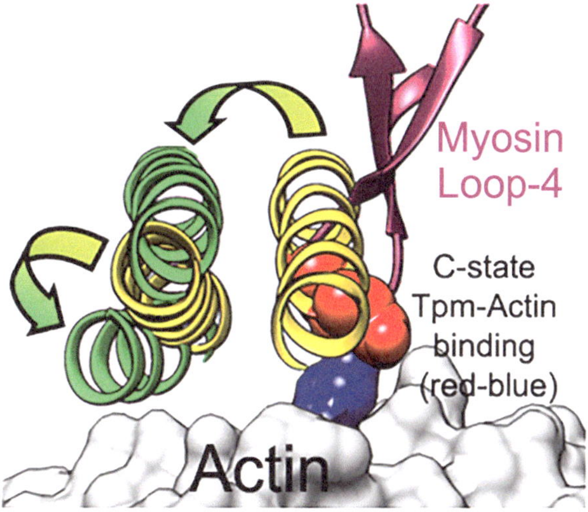

Fig. 3.11 Myosin Loop-4 causes displacement of tropomyosin from its C-state position (yellow) to the M-state (green). Note that myosin binding to actin disrupts electrostatic interactions between acidic residues on tropomyosin and lysine residues 326 and 328 on actin. (Figure from Doran et al. [8] with permission)

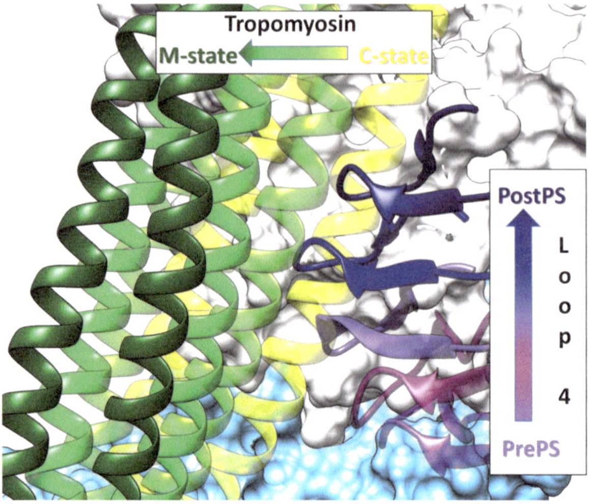

Fig. 3.12 Steered molecular dynamics targeting the pre-powerstroke myosin head to its post-powerstroke conformation on actin. During myosin's pre- to post-powerstroke transition on actin, Loop-4 sweeps upward toward the pointed end of actin. This myosin transition is depicted in the cartoon as if the Loop-4 myosin S1 extension is moving in 5 successive steps (colored in lavender to blue tones) over actin subunits [75]. Tropomyosin responds to the myosin S1 transition by translocating from its C-state (yellow coiled coil) to its M-state position (dark green) across the actin filament in corresponding steps (colored in yellow to dark green tones) [75]. (Cover illustration from J. Gen. Physiology 2024:156(5) with permission)

Perspective

The technological advances offered by the combination of cryo-EM and computational analysis have provided a robust picture of troponin-tropomyosin-linked thin filament regulation. Nevertheless, cryo-EM based on static, literally frozen, structures—even when supplemented by cutting-edge computation—cannot readily offer a complete molecular movie of real-time molecular dynamics occurring in vivo. In fact, X-ray fiber diffraction remains the only experimental tool currently available to assess the movement of thin filament components in working muscle. This was recognized in formative work by Kress et al. [36] and Poole et al. [70]. These and later studies by Koubassova and Bershitsky relied on low-resolution EM models to interpret the origins and dynamics of layer line changes on muscle activation [3, 35]. More recently, Koubassova et al. [34] have supplemented this approach by relying on cryo-EM-based atomic models of troponin-free thin filaments. We encourage continued modeling of muscle fiber diffraction based on ever-increasing resolution cryo-EM thin filament structure, particularly given that time-resolved fiber diffraction technology has advanced in parallel with the advent of time-resolved cryo-EM. If successful, the structural underpinning of thin filament steric regulation will have come full circle from the Hanson, Lowy, Huxley, Haselgrove, Parry, and Squire work to the present.

Acknowledgments This chapter summarizes material presented by Dr. Lehman as a "Distinguished Lecturer" at the University of Washington on April 25, 2024. Dr. Lehman's research work on thin filament regulatory mechanisms is supported by NIH grant R01-HL036153.

References

1. Baldo AP, Tardiff JC, Schwartz SD (2021) A proposed mechanism for the initial myosin binding event on the cardiac thin filament: a metadynamics study. J Phys Chem Lett 12(14): 3509–3513
2. Behrmann E, Müller M, Penczek PA, Mannherz HG, Manstein DJ, Raunser S (2012) Structure of the rigor actin-tropomyosin-myosin complex. Cell 150(2):327–338
3. Bershitsky SY, Koubassova NA, Ferenczi MA, Kopylova GV, Narayanan T, Tsaturyan AK (2017) The closed state of the thin filament is not occupied in fully activated skeletal muscle. Biophys J 112(7):1455–1461
4. Brown JH, Cohen C (2005) Regulation of muscle contraction by tropomyosin and troponin: how structure illuminates function. Adv Protein Chem 71:121–159
5. Brown JH, Kim KH, Jun G, Greenfield NJ, Dominguez R, Volkmann N, Hitchcock-DeGregori SE, Cohen C (2001) Deciphering the design of the tropomyosin molecule. Proc Natl Acad Sci USA 98(15):8496–8501
6. Cole HA, Perry SV (1975) The phosphorylation of troponin I from cardiac muscle. Biochem J 149(3):525–533
7. Doran MH, Pavadai E, Rynkiewicz MJ, Walklate J, Bullitt E, Moore JR, Regnier M, Geeves MA, Lehman W (2020) Cryo-EM and molecular docking shows myosin loop 4 contacts actin and tropomyosin on thin filaments. Biophys J 119(4):821–830
8. Doran MH, Rynkiewicz MJ, Pavadai E, Bodt SML, Rasicci D, Moore JR, Yengo CM, Bullitt E, Lehman W (2023a) Myosin loop-4 is critical for optimal tropomyosin repositioning on actin during muscle activation and relaxation. J Gen Physiol 155(2):e202213274
9. Doran MH, Rynkiewicz MJ, Rasicci D, Bodt SML, Barry ME, Bullitt E, Yengo CM, Moore JR, Lehman W (2023b) Conformational changes linked to ADP release from human cardiac myosin bound to actin-tropomyosin. J Gen Physiol 155(3):e202213267
10. Dror RO, Dirks RM, Grossman JP, Xu H, Shaw DE (2012) Biomolecular simulation: a computational microscope for molecular biology. Annu Rev Biophys 41:429–452
11. Ebashi S (1963) Third component participating in the superprecipitation of "natural actomyosin". Nature 200:1010
12. Ebashi S, Endo M, Ohtsuki I (1969) Control of muscle contraction. Q Rev Biophys 2(4): 351–384
13. Foster DB, Noguchi T, VanBuren P, Murphy AM, Van Eyk JE (2003) C-terminal truncation of cardiac troponin I causes divergent effects on ATPase and force: implications for the pathophysiology of myocardial stunning. Circ Res 93(10):917–924
14. Galińska A, Hatch V, Craig R, Murphy AM, Van Eyk JE, Wang CL, Lehman W, Foster DB (2010) The C terminus of cardiac troponin I stabilizes the Ca2+-activated state of tropomyosin on actin filaments. Circ Res 106(4):705–711
15. Geeves MA (2012) Thin filament regulation. In: Egelman EH, Goldman YE, Ostap EM (eds) Comprehensive biophysics, vol 4. Academic Press, Oxford, pp 251–267
16. Gordon AM, Homsher E, Regnier M (2000) Regulation of contraction in striated muscle. Physiol Rev 80(2):853–924
17. Gordon AM, Regnier M, Homsher E (2001) Skeletal and cardiac muscle contractile activation: tropomyosin "rocks and rolls". News Physiol Sci 16:49–55
18. Greenfield NJ, Huang YJ, Swapna GV, Bhattacharya A, Rapp B, Singh A, Montelione GT, Hitchcock-DeGregori SE (2006) Solution NMR structure of the junction between tropomyosin molecules: implications for actin binding and regulation. J Mol Biol 364(1):80–96

19. Hanson J, Lowy J (1964) The structure and periodicity of actin filaments and the origin of the axial periodicity in the I-substance of vertebrate striated muscle. Proc R Soc Lond B Biol Sci 160:449–460
20. Haselgrove JC (1972) X-ray evidence for a conformational change in actin-containing filaments of vertebrate striated muscle. Cold Spring Harb Symp Quant Biol 37:341–352
21. Hei B, Tardiff JC, Schwartz SD (2024) Human cardiac β-myosin powerstroke energetics: thin filament, Pi displacement, and mutation effects. Biophys J:S0006-3495(24)00451-X
22. Hitchcock-DeGregori SE, Barua B (2017) Chapter 9: Tropomyosin structure, function, and interactions: a dynamic regulator. In: Parry DAD, Squire JM (eds) Subcellular biochemistry vol 82: fibrous proteins: structures and mechanisms. Springer-Nature London, pp 253–284
23. Holmes KC, Lehman W (2008) Gestalt-binding of tropomyosin to actin filaments. J Muscle Res Cell Motil 29(6–8):213–219
24. Holmes KC, Popp D, Gebhard W, Kabsch W (1990) Atomic model of the actin filament. Nature 347(6288):44–49
25. Holthauzen LM, Corrêa F, Farah CS (2004) Ca2+-induced rolling of tropomyosin in muscle thin filaments: the alpha- and beta-band hypothesis revisited. J Biol Chem 279(15): 15204–15213
26. Hornos F, Feng HG, Rizzuti B, Palomino-Schätzlein M, Wierczorek D, Neira JL, Jin JP (2021) The muscle-relaxing C-terminal peptide from troponin I populates a nascent helix, facilitating binding to tropomyosin with a potent therapeutic effect. J Biol Chem 296:100228
27. Huxley HE (1971a) Cross-bridge movement and filament overlap. Biophys J 11:235a
28. Huxley HE (1971b) Structural changes during muscle contraction. Biochem J 125:85P
29. Huxley HE (1972) Structural changes in actin- and myosin-containing filaments during contraction. Cold Spring Harb Symp Quant Biol 37:361–376
30. Huxley HE (1970) Structural changes in muscle and muscle proteins during contraction. In: Gregory JG (ed) Proceedings of the 8th International Congress of Biochemistry, vol 4. Interlaken, p 23
31. Kabsch W, Mannherz HG, Suck D, Pai EF, Holmes KC (1990) Atomic structure of the actin: DNase I complex. Nature 347(6288):37–44
32. Kiani FA, Lehman W, Fischer S, Rynkiewicz MJ (2019) Spontaneous transitions of actin-bound tropomyosin toward blocked and closed states. J Gen Physiol 151(1):4–8
33. Klebl DP, McMillan SN, Risi C, Forgacs E, Virok B, Atherton JL, et al (2025) Swinging lever mechanism of myosin directly shown by time-resolved cryo-EM. Nature 642:519-526
34. Koubassova NA, Bershitsky SY, Ferenczi MA, Narayanan T, Tsaturyan AK (2017) Tropomyosin movement is described by a quantitative high-resolution model of X-ray diffraction of contracting muscle. Eur Biophys J 46(4):335–342
35. Koubassova NA, Bershitsky SY, Ferenczi MA, Tsaturyan AK (2008) Direct modeling of X-ray diffraction pattern from contracting skeletal muscle. Biophys J 95(6):2880–2894
36. Kress M, Huxley HE, Faruqi AR, Hendrix J (1986) Structural changes during activation of frog muscle studied by time-resolved X-ray diffraction. J Mol Biol 188(3):325–342
37. Kühlbrandt W (2014) The resolution revolution. Science 343(6178):1443–1444
38. Lehman SJ, Crocini C, Leinwand LA (2022) Targeting the sarcomere in inherited cardiomyopathies. Nat Rev Cardiol 19(6):353–363
39. Lehman W, Craig R, Vibert P (1994) Ca^{2+}-induced tropomyosin movement in limulus thin filaments revealed by three-dimensional reconstruction. Nature 368(6466):65–67
40. Lehman W, Hatch V, Korman V, Rosol M, Thomas L, Maytum R et al (2000) Tropomyosin and actin isoforms modulate the localization of tropomyosin strands on actin filaments. J Mol Biol 302(3):593–606
41. Lehman W, Li XE, Orzechowski M, Fischer S (2014) The structural dynamics of α-tropomyosin on F-actin shape the overlap complex between adjacent tropomyosin molecules. Arch Biochem Biophys 552–553:68–73
42. Lehman W, Medlock G, Li XE, Suphamungmee W, Tu AY, Schmidtmann A, Ujfalusi Z, Fischer S, Moore JR, Geeves MA, Regnier M (2015) Phosphorylation of Ser283 enhances the stiffness of the tropomyosin head-to-tail overlap domain. Arch Biochem Biophys 571:10–15

43. Lehman W, Pavadai E, Rynkiewicz MJ (2021) C-terminal troponin-I residues trap tropomyosin in the muscle thin filament blocked-state. Biochem Biophys Res Commun 551:27–32
44. Lehman W, Rynkiewicz MJ, Moore JR (2020) A new twist on tropomyosin binding to actin filaments: perspectives on thin filament function, assembly and biomechanics. J Muscle Res Cell Motil 41(1):23–38
45. Lehman W, Rynkiewicz MJ (2023) Troponin-I-induced tropomyosin pivoting defines thin-filament function in relaxed and active muscle. J Gen Physiol 155(7):e202313387
46. Lehman W, Vibert P, Uman P, Craig R (1995) Steric-blocking by tropomyosin visualized in relaxed vertebrate muscle thin filaments. J Mol Biol 251(2):191–196
47. Lehman W (2017) Switching muscles on and off in steps: the McKillop-Geeves three-state model of muscle regulation. Biophys J 112(12):2459–2466
48. Lehman W (2016) Thin filament structure and the steric blocking model. Compr Physiol 6(2): 1043–1069
49. Li XE, Holmes KC, Lehman W, Jung H, Fischer S (2010a) The shape and flexibility of tropomyosin coiled coils: implications for actin filament assembly and regulation. J Mol Biol 395(2):327–339
50. Li XE, Lehman W, Fischer S, Holmes KC (2010c) Curvature variation along the tropomyosin molecule. J Struct Biol 170(2):307–312
51. Li XE, Lehman W, Fischer S (2010b) The relationship between curvature, flexibility and persistence length in the tropomyosin coiled-coil. J Struct Biol 170(2):313–318
52. Li XE, Orzechowski M, Lehman W, Fischer S (2014) Structure and flexibility of the tropomyosin overlap junction. Biochem Biophys Res Commun 446(1):304–308
53. Li XE, Tobacman LS, Mun JY, Craig R, Fischer S, Lehman W (2011) Tropomyosin position on F-actin revealed by EM reconstruction and computational chemistry. Biophys J 100(4): 1005–1013
54. Lowy J, Vibert PJ (1972) Studies on low-angle x-ray pattern of a molluscan smooth muscle during tonic contraction and rigor. Cold Spring Harb Symp Quant Biol 37:353–360
55. Manning EP, Tardiff JC, Schwartz SD (2011) A model of calcium activation of the cardiac thin filament. Biochemistry 50(34):7405–7413
56. Marston S, Zamora JE (2020) Troponin structure and function: a view of recent progress. J Muscle Res Cell Motil 41(1):71–89
57. McKillop DF, Geeves MA (1993) Regulation of the interaction between actin and myosin subfragment 1: evidence for three states of the thin filament. Biophys J 65(2):693–701
58. McLachlan AD, Stewart M (1976) The 14-fold periodicity in alpha-tropomyosin and the interaction with actin. J Mol Biol 103(2):271–298
59. Memo M, Marston S (2013) Skeletal muscle myopathy mutations at the actin tropomyosin interface that cause gain- or loss-of-function. J Muscle Res Cell Motil 34(3–4):165–169
60. Metskas LA, Rhoades E (2015) Conformation and dynamics of the troponin I C-terminal domain: combining single-molecule and computational approaches for a disordered protein region. J Am Chem Soc 137(37):11962–11969
61. Metskas LA, Rhoades E (2016) Order-disorder transitions in the cardiac troponin complex. J Mol Biol 428(15):2965–2977
62. Moore JR, Campbell SG, Lehman W (2016) Structural determinants of muscle thin filament cooperativity. Arch Biochem Biophys 594:8–17
63. Murphy AM (2006) Heart failure, myocardial stunning, and troponin: a key regulator of the cardiac myofilament. Congest Heart Fail 12(1):32–38; quiz 39-40
64. Narolska NA, Piroddi N, Belus A, Boontje NM, Scellini B, Deppermann S, Zaremba R, Musters RJ, dos Remedios C, Jaquet K, Foster DB, Murphy AM, van Eyk JE, Tesi C, Poggesi C, van der Velden J, Stienen GJ (2006) Impaired diastolic function after exchange of endogenous troponin I with C-terminal truncated troponin I in human cardiac muscle. Circ Res 99(9):1012–1020
65. Orzechowski M, Li XE, Fischer S, Lehman W (2014) An atomic model of the tropomyosin cable on F-actin. Biophys J 107(3):694–699
66. Parry DA, Squire JM (1973) Structural role of tropomyosin in muscle regulation: analysis of the x-ray diffraction patterns from relaxed and contracting muscles. J Mol Biol 75(1):33–55

67. Pavadai E, Lehman W, Rynkiewicz MJ (2020) Protein-protein docking reveals dynamic interactions of tropomyosin on actin filaments. Biophys J 119(1):75–86
68. Perry SV (2003) What is the role of tropomyosin in the regulation of muscle contraction? J Muscle Res Cell Motil 24(8):593–596
69. Pirani A, Xu C, Hatch V, Craig R, Tobacman LS, Lehman W (2005) Single particle analysis of relaxed and activated muscle thin filaments. J Mol Biol 346(3):761–772
70. Poole KJ, Lorenz M, Evans G, Rosenbaum G, Pirani A, Craig R, Tobacman LS, Lehman W, Holmes KC (2006) A comparison of muscle thin filament models obtained from electron microscopy reconstructions and low-angle X-ray fibre diagrams from non-overlap muscle. J Struct Biol 155(2):273–284
71. Risi C, Eisner J, Belknap B, Heeley DH, White HD, Schröder GF, Galkin VE (2017) Ca(2+)-induced movement of tropomyosin on native cardiac thin filaments revealed by cryoelectron microscopy. Proc Natl Acad Sci USA 114(26):6782–6787
72. Risi C, Schäfer LU, Belknap B, Pepper I, White HD, Schröder GF, Galkin VE (2021a) High-resolution cryo-EM structure of the cardiac actomyosin complex. Structure 29(1):50–60.e4
73. Risi CM, Belknap B, Atherton J, Coscarella IL, White HD, Bryant Chase P, Pinto JR, Galkin VE (2024) Troponin structural dynamics in the native cardiac thin filament revealed by cryo electron microscopy. J Mol Biol 436(6):168498
74. Risi CM, Pepper I, Belknap B, Landim-Vieira M, White HD, Dryden K, Pinto JR, Chase PB, Galkin VE (2021b) The structure of the native cardiac thin filament at systolic Ca^{2+} levels. Proc Natl Acad Sci USA 118(13):e2024288118
75. Rynkiewicz MJ, Childers MC, Karpicheva OE, Regnier M, Geeves MA, Lehman W (2024) Myosin's powerstroke transitions define atomic scale movement of cardiac thin filament tropomyosin. J Gen Physiol 156(5):e202413538
76. Rynkiewicz MJ, Pavadai E, Lehman W (2022) Modeling human cardiac thin filament structures. Front Physiol 13:932333
77. Sevrieva I, Kampourakis T, Irving M (2024) Structural changes in troponin during activation of skeletal and heart muscle determined in situ by polarized fluorescence. Biophys Rev 16 (6):753-772
78. Solaro RJ, Kobayashi T (2011) Protein phosphorylation and signal transduction in cardiac thin filaments. J Biol Chem 286(12):9935–9940
79. Solaro RJ, Moir AJ, Perry SV (1976) Phosphorylation of troponin I and the inotropic effect of adrenaline in the perfused rabbit heart. Nature 262(5569):615–617
80. Sousa D, Cammarato A, Jang K, Graceffa P, Tobacman LS, Li XE, Lehman W (2010) Electron microscopy and persistence length analysis of semi-rigid smooth muscle tropomyosin strands. Biophys J 99(3):862–868
81. Squire JM, Morris EP (1998) A new look at thin filament regulation in vertebrate skeletal muscle. FASEB J 12(10):761–771
82. Tardiff JC (2011) Thin filament mutations: developing an integrative approach to a complex disorder. Circ Res 108(6):765–782
83. Tobacman LS (2021) Troponin revealed: uncovering the structure of the thin filament on-off switch in striated muscle. Biophys J 120(1):1–9
84. Vibert P, Craig R, Lehman W (1997) Steric-model for activation of muscle thin filaments. J Mol Biol 266(1):8–14
85. von der Ecken J, Müller M, Lehman W, Manstein DJ, Penczek PA, Raunser S (2015) Structure of the F-actin-tropomyosin complex. Nature 519(7541):114–117
86. Warren CM, Halas M, Feng HZ, Wolska BM, Jin JP, Solaro RJ (2021) NH_2-terminal cleavage of cardiac troponin I signals adaptive response to cardiac stressors. J Cell Signal 2(3):162–171
87. Yamada Y, Namba K, Fujii T (2020) Cardiac muscle thin filament structures reveal calcium regulatory mechanism. Nat Commun 11(1):153
88. Zhou X, Morris EP, Lehrer SS (2000) Binding of troponin I and the troponin I-troponin C complex to actin-tropomyosin. Dissociation by myosin subfragment 1. Biochemistry 39(5): 1128–1132

Chapter 4
Structural Biology of Actin: Interplay Between Molecular Assembly, Conformational Polymorphism and ATPase

Yuichiro Maéda, Toshiro Oda, Akihiro Narita, Yusuke Kanematsu, Mitsusada Iwasa, and Shuichi Takeda

Abstract Actin was first identified as a major muscle protein with two key activities: polymerization and ATPase. Although the significance of these activities in muscle contraction was unclear, subsequent cell biological studies revealed that actin is abundant also in non-muscle cells, where it drives dynamic remodeling through ATP-dependent treadmilling. Biochemical studies revealed that the polymerization coupled with ATP hydrolysis produces two distinct F-actin states: a stable ADP-P_i state and an unstable ADP state. The transition between the ADP-P_i and ADP states (i.e., P_i release) likely generates the free energy that drives treadmilling.

In this chapter, we introduce our structural biological studies of these states and the mechanisms underlying their formation. Our F-actin model showed that polymerization induces a rotation between its two rigid domains, shifting actin from a twisted G-form to a flat F-form. We also resolved the cryo-EM structure of cofilin-decorated F-actin, in which actin adopts a distinct C-form. Analysis of PDB data classified actin structures into four conformations: G-, F-, C-, and O-forms, each linked to specific functions—G-form for nucleotide exchange, F-form for ATP hydrolysis, and C-form for filament severing. High-resolution F-form structures further elucidated the ATP hydrolysis pathway and the basis for the stability of the ADP-P_i state.

Y. Maéda (✉) · S. Takeda
Graduate School of Informatics, Nagoya University, Nagoya, Japan
e-mail: ymaeda@cc.nagoya-u.ac.jp; takeda.shuichi.d0@f.mail.nagoya-u.ac.jp

T. Oda
Faculty of Health and Welfare, Tokai Gakuin University, Gifu, Japan
e-mail: toda375@tokaigakuin-u.ac.jp

A. Narita · M. Iwasa
Graduate School of Science, Nagoya University, Nagoya, Japan
e-mail: narita.akihiro.x8@f.mail.nagoya-u.ac.jp; iwasa.mitsusada.k1@a.mail.nagoya-u.ac.jp

Y. Kanematsu
Graduate School of Advanced Science and Engineering, Hiroshima University, Hiroshima, Japan
e-mail: ykanem@hiroshima-u.ac.jp

D. A. D. Parry, P. K. Luther (eds.), *Fibrous Proteins: Structures and Mechanisms*, Subcellular Biochemistry 113, https://doi.org/10.1007/978-3-032-05273-5_4

Despite these advances, key questions remain. Although the global structure of F-form actin is identical across nucleotide states, its properties differ: ATP/ADP-P_i states are stable and cofilin-resistant, whereas the ADP state is prone to depolymerization and cofilin-mediated severing. We suggest that each state should be characterized by the distinct nature of conformational fluctuations from F-form back to G-form.

Keywords Actin treadmilling · ATP hydrolysis · Conformational transition · Domain motion · Fiber diffraction · Crystallography · Cryo-EM · P_i release · QM/MM

Abbreviations

ABP	Actin-binding protein
B-end	(Fast-growing) Barbed end
C-form	Cofilin-bound actin conformation
F1A	Fragmin F1 domain-actin complex
F-form	F-actin conformation
G-form	G-actin conformation
GCF-space	A two-dimensional conformational space of actin defined by GC and GF axes
ID	Inner domain of actin
MD	Motor domain (of myosin)
OD	Outer domain of actin
O-form	Open (profilin-bound) actin conformation
P_i	Inorganic phosphate
P-end	(Slow-growing) Pointed end
QM/MM	Quantum mechanics/molecular mechanics calculations
SD1–SD4	Subdomain 1–4 of actin

Introduction

Actin is one of the most abundant proteins in our body [74]. Remarkably, it accounts for nearly 50% of the total ATP consumption in vertebrate cells [4, 11]. These facts, together with recent progress across various life sciences, including medical research, developmental biology, and neuroscience, highlight key roles of actin in a wide range of cellular functions, primarily organelle transport, maintenance of cellular architecture, and cell motility.

Many cellular roles of actin are mediated by the treadmilling mechanism, a molecular motion driven by ATPase activity; continuous cycles of polymerization and depolymerization generate unidirectional movement of actin filaments, producing force and motion. Treadmilling is an elegant integration of polymerization-depolymerization (molecular assembly-disassembly) and ATP hydrolysis (energy conversion).

Over the past two decades, structural biology has contributed to revealing that conformational polymorphism plays a significant role in actin treadmilling. A central task has been to establish a one-to-one correspondence between actin states and elementary processes. While such direct correspondence has been identified for some states, the F-form actin (the conformation of filament-forming actin) encompasses three nucleotide states (ATP-bound, ADP-P_i-bound, and ADP-bound), all sharing an identical global structure. This indicates that static global conformation alone is insufficient to define distinct functional states. We propose that ADP-P_i-F-form and ADP-F-form differ in their dynamic properties, with the ADP-F-form being significantly more prone to depolymerization than ATP- or ADP-P_i-F-form.

In this chapter, we discuss how we have elucidated F-form actin structures and how our concepts have evolved regarding the interplay between molecular assembly, conformational polymorphism, and energy transformation. On the other hand, the processes leading from the stable ADP-P_i-F-form actin to the unstable ADP-F-form actin, and further to disassembly into ADP-G-form actin, remain unclear. To characterize these species, it is essential to understand the nature of their conformational fluctuations.

Actin functions in concert with a large number of actin-binding proteins (ABPs), each specialized to provide local regulatory control. Therefore, in order to understand how actin works in specific cell types, detailed knowledge of ABP–actin interactions is essential. However, this chapter focuses primarily on actin free of accessory proteins with the exception of cofilin, an abundant ABP that plays a fundamental role in actin treadmilling.

We note that this chapter is not intended to provide a comprehensive review of actin studies, but rather to describe how our concepts have evolved based on our own work.

A Short History of Actin Study Before the Structural Elucidation of G-Actin at an Atomic Level

There are two major biological activities of actin, discovered shortly after the discovery of actin itself (Table 4.1). Actin was first isolated by Straub as a major muscle protein from a high salt extract of minced rabbit skeletal muscle [70]. The following year, Straub discovered that actin exists in two forms: at low salt concentrations, actin is in a monomeric form (G-actin), and when salt concentrations are raised, actin polymerizes into a filamentous form (F-actin) [71]. In 1950, Straub and Feuer further found that monomeric actin contains ATP in a bound form, which is hydrolyzed to ADP and inorganic phosphate (P_i) in association with polymerization [72].

Table 4.1 A short history of actin study until 1990

Year	Major progress	Reference
1942	Actin was isolated from rabbit skeletal muscle	[70]
1943	G-actin was shown to polymerize into F-actin	[71]
1950	Actin was found to have ATPase activity	[72]
1960s–1970s	Actin was identified and isolated from non-muscle cells	[1, 21]
1975	*Thermodynamics of the Polymerization of Protein* was published	[54]
1976	Actin treadmilling molecular motion was discovered	[80]
1980s–1990s	The treadmilling mechanism was characterized through biochemical studies.	[57, 61]
1990	Crystal structure of monomeric (G-form) actin was obtained	[30]

Despite the long-known polymerizability and ATPase activities of actin, the physiological significance of these activities remained unclear, as neither appeared to play major roles in muscle contraction. A significant twist occurred in the 1960s when actin was found to be a major protein in non-muscle cells (e.g., [1, 21]). Importantly, in 1976, Wegner found in vitro that actin performs the "treadmilling" molecular movement, which translocates the actin filament unidirectionally and generates force [80]. Later, it was confirmed that treadmilling occurs at the protruding cytoplasm in the lamellipodium of fibroblasts [78].

A fundamental understanding of protein function requires atomic-level structural information. In 1990, Holmes and colleagues reported the crystal structure of G-actin in complex with DNase I [30] (Fig. 4.1). The actin molecule consists of two domains—inner (ID) and outer (OD) domains—each comprising two subdomains: SD1/SD2 in OD and SD3/SD4 in ID. ATP binds at the cleft between ID and OD, at the center of the molecule.

How Actin Harnesses ATPase Activity for Treadmilling

Biochemical studies of actin treadmilling conducted in the 1980s and 1990s identified the step that is coupled to its energy source, namely, the transition from the stable ADP-P_i-bound F-actin to the less stable ADP-bound F-actin (i.e., the P_i release step). This transition is associated with a release of free energy, which plausibly represents the coupling of ATP hydrolysis and treadmilling [57, 61]. In this section, we explain how this conclusion was reached.

A cycle of treadmilling-derived molecular motion consists of the following states and steps:

1. ATP-G → ATP-F: A monomeric ATP-actin polymerizes to form the F-actin filament. This process does not require energy from ATP hydrolysis and is thought to be driven by thermal fluctuations resulting from molecular collisions.

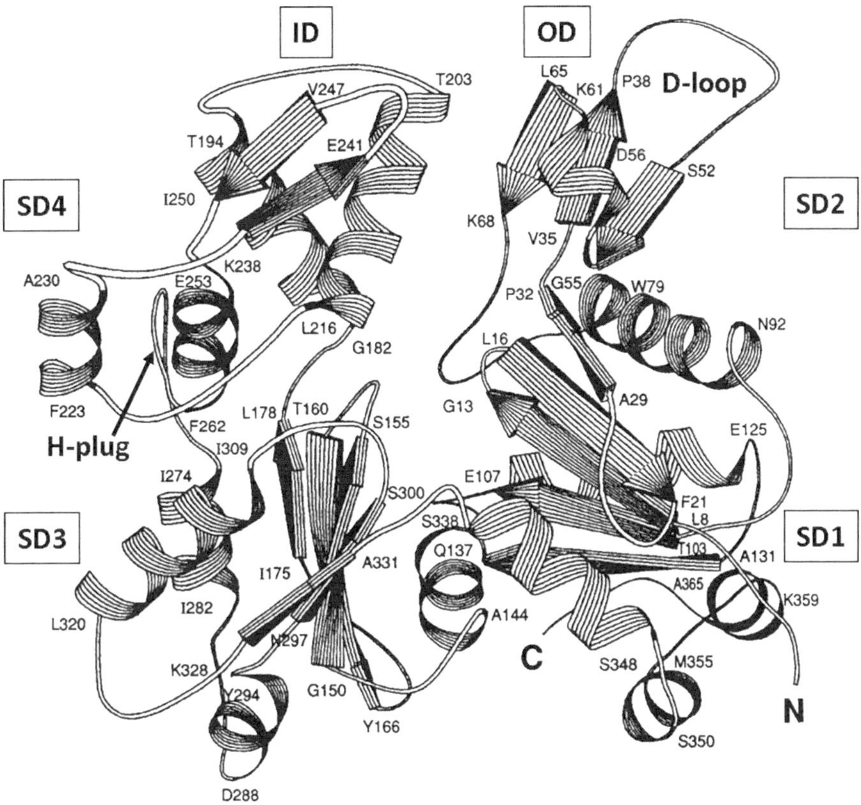

Fig. 4.1 Ribbon diagram of the atomic-level structure of monomeric actin (PDB ID: 1ATN; [30]). The molecule consists of two major domains, OD and ID, each further subdivided into two subdomains: OD into SD1 and SD2, and ID into SD3 and SD4, as indicated by the boxed labels. (This figure is reproduced with permission from Kabsch et al. [30])

2. ATP-F → ADP-P_i-F: ATP hydrolysis begins after polymerization (hydrolysis in G-actin is negligible) and proceeds at a rate of 0.3 s^{-1} [5], which is significantly slow compared to other ATPases (e.g., myosin hydrolyzes ATP at 20–200 s^{-1} [15]). The reaction is irreversible [7].
3. ADP-P_i-F → ADP-F: The release of P_i after hydrolysis is extremely slow (0.003 s^{-1}) from the interior subunits of the filament [6]. This reaction is reversible and occurs more rapidly at filament ends (>2 s^{-1}) [12].
4. ADP-F → ADP-G: Upon P_i release, F-actin becomes unstable, dissociates from the filament, and reverts to G-actin.
5. ADP-G → ATP-G: Nucleotide exchange from ADP to ATP occurs only in G-actin and does not take place in F-actin.

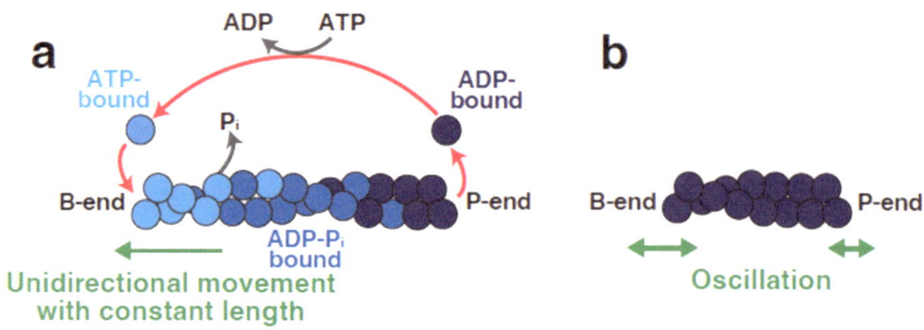

Fig. 4.2 Actin treadmilling. (**a**) Treadmilling occurs in the steady state in the presence of ATP and monomeric actin. (**b**) In the presence of ADP (without ATP) and monomeric actin, the positions of the two ends oscillate

Actin treadmilling F-actin is structurally and functionally polar: polymerization predominantly occurs at the barbed end (B-end), while depolymerization proceeds at the pointed end (P-end). At steady state, the polymerization of ATP-actin at the B-end is balanced by the depolymerization of ADP-actin at the P-end, resulting in an apparent movement of the filament without a net change in length—a process known as treadmilling [80] (Fig. 4.2a). This ATP hydrolysis-driven, directional filament movement generates forces that contribute to various cellular processes. Treadmilling also establishes a polar distribution of nucleotide states along the filament: ATP- or ADP-P_i-bound actin accumulates near the B-end, while ADP-bound actin predominates toward the P-end, reflecting the time elapsed since polymerization (Fig. 4.2a). Aged ADP—F-actin is more susceptible to spontaneous depolymerization and is preferentially severed by cofilin, thereby accelerating filament turnover.

Two sets of asymmetrical properties driving treadmilling Actin treadmilling is driven by a combination of two asymmetrical features: (1) kinetic differences between the two filament ends, and (2) distinct properties of actin depending on their bound nucleotides.

Dynamic B-end, less dynamic P-end Both polymerization and depolymerization occur more rapidly at the B-end than the P-end. After an ATP-actin monomer is incorporated at the filament end, ATP hydrolysis and subsequent P_i release follow. At the B-end, the next ATP-actin monomer is typically added before P_i is released from the preceding actin. In contrast, at the P-end, P_i is usually released before the next monomer is incorporated. These kinetic differences arise from the distinct steric environments at the two filament ends: the energy barriers associated with monomer addition may differ between the B- and P-ends.

Stable γ-phosphate-bound F-actin and unstable ADP-bound F-actin ATP-bound and ADP-P_i-bound F-actin are structurally and functionally very similar (see section “High-resolution crystal structures of F-form actin and the mechanism of actin ATP hydrolysis”). We collectively refer to these two species as γ-phosphate-bound F-actin. At B-end, the elongation rate is high for two reasons. First, due to the dynamic nature of the B-end, it is predominantly composed of γ-phosphate-bound F-actin (Fig. 4.2a). Second, γ-phosphate-bound F-form actin has a lower critical concentration for polymerization (C_c) than ADP-F-form actin. The rate of elongation is proportional to the difference $C_m - C_c$, where C_m is the monomer concentration. When $C_m > C_c$, elongation occurs; when $C_m < C_c$, shortening occurs. Since C_c is the inverse of the dissociation constant (K_d), γ-phosphate-bound F-actin is more stable in the filament form, whereas ADP-bound F-actin is unstable and more prone to depolymerization. Two additional conditions are required for actin treadmilling. First, ATP hydrolysis must occur within the filament, with negligible exchange of bound nucleotide. Second, P_i release must be sufficiently slow.

The P_i release step supplies energy for treadmilling In the presence of ADP (without ATP) and monomeric actin, ADP-bound G-actin can polymerize; however, the resulting actin filaments do not exhibit directional movement (Fig. 4.2b) [80]. In contrast to the ADP-containing system, polymerization in the presence of ATP leads to the formation of ADP-P_i-bound F-actin, followed by a transition to ADP-bound F-actin, the P_i release step. This transition, from the stable ADP-P_i-bound F-actin to the unstable ADP-bound F-actin, must be accompanied by a decrease in free energy. In conclusion, the free energy decrease associated with the P_i release step serves as the energy source for treadmilling.

Atomic Models for the Actin Filament Structure

In 2009, an atomic model of the actin filament was obtained, revealing that the shape of the actin molecule becomes flatter upon incorporation into the filament [51]. This also indicated, for the first time, that the actin molecule is better represented as comprising two connected rigid domains. The model elucidated details of inter-subunit contacts within the filament.

Previous works on actin filament modeling Determining the filament structure at atomic resolution was challenging, since actin filaments cannot be crystallized. As a first attempt to circumvent this problem, Holmes and his colleagues constructed a structural model of the actin filament by rigid-body fitting of actin monomers to an X-ray fiber diffraction diagram at 8.4 Å resolution [26], published back-to-back with the monomeric actin structure [30].

Their approach involved three key steps. First, in a glass capillary, highly concentrated and oriented actin filament sols were prepared. Second, from these oriented sols, X-ray fiber diffraction intensities were recorded. Third, the atomic model of monomeric actin was adjusted in silico until the calculated diffraction pattern matched the experimental data.

They successfully determined the orientation of actin subunits within the filament and greatly advanced actin research. However, during the modeling process, actin molecules were treated as four rigid subdomains that, while connected, were individually adjusted in their spatial arrangement. This assumption was based on their monomeric actin crystal structure, which appeared to show the molecule as composed of four distinct, rigid subdomains [30]. However, the outcome of this approach was difficult to interpret: accurate positioning of the subdomains within each subunit was complicated by their intricate connectivity, and inter-subunit interactions within the filament tended not to converge on a unique solution, as the subunits are not covalently linked. Later, the group refined their atomic model using techniques such as a directed mutation algorithm they had originally developed [36] and normal mode analysis based on the dihedral angles (φ, ψ) of the main chain [75]. These atomic models were widely accepted in the actin research community until about 2009.

Alternative actin filament model We reported a newly refined atomic model of the actin filament in 2009 [51]. The model was obtained using a similar approach to that of Holmes' lab, but with several technical modifications: we improved the quality of highly oriented actin filament sols and enhanced the resolution of the X-ray fiber diffraction intensities collected at SPring-8 beamlines, achieving data down to 3.3 Å resolution.

Here, we do not discuss methods for sample preparation or X-ray intensity acquisition, as cryo-EM has become the preferred approach for determining actin filament structures at atomic resolutions. Instead, we focus on the process of filament model building, as our efforts in this step have shaped our understanding of actin architecture: namely, that the actin molecule is better represented by two connected rigid domains, OD (SD1 + SD2) and ID (SD3 + SD4), rather than four individually connected rigid subdomains, which we explain below.

For the model-building step, we took a fundamentally different approach. We did not assume either a four-rigid-subdomain architecture or a two-rigid-domain model. Instead, we explored the conformational landscape stepwise by repeatedly applying normal mode analysis to calculate the lowest-frequency vibration modes of the actin molecule in the filament [17], combined with molecular dynamic simulations to optimize local contacts to optimize local inter-subunit contacts. This approach allowed for a broad and efficient search of the conformational space of the actin molecule. To validate the resulting model, we used an R-factor calculated from non-fitted areas, analogous to the free R-value used in crystallography.

The atomic model revealed that G-actin, which has a twisted conformation, becomes flatter upon incorporation into the filament. When the ID of F-actin is superimposed on that of G-actin (PDB ID: 1J6Z), the twisting angle between OD and ID is about 20° in G-actin, whereas it is nearly zero in F-actin, as calculated using the program DynDom [24]. This flattened conformation was later termed F-form, while the twisted conformation of G-actin was referred to as G-form [52]. This conformational transition supports the concept of actin architecture comprising two connected rigid domains (Fig. 4.4a–d).

Flattening of the actin molecule is essential for filament formation. As a result of the domain rotation, residues 61–65 of SD2 (OD) approach residues 200–208 and 241–247 of SD4 (ID), forming a ridge that encircles a hole, resembling a socket, into

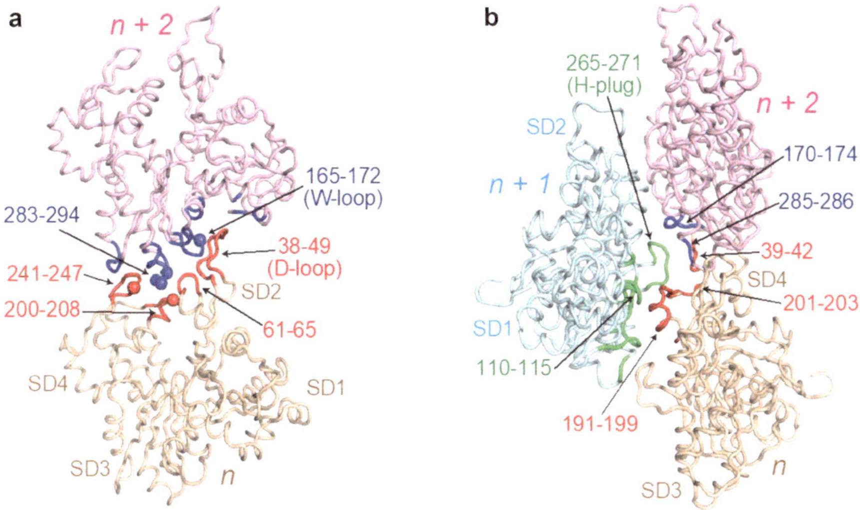

Fig. 4.3 Subunit contacts in the F-actin model (PDB ID: 2ZWH; [51]). Contact residues (within 10 Å of any other Cα atom) are highlighted. (**a**) Intra-strand interface. Residues critical for polymerization, Pro243 and Ala204 in (n), and Asp286-Asp288 (n + 2), as well as Tyr169 on the W-loop (n + 2), which interacts with the D-loop (n), are shown as spheres. See the text for details. (**b**) Inter-strand interface. This interface is primarily mediated by the H-plug (n + 1)

which residues 283–294 of SD3 (ID) from the upper subunit insert like a ball (Figs. 4.3a and 4.4f). The longitudinal (intra-strand) contacts are extensive and crucial for filament formation; mutations such as A204E/P243K [67] and D286A/V287A/D288A [83] result in non-polymerizable actin (see Fig. 4.3a for the mutated residues). This ball-and-socket-like interaction is further stabilized by a substantial interface between the D-loop (residues 39–48) of the lower subunit and the W-loop (residues 167–172) of the upper subunit, which becomes possible due to the flattening of the lower subunit. A later cryo-EM structure of F-actin revealed that the D-loop wraps around the side chain of Tyr169 on the W-loop [76].

Flattening also contributes to the diagonal (inter-strand) contacts, as these contact sites are distributed across both OD and ID; however, these interactions appear to be less extensive than intra-strand contacts. A prominent structure involved in inter-strand contacts is the H-plug (residues 265–271) of ID in actin (n + 1). The H-plug mediates a three-body interaction involving residues 39–42 and 201–203 of actin (n), and residue 286 of actin (n + 2), contributing to both the inter- and intra-strand stabilization (Figs. 4.3b and 4.4f).

Our 2009 F-actin model is highly consistent with the cryo-EM-based model [76] in terms of subunit conformation and inter-strand interactions. However, the side-chain structures were unclear, and water molecules were not included. As described in later sections, these details have been clarified by subsequent cryo-EM F-actin and F-form actin crystal structures (section "High-resolution crystal structures of F-form actin and the mechanism of actin ATP hydrolysis").

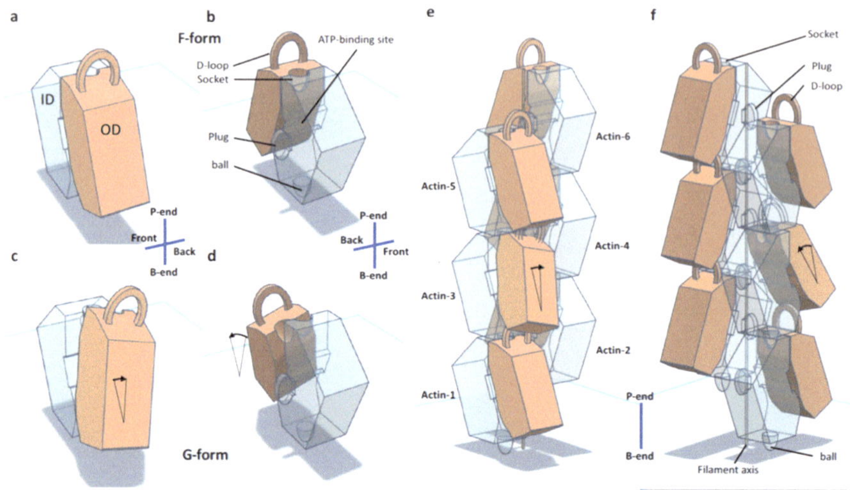

Fig. 4.4 Building block model of the actin filament. (**a**–**d**) Monomeric actin models. OD is shown as a solid beige body, and ID as a transparent light blue body. Panels a and b represent the filamentous F-form, while c and d represent the globular G-form. Only the plug, ball-and-socket interface, and D-loop, key regions for subunit contacts, are selectively presented. In F-form (**a**, **b**), the OD and ID are aligned in a flat conformation with no twist. In G-form (**c**, **d**), the OD is rotated by ~15° relative to the ID, compared to F-form. Note that in panel d, the socket is incomplete due to the twisted orientation of OD. (**e**–**f**) Six-subunit model of an actin filament. Only actin subunit 3 is shown in G-form; the others are in the F-form. This represents a hypothetical scenario in which a subunit within the filament spontaneously transitions toward G-form. For simplicity, the two strands are shown as parallel linear arrays rather than true helices, modeled here as a 2/1 helical symmetry rather than the actual 28/13 or 13/6 symmetry. Each subunit's ID contributes to inter- and intra-strand interactions around the filament axis, forming the filament core. In contrast, OD is positioned away from the axis, experiencing fewer constraints and allowing greater rotational freedom. However, this rotation is restricted when the D-loop contacts the SD3 of the upper subunit. Conversely, rotation of OD relative to ID can disrupt the D-loop–ID interaction, causing the D-loop to become disordered instead of folded (the D-loop of actin 3 is shown as transparent). All models were constructed using Tinkercad

Cofilin Binding and Conformational Fluctuations of Actin

Depolymerization of the actin filament is slow and constitutes a major rate-limiting factor in actin treadmilling. The actin-depolymerizing factor (ADF)/cofilin (hereafter cofilin) promotes actin filament disassembly and accelerates actin turnover [81]. Cofilin binds to the side of F-actin and exhibits a preference for ADP-F-actin over ADP-P_i-F-actin or ATP-F-actin [8, 62]. Cofilin binding is cooperative [40] and results in the formation of cofilin-binding clusters along the filament, reaching saturation at a 1:1 cofilin:actin molar ratio [14]. Upon binding to F-actin, cofilin induces filament severing [37, 39], which predominantly occurs at the pointed end side of the cofilin cluster [3]. We determined the 3.8 Å resolution cryo-EM structure of the actin filament decorated by cofilin (cofilactin) [73]. The actin subunit in cofilactin is distinct from the F-form and G-form, and is designated as the "C-form." Structural comparisons among the three forms have identified two rigid bodies. Subsequent studies have demonstrated that cofilin binding is largely dependent on various types of fluctuations of the actin filament.

Rigid bodies The conformational transition of actin subunits within cofilactin filaments from G-form is characterized by a 6.0° rotation of OD relative to ID around an axis, the GC axis. This axis is orthogonal to the GF axis, around which OD rotates during the G-to-F conformational transition (see Fig. 4.8). As a result, actin in cofilactin adopts a distinct structural state, termed the C-form [73] (see section "What is polymorphism? Why is it important?"). Notably, a major portion of ID and a significant region of SD1 remain unchanged by this transition and are thus defined as rigid bodies (ID and OD rigid bodies, respectively). The three actin forms—F, G, and C—can be distinguished by the relative orientation of these rigid bodies (Fig. 4.5a).

Cofilin/actin interfaces Cofilin binds two consecutive actin subunits along the same long-pitch helical strand [14] (Fig. 4.6a). If we define the B-end side subunit as the B-subunit and the P-end side subunit as the P-subunit, then cofilin's binding interface on the B-subunit (termed "F-site") is located on the OD rigid body (Fig. 4.6a). As a result, cofilin can bind actin subunits in any conformational state

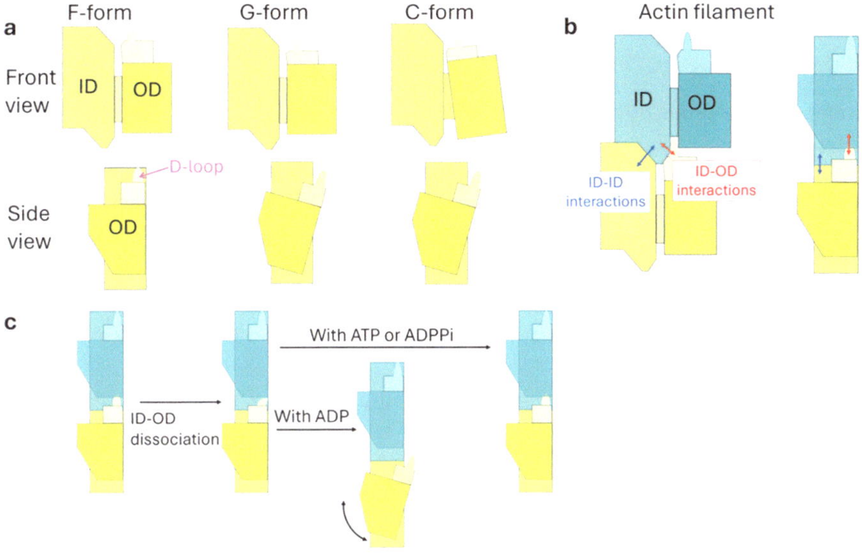

Fig. 4.5 (**a**) Schematic illustrations of the three actin forms. (**b**) Interactions between two subunits in one strand of the actin filament. (**c**) Model of actin filament destabilization with ADP. The F-form is stabilized by ATP or ADP-P_i. Even when the ID-OD interaction transiently dissociates, the subunit can maintain F-form, enabling the ID-OD interaction to readily re-form. However, with ADP, upon dissociation of ID-OD interactions, the actin subunit tends to adopt G-form and undergoes fluctuations

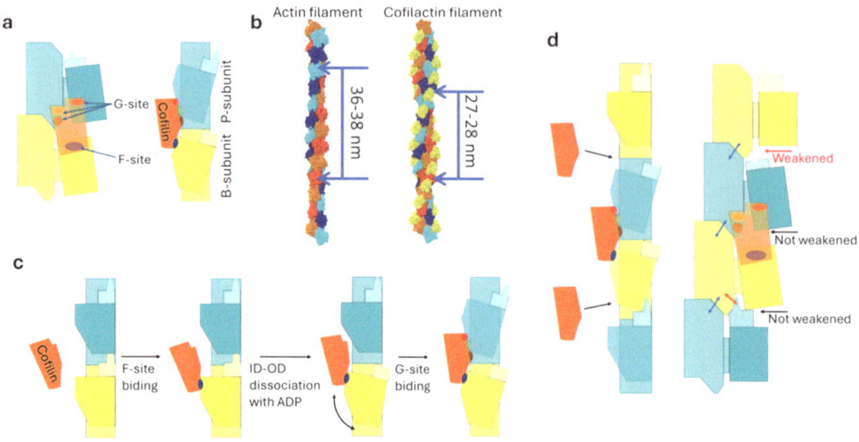

Fig. 4.6 (**a**) Schematic illustrations of actin-cofilin interfaces. (**b**) Cofilin binding alters the helical twist of the actin filament. (**c**) Schematic illustrations of the cofilin binding model. (**d**) Actin filament severing and cofilin binding cluster formation. Right panel: Cofilin binding weakens subunit-subunit interactions on the P-end side but not on the B-end side. Left panel: Additional cofilin binds more easily around previously bound cofilin because one of the two required subunits for binding is already in the C-form

via the F-site. However, the binding interface on the P-subunit (termed "G-site") spans both the ID and OD rigid bodies, requiring the actin subunit to adopt C-form for stable interaction (Fig. 4.6a).

Actin filament binding and severing Cofilin binding induces substantial structural changes in actin filaments, disrupting key inter-subunit interactions. The potential energy barrier associated with these structural transitions is likely to be significant, necessitating a multi-step binding process to minimize energetic costs. Initially, cofilin is proposed to bind to the B-subunit in F-form via the F-site (Fig. 4.6c). Cofilin is kept close to the G-site on the P-subunit, increasing the likelihood of inducing the structural transition of the P-subunit to C-form upon full binding. The probability of this occurring is much higher if the B-subunit is in ADP-bound form, rather than in ATP- or ADP-P_i-bound form. This is because the absence of the gamma phosphate weakens OD–ID interactions, increasing the probability of spontaneous transition to G-form (Fig. 4.5b, c). This transition is accompanied by an increase in filament twist (from $\Delta\phi \approx 27°$ to $\Delta\phi = 35.8°$) [40] (Fig. 4.6b). Following cofilin binding, subunit-subunit interactions weaken on the P-end side of the P-subunit, leading to filament severing (Fig. 4.6d). Cofilin binding also recruits additional cofilin molecules, which form a cofilin-binding cluster (Fig. 4.6d). Since actin filaments are double-stranded, a certain size of cofilin-binding cluster is required to sever both strands simultaneously [73].

This cofilin binding model is highly dependent on spontaneous dissociation of OD-ID interactions, particularly the flexibility of the D-loop, which plays a crucial role in OD-ID contacts. Mutations that reduce D-loop flexibility significantly weaken cofilin binding, supporting the idea that spontaneous D-loop dissociation is critical for cofilin-mediated structural transitions [38].

Helical twist and molecular fluctuations Cofilin binding increases the helical twist of actin filaments [40] (Fig. 4.6b). In the cofilin binding model (Fig. 4.6c), cofilin binds to the F-site of a subunit within an ADP filament, and the bound subunit shifts toward G-form (Fig. 4.6c). This shift results in steric hindrance between cofilin and a part of the adjacent P-subunit before attaching G-site with a canonical actin filament twist [50]. However, this steric collision is avoided when cofilin binds to a model actin filament composed of F-form subunits with a greater helical twist [50]. Consequently, cofilin preferentially binds to actin filaments that exhibit a more twisted helix than the canonical ones.

Mechanical tension significantly reduces cofilin affinity for actin filaments. When twisting fluctuations in actin filaments were analyzed under tension, they were found to be markedly suppressed [53]. A similar effect was observed when a rotational force was applied at the B-end by formin, an ABP that promotes B-end elongation while rotating; cofilin binding was inhibited, and the fraction of filaments adopting the more twisted form was diminished [44]. These findings suggest that cofilin binding is highly dependent on twisting fluctuations within actin filaments.

Cofilin binding is dependent on three key actin fluctuations (Fig. 4.7): (1) domain-level fluctuations (between ID and OD), which are enhanced in the ADP state; (2) D-loop fluctuations, which influence OD-ID dissociation; (3) twisting fluctuations of the filament. Notably, these fluctuations regulate protein function not

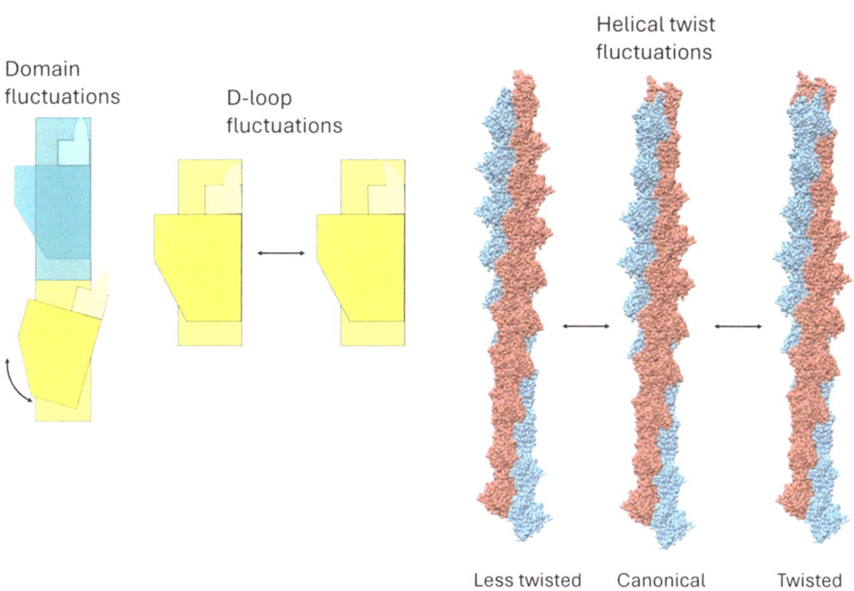

Fig. 4.7 Three types of fluctuations are important for cofilin binding

through structural changes but through the degree of the fluctuations. This fluctuation-function relationship may represent a fundamental principle in protein regulation across diverse biological systems.

What Is Polymorphism? Why Is It Important?

As described in sections "Atomic models for the actin filament structure" and "Cofilin binding and conformational fluctuations of actin", actin undergoes conformational changes upon filament formation (G-form to F-form) and cofilin binding (F-form to C-form). These observations prompted us to systematically investigate the polymorphism of actin—how many distinct conformations can actin adopt, and how is this polymorphism related to its cellular functions. Here, we summarize key findings on actin polymorphism based on our structural analyses [52].

Rigid bodies and domain movement As mentioned, the actin molecule comprises two major domains, OD and ID. During conformational transitions, the core of each domain retains its structure and behaves as a rigid body. Only their relative orientation changes. This contrasts with a homogeneously elastic model in which the entire protein molecule deforms uniformly. By analyzing over 200 published actin structures at atomic resolution, we confirmed that actin's global conformations can be effectively described by the orientation of OD relative to ID [52].

The orientation of the OD was determined as follows: First, the ID of a test structure (ID (test)) was aligned to the ID of a reference structure, ID (ref) (PDB: 1J6Z). Then, OD (test) was translated so that its center of mass coincided with that of OD (ref). Finally, the translated OD (test) was rotated around its center of mass to align with OD (ref), and the resulting rotation matrix was calculated. Although three Euler angles were derived from the matrix, two of them, rotations around the GC axis and the GF axis, shown in Fig. 4.8, are sufficient to describe the global conformation of actin. The GF axis represents the degree of flattening (propeller motion), whereas the GC axis corresponds to cleft closure (scissor motion) (Fig. 4.8). Because these two axes are approximately perpendicular to each other, using them to define OD orientation provides an intuitive description of the actin conformation defined by two rigid bodies. This two-dimensional conformational space is hereafter referred to as GCF-space.

The rotation of OD during the G-to-F conformational transition is associated with bends in the main chain at residues 141–142 and 336–337 [51]. In the G-to-C transition, the rotation of OD pivots around residues 137 and 336 [73]. In both transitions, the main chain segments linking OD/ID, residues 137–142 and 336–337, constitute a common hinge, although the directions of rotation are perpendicular to each other in the two transitions. Thus, actin conformational transitions can be well approximated using only two rotational angles, with negligible translation. Principal component analysis (PCA) has been proposed for classifying actin conformations

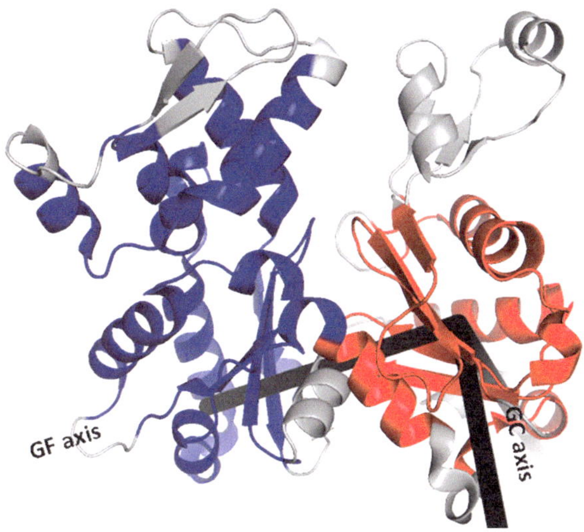

Fig. 4.8 Definition of the actin core domains. Ribbon diagram of the crystal structure of actin in G-form (PDB ID: 1J6Z). The core of the outer domain (OD) is colored red, and the core of the inner domain (ID) is colored blue. The thick black lines represent the three Cartesian axes used to describe the orientation of the OD core. Rotation of OD around the GF axis induced conformational transitions between G- and F-forms, while rotation around the GC axis leads to transitions between G- and C-forms

[82]. While PCA clearly distinguishes subtypes within G-form, it does not effectively separate F-form from G-form, making interpretation less intuitive.

Four distinct conformations: G-, F-, C- and O-forms More than 200 actin conformations are plotted in the GCF space shown in Fig. 4.9, where global conformations are classified into four groups [52]. G-form (black) forms a large cluster that includes many crystal structures of monomeric actin bound to actin-binding proteins (ABPs) or actin-binding small molecules. Within this large cluster, data points (specified by OD orientations) form subgroups dependent on the bound ABPs or small molecules. The remaining three groups, F-form (blue), O-form (orange), and C-form (red), are well separated from the G-form cluster. F-form includes subunits in F-actin (e.g., PDB ID: 2ZWH, 5JLF, 6DJM) and fragmin-actin complexes (e.g., 7W50) (see section "High-resolution crystal structures of F-form actin and the mechanism of actin ATP hydrolysis"). O-form includes profilin-actin complexes (1HLU and 3UB5), in which the nucleotide-binding cleft is widely open. Other profilin-actin complexes, aside from these two, fall within the G-form cluster. C-form includes cofilin-decorated F-actin filaments (5YU8) and the twinfilin-actin complex (3DAW). Twinfilin contains the same ADF-H domain as cofilin, which binds actin. In C-form, the cleft is more closed, in the direction opposite to that of O-form.

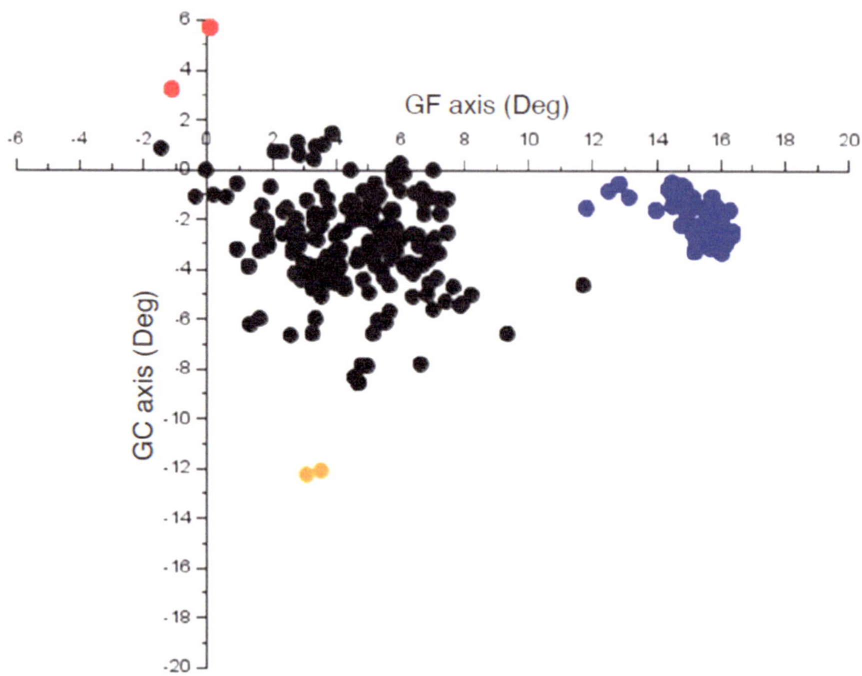

Fig. 4.9 Orientations of the OD core in a 2D space defined by the GF and GC axes. Each dot represents a PDB-deposited 3D structure of actin: F-form in blue, C-form in red, O-form in orange, and G-form in black

The stability of each actin form and the possibility of conformational transitions were evaluated using molecular dynamics simulations, which examined whether an isolated actin molecule in one form could spontaneously transition to another form under ambient conditions through thermal fluctuations [52]. No transitions from G-form to either C-form or F-form were observed within a 500 ns timescale. In contrast, simulations initiated from C-form showed rapid conversion to G-form. For the isolated F-form, some trajectories reverted to G-form, while others remained in F-form. Simulations that started from G-form resulted in trajectories that explored the entire G-form conformational cluster during a 500 ns timescale. These observations suggest an underlying energy landscape in which G-form corresponds to a stable state, C-form represents an unstable state, and F-form behaves as a semi-stable state with a possible energy barrier separating it from G-form. Within G-form cluster, multiple sub-conformations form distinct clusters that are accessible to each other via thermal fluctuations.

Actin conformation is shaped by both external interactions and internal structural preferences. The orientation of OD is confined to a few discrete regions—corresponding to F-, C-, and O-forms—due to the stability of ID/OD interactions in those areas. Selection among these regions is governed by factors such as binding partners, species-specific traits, or steric constraints, indicating that conformation is initially determined by the nature of external interactions. Once a region is selected, the specific OD orientation is further refined by the local energy minimum intrinsic to the actin molecule. This reflects an autonomous aspect of actin's structural behavior. For example, F-form is adopted both in filaments and in complex with fragmin domain-1 (see section "High-resolution crystal structures of F-form actin and the mechanism of actin ATP hydrolysis"), despite differing binding contexts—highlighting that actin can stabilize the same conformation independently of its partner.

Polymorphism and function How are distinct actin conformations linked to specific functions? To understand the mechanisms underlying each function, it is essential to decompose the function into elementary processes or reactions. For example, the ATPase cycle consists of ATP binding, ATP hydrolysis, P_i release, and ADP/ATP exchange. For each process, the initial state from which it is initiated should be identified. In this way, one-to-one correspondence between an initial state and a process or reaction can be established.

As we have shown, G-form is competent for ADP/ATP exchange (section "How actin harnesses ATPase activity for treadmilling"), F-form for ATP hydrolysis (sections "How actin harnesses ATPase activity for treadmilling" and "Atomic models for the actin filament structure"), and C-form for filament severing (section "Cofilin binding and conformational fluctuations of actin"). Each conformation thus serves as the initial state for a distinct process. However, as we will discuss in section "High-resolution crystal structures of F-form actin and the mechanism of actin ATP hydrolysis", even within the same F-form conformation, at least three distinct functional states can be distinguished depending on the bound nucleotide: ATP-F is competent for ATP hydrolysis, ADP-P_i-F for P_i release, and ADP-F for depolymerization. Moreover, only ADP-F is competent for cofilin binding, whereas ATP-F and ADP-P_i are not. These notions indicate that conformation alone is not sufficient to uniquely define the process initiated from a given state. We propose that each state should be characterized not only by its global conformation but also by its dynamic properties, such as conformational fluctuations (see section "Concluding remarks and future directions" for further discussions).

It is worth noting that F-form is substantially different from C-form. Unlike C-form, F-form likely has an energy barrier separating it from G-form, with an energy minimum that makes it an autonomous state. More importantly, F-form is a stable conformation as long as the actin subunit resides in the filament interior. In this environment, inter-subunit interactions constrain the actin conformation so that the influence of the bound nucleotide on the global conformation is effectively suppressed. Following P_i release, the ADP-F-form becomes unstable, allowing the actin subunit to overcome the energy barrier and transition from F-form to G-form, resulting in depolymerization.

High-Resolution Crystal Structures of F-Form Actin and the Mechanism of Actin ATP Hydrolysis

High-resolution crystal structures of F-form actin at 1.15 Å resolution were recently determined, providing detailed insights into the mechanism of ATP hydrolysis in actin [31]. This section outlines the background that led to the structural determination and highlights the mechanistic findings derived from these structures.

As noted in section "Atomic models for the actin filament structure", the atomic structure of G-form actin, an inactive conformation with negligible ATPase activity, was resolved over three decades ago [30]. Subsequent G-actin structures with resolved water molecules provided a basis for investigating the structural mechanism of actin ATP hydrolysis. Notably, Almo and colleagues identified a water molecule (W1) likely to perform a nucleophilic attack on ATP and highlighted the involvement of the active site residues Gln137 and His161 [77]. As introduced in section "Cryo-EM study of the actin filament", recent advancements in cryo-EM have significantly improved the resolution of actin filament structures. Current cryo-EM reconstructions, reaching ~2.2 Å resolution, allow for the interpretation of water molecule densities [55, 65]. Nevertheless, these cryo-EM models are not yet sufficiently accurate for quantum chemical analyses (see section "ATP hydrolysis reaction mechanism of actin: comparison with motor proteins"). To achieve the resolution required for such analysis, X-ray crystallography was employed. Because actin filaments are unsuitable for crystallization due to their variable lengths, an alternative strategy was employed: capturing the active F-form conformation by using fragmin, a protein that stabilizes actin in the F-form.

Fragmin as a tool for stabilizing the F-form of actin The use of fragmin in this context was not a preconceived strategy but rather a fortunate discovery. Fragmin, a member of the gelsolin family isolated from the slime mold *Physarum polycephalum*, exhibits characteristic activities such as Ca^{2+}-dependent actin filament severing and barbed end capping [2, 20, 25]. In addition to these common features, fragmin shows unique characteristics. It forms a ternary complex with actin and a kinase, leading to phosphorylation of actin residues Thr202 and Thr203, which likely suppresses polymerization [13, 16]. Moreover, actin has been reported to exhibit high dynamism in the presence of fragmin [22]. This unique behavior sparked renewed interest in fragmin-induced actin dynamics. Although further functional insights were not obtained, a parallel structural study using fragmin unexpectedly enabled the successful crystallization of F-form actin, leading to a major breakthrough.

Fragmin contains three domains (F1–F3), with the N-terminal domain (F1) responsible for actin filament severing and capping, as is its counterpart (G1) in gelsolin [47]. The crystal structure of F1–actin complex (F1A) revealed that while F1 binds actin in a manner similar to G1 in "G1A" [42], unexpectedly, the actin conformation in F1A adopts F-form [31] rather than G-form seen in G1A crystals (Fig. 4.10a). F-form stabilization is mediated by FNEX (Fragmin N-terminal

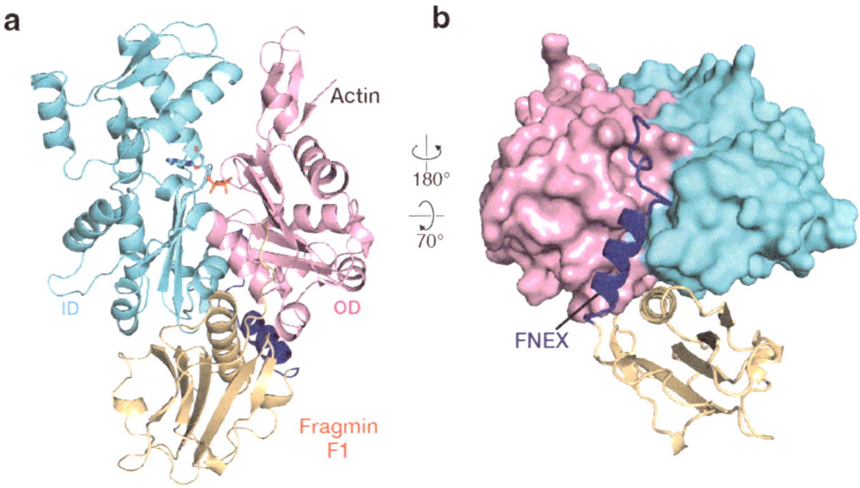

Fig. 4.10 Crystal structure of F-form actin bound to fragmin F1. (**a**) Overall structure (PDB ID: 7W4Z). The ID and OD of actin are shown in cyan and pink, respectively. The domain core of fragmin F1 is colored orange, with FNEX shown in blue. Here, the bound AMPPNP molecule at the center of actin is depicted as sticks. (**b**) Actin is shown in surface representation from the back and slightly lower (P-end) side to illustrate the binding mode of FNEX. FNEX fits snugly between the flattened ID and OD of actin, thereby stabilizing F-form

EXtension), a 30-residue N-terminal sequence absent in vertebrate gelsolins. FNEX fits snugly between ID and OD on the back surface (the side facing to filament axis) of actin (Fig. 4.10b). This configuration is only possible when actin is in the flat F-form conformation, suggesting that FNEX contributes to preventing the dissociation of the terminal actin when fragmin caps the barbed end.

High-resolution crystal structures of F-form actin in three nucleotide states Using a fragmin-mediated crystallization system, F-form actin structures were determined in three different nucleotide states—AMPPNP (pre-hydrolysis), ADP-P_i (post-hydrolysis), and ADP (after P_i release)—each coordinated with Mg^{2+}, the predominant cytoplasmic divalent cation (Fig. 4.11). Except for the D-loop, which is disordered due to the absence of filament contacts, the overall structure, including the active site, is nearly identical to that of actin subunits in cryo-EM filament structures [10, 43]. The bound nucleotide does not affect the overall actin structure. These structures provide precise atomic information for key components of the hydrolysis reaction, including active site residues, nucleotides, Mg^{2+} ions, and associated water molecules (Fig. 4.12a–c). The following important insights were obtained from these high-resolution structures [31]:

1. G-to-F conformation transition triggers ATP hydrolysis.
   The conformational transition from G-form to F-form primarily induces structural changes in two active site residues, Gln137 and His161 [10, 31, 43]. The importance of these residues has been confirmed through mutant analyses [27, 28]. In

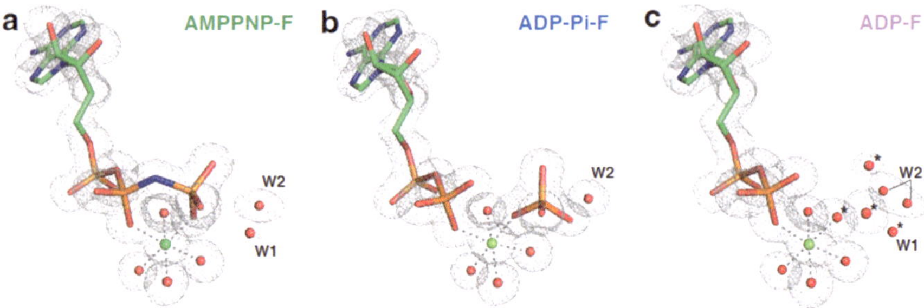

Fig. 4.11 Actin-bound nucleotide, Mg^{2+} (with coordinated waters), and lytic water (W1) and helping water (W2) in AMPPNP (**a**; PDB ID: 7W4Z), ADP-P_i (**b**; 7W50), and ADP (**c**; 7W51) states, with the omit map contoured at 3.5 σ. Asterisks in panel c denote water molecules replacing P_i in the ADP state

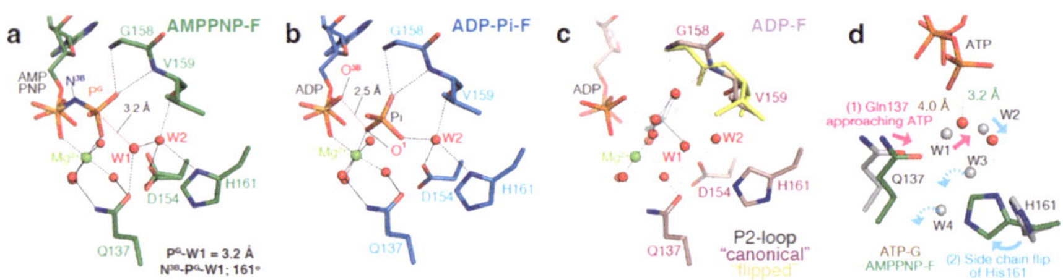

Fig. 4.12 Active site structures. (**a**–**c**) Nucleotide, Mg^{2+}, key residues, and water molecules are shown for the AMPPNP (**a**), ADP-P_i, and ADP (**c**) states. In the AMPPNP state, the lytic water molecule W1 is positioned 3.2 Å from the γ-phophorous (P^G) of AMPPNP, forming a P^G-N^{3B}-W1 angle of 161°, suitable for nucleophilic attack. After hydrolysis, the resulting inorganic phosphate (P_i) forms a hydrogen bond with the O^{3B} atom of ADP at a distance of 2.5 Å. P_i is further stabilized by hydrogen bonds with residues in P1- and P2-loops, accounting for the remarkable stability of the ADP-P_i state. Upon P_i release, P_i (shown as a gray stick) is replaced by water molecules. Notably, the P2-loop adopts two conformations. In the non-canonical conformation (yellow), the main chain oxygen of Gly158 flips toward the nucleotide, likely compensating for the absence of P_i. (**d**) Activation of hydrolysis by the G-to-F transition. Compared to the Mg-ATP–G-actin structure (PDB ID: 4B1Y; gray), (1) the closer positioning of Gln137 to ATP and (2) the flipping of His161 toward the nucleotide lead to a rearrangement of water molecules. As a result, W1 is positioned within reach of the P^G and forms a hydrogen bond with W2, which participates in a proton transfer ring (see section "ATP hydrolysis reaction mechanism of actin: comparison with motor proteins")

the AMPPNP-bound F-form structure, a rearrangement of water molecules occurs in the active site. Water molecule W1 is positioned closer to the P^G atom (3.2 Å in F-form vs. 4.0 Å in G-form; (PDB ID: 4B1Y; [46])), favoring nucleophilic attack (Fig. 4.12d). W1 forms a hydrogen bond with a second water molecule (W2), which is essential for catalysis (see also section "ATP hydrolysis reaction mechanism of actin: comparison with motor proteins").

2. Stabilization of the ADP-P_i state.
   The ADP-P_i-bound structure shows minimal differences from the pre-hydrolysis structure. The P_i cleaved from ATP is stabilized by multiple hydrogen bonds, primarily involving the backbone amides of the P1- and P2-loop residues. These interactions appear stronger than in the pre-hydrolysis structure. In addition, a short hydrogen bond (2.5 Å) between ADP and P_i further contributes to P_i stabilization. These findings are consistent with the very slow P_i release from ADP-P_i filaments [6].
3. Instability of the ADP-bound F-form.
   The overall structure remains largely unchanged even after P_i release. The position previously occupied by P_i is filled by water molecules (Figs. 4.11c and 4.12c), which will not prevent P_i reuptake [7]. However, this substitution weakens the linkage between the ID and OD, likely contributing to the reduced stability of the ADP-bound actin filament [60]. This effect may be further enhanced by dual conformations in Gly158-Val159 of the P2-loop.

ATP Hydrolysis Reaction Mechanism of Actin: Comparison with Motor Proteins

The ATP hydrolysis reaction pathway in actin was elucidated using high-resolution crystal structures (1.15 Å) described above in combination with hybrid quantum mechanics/molecular mechanics (QM/MM) calculations [31].

ATP hydrolysis occurs at the nucleotide-binding site, known as the "active site," located deep within the central cleft of actin, between OD and ID (Fig. 4.13a). The site is highly enclosed, with three phosphate moieties of ATP positioned between two phosphate-binding loops: P1-loop from OD and P2-loop from ID (Fig. 4.13b).

QM/MM calculations Currently, no experimental method directly reveals the atomic-scale reaction path proceeding inside proteins. Provided that high-resolution structures are available, QM/MM calculations are powerful and reliable approaches to achieve precise information (positions and occupancies) about water molecules trapped in the ATP binding site, and the initial (reactant), intermediate, and final (product) structures of the reaction. In QM/MM calculations, atoms whose electrons move and are involved in chemical reactions should be treated by quantum mechanics, while the other atoms should be treated by classical mechanics for computational efficiency.

The QM/MM calculations indicated that ATP hydrolysis reaction in actin proceeds through four steps, involving distinct intermediates (IMs) and transition states (TSs) along the reaction path from the ATP-bound (reactant) to the ADP-P_i-bound (product) F-form (Fig. 4.14a). Atomic positions at key stages are shown in Fig. 4.14b–d.

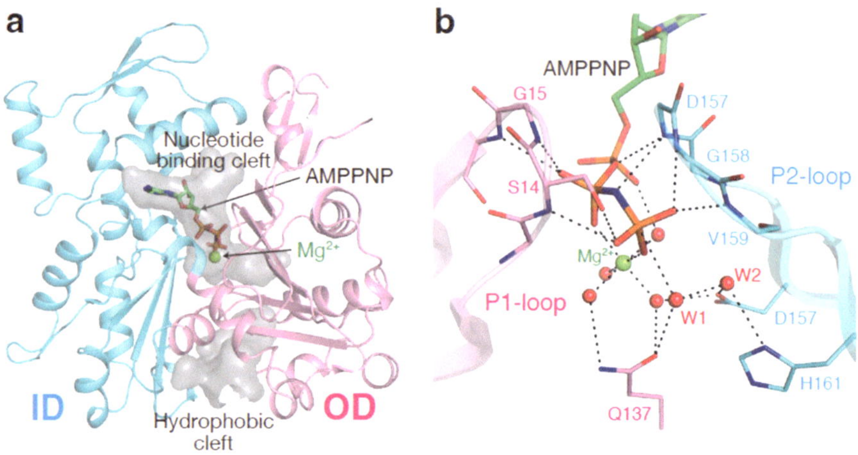

Fig. 4.13 ATP-binding site of actin (PDB ID: 7W4Z). (**a**) Nucleotide-binding cleft. The inner domain (ID) and outer domain (OD) are shown in cyan and pink, respectively. (**b**) Nucleotide-binding loops. The triphosphate moiety of AMPPNP is clamped primarily by the backbone nitrogen atoms of two actin-specific nucleotide-binding motifs: the P1-loop (in OD) and the P2-loop (in ID). W1 and W2 denote the lytic and helping water molecules, respectively

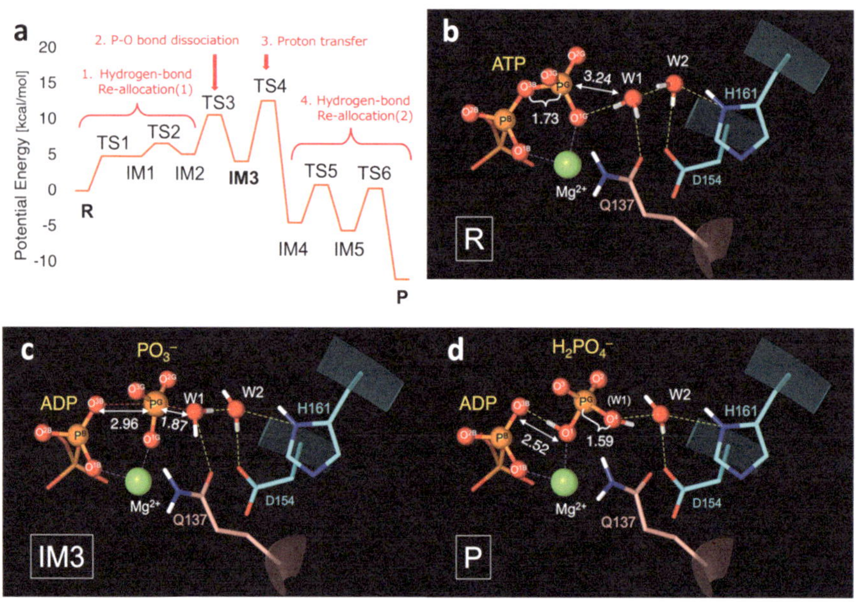

Fig. 4.14 ATP hydrolysis reaction pathway of F-form actin, simulated using QM/MM calculations. (**a**) Potential energy profiles along the reaction pathway. (**b–d**) Atomic configurations of the catalytic site and selected interatomic distances in states R (**b**), IM3 (**c**), and P (**d**). In b–d, phosphorus (P), oxygen (O), protons (H^+), and magnesium ions (Mg^{2+}) are colored orange, red, white, and light green, respectively

Step 1: Hydrogen bond rearrangement I (R → IM2)

Water molecules W1 and W2 rotate and shift slightly, rearranging the hydrogen bond network. This step incurs an energy cost of 5 kcal/mol and initiates the reaction.

Step 2: PO-bond cleavage (IM2 → IM3)

The P^G–O^{3B} bond is cleaved, producing a planar PO^{3-} (metaphosphate) intermediate. The dissociation occurs upon the approach of W1 to PO^{3-}, but prior to its nucleophilic attack.

Step 3: Proton transfer and nucleophilic attack (IM3 → IM4)

Three events occur nearly simultaneously: W1 splits into OH^- and H^+ (proton); the OH^- attacks the metaphosphate (the nucleophilic attack); and the proton dissociated from W1 is transferred to W2, in concert with another proton transfer from W2 to the cleaved phosphate resulting in the formation of doubly protonated $H_2PO_4^-$. This step is the rate-limiting step and is associated with the highest energy barrier.

Step 4: Hydrogen bond rearrangement II (IM4 → P)

The transferred proton further stabilizes the ADP–P_i product by forming a strong hydrogen bond between ADP and the cleaved phosphate.

QM/MM calculations provide mechanistic explanations for the following functional properties: First, how the conformational transition from G-form to F-form induces ATP hydrolysis (see section "High-resolution crystal structures of F-form actin and the mechanism of actin ATP hydrolysis"). Second, why the actin ATP hydrolysis reaction is irreversible [7]. The stabilization energy of Step 4 is substantial ($-12 \sim -15$ kcal/mol), making the kinetic barrier for the reverse reaction too high to overcome under ambient conditions. In contrast, in motor proteins such as F_1-ATPase, the stabilization energy is -3 kcal/mol, allowing the reaction to proceed reversibly. This step likely determines whether ATP synthesis, the reverse reaction of hydrolysis, can occur. Third, why the ATP hydrolysis reaction proceeds slowly ($0.3\ s^{-1}$) [5]. The proton transfer ring is composed of only two water molecules and is not uniformly curved. These geometric constraints elevate the energy barrier at the transition state (TS4), resulting in a slow reaction rate.

Mechanistic comparison with other ATPases To date, the hydrolysis reaction pathways have been comprehensively determined only for F_1-ATPase [23], kinesin [41], myosin [33], and actin [31] among ATPase and GTPase proteins [32, 34]. Why have so few ATP hydrolysis mechanisms been elucidated? This is primarily due to the generally insufficient quality of structural data required for QM/MM simulations. The following discussion compares the ATP hydrolysis mechanism of actin with those of motor proteins.

The ATP hydrolysis mechanism in actin is nearly identical to that in motor proteins. The shared mechanism is characterized by three key features: First, the reaction is dissociative, with PO bond disruption occurring before PO_3-–W1 bond formation. Second, γ-phosphate is the destination of the proton extracted from W1. Third, a two-water system, where the proton transfer ring consists of at least two water molecules [68].

Table 4.2 Comparison of ATP hydrolysis mechanisms between actin and motor proteins

a. The common catalytic strategy:				
	F1-ATPase	**kinesin**	**myosin**	**F-form actin**
QM/MM calculations	[23]	[41]	[33]	[31]
1st step: HBR-1	√	**Not described**	**Not described**	√
2nd step: P^G-O^{3B} Dissociation	√	√	√	√
3rd step: Proton Transfer O^{W1}-H bond cleavage Nucleophilic attack (→HPO_4^{2-}) Proton dissociation Proton transfer to γ-phosphate (→$H_2PO_4^-$)	√	√	√	√
4th step: HBR-2	√	**Not described**	√	√
b. Structural elements in the catalytic site:				
Phosphate-binding loop	P-loop	P-loop	P-loop	**P1-, P2-loops**
Switch-1 loop	√	√	√	**No**
Switch-2 loop	√	√	√	**No**
c. Quality of structural information:				
Initial structure (***R***) is experimentally available	Yes β_{DP} of **2JDI** (1.9 Å)	No Derived from **3HDQ** (2.2 Å)	No Derived from **1VOM** (1.9 Å)	Yes **7W4Z** (1.15 Å)
Final structure (***P***) is experimentally available	No	No	No Later 4PJM (2015)	Yes **7W50** (1.15 Å)

In each motor protein, ATP hydrolysis proceeds through four steps, following the same sequences as in actin. However, Steps 1 and 4 remain undetermined for kinesin and myosin (Table 4.2a). This is because structural information was unavailable for both the initial (reactant) and final (product) states of the reaction in these proteins (Table 4.2c). To properly identify Steps 1 and 4, precise coordinates and occupancies of individual water molecules in the ATP binding site are essential. It is widely believed that the ATP hydrolysis reaction mechanism is shared across all ATPase proteins. Our findings in actin not only confirm that the ATP hydrolysis pathway is common to actin and motor proteins, but also define the shared mechanism. This is made possible by the high-quality structural information obtained from actin.

The ATP hydrolysis mechanism of actin exhibits distinct features when compared to motor proteins. Here, key structural differences are discussed.

The architecture of the ATP-binding site differs significantly. In motor proteins, the phosphate moiety of ATP interacts with the P-loop. In contrast, actin lacks a conventional P-loop; instead, it possesses two distinct motifs, P1-loop and P2-loop.

More importantly, the motor proteins feature switch I and switch II loops, which undergo conformational changes upon ATP hydrolysis. Actin, however, lacks both of these structural elements (Table 4.2b).

The ATPase mechanism in actin, particularly the process of P_i release, differs substantially from that of other ATPases, including the three motor proteins discussed above. NDP-P_i-bound state is generally unstable in most ATPase and GTPase (collectively referred to as NTPase) proteins, and only a limited number of crystal structures have been reported. In F-actin, ADP-P_i-bound state is stable and can be readily captured in crystal structures in the presence of fragmin F1, which traps actin in F-form. In addition, the overall conformation of F-form actin remains unchanged regardless of the bound nucleotides—ATP, ADP–P_i, or ADP. This contrasts sharply with many other NTPases, where P_i release is accompanied by substantial conformational changes.

Further comparison with a focus on the P_i release mechanism To better understand the distinctions between the ATPase processes of actin and those of motor proteins, we propose a common framework applicable to all NTPase systems. First, the NTP hydrolysis and subsequent P_i release should be considered as two sequential but mechanistically distinct steps. That is, the NDP-P_i-bound state is not merely a transient intermediate, but rather the final product of the hydrolysis reaction and, simultaneously, the starting point of the P_i release process. Second, the global conformation of the protein is expected to remain unchanged before and after the hydrolysis step in all NTPase reactions.

The ATPase cycle of myosin serves as a representative example among motor proteins. Crystal structures of the ADP-P_i-bound motor domain (MD) have been reported for myosin VI (PDB ID: 4PJK; [35]) and bovine cardiac muscle myosin (5N6A; [59]). These structures illustrate the transition from ATP hydrolysis to P_i release in actomyosin complexes. ATP hydrolysis occurs while the MD is dissociated from actin. The resulting ADP-P_i-bound MD then binds actin via electrostatic interactions, which are reinforced by stereospecific contacts to stabilize the complex in the P_i release (PiR) state. At this point, switch II is thought to shift to open a pathway for P_i release. Finally, P_i is released from the MD along with the lever arm swinging, which is coupled to the power stroke of the myosin cross-bridge [35, 66]. In summary, the process proceeds as follows: P_i cleavage → MD property change → actin binding → conformational shift, including switch II → P_i release. This sequence suggests that restricting MD conformational changes allows stable capture of the ADP-P_i-bound state.

Similar observations have been reported for small GTPases. When conformational changes of these proteins are restricted, the GDP-P_i-bound state can be stably captured. For example, Arl2 adopts a stable GDP-P_i-bound form upon forming a complex with another protein (1KSH; [19]); Rab11a is trapped in this state due to a point mutation (1OIX; [58]); and the RhoGAP–RhoA complex adopts this conformation within the pre-formed crystal lattice (6R3V; [45]). Notably, small GTPases also possess switch I and switch II loops at their nucleotide-binding sites.

In NTPase proteins equipped with switch I and switch II loops at the nucleotide-binding site, P_i release is generally mediated by a conformational shift of switch II, likely triggered by the completion of ATP hydrolysis through currently unknown mechanisms. In contrast, actin lacks both switch I and switch II loops, as well as any associated triggering mechanism. As a result, P_i release in actin does not involve major conformational changes and proceeds much more slowly. ADP-P_i-bound actin is a stable intermediate in the actin filament, and P_i is released on a timescale of several minutes on average [12]. We suggest that P_i release in actin occurs through a stochastic process driven by spontaneous conformational fluctuations between F-form and G-form.

Cryo-EM Study of the Actin Filament

Actin filaments have been a major target of electron microscopy for a long time [18]. The first high-resolution structure obtained by cryo-electron microscopy was published in 2015 [76]. Since then, several actin filament structures have been reported, including those from yeast [69], plants [64], and the malaria parasite [63]. All of these structures are highly similar, confirming the strong evolutionary conservation of actin among eukaryotes.

The nucleotide-dependent conformational changes of the D-loop were initially reported [43]. However, the electron density of the D-loop is often weak, making it difficult to model. Recent high-resolution structures at resolutions better than 3 Å [55, 65] have shown almost no nucleotide-dependent differences, including in the D-loop, making it challenging to explain how nucleotide binding influences actin filament dynamics. The very limited structural differences observed between nucleotide states are consistent with high-resolution F-form crystal structures [31].

The stabilization of actin filaments by ADP-P_i and ATP is most likely due to F-form stabilization mediated by the inorganic phosphate or the γ-phosphate, as observed in F-form crystal structures [31]. When OD–ID interactions spontaneously dissociate, actin subunits can maintain F-form in the presence of ADP-P_i or ATP. However, in the ADP state, actin subunits tend to adopt the G-form with the loss of OD–ID interactions, leading to filament destabilization (Fig. 4.5c). The dependence of depolymerization on the spontaneous dissociation of OD–ID interactions is supported by findings showing that reducing the D-loop flexibility decreases depolymerization rates [38].

The structures of the two ends of the actin filament are critical for understanding filament dynamics, as polymerization and depolymerization occur exclusively at the ends. The first structural determinations of filament ends were performed for the capped barbed end by CapZ [48] and then for the free pointed end [49], but the resolution was limited. A high-resolution structure of the free filament ends was published in 2023 [9], revealing a key finding: subunits at the pointed end adopt the G-form. This observation is plausible given the nature of subunit-subunit interactions within the strand. Since the OD of subunits exposed at the pointed end is not

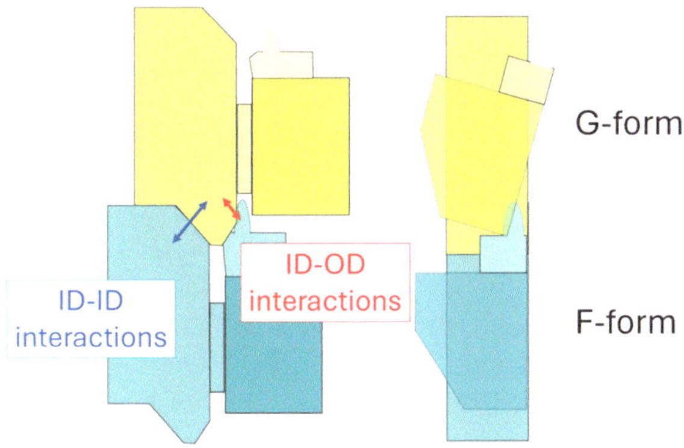

Fig. 4.15 Schematic illustration of the pointed end. The terminal subunit at the pointed end can adopt any form without disrupting ID-OD and ID-ID interactions with the adjacent barbed end side subunit

involved in ID–OD or ID–ID interactions, these subunits can adopt various conformations without disrupting intra-strand interactions, and the G-form being favored as the ground state (Fig. 4.15).

Additionally, a high-resolution structure of the phalloidin-bound barbed end was published in 2023 [56]. Surprisingly, a tunnel facilitating phosphate release was identified in the terminal subunit at the barbed end subunit, explaining why phosphate release occurs rapidly at the barbed end. This discovery led to the identification of key residues involved in phosphate release, although the detailed mechanism of phosphate release remains unclear except at the terminal subunit at the barbed end. The reasons why depolymerization and polymerization rates are significantly slower at the pointed end also remain unclear. Further research is needed to fully understand these phenomena.

Concluding Remarks and Future Directions

The central question in actin treadmilling is how the energy source, ATP hydrolysis, drives the process. The entire treadmilling cycle can be divided into two phases: the filament assembly pathway and the filament disassembly pathway.

In the filament assembly pathway, first, polymerization is coupled with a conformational transition from G- to F-form via a domain motion (section "Atomic models for the actin filament structure"). Second, this conformational change triggers ATP hydrolysis (section "High-resolution crystal structures of F-form actin and the mechanism of actin ATP hydrolysis"). Third, because ADP-P_i binding stabilizes F-form (section "ATP hydrolysis reaction mechanism of actin: comparison with motor proteins"), ADP-P_i-bound F-actin exposed at the B-end promotes asymmetric elongation due to its lower critical concentration (section "How actin harnesses ATPase activity for treadmilling"). Polymerization still occurs in the presence of ADP alone (without ATP) but does not induce treadmilling movement in the steady state. As described in this chapter, structural biology studies have significantly advanced our understanding of the filament assembly pathway.

The filament disassembly pathway begins with P_i release from F-form actin located within the filament interior. Since actin lacks a structural element facilitating P_i release, such as the switch II motif in myosin (section "ATP hydrolysis reaction mechanism of actin: comparison with motor proteins"), P_i release proceeds slowly, thereby allowing sufficient time for filament elongation. Upon P_i release, F-form actin becomes unstable (section "High-resolution crystal structures of F-form actin and the mechanism of actin ATP hydrolysis") and stochastically depolymerizes into G-form, thereby providing additional time for elongation. On the other hand, there are mechanisms that accelerate both P_i release and filament disassembly. P_i release occurs much more rapidly at the B-end, and ADP-bound F-form actin recruits cofilin, whose binding promotes filament disassembly.

Key features characteristic of the filament disassembly pathway include the following: First, actin molecules within the filament adopt the filament-specific F-form conformation and retain it until depolymerization occurs. Second, changes in bound nucleotides, i.e., cleavage of γ-phosphate and the Pi release, do not affect the global conformation but destabilize F-form actin upon P_i loss. Third, actin employs a disassembly mechanism based on enhanced instability induced by P_i release. Fourth, this mechanism is slow, which is well-suited to allow sufficient time (on the order of minutes) for filament elongation. Interestingly, this disassembly mechanism is shared by other NTPase proteins such as GroEL, which form large complexes for physiological functions. In contrast to motor proteins equipped with switch II, like myosin, where P_i release is linked to power strokes, P_i release in these NTPases appears to enhance conformational instability, preparing for the collapse of large macromolecular assemblies.

Although the destabilization of F-form upon P_i-release (Fig. 4.12b, c) has contributed to our understanding of its significance, the pathway of actin filament disassembly remains incompletely understood. For example, what is the precise mechanism of P_i release? Why does ADP-F depolymerize more readily than ADP-P_i-F? Why does ADP-F have a higher affinity for cofilin than either ATP-F or ADP-P_i-F, despite sharing the same global conformation? To address these and related questions, it is crucial to understand the dynamic structural properties of actin within the filament, particularly the conformational fluctuations of F-form actin throughout the disassembly pathway.

Recent studies of the P_i release mechanism using cryo-EM and molecular dynamics simulations have provided new insights. The slow P_i release may be explained either by the "backdoor," a proposed P_i exit gate, that remains closed in filament interior subunits [56], or by a high-energy barrier for P_i dissociation from the nucleotide-coordinated cation (typically Mg^{2+}) [79]. In actin, P_i is held in place by the P1-loop (OD) and P2-loop (ID), so fluctuations between these domains likely facilitate P_i release. This interpretation aligns with observations of faster P_i release near filament ends [12, 29], where greater thermal motion is expected. Note that rapid P_i release from the terminal subunit at the barbed end may result from a constitutively open backdoor [56]. To gain deeper insight into the P_i release mechanism, it is essential to characterize the dynamic properties of actin.

A promising approach to observe both domain and local fluctuations in actin is solid-state NMR. During our investigation of the actin ATP hydrolysis reaction pathway, we realized that actin is especially well-suited for ATPase studies compared to other ATPases such as motor proteins (section "ATP hydrolysis reaction mechanism of actin: comparison with motor proteins"): (1) its molecular weight (~43 kDa) is not too large; (2) the active site has a relatively simple structure without switch I and switch II loops and does not undergo nucleotide-dependent conformational changes; (3) the ATP-binding site is closed and buried (Fig. 4.12a), reducing susceptibility to electrostatic disturbances from solvent water; (4) high-resolution structural coordinates are available; and (5) the fragmin F1 domain provides a unique system to stabilize a single actin molecule in F-form.

References

1. Adelman MR, Taylor EW (1969) Further purification and characterization of slime mold myosin and slime mold actin. Biochemistry 8:4976–4988
2. Ampe C, Vandekerckhove J (1987) The F-actin capping proteins of Physarum polycephalum: cap42(a) is very similar, if not identical, to fragmin and is structurally and functionally very homologous to gelsolin; cap42(b) is Physarum actin. EMBO J 6:4149–4157
3. Andrianantoandro E, Pollard TD (2006) Mechanism of actin filament turnover by severing and nucleation at different concentrations of ADF/cofilin. Mol Cell 24:13–23
4. Bernstein BW, Bamburg JR (2003) Actin-ATP hydrolysis is a major energy drain for neurons. J Neurosci 23:1–6
5. Blanchoin L, Pollard TD (2002) Hydrolysis of ATP by polymerized actin depends on the bound divalent cation but not profilin. Biochemistry 41:597–602
6. Carlier M-F, Pantaloni D (1986) Direct evidence for ADP-Pi-F-actin as the major intermediate in ATP-actin polymerization. Rate of dissociation of Pi from actin filaments. Biochemistry 25: 7789–7792
7. Carlier M-F, Pantaloni D, Evans JA, Lambooy PK, Korn ED, Webb MR (1988) The hydrolysis of ATP that accompanies actin polymerization is essentially irreversible. FEBS Lett 235:211–214
8. Carlier M-F, Laurent V, Santolini J, Melki R, Didry D, Xia G-X, Hong Y, Chua N-H, Pantaloni D (1997) Actin depolymerizing factor (ADF/cofilin) enhances the rate of filament turnover: implication in actin-based motility. J Cell Biol 136:1307–1322
9. Carman PJ, Barrie KR, Rebowski G, Dominguez R (2023) Structures of the free and capped ends of the actin filament. Science 380:1287–1292
10. Chou SZ, Pollard TD (2019) Mechanism of actin polymerization revealed by cryo-EM structures of actin filaments with three different bound nucleotides. Proc Natl Acad Sci USA 116: 4265–4274
11. Daniel JL, Molish IR, Robkin L, Holmsen H (1986) Nucleotide exchange between cytosolic ATP and F-actin-bound ADP may be a major energy-utilizing process in unstimulated platelets. Eur J Biochem 156:677–684
12. Fujiwara I, Vavylonis D, Pollard TD (2007) Polymerization kinetics of ADP- and ADP-Pi-actin determined by fluorescence microscopy. Proc Natl Acad Sci USA 104:8827–8832
13. Furuhashi K, Hatano S (1990) Control of actin filament length by phosphorylation of fragmin-actin complex. J Cell Biol 111:1081–1087
14. Galkin VE, Orlova A, Kudryashov DS, Solodukhin A, Reisler E, Schröder GF, Egelman EH (2011) Remodeling of actin filaments by ADF/cofilin proteins. Proc Natl Acad Sci USA 108: 20568–20572

15. Geeves MA, Fedorov R, Manstein DJ (2005) Molecular mechanism of actomyosin-based motility. Cell Mol Life Sci 62:1462–1477
16. Gettemans J, De Ville Y, Vandekerckhove J, Waelkens E (1992) Physarum actin is phosphorylated as the actin-fragmin complex at residues Thr203 and Thr202 by a specific 80 kDa kinase. EMBO J 11:3185–3191
17. Go N, Noguti T, Nishikawa T (1983) Dynamics of a small globular protein in terms of low-frequency vibrational modes. Proc Natl Acad Sci USA 80:3696–3700
18. Hanson J, Lowy J (1963) The structure of F-actin and of actin filaments isolated from muscle. J Mol Biol 6:46–60
19. Hanzal-Bayer M, Renault L, Roversi P, Wittinghofer A, Hillig RC (2002) The complex of Arl2-GTP and PDE delta: from structure to function. EMBO J 21:2095–2106
20. Hasegawa T, Takahashi S, Hayashi H, Hatano S (1980) Fragmin: a calcium ion sensitive regulatory factor on the formation of actin filaments. Biochemistry 19:2677–2683
21. Hatano S, Oosawa F (1966) Isolation and characterization of plasmodium actin. Biochim Biophys Acta 127:488–498
22. Hatano S, Hasegawa T, Sugino H, Ozaki K (1982) Physical properties of fragmin, a Ca^{++}-sensitive regulatory protein of actin polymerization isolated from Physarum plasmodium. In: Kakiuchi, Hidaka and Means (eds) Calmodulin and intracellular Ca^{2+} receptors. Plenum Publishing Corporation, New York, pp 403–420
23. Hayashi S, Ueno H, Shaikh AR, Umemura M, Kamiya M, Ito Y, Ikeguchi M, Komoriya Y, Iino R, Noji H (2012) Molecular mechanism of ATP hydrolysis in F1-ATPase revealed by molecular simulations and single-molecule observations. J Am Chem Soc 134:8447–8454
24. Hayward S, Kitao A, Berendsen HJ (1997) Model-free methods of analyzing domain motions in proteins from simulation: a comparison of normal mode analysis and molecular dynamics simulation of lysozyme. Proteins 27:425–437
25. Hinssen H (1981) An actin-modulating protein from Physarum polycephalum. I. Isolation and purification. Eur J Cell Biol 23:225–233
26. Holmes KC, Popp D, Kabsch W, Gebhard W (1990) Atomic model of the actin filament. Nature 347:44–49
27. Iwasa M, Maeda K, Narita A, Maéda Y, Oda T (2008) Dual roles of Gln137 of actin revealed by recombinant human cardiac muscle alpha-actin mutants. J Biol Chem 283:21045–21053
28. Iwasa M, Takeda S, Narita A, Maéda Y, Oda T (2023) Mutagenic analysis of actin reveals the mechanism of His161 flipping that triggers ATP hydrolysis. Front Cell Dev Biol 11:1105460
29. Jégou A, Niedermayer T, Orbán J, Didry D, Lipowsky R, Carlier M-F, Romet-Lemonne G (2011) Individual actin filaments in a microfluidic flow reveal the mechanism of ATP hydrolysis and give insight into the properties of profilin. PLoS Biol 9:e1001161
30. Kabsch W, Mannherz HG, Suck D, Pai EF, Holmes KC (1990) Atomic structure of the actin: DNase I complex. Nature 347:37–44
31. Kanematsu Y, Narita A, Oda T, Koike R, Ota M, Takano Y, Moritsugu K, Fujiwara I, Tanaka K, Komatsu H et al (2022) Structures and mechanisms of actin ATP hydrolysis. Proc Natl Acad Sci USA 119:e2122641119
32. Khrenova MG, Mulashkina TI, Stepanyuk RA, Nemukhin AV (2024) Modeling of enzyme-catalyzed P-O bond cleavage in the adenosine triphosphate molecule. Mendeleev Commun 34:1–7
33. Kiani FA, Fischer S (2014) Catalytic strategy used by the myosin motor to hydrolyze ATP. Proc Natl Acad Sci USA 111:E2947–E2956
34. Kiani FA, Fischer S (2016) Comparing the catalytic strategy of ATP hydrolysis in biomolecular motors. Phys Chem Chem Phys 18:20219–20233
35. Llinas P, Isabet T, Song L, Ropars V, Zong B, Benisty H, Sirigu S, Morris C, Kikuti C, Safer D et al (2015) How actin initiates the motor activity of myosin. Dev Cell 33:401–412
36. Lorenz M, Popp D, Holmes KC (1993) Refinement of the F-actin model against X-ray fiber diffraction data by the use of a directed mutation algorithm. J Mol Biol 234:826–836

37. Maciver SK, Zot HG, Pollard TD (1991) Characterization of actin filament severing by actophorin from Acanthamoeba castellanii. J Cell Biol 115:1611–1620
38. Matsuzaki M, Fujiwara I, Kashima S, Matsumoto T, Oda T, Hayashi M, Maeda K, Takiguchi K, Maéda Y, Narita A (2020) D-loop mutation G42A/G46A decreases actin dynamics. Biomolecules 10:736
39. McCullough BR, Grintsevich EE, Chen CK, Kang H, Hutchison AL, Henn A, Cao W, Suarez C, Martiel JL, Blanchoin L et al (2011) Cofilin-linked changes in actin filament flexibility promote severing. Biophys J 101:151–159
40. McGough A, Pope B, Chiu W, Weeds A (1997) Cofilin changes the twist of F-actin: implications for actin filament dynamics and cellular function. J Cell Biol 138:771–781
41. McGrath MJ, Kuo IF, Hayashi S, Takada S (2013) Adenosine triphosphate hydrolysis mechanism in kinesin studied by combined quantum-mechanical/molecular-mechanical metadynamics simulations. J Am Chem Soc 135:8908–8919
42. McLaughlin P, Gooch J, Mannherz H, Weeds A (1993) Structure of gelsolin segment 1-actin complex and the mechanism of filament severing. Nature 364:685–692
43. Merino F, Pospich S, Funk J, Wagner T, Kullmer F, Arndt HD, Bieling P, Raunser S (2018) Structural transitions of F-actin upon ATP hydrolysis at near-atomic resolution revealed by cryo-EM. Nat Struct Mol Biol 25:528–537
44. Mizuno H, Tanaka K, Yamashiro S, Narita A, Watanabe N (2018) Helical rotation of the diaphanous-related formin mDia1 generates actin filaments resistant to cofilin. Proc Natl Acad Sci USA 115:E5000–5007
45. Molt RW Jr, Pellegrini E, Jin Y (2019) A GAP-GTPase-GDP-Pi intermediate crystal structure analyzed by DFT shows GTP hydrolysis involves serial proton transfers. Chemistry 25:8484–8488
46. Mouilleron S, Wiezlak M, O'Reilly N, Treisman R, McDonald NQ (2012) Structures of the Phactr1 RPEL domain and RPEL motif complexes with G-actin reveal the molecular basis for actin binding cooperativity. Structure 20:1960–1970
47. Nag S, Larsson M, Robinson RC, Burtnick LD (2013) Gelsolin: the tail of a molecular gymnast. Cytoskeleton 70:360–384
48. Narita A, Takeda S, Yamashita A, Maéda Y (2006) Structural basis of actin filament capping at the barbed-end: a cryo-electron microscopy study. EMBO J 25:5626–5633
49. Narita A, Oda T, Maéda Y (2011) Structural basis for the slow dynamics of the actin filament pointed end. EMBO J 30:1230–1237
50. Narita A (2020) ADF/cofilin regulation from a structural viewpoint. J Muscle Res Cell Motil 41: 141–151
51. Oda T, Iwasa M, Aihara T, Maéda Y, Narita A (2009) The nature of the globular-to fibrous-actin transition. Nature 457:441–445
52. Oda T, Takeda S, Narita A, Maéda Y (2019) Structural polymorphism of actin. J Mol Biol 431(17):3217–3228
53. Okura K, Matsumoto T, Narita A, Tatsumi H (2023) Mechanical stress decreases the amplitude of twisting and bending fluctuations of actin filaments. J Mol Biol 435:168295
54. Oosawa F, Asakura S (1975) Thermodynamics of the polymerization of protein. Academic Press, London
55. Oosterheert W, Klink BU, Belyy A, Pospich S, Raunser S (2022) Structural basis of actin filament assembly and aging. Nature 611:374–379
56. Oosterheert W, Blanc FEC, Roy A, Belyy A, Sanders MB, Hofnagel O, Hummer G, Bieling P, Raunser S (2023) Molecular mechanisms of inorganic-phosphate release from the core and barbed end of actin filaments. Nat Struct Mol Biol 30:1774–1785
57. Pantaloni D, Clainche CL, Carlier M-F (2001) Mechanism of actin-based motility. Science 292: 1502–1506
58. Pasqualato S, Cherfils J (2005) Crystallographic evidence for substrate-assisted GTP hydrolysis by a small GTP binding protein. Structure 13:533–540
59. Planelles-Herrero VJ, Hartman JJ, Robert-Paganin J, Malik FI, Houdusse A (2017) Mechanistic and structural basis for activation of cardiac myosin force production by omecamtiv mecarbil. Nat Commun 8:190

60. Pollard TD (1984) Polymerization of ADP-actin. J Cell Biol 99:769–777
61. Pollard TD, Blanchoin L, Mullins RD (2000) Molecular mechanisms controlling actin filament dynamics in nonmuscle cells. Annu Rev Biophys Biomol Struct 29:545–576
62. Pollard TD, Borisy GG (2003) Cellular motility driven by assembly and disassembly of actin filaments. Cell 112:453–465
63. Pospich S, Kumpula EP, von der Ecken J, Vahokoski J, Kursula I, Raunser S (2017) Near-atomic structure of jasplakinolide-stabilized malaria parasite F-actin reveals the structural basis of filament instability. Proc Natl Acad Sci USA 114:10636–10641
64. Ren Z, Zhang Y, Zhang Y, He Y, Du P, Wang Z, Sun F, Ren H (2019) Cryo-EM structure of actin filaments from Zea mays pollen. Plant Cell 31:2855–2867
65. Reynolds MJ, Hachicho C, Carl AG, Gong R, Alushin GM (2022) Bending forces and nucleotide state jointly regulate F-actin structure. Nature 611:380–386
66. Robert-Paganin J, Pylypenko O, Kikuti C, Sweeney HL, Houdusse A (2020) Force generation by myosin motors: a structural perspective. Chem Rev 120:5–35
67. Rould MA, Wan Q, Joel PB, Lowey S, Trybus KM (2006) Crystal structures of expressed non-polymerizable monomeric actin in the ADP and ATP states. J Biol Chem 281:31909–31919
68. Scheidig AJ, Burmester C, Goody RS (1999) The pre-hydrolysis state of p21(ras) in complex with GTP: new insights into the role of water molecules in the GTP hydrolysis reaction of ras-like proteins. Structure 7:1311–1324
69. Stevenson SR, Tzokov SB, Lahiri I, Ayscough KR, Bullough PA (2025) Cryo-EM reconstruction of yeast ADP-actin filament at 2.5 Å resolution. A comparison with vertebrate F-actin. Structure 33:435–442
70. Straub F (1942) Actin. Stud Inst Med Chem Univ Szeged 2:3–15
71. Straub F (1943) Actin, ii. Stud Inst Med Chem Univ Szeged 3:23–37
72. Straub F, Feuer G (1950) Adenosinetriphosphate the functional group of actin. Biochim Biophys Acta 4:455–470
73. Tanaka K, Takeda S, Mitsuoka K, Oda T, Kimura-Sakiyama C, Maéda Y, Narita A (2018) Structural basis for cofilin binding and actin filament disassembly. Nat Commun 9:1860
74. Tarnutzer K, Siva Sankar D, Dengjel J, Ewald CY (2023) Collagen constitutes about 12% in females and 17% in males of the total protein in mice. Sci Rep 13:4490
75. Tirion MM, Ben-Avraham D, Lorenz M, Holmes KC (1995) Normal modes as refinement parameters for the F-actin model. Biophys J 68:5–12
76. von der Ecken J, Müller M, Lehman W, Manstein DJ, Penczek PA, Raunser S (2015) Structure of the F-actin-tropomyosin complex. Nature 519:114–117
77. Vorobiev S, Strokopytov B, Drubin DG, Frieden C, Ono S, Condeelis J, Rubenstein PA, Almo SC (2003) The structure of nonvertebrate actin: implications for the ATP hydrolytic mechanism. Proc Natl Acad Sci USA 100:5760–5765
78. Wang YL (1985) Exchange of actin subunits at the leading edge of living fibroblasts: possible role of treadmilling. J Cell Biol 101:597–602
79. Wang Y, Wu J, Zsolnay V, Pollard TD, Voth GA (2024) Mechanism of phosphate release from actin filaments. Proc Natl Acad Sci USA 121:e2408156121
80. Wegner A (1976) Head to tail polymerization of actin. J Mol Biol 108:139–150
81. Wioland H, Guichard B, Senju Y, Myram S, Lappalainen P, Jégou A, Romet-Lemonne G (2017) ADF/Cofilin accelerates actin dynamics by severing filaments and promoting their depolymerization at both ends. Curr Biol 27:1956–1967
82. Xue B, Leyrat C, Grimes JM, Robinson RC (2014) Structural basis of thymosin-beta4/profilin exchange leading to actin filament polymerization. Proc Natl Acad Sci USA 111:E4596–E4605
83. Zahm JA, Padrick SB, Chen Z, Pak CW, Yunus AA, Henry L, Tomchick DR, Chen Z, Rosen MK (2013) The bacterial effector VopL organizes actin into filament-like structures. Cell 155:423–434

Chapter 5
Structural and Functional Elements of the Epidermal Appendages in Birds and Reptiles: Conformations and Modes of Assembly of the Constituent β-Filaments and Keratin Intermediate Filaments (IF)

David A. D. Parry

This chapter is dedicated to two outstanding scientists—Bruce Fraser and John Squire. Both were greatly valued friends and research colleagues over a span of more than 50 years, and each played a leadership role in their own field of research—keratin and collagen (Bruce Fraser) and muscle (John Squire). Both passed away in recent years—Bruce Fraser on 16 June 2019 and John Squire on 31 January 2021. My career owes much to both of them but in quite different ways. With Bruce, I co-authored 40 research papers over a span of 53 years, mostly on keratin. It was an area of research that he inspired me to contribute to over an extended period, as he did also with other fibrous proteins. With John, I co-organised five workshops in Alpbach, Austria, on coiled-coils, collagen and co-proteins, and we also co-edited four books, including the first edition of this volume. John and I also wrote a paper on the regulation of vertebrate skeletal muscle that has become increasingly important with time. My sincere thanks went to both of them—it was fun, it was productive and it was a privilege to collaborate with them.

Abstract Much is now known about the structures of the β-filaments and the intermediate filaments that together constitute the bulk of the avian and reptilian (sauropsid) appendages (claws, scales, feathers and beaks). New sequence data from the Rhynchocephalia (the tuatara), the last branch of the phylogenetic classification of the sauropsids to be studied, has confirmed that all members of the sauropsids are based on common structures. In addition, an examination of the sequence data has revealed that the β-filaments in the lepidosaurs (lizards, snakes and tuatara) contain a chain that is likely to be a structural component of two separate filaments, thereby providing a unique feature that could facilitate ordered filament aggregation.

D. A. D. Parry (✉)
School of Natural Sciences, Massey University, Palmerston North, New Zealand
e-mail: d.parry@massey.ac.nz

D. A. D. Parry, P. K. Luther (eds.), *Fibrous Proteins: Structures and Mechanisms*, Subcellular Biochemistry 113, https://doi.org/10.1007/978-3-032-05273-5_5

Similarly, a Type II IF chain (K80) in lizards appears capable of forming an interaction in inter-filament space that would link adjacent IF through tail-tail interactions. A Type II IF chain in zebra finch (K78LT) also seems likely to play a similar role. More details of the surface lattice structure in the IF have also been obtained, as has information on the lateral packing of the protofilaments in the IF. Consequently, an increasingly detailed picture has emerged of the structure and assembly of the filamentous structures that comprise the corneous appendages in the sauropsids.

Keywords β-Filaments · Corneous β-proteins · Intermediate filaments · Inter-filament linkages · Lateral organisation in IF · Surface lattice structure · Sequence repeats · Rhynchocephalia · Squamates · Lepidosaurs

Abbreviations

CBP	Corneous β-protein
DST	Disulpho-succinimidyl-tartrate
EDC	Epidermal differentiation complex
HGT	High glycine-tyrosine proteins
HS	High-sulphur proteins
IF	Intermediate filament
KAP	Keratin-associated protein
STEM	Scanning transmission electron microscopy
UHS	Ultra-high sulphur proteins

Introduction

Ultrastructural studies on the scales, claws, beaks and feathers of the sauropsids (reptiles and birds) have revealed a fibril-matrix composite [1–3, 11, 50] designed to provide the necessary mechanical attributes to the animal's appendages so that they may function and thrive in a hostile environment. This ultrastructure has some similarities to that characterised in the mammalian hard α-keratins, such as quill and hair, where intermediate filaments (IF) of well-defined structure [27, 73] exist in a matrix comprised of several families of keratin-associated proteins (KAPs). These are characterised by high contents of cysteine residues (high sulphur (HS) and ultra-high sulphur (UHS) proteins) and also by glycine and tyrosine residues (HGT proteins) [14, 24, 25, 65, 73]. Some of the matrix proteins are known to contain sequence regularities, and these are expected to give rise to elements of local secondary structure. However, none of the matrix proteins would be expected to have a unique tertiary conformation and, in all probability, would adopt a structure consistent with its local environment [28].

However, the predominant filaments in the corneous appendages are based on a β-structure and have diameters of only 3.4 nm. These β-filaments are of pre-eminent importance in providing the scales, claws, feathers and beaks with their appropriate tensile properties. The inter-filament matrix is formed from both terminal domains of the corneous β-proteins (CBP) that form the β-filaments and the proteins of the epidermal differentiation complex (EDC) [69]. Together, these endow the appendages with their strength, flexibility and ability to resist compression and torsion [16, 23, 41, 57, 71]. Keratins that form IF are found in these tissues too, but they are believed to be far fewer in number than the CBP forming the β-keratin filaments and are therefore of lesser significance as regards their contribution to the tissue's mechanical attributes. It is important to note, however, that the relative percentages of the constituent IF and β-filaments have not yet been determined quantitatively for any corneous appendage but various strands of evidence indicate that the number of β-filaments far exceeds those of the IF (see, for example, [9, 39, 40, 49, 53, 77, 78]). In addition, the precise localisation of both sets of filaments has not yet been ascertained for any of the appendages in the sauropsids, though there are, once again, indications that the β-filaments form the core of the structure whilst the IF are more peripherally located in discrete layers. More data, however, will be required to confirm both of these points.

Much is now known about the primary, secondary and tertiary structures of both sets of filaments. Firstly, hundreds of sequences of the CBPs have been published, and, more recently, the database of sequences of the IF proteins has grown considerably (see Ehrlich et al. [9]). Alibardi et al. [4] have subsequently defined different regions of the CBPs and produced specific antibodies against various epitopes that are useful in structural/biological studies. Well-determined secondary and tertiary structures have been published for both the β-filaments (see Fraser and Parry [25], Parry [57]) and the IF in hard α-keratins (but not for those in the corneous appendages) [24]. While the reader is referred to the full details of these conformations in the reviews noted above, it is suffice here to outline their main features and, in the following text, to concentrate on the structural aspects of both sets of filaments that have more recently been established.

The sequences of the keratin chains that form the β-filaments in the archelosaurs (birds, crocodiles and turtles) comprise four segments (segment A, a largely conserved 34-residue segment with a high propensity to form a β-structure, and segments C and D). In the lepidosaurs (lizards, snakes and the tuatara, see section "Corneous β-Proteins and Assembly of β-Filaments"), the sequences have five segments (segments A and B, a conserved 34-residue with a high propensity to form a β-structure, and segments C and D) (Fig. 5.1). The sequences and composition characteristics of segments A, B, C and D are described in detail by Fraser and Parry [25], but the key observations are as follows, though a degree of variation from the stated norms has been observed in a few cases. Segment A in the archelosaurs is short (22–26 residues), and the sequences display high homology with one another. In the squamates, the sequences in segment A (18–25 residues in the lizards and snakes) are relatively rich in cysteine, proline and serine residues and are also

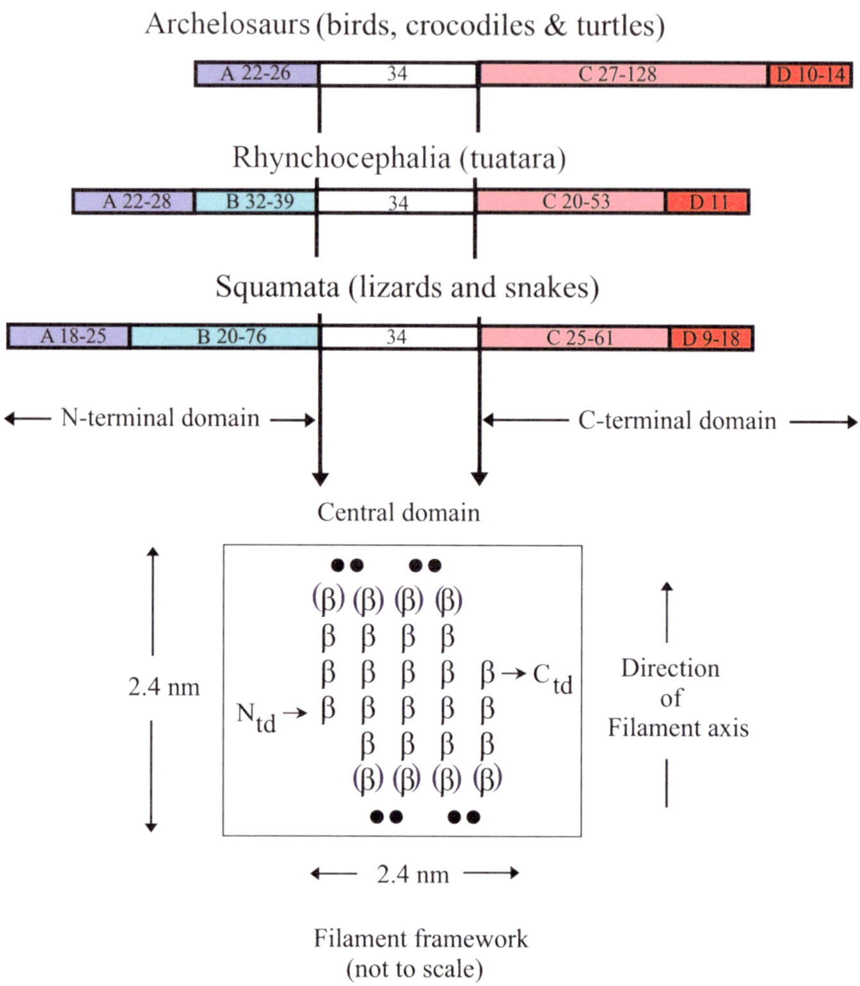

Fig. 5.1 Sequence comparisons from representatives of the sauropsids reveal a central homologous 34-residue domain that is believed to form a right-handed, antiparallel twisted β-sheet with three central strands and two partial outer ones. The head (N-terminal domain) is characterised by an A domain in the archelosaurs and an A and B domain in the lepidosaurs. In contrast, the tail (C-terminal domain) contains a C and a D domain in both the archelosaurs and the lepidosaurs. The characteristics of the A, B, C and D subdomains are noted in the text. (Reproduced from Fig. 4 of Parry et al. [63] with permission of Elsevier)

homologous to one another. They are, nonetheless, different from those in the archelosaurs.

Segment B (a feature of the squamates but not the archelosaurs) is rich in glycine and tyrosine residues, and these residues are commonly found in repeats 2–7 residues in length. In both the archelosaurs and the squamates, segment C has an elevated content of glycine, serine and tyrosine residues, and this segment too often contains short sequence repeats. In contrast, segment D in both the archelosaurs and squamates is rich in cysteine and sometimes arginine residues [22]. The key structural feature of all CBPs, however, is the homologous length of sequence (34 residues) that is found in all sauropsids. Using sequence analysis techniques, it was shown that this region consisted of out-of-phase periodicities of about eight residues in the propensity of the residues to form β- and β-turn structures, thereby indicating an antiparallel β-sheet conformation consistent with earlier infrared data [15, 22]. X-ray fibre diffraction and model building [12, 13, 25, 34] have shown that the 34-residue length of sequence will form a right-handed, twisted, antiparallel β-sheet comprising three inner strands and two partial outer ones. This sheet, in turn, assembles in an antiparallel manner with a similar sheet in a second molecule. These are related by a perpendicular dyad axis of symmetry. The role played by the apolar residues in the assembly and stabilisation of the β-sandwich thus formed is evident in the molecular dynamics simulations of Calvaresi et al. [6]. In turn, the β-sandwiches assemble axially with a left-handed four-fold screw symmetry (axial rise 2.4 nm per sandwich, pitch length 9.6 nm) (Fig. 5.2). The fully assembled filaments have diameters of 3.4 nm, and the N- and C-terminal domains are located on the surface of the filament.

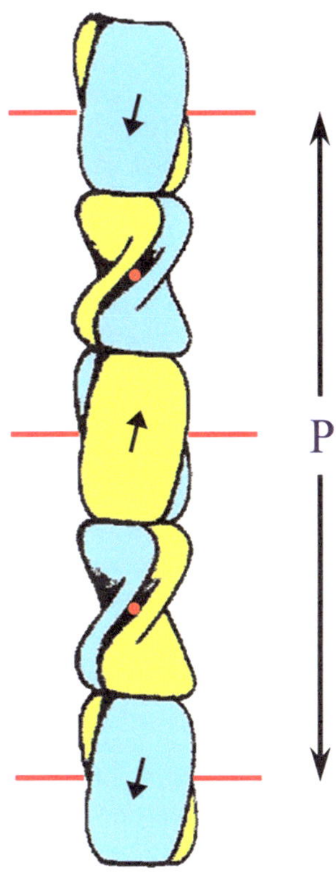

Fig. 5.2 The β-sandwiches (coloured blue and yellow at the same axial coordinate) are formed from the antiparallel interaction of a pair of β-sheets. Each of the two constituent β-sheets displays a right-handed twist and is formed from the conserved 34-residue sequence present in each of the pair of molecules involved. The perpendicular dyad axes of rotation in the filament are marked by a red line or by a solid red circle. These are found at axial intervals of 2.4 nm. The pitch length (P) of the helix with left-handed four-fold screw symmetry is 9.6 nm, and the diameter of the filament thus formed is 3.4 nm. (Reproduced from Fig. 5 of Parry [57] with permission of Springer)

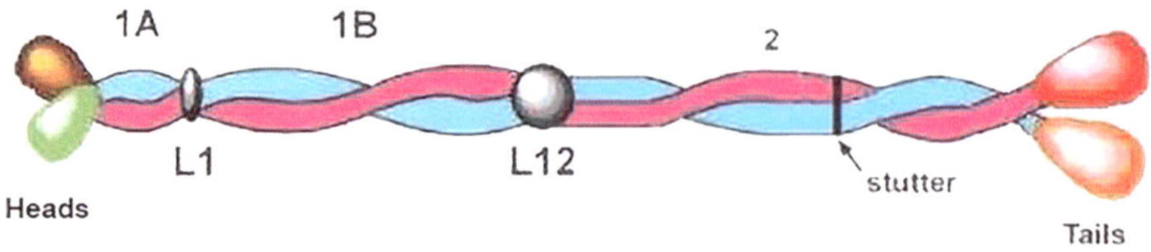

Fig. 5.3 IF molecules are in-register parallel heterodimers formed from a Type I and a Type II chain. The central rod domain of the molecules (1A-L1-1B-L12-2) is largely characterised by a heptad substructure favouring an α-helical, left-handed coiled-coil conformation. The rod is enclosed by an N-terminal (head) and a C-terminal (tail) domain. (Reproduced from Fig. 5.3 of Fraser and Parry [24] with permission of Springer)

IF structure is well characterised for the hard α-keratins of quill and hair but thus far there are no comparable studies for the IF in the corneous appendages. Nonetheless, the marked sequence homologies that exist between their respective IF proteins (see later) strongly suggest that the filamentous structures adopted will be either identical or very closely similar. Hence, the IF structure for the hard α-keratins will be described here briefly [24]. The IF sequences in both the sauropsid and mammalian hard keratins can be subdivided into an N-terminal cysteine-rich head domain, a central α-helical-rich rod domain and a cysteine-rich C-terminal tail domain (Fig. 5.3). The rod domain is comprised of coiled-coil segments, each with a heptad substructure. These are known as segments 1A (35 residues), 1B (101 residues) and 2 (148 residues). Linkers, generally non-α-helical, are termed L1 and L12, and join 1A to 1B and 1B to 2, respectively. Importantly, two closely related but nonetheless distinguishable IF chain types with the sequence features noted above have been identified. These are known as Type I and Type II IF chains. There is now abundant evidence that each IF molecule is a Type I/Type II heterodimer with the chains lying parallel (rather than antiparallel) to one another and that these lie in axial register (Fig. 5.3). One of the distinguishing features between the chains is the conserved but different length of the linker L12 (16 residues in the Type I chains and 17 residues in the Type II chains).

Using chemical crosslinking between spatially adjacent lysine residues in intact filaments and the subsequent characterisation of the residues involved, it has been possible to distinguish experimentally three unique modes of axial alignment within the IF [66, 67, 74]. These are known as A_{11}, A_{12} and A_{22} and correspond respectively to an overlap of antiparallel 1B segments, an antiparallel near-complete overlap of a pair of molecules, and an antiparallel overlap of 2 segments (Fig. 5.4). This result confirmed the model proposed by Crewther et al. [7], who noted that the periodicities observed in the linear distribution of the acidic and basic residues in the 1B and 2 segments [54, 58, 62] would allow the maximisation of inter-molecular ionic interactions to occur if the three arrangements of the two segments (1B–1B, 1B–2 and 2–2) were largely overlapped in an antiparallel manner.

It has also proved possible to characterise the surface lattice structure of the IF in the hard α-keratins of quill and hair using the positions of the meridional and

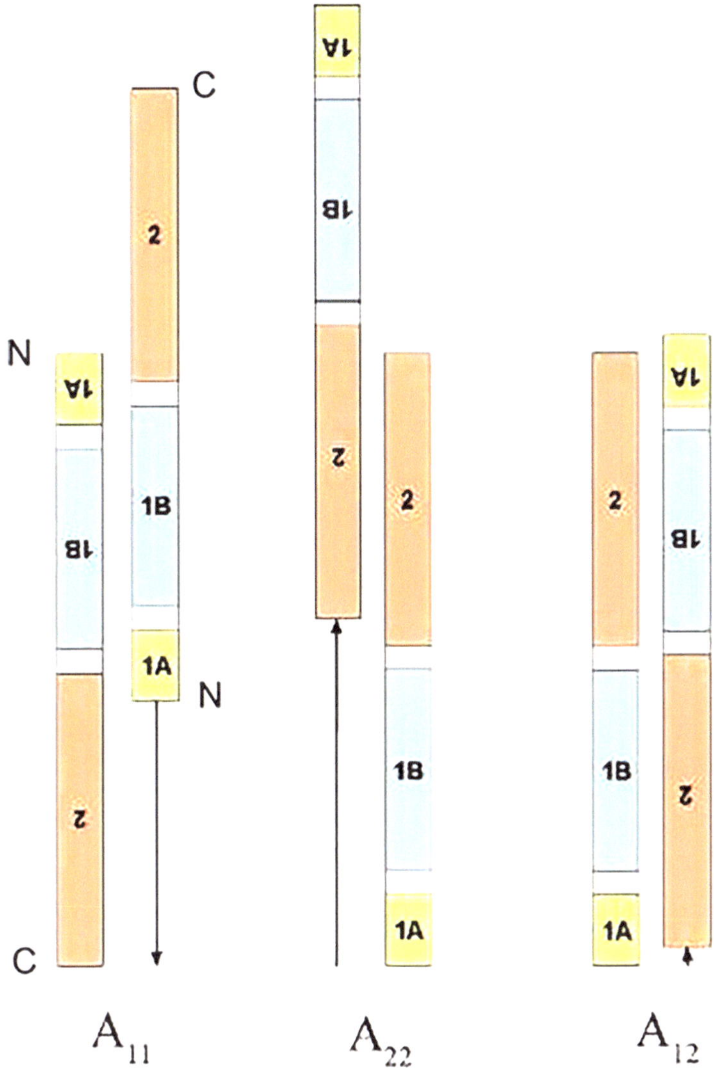

Fig. 5.4 The intermolecular staggers of the molecules in the IF have been deduced from DST crosslinking data [66, 67, 74] and have thereby confirmed the modes of assembly initially proposed by Crewther et al. [7]. The three (and only three) modes are termed A_{11}, A_{22} and A_{12}, and correspond respectively to antiparallel overlaps of the 1B segments, the 2 segments and the molecules as a whole. (Reproduced from Fig. 5.4 of Fraser and Parry [24] with permission of Springer)

near-meridional reflections (Fig. 5.5: [17, 18, 35–37]). The surface lattice thus characterised revealed a helical dislocation in the packing of the protofilaments, each of which is formed from a regular axial arrangement of a pair of antiparallel

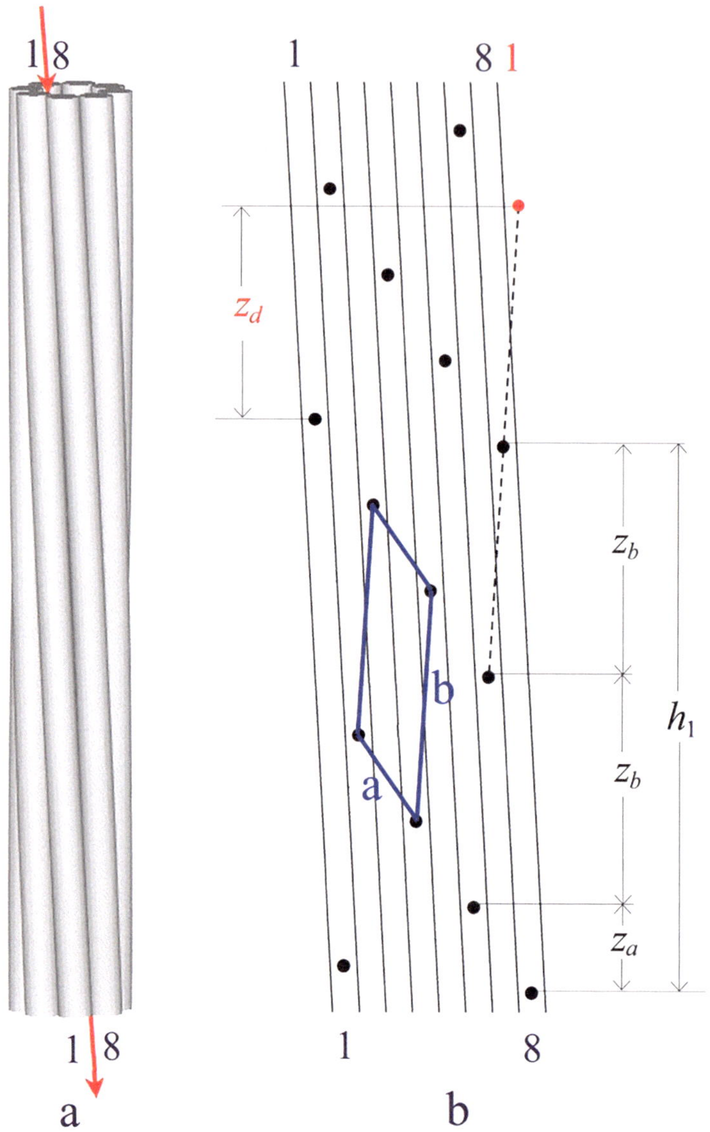

Fig. 5.5 Schematic diagram illustrating the helical dislocation for oxidised trichocyte keratin IF. (**a**) Three-dimensional view of an IF with eight helical four-chain protofilaments on a ring of radius 2.99 nm. The interface (or seam) between neighbouring protofilaments is generally identical, but that between protofilaments 1 and 8 are disrupted as a result of a helical dislocation (see red arrows), (**b**) the unit cell of the limited surface lattice is shown in blue. Black dots indicate equivalent points in the protofilaments. z_a and z_b are the axial projections of the unit cell lattice dimensions a and b, and z_d is the axial projection of the dislocation along a helical path. z_d is equivalent to $2z_b$ - $3z_a$. For a regular helical structure, the expected position of the reference point in the protofilament is shown in red on the right-hand side of the figure but is displaced from it in the structure by the dislocation z_d, as shown by the arrows. (Reproduced from Fig. 2 of Fraser and Parry [29] with permission of Elsevier)

molecules. From the data noted above, plus parameters determined from fibre X-ray diffraction studies, a detailed picture of the structure of the intact IF at low resolution has been determined.

This chapter will primarily consider data that have become available for the proteins constituting the corneous appendages since 2017 (see review of Fraser and Parry [24, 25] in the first edition of this book) and the interpretation of these data in terms of the structure and assembly of both filament types. It will also highlight those areas of research that require further investigation.

Sequences of Corneous β-Proteins and Keratin Intermediate Filament Chains

Secondary structure, as well as elements of tertiary structure and function, can often be deduced from a knowledge of the amino acid sequences of the proteins that constitute the tissue. In this case, the structural proteins that form claws, scales, beaks and feathers in the sauropsids (birds and reptiles) have been studied in considerable detail, and sets of sequence data relating to each filament type have been obtained. The first set relates to the CBPs that form the filaments with diameters of about 3.4 nm, and the second set to the α-helix-rich keratin proteins that form the IF with diameters of 7–10 nm.

Corneous β-Proteins

Several hundred sauropsid β-sequences have now been published (see, for example, [8, 39, 42–44, 51, 52]; and those listed in the review by Ehrlich et al. [9]). However, one major omission was evident and that related to the absence of sequence data for the Rhynchocephalia, a branch of the phylogenetic classification of birds and reptiles ([70]; Fig. 5.6).

The sole member of the Rhynchocephalia—the tuatara (*Sphenodon punctatus*)—evolved from other lepidosaurs at the time of the Triassic period (Fig. 5.7). Interestingly, the tuatara represents a link to the now extinct stem reptiles from which the dinosaurs evolved. This sequence gap has recently been filled and tuatara corneous β-protein sequences have been reported by Parry et al. [63]. Consequently, it has now become possible to analyse the similarities and differences of these sequences with the other sauropsid chains, as well as with humans [60]. The results show that the 37 partial tuatara sequences recorded all contain the same homologous sequence 34-residues long that is conserved across the entire range of the CBPs in the sauropsids. Simply put, it can be concluded that the β-filamentous structures proposed earlier for the birds, crocodiles, lizards and snakes is equally applicable to the tuatara. While this is not a surprise, of course, it is pleasing to see this gap finally filled with hard data.

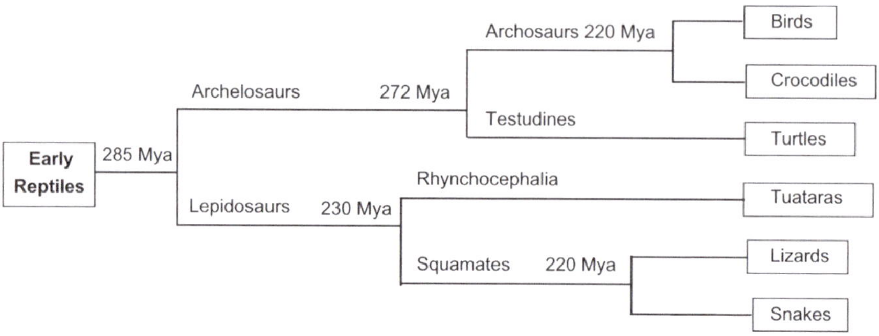

Fig. 5.6 Phylogenetic classification of the sauropsids with branching times indicated in millions of years (Mys). The expectation is that the archelosaurs, i.e. the birds and crocodiles (archosaurs) and turtles (testudines), would show greater similarity between themselves than with the lepidosaurs, i.e. lizards and snakes (squamates) and the tuatara (rhynchocephalia). Likewise, the lepidosaurs would also show greater similarity between themselves than with the archelosaurs. (Reproduced from Fig. 1 of Parry et al. [63] with permission of Elsevier)

Fig. 5.7 The tuatara (*Sphenodon punctatus*), which evolved from the lizards and snakes at the time of the Triassic period, is a link to the now extinct stem reptiles from which the dinosaurs evolved

Type I and Type II Keratin IF Chains

In contrast to the CBPs, the sequences reported for the Type I and Type II IF proteins have been fewer in number. However, that is now beginning to change. The genome sequences of green anole lizard (*Anolis carolinensis*; [72]), red jungle fowl chicken (*Gallus gallus*), Chinese alligator (*Alligator sinensis*), painted turtle (*Chrysemys picta bellii*) and the venomous pit viper (*Protobothrops mucrosquamatus*) (see

Table 5.1 Numbers of Type I and Type II keratin IF chains in sauropsids

		Total chains		Fits Criteria		Hard		Epid		Epit	
Common name	Species	I	II	I	II	I	II	I	II	I	II
Common wall lizard[1]*	*Podarcis muralis*	19	17	16	12	3	2	9	7	3	2
Green anole lizard[2]	*Anolis carolinensis*	19	20	nd	nd	nd	nd	nd	nd	nd	nd
Green anole lizard[1]	*Anolis carolinensis*	19	21	15	18	3	4	8	10	4	4
Tuatara[1]	*Sphenodon punctatus*	15	17	13	12	3	2	7	6	3	4
Red jungle fowl[2]	*Gallus gallus*	15	15	nd	nd	nd	nd	nd	nd	nd	nd
Chinese alligator[2]	*Alligator sinensis*	21	20	nd	nd	nd	nd	nd	nd	nd	nd
Painted turtle[2]	*Chrysemys picta bellii*	12	15	nd	nd	nd	nd	nd	nd	nd	nd
Venomous pit viper[2]	*Protobothrops mucrosquamatus*	14	14	nd	nd	nd	nd	nd	nd	nd	nd
Tibetan frog[2]	*Xenopus tropicalis*	23	14	nd	nd	nd	nd	nd	nd	nd	nd

*One Type I and one Type II chain have not been classified as hard, epidermal or epithelial since they have at least one unusual feature (discussed later). The term "nd" means "not determined". Superscript 1 refers to Parry and Winter [61], and 2 refers to Ehrlich et al. [9]. Hard, Epid and Epit refer to hard, epidermal and epithelial keratin chains

Ehrlich et al. [9] for details) have been investigated to ascertain whether keratin genes were present. They were indeed and multiple sequences of the Type I and Type II IF chains (hair-like, epidermal and epithelial) have now been published for each species (Table 5.1). In addition, Ehrlich et al. [9] have reported that genes that encoded cysteine-rich keratins (thereby implying the presence of both Type I and Type II hard keratin chains) have been found in the Tibetan frog (*Nanorana parkeri*), clawed frog (*Xenopus laevis* and *Xenopus tropicalis*), zebrafish (*Danio rerio*), kiwi (*Apteryx australis mantelli*), ostrich (*Struthio camelus australis*), zebra finch (*Taeniopygia guttata*), American alligator (*Alligator mississipiensis*), green sea turtle (*Chelonia mydas*), Burmese python (*Python bivittatus*) and the platypus (*Ornithorhynchus anatinus*). It is important to emphasise, however, that not all the keratin chains described above (chicken, alligator, turtle, lizard and snake), as well as those present in the cysteine-rich HAS (hard acidic sauropsid-specific) and HBS (hard basic sauropsid-specific) proteins identified by Ehrlich et al. [9], will necessarily give rise to chains capable of assembly into functional IF. This followed the work of Parry and Winter [61], who defined a structural protocol to identify those IF chains that would indeed exhibit the necessary assembly characteristics. This included, for example, the requirement that the lengths of the coiled coil segments 1A, 1B and 2 be conserved (35, 101 and 148 residues, respectively), that the central linker L12 be 16 residues in Type I chains and 17 residues in Type II chains, and that the heptad/hendecad/stutter substructure in the sequences of the coiled-coil segments be maintained. To illustrate this point, Parry and Winter [61] used the genome sequences of tuatara, the European common wall lizard and the green anole lizard to ascertain those chains consistent with these criteria and which were therefore deemed capable of forming viable IF. The results showed that about 80% of the chains identified as having the above characteristics were likely to assemble into viable IF (Table 5.1). The remaining 20% were termed IF-like chains, as some of the coiled-coil segments were missing significant numbers of residues. Table 5.1 also illustrates that three Type I and two Type II "hair-like" keratin chains exist in tuatara, three Type I and four Type II chains in the green anole lizard, and three Type I and two Type II chains in the European common wall lizard. Comparable numbers were listed for the epidermal and epithelial chains.

It was considered that small sequence errors in the genome assembly might have been the reason why it was necessary to designate some of the chains as IF-like rather IF, and that if these minor genome sequence errors were corrected then these chains too might indeed be genuine IF ones. However, the possibility that some IF-like chains may represent pseudogenes in the genome was also noted.

Assembly Studies

Proteins lying between the β-filaments consist of the head and tail domains of the CBPs and the proteins of the EDC. Indirect linkages between adjacent filaments via these proteins or protein segments must therefore occur. However, it has long been speculated that direct linkages might also occur between filaments. The first evidence that this is indeed a feature of assembly in at least half of the sauropsids (the lepidosaurs) is outlined in section "Assembly of β-Filaments".

The question also arises as to whether similar direct linkages occur between IF, in addition to the indirect ones made through the proteins of the EDC proteins in the sauropsids (and the KAP proteins in the matrix of the mammals). Once again, recent sequence data suggests that this is highly likely. Details of the modelling for the IF are discussed in section "Assembly of IF".

Assembly of β-Filaments

Earlier sequence data had revealed the presence of a minor chain component in lizards and snakes that had a molecular weight approximately twice that of the usual complement of β-corneous proteins. Furthermore, each of these chains contained four 34-residue repeats rather than just one (Fig. 5.8: King Cobra (*Ophiophagus Hannah*) and Burmese python (*Python bivittatus*), [43]; Japanese gecko (*Gekko japonicus*), [52]; green anole lizard (*Anolis carolinensis*), [8]). New sequence data revealed that a similar chain was also present in tuatara (*Sphenodon punctatus*) [63], thereby indicating the existence of a feature common to all lepidosaurs. Fraser and Parry [30] proposed that its commonality implied a structural and/or functional role in vivo, and subsequent modelling indicated that just six families of structure were

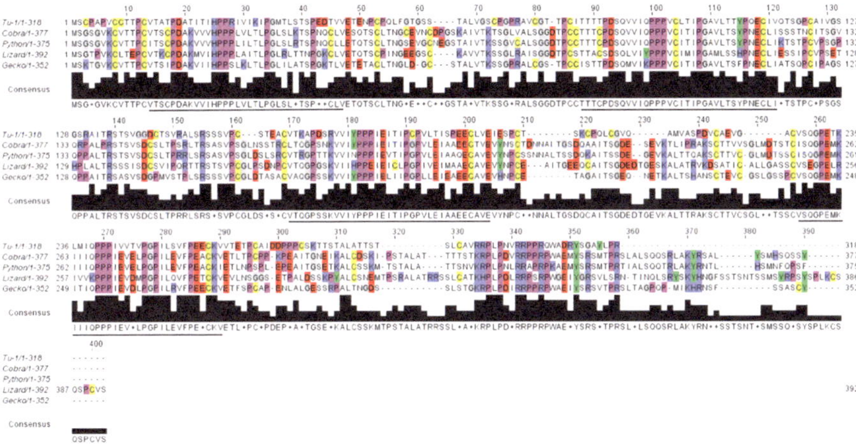

Fig. 5.8 A specialised chain present in each branch of the sauropsids contains four 34-residue repeats (residues 14–47, 88–121, 170–203 and 257–290). Five representative sequences are included here for comparison purposes—tuatara (*Sphenodon punctatus*), King cobra (*Ophiophagus hannah*), Burmese python (*Python bivittatus*), green anole lizard (*Anolis carolinensis*) and Japanese gecko (*Gekko japonicus*). Each of the four 34-residue repeats is underlined by a black bar. (Reproduced from Fig. 2 of Fraser and Parry [30] with permission of Elsevier)

possible: in three of them, all four 34-residue repeats existed within a single filament, and in the three others the repeats were found in either two, three or four filaments, respectively. The latter three cases implied the presence of direct inter-filament linkages. The favoured model was one in which a β-sandwich in filament 1 (comprising the first two 34-residue repeats) was connected via a centrally located sequence (designated the middle linker) to a β-sandwich in filament 2 (comprising the last two 34-residue repeats).

More detailed investigations by Parry [56] considered each of the six families of structure noted above and the possible members within each class (a total of 20 potential arrangements). The topology of the three linking sequences (designated the first, middle and last linkers that occur between the first two 34-residue repeats, the middle two 34-residue repeats and the last two 34-residue repeats, respectively) and the chemical characteristics of these linkers were studied in an effort to discriminate between the 20 arrangements.

As regards the sequence characteristics of the linkers, it was observed that the middle linker exhibited special features that were noticeably different from those present in the first and last ones. For example, the middle linker was the longest of the three linkers and exhibited the least variation in length, it was the least acidic but the most basic, it was the most proline-rich, and it contained the highest percentage of conserved residues. Taken together, these indicate a unique role for the middle linker [30]. Further analyses [56] also showed that the number of serine residues in the middle linker was approximately twice that of those in the first and last linker and, furthermore, that the serine residues in the middle linker (and only that linker) had a particular substructure of the form $(S\text{-}X)_{2,3}$ separated by sequences containing proline and cysteine residues. These observations were consistent with the middle linker possibly adopting a β-like structure and having a special functional role in vivo. Further, the first and second 34-residue segments in this chain were shown to be the most similar to one another, as were the third and fourth.

The topology maps indicated that eight of the 20 arrangements would lead to the expectation of similar structures for all three linkers, a feature inconsistent with the sequence data noted above. Similarly, four other arrangements would lead to the expectation of a unique structure for either the first or the last linker, and two would give rise to physically improbable linker lengths. Six arrangements, however, were consistent with a unique role for the middle linker.

Thus, using all the sequence analysis data and the topological criteria, Parry [56] was able to conclude that the most likely model was indeed one in which the first and second 34-residue repeats formed a β-sandwich in filament 1, that the third and fourth 34-residue repeats formed a β-sandwich in filament 2, and that these were physically joined via the middle linker (Fig. 5.9). This model, if proven correct, would be the first example of direct filament-filament interactions and therefore an important observation on one of the potential modes of assembly of the β-filaments. At this stage, however, a chain capable of playing a similar role for the archelosaurs has yet to be identified.

Assembly of IF

In determining the IF sequences in tuatara, an effort was made to ascertain whether the equivalent classes of matrix proteins found in mammals (HS, UHS and HGT) were also present in tuatara [60]. The answer is clear—they are not. There are, of course, many other proteins present in the EDC that might play a comparable role but these are clearly different in some important respects from those played by the HS, UHS and HGT proteins in the mammals.

In addition to potential matrix interactions that may be involved in filament assembly it is of considerable interest to ascertain whether direct IF-IF interactions might also occur. Recent observations suggest that they might indeed be present in at least some of the sauropsids. A study of the genome of the European common wall lizard (*Podarcis muralis*) and the green anole lizard (*Anolis carolinensis*) has revealed a keratin-like Type II IF chain termed K80 [61]. It has a one residue insertion at the boundary between linker L12 and segment 2 but is otherwise compatible with the designation of a keratin IF chain. For both lizards, the K80 chain has a very extensive C-terminal domain (869 residues in the common wall lizard and 916 residues in the green anole lizard) characterised by a remarkable substructure [61]. In the European common wall lizard, the sequence contains six tandem repeats of a 51-residue motif enclosed by 15 and 16 tandem repeats of a 16-residue motif (Table 5.2). In the green anole lizard, the sequence contains four tandem repeats of a highly related 55/56-residue motif enclosed by tandem repeats of a very similar 16-residue motif. The consensus 16 residue motifs are homologous

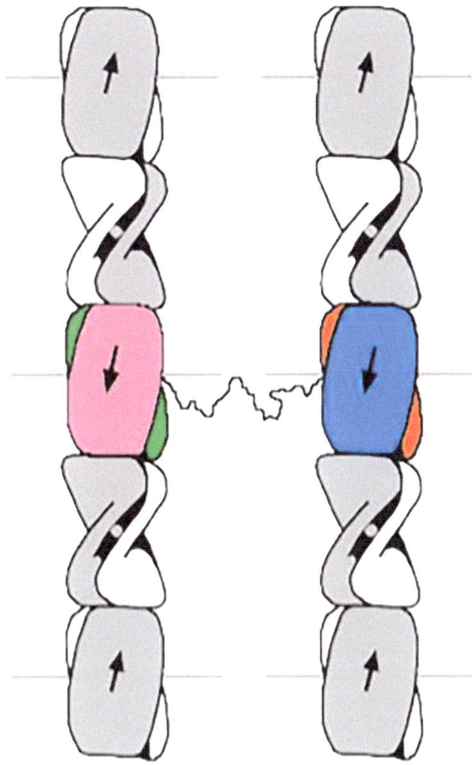

Fig. 5.9 In the chains with four 34-residue repeats, it is proposed that the first two repeats (pink and green) form a β-sandwich in one filament and that the third and fourth repeats (blue and orange) form a β-sandwich in a different filament. The region formed by the sequence between repeats 2 and 3, known as the middle linker, thereby forms a physical connection between filaments. (Reproduced from Fig. 4 of Parry [56] with permission of Elsevier)

within the same species, and, furthermore, they are homologous between the two species. There is a degree of degeneracy in the 16-residue repeats N-terminal to the 51-residue repeats in the European common wall lizard, as there is indeed in the 16-residue repeats C-terminal to the 55/56 residue sequences in the green anole lizard. Nonetheless, the conserved features strongly suggest a common structural and functional role (Table 5.2). Unfortunately, there are no similar 51–56 residue motifs in the structural database that might indicate the likely tertiary structure and function of this particular region in the K80 chain in the two species studied here. However, the 16 residue sequences do exhibit an interesting substructure. In particular, the serine residues are often two residues apart, and predominately occupy positions 7, 9 and 11 as well as 14, 16, 2 and 4. Equally important is the observation that the remaining positions 6, 8, 10 and 12 as well as 13, 15, 1, 3 and 5 are commonly occupied by charged residues. Thus, each 16-residue repeat comprises a pair of segments, each with a two-residue repeat, which, it has been speculated, may adopt a β-sheet-like conformation with one face dominated by serine residues and the other by charged residues [61].

Kajava et al. [48] has shown that five principal β-structural folds exist, and that some of these involve more than a single chain. One role for such structures is in the assembly of networks [46]. It can be speculated that such a role may occur here also

Table 5.2 Consensus sequences of the 16-residue and 51/56-residue motifs in the C-terminal domain of the K80 chains in the European common wall lizard and the green anole lizard

(a)	R	S	H	S	N/G	R/E	S	G/E	G/S	ap	X	E	G	S	ap	S
(b)	X	S/G	S/H	S/R	G	ap	S	E/D	S	V/G	S/ap	S	R	S	Q/L	S
(c)	R	S	H/S	S	E/S	E/R	S	G/P	E/G	ap	S/ap	E/L	G/S	S/R	S/L	S
(d)	R	S	H/S	S	D/E	R/D	S	E/R	G	F/S	S	E	G/S	S	L	S
(e)	E	P	I	V	S	R	T	N	S	A	E	D	G	G	V	P
(f)	D/E	P	V	V	S	R	A	R	S	A	E	E/D	G	G	V	P
(e)	V	G	G	I	S	R	S	N	S	R	A	S	S	Y	V/A	S
(f)	V	G	G	I	S	R	S	Q	S	G	T	S	T	D	ap	S
(e)	G	G	S	H/L	S	R	N/I	E/Q	S/N	D	V/S	S	H	G	S/N	E
(f)	G	G	S	L	S	R	S	Q	S	D	E	N	K	F	H	ap
(e)	G	–	–	–	–	Y	A									
(f)	S	X	X	S	X	Y	T									

Ap refers to apolar residues. (a) and (c) refer to the 16-residue consensus repeats that lie N-terminal and C-terminal, respectively, to the 51-residue repeat in the European common wall lizard, whereas (b) and (d) refer to the 16-residue consensus repeat that lies N-terminal and C-terminal, respectively, to the 55/56-residue repeat in the green anole lizard. (e) and (f) refer to the 51 and 55/56 residue repeats that lie between the 16-residue repeats in the European common wall lizard and the green anole lizard, respectively

for the IF. Thus, the K80 chain would form part of an IF as usual whilst its protruding C-terminal domain with its array of repeating motifs would likely interact with a similar array from a K80 molecule lying in an adjacent IF ([61]; Fig. 5.10). It follows that this situation, if confirmed, would represent the first example of direct linkages/crossbridges occurring between molecules in different IF.

Although this conclusion pertains specifically to two lizards the question must be asked as to whether this attribute is a more general feature of IF assembly. Some preliminary studies suggest that it might be the latter. For example, the Type II K78LT chain in zebra finch has many similarities to the K80 chains. It too has an extensive C-terminal tail domain characterised by 27 tandem repeats of a 54-residue motif as well as multiple tandem repeats of a 38-residue motif. In the C-terminal domains of nestin IF chains, there are multiple 22-residue repeats in human and 44-residue repeats in hamster and rat. Steinert et al. [68] have proposed that a minor component of mammalian nestin in Types III and IV IF is important in allowing the regulation of dynamic assembly of IF in cells. Additionally, they proposed that these nestin sequences might act as crossbridges between IF. Plectin has also been shown to link IF to one another (as well as to other filamentous structures). However, the

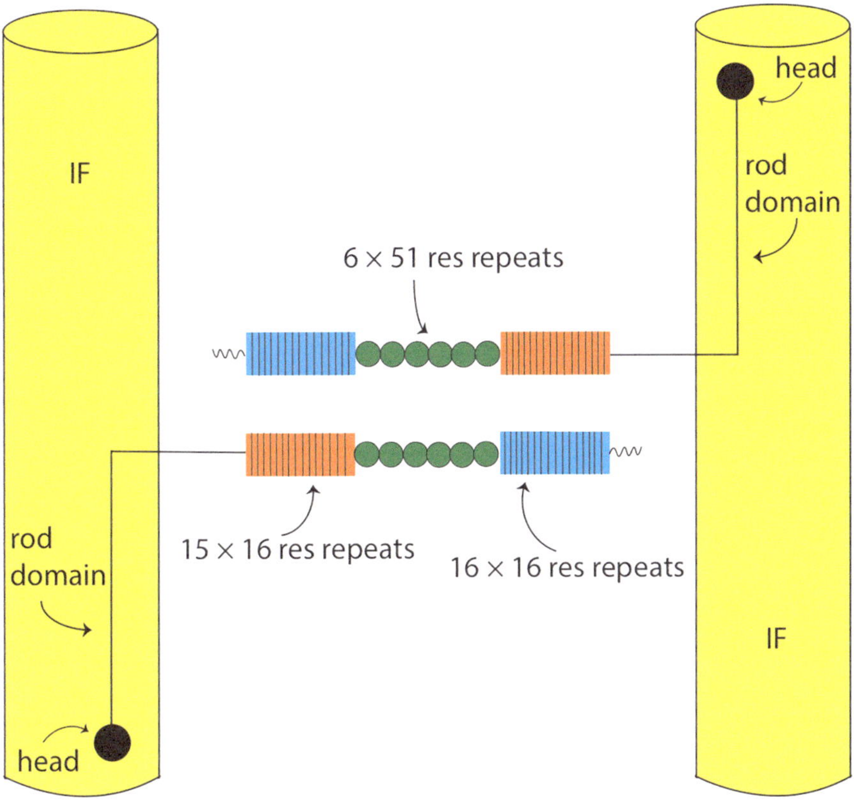

Fig. 5.10 Model derived for the K80 chain present in the European common wall lizard. The large C-terminal (tail) domain is characterised by multiple 16- and 51-residue repeats that are believed to assemble with a similar region emanating from a neighbouring IF. The IF are coloured yellow, the 16-residue repeats are coloured orange and light blue, and the 51-residue repeats are represented as dark green spheres. It has been postulated that the 16-residue repeat regions adopt a β-like conformation and that the subsequent pairing of the β-sheets facilitates assembly of IF into larger groupings. (Reproduced from Fig. 4 of Parry and Winter [61] with permission of Elsevier)

plectin molecule is not a constituent of the IF, and the interactions made between plectin and the IF proteins are made through their external surfaces [76].

Structural Data

The reader is referred to reviews by Fraser and Parry [24, 25, 27] and Parry [57], which describe in some detail the tertiary structures of both the β-filaments and the IF. Since the time of the earlier reviews, new data have been reported and these are noted below. In section "IF in Axial Projection", the lateral structure of hard keratin

IF has been studied using the X-ray diffraction intensity distribution in the 1 nm equatorial region, and this has revealed the projected arrangement of the constituent protofilaments in the IF. Further, in section "IF Surface Lattice Structure", the surface lattice structure of the IF has been reinvestigated using a 1-D Patterson function. The method used was different from those previously used, and this has allowed independent confirmation to be made of the proposed surface lattice structure.

IF in Axial Projection

In hard α-keratin IF, the protofilaments are constructed axially from pairs of antiparallel molecules aligned using the A_{11} and A_{22} modes of assembly. Laterally, these are assembled using the A_{12} mode [66, 67]. Previously published STEM data indicate that there are 32 chains in toto in cross-section [47, 75], thereby implying the existence of eight protofilaments.

The interpretation of the electron microscope images of transverse sections of wool or quill, however, has proved problematical as regards the disposition of those eight protofilaments within the IF cross-section. When stained with osmium the IF often displayed a central core in addition to a peripheral ring of density, thus giving rise to a so-called "ring-core" structure. This implies that one protofilament is located centrally and that the remaining seven lie peripherally at the same radial coordinate in the IF, i.e. a so-called (7 + 1) structure (Fig. 5.11). Conversely, when stained with potassium permanganate, the appearance of the IF lacks the central core and shows only a "ring-like" structure [64]. This would indicate that all eight protofilaments have a common radial coordinate in the IF, i.e. an (8 + 0) structure. Consequently, electron microscopy has thus far been unable to determine which of the two possible arrangements best represents the in vivo situation.

Recently, however, this problem has been tackled from a different standpoint by Fraser and Parry [26], who used data from the equatorial region of the fibre X-ray diffraction patterns. Calculation of the Fourier transform of a single coiled-coil molecule in a matrix of electron density equal to that of the mean value of the coiled-coil results in a uniform peak on the equator at a spacing of about 1 nm. However, the experimental data for the same region reveals fine structure present in the intensity distribution (Fig. 5.12a). This has been explored in terms of small packing differences that will almost certainly occur between constituent molecules. Such differences can be defined in terms of the distances a, b and c between nearest neighbours and the inner and outer radii (*ir* and *or*, respectively) on which the molecules are centred (Fig. 5.12b). If $a = b = c = 1.92$ nm, then the positions of the peaks in the fine structure can be reproduced satisfactorily for the (8 + 0) model (Fig. 5.12c) but their intensities cannot. However, allowing small and realistic variations in these parameters to occur ($a = 1.81$ nm, $b = 2.09$ nm and $c = 2.07$ nm) both the positions and the intensities can be reproduced accurately (Fig. 5.12d). Allowing similar variations in the parameters for the (7 + 1) model does

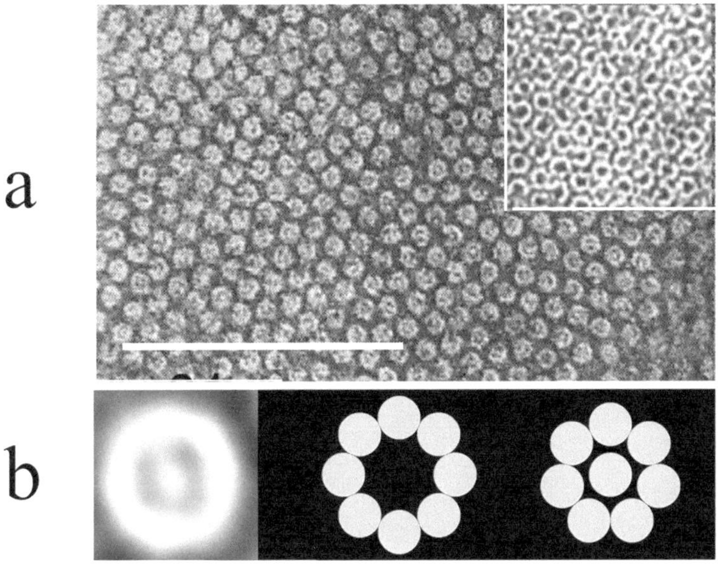

Fig. 5.11 (**a**) Transmission electron micrograph of a transverse section from the paracortex of Merino wool showing a quasi-hexagonal array of IF with diameters of 7–10 nm [10]. Many of the IF in this osmium-stained preparation have an apparent "ring-core" substructure. Magnification bar 100 nm. The inset shows a cross-section from porcupine quill post-stained with potassium permanganate [64]. Here, the IF have a "ring-like" appearance. (**b**) left, an average appearance of the IF in (a) achieved by template matching; centre, schematic representation of the (8 + 0) arrangement of four-chain protofilaments, which forms the basis of the structure proposed for the oxidised structure of keratin IF and which is also believed to be identical or near identical to the IF in the corneous appendages; right, schematic representation of the (7 + 1) arrangement of protofilaments. Each protofilament is represented by a grey disc. (Reproduced from Fig. 1 of Fraser and Parry [26] with permission of Elsevier)

not, however, result in a similar degree of fit to the data that was achieved for the (8 + 0) structure.

If the electron density is cylindrically averaged for the (8 + 0) arrangement, there will be a maximum originating from the position of the protofilaments. The ring will have a radius given by the harmonic mean (*hr*) of the values of *ir* (2.36 nm) and *or* (4.06 nm), where *hr* is given by the expression

$$hr = (2\,ir.or)/(ir + or)$$

This equates to 2.99 nm, a value which fits within the observed range of 2.9 nm (dehydrated) to 3.0 nm (hydrated) [32, 33]. Figure 5.13 illustrates a composite axial (partial) and radial projection of the favoured model.

In the (8 + 0) arrangement of protofilaments, the sizes of the terminal domains would permit them to fit within the core region, and if this occurs, this would likely be stabilised by many disulphide bonds. The high content of apolar residues and the corresponding lack of many charged residues in both the head and tail domains are

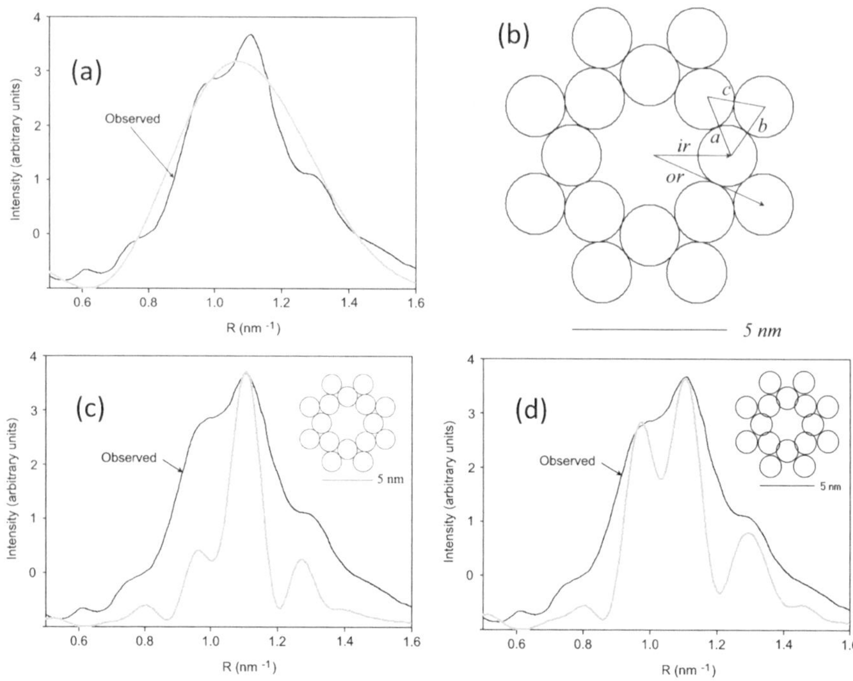

Fig. 5.12 (**a**) Calculated intensity profile for a two-stranded coiled coil rope of pitch length – 14 nm and major helix radius of 0.5 nm (grey) in the 1 nm equatorial region compared to that observed for oxidised hard keratin IF (black). (**b**) A simple model for the 16 molecules in the IF is defined by an inner radius (*ir*), an outer radius (*or*) and the separations of nearest neighbours *a*, *b* and *c*. Each white circle represents a single two-stranded IF molecule. (**c**) Calculated intensity distribution for the model in (b) with $a = b = c = 1.92$ nm. The positions of the maxima in reciprocal space correspond to the observed values, but the relative intensities are mismatched. The lateral arrangement of the 16 molecules, each shown as a white disc, is shown in the inset. (**d**) Calculated intensity distribution for the model in (b) with $a = 1.81$ nm, $b = 2.09$ nm and $c = 2.07$ nm. Not only do the positions of the maxima correspond to those observed, but the intensities are also closely matched. The lateral arrangement of the 16 molecules, each shown as a white disc, is shown in the inset. (Reproduced from Fig. 2 of Fraser and Parry [26] with permission of Elsevier)

therefore consistent with one or both regions lying along the axis of the IF and therefore shielded from water. This would be an energetically favourable situation. Furthermore, any apolar, disulphide-bonded terminal domains lying along the core of the IF would likely exclude some stains employed in electron microscopy (such as the osmiophilic ones) and therefore give rise to the appearance of a central core. This would explain the "ring-core" structure previously noted. Also, the fact that there is a high degree of sequence conservation in a major portion of the terminal domains argues that they might play an assembly role in vivo. Again, it was worth noting that the structural studies have all been carried out on hard keratin IF rather than those

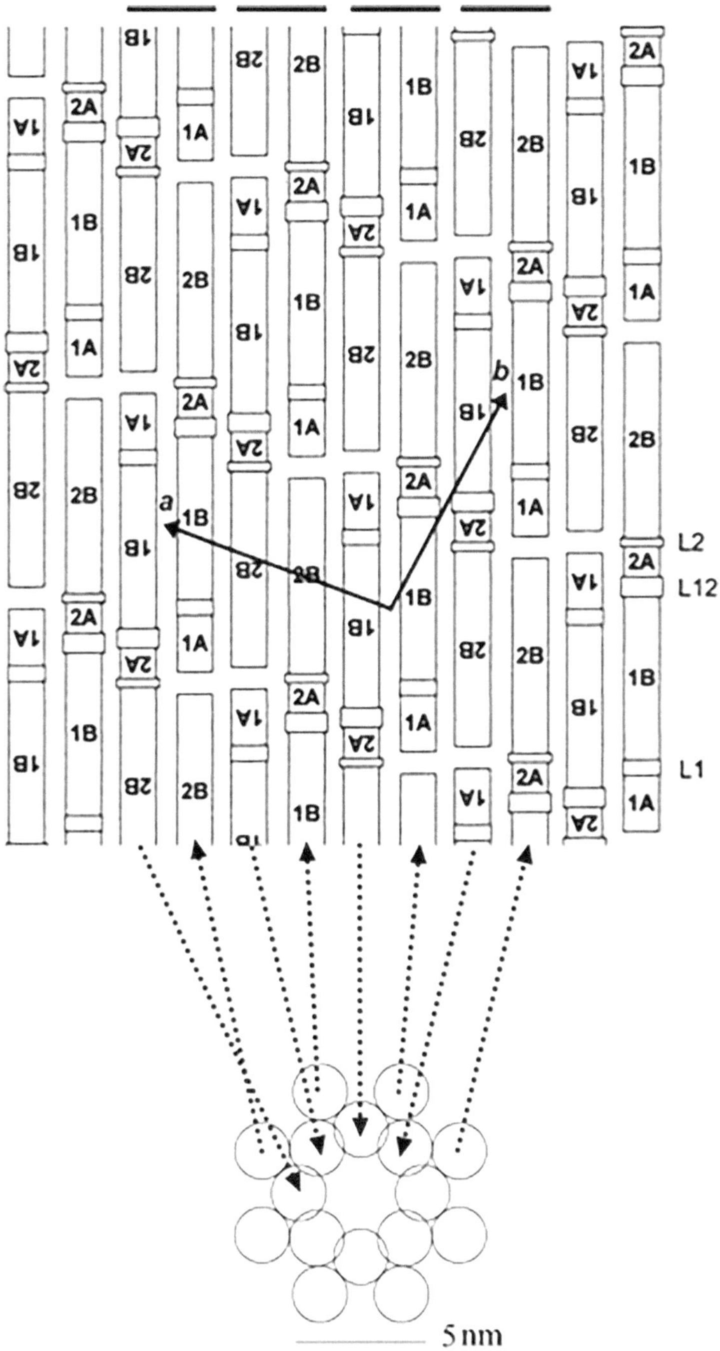

Fig. 5.13 An axial projection (upper) and a radial projection (lower) of the favoured model for the distribution of protofilaments in the oxidised trichocyte IF. An identical or near identical model is believed appropriate for the IF in the corneous appendages. The vectors *a* and *b* define the surface lattice previously described (see text). Upper: Partial radial projection of a simple (8 + 0) model.

from the corneous appendages. The strong similarity, however, in the chain sequences from both sources strongly implies a common or a very closely related structure, hence their inclusion in this review.

IF Surface Lattice Structure

Recent work using a 1-D Patterson approach that utilises the X-ray diffraction *intensities* of the meridional reflections [29] has independently confirmed the surface lattice parameters proposed earlier using the *positions* of the meridional and near-meridional reflections (see Table 5.3) ([17, 18, 35–37]; section "Introduction"). It is pertinent to note that the surface lattice parameters are the same within the experimental error for the hard keratins in the reduced state and for the epidermal keratin IF. However, the hard keratin IF in their (functional) oxidised state has quite different parameters (Table 5.3) following the molecular slippage that occurs between the reduced and oxidised forms [24, 27, 74].

Missing Data

Studies on the structures and mechanics of the corneous appendages have highlighted the lack of quantitative data in three particular areas of research. The first relates to the relative proportions of the IF and β-keratin filaments in claws, scales, beaks and feathers, and the second to the spatial locations of these two filament types. The third relates to the likely variation that occurs in the matrix proteins as regards heterogeneity, size and composition. These topics are discussed in sections "Relative Contents of the β-Filaments and the Intermediate Filaments, Spatial Location of the Filaments, and Protein-Protein Interactions and Their Relationship to the Physical-Chemical Properties of the Corneous Appendages", respectively.

Table 5.3 Surface lattice parameters for keratin IF (in nm)

	Hard IF (reduced)	Hard IF (oxidised)	Epidermal IF
z_a	11.06,11.23	7.42, 7.36, 7.24, 7.27	11.93
z_b	16.60, 16.85	19.79,19.82, 20.03, 19.86	16.64
$z_a + 2z_b$	44.26, 44.92	47.00, 47.06, 47.30, 46.99	45.21

Data from [17, 19–21, 29, 38, 55, 59]

Fig. 5.13 (continued) Four of the eight protofilaments are marked by a black bar. Each molecule consists of a Type I and a Type II chain, and these chains lie in axial register and are parallel to one another (rather than antiparallel). It is assumed that the molecules lie parallel to the fibre axis. Lower: Pairs of antiparallel coiled-coil IF molecules form the protofilaments. (Reproduced from Fig. 3 of Fraser and Parry [26] with permission of Elsevier)

Relative Contents of the β-Filaments and the Intermediate Filaments

The relative percentages of the β-filaments and the intermediate filaments in the various corneous appendages have not, as yet, been reported. It is likely, however, that while such figures will vary with appendage, they will, nonetheless, only show relatively small variations in absolute terms between the different tissues. Such data are important since they relate directly to the interpretation of the mechanical properties exhibited by the individual corneous appendages. However, it has been noted that the number of β-genes expressed far exceeds those for the IF proteins, a fact indicating that the β-filaments will predominate in vivo. Also, data from fibre X-ray diffraction experiments, whilst clearly identifying the β-structure present (see, for example, Fraser and MacRae [12, 13]), failed to record the presence of any component characteristic of IF. This result indicates that the percentage of α-material (IF) must be very small in comparison to that in the β-filaments. Alternatively, it could be construed that the IF, unlike the β-filaments, lacks regular orientation, thereby smearing out the diffraction intensity and rendering it difficult to recognise. Likewise, infrared studies of feather keratin [31] showed bands expected of a β-structure but did not find evidence that would be expected of an α-component. This result also strongly indicates that the α-content in the corneous appendages must be low, probably just a few percent and possibly even lower. However, experimental data are needed if these ideas are to be put on a firm basis.

Spatial Location of the Filaments

A second point of concern is that it is not yet known whether the two types of filaments are (a) spatially co-located, (b) in spatially diverse areas of the tissue or (c) in some mixture of the two. The first (a) and third possibilities (c), however, would seem improbable since no electron micrographs have, to date, shown the presence of both the 7–10 nm diameter IF and the 3.4 nm β-keratin filaments co-existing in the same localised region (see, for example, Filshie and Rogers [11], Brush [5]). If the β-keratin filaments do indeed dominate in terms of numbers (see previous section), then it would seem likely that they would form the core of the corneous appendage and that the IF would be primarily localised around the periphery, probably in sheets. The limited data at hand, therefore, suggests that the filaments will exist in spatially diverse areas of the various appendages, but, once again, the field would benefit from some firm experimental data, most likely from transmission electron microscopy.

Protein-Protein Interactions and Their Relationship to the Physical-Chemical Properties of the Corneous Appendages

The corneous appendages of the sauropsids show considerable variation in their mechanical attributes, as, for example, in those epitomised by feather and claw. An interesting question that must now be posed is whether this can be related to the protein families present and the interactions that they make with one another. While microheterogeneity is known to occur in the CBP and IF proteins from the various species within the sauropsid classification, the most significant differences are likely to occur in the interfilamentous material, i.e. the matrix, formed by the terminal domains of the CBP and the proteins of the EDC. A comparison with mammalian wool/hair is pertinent here, where it is well established that the matrix proteins are extremely heterogeneous with regard to size (molecular weight) and composition [14]. Also, considerable variations in type and proportion of the matrix proteins occur between species and with the nutritional state experienced by the animal. One example is that of the HGT content of the matrix, which varies from just 1% in horn and nail to about 30% in echidna quill. It has also been established that some of the proteins in the mammalian matrix show limited sequence regularity and are therefore believed to contain elements of regular secondary structure. However, it is thought that none of these proteins is likely to adopt a unique tertiary conformation, and, to a large extent, this would most likely be defined by the protein's local environment.

Comparative studies of the matrix proteins constituting the corneous appendages are presently limited, so it remains a matter of speculation as to whether or not similar features are also a characteristic of the sauropsid structures. However, the first research aimed at addressing this problem has recently been reported by Holthaus et al. [45], who studied the organisation of the EDC in the common wall lizard (*Podarcis muralis*). Interestingly, they showed that the cornification proteins are different from those reported in the homologous proteins present in other reptiles. In addition, they characterised (a) sixteen members of a family of proteins with >50% cysteine residue content and which were further characterised by a PCCC motif. Glutamine and lysine residues were also conserved as potential sites for transglutamination, (b) a family of four loricrin proteins with long stretches of glycine and serine residues interspersed with short regions 4–7 residues long that commonly include valine and isoleucine residues, (c) two CBPs with long C-terminal domains characterised by about 50% glycine residue content, and (d) a family of five epidermal differentiation proteins with a C-terminal glutamine motif. For these proteins as a whole, glycine and cysteine residues account for more than 50% of the total number of residues. While it was likely that the proteins of the EDC and the mammalian matrix would be clearly distinguishable, as now appears to be the case, it would be both valuable and informative to extend the study of the EDC proteins to different corneous appendages and to different species. As noted previously, the equivalent protein families that form the matrix in the mammalian hard keratins (KAPs) are known to differ significantly in this regard. The work of Holthaus et al. [45] represents an important first step in tackling this problem.

A closely related area ripe for investigation is to determine experimentally the nature of the interactions occurring between the CBP and the proteins of the EDC, as well as those between the various constituents of the EDC. The very high content of cysteine residues in the cornified proteins of the common wall lizard strongly indicates that the formation of multiple crosslinks through disulphide bonding will be a feature. It would seem certain that such links in the sauropsids will play a significant mechanical role, particularly in the toughest appendages, as they have been shown to do in hair. In the latter case, such bonds were primarily (but not totally) provided by the HS and UHS proteins in the mammalian matrix. It can be speculated that the high glycine residue content of the corneous proteins would also likely endow the proteins with a substantial degree of flexibility. Studies of the type described here would provide a valuable insight into those features of the constituent proteins that have special importance in determining the mechanical attributes of the corneous appendages.

Summary

Recent sequence data, allied to structural modelling, have revealed several new aspects of the filamentous structures of both the β-filaments and IF, together with their potential modes of assembly in vivo. For example, the lack of sequence data for the sole member of the Rhynchocephalia—the tuatara—has finally been filled, and it can now be said that all members constituting the phylogenetic classification encompassing the birds and reptiles exhibit the same structural features. In addition, it has now been suggested that the β-filaments in the lepidosaurs are likely to be linked directly to one another through a specialised chain, albeit a minor component, that is incorporated directly in two neighbouring filaments. More work, however, needs to be done to clarify the extent to which this feature (or a related one) is a feature of all sauropsids. Likewise, it has been proposed that another minor component is present in IF and that the large C-terminal tail domains of these proteins, characterised by multiple sequence repeats, interact and thereby facilitate the assembly of filaments into larger aggregates. The possibility that multiple repeats in the tail of nestin are intimately involved in the regulation of dynamic assembly of IF in mammalian cells has also been suggested, as has the concept that such repeats might act as crossbridges between IF. Plectin has also been shown to link IF through interactions made through their external surfaces [76].

The arrangement of the eight protofilaments that constitute the IF in hard α-keratin has long been a matter of debate. Over many years, electron microscopy data failed to distinguish clearly between a (7 + 1) "ring-core" and a (8 + 0) "ring" arrangement. Recently, however, modelling of both structures against the fine structure of the 1 nm equatorial region of the fibre X-ray diffraction patterns has shown that the (8 + 0) model provides a significantly better fit to the observed intensity distribution than can be achieved for the (7 + 1) structure. In addition, the fact that both the head and tail domains are largely apolar in character, that these domains display a significant amount of sequence conservation and that their positioning in the central disulphide-bonded core would shield them from water and thus lead to an energetically favourable situation all suggests that the (8 + 0) model is likely to be the correct one. These structural studies have direct relevance for the IF in the corneous appendages since their constituent proteins are strongly homologous to those found in wool and hair.

Finally, the surface lattice structure of the IF has been determined using a different technique from those previously employed. It has, however, allowed confirmation of the parameters derived earlier.

Thus, while much is now known at low resolution about both the β-filaments and the IF in claw, beak, feather and scales, it may be speculated that high-resolution electron microscopy, for example, will reveal these structures in even greater detail over the next ten years or so. Further, more will be understood about the various modes of filament assembly, which are only now beginning to be fully appreciated. Exciting times are ahead.

References

1. Alexander NJ (1970) Comparison of alpha and beta keratins in reptiles. Z Zellforsch Mikrosk Anat 110:153–165
2. Alibardi L, Dalla Valle L, Toffolo V, Toni M (2006) Scale keratin in lizard epidermis reveals amino acid regions homologous with avian and mammalian proteins. Anat Rec 288A:734–752
3. Alibardi L, Dalla Valle L, Toni M (2009a) Cell biology of adhesive setae in gecko lizards. Zoology 112:403–424
4. Alibardi L, Dalla Valle L, Nardi A, Toni M (2009b) Evolution of hard proteins in the sauropsid integument in relation to the cornification of skin derivatives in amniotes. J Anat 214:560–586
5. Brush HA (1983) Self-assembly of avian ϕ-keratins. J Protein Chem 2:63–75
6. Calvaresi M, Eckhart L, Alibardi L (2016) The molecular organization of the beta-sheet region in corneous beta-proteins (beta-keratins) of sauropsids explains its stability and polymerization into filaments. J Struct Biol 194:282–291
7. Crewther WG, Dowling LM, Steinert PM, Parry DAD (1983) Structure of intermediate filaments. Int J Biol Macromol 5:267–274
8. Dalla Valle L, Nardi GB, Bonazza G, Zuccal C, Emera D, Alibardi L (2010) Forty keratin-associated beta-proteins (β-proteins) form the hard layers of scales, claws, adhesive pads in the green anole lizard, *Anolis carolinensis*. J Exp Zool 314B:11–32
9. Ehrlich F, Lachner J, Hermann M, Tschachler E, Eckhart L (2020) Convergent evolution of cysteine-rich keratins in hard skin appendages of terrestrial vertebrates. Mol Biol Evol 37:982–993
10. Filshie BK, Rogers GE (1961) The fine structure of α-keratin. J Mol Biol 3:784–786
11. Filshie BK, Rogers GE (1962) An electron microscope study of the fine structure of feather keratin. J Cell Biol 13:1–12
12. Fraser RDB, MacRae TP (1959) Molecular organization in feather keratin. J Mol Biol 1:387–397
13. Fraser RDB, MacRae TP (1963) Structural organization in feather keratin. J Mol Biol 7:272–280

14. Fraser RDB, MacRae TP (1973) Conformation in fibrous proteins and related synthetic polypeptides. Academic Press, London
15. Fraser RDB, MacRae TP (1976) The molecular structure of feather keratin. In: Proceedings on 11th International Ornithological Congress. Canberra, Australia, pp 443–451
16. Fraser RDB, MacRae TP (1980) Molecular structure and mechanical properties of keratin. In: Vincent JFV, Currey JD (eds) The mechanical properties of biological materials. Society for Experimental Biology, London, UK, pp 211–246. S.E.B. Symp. No. 34
17. Fraser RDB, MacRae TP (1983) The structure of the α-keratin microfibril. Biosci Rep 3:517–525
18. Fraser RDB, MacRae TP (1985) Intermediate filament structure. Biosci Rep 5:573–579
19. Fraser RDB, Macrae TP (1988) Surface lattice in α-keratin filaments. Int J Biol Macromol 10:573–579
20. Fraser RDB, Parry DAD (2007) Structural changes in the trichocyte intermediate filaments accompanying the transition from the reduced to the oxidized form. J Struct Biol 159:36–45
21. Fraser RDB, Parry DAD (2008) Molecular packing in the feather keratin filament. J Struct Biol 162:1–13
22. Fraser RDB, Parry DAD (2011) The structural basis of the filament-matrix texture in the avian/reptilian group of hard β-keratins. J Struct Biol 173:391–405
23. Fraser RDB, Parry DAD (2014) Amino acid sequence homologies in the hard keratins of birds and reptiles, and their implications for molecular structure and physical properties. J Struct Biol 188:213–224
24. Fraser RDB, Parry DAD (2017a) Structural transition of trichocyte keratin intermediate filaments during development in the hair follicle. In: Parry DAD, Squire JM (eds) Subcellular biochemistry: Fibrous proteins: structures and mechanisms, vol 82. Springer International Publishing Switzerland, pp 131–149
25. Fraser RDB, Parry DAD (2017b) Filamentous structure of hard β-keratin in the epidermal appendages of birds and reptiles. In: Parry DAD, Squire JM (eds) Subcellular biochemistry: Fibrous proteins: structures and mechanisms, vol 82. Springer International Publishing Switzerland, pp 231–252
26. Fraser RDB, Parry DAD (2017c) Intermediate filament structure in fully differentiated (oxidised) trichocyte keratin. J Struct Biol 200:45–53
27. Fraser RDB, Parry DAD (2018a) Structural hierarchy of trichocyte keratin intermediate filaments. In: Plowman J, Harland D, Deb-Choudhury S (eds) The hair fibre: proteins, structure and development, Advances in Experimental Medicine and Biology 1054. Springer, pp 57–70
28. Fraser RDB, Parry DAD (2018b) Trichocyte keratin-associated proteins (KAPs). In: Plowman J, Harland D, Deb-Choudhury S (eds) The hair fibre: proteins, structure and development, Advances in Experimental Medicine and Biology 1054. Springer, pp 71–86
29. Fraser RDB, Parry DAD (2018c) Direct evidence supporting the existence of a helical dislocation in protofilament packing in the intermediate filaments of oxidized trichocyte keratin. J Struct Biol 204:491–497
30. Fraser RDB, Parry DAD (2020) Lepidosaur β-keratin chains with four 34-residue repeats: modelling reveals a potential filament-crosslinking role. J Struct Biol 209:107413
31. Fraser RDB, Suzuki E (1965) Polypeptide chain conformation in feather keratin. J Mol Biol 14:279–282
32. Fraser RDB, MacRae TP, Miller A (1965) X-ray diffraction pattern of α-fibrous proteins. J Mol Biol 14:432–442
33. Fraser RDB, MacRae TP, Parry DAD (1967) The substructure of the α-keratin microfibril. In: Crewther WG (ed) Symposium on Fibrous Proteins (Australia 1967). Butterworths, Australia, pp 279–286
34. Fraser RDB, MacRae TP, Parry DAD, Suzuki E (1971) The structure of feather keratin. Polymer 12:35–56
35. Fraser RDB, MacRae TP, Suzuki E (1976) Structure of the α-keratin microfibril. J Mol Biol 108:435–452

36. Fraser RDB, MacRae TP, Parry DAD, Suzuki E (1986) Intermediate filaments in α-keratins. Proc Natl Acad Sci USA 83:1179–1183
37. Fraser RDB, MacRae TP, Parry DAD (1990) The three-dimensional structure of IF. In: Goldman RD, Steinert PM (eds) Cellular and molecular biology of intermediate filaments. Plenum, New York, pp 205–231
38. Fraser RDB, Steinert PM, Parry DAD (2003) Structural changes in trichocyte keratin intermediate filaments during keratinization. J Struct Biol 142:266–271
39. Greenwold MJ, Sawyer RH (2013) Molecular evolution and expression of Archosaurian β-keratins: diversification and expansion of Archosaurian β-keratins and the origin of feather β-keratins. J Exp Zool 320:393–405
40. Greenwold MJ, Bao W, Jarvis ED, Hu H, Li C, Gilbert MTP, Zhang G, Sawyer RH (2014) Dynamic evolution of the alpha (α) and beta (β) keratins has accompanied integument diversification and the adaption of birds into novel lifestyles. BMC Evol Biol 14:249
41. Harris B (1980) The mechanical behaviour of composite materials. In: Vincent JFV, Currey JD (eds) The mechanical properties of biological materials. Society for Experimental Biology, London, pp 37–74. S.E.B. Symp. No. 34.
42. Holthaus KB, Strasser B, Sipos W, Schmidt HA, Mlitz V, Sukseree S, Weissenbacher A, Tschachler E, Alibardi L, Eckhart L (2016) Comparative genomics identifies epidermal proteins associated with the evolution of the turtle shell. Mol Biol Evol 33:726–737
43. Holthaus KB, Mlitz V, Strasser B, Tschachler E, Alibardi L, Eckhart L (2017) Identification and comparative analysis of the epidermal differentiation complex in snakes. Sci Rep 7:45338
44. Holthaus KB, Eckhart L, Dalla Valle L, Alibardi L (2018) Review: evolution and diversification of corneous beta-proteins, the characteristic epidermal proteins of reptiles and birds. J Exp Zool (Mol Dev Evol) 330:438–453
45. Holthaus KB, Sachslehner AP, Steinbinder J, Eckhart L (2024) Epidermal differentiation genes of the common wall lizard encode proteins with extremely biased amino acid contents. Genes 15:1136
46. Hughes MP, Sawaya MR, Boyer DR, Goldschmidt L, Rodriguez JA, Cascio D, Chong L, Gonen T, Eisenberg DS (2018) Atomic structures of low-complexity protein segments reveal kinked ß-sheets that assemble networks. Science 359:698–701
47. Jones LN, Simon N, Watts NR, Booey FP, Steven AC, Parry DAD (1997) Intermediate filament structure: hard alpha-keratin. Biophys Chem 68:83–93
48. Kajava AV, Squire JM, Parry DAD (2006) ß-structures in fibrous proteins. Adv Protein Chem 73:1–15
49. Lachner J, Ehrlich F, Mlitz V, Hermann M, Alibardi L, Tschachler E, Eckhart L (2019) Immunolocalization and phylogenetic profiling of the feather protein with the highest cysteine content. Protoplasma 256:1257–1265
50. Landmann L (1979) Keratin formation and barrier mechanisms in the epidermis of Natrix natrix (Reptilia, serpentes): an ultrastructural study. J Morphol 162:93–126
51. Li YI, Kong L, Ponting CP, Haerty W (2013) Rapid evolution of beta-keratin genes contribute to phenotypic differences that distinguish turtles and birds from other reptiles. Genome Biol Evol 5:923–933
52. Liu Y, Zhou Q, Wang Y, Luo L, Yang J, Yang L, Liu M, Li YR, Qian TM, Zheng Y, Li M, Li J, Gu Y, Han Z, Xu M, Wang Y, Zhu C, Yu B, Yang Y, Ding F, Jiang J, Yang H, Gu X (2015) *Gekko japonicus* genome reveals evolution of adhesive toe pads and tail regeneration. Nat Commun 6:10033
53. Ng CS, Wu P, Fan WL, Yan J, Chen CK, Lai YT, Wu SM, Mao CT, Chen JJ, Lu MYJ, Ho MR, Widelitz RB, Chen CF, Chuong CM, Li WH (2014) Genomic organization, transcriptomic analysis, and functional characterization of avian α- and β-keratins in diverse feather forms. Genome Biol Evol 6:2258–2273
54. Parry DAD (1979) Determination of structural information from the amino acid sequences of fibrous proteins. In: Parry DAD, Creamer LK (eds) Fibrous proteins: scientific, industrial and medical aspects, vol 1. Academic Press, London, pp 393–427

55. Parry DAD (1999) Structural features of IF proteins. In: Kreis T, Vale R (eds) Guidebook to the cytoskeletal and motor proteins, Second edition, pp 285–291
56. Parry DAD (2020) Structure and topology of the linkers in the conserved lepidosaur β-keratin chain with four 34-residue repeats support an interfilament role for the central linker. J Struct Biol 212:107599
57. Parry DAD (2021) Structures of the β-keratin filaments and keratin intermediate filaments in the epidermal appendages of birds and reptiles (sauropsids). Genes 12:591
58. Parry DAD, Fraser RDB (1985) Intermediate filament structure. 1. Analysis of IF protein sequence data. Int J Biol Macromol 7:203–213
59. Parry DAD, Steinert PM (1999) Intermediate filaments: molecular architecture, assembly, dynamics and polymorphism. Q Rev Biophys 32:99–187
60. Parry DAD, Winter DJ (2021a) Keratin intermediate filament chains in tuatara (Sphenodon punctatus): a comparison of tuatara and human sequences. J Struct Biol 213:107706
61. Parry DAD, Winter DJ (2021b) Keratin intermediate filament chains in the European common wall lizard (Podarcis muralis) and a potential keratin filament crosslinker. J Struct Biol 213: 107793
62. Parry DAD, Crewther WG, Fraser RDB, MacRae TP (1977) Structure of α-keratin: structural implication of the amino acid sequences of the Type I and Type II chain segments. J Mol Biol 113:449–454
63. Parry DAD, Fraser RDB, Alibardi L, Rutherford K, Gemmell N (2019) Molecular structure of sauropsid β-keratins from tuatara (Sphenodon punctatus). J Struct Biol 207:21–28
64. Rogers GE, Filshie BK (1962) Electron staining and fine structure of keratins. In: Proceedings of International Congress Electron Microscopy, 5th Philadelphia, vol 2, O-2. Academic Press, New York
65. Rogers MA, Langbein L, Praetzel-Wunder S, Winter H, Schweizer J (2006) Human hair keratin-associated proteins (KAPs): sequence regularities and structural implications. Int Rev Cytol 251:209–263
66. Steinert PM, Marekov LN, Fraser RDB, Parry DAD (1993a) Keratin intermediate filament structure: crosslinking studies yield quantitative information on molecular dimensions and mechanism of assembly. J Mol Biol 230:436–452
67. Steinert PM, Marekov LN, Parry DAD (1993b) Conservation of the structure of keratin intermediate filaments: molecular mechanism by which different keratin molecules integrate into pre-existing keratin intermediate filaments during differentiation. Biochemistry 32:10046–10056
68. Steinert PM, Chou Y-H, Prahlad V, Parry DAD, Marekov LN, Wu KC, Jang S-I, Goldman RD (1999) A high molecular weight intermediate filament-associated protein in BHK-21 cells is nestin, a type VI intermediate filament protein: limited co-assembly in vitro to form heteropolymers with type III vimentin and type IV α-internexin. J Biol Chem 274:9881–9890
69. Strasser B, Mlitz V, Hermann M, Rice RH, Eigenheer RA, Alibardi L, Tschachler E, Eckhart L (2014) Evolutionary origin and diversification of epidermal barrier proteins in amniotes. Mol Biol Evol 31:3194–3205
70. Strasser B, Mlitz V, Hermann M, Tschachler E, Eckhart L (2015) Convergent evolution of cysteine-rich proteins in feathers and hair. BMC Evol Biol 15:82
71. Taylor AM, Bonser RHC, Farrent JW (2004) The influence of hydration on the tensile and compressive properties of avian keratinous tissues. J Mater Sci 39:939–942
72. Vandebergh W, Bossuyt F (2012) Radiation and functional diversification of alpha keratins during early vertebrate evolution. Mol Biol Evol 29:995–1004
73. Wang B, Yang W, McKittrick J, Meyers MA (2016) Keratin: structure, mechanical properties, occurrence in biological organisms and efforts in bioinspiration. Prog Mater Sci 76:229–318
74. Wang H, Parry DAD, Jones LN, Idler WW, Marekov LN, Steinert PM (2000) In vitro assembly and structure of trichocyte keratin intermediate filaments: a novel role for stabilization by disulfide bonding. J Cell Biol 151:1459–1468

75. Watts NR, Jones LN, Cheng N, Wall JS, Parry DAD, Steven AC (2002) Cryo-electron microscopy of trichocyte (hard α-keratin) intermediate filaments reveals a low-density core. J Struct Biol 137:109–118
76. Wiche G (1998) Role of plectin in cytoskeleton organization and dynamics. J Cell Sci 111: 2477–2486
77. Wu P, Ng CS, Yan J, Lai YC, Chen CK, Lai YT, Wu SM, Chen JJ, Luo W, Widelitz RB, Li WH, Chuong CM (2015) Topographical mapping of α- and β-keratins on developing chicken skin integument: functional interaction and evolutionary perspectives. Proc Natl Acad Sci USA: E6770–E6779
78. Yang J, Qu Y, Huang Y, Lei F (2018) Dynamic transcriptome profiling towards understanding the morphogenesis and development of diverse feathers in domestic duck. Genomics 19:391

Chapter 6
Insights into the Structure of Intermediate Filaments

Matthias Eibauer and Ohad Medalia

Abstract Intermediate filaments (IFs) are central to the mechanical integrity of metazoan cells and play critical roles in various fundamental cellular and multicellular processes, including cell motility, signal transduction, and wound healing. To perform their functions, IF proteins self-assemble into nanoscale biopolymers, each exhibiting unique properties that are finely tuned to their specific roles across different tissue types. However, the 3D structure of IFs has remained largely unresolved due to their intrinsic flexibility and polymorphism. This chapter reviews recent advances in the structural analysis of IFs, with a focus on vimentin IFs (VIFs), which are featuring a helical tube with a central luminal fiber. We discuss how AlphaFold-based modeling, chemical cross-linking data, and cryo-electron microscopy (cryo-EM) reconstructions have been integrated to generate a detailed structural model of VIFs, highlighting key features such as the helical symmetry of the filaments and the nature of the luminal fiber. Additionally, we explore potential sources of IF polymorphism and their implications for the analysis of IF structures.

Keywords Cryo-EM · Helical symmetry · Intermediate filaments · Structural determination · Vimentin

Abbreviations

Cryo-EM	Cryo-electron microscopy
Cryo-ET	Cryo-electron tomography
GFAP	Glial fibrillary acidic protein
IF	Intermediate filament
KIF	Keratin intermediate filament
MEFs	Mouse embryonic fibroblasts
MPL	Mass per length

M. Eibauer (✉) · O. Medalia (✉)
Department of Biochemistry, University of Zurich, Zurich, Switzerland
e-mail: m.eibauer@bioc.uzh.ch; omedalia@bioc.uzh.ch

D. A. D. Parry, P. K. Luther (eds.), *Fibrous Proteins: Structures and Mechanisms*, Subcellular Biochemistry 113, https://doi.org/10.1007/978-3-032-05273-5_6

NF	Neurofilament
NTE	N-terminal end
STEM	Scanning transmission electron microscopy
VIF	Vimentin intermediate filament

Introduction

IFs are fundamental components of the metazoan cytoskeleton, which contribute significantly to the mechanical resilience and elasticity of cells and tissues due to their distinct viscoelastic properties [29, 33, 57, 70]. IFs are biopolymers that confer cell-type-specific mechanical properties and participate in diverse cellular processes, such as cell motility [86], signal transduction [59], and the epithelial-mesenchymal transition [48]. IFs are encoded by 70 human genes [77], including vimentin, keratins, and neurofilaments, and are implicated in over 70 pathologies [54].

IFs are classified into five primary types based on their sequence homologies [19]. The first four types (I–IV) are cytoplasmic, whereas type V IFs reside in the nucleus. The cytoplasmic IF proteins assemble into 10 nm thick filaments although nuclear lamins assemble into thinner filaments around 4 nm in diameter. Among these types, types I and II are classified as acidic and neutral-basic keratins, respectively [32]. These keratins assemble into heteropolymer filaments, which are commonly found in epithelial cells. In humans, 54 distinct keratins are expressed, with their expression being cell type and stage of differentiation dependent. Type III IFs are composed of homopolymers of vimentin [62], desmin [1], peripherin [45], or glial fibrillary acidic protein (GFAP) [30]. Vimentin is predominantly expressed in fibroblasts, but it is also expressed in endothelial cells, the eye lens epithelium, and the dendritic reticulum cells of lymphoid follicles. Desmin is the major IF protein of smooth, skeletal, and cardiac muscle, while peripherin is found mainly in neurons of the peripheral nervous system. GFAP is located in astrocytes and glial cells. Type IV IFs include those expressed in the nervous system either as complex heteropolymers, such as the neurofilament triplet proteins (NF-L, NF-M, and NF-H), or as homopolymers of alpha-internexin. Nestin, another Type IV protein, does not form IFs autonomously; rather, it requires association with other IF family members, such as vimentin or desmin. Type V IFs include the nuclear lamins (lamin A/C, B1, and B2), which are the only assembled cytoskeleton element in the nucleus [78, 80].

Despite their importance as fundamental components of the cytoskeleton, the 3D structure of IFs has largely remained unresolved due to their intricate and flexible architecture, large intrinsically disordered head and tail domains, and their general structural heterogeneity and plasticity. Recently, the first single particle cryo-EM structure of an IF was determined [12], providing the first glimpse into the structure of VIFs at ~7 Å resolution. Although this structure lacks atomic detail, it yields insights into the building principles of IFs and sheds light on the structural organization of the low-complexity head and tail domains within a tubular helical assembly formed of alpha-helical coiled-coil structures.

In this chapter, we outline key aspects of the general building plan of IFs, highlight the structural insights derived from the VIF structure, and explore how these principles may be extended to other IF systems. Particular emphasis is placed on the challenges of reconstructing IFs by cryo-EM, including the determination of helical symmetry and potential causes of the pronounced structural polymorphism observed in these filaments.

Elementary Dimers

The domain architecture of IF proteins is highly conserved [24, 57]. It features a central rod domain flanked by low-complexity head and tail domains (Fig. 6.1a). The rod domain consists of two extended alpha-helical segments [76], termed as 1A-1B and 2A-2B segments, which are connected by the non-helical, flexible linker L12. The helical segments can be further subdivided into their individual alpha-helical domains (1A, 1B, 2A, 2B) [60]. Highly conserved amino acid sequences (Fig. 6.1a, indicated by hatching) among IF proteins are present in the N-terminal region of the 1A helix and the C-terminal region of the 2B helix [28, 47].

In solution, IF proteins readily assemble into parallel, in-register dimers [26]. Within these dimers, the 1A-1B and 2A-2B helices form coiled-coil structures [13] driven by the characteristic heptad repeat patterns in their amino acid sequences [85]. The atomic structure of IF dimers have been elucidated in detail within the coiled-coil regions, based on numerous X-ray crystallography studies [2, 6, 13, 52, 56, 76]. For example, Fig. 6.1b shows the X-ray structure of the vimentin helix 1B coiled-coil dimer, while Fig. 6.1c shows the structure of the vimentin helix 2B coiled-coil dimer. These and additional overlapping X-ray structures enabled the construction of an atomic model of the complete vimentin dimer [76], which was subsequently refined through molecular dynamics simulations (Fig. 6.1d) [60].

The atomic structure of IF dimers can also be predicted with AlphaFold [34]. In Fig. 6.1e-g models of IF dimers are shown, namely the homodimers of vimentin and GFAP, and the heterodimer of keratin 5/15. These models recapitulate previous results, showing the alpha-helical 1A-1B and 2A-2B coiled-coil dimers, the flexible L12 linkers that connect them, and the low-complexity head and tail domains that emanate from the terminals of the central coiled-coil region. In detail, the models show that the 1A helices have the tendency to separate and to form a V-shaped structure (Fig. 6.1e, bracket labelled 1A), confirmed by previous X-ray crystallography structural analysis [47, 69, 76]. Additionally, the models indicate a potential transition region between the 1A and 1B helices, where the coiled-coil structure uncoils into a linker between these two regions (Fig. 6.1f-g, arrowheads), a feature which has also been described by molecular dynamics simulation [60]. This structural property could accommodate changes in the position of the 1A helices relative to the 1B helices. Finally, the models show that the 2A helices exhibit a modulation of their coiled-coil parameters compared to the 2B helices [52, 60], so that the 2A helices almost run parallel to each other (Fig. 6.1e, bracket labelled 2A), before transitioning into a typical coiled-coil structure adopted by the 2B helices.

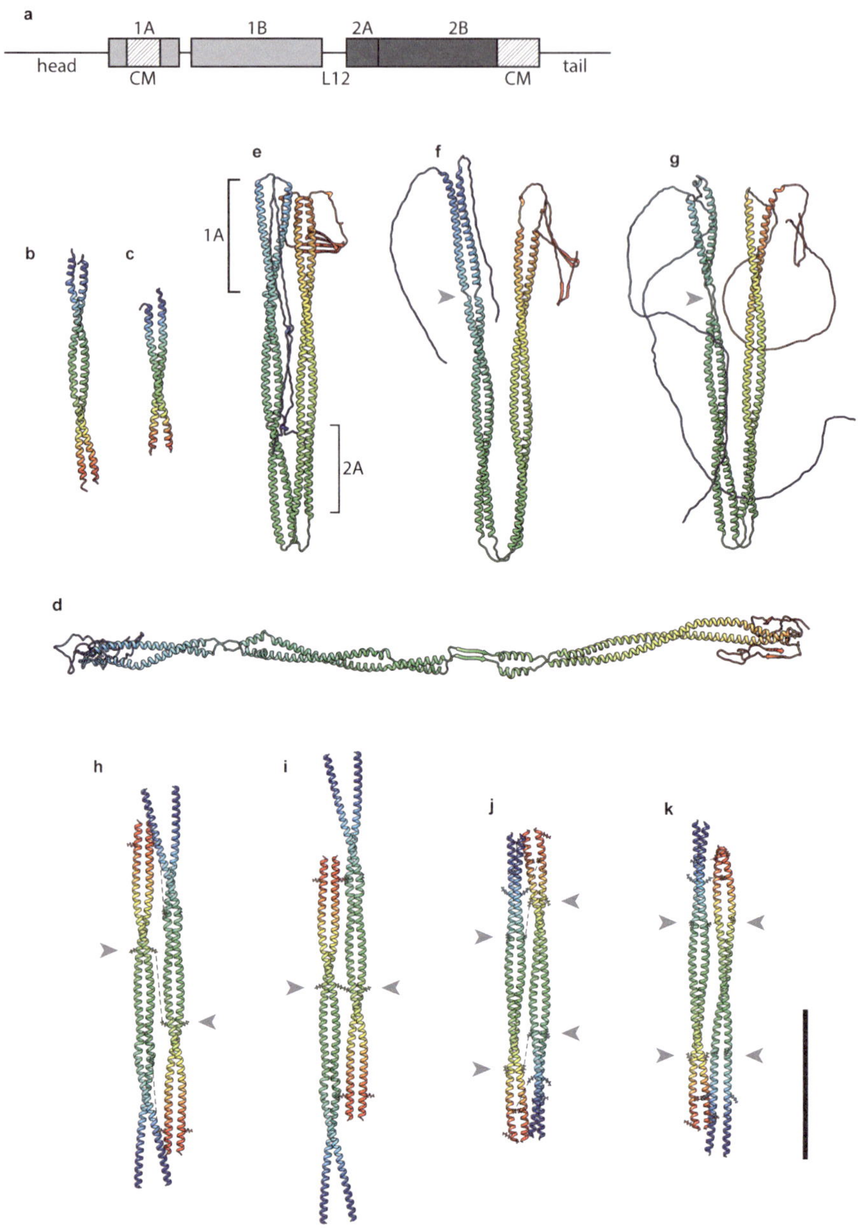

Fig. 6.1 Elementary dimers and tetramer interactions in IFs. (**a**) Conserved domain structure of IFs. The highly conserved regions in the 1A and 2B domains are indicated by hatching (CM, consensus motif [47]). (**b**) X-ray structure 3UF1 [2] of the vimentin 1B coiled-coil dimer. All models are colored from N-terminus (blue) to C-terminus (red). (**c**) X-ray structure 1GK4 [76] of the vimentin 2B coiled-coil dimer. (**d**) Model of the vimentin dimer based on X-ray crystallography and molecular dynamics simulation [60]. (**e**) AlphaFold model of the vimentin homodimer. The positions of helices 1A and 2A are indicated by brackets. (**f**) AlphaFold model of the GFAP

In the full-length AlphaFold models, the conservation of IF domain structure can be clearly seen, recapitulating that the rod domains of IFs consist of alpha-helical coiled-coil structures, which are defined by their heptad repeat pattern, but also that the head and tail domains can vary significantly in length between different IF [13]. Combining all these experimental, simulation, and modelling results, we have a detailed and well-founded understanding of the structure of elementary IF dimers.

Tetramer Interactions

To address the question of how IF dimers are packed into a mature, fully polymerized IF, a wealth of information and experimental results based on cross-linking experiments, was established by the pioneering effort of Steinert, Parry, and coworkers [18, 57, 71–73]. The spatial distributions of specific cross-links were used to infer the architectures of different IFs, and how the elementary dimers are packed into these structures. Based on their work, it became clear that there is a general building plan of IFs. For all the analyzed IFs, this group consistently demonstrated, that there are three basic modes of interaction occurring between the coiled-coil protein domains within fully-assembled IFs. These basic modes of interaction are termed the A11, A22, and A12 binding modes. These interactions occur between anti-parallel coiled-coil dimers in the mature filaments. In the A11 mode, two anti-parallel dimers are laterally aligned and displaced relative to each other, with a stagger that is IF specific, such that their 1B domains are brought into register in the resulting tetrameric assembly. Similarly, in the A22 mode two anti-parallel dimers are staggered, such that their 2B domains are aligned in the resulting tetrameric complex. Finally, in the A12 mode a tetrameric unit is formed from two anti-parallel dimers which are laterally aligned and registered with only minimal staggering.

The information from cross-linking experiments can be employed as spatial restraints during AlphaFold modelling [68] of extended tetrameric units of the IF rod domain. In this context, it is important to note that the precision of these models is typically very high for the coiled-coil dimers, and for the prediction of their anti-parallel orientation in tetrameric IF assemblies. However, the relative stagger between two of such dimers typically show more degrees of freedom in the AlphaFold modelling, meaning that differently ranked predictions primarily vary in the stagger between the coiled-coils. Therefore, the results from cross-linking

Fig. 6.1 (continued) homodimer. (**g**) AlphaFold model of the keratin 5/15 heterodimer. In (**f**) and (**g**), the transition regions between helices 1A and 1B are indicated by arrowheads. (**h**) AlphaFold model of the 1A-1B vimentin tetramer. Cross-linked lysine side chains are shown (arrowheads), and connected by dashed lines. (**i**) Cross-link correction of the 1A-1B vimentin tetramer model. (**j**) AlphaFold model of the 2A-2B vimentin tetramer. (**k**) Cross-link correction of the 2A-2B vimentin tetramer model. Scale bar is 10 nm experiments are crucial to reduce the number of different stagger conformations between two anti-parallel dimers.

This consideration is visualized in Fig. 6.1h-k. In Fig. 6.1h, the highest ranked AlphaFold model of a vimentin 1A-1B tetramer is depicted. As shown previously, based on the model of the vimentin dimer (Fig. 6.1e), this tetrameric model recapitulates the V-shaped structure of the 1A helices [47, 76], correctly predicts the anti-parallel orientation of the coiled-coil dimers [76], and reaches an overall high precision, as reported by the AlphaFold pLDDT [34] confidence score. However, the predicted relative stagger between the dimers places the cross-linked lysine residues ~46 Å apart from each other (Fig. 6.1h, dashed lines). This distance is significantly larger than the upper limit of ~27 Å, which would been expected from such cross-linking experiments [61]. In turn, the distance between the cross-linked residues can be minimized to create a cross-link corrected AlphaFold model [68]. After applying this correction to the tetrameric 1A-1B model (Fig. 6.1i), it becomes clear that the 1B helices are in register, as expected based on the their A11 mode of interaction. In Fig. 6.1j-k, a similar modelling and cross-linking based stagger correction is shown for a tetramer of the 2A-2B helices of vimentin.

Helical Symmetry

The results of the cross-linking experiments were used to infer how coiled-coil dimer interactions form a repeating unit within IFs [18], and how this unit is distributed with helical symmetry in the 3D architecture of mature IFs. The observation that IFs exhibit helical symmetry is highly significant, as helical symmetry is a common building principle of filamentous structures in living organisms, including DNA, alpha-helices, coiled-coils [85], actin filaments [20, 55] microtubules [89], bacterial pili [22], filamentous viruses [4], and amyloids [17]. These polymers are all helical, which facilitates cryo-EM data analysis and 3D reconstruction [11, 23]. Furthermore, this symmetry simplifies the conceptual understanding of IF structure. Recent studies on VIFs [12] and keratin IFs [84] (KIFs) have combined established cryo-EM methodologies with newly developed techniques to measure the helical symmetry of IFs and to use this information to reconstruct their 3D structures.

Helical symmetry can be described using two parameters [9, 23]. These are the helical rise h_r and the helical twist h_t, which define how copies of the asymmetric unit are translated along the filament axis by the helical rise and rotated around the filament axis by the helical twist, to form the helical assembly. This concept is illustrated in Fig. 6.2a, which shows how the asymmetric unit of a helical assembly is translated and rotated by h_r and h_t, respectively. Alternatively, helical symmetry can be parameterized by the helical pitch P, which is the axial distance required to complete a full helix turn of 360°, and the number of asymmetric units n_{asu}, which are contained within the pitch length. Between these two parameter sets, the following transformation exists, that is $n_{asu} = 360°/h_t$ and $P = h_r * n_{asu}$. The first formulation offers an intuitive description of the construction of a helical filament from its asymmetric unit, whereas in the second formulation, helical symmetry can be

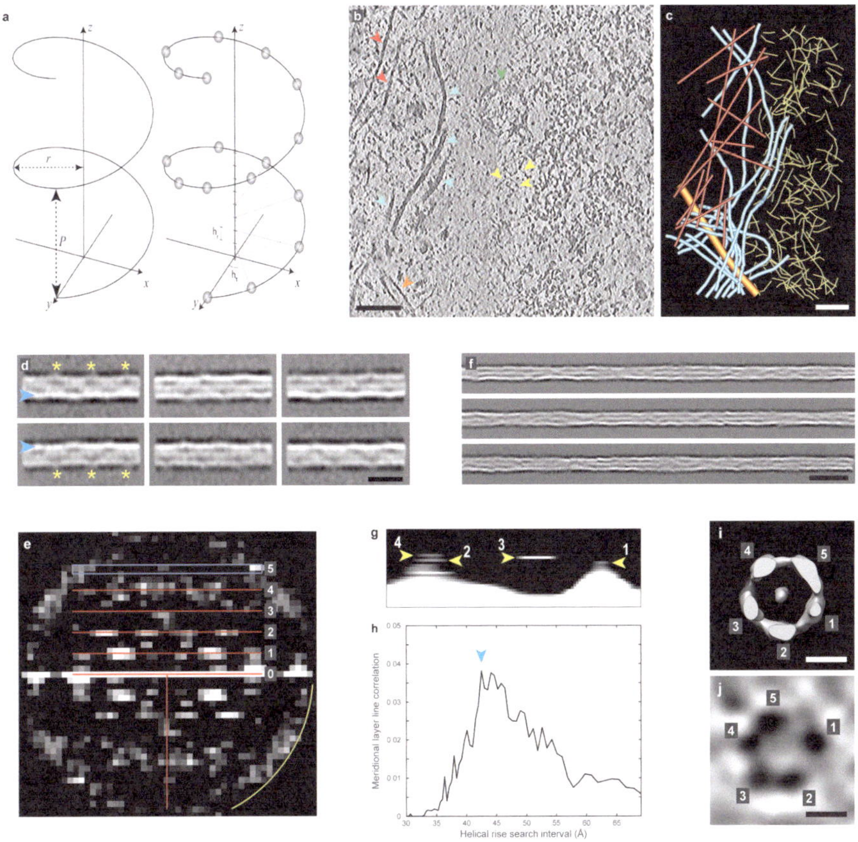

Fig. 6.2 Helical symmetry of VIFs. (**a**) A continuous helical path (depicted as a solid helical curve) with radius r completes a 360° rotation over the length of the helical pitch *P*. In helical symmetry, the asymmetric unit (illustrated as spheres on the right side of the figure) is placed along the continuous helical path. Therefore, the asymmetric unit is successively translated by the helical rise h_r and rotated by the helical twist h_t around the helical axis (*z*-axis) to generate the helical filament. (Adapted from [9], with permission from Elsevier). (**b**) Slice (~1 nm thick) through a tomogram of a cryo-FIB milled MEF, recorded in a region around the nuclear envelope. The following cellular structures are indicated: Red arrowheads, actin filaments; cyan arrowheads, VIFs; orange arrowhead, microtubule; green arrowhead, nuclear pore complex; and yellow arrowheads, lamin filaments. Scale bar is 100 nm. (**c**) Segmentation of the biopolymers contained in (**b**). Scale bar is 100 nm. (**d**) Class averages of VIFs, showing different views of a helical pattern. The yellow asterisks mark a repeating feature with a repeat distance of around 180 Å. Scale bar is 180 Å. (**e**) Combined power spectrum of the VIF class averages. Layer lines are indicated with horizontal red lines, the meridian by a vertical red line. The yellow arc represents 1/26 Å. (**f**) Gallery of three montaged and computationally straightened VIFs. Scale bar is 35 nm. (**g**) Zoom into the pitch region of the combined power spectrum of VIF filament montages. The yellow arrowheads, labelled from 1 to 4, indicate the position of the following layer lines: 1/207.4 Å, 1/195.9 Å, 1/185.6 Å, and 1/176.3 Å. (**h**) Helical rise candidates were scored by the similarity of the layer line with respect to a meridional reflection. The layer line connected to a helical rise value of 42.5 Å (indicated with the blue arrowhead) exhibits the highest similarity within the search range. (**i**) Subtomogram averaging [66] shows a pentameric VIF cross-section. Scale bar is 5 nm. (**j**) Pentameric cross-section of an individual VIF, extracted from a tomogram of a cryo-FIB milled MEF. Scale bar is 5 nm

imagined as a 3D packing problem, which describes how n_{asu} copies of the asymmetric unit are packed into the helical pitch P of the filament.

To determine the helical symmetry of VIFs [12], cryo-electron tomography (cryo-ET) of mouse embryonic fibroblasts (MEFs) was employed (Fig. 6.2b-c). As a prerequisite for cryo-ET [3, 43], MEFs cultured directly on electron microscopy grids were thinned to 200–300 nm using two different approaches. In the first method, MEFs were detergent-treated [80] prior to vitrification using a protocol that preserves the VIF network. In the second approach, MEFs were vitrified directly and subsequently thinned using cryo-focused ion beam (cryo-FIB) milling [44, 63, 64].

First, the VIFs were detected in the cryo-tomograms of the detergent-treated MEFs and subdivided with standard helical processing methods into short segments [23, 46, 83]. Subsequently, these segments were classified according to their structural patterns [65] and finally combined into 2D class averages (Fig. 6.2d). As a result of this procedure, the class averages exhibit an improved signal-to-noise ratio and resolution compared to the initially extracted filament segments. The combined power spectrum of these class averages shows a clear layer line spectrum (Fig. 6.2e), which is a unique consequence and proof of the helical symmetry of the observed filaments [9].

The resolution of such a power spectrum depends, among other factors (see section about polymorphism of IFs below), on the length of the analyzed filament segments. However, the filament segments need to be kept relatively short (typically around 30 nm) to compensate for the significant bending and flexibility of IFs [31, 33]. To address this issue, a method termed filament montaging was developed [84]. Here, the initially extracted segments are replaced by their respective 2D class averages at the position of their original extraction along a filament, and subsequently montaged to form extended stretches of IFs, typically between 100 and 300 nm in length. Finally, these filament montages were computationally straightened [10] (Fig. 6.2f).

This approach allows the analysis of long-range repeating patterns along highly flexible filaments [84]. Based on the combined power spectrum of montaged VIFs, the helical pitch and helical rise of VIFs was measured [12]. Here, it was observed that the previously detected layer lines split into a spectrum of fine layer lines, as shown in Fig. 6.2g for the layer lines which are related to the helical pitch of VIFs. The positions of these layer lines are connected to a range of possible helical pitch values, including 207.4 Å, 195.9 Å, 185.6 Å, and 176.3 Å (Fig. 6.2g, arrowheads). In order to evaluate the layer line spectrum related to the helical rise of VIFs, the layer lines were scored with respect to their similarity to a meridional reflection, which indicates the helical rise of the underlying helical assembly. According to this measurement, the most likely helical rise of VIFs is 42.5 Å (Fig. 6.2h, arrowhead).

Since there are still multiple possible solutions of helical symmetry arising from the different combinations of helical pitch and helical rise values with lower meridional similarity scoring, this ambiguity was resolved by analyzing the shape of the cross-section of the filaments [12], and choosing the helical symmetry solution which corresponds to the observed filament cross-section. Based on this analysis, conducted with cryo-tomograms recorded from cryo-FIB milled MEFs (Fig. 6.2b-c), it turned out that the cross-section of VIFs displays a pentameric shape (Fig. 6.2i-j). Combining these observations together with the power spectrum analysis, a helical rise of 42.5 Å and a helical pitch of 207.4 Å is the most likely helical symmetry solution. This solution corresponds to a packing of ≈5 asymmetric units in the helical pitch length of VIFs, which in turn explains the pentameric cross-section of VIFs.

Map and Model of Vimentin Filaments

Using these helical symmetry parameters in combination with single-particle cryo-EM of in vitro polymerized VIFs, a 3D electron density map of VIFs at sub-nanometer resolution was reconstructed [12] (Fig. 6.3a). In cross-section (Fig. 6.3b), the structure reveals that VIFs consist of a pentameric outer tube,

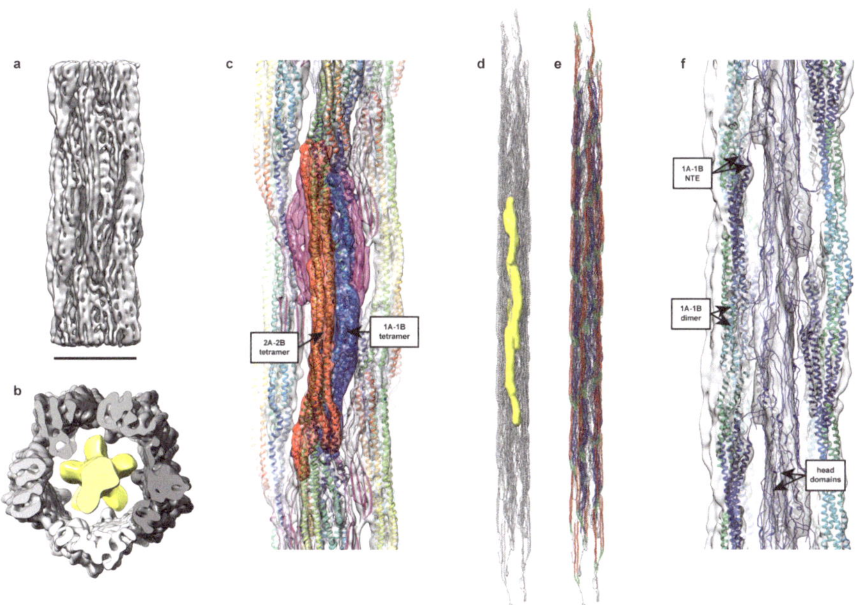

Fig. 6.3 Structure of VIFs. (**a**) Sub-nanometer resolution VIF structure, which was reconstructed with helical single particle cryo-EM. Scale bar is 10 nm. (**b**) Cross-section of the VIF structure. The pentameric outer tube (diameter around 10 nm) is rendered in different grey tones, while the central luminal fiber is depicted in yellow. (**c**) The octameric repeating unit of the VIF structure is assembled from two tetrameric densities, colored red and blue. The octamers are laterally connected by the vimentin tail domains (magenta-colored densities). The VIF atomic model is docked into the electron density map. The two tetrameric units are formed from two anti-parallel 2A-2B dimers (red helices) and two anti-parallel 1A-1B dimers (blue helices), respectively. (**d**) Atomic model of VIFs (chains are colored grey) with a length of around 190 nm. A single A11 tetramer is shown as yellow density. (**e**) Atomic model of VIFs, with helix 1A-1B shown in blue, helix 2A-2B shown in red, and the conserved regions within helices 1A and 2B highlighted in green. (**f**) The N-terminal ends (NTEs) of each 1A-1B dimer position the head domains to protrude into the lumen of the filament, forming the luminal fiber (blue chains)

approximately 10 nm in diameter, which surrounds a central luminal fiber (Fig. 6.3b, yellow density).

Based on this map, it is evident that the outer tube is composed of repeating octameric units (Fig. 6.3c), which are interconnected both laterally and longitudinally along the filament. Each of these octameric densities is assembled from two tetrameric subunits (Fig. 6.3c, red and blue densities), which differ in length and shape, allowing them to be clearly distinguished in the map. In concordance with previous literature, these longitudinally connected octameric units are referred to as protofibrils [58], which are defined as structural elements containing eight chains in cross-section. Since the 3D reconstruction reveals five protofibrils, the observed state of VIFs is composed of 40 chains in cross-section [49]. A second cryo-EM map, derived from VIFs polymerized from tailless vimentin [26], enabled localization of the tail domains. These tail domains are situated between the protofibrils and contribute to their lateral connections (Fig. 6.3c, magenta densities).

An atomic model of VIFs (Fig. 6.3c) was derived by integrating multiple lines of evidence [12]. This included the position of the tail domains, AlphaFold-based modeling of the vimentin 1A-1B and 2A-2B tetramers (Fig. 6.1h, j), incorporation of cross-linking data as spatial restraints for the lateral stagger of antiparallel dimers (Fig. 6.1i, k), rigid and flexible docking [79] of these models into the octameric densities observed in the cryo-EM map (Fig. 6.3c, red and blue helices), together with the measured helical symmetry of the filament. This analysis shows that the complete filament assembly can be constructed from the application of helical symmetry to an A11 tetramer (Fig. 6.3d). Therefore, it can be assumed that the A11 tetramer is the asymmetric unit of VIFs. While VIFs are constructed solely from A11 tetramers, the additional A22 and A12 modes of interaction between the coiled-coil dimers are realized within the repeating octameric units forming mature VIFs.

This model shows a further fascinating property of VIFs, namely that the helical geometry of the filaments intertwines the conserved regions in the 1A and 2B helices, which are situated at opposite sides of the rod domain in the elementary IF dimers [28, 47, 76]. However, in the mature filaments, these helices are situated in close proximity [25, 39, 69] (Fig. 6.3e, green helices). This interlock region is essential for the correct polymerization of VIFs [47]. It was previously shown based on mutation studies and near-UV circular dichroism [21, 47] that a specific mutation in this region (Y117L) leads to the failure of the filaments to elongate. Furthermore, the interlock region is a hot-spot for disease-associated mutations in other IFs [54], including Alexander disease [87], Charcot-Marie-Tooth [75] disease, and Epidermolysis bullosa simplex [7]. Since the amino acid sequences in the interlock region are highly conserved among IFs, it appears reasonable to speculate that the formation of an interlock region between the 1A and 2B helices is a general feature of IF architecture. The interaction between helices 1A and 2B in IFs has also been previously described based on X-ray crystallography and cross-linking data [25, 69, 71–73] and has been referred to as head-to-tail interaction and ACN binding mode.

The structure of VIFs suggests a further fascinating property of IFs, namely that the low-complexity head domains protrude into the lumen of the filament tube and form the luminal fiber (Fig. 6.3f). This observation is consistent with prior research indicating that the isolated, low-complexity head domains of vimentin, as well as the head domains of other IFs such as neurofilaments, can self-assemble into amyloid-like fibrillar structures [37, 91]. These assemblies are stabilized through transient cross-beta interactions, as demonstrated by X-ray diffraction [37] and NMR studies [50, 90].

While the head domains vary significantly in length between different IFs [13] (Fig. 6.1e-g), these domains still show similar characteristics compared among IFs, which were demonstrated by numerical simulations of their solvent accessibility [36]. These calculations demonstrated that the head domains of IFs in general are solvent inaccessible and are therefore likely to protrude into the lumen of the filaments. In contrast, this analysis showed that the low-complexity tail domains of IFs are typically solvent accessible. Consequently, as shown in the VIF structure [12] the tail domains are located between the protofibrils, thereby forming parts of the outer filament surface. This behavior of the tail domains is particularly pronounced in neurofilaments [81], where they are proposed to form a polymer brush surrounding the filaments [92]. In this context, the tail domains play a crucial role in maintaining proper spacing and organizing neurofilaments into bundles within axons [81].

Polymorphism

While the VIF model provides new insights into the 3D structure of IFs, the resolution of the corresponding 3D electron density map [12], and therefore the accuracy of the model, remains limited. Further improvements are required to achieve a more precise representation and deeper understanding of IF structure, polymerization, and function. One major challenge is that the resolution of 2D class averages of filament segments (Fig. 6.2d), particularly for VIFs [12] and KIFs (e.g., keratin 5/15 filaments [84]), remains restricted to approximately 20–10 Å. In contrast, class averages of actin filaments or microtubules frequently reach atomic resolution, allowing for the reconstruction of high-resolution 3D structures [55, 89]. As the analyzed VIF samples also contain a substantial number of actin filaments, these can be used as internal controls to assess micrograph quality [84]. When processed using the same single-particle helical reconstruction pipelines applied to IFs, actin filaments typically reach resolutions around 4 Å. This outcome confirms both the high quality of the micrographs and the utility of the image processing pipeline. However, it also highlights the unique challenges posed by IFs, namely their pronounced flexibility and structural polymorphism, which currently prevent atomic-resolution reconstruction.

Polymorphism of IFs was precisely documented and quantified with a series of STEM mass-per-length (mpl) measurements performed along different IFs [26, 74]. Importantly, the mpl is directly connected to the helical rise of the analyzed filaments [8] ($h_r = m/\text{mpl}$, with m is the mass of the asymmetric unit of the helical assembly). These measurements show typically three mpl peaks [26, 74], in case of VIFs centered around 25.0, 36.5, and 51.0 kDa/nm. With m set to the mass of the human vimentin tetramer of 214.6 kDa, these values translate into a helical rise of 85.8, 58.8, and 42.1 Å, respectively, with the third helical rise value in agreement with the helical rise of the pentameric VIF structure [12] (Fig. 6.2h). In their impressive work, Harald Herrmann and coworkers show [26] that depending on the chosen in vitro polymerization protocol individual VIFs can tune within a length of ~300 nm either through an extremely wide range of mpl values (in the following expressed as helical rise values) from 107.3 to 35.8 Å, when polymerized by dilution into filament buffer, or in contrast, show a drastically reduced mpl fluctuation within a range between 65.0 and 52.3 Å, when polymerized by dialysis against filament buffer.

There are multiple plausible mechanisms of how such an extended structural polymorphism could occur in IFs. First of all, as shown in fluorescence-recovery-after-photobleaching experiments [53, 88], IF subunits can detach and attach laterally along the filaments in a dynamic manner. This would introduce local variations to the filament structure and create a multitude of different structural states along the filaments. In addition, experimental evidence suggests that the low-complexity head domains of IFs self-assemble into filamentous structures [37], which are stabilized by transient, cross-beta interactions [90]. Possibly, such structures like the luminal fiber could oscillate through a continuum of structural states, varying from predominately intrinsically disordered to largely compacted and structurally stable. Since the luminal fiber is located at the center of the filament tube (Fig. 6.3f), changes in its conformation [40, 51] could directly feedback to the structure of the outer filament tube, resulting in variation in helical symmetry, subunit packing, and diameter. It has also been shown that such cross-beta structural motifs are highly sensitive to urea concentration [35]. The potentially associated structural changes in the luminal fiber could explain why different in vitro polymerization protocols result in a different variability of mpl, respectively helical rise values.

Another source of structural polymorphism in IFs may arise from post-translational modifications [14, 67]. Multiple kinases are known to phosphorylate IFs, and the low-complexity head and tail domains contain multiple phosphorylation sites [15]. Structural modifications through phosphorylation may drastically alter the conformation of the low-complexity domains [51], for example, through a gradual expansion or destabilization of the luminal fiber, until complete filament disassembly, as a function of the number of added phosphate groups. Based on force-strain curves measured with optical traps, it was shown that phosphorylation of VIFs softens the filaments [41]. It is conceivable that phosphorylation of the luminal fiber may lead to a reduced electrostatic coupling of the luminal fiber with the outer filament tube [41], which would soften the filaments and increase their flexibility.

Phosphorylation of the tail domains, which are situated between the protofibrils in VIFs and form their lateral connections, may induce structural reorganization or even lead to their detachment. Such modifications could locally relieve the constraints on filament diameter [26], promoting filament opening and increasing the spacing between the protofibrils. Consistent with this, experimental evidence shows that cells stably transfected with tailless vimentin exhibit VIFs with an increased diameter.

VIFs and KIFs are involved in force-related cellular processes [29]. It is therefore reasonable to assume that their structures are locally altered through applied forces. These forces occur in the cellular environment through stretching or bending of the filaments. The effect of forces on the structure of VIFs and KIFs was measured experimentally with optical tweezers [5, 38, 42]. The authors found that for KIFs (e.g., keratin 8/18 filaments) the structural changes upon repeated force loading lead to filament elongation and can be explained by the sliding of keratin subunits relative to each other. In contrast, VIFs keep their initial length when repeatedly stretched and relaxed but show structural changes related to the uncoiling of the alpha-helices in the filament structure. Subunit sliding within the observed filaments would lead to local changes in the lateral registration and relative staggering of the anti-parallel coiled-coil building blocks and the head-to-tail overlap in the interlock region. Such a sliding mechanism could explain the superposition of multiple discrete helical pitch and rise values in the averaged power spectra. Local stretching and eventually uncoiling of the alpha-helices in the structure could lead to a random blurring effect, which would be convolved onto the filament segments and distort their exact helical symmetry.

All the aforementioned chemical and mechanical processes may contribute to a structural continuum of polymorphic states within IFs. Given the low signal-to-noise ratio of cryo-EM images and limited dataset size, classifying this continuum into sufficiently homogeneous structural classes remains challenging. This likely accounts for the difficulty in achieving higher resolution in the 2D class averages and ultimately 3D reconstructions of IFs.

Discussion and Outlook

Reflecting on the remarkable classical image (dating back to 1961) of a cross-section of macrofibrils in the paracortex of fine Merino wool [16], which reveals a quasi-hexagonally packed bundle of alpha-keratin fibers, already clearly indicated that the architecture of these IFs adopts a ring-core structure. In this chapter, we aimed to outline the thought process that led us from this classical view of IFs to the refined model of a helical tube containing a luminal fiber.

A review of the literature suggests the coexistence of two generally accepted models describing the structure of fully polymerized, mature IFs [82]. In the first model (the helical lattice model [57]), IFs are typically represented in an abstract form as a helical lattice [71–73], in which alternating rows of head-to-tail connected IF dimers and their anti-parallel counterparts are axially shifted to form A11, A22, and A12 interactions between neighboring anti-parallel dimers. In the second model (the unit-length filament model [26, 27], IFs are depicted as being assembled from cylindrical building blocks composed of laterally aligned A11 tetramers

[29, 81]. These unit-length filaments are merging in a staggered manner along the filament axis and then radially compacting to form a mature IF [82], which is characterized by alternating bands of A11 and A22 interactions between neighboring anti-parallel dimers.

Both models typically are calibrated to contain a number of 32 chains (comprising 4 protofibrils) in the cross-section of a mature IF. Consequently, both models are consistent with the most prominent mpl peak [26], which, for example, in VIFs is ~37 ± 4 kDa/nm, and accounts for ~76% of the mpl measurements [74], whereas the pentameric VIF model explains the secondary mpl peak at ~51 kDa/nm and accounts for ~7% of the mpl measurements [74]. Additionally, both models predict a close spatial proximity or overlap of the highly conserved regions within the 1A and 2B helices, corresponding to the ACN interaction [71]. These models do not provide explicit descriptions of the position or structure of the head and tail domains (except in the case of neurofilament tail domains [81]).

However, it is evident that these models also exhibit clear differences. Since the unit-length filament model is non-helical, neighboring A11 tetramers are not axially shifted relative to one another (i.e., the helical rise is zero), and the resulting filament geometry does not give rise to A12 interactions between adjacent anti-parallel dimers. It is, however, plausible that during filament maturation, particularly during the radial compaction step, the laterally aligned A11 tetramers within unit-length filaments undergo axial shifts, thereby introducing helical rise and enabling A12 interactions. Nonetheless, based on the current cryo-EM data, the assembly pathway leading to a specific mature filament architecture cannot be resolved. What can be observed is the final product of what is likely an extraordinarily complex polymerization and maturation process. The VIF reconstruction clearly exhibits helical symmetry. A central feature of this structure is an octameric repeating unit composed of two laterally aligned but structurally distinct tetramers, each formed by two anti-parallel dimers involving the 1A-1B and 2A-2B domains. This helical assembly naturally realizes A11, A22, and A12 interactions between adjacent anti-parallel dimers. Thus, the VIF model is consistent with the helical lattice model previously proposed based on cross-linking data.

If we were to speculate and extend insights from the pentameric VIF model to a more general model of IF architecture, we propose that the octameric repeating unit represents a fundamental structural feature shared across different IF types. In the case of pentameric VIFs, this octamer is relatively confined to the diameter of a protofibril and is oriented almost parallel to the filament axis. However, it is plausible that in other VIF polymorphs or IF structures, the octameric repeating unit spans a wider azimuthal angular range and adopts a different tilt relative to the filament axis. Such structural variation would effectively increase the helical rise and helical twist angle, thereby decreasing the number of asymmetric units within one helical pitch and reducing the mpl, and consequently, the number of chains in the filament cross-section. Nonetheless, other structural features inferred from the pentameric VIF model would remain unaffected. These include the formation of A11, A22, and A12 interactions within the octamer, the identity of the asymmetric unit as an A11 tetramer, the assembly of the interlock region, and the emergence of the luminal fiber.

References

1. Agnetti G, Herrmann H, Cohen S (2022) New roles for desmin in the maintenance of muscle homeostasis. FEBS J 289:2755–2770. https://doi.org/10.1111/febs.15864
2. Aziz A et al (2012) The structure of vimentin linker 1 and rod 1B domains characterized by site-directed spin-labeling electron paramagnetic resonance (SDSL-EPR) and X-ray crystallography. J Biol Chem 287:28349–28361. https://doi.org/10.1074/jbc.M111.334011
3. Beck M, Baumeister W (2016) Cryo-electron tomography: can it reveal the molecular sociology of cells in atomic detail? Trends Cell Biol 26:825–837. https://doi.org/10.1016/j.tcb.2016.08.006
4. Bharat TA et al (2012) Structural dissection of Ebola virus and its assembly determinants using cryo-electron tomography. Proc Natl Acad Sci USA 109:4275–4280. https://doi.org/10.1073/pnas.1120453109
5. Block J et al (2018) Viscoelastic properties of vimentin originate from nonequilibrium conformational changes. Sci Adv 4:eaat1161. https://doi.org/10.1126/sciadv.aat1161
6. Chernyatina AA, Nicolet S, Aebi U, Herrmann H, Strelkov SV (2012) Atomic structure of the vimentin central alpha-helical domain and its implications for intermediate filament assembly. Proc Natl Acad Sci USA 109:13620–13625. https://doi.org/10.1073/pnas.1206836109
7. Cummins RE et al (2001) Keratin 14 point mutations at codon 119 of helix 1A resulting in different epidermolysis bullosa simplex phenotypes. J Invest Dermatol 117:1103–1107. https://doi.org/10.1046/j.0022-202x.2001.01508.x
8. Davies DB, Saenger W, Danyluk SS, Federation of European Biochemical Societies. & North Atlantic Treaty Organization (1982) Scientific Affairs Division. Structural molecular biology : methods and applications. Plenum Press
9. Diaz R, Rice WJ, Stokes DL (2010) Fourier-Bessel reconstruction of helical assemblies. Methods Enzymol 482:131–165. https://doi.org/10.1016/S0076-6879(10)82005-1
10. Egelman EH (1986) An algorithm for straightening images of curved filamentous structures. Ultramicroscopy 19:367–373. https://doi.org/10.1016/0304-3991(86)90096-3
11. Egelman EH (2010) Reconstruction of helical filaments and tubes. Methods Enzymol 482:167–183. https://doi.org/10.1016/S0076-6879(10)82006-3
12. Eibauer M et al (2024) Vimentin filaments integrate low-complexity domains in a complex helical structure. Nat Struct Mol Biol 31:939–949. https://doi.org/10.1038/s41594-024-01261-2
13. Eldirany SA, Lomakin IB, Ho M, Bunick CG (2021) Recent insight into intermediate filament structure. Curr Opin Cell Biol 68:132–143. https://doi.org/10.1016/j.ceb.2020.10.001
14. Eriksson JE, Opal P, Goldman RD (1992) Intermediate filament dynamics. Curr Opin Cell Biol 4:99–104. https://doi.org/10.1016/0955-0674(92)90065-k
15. Eriksson JE et al (2004) Specific in vivo phosphorylation sites determine the assembly dynamics of vimentin intermediate filaments. J Cell Sci 117:919–932. https://doi.org/10.1242/jcs.00906
16. Filshie BK, Rogers GE (1961) The fine structure of alpha-keratin. J Mol Biol 3:784–786. https://doi.org/10.1016/s0022-2836(61)80084-3
17. Fitzpatrick AWP et al (2017) Cryo-EM structures of tau filaments from Alzheimer's disease. Nature 547:185–190. https://doi.org/10.1038/nature23002
18. Fraser RD, Parry DA (2006) The three-dimensional structure of trichocyte (hard alpha-) keratin intermediate filaments: the nature of the repeating unit. J Struct Biol 155:375–378. https://doi.org/10.1016/j.jsb.2005.12.014
19. Fuchs E, Weber K (1994) Intermediate filaments: structure, dynamics, function, and disease. Annu Rev Biochem 63:345–382. https://doi.org/10.1146/annurev.bi.63.070194.002021
20. Galkin VE, Orlova A, Vos MR, Schroder GF, Egelman EH (2015) Near-atomic resolution for one state of F-actin. Structure 23:173–182. https://doi.org/10.1016/j.str.2014.11.006
21. Georgakopoulou S, Moller D, Sachs N, Herrmann H, Aebi U (2009) Near-UV circular dichroism reveals structural transitions of vimentin subunits during intermediate filament assembly. J Mol Biol 386:544–553. https://doi.org/10.1016/j.jmb.2008.12.053

22. Gong M, Makowski L (1992) Helical structure of P pili from Escherichia coli. Evidence from X-ray fiber diffraction and scanning transmission electron microscopy. J Mol Biol 228:735–742. https://doi.org/10.1016/0022-2836(92)90860-m
23. He S, Scheres SHW (2017) Helical reconstruction in RELION. J Struct Biol 198:163–176. https://doi.org/10.1016/j.jsb.2017.02.003
24. Herrmann H, Aebi U (2004) Intermediate filaments: molecular structure, assembly mechanism, and integration into functionally distinct intracellular scaffolds. Annu Rev Biochem 73:749–789. https://doi.org/10.1146/annurev.biochem.73.011303.073823
25. Herrmann H, Aebi U (2016) Intermediate filaments: structure and assembly. Cold Spring Harb Perspect Biol 8. https://doi.org/10.1101/cshperspect.a018242
26. Herrmann H et al (1996) Structure and assembly properties of the intermediate filament protein vimentin: the role of its head, rod and tail domains. J Mol Biol 264:933–953. https://doi.org/10.1006/jmbi.1996.0688
27. Herrmann H, Haner M, Brettel M, Ku NO, Aebi U (1999) Characterization of distinct early assembly units of different intermediate filament proteins. J Mol Biol 286:1403–1420. https://doi.org/10.1006/jmbi.1999.2528
28. Herrmann H et al (2000) The intermediate filament protein consensus motif of helix 2B: its atomic structure and contribution to assembly. J Mol Biol 298:817–832. https://doi.org/10.1006/jmbi.2000.3719
29. Herrmann H, Bar H, Kreplak L, Strelkov SV, Aebi U (2007) Intermediate filaments: from cell architecture to nanomechanics. Nat Rev Mol Cell Biol 8:562–573. https://doi.org/10.1038/nrm2197
30. Hol EM, Pekny M (2015) Glial fibrillary acidic protein (GFAP) and the astrocyte intermediate filament system in diseases of the central nervous system. Curr Opin Cell Biol 32:121–130. https://doi.org/10.1016/j.ceb.2015.02.004
31. Hu J et al (2019) High stretchability, strength, and toughness of living cells enabled by hyperelastic vimentin intermediate filaments. Proc Natl Acad Sci USA 116:17175–17180. https://doi.org/10.1073/pnas.1903890116
32. Jacob JT, Coulombe PA, Kwan R, Omary MB (2018) Types I and II keratin intermediate filaments. Cold Spring Harb Perspect Biol 10. https://doi.org/10.1101/cshperspect.a018275
33. Janmey PA, Euteneuer U, Traub P, Schliwa M (1991) Viscoelastic properties of vimentin compared with other filamentous biopolymer networks. J Cell Biol 113:155–160. https://doi.org/10.1083/jcb.113.1.155
34. Jumper J et al (2021) Highly accurate protein structure prediction with AlphaFold. Nature 596: 583–589. https://doi.org/10.1038/s41586-021-03819-2
35. Kakeshpour T et al (2021) A lowly populated, transient β-sheet structure in monomeric Aβ identified by multinuclear NMR of chemical denaturation. Biophys Chem 270:106531. https://doi.org/10.1016/j.bpc.2020.106531
36. Kornreich M, Avinery R, Malka-Giborb E, Laser-Azogui A, Beck R (2015) Order and disorder in intermediate filament proteins. FEBS Lett 589:2464–2476. https://doi.org/10.1016/j.febslet.2015.07.024
37. Lin Y et al (2016) Toxic PR poly-dipeptides encoded by the C9orf72 repeat expansion target LC domain polymers. Cell 167:789–802 e712. https://doi.org/10.1016/j.cell.2016.10.003
38. Lorenz C, Forsting J, Style RW, Klumpp S, Koster S (2023) Keratin filament mechanics and energy dissipation are determined by metal-like plasticity. Matter 6:2019–2033. https://doi.org/10.1016/j.matt.2023.04.014
39. Kapinos LE et al (2010) Characterization of the head-to-tail overlap complexes formed by human lamin A, B1 and B2 "half-minilamin" dimers. J Mol Biol 396:719–731. https://doi.org/10.1016/j.jmb.2009.12.001
40. Kato M, Zhou XM, McKnight SL (2022) How do protein domains of low sequence complexity work? RNA 28:3–15. https://doi.org/10.1261/rna.078990.121
41. Kraxner J et al (2021) Post-translational modifications soften vimentin intermediate filaments. Nanoscale 13:380–387. https://doi.org/10.1039/d0nr07322a

42. Lorenz C et al (2019) Lateral subunit coupling determines intermediate filament mechanics. Phys Rev Lett 123:188102. https://doi.org/10.1103/PhysRevLett.123.188102
43. Lucic V, Forster F, Baumeister W (2005) Structural studies by electron tomography: from cells to molecules. Annu Rev Biochem 74:833–865. https://doi.org/10.1146/annurev.biochem.73.011303.074112
44. Mahamid J et al (2016) Visualizing the molecular sociology at the HeLa cell nuclear periphery. Science 351:969–972. https://doi.org/10.1126/science.aad8857
45. Manco C et al (2025) Peripherin, a new promising biomarker in neurological disorders. Eur J Neurosci 61:e70030. https://doi.org/10.1111/ejn.70030
46. Martins B et al (2021) Unveiling the polarity of actin filaments by cryo-electron tomography. Structure. https://doi.org/10.1016/j.str.2020.12.014
47. Meier M et al (2009) Vimentin coil 1A-A molecular switch involved in the initiation of filament elongation. J Mol Biol 390:245–261. https://doi.org/10.1016/j.jmb.2009.04.067
48. Mendez MG, Kojima S, Goldman RD (2010) Vimentin induces changes in cell shape, motility, and adhesion during the epithelial to mesenchymal transition. FASEB J 24:1838–1851. https://doi.org/10.1096/fj.09-151639
49. Mucke N et al (2018) Assembly kinetics of Vimentin tetramers to unit-length filaments: a stopped-flow study. Biophys J 114:2408–2418. https://doi.org/10.1016/j.bpj.2018.04.032
50. Murray DT et al (2017) Structure of FUS protein fibrils and its relevance to self-assembly and phase separation of low-complexity domains. Cell 171:615–627 e616. https://doi.org/10.1016/j.cell.2017.08.048
51. Newcombe EA, Delaforge E, Hartmann-Petersen R, Skriver K, Kragelund BB (2022) How phosphorylation impacts intrinsically disordered proteins and their function. Essays Biochem 66:901–913. https://doi.org/10.1042/Ebc20220060
52. Nicolet S, Herrmann H, Aebi U, Strelkov SV (2010) Atomic structure of vimentin coil 2. J Struct Biol 170:369–376. https://doi.org/10.1016/j.jsb.2010.02.012
53. Noding B, Herrmann H, Koster S (2014) Direct observation of subunit exchange along mature vimentin intermediate filaments. Biophys J 107:2923–2931. https://doi.org/10.1016/j.bpj.2014.09.050
54. Omary MB, Coulombe PA, McLean WHI (2004) Mechanisms of disease: intermediate filament proteins and their associated diseases. New Engl J Med 351:2087–2100. https://doi.org/10.1056/NEJMra040319
55. Oosterheert W, Klink BU, Belyy A, Pospich S, Raunser S (2022) Structural basis of actin filament assembly and aging. Nature 611:374–379. https://doi.org/10.1038/s41586-022-05241-8
56. Pang AH, Obiero JM, Kulczyk AW, Sviripa VM, Tsodikov OV (2018) A crystal structure of coil 1B of vimentin in the filamentous form provides a model of a high-order assembly of a vimentin filament. FEBS J 285:2888–2899. https://doi.org/10.1111/febs.14585
57. Parry DA, Steinert PM (1999) Intermediate filaments: molecular architecture, assembly, dynamics and polymorphism. Q Rev Biophys 32:99–187. https://doi.org/10.1017/s0033583500003516
58. Parry DA, Marekov LN, Steinert PM (2001) Subfilamentous protofibril structures in fibrous proteins: cross-linking evidence for protofibrils in intermediate filaments. J Biol Chem 276:39253–39258. https://doi.org/10.1074/jbc.M104604200
59. Perlson E et al (2005) Vimentin-dependent spatial translocation of an activated MAP kinase in injured nerve. Neuron 45:715–726. https://doi.org/10.1016/j.neuron.2005.01.023
60. Qin Z, Kreplak L, Buehler MJ (2009) Hierarchical structure controls nanomechanical properties of vimentin intermediate filaments. PLoS One 4:e7294. https://doi.org/10.1371/journal.pone.0007294
61. Rappsilber J (2011) The beginning of a beautiful friendship: cross-linking/mass spectrometry and modelling of proteins and multi-protein complexes. J Struct Biol 173:530–540. https://doi.org/10.1016/j.jsb.2010.10.014

62. Ridge KM, Eriksson JE, Pekny M, Goldman RD (2022) Roles of vimentin in health and disease. Genes Dev 36:391–407. https://doi.org/10.1101/gad.349358.122
63. Rigort A et al (2010) Micromachining tools and correlative approaches for cellular cryo-electron tomography. J Struct Biol 172:169–179. https://doi.org/10.1016/j.jsb.2010.02.011
64. Rigort A et al (2012) Focused ion beam micromachining of eukaryotic cells for cryoelectron tomography. Proc Natl Acad Sci USA 109:4449–4454. https://doi.org/10.1073/pnas.1201333109
65. Scheres SH (2012) RELION: implementation of a Bayesian approach to cryo-EM structure determination. J Struct Biol 180:519–530. https://doi.org/10.1016/j.jsb.2012.09.006
66. Scheres SHW, Melero R, Valle M, Carazo JM (2009) Averaging of electron subtomograms and random conical tilt reconstructions through likelihood optimization. Structure 17:1563–1572. https://doi.org/10.1016/j.str.2009.10.009
67. Sihag RK, Inagaki M, Yamaguchi T, Shea TB, Pant HC (2007) Role of phosphorylation on the structural dynamics and function of types III and IV intermediate filaments. Exp Cell Res 313: 2098–2109. https://doi.org/10.1016/j.yexcr.2007.04.010
68. Stahl K et al (2024) Modelling protein complexes with crosslinking mass spectrometry and deep learning. Nat Commun 15:7866. https://doi.org/10.1038/s41467-024-51771-2
69. Stalmans G et al (2020) Addressing the molecular mechanism of longitudinal Lamin assembly using chimeric fusions. Cells 9:1633. https://doi.org/10.3390/cells9071633
70. Steinert PM, Steven AC, Roop DR (1985) The molecular biology of intermediate filaments. Cell 42:411–420. https://doi.org/10.1016/0092-8674(85)90098-4
71. Steinert PM, Marekov LN, Parry DA (1993a) Diversity of intermediate filament structure. Evidence that the alignment of coiled-coil molecules in vimentin is different from that in keratin intermediate filaments. J Biol Chem 268:24916–24925
72. Steinert PM, Marekov LN, Parry DA (1993b) Conservation of the structure of keratin intermediate filaments: molecular mechanism by which different keratin molecules integrate into preexisting keratin intermediate filaments during differentiation. Biochemistry 32:10046–10056. https://doi.org/10.1021/bi00089a021
73. Steinert PM, Marekov LN, Fraser RD, Parry DA (1993c) Keratin intermediate filament structure. Crosslinking studies yield quantitative information on molecular dimensions and mechanism of assembly. J Mol Biol 230:436–452. https://doi.org/10.1006/jmbi.1993.1161
74. Steven AC, Hainfeld JF, Trus BL, Wall JS, Steinert PM (1983) The distribution of mass in heteropolymer intermediate filaments assembled in vitro. Stem analysis of vimentin/desmin and bovine epidermal keratin. J Biol Chem 258:8323–8329
75. Stone EJ, Kolb SJ, Brown A (2021) A review and analysis of the clinical literature on Charcot-Marie-Tooth disease caused by mutations in neurofilament protein L. Cytoskeleton 78:97–110. https://doi.org/10.1002/cm.21676
76. Strelkov SV et al (2002) Conserved segments 1A and 2B of the intermediate filament dimer: their atomic structures and role in filament assembly. EMBO J 21:1255–1266. https://doi.org/10.1093/emboj/21.6.1255
77. Szeverenyi I et al (2008) The human intermediate filament database: comprehensive information on a gene family involved in many human diseases. Hum Mutat 29:351–360. https://doi.org/10.1002/humu.20652
78. Tenga R, Medalia O (2020) Structure and unique mechanical aspects of nuclear Lamin filaments. Curr Opin Struct Biol 64:152–159. https://doi.org/10.1016/j.sbi.2020.06.017
79. Trabuco LG, Villa E, Mitra K, Frank J, Schulten K (2008) Flexible fitting of atomic structures into electron microscopy maps using molecular dynamics. Structure 16:673–683. https://doi.org/10.1016/j.str.2008.03.005
80. Turgay Y et al (2017) The molecular architecture of lamins in somatic cells. Nature 543:261–264. https://doi.org/10.1038/nature21382
81. van Asperen JV, Kotaich F, Caillol D, Bomont P (2024) Neurofilaments: novel findings and future challenges. Curr Opin Cell Biol 87:102326. https://doi.org/10.1016/j.ceb.2024.102326

82. Vermeire PJ et al (2021) Molecular interactions driving intermediate filament assembly. Cells 10. https://doi.org/10.3390/cells10092457
83. Wagner T et al (2019) SPHIRE-crYOLO is a fast and accurate fully automated particle picker for cryo-EM. Commun Biol 2:218. https://doi.org/10.1038/s42003-019-0437-z
84. Weber MS et al (2021) Structural heterogeneity of cellular K5/K14 filaments as revealed by cryo-electron microscopy. elife 10. https://doi.org/10.7554/eLife.70307
85. Woolfson DN (2023) Understanding a protein fold: the physics, chemistry, and biology of alpha-helical coiled coils. J Biol Chem 299:104579. https://doi.org/10.1016/j.jbc.2023.104579
86. Wu HY et al (2022) Vimentin intermediate filaments and filamentous actin form unexpected interpenetrating networks that redefine the cell cortex. Proc Natl Acad Sci USA 119: e2115217119. https://doi.org/10.1073/pnas.2115217119
87. Yang AW, Lin NH, Yeh TH, Snider N, Perng MD (2022) Effects of Alexander disease-associated mutations on the assembly and organization of GFAP intermediate filaments. Mol Biol Cell 33:ar69. https://doi.org/10.1091/mbc.E22-01-0013
88. Yoon M, Moir RD, Prahlad V, Goldman RD (1998) Motile properties of vimentin intermediate filament networks in living cells. J Cell Biol 143:147–157. https://doi.org/10.1083/jcb.143.1.147
89. Zhang R, LaFrance B, Nogales E (2018) Separating the effects of nucleotide and EB binding on microtubule structure. Proc Natl Acad Sci USA 115:E6191–E6200. https://doi.org/10.1073/pnas.1802637115
90. Zhou XM et al (2021) Transiently structured head domains control intermediate filament assembly. Proc Natl Acad Sci USA 118:e2022121118. https://doi.org/10.1073/pnas.2022121118
91. Zhou XM, Kato M, Mcknight SL (2023) How do disordered head domains assist in the assembly of intermediate filaments? Curr Opin Cell Biol 85:102262. https://doi.org/10.1016/j.ceb.2023.102262
92. Zhulina EB, Leermakers FA (2010) The polymer brush model of neurofilament projections: effect of protein composition. Biophys J 98:462–469. https://doi.org/10.1016/j.bpj.2009.10.033

Chapter 7
Intermediate Filaments: Structure-Mechanics Crosstalk

Sarah Köster and Harald Herrmann

Abstract Intermediate filaments (IFs) possess unique mechanical properties that distinguish them from actin filaments and microtubules. In particular, they exhibit high flexibility, pronounced extensibility, and complete stability during biochemical extractions from cells and tissues. These characteristics stem from their molecular structure, which is typical of fibrous proteins. A defining feature is the central ~300 amino acid long α-helical segment with a distinct hydrophobic sequence pattern, facilitating the formation of a parallel coiled-coil dimer. Under low ionic strength conditions, two such dimers interact via their basic amino-terminal domains with the acidic coiled-coil domains to form distinct, rod-like tetrameric complexes. Upon addition of salt, the tetramers first assemble laterally into full-width, unit-length filaments, which then anneal longitudinally into micrometer-long filaments with a characteristic, 10-nm diameter. Advanced experimental techniques enable us to measure piconewton forces and micrometer length scales. By combining, for example, optical tweezers or atomic force microscopy with sophisticated data analysis and numeric modeling, we have deepened our understanding of the structure-mechanics relationship in IFs, including their force-extension behavior and the low bending rigidity. These findings enable us to hypothesize about the mechanical roles of these filaments within the living cell and speculate about biomimetic, synthetic materials.

Keywords Cytoskeleton · Vimentin · Keratin · Molecular structure · Stress-strain relationship · Bending rigidity · Persistence length · α helix · Coiled-coil

S. Köster (✉)
University of Göttingen, Institute for X-Ray Physics, Göttingen, Germany
e-mail: sarah.koester@uni-goettingen.de

H. Herrmann
Friedrich-Alexander University Erlangen-Nürnberg, Department of Physics, Erlangen, Germany
e-mail: harald.herrmann-lerdon@fau.de

D. A. D. Parry, P. K. Luther (eds.), *Fibrous Proteins: Structures and Mechanisms*, Subcellular Biochemistry 113, https://doi.org/10.1007/978-3-032-05273-5_7

Abbreviations

AFM	Atomic force microscope/microscopy
EM	Electron microscopy
EMT	Epithelial-to-mesenchymal transition
GFAP	Glial fibrillary acidic protein
IF	Intermediate filament
MD	Molecular dynamics
SAXS	Small-angle X-ray scattering
STEM	Scanning transmission electron microscopy
ULF	Unit-length filament
WAXS	Wide-angle X-ray scattering
WLC	Worm-like chain

Introduction

Biological cells and tissues are determined by their mechanical properties as much as by biochemical processes. For example, muscle is highly contractile, skin is extensible, and immune cells exhibit high migratory activity and extensive shape changes as they move through the extracellular matrix. The most important determinant of cell mechanics is the cytoskeleton, a dense, intracellular network of filamentous proteins: actin filaments, microtubules, and intermediate filaments (IFs). Actin filaments and microtubules share a number of common features, including dynamic assembly and disassembly of globular actin monomers or tubulin dimers and genetic conservation across cell types and even species. The structural polarity of both types of filaments enables them to serve as "tracks" for motor proteins such as myosin and kinesin, respectively. Furthermore, actin filaments and microtubules are virtually inextensible [49]. IFs, by contrast, comprise a large family of over 70 different proteins in humans that are expressed in a cell- and tissue-type-specific manner, including lamins, which will not be discussed here [15, 46]. For example, desmin is found in muscle (see part B of this book), keratin in skin and epidermal appendages (see Chap. 5), and vimentin in cells of mesenchymal origin, such as fibroblasts. Due to their enormous variety, IF proteins have been categorized into six "types" or sequence-homology groups, according to expression patterns and sequence homology (see Table 7.1) [13, 27].

Despite significant differences in their amino acid sequences, i.e., their primary protein structure, IF proteins share a common structural building plan that justifies the definition of a large protein family. Namely, they all exhibit a ~300 amino acid-long central α-helical segment, and in vitro IF proteins assemble in a strictly hierarchical manner into long filaments (see section "Lateral and Longitudinal Assembly"). The biological and biochemical functions of IFs are versatile—from acting as signaling modulators to their involvement in cell death [64]. This range of

Table 7.1 The six classes of IF proteins according to amino acid sequence homology of the α-helical rod domain along with examples of tissues where they are expressed. We include IF-related proteins that do not fit into any of the six classes as type X according to Guzenko et al. [25]

Type	IF proteins	Expression in
I	Acidic keratins	Cytokeratins in epithelial cells; hair keratins in hair, nails, horns
II	Basic keratins	
III	Desmin, vimentin, glial fibrillary acidic protein (GFAP), peripherin	Muscle, fibroblasts, blood vessels, astrocytes, peripheral neurons
IV	Neurofilament proteins, α-internexin	Neurons, muscle
V	Nuclear lamins	Nuclear lamina
VI	Phakinin	Eye lens
X	Filensin, nestin, synemin, syncoilin	IF-related proteins

functions explains the importance of IFs in physiological and pathological contexts. In fact, IFs are associated with many genetic diseases, most of which are caused by single point mutations, such as is the case for Alexander disease and certain cardiomyopathies [35, 63].

This chapter will focus on the mechanical properties of IFs. It will begin with the structural basics, by unraveling the assembly pathway from the rod-shaped α-helical tetramers to micrometer-long filaments and will shed light on the physical mechanisms that lead to this unique assembly and resulting filament architecture. Next, modern biophysical methods to investigate single IFs will be presented along with the current state-of-the-art of in vitro IF mechanics. Third, we will look into theoretical and computational models that help to understand the link between IF structure and mechanics. Most in vitro work on IFs has been performed on vimentin, which is a highly abundant IF protein in human and which is furthermore evolutionary highly conserved from shark to man [74]. Early on, protocols for purification of vimentin from eye lenses and cell cultures have been developed. Today, the different proteins are usually expressed recombinantly in bacteria, which allows for direct comparison of the properties of vimentin from different species such as frog and human [30]. Recombinant expression [6, 34] has the advantage that the protein can be produced in a very pure form, free of posttranslational modifications, and directed mutagenesis can be used to introduce desired mutations, e.g., disease mutations or phosphomimetic mutations. Therefore, we will focus on vimentin in this chapter and will explicitly mention when other IFs are involved.

Structure of Intermediate Filaments

Organization of the Coiled Coils Within the Filament

The structural building plan of IF proteins exhibits a central domain of three α-helical segments connected by non-α-helical linkers L1 and L12 (Fig. 7.1a).

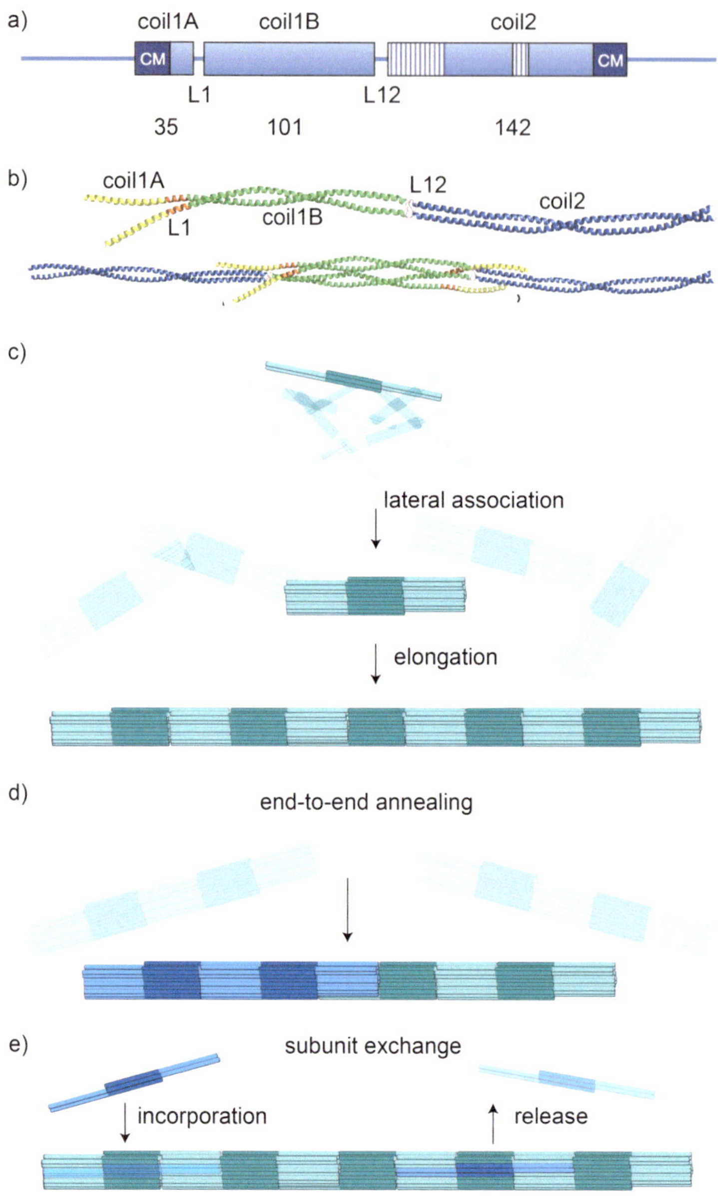

Fig. 7.1 (**a**) Domain structure of vimentin. The boxes represent α-helical segments coil 1A, coil 1B, and coil 2 as separated by non-α-helical linkers L1 and L12; striped parts of the boxes represent regions with hendecad repeats instead of standard heptad repeat patterns; CM: conserved amino acid motifs. The numbers indicate amino acids per segment. (**b**) Model of the structure of a vimentin coiled-coil dimer and the half-staggered, antiparallel tetramer. (Reprinted with permission from Ref. [10]). (**c**) Schematic of the assembly pathway of vimentin from tetramers to filaments. (**d**) End-to-end annealing of two extended filaments; this process is in stark contrast to classical "polymerization," where subunits are added one-by-one. (**e**) Subunit exchange; the massively parallel arrangement of tetramers in the filament provides the possibility for exchange of these subunits without filament breaking. The regular segmentation of the filament reflects a 22 nm-repeat seen early on by glycerol spraying/low-angle rotary metal shadowing EM [29]

This so-called rod domain is flanked by intrinsically disordered N-terminal ("head") and C-terminal ("tail") domains of different lengths for the individual IF proteins. Due to a heptad repeat amino acid pattern (***abcdefg***), in which ***a*** and ***d*** contain apolar residues, IF proteins of one kind usually form coiled-coil dimers with each other. Keratins, however, form obligatory heterodimers from one type I and one type II protein each. Moreover, IF proteins of type III and those of type IV can form heterodimers [12, 25]. In fact, these extended coiled coils serve as the fundamental elements necessary for the generation of an IF.

IFs are notoriously insoluble in buffers of physiological ionic strength. Therefore, in order to obtain free monomers for the purification of IF proteins after isolation from tissues by extraction with solutions of high concentrations of salt and anionic detergent, the crude filament pellets obtained by centrifugation must be denatured with a strong chaotropic agent such as urea at high concentration (8 M). Then, to assemble purified IF proteins into filaments, the chaotropes must be removed. This is conveniently done by dialysis. When IF proteins are dialyzed directly into buffers of physiological ionic strength, very regular filaments are formed. However, to be able to study the assembly process in vitro, one needs to generate soluble, assembly-competent complexes, which is done by dialysis into buffers of very low ionic strength [57]. Analytical ultracentrifugation has revealed that at 6 M urea, vimentin is present as a dimer, which associates further into a tetrameric complex at about 5 M urea. Further removal of the urea does not alter the molecular size of the complexes [30]. Hence, the position of the head domains within a tetramer is the consequence of the folding events during renaturation. By both chemical crosslinking and electron microscopy (EM), this tetramer was defined as a uniform complex of two approximately half-staggered, antiparallel dimers of about 60-nm length (for review see [28, 65]). Eventually, an atomic model for the tetrameric complex of vimentin was achieved by X-ray crystallization of overlapping rod segments (Fig. 7.1b) [10]. However, further molecular structural information on the next level of IF organization has not been obtained, except for cryo-electron tomography data on in vitro assembled vimentin filaments that revealed four protofibrillar strands exhibiting a right-handed supertwist with a 96 nm helical pitch [24].

Recent work on the structure of a complete IF combined cryo-focused ion-beam milling, cryo-EM, and tomography (see Chap. 6). Thereby, a 3D structure of vimentin IFs as found in cells is obtained for the first time. From these data a model is derived that depicts 40 vimentin monomers, and thus five octamers, arranged in parallel, with the intrinsically unstructured head domains positioned in the center of the filaments [18]. Indeed, the existence of a small "hollow core" of IFs has been proposed early on from electron microscopic images. This finding is also in line with previous work that revealed a lower electron and mass density in the center of the filament [9]. According to this model, the non-α-helical tail domains should form the lateral links between the protofibrils within the filament. The current structural resolution is 7.2 Å and higher resolution cryo-EM data will be necessary to derive a definitive model.

In a mechanistic view, tetramers appear to be the true subunits for filament assembly in vitro, because dimers cannot be obtained in solution without denaturing

Table 7.2 Size and charge distribution of the head, rod, and tail domain of four different IF proteins; the numbers of *acidic (negatively charged)* and *basic (positively charged)* amino acids are listed in red and blue, respectively; the numbers of the *histidines* are in green

protein	head (size / charge)	rod (size / charge)	tail (size / charge)
Vimentin	96 / 4 12 0	308 / 67 47 3	62 / 12 6 3
Desmin	101 / 3 10 0	308 / 65 48 4	61 / 11 7 2
Keratin 8	84 / 2 6 0	311 / 60 47 1	88 / 4 8 1
Keratin 18	73 / 0 6 0	311 / 58 44 5	46 / 9 6 0

agents. Notably, IFs can be disassembled into tetramers by incubation in low ionic strength solutions, and these tetramers are productive precursors for IFs in another salt-induced assembly cycle [59]. This observation cannot be explained with the above study that puts the entire head domains, without structural relation to the tetramers, into the center of the filament, which is envisioned as a hollow tube formed by five protofibrils [18]. Instead, one may suppose that parts of the basic head domain, which contains in total 12 charged arginines and no acidic amino acid (see Table 7.2), permanently and strongly connect neighboring dimers, which are highly acidic, to distinct complexes, the topology of which has been demonstrated by chemical crosslinking of vimentin tetramers [84]. In any case, because of its observed stability against dissociation in low ionic strength buffers, the intra-tetrameric interaction of two dimers is apparently stronger than the lateral association of two tetramers in octameric complexes, which are the modules of the so-called protofibrils in the mature filament. It is also apparently stronger than the protofibril-protofibril interaction within filaments, which can be already disrupted by incubation of IF with 10 mM sodium phosphate buffer [1]. This hierarchy of binding strengths between subunits—dimers to tetramers, tetramers to octamers, and octamers to filaments—reflects the hierarchy of interactions that occur during filament assembly and is certainly important for the mechanical properties of the filament during stretching.

Lateral and Longitudinal Assembly

Vimentin tetramers are stable under low ionic strength conditions and remain perfectly assembly-competent for days when stored refrigerated. Assembly is simply started by an increase of the ionic strength of the solution, e.g., by adding sodium or potassium chloride to a physiological concentration. As directly shown by time-lapse EM, the tetramers laterally associate within seconds into open complexes of 60 nm lengths, termed unit-length filaments (ULFs) [30]. This is a common feature of all IF proteins, although the optimal ionic conditions may differ for different candidates. For a mechanism that triggers assembly, we propose that some of the basic residues of the head domains are released from the acidic rod by the added monovalent ions, thereby releasing the head domain from an assembly blockade. Interestingly, the assembly of IFs can also be triggered by the addition of certain short head peptides, without any change in the ionic strength or pH value [33]. Thereby, parts of the head domains are released from being blocked within the tetramers, and tetramers are set free to start associating laterally.

For kinetic measurements, the ionic strength is raised in a "kick-start" mode by addition of a salt solution. Mass measurements by scanning transmission electron microscopy (STEM) revealed different numbers of molecules per cross-section in individual filaments, i.e., 8, 10, and 12 tetramers [30, 31]. With respect to the hypothesis that IFs are made from 8-meric protofibrils, these filaments are therefore supposed to be composed of 4, 5, and 6 protofibrils. Notably, measurements of single IFs revealed that segments with different mass can occur along one single filament. In stark contrast, this polymorphism does not occur when the salt concentration is slowly raised for tetramers by dialysis against the corresponding salt solution. Obviously, a slow increase in the ionic strength over time allows for more control over the lateral association of tetramers. This results in filaments that exhibit eight tetramers per cross-section [30, 82]. In mechanical stretching experiments (see section "Stretching Intermediate Filaments") only the dialyzed filaments, but not the "kick-started" filaments, are stable against breaking under strain. This experimental observation suggests that the nonuniformity of the latter provides weak points in the filaments. Continuous flow microfluidic chambers were used as an approach to perform in situ measurements of the assembly process: while tetrameric protein solution and assembly buffer are mixed by diffusion in a co-flow geometry, small-angle X-ray scattering (SAXS) data have been recorded, thus capturing the different stages of the assembly [8]. Notably, the paper by Steven and Steinert [82] demonstrated that vimentin IFs isolated from cultured cells and directly subjected to mass determination by STEM did not differ from those that had been reassembled after a round of purification of monomers in urea and corresponding reassembly after removal of urea. This indicated that IFs can be reliably assembled in vitro, see Fig. 7.4a for an example electron micrograph.

A key segment of the IF protein for tetramer formation and ULF assembly is the positively charged, approximately 100 amino acid long amino-terminal domain ("head"). This has been concluded from experiments with "headless" vimentin dimers. Headless vimentin forms dimers, but these dimers do not associate to tetramers during reconstitution into low ionic strength conditions. Under filament assembly conditions, however, headless dimers do laterally anneal to tetrameric rods of approximately 65 nm length as revealed by glycerol spraying/low angle rotary metal shadowing EM. Since these rods are slightly longer than tetramers of intact vimentin under low salt conditions, we propose that the coil 2 dimer contains salt-inducible binding sites to interact specifically with a second coil 2 dimer. No further assembly or aggregation occurs, neither laterally nor longitudinally [58].

The tail domain seems less important for assembly. It has always been assumed that most of it stands out of the filament, as judged from rotary shadowing EM of neurofilaments, and it was suggested that they are probably engaged in the interaction with other cellular factors [27, 28]. Interestingly, the consequence of its removal is that tailless vimentin assembles into six-stranded filaments compared to the four-stranded wildtype IFs, and thus the absence of the tail alters the overall design of the IF [30]. Interestingly, a mutant vimentin with only approximately half of the tail removed assembles into filaments containing both four and five protofibrils along its length. These results clearly show that the tail domain is not needed to connect individual protofibrils with one another. However, the tail domains apparently play a role in regulating the interactions and the number of protofibrils in a filament.

The key to the IF formation by longitudinal annealing of two ULFs is the head-to-tail overlap reaction of two single dimers with their amino- and carboxy-terminal rod segments, respectively. This longitudinal interaction was clearly demonstrated by chemical crosslinking analysis of vimentin and keratins and has been termed A_{CN} overlap (for review, see [65]). Like the other interactions, except those of the dimers within the tetramer, the head-to-tail overlap is released, when the filaments are exposed to low ionic strength buffers. The head-to-tail interaction can be directly affected and vimentin assembly prevented by point mutations in the consensus motifs of coil 1A and coil 2 (see Fig. 7.1a), which either arrest the ULF at that stage completely (mutants Y117L and Y400L) or do so in a temperature-specific fashion (mutant K139C) [10, 54, 70]. The latter vimentin mutant stays at the ULF stage at room temperature but assembles perfectly fine at 37 °C. Thus, the physical interactions of the rod end domains that mediate filament elongation are absolutely essential for the structural integrity of IFs, certainly with consequences for filament mechanics. The modular mode of assembly, dimer strand by dimer strand, also explains why a 16-mer ULF can connect to a 48-mer ULF. It furthermore suggests that individual protofibrils do not necessarily run continuously through a filament, at least not in filaments assembled in vitro. In addition, the dimeric ends of a protofilament could in principle contact the dimeric ends of two incoming protofibrils. In any way, because of this polymorphism, it is not recommended to use kick-started IFs for mechanical measurements.

The complete in vitro assembly pathway, as shown in Fig. 7.1c, can be described by a step growth process [50, 69]. Furthermore, IFs display additional assembly and disassembly features. First, not only single ULFs can associate with existing filaments, but two filaments of any length may end-to-end anneal, see in Fig. 7.1d [14, 87]. Moreover, even two types of IFs, such as vimentin and desmin IFs, can longitudinally anneal, both seen in vitro and in cells that contain both IF proteins such as BHK-21 cells [29, 85]. This ability to form "mixed" IFs reflects the fact that the sequences for the head-to-tail overlap which are instrumental in the longitudinal annealing process, are conserved in both proteins. These are indeed the so-called IF consensus domains, which are evolutionarily highly conserved from human vimentin to hair keratins [32]. Nonetheless, the longer the filaments are, the less likely they are to anneal, due to slower translational and rotational diffusion and a decrease in the number of available filament ends [41, 69, 79].

Second, because of the massively parallel arrangement of monomers within a filament, subunits may be "exchanged," meaning that they may leave the filament and be replaced by soluble subunits (see Fig. 7.1e). This process has been shown both in cells and in vitro [14, 62]. It is currently a matter of debate which size (tetramers or larger) these subunits have. Subunit exchange in vimentin is slow, i.e., about 10% of the protein within 8 h, reflecting the strong lateral interactions of the fibrous sub-complexes. Thus, IFs are very stable structures complementing the dynamically assembling and disassembling actin filaments and microtubules. Furthermore, since the kick-start mode of assembly incorporates segments with a different number of molecules per cross-section within a single filament, in the extreme 16 and 48 [30], the fragmentation of filaments over time is a potential concern for mechanical experiments. Indeed, such fragmentation has been documented in vitro using fluorescently labeled vimentin [83].

Interactions Within Filaments

Electrostatic and Hydrophobic Interactions

As described in section "Lateral and Longitudinal Assembly", IF proteins self-assemble in a hierarchical manner to form filaments of approximately 10 nm diameter. In general, IFs are highly negatively charged, as is typical for intracellular filaments [39]. For example, the linear charge density of vimentin, desmin, and keratin 8/18 is -19 *e*/nm, -6 *e*/nm, and -3.2 *e*/nm, respectively [26]. Table 7.2 shows the charge distribution for the head, the rod, and the tail domains for these four IF proteins in the monomer. Interestingly, the intrinsically disordered parts of the head domains are overall positively charged. We propose that these positive charges mediate the interaction of two dimers via acidic residues on their rod domains, thereby conferring exceptional stability to the tetramer complex. Notably, many of these charged residues are located on the surface of the coiled coil, particularly in the *b*, *c*, and *f* positions of the heptad repeat pattern, and are therefore free for ionic interactions with other molecules.

In general, the differences in total charge and its distribution along the molecule are not very large between the four example proteins. However, the head domains of vimentin and desmin stand out because there is no acidic residue within the intrinsically disordered part. The acidic charges listed in Table 7.2 for the head (4 for vimentin, 3 for desmin) are actually located in a 20-amino acid pre-coil region that most likely adopts an α-helical structure [28]. The heads of keratins 8 and 18 are less basic. In the rod domain, acidic residues dominate, with 20 more acidic than basic residues in vimentin, 17 in desmin, and 13 and 14 in keratin 8 and 18, respectively. Consequently, for in vitro studies the ionic environment of the buffer plays an important role for intermolecular interactions; e.g., monovalent cations may screen charges on the filament and di- or multivalent ions may lead to counterion condensation [16, 17]. The pH of the buffer changes the protein's overall charge via the histidine side groups [75]. Histidine is the only amino acid that changes its charge in the physiological range, i.e., it is positively charged at pH 5 but neutral and hydrophobic at pH 7.4. In the tetrameric complex, however, some of these charges are involved in the interaction of the two coiled-coil dimers via salt bridges in the *e* and *g* positions, and the charge presented to the surrounding solvent is specific for the individual protein complexes. These assumptions reflect our knowledge of the specific assembly behaviors of vimentin, desmin, and keratins, particularly with respect to their differential response to the ionic conditions used for in vitro assembly. Within a cell, the situation is considerably more complex; however, local differences in ion concentrations and pH may also occur and influence the charge interactions within IFs. In any case, we believe that the tetrameric complex is also an elementary intermediate for the assembly process in cells.

Posttranslational Modifications

IF proteins are subject to posttranslational modifications, introduced by a multitude of specific enzymes, such as phosphorylation, glycosylation, sumoylation, acetylation, prenylation, ubiquitylation, and transamidation [81]. The single cysteine of vimentin can be glutathionylated or nitrosylated, as seen in rheumatoid tissues [40]. However, the most important posttranslational modification for the dynamics of the assembly-disassembly cycle of IFs is phosphorylation. Unlike actin filaments and microtubules, which assemble and disassemble in a highly dynamic, ATP- and GTP-dependent manner, respectively, IFs are very stable structures in cells. In addition, they typically form stable "cages" around the nucleus [13, 86] that must be disassembled in many cell types for proper cell division to occur. Phosphorylation adds two additional charges to certain amino acids, notably serine and threonine. In vimentin, for example, these extra charges are added to one of the 31 hydroxyl residues of the head domain. These are all located near one of the regularly spaced arginines in the head that are an essential element of recognition sites for various kinases. Of course, these sites must be accessible to the enzymes for modification to occur. As a result of phosphorylation, the filament structure is destabilized and eventually massive disassembly reactions occur [37].

Experimental Investigations of Intermediate Filament Mechanics

The mechanical properties of matter determine the way it responds to external forces. Simple materials are either elastic (solid-like) or viscous (fluid-like). Figure 7.2a,b shows that for a purely elastic material the strain (ε, red line) precisely follows the applied stress (σ, gray line) and for a purely viscous material, the strain linearly increases in response to a constant strain, so that the strain rate is constant. Upon

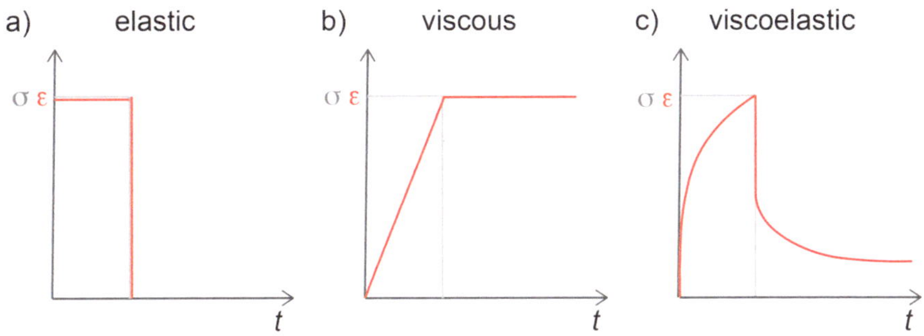

Fig. 7.2 Strain (ε) response (red) of (**a**) an elastic (solid-like), (**b**) a viscous (liquid-like), and (**c**) a viscoelastic material to a short, constant stress pulse σ (gray), plotted versus time

release of the stress, the stress remains constant and the material does not relax back to its original shape. By contrast, complex, biological materials are referred to as viscoelastic, because they combine both properties, typically on different time scales, see Fig. 7.2c. Such viscoelasticity is a hallmark of biological cells and tissues and of utmost importance for their function. An illustrative example is provided by immune cells, which are capable of extreme deformation in order to squeeze through dense layers of connective tissue and endothelial cells, while maintaining their shape and protecting themselves from external influences and damage.

Simple solids display a linear relationship between the force F and the extension Δx with the proportionality constant k, as we know it from a Hookean spring

$$F = k\Delta x. \tag{7.1}$$

More complex materials are characterized by nonlinearities in the $F(x)$ curve; additionally, the yield point, i.e., the point of the curve, where the material breaks, is of interest. The very special mechanical properties of IFs were demonstrated early, when Janmey and colleagues showed that vimentin networks can be sheared to high strains and bear high stresses, whereas actin or microtubule networks yield at fairly low strains or stress, respectively [38], see Fig. 7.3.

While these experiments were performed on fully entangled networks of IFs, thus probing a convolution of the filament properties and the network properties, i.e., the interactions between the filaments, it is more straightforward to probe specifically the response of the filaments to deformation modes on a single filament level. In general, we may distinguish three such modes—bending, stretching, and twisting—and different experimental methods exist to probe each of them.

Bending Intermediate Filaments

From a physical point of view and on a mesoscopic length scale, IFs are long and thin objects with a high aspect ratio. This is certainly the case for in vitro assembled vimentin filaments, which can be tens of micrometers long and about 10 nm in

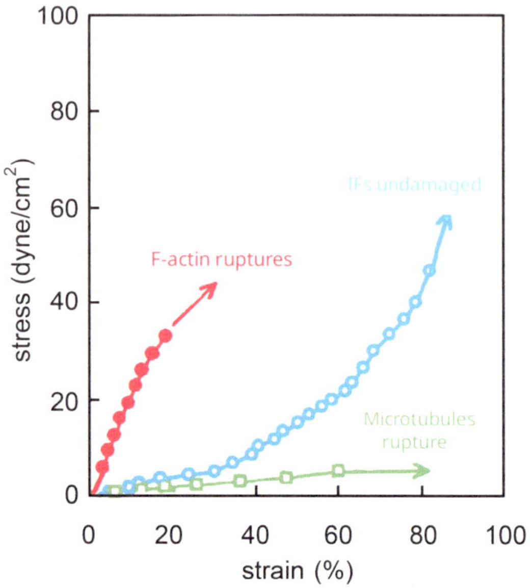

Fig. 7.3 Rheology data (stress against strain) for networks of the different cytoskeletal filaments. In comparison to actin and microtubule networks, vimentin networks resist high strains and stresses. (Data adapted from Janmey et al. [38])

diameter. IFs share this property with polymers, a class of macromolecules that assemble via one-by-one addition of their building blocks, the monomers. Note, however, that the assembly process of IFs (see section "Lateral and Longitudinal Assembly") is not a polymerization as it follows a hierarchical pathway rather than being rooted in the addition of individual monomers (or subunits) of the same kind. In a coarse-grained manner, and disregarding molecular details, we can therefore describe the filaments using polymer models that have originally been developed for synthetic macromolecules. A suitable description for so-called semiflexible polymers, where the contour length and the persistence length are on the same order of magnitude, is the worm-like chain model (WLC, see section "Worm-Like Chain Model"). The persistence length L_P is a measure of the bending rigidity of the filament (see Eq. (7.4) in section "Worm-Like Chain Model"). Of all cytoskeletal filaments, IFs have the smallest persistence length, on the order of 1 μm, see Table 7.3 for example values for different types of IFs. The numerical values differ, depending on the measurement method. However, all values are on the same order and about tenfold smaller than for actin filaments.

Probably the most direct way of determining the persistence length of filaments, and thus their propensity to bend, is by letting them thermally fluctuate to observe how much they curve. This phenomenon is routed in Brownian motion, which all matter experiences and which becomes noticeable and influential on small length scales such as those of a cell. The reason is that the thermal energy at room temperature ($k_BT \sim 4 \times 10^{-21}$ J) is on the same order of magnitude as the cellular energy unit, i.e., the hydrolysis of one ATP molecule which is around 20 k_BT [55]. This thermal energy is fully sufficient to excite fluctuations in the biopolymer.

Table 7.3 Values measured for the persistence length of different types of IFs

IF	L_P (μm)	Method	References
Keratin 8/18	0.65 0.3–0.48	Rheology EM	[47, 66]
Desmin	0.1–1 0.9	Quasi-elastic light scattering Rheology	[36, 78]
Vimentin	1 0.4–0.5 2–2.1	AFM, EM Rheology Fluorescence microscopy	[48, 57, 61, 78]
Neurofilaments	0.05–0.2 0.2	AFM Rheology	[2, 48]

By measuring tangents along their contour and using Eq. (7.4) to fit the data, or alternatively by measuring the distance between the two end-points of the polymer and using Eq. (7.5), the persistence length is determined. The procedure is explained in more detail in section "Worm-Like Chain Model". Formally, the persistence length is defined as the length over which correlations in the direction of the tangent are lost.

Imaging can be performed using atomic force microscopy (AFM, see section "Stretching Intermediate Filaments" and Fig. 7.5a) with the advantage of high spatial resolution, and no labeling of the filaments is necessary [57]. A drawback of this method is that great care needs to be taken so as to deposit the filaments on the substrate in equilibrium so that exactly this equilibrium conformation of the chain is being imaged. For example, filaments on mica, which strongly interact with the filaments, are observed to have an apparent persistence length of $L_P = 0.3$ μm, while on glass, $L_P = 1$ μm [57]. Figure 7.4b depicts examples of such AFM images of vimentin filaments. In addition, when observed by cryo-electron tomography, in which objects are trapped in solution, vimentin IFs appeared to be more straight and therefore stiffer than when attached to a support [24]. Assuming the ergodic hypothesis, it is not necessary to image the filaments dynamically, i.e., capture multiple temporal conformations per filament, as long as a statistically large enough number of filaments is being included in the analysis.

With the development of methods to directly fluorescently label IF proteins on cysteine residues [61, 87], dynamic imaging of the filaments became possible, in which case surface interactions are entirely avoided, see Fig. 7.4c. Due to the small persistence length of the filaments, they fluctuate fast and imaging is typically performed in two dimensions, omitting the z-direction. However, since fluctuations in the microscopy plane (x-y) and perpendicular to it are decoupled, a projection in the microscopy plane yields the desired results with an equilibrium persistence length of 2 μm [61].

At first glance it may be surprising that an IF with a similar diameter as an actin filament has a persistence length of one order smaller. Both filament types can, in first approximation, be described as "long, thin cylinders"; however, when regarding the details of the molecular structure (Section "Structure of Intermediate Filaments"), important differences persist. Actin filaments are built up from rather

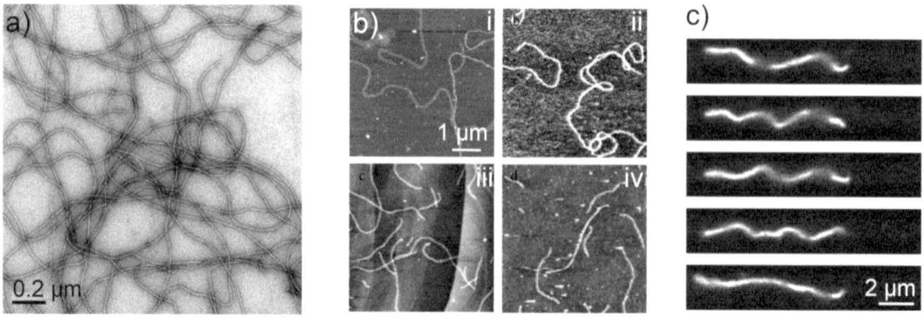

Fig. 7.4 (**a**) Electron micrograph of in vitro assembled vimentin filaments. (**b**) AFM data (height images) of individual vimentin filaments on different substrates: (i) on mica in air; (ii) on mica in liquid; (iii) on highly oriented pyrolytic graphite in liquid; (iv) on hydrophilic glass in liquid; height scale from 0 (black) to 20 nm (white). (Reprinted from: J. Mol. Biol. 335, Mücke et al., Assessing the Flexibility of Intermediate Filaments by Atomic Force Microscopy. Copyright (2004) with permission from Elsevier [57]). (**c**) Temporal (conformational) snapshots a vimentin filament suspended in buffer and confined in a microchannel of different widths $d = 1.5$. μm, fluorescence microscopy. (Reprinted figure with permission from Nöding and Köster, Phys. Rev. Lett. 108, 088101, 2012. Copyright (2012) by the American Physical Society)

compact, globular monomers that line up in a double-helical structure. Thus, the filament core is actually "filled" with peptide chains. IFs exhibit a much more open structure. The subunits (Section "Organization of the Coiled Coils Within the Filament") are held together by electrostatic and hydrophobic interactions (Section "Interactions Within Filaments"), which leave space for water molecules in between. In fact, SAXS data [9] as well as recent cryo-EM data [18] show that the core of vimentin filaments is less dense than the shell.

Stretching Intermediate Filaments

While the experimental setup to study the bending of filaments is conceivably straightforward and the deformation occurs spontaneously, the response to stretching deformation requires sophisticated experimental tools beyond mere microscopy that have become available recently. Mainly three methods are capable of manipulating filaments in a controlled manner on the relevant length (nanometers to micrometers) and forces (piconewton to nanonewton) scales: AFM, optical tweezers, and magnetic tweezers. By means of mechanical probes, optical forces, or magnetic forces, respectively, these state-of-the-art methods allow for precise determination of force-distance data while pulling on filaments.

The AFM was introduced in 1986 by Binning, Quate, and Gerber [4]. The working principle is elegant and seemingly simple (see Fig. 7.5a), although the technical implementation is extremely challenging due to the small length scales and forces. A cantilever with a sharp tip is used to approach a surface. In imaging mode (see section "Bending Intermediate Filaments") the tip scans the sample, typically in

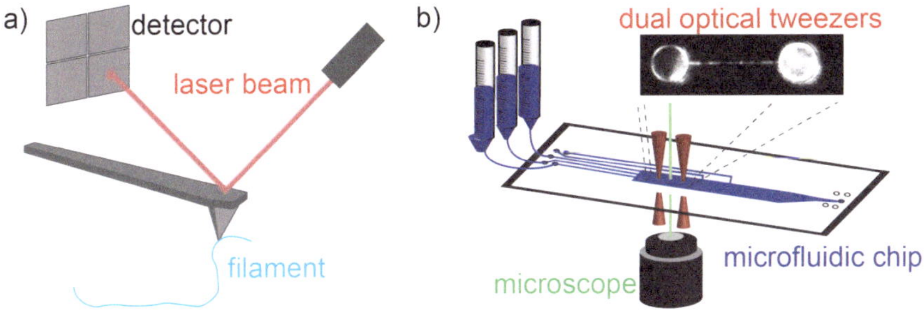

Fig. 7.5 Working principle of (**a**) AFM and (**b**) optical tweezers in combination with microscopy and microfluidics. (b adapted figure with permission from Block et al., Phys. Rev. Lett. 118, 048101, 2017. Copyright (2017) by the authors, licensed under a Creative Commons Attribution (CC BY) License [5])

an *x*-*y*-grid, and the instrument generates two-dimensional topographic images of the surface. The cantilever acts as a Hookean spring and, therefore, may be used in force-spectroscopy mode as well. Here, the deflection of the cantilever is measured by a laser beam that is reflected from the back of the cantilever and detected by a four-quadrant diode. With the known stiffness of the cantilever, which can be determined by measuring its equilibrium fluctuations and using the equipartition theorem [22], the force needed to deflect the cantilever is calculated, via Hooke's law (see Eq. (7.1)).

The filament is placed on a substrate to which is either adheres along its length, by physisorption. An elegant assay combining imaging and lateral stretching of filaments was developed by Kreplak and Aebi and applied to desmin filaments, neurofilaments, and keratin 5/14 filaments. The IFs were adsorbed on either mica or graphite substrates, first imaged by AFM in unstretched conditions, then laterally pulled by the AFM tip by increasing the force to 30–40 nN until they ruptured and imaged again (see sketch in Fig. 7.6a). From comparing the length before pulling and just after rupturing, a maximum strain up to 3.6 was determined at a pulling force of up to 3.5 nN [44, 45]. The strain ε is defined as

$$\varepsilon = \frac{L - L_0}{L_0} = \frac{\Delta L}{L_0}, \tag{7.2}$$

with the original length (before pulling) L_0 and the change in length upon pulling ΔL. Thus, a strain of 3.6. corresponds to an extension to 4.6-fold the original length [44], which by far extends the extensibility of the other cytoskeletal filament types, actin filaments and microtubules. The great advantage of this setup is the possibility to image the filaments at high resolution before and after stretching them. The example in Fig. 7.6b shows that a filament segment of original length 280 nm was stretched to 950 nm before it ruptured, resulting in a maximum strain of 2.4. Using this method, it was observed that the enormous elongation is accompanied by filament thinning [45].

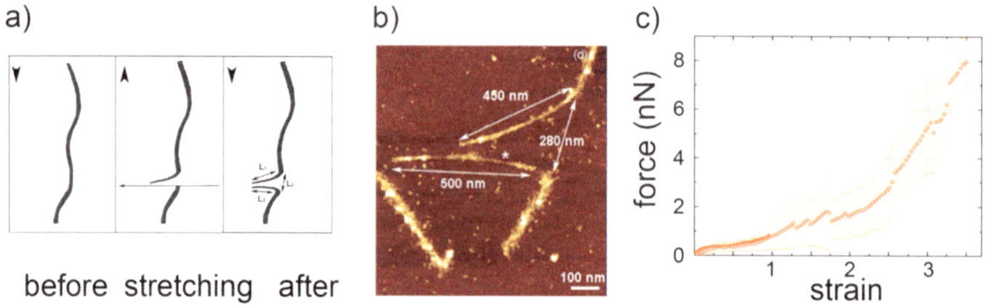

Fig. 7.6 AFM measurements to determine the stretchability of IFs. (**a**) Sketch of an AFM experiment, where the filament is laterally stretched until rupture and imaged before and after the stretching. (**b**) Example result of such an experiment (AFM height image). (a,b reprinted from: J. Mol. Biol. 354, Kreplak et al., Exploring the mechanical behavior of single intermediate filaments, 569–577. Copyright (2005) with permission from Elsevier [44]). (**c**) AFM experiment, where the filament is pulled upwards in z-direction up to a pulling force of 8 nN. (Adapted figure with permission from Block et al., Phys. Rev. Lett. 118, 048101, 2017. Copyright (2017) by the authors, licensed under a Creative Commons Attribution (CC BY) License [5])

However, in this assay the filament is attached to a substrate along its contour, and this interaction may influence the measurements, similar to the determination of filament fluctuations and the persistence length from adsorbed filaments, as discussed in section "Bending Intermediate Filaments". This difficulty can be overcome by pulling the filament that is adsorbed to the substrate with one end vertically and thereby stretching it, see Fig. 7.5a. Some modern AFMs have a large travel range and enable such experiments. Figure 7.6c shows a vimentin filament that is extended to a strain of 3.5 without rupturing [5]. In this assay, the force needed to extend the filament this far was recorded and amounted to 8 nN. Hence, these filaments can bear enormous forces compared to the typical forces generated by molecular motors, which are in the few-pN-range.

While it is highly interesting to study the extension of filaments at such high forces, the degree of control within this AFM assay is somewhat limited. The filament is pulled upwards, i.e., perpendicularly to the image plane of a microscopy that most bio-AFMs have integrated from the bottom of the setup. Therefore, concomitant imaging is not possible and the question of whether a single filament, or a filament bundle is being pulled on, can only be answered by comparing the results to data derived by more controlled experiments, such as optical tweezer measurements. Here, optical tweezer setups (Fig. 7.5b), where the filament is freely suspended in liquid between two beads, provide all the control needed: the beads have a radius of a few micrometers and are functionalized for a single filament to attach. This process is controlled via fluorescence microscopy. One bead is then moved by one optical tweezer to stretch the filament, and for every time point the length of the filament is determined via the distance of the beads. Concomitantly, the force is measured using the second optical tweezer: when the filament is stretched, the bead is displaced slightly from the center of the focused laser beam, and this displacement can be transformed into a force by applying Hooke's law (see

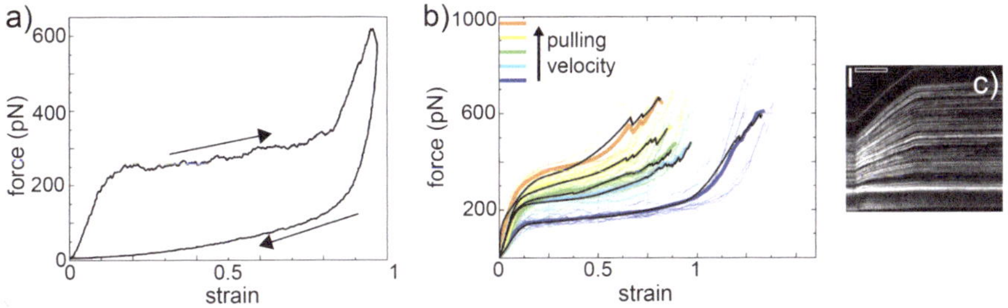

Fig. 7.7 (**a**) A typical stretch-relax curve of a vimentin IF. Three regimes are visible, additionally a strong hysteresis. After one full cycle, the filament returns to its original length. (**b**) The force-strain behavior of vimentin filaments depends on the pulling speed. For slow pulling (blue) the filament can be extended much more than for fast pulling (orange, about 100x faster). (Reprinted figure with permission from Block et al., Phys. Rev. Lett. 118, 048101, 2017. Copyright (2017) by the authors, licensed under a Creative Commons Attribution (CC BY) License [5]). (**c**) Kymograph of a vimentin filament that is being stretched. The vertical axis shows the extension of the filament (scale bar 5 μm), and the horizontal axis shows the time (scale bar 15 s). The white lines (fluorescence microscopy) show the unevenly labeled filament. The filament is oriented vertically in the image, with the moving bead at the top and the force-measuring tweezer at the bottom

Eq. (7.1)) and using the previously determined stiffness k of the optical tweezer. The experimental procedure is described in detail in Schepers et al. [77].

This setup offers a high degree of control, both in terms of force-strain data and in terms of imaging. The typical force-strain curve of a vimentin filament shows several very interesting features (see Fig. 7.7a): first, there are three regimes, a low strain (up to about 10–20%) regime, where the force increases linearly with strain (the elastic, Hookean behavior), followed by a flat regime, where little force is required to extend the filament further, and then a stiffening regime with a much steeper data curve. Second, the stretching curve and the relaxation curve differ considerably, leading to a high degree of hysteresis, and therefore energy dissipation, which is proportional to the area between the two curves. In vimentin filaments, 70–80% [6] of the input energy is dissipated, in keratin filaments around 50% [53]. Third, despite this large amount of dissipated energy, vimentin filaments return to their original length when the force is back to zero. Thus, we do not observe plastic elongation at the accessible resolution. The force-strain relationship is highly loading rate dependent: at low pulling velocities (0.05 μm/s), the flat region extends to high strains. By contrast for 100-fold faster pulling, there is almost no flat region. Merely a “shoulder” is observed (Fig. 7.7b). Figure 7.7c shows a typical kymograph of a filament that is being stretched. How such intriguing behavior may be theoretically modeled and explained is presented in section “Theoretical Models”. The exact course of the force-strain data and in particular the slope in the beginning, which informs us about the stiffness of the filaments, depends sensitively on the electrostatic interactions within the filament. This is obvious when varying the ion type and concentration in the buffer and when altering the pH [75]: monovalent ions screen electrostatic repulsion within the filaments, thus increasing the charge interactions and stiffening the filaments; divalent ions lead to counterion condensation and likewise increased interactions. A lower pH renders histidines positively charged, also stiffening the filaments. These experimental findings indicate that electrostatic interactions are involved in determining the mechanical properties of single IFs. This hypothesis is supported by observation of partly phosphorylated vimentin filaments: up to 10% of phosphorylated (or phosphomimetic) protein may be incorporated during assembly with intact filaments forming. Upon stretching, however, the filaments are softer the more phosphorylated protein is in the filament [43]. This finding is in line with the observation that in the cell, pronounced phosphorylation of IFs leads to disassembly of the filaments and is a prerequisite for cell division of many different cell types [19, 37].

The optical tweezer setup provides the possibility of cycling the filament length between a "relaxed" and an "extended" state, because the two plastic beads to which the filament is attached are captured with the laser beams very much like with mechanical tweezers. Above, we explained that upon stretching and relaxation, vimentin filaments return to their original length. Interestingly, when pulling again, the filament is softer, showing strain memory [6]—and this softening behavior is observed even if between the pulling cycles an extended recovery time is included [21]. Different IF proteins differ in their amino acid sequence and, therefore, in their charge and hydrophobicity pattern, see section "Structure of Intermediate Filaments". A particularly interesting "pair" of IFs, usually not expressed in one cell type, is vimentin and keratin, as they are involved in the epithelial-to-mesenchymal transition (EMT). Epithelial cells express keratins and are strongly connected to their neighbors via cell-cell junctions, which anchor keratin IFs. Complex signaling activities during EMT lead to a downregulation of keratin and an upregulation of vimentin. The cells become migratory and may travel to distant locations in the organism. EMT is an important process during embryogenesis and wound healing. Of particular concern is EMT during cancer cell metastasis. On the single filament level, vimentin and keratin behave strikingly different. As discussed above, vimentin filaments become softer with each pulling cycle, but retain their original length, whereas keratin filaments become longer with each cycle, but keep their stiffness [53]. This finding is all the more fascinating as the molecular structure of the filaments is very similar, and the main difference lies in the amino acid sequences of the two types of proteins, as exemplified already by their different charge patterns with respect to the head and tail domains (see Table 7.2).

In the cell, IFs form bundles and networks and interact with the other cytoskeletal filament types, i.e., actin filaments and microtubules. Optical tweezers provide the possibility to probe the interaction between two filaments directly and in a quantitative manner. For such measurements, two filaments are bound to two plastic beads each and four optical tweezers steer these beads. In a crossed-filament geometry, one filament is moved with respect to the other one (see Fig. 7.8) [52], and the tweezers serve to measure the forces. Interestingly, vimentin IFs interact with one another [76] as well as with microtubules [73], and the forces reach up to tens of pN, thus lie in a range relevant for the cell. These interactions can be tuned by manipulating the electrostatic and hydrophobic conditions. Thus, interactions within the filaments, that lead to filament assembly (see section "Interactions Within Filaments") as well

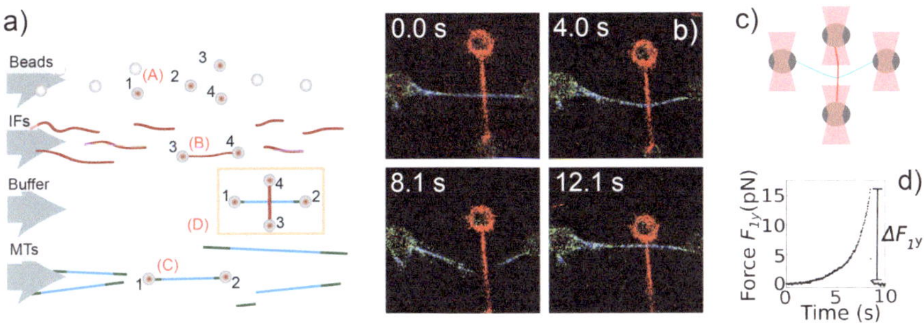

Fig. 7.8 (**a**) Experimental protocol for a filament-crossing experiment, using quadruple optical tweezers. The laminar inflow into the microfluidic channel is denoted by gray arrows. The optical tweezers are used to capture beads 1 to 4 and moved through the device according to the letters A to D. (**b**) Temporal snapshots of microscopy recordings during the interaction experiment. (**c**) Sketch of the bead movement; the top and bottom bead (attached to the vimentin IF, red) are moved down and up. (**d**) Force-time recording with the maximum force (breaking force) ΔF_{1y}. (Adapted figure with permission from Schaedel et al., Nat. Commun. 12, 3799. 2021. Copyright (2021) by the authors, licensed under a Creative Commons Attribution (CC BY) License [73])

as between filaments, are governed by electrostatic and hydrophobic forces. Whether or not such direct interaction leads to synergistic effects in mixed cytoskeletal networks remains a matter of debate [20, 23, 67, 80].

Experiments to study structural phenomena of IFs while stretching them, such as the transition of the α helices in the protein rod region to β sheets, are rare. The reason is that combining a controlled optical tweezers, magnetic tweezers, or AFM setup with any kind of spectroscopy method that would probe the molecular structure of the protein is a real challenge. Typically, spectroscopy methods require a considerable amount of material and cannot be performed on a single filament, because the signal would be too weak. Fibers made from hydrogels of vimentin can be strained and show the typical signature of an a α-β transition, where α-helical protein structures transition into β sheets, as shown by wide-angle X-ray scattering (WAXS) [68]. The α-β transition was also reproduced in simulations of vimentin dimers [71, 72]. Interestingly, the simulated data (force versus strain) of the pulled dimers share the same signature with three regimes (linear, plateau, stiffening) as in experiments of fully assembled filaments, even including the force ranges. This latter point is surprising given that a regular vimentin filament consists of 16 dimers, and in a parallel configuration, the pulling forces should be additive. It is unlikely that the attachment of the filament to the bead is only via a dimer. Microscopy images very clearly show that the filaments are wrapped around the beads.

Twisting Intermediate Filaments

Whereas filament bending and stretching can be investigated in a quantitative manner using modern experimental methods, twisting filaments is all the more challenging. There is indication that vimentin IFs possess a helical turn, similar to actin filaments and microtubules, from filament shape fluctuation analysis [7] and from electron tomographic structural analysis [18]. How much a filament would resist being twisted and what its torsional stiffness might be, however, remains to be investigated. Modern methods using high-end magnetic tweezer setups may potentially provide a suitable approach here [56].

Theoretical Models

Worm-Like Chain Model

The persistence length L_P of IFs is on the order of few micrometers, see Table 7.3 [57, 61] and therefore on the same order of magnitude as the contour length L. This categorizes them as "semiflexible filaments" and distinguishes them from stiff polymers that are mostly straight in force-free conditions (e.g., microtubules) on the one hand, and from flexible polymers that are best described as random coils (e.g., synthetic polymers, DNA) on the other hand.

The WLC model was first introduced by Kratky and Porod as a theoretical description of scattering data from polymer solutions [42] and can be applied to biopolymers, which are imaged by scanning probe or light microscopy, in a straightforward manner. In contrast to random coil models that describe flexible polymers (e.g., freely jointed chain, freely rotating chain), a WLC-polymer is not described by individual "chain segments" with coordinates, but by a continuous function that describes the spatiotemporal shape of the filaments, either in terms of arc length s or in an y-x coordinate system, i.e., using Monge representation. The key point of the coarse-grained model is that only the bending energy, described by the bending rigidity κ, enters the Hamiltonian (energy function) to describe the polymer:

$$H = \frac{\kappa}{2} \int_0^L \left(\frac{d\vec{t}}{ds} \right)^2 ds, \tag{7.3}$$

with the tangent vector $\vec{t}$. To quantify the propensity to thermally bend, the persistence length L_P is determined by correlating tangents on the contour (see Fig. 7.9a):

$$\left\langle \vec{t_1} \cdot \vec{t_2} \right\rangle = \left\langle \vec{t}\,(s) \cdot \vec{t}\,(s+l) \right\rangle = \langle \cos\theta \rangle = \exp\left(-\frac{l}{L_P} \right). \tag{7.4}$$

Equation (7.4) shows that the persistence length is the characteristic decay length of the correlations, i.e., when moving along the filaments, the correlation is decayed

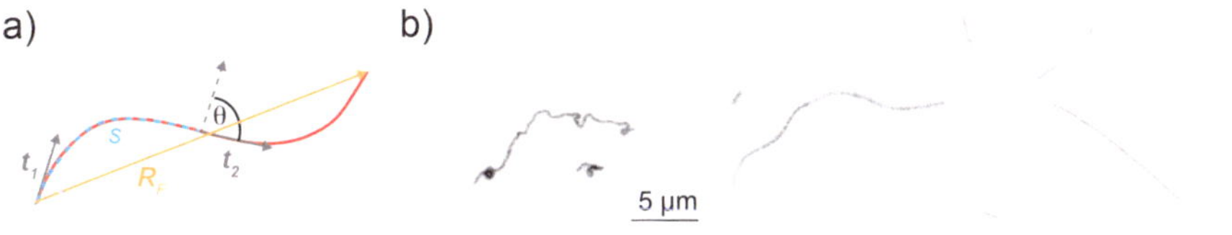

Fig. 7.9 The WLC model. (**a**) Schematic explaining Eqs. (7.4) and (7.5). The contour of the filament is shown in red, the arc length s in blue and dashed; furthermore, two tangent vectors t_1 and t_2 (gray), the angle θ between them (black) and the end-to-end distance R_E (orange) are shown. (**b**) Cytoskeletal filaments of different lengths are shown. Each of the filaments is about 20 μm long. From left to right: vimentin IF ($L_P \sim 1$ μm), actin filament ($L_P \sim 15$ μm), microtubule ($L_P \sim 1$ mm)

to $1/e$ after a distance $l = L_P$. An alternative way to determine the persistence length is by measuring the end-to-end distances R_E with

$$\overline{R_E^2} = 2L_P^2\left(\frac{L}{L_P} - 1 + \exp(-l/L_P)\right). \tag{7.5}$$

While the approach using Eq. (7.4) is statistically more accurate as it takes into account the complete contour length, the one using Eq. (7.5) does not require determining the contour accurately, but only the two ends of the filament, and is therefore in some cases easier to apply. Figure 7.9b illustrates that the different persistence lengths of different filament types (from left to right: vimentin IF, actin filament, microtubule), and thereby the bending rigidity, can be distinguished, in a qualitative manner, merely by microscopic imaging.

Two-State Models

It should be noted that the worm-like chain model per se does not include extensibility of the filament. Nevertheless, it is well suited to describe thermal equilibrium fluctuations of IFs. However, in contrast to actin filaments that break upon applied strains of a few piconewton, IFs are very extensible. In order to understand this phenomenon and the special mechanical properties of IFs, let us recall the molecular structure of the filaments with a massively parallel and serial array of α-helical monomers (see Fig. 7.1b). Experimentally [68] and in simulations [71, 72], an α-β transition has been shown, which is accompanied by an elongation, which can be estimated from structural data [5, 11, 60]. The simplest model to describe the nonlinear force-extension curves measured for vimentin and the loading-rate dependence associated with it is therefore a two-state model. It includes a polymer model (such as WLC or freely jointed chain) and an elastic element (accounting for the experimental realization of the optical tweezer setup as well as elastic contributions to the elongation) in series and takes two states into account, a shorter, α-helical one and an extended, longer one (see sketch in Fig. 7.10a). The α helices switch

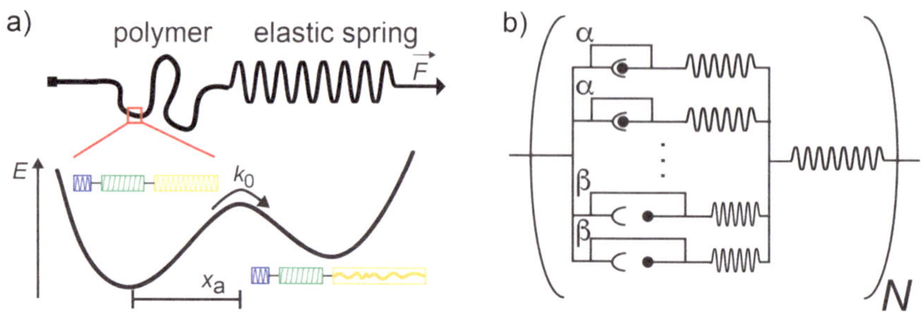

Fig. 7.10 (**a**) Two-state model to describe the force-strain behavior of individual IFs. The model combines a polymer model and an elastic element in series and includes a double-well potential with the α-helical state (left minimum) and an extended state (right minimum). (Adapted figure with permission from Block et al., Phys. Rev. Lett. 118, 048101, 2017. Copyright (2017) by the authors, licensed under a Creative Commons Attribution (CC BY) License [5]). (**b**) Sketch of a Monte Carlo model with a high number of vimentin monomers in parallel and in series. The monomers can be in closed (α) or open (β) state. In this case, all monomers within a ULF are coupled, leading to cooperative behavior. (Adapted figure with permission from Block et al., Sci. Adv. 4, eaat1161, 2018. Copyright (2018) by the authors, licensed under a Creative Commons Attribution (CC BY) License [6])

stochastically and in a force-dependent manner into the extended state following Bell-Evans kinetics [3]. Such a comparatively simple model describes force-strain data of single vimentin filaments very well, including the loading-rate dependence of the behavior, as shown in Fig. 7.7b, and mesoscopic filament elongation by several micrometers can be traced to monomer elongation, due to unfolding of the α helices by a few nanometers each.

Monte-Carlo Simulations

However, simple two-state models cannot describe the rich mechanical behavior of single IFs upon stretching and relaxing and when performing several such stretch-relax cycles [6]. Stochastic and numerical modeling, taking into account the peculiar structure of IFs when fully assembled, however, sheds light on these interesting phenomena. The massively parallel arrangement of monomers within a ULF can be reproduced in silico by representing each monomer by an element consisting of an elastic (Hookean) spring and a "switch" that transitions between a closed (α) and an open (β) state, where, however, the β state does not necessarily have to be a β sheet (see sketch in Fig. 7.10b). The opening is force dependent and due to the parallel arrangement, once the first element (monomer) is open, the pulling force has to be borne by the *N*-1 remaining elements; thus the force per element increases, leading to an avalanche-like opening of the elements. Once all elements within one ULF are open, the ULF, and therefore the filament, elongates. This process occurs in all ULFs that are placed in series for the model. The strength of this model is that it captures the softening of the filament with each pulling cycle as well: upon relaxation, the elements close in a force-independent manner. As soon as the first element within a ULF is closed, the ULF shortens. Therefore, with this model, the short, original filament length is reached at zero force, but it is likely that some elements are still in the longer β state, and thus a new stretching of the filament becomes easier. The lateral coupling of the elements (e.g., dimers or tetramers) within one ULF leads to cooperative mechanical behavior of the filaments shown by the characteristic, nonlinear force-strain curve.

This basic model to describe the mechanics of single IFs under strain can be adapted to specific experimental conditions. As discussed above, there is evidence that electrostatic and hydrophobic interactions within the filament, e.g., between the monomers, dimers, or tetramers, exist. In the model, the stiffness of the α-helical protein structures as well as the lateral coupling between elements can be tuned and the experimental findings are confirmed: stiffer α helices lead to stiffer filaments [75] and decreased lateral coupling, as in phosphorylated filaments, leads to softer filaments [43]. By releasing the coupling within each ULF and thus diminishing the cooperative behavior upon stretching, the plateau-like region in the force-strain curve disappears from the simulation, which is in agreement with stretched keratin 8/18 filaments [51], and the strikingly different behavior of vimentin and keratin may be thereby explained. These two filament types, which both play a major role in the EMT and are thus crucial for important biological processes, have "opposite" material properties: keratin elongates, whereas vimentin softens [53]. Both filament types dissipate a large amount of the input energy, and by using the model, this observation can be explained in the following way: in keratin IFs, subunits within a ULF can slide against each other. Note that the model does not specify what a "subunit" is, and it could be a tetramer or a larger entity. This sliding of subunits dissipates energy by viscous friction, and upon relaxation the filament ends up in a slight elongated state, as there is no "restoring force" pulling the subunits back entirely. By contrast, in vimentin, where subunit sliding is impossible due to the ULF coupling, the energy is dissipated via nonequilibrium folding and unfolding of the α helices in the monomers. Thus, these Monte Carlo simulations provide a versatile toolbox to reproduce the mechanical behavior of complex protein filaments in silico. However, ultimately one will need coarse-grained molecular dynamics (MD) simulations, possibly set up in a hybrid manner of coarse-grained coiled-coils and fully atomistic heads.

Conclusion and Future Perspectives

Cytoplasmic IFs share two important structural features that encode their intriguing mechanical behavior: (i) the mostly α-helical secondary structure of the monomers resulting in the formation of extended coiled coils and (ii) the massive parallel arrangement of the rod-like dimers. Although each of the human IF proteins displays its unique amino acid sequence (primary structure), the resulting dimeric structures represent coiled coils of virtually identical conformation. In addition, the assembly pathways are conserved. Differences concern the intrinsically unstructured head and tail regions and the number of monomers across the filament diameter. Together, these features lead to several properties that clearly distinguish IFs from actin filaments and microtubules:

- IFs are more *flexible* with a ten-times smaller persistence length, despite a similar diameter compared to actin filaments, because their core is less dense than the "shell" of the filament made by independent protofibrils. Hence, IFs are open structures.
- IFs are much more *extensible* and can be elongated to several times their original length until filament failure takes place, due to their special structure with a large number of α-helical, strongly linked monomers, dimers, and tetramers arranged in parallel and in series. The α helices can elongate or slide against each other when stress is applied.
- In cells, as well as in vitro, IFs have a *slow turnover* unless dynamics are initiated by posttranslational modifications. Turnover occurs not only at the ends of the filaments but also from any position, due to the many tetramers arranged in parallel.
- IFs are *nonpolar*, because the smallest soluble subunit—the tetramer—consists of two antiparallel, half-staggered coiled-coil dimers; therefore, IFs do not serve as tracks for molecular motors.
- IFs are expressed in a *cell-type-specific* manner and are evolutionarily conserved across species. It remains to be investigated if they each impose their own "mechanical footprint" on the cell type and tissue.

Therefore, IFs provide a prime example of structure-mechanics relationships. In a next step, this research will lead to a better understanding of the mechanical nature and the biological function of cell structure. In particular, research on the mechanics of IF is interesting and relevant for three reasons. First, from a biological and biomedical point of view, understanding IFs in health and disease is highly desirable. Single point mutations have been demonstrated to lead to severe problems in assembly and mechanics in tissues and organs. In addition, IF proteins seem to play an important role in EMT, which is involved in cancer metastasis and wound healing, the details of which are still not known. Moreover, IFs form a cage around the cell nucleus and thus provide a mechanical protection against damage of the genetic material. Second, from a soft matter physics point of view, these flexible and extensible filaments form an interesting and unique class of biopolymers. And third, it will be interesting to think about how the physical principles underlying IF mechanics and rendering them "high-performance biomaterials" may be used in man-made, synthetic, biomimetic materials. Nature has already explored this, using keratins in combination with associated proteins, to make materials as diverse as hair and baleen. It would be very interesting to transfer these principles to synthetic materials.

Acknowledgments We would like to thank all the past and present members of our research groups for all their creative ideas and fruitful discussions. This work was funded by the German Research Foundation (Deutsche Forschungsgemeinschaft, DFG), project number 429958739—HE 1853/15-1 and KO 3752/7-1.

Compliance with Ethical Standards There are no conflicts to declare.

References

1. Aebi U, Fowler WE, Rew P, Sun TT (1983) The fibrillar substructure of keratin filaments unraveled. J Cell Biol 97:1131–1143. https://doi.org/10.1083/jcb.97.4.1131
2. Beck R, Deek J, Choi MC et al (2010) Unconventional salt trend from soft to stiff in single neurofilament biopolymers. Langmuir 26:18595–18599. https://doi.org/10.1021/la103655x
3. Bell GI (1978) Models for the specific adhesion of cells to cells. Science 200:618–627. https://doi.org/10.1126/science.347575
4. Binnig G, Quate CF, Gerber C (1986) Atomic force microscope. Phys Rev Lett 56:930–933. https://doi.org/10.1103/PhysRevLett.56.930
5. Block J, Witt H, Candelli A et al (2017) Nonlinear loading-rate-dependent force RESPONSE of individual vimentin intermediate filaments to applied strain. Phys Rev Lett 118:048101–048101. https://doi.org/10.1103/PhysRevLett.118.048101
6. Block J, Witt H, Candelli A et al (2018) Viscoelastic properties of vimentin originate from nonequilibrium conformational changes. Science. Advances 4:eaat1161. https://doi.org/10.1126/sciadv.aat1161
7. Bouzar L, Michael Müller M, Messina R et al (2019) Helical superstructure of intermediate filaments. Phys Rev Lett 122:098101–098101. https://doi.org/10.1103/PhysRevLett.122.098101
8. Brennich ME, Nolting JF, Dammann C et al (2011) Dynamics of intermediate filament assembly followed in micro-flow by small angle X-ray scattering. Lab Chip 11:708–716. https://doi.org/10.1039/c0lc00319k
9. Brennich ME, Bauch S, Vainio U et al (2014) Impact of ion valency on the assembly of vimentin studied by quantitative small angle X-ray scattering. Soft Matter 10:2059–2068. https://doi.org/10.1039/C3SM52532E
10. Chernyatina AA, Nicolet S, Aebi U et al (2012) Atomic structure of the vimentin central -helical domain and its implications for intermediate filament assembly. Proc Natl Acad Sci 109:13620–13625. https://doi.org/10.1073/pnas.1206836109
11. Chernyatina AA, Guzenko D, Strelkov SV (2015) Intermediate filament structure: the bottom-up approach. Curr Opin Cell Biol 32:65–72. https://doi.org/10.1016/j.ceb.2014.12.007
12. Chernyatina AA, Hess JF, Guzenko D et al (2016) How to study intermediate filaments in atomic detail. In: Methods in enzymology. Academic Press Inc, pp 3–33
13. Coelho-Rato LS, Parvanian S, Andrs Salajkova S et al (2024) Intermediate filaments at a glance. J Cell Sci 137:jcs261386. https://doi.org/10.1242/jcs.261386
14. Çolakoğlu G, Brown A (2009) Intermediate filaments exchange subunits along their length and elongate by end-to-end annealing. J Cell Biol 185:769–777. https://doi.org/10.1083/jcb.200809166
15. Coulombe PA, Wong P (2004) Cytoplasmic intermediate filaments revealed as dynamic and multipurpose scaffolds. Nat Cell Biol 6:699–706. https://doi.org/10.1038/ncb0804-699
16. Dammann C, Köster S (2014) Dynamics of counterion-induced attraction between vimentin filaments followed in microfluidic drops. Lab Chip 14:2681–2687. https://doi.org/10.1039/c3lc51418h

17. Dammann C, Herrmann H, Köster S (2016) Competitive counterion binding regulates the aggregation onset of vimentin intermediate filaments. Israel J Chem 56:614–621. https://doi.org/10.1002/ijch.201400153
18. Eibauer M, Weber MS, Kronenberg-Tenga R et al (2024) Vimentin filaments integrate low-complexity domains in a complex helical structure. Nat Struct Mol Biol 31:939–949. https://doi.org/10.1038/s41594-024-01261-2
19. Eriksson JE, Brautigan DL, Vallee R et al (1992) Cytoskeletal integrity in interphase cells requires protein phosphatase activity. Proc Natl Acad Sci USA 89:11093–11097. https://doi.org/10.1073/pnas.89.22.11093
20. Esue O, Carson AA, Tseng Y, Wirtz D (2006) A direct interaction between actin and vimentin filaments mediated by the tail domain of vimentin. J Biol Chem 281:30393–30399. https://doi.org/10.1074/jbc.M605452200
21. Forsting J, Kraxner J, Witt H et al (2019) Vimentin intermediate filaments undergo irreversible conformational changes during cyclic loading. Nano Lett 19:7349–7356. https://doi.org/10.1021/acs.nanolett.9b02972
22. Gavara N (2017) A beginner's guide to atomic force microscopy probing for cell mechanics. Microsc Res Tech 80:75–84. https://doi.org/10.1002/jemt.22776
23. Golde T, Huster C, Glaser M et al (2018) Glassy dynamics in composite biopolymer networks. Soft Matter 14:7970–7978. https://doi.org/10.1039/C8SM01061G
24. Goldie KN, Wedig T, Mitra AK et al (2007) Dissecting the 3-D structure of vimentin intermediate filaments by cryo-electron tomography. J Struct Biol 158:378–385. https://doi.org/10.1016/j.jsb.2006.12.007
25. Guzenko D, Chernyatina AA, Strelkov SV (2017) Crystallographic studies of intermediate filament proteins. In: Parry DAD, Squire JM (eds) Fibrous proteins: structures and mechanisms. Springer International Publishing, Cham, pp 151–170
26. Hémonnot CYJ, Mauermann M, Herrmann H, Köster S (2015) Assembly of simple epithelial keratin filaments: deciphering the ion dependence in filament organization. Biomacromolecules 16:3313–3321. https://doi.org/10.1021/acs.biomac.5b00965
27. Herrmann H, Aebi U (2000) Intermediate filaments and their associates: multi-talented structural elements specifying cytoarchitecture and cytodynamics. Curr Opin Cell Biol 12:79–90. https://doi.org/10.1016/S0955-0674(99)00060-5
28. Herrmann H, Aebi U (2004) Intermediate filaments: molecular structure, assembly mechanism, and integration into functionally distinct intracellular scaffolds. Annu Rev Biochem 73:749–789. https://doi.org/10.1146/annurev.biochem.73.011303.073823
29. Herrmann H, Aebi U (2016) Intermediate filaments: structure and assembly. Cold Spring Harb Perspect Biol 8:a018242–a018242. https://doi.org/10.1101/cshperspect.a018242
30. Herrmann H, Häner M, Brettel M et al (1996) Structure and assembly properties of the intermediate filament protein vimentin: the role of its head, rod and tail domains. J Mol Biol 264:933–953. https://doi.org/10.1006/jmbi.1996.0688
31. Herrmann H, Häner M, Brettel M et al (1999) Characterization of distinct early assembly units of different intermediate filament proteins. J Mol Biol 286:1403–1420. https://doi.org/10.1006/jmbi.1999.2528
32. Herrmann H, Strelkov SV, Feja B et al (2000) The intermediate filament protein consensus motif of helix 2B: its atomic structure and contribution to assembly. J Mol Biol 298:817–832. https://doi.org/10.1006/jmbi.2000.3719
33. Herrmann H, Wedig T, Porter RM et al (2002) Characterization of early assembly intermediates of recombinant human keratins. J Struct Biol 137:82–96. https://doi.org/10.1006/jsbi.2002.4466
34. Herrmann H, Kreplak L, Aebi U (2004) Isolation, characterization, and in vitro assembly of intermediate filaments. In: Methods in cell biology. Elsevier, pp 3–24
35. Herrmann H, Cabet E, Chevalier NR et al (2020) Dual functional states of R406W-desmin assembly complexes cause cardiomyopathy with severe intercalated disc derangement in

humans and in knock-in mice. Circulation 142:2155–2171. https://doi.org/10.1161/CIRCULATIONAHA.120.050218
36. Hohenadl M, Storz T, Kirpal H et al (1999) Desmin filaments studied by quasi-elastic light scattering. Biophys J 77:2199–2209. https://doi.org/10.1016/S0006-3495(99)77060-8
37. Inagaki M, Matsuoka Y, Tsujimura K et al (1996) Dynamic property of intermediate filaments: Regulation by phosphorylation. Bioessays 18:481–487. https://doi.org/10.1002/bies.950180610
38. Janmey PA, Euteneuer U, Traub P, Schliwa M (1991) Viscoelastic properties of vimentin compared with other filamentous biopolymer networks. J Cell Biol 113:155–160. https://doi.org/10.1083/jcb.113.1.155
39. Janmey PA, Slochower DR, Wang YH et al (2014) Polyelectrolyte properties of filamentous biopolymers and their consequences in biological fluids. Soft Matter 10:1439–1449. https://doi.org/10.1039/c3sm50854d
40. Kaus-Drobek M, Mücke N, Szczepanowski RH et al (2020) Vimentin S-glutathionylation at Cys328 inhibits filament elongation and induces severing of mature filaments in vitro. FEBS J 287:5304–5322. https://doi.org/10.1111/febs.15321
41. Kirmse R, Portet S, Mücke N et al (2007) A quantitative kinetic model for the in vitro assembly of intermediate filaments from tetrameric vimentin. J Biol Chem 282:18563–18572. https://doi.org/10.1074/jbc.M701063200
42. Kratky O, Porod G (1949) Röntgenuntersuchung gelöster Fadenmoleküle. Recueil des Travaux Chimiques des Pays-Bas 68:1106–1122. https://doi.org/10.1002/recl.19490681203
43. Kraxner J, Lorenz C, Menzel J et al (2021) Post-translational modifications soften vimentin intermediate filaments. Nanoscale 13:380–387. https://doi.org/10.1039/d0nr07322a
44. Kreplak L, Bär H, Leterrier JF et al (2005) Exploring the mechanical behavior of single intermediate filaments. J Mol Biol 354:569–577. https://doi.org/10.1016/j.jmb.2005.09.092
45. Kreplak L, Herrmann H, Aebi U (2008) Tensile properties of single desmin intermediate filaments. Biophys J 94:2790–2799. https://doi.org/10.1529/biophysj.107.119826
46. Lazarides E (1982) Intermediate filaments: a chemically heterogeneous, developmentally regulated class of proteins. Annu Rev Biochem 51:219–250. https://doi.org/10.1146/annurev.bi.51.070182.001251
47. Lichtenstern T, Mücke N, Aebi U et al (2012) Complex formation and kinetics of filament assembly exhibited by the simple epithelial keratins K8 and K18. J Struct Biol 177:54–62. https://doi.org/10.1016/j.jsb.2011.11.003
48. Lin YC, Yao NY, Broedersz CP et al (2010) Origins of elasticity in intermediate filament networks. Phys Rev Lett 104:058101–058101. https://doi.org/10.1103/PhysRevLett.104.058101
49. Liu X, Pollack GH (2002) Mechanics of F-actin characterized with microfabricated cantilevers. Biophys J 83(5):2705–2715
50. Lopez CG, Saldanha O, Huber K, Köster S (2016) Lateral association and elongation of vimentin intermediate filament proteins: a time-resolved light-scattering study. Proc Natl Acad Sci 113:11152–11157. https://doi.org/10.1073/pnas.1606372113
51. Lorenz C, Forsting J, Schepers AV et al (2019) Lateral subunit coupling determines intermediate filament mechanics. Phys Rev Lett 123:188102–188102. https://doi.org/10.1103/PhysRevLett.123.188102
52. Lorenz C, Schepers AV, Köster S (2022) Quantifying the interaction strength between biopolymers. In: Gennerich A (ed) Optical Tweezers. Springer US, New York, pp 701–723
53. Lorenz C, Forsting J, Style RW et al (2023) Keratin filament mechanics and energy dissipation are determined by metal-like plasticity. Matter 6:S2590238523002035. https://doi.org/10.1016/j.matt.2023.04.014
54. Meier M, Pauline Padilla G, Herrmann H et al (2009) Vimentin coil 1A-A molecular switch involved in the initiation of filament elongation. J Mol Biol 390:245–261. https://doi.org/10.1016/j.jmb.2009.04.067
55. Milo R, Jorgensen P, Moran U et al (2010) BioNumbers—the database of key numbers in molecular and cell biology. Nucleic Acids Res 38:D750–D753. https://doi.org/10.1093/nar/gkp889

56. Mosconi F, Allemand JF, Bensimon D, Croquette V (2009) Measurement of the torque on a single stretched and twisted DNA using magnetic tweezers. Phys Rev Lett 102:078301. https://doi.org/10.1103/PhysRevLett.102.078301
57. Mücke N, Kreplak L, Kirmse R et al (2004a) Assessing the flexibility of intermediate filaments by atomic force microscopy. J Mol Biol 335:1241–1250. https://doi.org/10.1016/j.jmb.2003.11.038
58. Mücke N, Wedig T, Bürer A et al (2004b) Molecular and biophysical characterization of assembly-starter units of human vimentin. J Mol Biol 340:97–114. https://doi.org/10.1016/j.jmb.2004.04.039
59. Mücke N, Kämmerer L, Winheim S et al (2018) Assembly kinetics of vimentin tetramers to unit-length filaments: a stopped-flow study. Biophys J 114:2408–2418. https://doi.org/10.1016/j.bpj.2018.04.032
60. Nicolet S, Herrmann H, Aebi U, Strelkov SV (2010) Atomic structure of vimentin coil 2. J Struct Biol 170:369–376. https://doi.org/10.1016/J.JSB.2010.02.012
61. Nöding B, Köster S (2012) Intermediate filaments in small configuration spaces. Phys Rev Lett 108. https://doi.org/10.1103/PhysRevLett.108.088101
62. Nöding B, Herrmann H, Köster S (2014) Direct observation of subunit exchange along mature vimentin intermediate filaments. Biophys J 107:2923–2931. https://doi.org/10.1016/j.bpj.2014.09.050
63. Omary MB, Coulombe PA, McLean WHI (2004) Intermediate filament proteins and their associated diseases. N Engl J Med 351:2087–2100. https://doi.org/10.1056/NEJMra040319
64. Pallari H-M, Eriksson JE (2006) Intermediate filaments as signaling platforms. Sci STKE 2006: pe53. https://doi.org/10.1126/stke.3662006pe53
65. Parry DAD, Steinert PM (1999) Intermediate filaments: molecular architecture, assembly, dynamics and polymorphism. Q Rev Biophys 32:99–187. https://doi.org/10.1017/S0033583500003516
66. Pawelzyk P, Mü Cke N, Herrmann H et al (2014) Attractive interactions among intermediate filaments determine network mechanics in vitro. PLoS One 9:e93194–e93194. https://doi.org/10.1371/journal.pone.0093194
67. Petitjean II, Tran QD, Goutou A et al (2024) Reconstitution of cytolinker-mediated crosstalk between actin and vimentin. Eur J Cell Biol 103:151403. https://doi.org/10.1016/j.ejcb.2024.151403
68. Pinto N, Gillis TE, Fudge DS et al (2014) Self-assembly enhances the strength of fibers made from vimentin intermediate filament proteins. Biomacromoleules 15:574–581. https://doi.org/10.1021/bm401600a
69. Portet S, Mücke N, Kirmse R et al (2009) Vimentin intermediate filament formation: in vitro measurement and mathematical modeling of the filament length distribution during assembly. Langmuir 25:8817–8823. https://doi.org/10.1021/la900509r
70. Premchandar A, Mücke N, Poznański J et al (2016) Structural dynamics of the vimentin coiled-coil contact regions involved in filament assembly as revealed by hydrogen-deuterium exchange. J Biol Chem 291:24931–24950. https://doi.org/10.1074/jbc.M116.748145
71. Qin Z, Kreplak L, Buehler MJ (2009a) Nanomechanical properties of vimentin intermediate filament dimers. Nanotechnology 20:425101–425101. https://doi.org/10.1088/0957-4484/20/42/425101
72. Qin Z, Kreplak L, Buehler MJ (2009b) Hierarchical structure controls nanomechanical properties of vimentin intermediate filaments. PLoS One 4:e7294. https://doi.org/10.1371/journal.pone.0007294
73. Schaedel L, Lorenz C, Schepers AV et al (2021) Vimentin intermediate filaments stabilize dynamic microtubules by direct interactions. Nat Commun 12:3799–3799. https://doi.org/10.1038/s41467-021-23523-z
74. Schaffeld M, Herrmann H, Schultess J, Markl J (2001) Vimentin and desmin of a cartilaginous fish, the shark Scyliorhinus stellaris: Sequence, expression patterns and in vitro assembly. Eur J Cell Biol 80:692–702. https://doi.org/10.1078/0171-9335-00206

75. Schepers AV, Lorenz C, Köster S (2020) Tuning intermediate filament mechanics by variation of pH and ion charges. Nanoscale 12:15236–15245. https://doi.org/10.1039/D0NR02778B
76. Schepers AV, Lorenz C, Nietmann P et al (2021) Multiscale mechanics and temporal evolution of vimentin intermediate filament networks. Proc Natl Acad Sci 118:e2102026118–e2102026118. https://doi.org/10.1073/pnas.2102026118
77. Schepers AV, Kraxner J, Lorenz C, Köster S (2022) Mechanics of single vimentin intermediate filaments under load. In: Gennerich A (ed) Optical Tweezers. Springer US, New York, pp 677–700
78. Schopferer M, Bär H, Hochstein B et al (2009) Desmin and vimentin intermediate filament networks: their viscoelastic properties investigated by mechanical rheometry. J Mol Biol 388: 133–143. https://doi.org/10.1016/j.jmb.2009.03.005
79. Schween L, Mücke N, Portet S, et al (2022) Dual-wavelength stopped-flow analysis of the lateral and longitudinal assembly kinetics of vimentin. https://doi.org/10.1101/2022.05.05.490764
80. Shen Y, Wu H, Lu PJ et al (2021) Effects of vimentin intermediate filaments on the structure and dynamics of in vitro multicomponent interpenetrating cytoskeletal networks. Phys Rev Lett 127:108101–108101. https://doi.org/10.1103/physrevlett.127.108101
81. Snider NT, Omary MB (2014) Post-translational modifications of intermediate filament proteins: mechanisms and functions. Nat Rev Mol Cell Biol 15:163–177. https://doi.org/10.1038/nrm3753
82. Steven AC, Wall J, Hainfeld J, Steinert PM (1982) Structure of fibroblastic intermediate filaments: analysis by scanning transmission electron microscopy. Proc Natl Acad Sci USA 79:3101–3105. https://doi.org/10.1073/pnas.79.10.3101
83. Tran QD, Sorichetti V, Pehau-Arnaudet G et al (2023) Fragmentation and entanglement limit vimentin intermediate filament assembly. Phys Rev X 13:011014. https://doi.org/10.1103/PhysRevX.13.011014
84. Vermeire P-J, Lilina AV, Hashim HM et al (2023) Molecular structure of soluble vimentin tetramers. Sci Rep 13:8841. https://doi.org/10.1038/s41598-023-34814-4
85. Wickert U, Mücke N, Wedig T et al (2005) Characterization of the in vitro co-assembly process of the intermediate filament proteins vimentin and desmin: mixed polymers at all stages of assembly. Eur J Cell Biol 84:379–391. https://doi.org/10.1016/j.ejcb.2005.01.004
86. Windoffer R, Beil M, Magin TM, Leube RE (2011) Cytoskeleton in motion: the dynamics of keratin intermediate filaments in epithelia. J Cell Biol 194:669–678. https://doi.org/10.1083/jcb.201008095
87. Winheim S, Hieb AR, Silbermann M et al (2011) Deconstructing the late phase of vimentin assembly by total internal reflection fluorescence microscopy (TIRFM). PLoS One 6:e19202–e19202. https://doi.org/10.1371/journal.pone.0019202

Chapter 8
The Fibrillar Collagens of Bony Fish: Genes, Proteins, and Biomedical Applications

Ian P. Smith and Jordi Bella

Abstract Fibrillar collagens are the most abundant structural proteins in vertebrates, forming the backbone of connective tissues such as skin, bone, tendon, and cartilage. In bony fish (teleosts), fibrillar collagens exhibit unique genetic and biochemical properties that reflect a complex evolutionary history of this taxonomic group and its adaptation to diverse aquatic environments—from tropical to polar habitats. This review summarises the current understanding of the genetic organisation and biochemical characteristics of fibrillar collagens in bony fish. They show significant differences in amino acid composition to mammalian collagens, especially in cold-adapted species, where collagens display lower thermal stability and reduced hydroxyproline content relative to their mammalian counterparts.

Advances in genomic, transcriptomic, and proteomic profiling have provided new perspectives on the molecular diversity and tissue-specific roles of collagen chains in teleosts. Furthermore, the biomedical potential of fish-derived collagens is receiving growing attention, particularly in biomaterials, wound healing, tissue engineering, and drug delivery systems, owing to their biocompatibility, low immunogenicity, and ease of extraction from byproducts of the fishing industry.

By looking at molecular, structural, and applied perspectives, this review highlights the relevance of bony fish collagens as a subject of fundamental biological interest and as a valuable resource for biotechnological and biomedical innovation.

Keywords Fibrillar collagen · Bony fish (teleosts) · Gene and genome duplication · Thermal stability · Biomedical applications · Tissue engineering

I. P. Smith · J. Bella (✉)
Division of Cell Matrix Biology and Regenerative Medicine, School of Biological Sciences, Faculty of Biology, Medicine and Health, University of Manchester, Manchester, UK
e-mail: jordi.bella@manchester.ac.uk

D. A. D. Parry, P. K. Luther (eds.), *Fibrous Proteins: Structures and Mechanisms*, Subcellular Biochemistry 113, https://doi.org/10.1007/978-3-032-05273-5_8

Introduction

Collagens are the most abundant proteins of the animal kingdom, responsible for maintaining the structural integrity and correct organisation of the different tissues and organs [77, 110]. Collagens are found in the extracellular matrix of connective tissues such as skin, tendons, ligaments, bones, and cartilage, where they provide mechanical strength and help such tissues to maintain their shape and resist stretching [51, 105, 140]. The importance of collagens is highlighted by the multiple connective tissue disorders caused by mutations in collagen genes and the fibroproliferative diseases that result from excessive collagen deposition in the extracellular matrix [80, 88, 109, 110]. Beyond structural support, collagens also play a key physiological role through interaction with cell surface receptors and other extracellular matrix proteins [2, 11, 42, 62, 66, 78, 79, 114]. Cell-collagen interactions are key in processes like cell signalling, wound healing, or tissue remodelling and repair [35, 131, 141, 163]. Without collagens, tissues would lack mechanical stability, resilience, and durability, which would result in weakened bones, fragile skin, and impaired movement.

Collagen Genes, Chains, and Types: Fibrillar and Non-fibrillar Collagens

Collagens are hallmark metazoan proteins and are found both in vertebrates and invertebrates. The number of collagen proteins and their respective genes vary significantly across the animal kingdom. Humans and mammals in general have 28 different types of collagens, built from at least 46 polypeptide chains which are coded by 45 genes [57, 77, 128]. However, other vertebrate groups have different numbers of collagen genes and proteins. The number of collagen genes in invertebrates shows large variability. The nematode worm *Caenorhabditis elegans* has 181 collagen genes that code for various types of collagen proteins. Most of these genes (173) are unique to nematodes and are probably involved in the formation of the cuticle, the outer layer that surrounds the worm [149]. At the other end, the genome of the fruit fly *Drosophila melanogaster* only contains four collagen-encoding genes [127].

All collagen chains are multi-domain proteins and contain at least one instance of a characteristic collagen domain, easily recognisable at the primary structure level by a repetitive sequence in which Gly residues appear at every third position, forming a (Gly-X-Y)$_n$ pattern where X is often proline and Y is frequently its posttranslational modification, hydroxyproline. Three collagen chains must come together so that this repetitive (Gly-X-Y)$_n$ sequence is folded in a defining triple-helical structure, the collagen triple helix (reviewed in [5]). Thus, the minimal functional unit for all collagen types must be a trimeric protein. Once formed, trimeric collagen proteins form supramolecular structures such as cable-like fibres, mesh-like networks, or beaded filaments that define their biological function [57, 77, 128]. Depending on their supramolecular structure, collagens are broadly classified into two major groups, fibrillar and non-fibrillar. Fibrillar collagens are the most abundant [6, 39]. They form long insoluble fibres that show a periodic structure due to regular staggering of collagen triple-helical molecules. Fibrillar collagens are primarily structural, and their long fibres provide tensile strength to tissues [76]. Other collagens form networks that contribute to tissue organisation and cellular interactions or establish different forms of filaments or microfibrils that connect with other collagens or other components of the extracellular matrix.

In vertebrates, collagen types are designated with Roman numerals I to XXVIII. Each collagen chain is named with the Greek letter α, a numeral indicating the chain number, and the collagen type to which it belongs, such as α1(I) or α3 (V) [57, 128]. The genes coding for these chains are named accordingly, such as *col1a1* or *col5a3* for the examples above, where the first number refers to the collagen type and the second to the chain number. Type I collagen, the most abundant and best studied of all collagens, is a fibrillar type that in mammals is encoded by the *col1a1* and *col1a2* genes and is expressed in almost all connective tissues, including bone, skin, tendons, cornea, and vasculature [67].

Collagens as Biomaterials

The central role of fibrillar collagens in organising the extracellular matrix has resulted in significant interest in their use as biomaterials for a variety of applications in the medical, pharmaceutical, and cosmetic industries, as well as in research laboratories. Type I collagen is by far the most used in these applications. Fibrillar collagens, especially those of mammalian origin, are thought to be biocompatible with minimal risks of causing adverse reactions. Fibrillar collagens in general are known to promote cell growth and therefore are attractive components for tissue remodelling. Their cell-adhesion properties allow cells to attach, proliferate, and differentiate on their surface. Thus, fibrillar collagens are key biomaterials for wound healing, tissue engineering, and regenerative medicine. Type I collagen can be processed into various forms such as hydrogels, sponges, sheets, fibres, and 3D scaffolds, allowing for flexible designs tailored to specific medical applications, drug delivery systems, wound dressings, surgical implants, and cell substrates for research applications. Type I collagen is biodegradable, an attractive quality in tissue engineering or wound healing applications, where the initial scaffold material is degraded as new collagen and new tissue are produced by cells, thereby not interfering in the final remodelling process.

Collagens for biotechnological applications are mainly sourced from animal sources such as hides or bones of cattle, horses, or pigs [17], and type I collagen is the most abundant in those tissues. Animal-derived collagen is therefore cheap, as it is obtained from byproducts of the meat industry. On the other hand, animal-derived collagen has a few shortcomings such as some degree of molecular heterogeneity or batch-to-batch variability, which is an important consideration for most clinical applications. Although there is great interest in developing synthetic collagen-like alternatives to animal-derived collagen [16], progress is slow and the cost of completely replacing animal derived collagen in biotechnological applications is currently prohibitive [21].

Marine Collagens

Marine organisms are getting increased attention as alternative sources for biotechnological collagen. For a general overview on marine collagens and their use in the development of new biomaterials, the reader is directed to an excellent 2021 review [43]. Marine collagen products, from tissue-extracted collagen to gelatine or collagen peptides, show interesting properties in comparison to the more traditional mammalian collagen sourced from farm animals [29, 43, 44]. Collagens from bony fish (teleosts) have attracted most attention to date given their economic advantages, as they can be extracted from waste products from the fishing industry, and they are the focus of this review. However, more recently, invertebrate marine collagens have also gained interest [30, 45, 46, 122, 136, 143].

When compared to mammalian collagens, marine collagens show several key differences:

- Amino acid composition: lower imino acid content. The collagen domains of marine collagens generally contain lower proportions of proline and hydroxyproline when compared to the same domains from mammalian collagens (see section "Amino Acid Sequence Preferences in Fish Collagens and Their Relation to Thermal Stability").
- Lower thermal stability. This lower imino acid content typically results in a lower denaturation temperature. Marine collagens may denature at temperatures just above room temperature or lower, depending on their natural habitat, whereas mammalian collagens are stable at higher temperatures. This difference can be critical when designing products for specific thermal environments.
- Marine organisms show a wide variety of fibrillar collagen genes. Many of these collagens show significant differences in domain architecture and amino acid composition with mammalian collagens. Bony and cartilaginous fish have similar collagens to other vertebrates but often have additional genes for some collagen types while they may completely lack others.
- Marine collagens often exhibit fewer intermolecular cross-links, which can lead to differences in fibril formation and mechanical strength. As a result, marine collagen-based materials might be less robust or have different elasticity compared to those made from mammalian collagens.
- Collagen extraction from marine tissues. The processes for extraction of collagen from marine tissues can lead to variations in molecular weight distributions, influencing properties like viscosity, gelation behaviour, and mechanical

performance. Marine collagen is usually extracted under milder conditions such as lower temperatures and less acidic pH levels to prevent denaturation (due to their lower thermal stability as mentioned above), whereas mammalian collagens can often withstand harsher extraction processes.
- Enhanced solubility. In many cases, marine collagens show higher solubility in aqueous solutions, which can be beneficial for applications like injectable formulations, drug delivery systems, or hydrogels.
- Marine collagens are often reported to have lower antigenicity. This makes them attractive for biomedical applications, as they may elicit a lesser immune response compared to collagens derived from mammals.
- Safety and cultural acceptance. In addition to lower zoonotic disease risk, marine collagens are more acceptable to populations with religious or cultural restrictions regarding porcine or bovine products.

Each of these differences can be potentially exploited for specific biotechnological applications.

Bony Fish Fibrillar Collagens and Their Genes

Early Years

Research on collagen started early in the twentieth century. Most of the initial work was carried out on mammals, trying to understand the structure and properties of the main fibrillar protein from connective tissues such as skin, bone, or tendons, which we now know as type I collagen. By the mid-1950s, scientists started to investigate how fish collagen differs from that of the terrestrial animals, and fish collagen fibrils with the characteristic banding pattern of collagen were observed under the electron microscope [4, 58, 59, 68]. The largely forgotten term *ichthyocol*, referring to collagen-based extracts from fish swim bladder, dates from that period. Methods were developed for the isolation of collagen from fish skins, bones, scales, swim bladders, and other byproducts of the fishing industry [49, 50, 161, 162]. Early studies on the amino acid composition of fish collagen and gelatine already noticed a reduced thermal stability compared to mammalian collagens, and related it to a lower content of hydroxyproline [34, 60, 121]. Fish collagen (ichthyocol) was used in early studies of the transition between denatured collagen (gelatine) and folded collagen [28, 64, 155]. All the work carried out during this period related exclusively to type I collagen, the main component in the connective tissues of fish from which it was extracted. A key difference between fish and mammalian collagens was discovered in 1964, when three non-identical α chains were identified chromatographically in codfish skin collagen [119, 120] (see section "Multiple Fibrillar Collagen Genes in Bony Fish Genomes").

Zebrafish Fibrillar Collagen Genes

As multiple collagen types were being discovered in mammalian connective tissues, interest in fish collagens shifted from a mainly biochemical focus to the genetic identification and functional characterisation of these collagens, particularly in zebrafish (*Danio rerio*), a model organism for the study of developmental biology in vertebrates [31, 154]. Transparent zebrafish embryos are ideally suited for the study of gene function by introducing mutations and analysing the resulting phenotypes [135]. Many of these mutants affect embryonic development and zebrafish physiology, or closely mimic human skeletal diseases such as osteogenesis imperfecta. Thus, several studies have identified fibrillar collagen genes in zebrafish embryos, investigated their expression patterns under different conditions, and explored the effects of mutations of these genes on embryonic development [47, 53, 107, 117, 151, 153, 159]. Sequencing of the zebrafish genome [70] has helped to complete the identification of the zebrafish genes coding for fibrillar collagen types I, II, V, XI, XXIV, and XXVII (Table 8.1). Three additional fibrillar collagen genes have been identified in the zebrafish genome (*dkey-21p1.3*, *dkey-61 l1.4*, *ch211-196i2.1*), although their biological function is largely unknown. Notably, there is no gene coding for type III collagen in the zebrafish genome (Table 8.1).

Updated information on the zebrafish genome is maintained by the Zebrafish Information Network (ZFIN), the knowledge base for the model organism *Danio rerio* [14]. The reader is directed to their websites for curated, up-to-date pages on each fibrillar collagen gene in zebrafish. ZFIN coordinates the naming of zebrafish genes. Throughout this review, we have followed ZFIN recommendations for the naming of fibrillar collagen genes from zebrafish and other bony fish.

Multiple Fibrillar Collagen Genes in Bony Fish Genomes

Since the publication of the zebrafish genome, many additional genomes and transcriptomes for bony fish have been published and their data have become publicly available. Zebrafish fibrillar collagen sequences have been used for the identification and annotation of genes for the different fibrillar collagen types in the genomes of other bony fish. The biological function of these genes has been largely inferred from their orthologs in humans and mammals, but very limited actual experimental work has been carried out for most fish collagen types other than type I.

A significant difference between fish and mammalian collagens is the existence of multiple genes for most collagen types (Table 8.2). This multiplicity was first discovered in type I collagen before the genomic era. In mammals, type I collagen is mainly a heterotrimer made from two polypeptide chains α1(I) and α2(I) in a 2:1 stoichiometry $\alpha1(I)_2\alpha2(I)$. These chains are coded by their respective genes *col1a1* and *col1a2*. The only exception is a homotrimer $\alpha1(I)_3$ version, which has been

Table 8.1 Fibrillar collagen genes in zebrafish, chromosomal location (Chr.), and accession codes in several sequence databases

Gene name	Chr.	ZFIN gene	Ensembl gene (GRCz11)	NCBI gene	NCBI protein
col1a1a	3	ZDB-GENE-030131-9102	ENSDARG00000012405	337158	AAH63249
col1a1b	12	ZDB-GENE-030131-4400	ENSDARG00000035809	325675	NP_958886.1
col1a2	19	ZDB-GENE-030131-8415	ENSDARG00000020007	336471	NP_892013.2
col2a1a	8	ZDB-GENE-980526-192	ENSDARG00000069093	562496	NP_571367.1
col2a1b	11	ZDB-GENE-050302-9	ENSDARG00000011407	503730	NP_001268407.1
col5a1	21	ZDB-GENE-041105-6	ENSDARG00000012593	100334431	XP_021324673.1
col5a2a	9	ZDB-GENE-030616-13	ENSDARG00000031678	564821	NP_001139254.1
col5a2b	6	ZDB-GENE-110406-2	ENSDARG00000024847	100006366	XP_001920005.4
col5a3a	3	ZDB-GENE-030131-968	ENSDARG00000098294	100330523	NP_001177685.1
col5a3b	1	ZDB-GENE-110728-4	ENSDARG00000101816	100463515	NP_001177687.1
col11a1a	24	ZDB-GENE-070209-167	ENSDARG00000026165	565402	NP_001077313.1
col11a1b	2	ZDB-GENE-070912-607	ENSDARG00000009014	555202	NP_001171883.1
col11a2	19	ZDB-GENE-000208-23	ENSDARG00000012422	561679	NP_001073461.2
col24a1	23	ZDB-GENE-110411-128	ENSDARG00000095843	100332483	XP_002666614.2
col27a1a	16	ZDB-GENE-100119-1	ENSDARG00000034373	100004688	NP_001156766.1
col27a1b	5	ZDB-GENE-100119-2	ENSDARG00000099930	560145	NP_001074044.1
si:dkey-21p1.3	8	ZDB-GENE-081105-162	ENSDARG00000002295	101883345	XP_021334064.2
si:dkey-61l1.4	5	ZDB-GENE-070705-483	ENSDARG00000092567	100004431	XP_021332163.1
si:ch211-196i2.1	21	ZDB-GENE-091204-150	ENSDARG00000007912	564099	XP_009293536.1

identified in small amounts in skin and embryonic tissues [75, 152], and is also found in some pathological conditions [15, 100]. The possibility of a third α chain in fish type I collagens, α3(I), was first introduced from the observation by Karl Piez of three separate protein peaks in the analysis of denatured cod skin collagen by carboxymethylcellulose chromatography [119, 120]. Amino acid analysis of these

Table 8.2 Numbers of fibrillar collagen genes in mammals and in representative groups of ray-finned fish (class Actinopterygii)

Taxonomic group	Representative species	I	II	III	V	XI	XXIV	XXVII	Other
Mammals	*Homo sapiens*	2	1	1	3	2	1	1	0
Danionidae	*Danio rerio* (zebrafish)	3	2	0	5	3	1	2	3
Cyprinidae	*Carassius gibelio* (crucian carp)	6	3	0	9	6	2	4	3
Siluriformes	*Ictalurus punctatus* (channel catfish)	3	1	0	5	3	1	2	3
Salmoniformes	*Salmo salar* (Atlantic salmon)	6	4	0	9	6	2	3	4
Gadiformes	*Gadus morhua* (Atlantic cod)	3	2	0	5	3	1	2	3
Perciformes	*Gasterosteus aculeatus* (stickleback)	3	2	0	5	3	1	2	2
Beloniformes	*Oryzias latipes* (medaka)	3	2	0	5	3	1	2	2
Tetraodontiformes	*Takifugu rubripes* (pufferfish)	3	2	0	5	3	1	2	2
Cichliformes	*Oreochromis niloticus* (Nile tilapia)	3	2	0	5	3	1	2	3
Scombriformes	*Thunnus albacares* (yellowfin tuna)	3	2	0	5	3	1	2	3
Osteoglossomorpha	*Sclerophages formosus* (bonytongue)	3	2	0	6	3	1	2	1
Elopomorpha	*Anguilla anguilla* (European eel)	4	2	0	4	3	1	1	3
Chondrostei	*Polyodon spathula* (paddlefish)	4	2	2	4	4	1	2	1
Holostei	*Lepisosteus oculatus* (spotted gar)	2	1	1	3	2	1	1	1
Cladistia	*Polypterus senegalus* (gray bichir)	2	1	1	3	2	1	1	1

peaks indicated three closely related, though not identical, protein chains, which were named accordingly α1, α2, and α3. The equal size of these peaks suggested an α1α2α3 heterotrimer [119, 120]. Subsequent work by Kimura and co-workers showed that a third α3(I) collagen chain was present in the skins of many bony fish but not in their swim bladders, indicating a tissue-specific heterotrimeric chain composition of fish type I collagen, α1(I)α2(I)α3(I) in skin, and $\alpha1(I)_2\alpha2(I)$ in muscle [84, 85, 101, 102, 106, 124]. The fish-specific α1(I)α2(I)α3(I) chain composition was identified in most groups of bony fish, including sturgeons [83, 124, 132, 164].

Early biochemical characterisation of the α3(I) chain from fish skin type I collagens showed that it was much closer to α1(I) than to α2(I), and it was proposed that the gene coding for α3(I) could have arisen in the past as a duplication of an ancestral fish *col1a1* gene [83, 164]. The first sequence analysis at the cDNA level

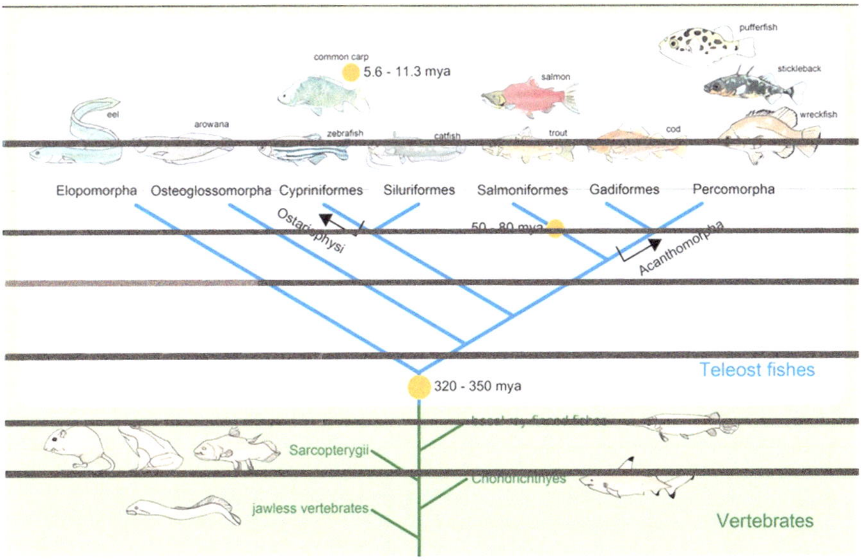

Fig. 8.1 Diagram showing a simplified phylogeny of teleost fish, reproduced from [55] with permission. A whole duplication event (WGD) occurred in the teleost group around 320–350 million years ago (mya), after they had split from the basal ray-finned fish (chondrostei, holostei, cladistia). Additional WGDs have occurred in the Salmoniformes around 80 mya, and in a closely related ancestor of the common carp around 11 mya. The orange dots represent WGD events in the teleost lineage. The degree of multiplicity in the numbers of collagen genes in living bony fish is consistent with the WGD events depicted in this diagram

confirmed the close resemblance between α1(I) and α3(I) [132], and further cDNA sequencing of type I collagen genes in rainbow trout and zebrafish also demonstrated that α3(I) is phylogenetically close to α1(I), supporting the hypothesis of gene duplication [108, 133]. The gene coding for α3(I) has been identified in multiple fish genomes and is now considered the result of an ancient genome-wide duplication in the ray fish lineage after it diverged from the lobe fin lineage, around 320–350 million years ago [1, 55] (Fig. 8.1). The recommended naming for the α1(I) and α3 (I) genes in zebrafish is *col1a1a* and *col1a1b,* respectively (ZFIN Zebrafish Nomenclature Conventions), and the same naming scheme has been adopted for other fish genomes. Table 8.3 lists the type I collagen genes from currently available bony fish genomes, and provides links to gene and protein entries for type I collagen in major sequence databases. The *col1a1b* gene is present in all bony fish with the exception of the polypteridae (bichirs and reedfish) and the holostei (bowfins and gars), all of them considered relic species and part of the basal ray-finned fish lineage from which modern teleost fish diverged before the ancient whole genome duplication (WGD) event mentioned above (Fig. 8.1).

Due to this ancient WGD event, all teleostei show duplicates of many of their fibrillar collagen genes in addition to the *col1a1b* gene (Table 8.2). For instance, mammalian type II collagen is coded by a single gene *col2a1*, whereas zebrafish and

Table 8.3 Collagen type I genes from currently available bony fish genomes, including links to gene and protein entries in major sequence databases

NCBI gene	Gene	Chromosome	NCBI gene	NCBI protein	aa	exons	Uniparc	Ensembl gene	Ensembl protein	Uniprot
Erpetoichthys calabaricus	col1a1	14	LOC114664453	XP_051792335.1	1455	51	UPI0022344DF3	ENSECRG00000012828	ENSECRP00000019185.1*	A0A8C4 SSU9*
	col1a2	13	LOC114663378	XP_051791432.1	1351	50	UPI0022348C9E	ENSECRG00000013418	ENSECRP00000019987.1*	A0A8C4SP70*
Polypterus senegalus	col1a1	17	LOC120517638	XP_039595995.1*	1248*	43*	UPI001963FB8D*	ENSSTSG00000024652	ENSSTSP00000056813.1*	ND
	col1a2	15	LOC120515712	XP_039592895.1	1359	51	UPI001963AFE1	ENSSTSG00000019643	ENSSTSP00000045317.1	ND
Acipenser ruthenus	col1a1a	33	LOC117433526	XP_033911740.1	1452	51	UPI00145AF3FD	ND		ND
	col1a1b	28	LOC117434956	XP_033913598.1	1457	51	UPI00145A5173	ND		ND
	col1a2a	3	LOC117394450	XP_058868361.1	1349	50	UPI002740A6F3	ND		ND
	col1a2b	4	LOC121313553	XP_033855900.2*	1317*	*49	UPI0027413949*	ND		ND
Polyodon spathula	col1a1a	28	LOC121301872	XP_041087483.1	1452	51	UPI001B7E8B60	ENSPAHG00005018776	ENSPAHP00005026611.1	ND
	col1a1b	12	LOC121324653	XP_041122692.1	1467	51	UPI001B7DD656	ENSPAHG00005011627	ENSPAHP00005016705.1*	ND
	col1a2a	4	LOC121315031	XP_041104831.1	1357	51	UPI001B7EC1D7	ENSPAHG00005001160	ENSPAHP00005001771.1	ND
	col1a2b	3	LOC121313553	XP_041102168.1	1357	51	UPI001B7DDFBE	ENSPAHG00005008426	ENSPAHP00005012196.1*	ND
Amia calva	col1a1	3	LOC136744273	XP_066554840.1	1450	51	UPI0034649353	ND	ND	ND
	col1a2	10	LOC136760191	XP_066571579.1	1350	50	UPI003463A7FF	ND	ND	ND
Lepisosteus oculatus	col1a1	28	LOC102682998	XP_006638297.1	1450	51	UPI0003EAADA0	ENSLOCG00000013046	ENSLOCP00000016069.1*	ND
	col1a2	10	LOC102694884	XP_069050866.1	1351	50	UPI0003EADBFF	ENSLOCG00000010835	ENSLOCP00000013346.1*	ND
Anguilla anguilla	col1a1a	17	LOC118216611	XP_035253836.1	1451	51	UPI0015ACED39	ENSAANG00010021368	ENSAANP00010054619.1	A0A9D3LM48
	col1a1b	2	LOC118220696	XP_035260697.1	1447	51	UPI0015ADFD44	ND		ND
	col1a2a	8	LOC118233663	XP_035285365.1	1357	51	UPI0015B1F361	ENSAANG00010022115	ENSAANP00010056806.1*	ND
	col1a2b	1	LOC118214884	XP_035251069.1	1351	50	UPI0015AAC00D	ENSAANG00010013192	ENSAANP00010032232.1*	ND
Anguilla rostrata	col1a1a	17	LOC135244222	XP_064172601.1	1450	51	UPI0030D58238	ENSAROG00010020319	ENSAROP00010038181.2*	ND
	col1a1b	2	LOC135247826	XP_064177791.1	1447	51	UPI0030D02954	ENSAROG00010023101	ENSAROP00010042597.2	ND
	col1a2a	8	LOC135261713	XP_064204340.1	1353	50	UPI0030D5E36E	ENSAROG00010009943	ENSAROP00010018790.2*	ND
	col1a2b	1	LOC135260851	XP_064202565.1	1349	50	UPI0030CF0734	ENSAROG00010004852	ENSAROP00010069727.1	ND
Anguilla japonica	col1a1a	17	ND	BAP63924.1	1450	51	UPI00050D8AE3	ENSAJAG00030029407	ENSAJAP00030054894.1	A0A090B2K5
	col1a1b	2	ND	BAP63926.1	1446	51	UPI00050D8331	ENSAJAG00030025565	ENSAJAP00030048370.1*	A0A090AZH6
	col1a2a	8	ND	BAP63925.1	1358	49	UPI00050D684C	ENSAJAG00030023711	ENSAJAP00030044788.1	A0A090AX23
	col1a2b	1	ND	ND	1351	50	ND	ENSAJAG00030028866	ENSAJAP00030054025.1	ND
Conger conger	col1a1a	16	LOC133113963	XP_061079118.1	1449	51	UPI002A5B0EB4	ENSRRGG00005027144	ENSRRGP00005062889.1	ND
	col1a1b	2	LOC133121223	XP_061086408.1*	1396*	48*	UPI002A5ACE77*	ENSRRGG00005026945	ENSRRGP00005062007.1*	ND
	col1a2b	9	LOC133136668	XP_061110298.1	1351	50	UPI002A5A4ABF	ENSRRGG00005007662	ENSRRGP00005015070.1*	ND
	col1a2a	1	LOC133126844	XP_061095255.1	1349	50	UPI002A59D888	ENSRRGG00005013190	ENSRRGP00005030964.1*	ND
Megalops cyprinoides	col1a1a	19	LOC118794176	XP_036408262.1	1450	51	UPI0018649016	ENSMCNG00005013441	ENSMCNP00005033731.1*	ND

	col1a1b	1	LOC118793509	XP_036407625.1	1449	51	UPI0018646B89	ENSMCNG00005005168	ENSMCNP00005011838.1*	ND
	col1a2a	21	LOC118796578	XP_036411440.1	1361	51	UPI00186504C9	ENSMCNG00005024455	ENSMCNP00005064183.1*	ND
	col1a2b	10	LOC118784937	XP_036395323.1	1353	50	UPI00186559E1	ENSMCNG00005021257	ENSMCNP00005054997.1*	ND
Brienomyrus brachyistius	col1a1a	22	LOC125717923	XP_048847252.1	1450	51	UPI0020B2C6DE	ENSBBCG00000028625	ENSBBCP00000039858.1	ND
	col1a1b	5	LOC125742925	XP_048871344.1	1452	51	UPI0020B36D34	ENSBBCG00000025387	ENSBBCP00000039824.1	ND
	col1a2	23	LOC125718755	XP_048848747.1	1355	50	UPI0020B1F254	ENSBBCG00000015231	ENSBBCP00000025148.1	ND
Paramormyrops kingsleyae	col1a1a	Unplaced	LOC111855986	XP_023691321.1	1447	51	UPI000CD63770	ENSPKIG00000017673	ENSPKIP00000040891.1	A0A3B3TF17
	col1a1b	Unplaced	LOC111848653	XP_023676618.1	1449	51	UPI000CD6500E	ENSPKIG00000021151	ENSPKIP00000003901.1*	A0A3B3 QEQ9*
	col1a2	Unplaced	LOC111843899	XP_023667631.1	1354	50	UPI000CD64FA1	ENSPKIG00000001264	ENSPKIP00000014068.1*	A0A3B3 R6R1*
Scleropages formosus	col1a1a	8	LOC108941816	XP_018620194.1	1448	51	UPI0008782A8F	ENSSFOG00015020268	ENSSFOP00015031767.2	A0A8C9SH96
	col1a1b	25	LOC108928208	XP_018597555.1	1448	51	UPI0008791E95	ENSSFOG00015004978	ENSSFOP00015007726.2*	A0A1W4ZA01
	col1a2	23	LOC108922001	XP_018587334.1	1358	51	UPI00087864A1	ENSSFOG00015012447	ENSSFOP00015064451.1*	A0A1W4Y6I4
Alosa alosa	col1a1a	6	LOC125296068	XP_048101686.1	1443	51	UPI002015251E	ENSAAOG00005015007	ENSAAOP00005030636.1	ND
	col1a1b	22	LOC125287101	XP_048088557.1	1448	51	UPI00201534ED	ENSAAOG00005001806	ENSAAOP00005003229.1	ND
	col1a2	20	LOC125285020	XP_048085161.1	1349	50	UPI0020154CF4	ENSAAOG00005006015	ENSAAOP00005012035.1*	ND
Alosa sapidissima	col1a1a	3	LOC121705000	XP_041941592.1	1443	51	UPI001C09B2B5	ENSASDG00000004821	ENSASDP00000023483.1	ND
	col1a1b	24	LOC121699581	XP_041937770.1	1448	51	UPI001C091D2E	ENSASDG00000022155	ENSASDP00000062545.1	ND
	col1a2	21	LOC121695225	XP_041931796.1	1349	50	UPI001C093DF9	ENSASDG00000006124	ENSASDP00000020814.1*	ND
Clupea harengus	col1a1a	Unplaced	LOC122130930	XP_042561723.1	1443	51	UPI0006444C51	ENSCHAG00020001904	ENSCHAP00020003719.1	A0A8M1 KHC8
	col1a1b	23	LOC105909513	XP_031416510.1	1453	50	UPI0012ABD57B	ENSCHAG00020008521	ENSCHAP00020018395.1*	A0A6P8EXC7
	col1a2	19	LOC105902040	XP_012685017.1	1349	50	UPI000644328D	ENSCHAG00020017383	ENSCHAP00020039279.1*	A0A6P3W0U2
Sardina pilchardus	col1a1a	3	LOC134077254	XP_062388734.1	1443	51	UPI002E12DFD3	ENSSPIG00015012108	ENSSPIP00015028242.1*	ND
	col1a1b	22	LOC134070326	XP_062382626.1	1447	51	UPI002E1135F2	ENSSPIG00015001415	ENSSPIP00015001624.1	ND
	col1a2	24	LOC134073018	XP_062385933.1	1352	50	UPI002E15983A	ENSSPIG00015021780	ENSSPIP00015052709.1*	ND
Denticeps clupeoides	col1a1a	7	LOC114793647	XP_028841563.1	1446	51	UPI0010A4CB64	ENSDCDG00010025392	ENSDCDP00010049907.1	A0A8C4C7W2
	col1a1b	8	LOC114796062	XP_028845768.1	1447	51	UPI0010A3418B	ENSDCDG00010017990	ENSDCDP00010033381.1	A0AAY4 CLK4
	col1a2	20	LOC114769999	XP_028819400.1	1353	50	UPI0010A4C6D5	ENSDCDG00010030563	ENSDCDP00010052789.1*	ND
Engraulis encrasicolus	col1a1a	2	LOC134438270	XP_063043863.1	1444	51	UPI002FD4E6D8	ENSPGJG00000020560	ENSPGJP00000048598.1	ND
	col1a1b	17	LOC134466835	XP_063076789.1	1446	51	UPI002FD709D8	ENSPGJG00000015707	ENSPGJP00000037162.1	ND
	col1a2	20	LOC134436327	XP_063041556.1	1352	50	UPI002FCF7C5D	ENSPGJG00000020732	ENSPGJT00000050862.1*	ND
Chanos chanos	col1a1a	16	LOC115829318	XP_030649248.1	1448	51	UPI0011F2083C	ND		A0A6J2WY88
	col1a1b	5	LOC115811462	XP_030629529.1	1447	51	UPI0011F246D7	ND		A0A6J2VB72
	col1a2	12	LOC115825069	XP_030644688.1	1352	50	UPI0011F3F489	ND		A0A6J2WK72
Astyanax mexicanus	col1a1a	19	LOC103041791	XP_007232286.3	1447	51	UPI0020CB0A41	ENSAMXG00025011865	ENSAMXP00025015007.1	ND

(continued)

Table 8.3 (continued)

NCBI gene	Gene	Chromosome	NCBI gene	NCBI protein	aa	exons	Uniparc	Ensembl gene	Ensembl protein	Uniprot
	col1a1b	15	LOC103043248	XP_022525013.2	1446	51	UPI0020CAAB74	ENSAMXG00025024877	ENSAMXP00025040184.1	ND
	col1a2	3	LOC103037264	XP_049331372.1	1351	50	UPI0020CB242E	ENSAMXG00025024157	ENSAMXP00025038464.1*	ND
Hoplias malabaricus	col1a1a	13	LOC136665396	XP_066499104.1	1447	50	UPI003461E82D	ND		ND
	col1a1b	3	LOC136690654	XP_066518689.1	1445	51	UPI0034633AB8	ND		ND
	col1a2	6	LOC136699370	XP_066530157.1	1351	50	UPI003461E0F3	ND		ND
Colossoma macropomum	col1a1a	Unplaced	LOC118821406	XP_036445890.1	1447	51	UPI001863ACD2	ND		ND
	col1a1b	Unplaced	LOC118809800	XP_036429267.1	1446	51	UPI001865177F	ND		ND
	col1a2	Unplaced	LOC118805569	XP_036422339.1	1351	50	UPI001863FA02	ND		ND
Pygocentrus nattereri	col1a1a	14	LOC108412936	XP_017540674.2	1447	51	UPI0018912146	ENSPNAG00000003407	ENSPNAP00000010420.2*	ND
	col1a1b	13	LOC108444293	XP_017580869.1	1448	51	UPI0008147DE4	ENSPNAG00000012127	ENSPNAP00000005928.2	A0A3B4C2C3
	col1a2	27	LOC108425881	XP_017550401.1	1351	50	UPI0008143979	ENSPNAG00000029588	ENSPNAP00000037653.1*	ND
Electrophorus electricus	col1a1a	1	LOC113569930	XP_026853844.2	1448	51	UPI0015D06D84	ENSEEEG00000004027	ENSEEEP00000007693.2	A0A4W4 E9W2
	col1a1b	14	LOC113574298	XP_026860911.2	1450	51	UPI0015D07636	ENSEEEG00000003885	ENSEEEP00000007721.2	A0A4W4E911
	col1a2	8	LOC113578661	XP_026867852.2	1352	50	UPI0015CFCA59	ENSEEEG00000021238	ENSEEEP00000045876.1*	ND
Hemibagrus wyckioides	col1a1a	LG02	LOC131370223	XP_058273421.1	1449	51	UPI001C7AFD4A	ENSHWYG000000192 60	ENSHWYP00000041591.1	A0A9D3P5F5
	col1a1b	LG13	LOC131363742	XP_058262567.1	1446	51	UPI00266B7D08	ENSHWYG0000000743 6	ENSHWYP00000018806.1	ND
	col1a2	LG24	LOC131345142	XP_058233872.1	1352	50	UPI00266B81FE	ENSHWYG0000001473 5	ENSHWYP00000034736.1	ND
Trichomycterus rosablanca	col1a1a	22	LOC134300158	XP_062840896.1	1449	51	UPI002F352DCE	ND		ND
	col1a1b	10	LOC134321199	XP_062858879.1	1447	51	UPI002F356539	ND		ND
	col1a2	2	LOC134332420	XP_062870135.1	1352	49	UPI002F356FDB	ND		ND
Neoarius graeffei	col1a1a	20	LOC132868675	XP_060757739.1	1449	51	UPI00298C8E27	ENSNGEG00005015743	ENSNGEP00005024040.1	ND
	col1a1b	14	LOC132898066	XP_060795426.1	1442	51	UPI00298D52A7	ENSNGEG00005019068	ENSNGEP00005029385.1	ND
	col1a2	16	LOC132900576	XP_060798734.1	1353	50	UPI00298C8F02	ENSNGEG00005032894	ENSNGEP00005065744.1	ND
Tachysurus fulvidraco	col1a1a	15	LOC113638083	XP_026994881.2	1449	51	UPI001FEE0B04	ENSTFUG00005008838	ENSTFUP00005021170.1	ND
	col1a1b	8	LOC113653604	XP_047673196.1	1447	51	UPI001FEE23D7	ENSTFUG00005009682	ENSTFUP00005022133.1	ND
	col1a2	24	LOC113661410	XP_047663913.1	1352	50	UPI001FEF43ED	ENSTFUG00005034539	ENSTFUP00005059655.1	ND
Tachysurus vachellii	col1a1a	18	LOC132861286	XP_060748721.1	1449	51	UPI0029485E8D	ENSYVCG00005004823	ENSYVCP00005015594.1	A0AA88M604
	col1a1b	2	LOC132841535	XP_060719869.1	1446	51	UPI00296AB511	ENSYVCG00005005883	ENSYVCP00005015189.1	ND
	col1a2	14	LOC132864293	XP_060753644.1	1352	50	UPI00296AAADD	ENSYVCG00005003924	ENSYVCP00005013278.1	ND
Clarias gariepinus	col1a1a X2	16	LOC128545027	XP_053371332.1	1449	51	UPI00234D73E5	ENSCGSG00000014730	ENSCGSP00000035648.1	ND

	col1a1b	14	LOC128541637	XP_053368126.1	1444	51	UPI00234CFB62	ENSCGSG00000012623	ENSCGSP00000030982.1	ND
	col1a2	3	LOC128518989	XP_053348445.1	1352	50	UPI00234D849B	ENSCGSG00000021321	ENSCGSP00000053908.1	ND
Clarias fuscus	col1a1a	19	ND	ND	1449	51	ND	ENSCFSG00010006049	ENSCFSP00010013787.1	ND
	col1a1b	17	ND	ND	1289*	40*	ND	ENSCFSG00010003063	ENSCFSP00010010218.1*	ND
	col1a2	12	ND	ND	1352	50	ND	ENSCFSG00010023110	ENSCFSP00010056842.1	ND
Ictalurus furcatus	col1a1a X2	2	LOC128600534	XP_053469073.1	1449	51	UPI00234FFEFF	ENSIFRG00005009161	ENSIFRP00005027040.1*	ND
	col1a1b	13	LOC128616544	XP_053495105.1	1449	51	UPI00235098A9	ENSIFRG00005007827	ENSIFRP00005019197.1	ND
	col1a2	24	LOC128600351	XP_053468726.1	1352	50	UPI00235052CE	ENSIFRG00005014099	ENSIFRP00005040798.1	ND
Ictalurus punctatus	col1a1a X2	2	LOC108279263	XP_017348858.2	1449	51	UPI002353B2DC	ENSIPUG00015016008	ENSIPUP00015035496.1	A0A2D0T1Q8
	col1a1b	13	LOC108273947	XP_017339145.1	1449	51	UPI0008029D4D	ENSIPUG00015018347	ENSIPUP00015037696.1	A0A2D0SA79
	col1a2	24	LOC108257075	XP_047006669.2	1352	50	UPI00235366BD	ENSIPUG00020014459	ENSIPUP00020029532.1	A0A979EH42
Pangasianodon gigas	col1a1a X1	LG13	PGIGA_ G00045920	MCI4385055.1	1450	51	UPI00219686E6	ND	ND	ND
	col1a1b	LG12	PGIGA_ G00033400	MCI4384012.1	1448	51	UPI00219F4854	ND	ND	ND
	col1a2	LG23	PGIGA_ G00140650	MCI4391979.1*	1332*	49*	UPI00219CBD6E*	ND	ND	ND
Pangasianodon hypophthalmus	col1a1a X2	13	LOC113528288	XP_026772561.1	1449	51	UPI000F009FD7	ENSPHYG00025013371	ENSPHYP00025032709.1	ND
	col1a1b	12	LOC113527704	XP_026771581.2	1448	51	UPI0012C7CF08	ENSPHYG00025017768	ENSPHYP00025042597.1	A0A5N5MJK3
	col1a2	23	LOC113536150	XP_026785766.1	1352	50	UPI000EFE5ECB	ENSPHYG00025016865	ENSPHYP00025040812.1	ND
Pangasius djambal	col1a1a	13	PDJAM_ G00045200	MCJ8739268.1	1449	51	UPI00219D6CF7	ND	ND	ND
	col1a1b	12	PDJAM_ G00033480	MCJ8738250.1	1448	51	UPI0021902910	ND	ND	ND
	col1a2	23	PDJAM_ G00139250	MCJ8746217.1	1349	50	UPI00219D4539	ND	ND	ND
Silurus meridionalis	col1a1a	14	LOC124396621	XP_046721734.1	1449	51	UPI00198A267B	ENSSMDG00005013243	ENSSMDP00005029634.1	A0A8T0AYI6
	col1a1b	7	LOC124388656	XP_046709509.1	1445	51	UPI0019A03850	ENSSMDG00005017895	ENSSMDP00005039990.1	A0A8T0BDF3
	col1a2	22	LOC124375842	XP_046690467.1	1352	50	UPI001EEA979D	ENSSMDG00005004298	ENSSMDP00005012060.1	ND
Silurus aristotelis	col1a1a	17	ND	ND	1449	51	ND	ENSSARG00005027328	ENSSARP00005035240.1	ND
	col1a1b	12	ND	ND	1445	51	ND	ENSSARG00005039640	ENSSARP00005064942.1	ND
	col1a2	21	ND	ND	1352	50	ND	ENSSARG00005032876	ENSSARP00005054612.1	ND
Myxocyprinus asiaticus	col1a1a1	13	LOC127450870	XP_051571265.1	1448	50	UPI002221B541	ENSMTCG00000024771	ENSMTCP00000054510.1	ND
		14	LOC127452223	XP_051573525.1	1448	50	UPI002223440D	ENSMTCG00000014610		ND

(continued)

Table 8.3 (continued)

NCBI gene	Gene	Chromosome	NCBI gene	NCBI protein	aa	exons	Uniparc	Ensembl gene	Ensembl protein	Uniprot
	col1a1a2								ENSMTCP00000029901.1	
	col1a1b1	36	LOC127427372	XP_051530903.1	1451	51	UPI002223D6FD	ENSMTCG00000031374	ENSMTCP00000065193.1	ND
	col1a1b2	32	LOC127422779	XP_051522495.1	1451	51	UPI002222FB0B	ENSMTCG00000020131	ENSMTCP00000041575.1	ND
	col1a2a	34	LOC127425309	XP_051527106.1	1352	50	UPI0022239887	ENSMTCG00000007939	ENSMTCP00000049758.1	ND
	col1a2b	30	LOC127421357	XP_051520340.1	1350	50	UPI002221818E	ENSMTCG00000035774	ENSMTCP00000073228.1*	ND
Xyrauchen texanus	col1a1a1	8	LOC127647399	XP_051987570.1	1447	50	UPI002241F8C9	ENSAUXG00000024249	ENSAUXP00000047386.1	ND
	col1a1a2	12	LOC127652258	XP_051994356.1	1448	50	UPI00224209CA	ENSAUXG00000007422	ENSAUXP00000013215.1	ND
	col1a1b1	40	LOC127633211	XP_051968096.1	1449	51	UPI002241EBEB	ENSAUXG00000018654	ENSAUXP00000035265.1	ND
	col1a1b2	33	LOC127626403	XP_051958101.1	1454	51	UPI002241FE57	ENSAUXG00000039837	ENSAUXP00000077391.1	ND
	col1a2b	26	LOC127619580	XP_051948423.1	1352	50	UPI002241ABAA	ENSAUXG00000027872	ENSAUXP00000058821.1	ND
	col1a2a	10	LOC127650655	XP_051992183.1	1350	50	UPI0022426704	ENSAUXG00000007886	ENSAUXP00000014496.1*	ND
Misgurnus anguillicaudatus	col1a1a	2	LOC129436408	XP_055050508.1	1447	51	UPI002434B815	ENSIRMG00000028813	ENSIRMP00000047674.1	ND
	col1a1b	17	LOC129420715	XP_055031722.1	1448	51	UPI002434F2B7	ENSIRMG00000021502	ENSIRMP00000039672.1	ND
	col1a2	12	LOC129455286	XP_055075952.1*	1352	50	UPI0024352D84*	ENSIRMG00000033192	ENSIRMP00000053298.1	ND
Paramisgurnus dabryanus	col1a1a	Unplaced	LOC135733939	XP_065108839.1	1447	51	UPI0031F373AE	ND	ND	ND
	col1a1b	Unplaced	LOC135734601	XP_065109521.1	1450	51	UPI0031F3F947	ND	ND	ND
	col1a2	Unplaced	LOC135781430	XP_065147983.1*	1270*	46*	UPI0031F38DD3*	ND	ND	ND
Triplophysa dalaica	col1a1a	7	LOC130426103	XP_056608623.1	1448	51	UPI0024DFA733	ENSTDLG00000019382	ENSTDLP00000039868.1	ND
	col1a1b	17	LOC130438903	XP_056627160.1	1452	51	UPI0024DF69ED	ENSTDLG00000001934	ENSTDLP00000002944.1	ND
	col1a2	25	LOC130415243	XP_056596782.1	1352	50	UPI0024DFEE3E	ENSTDLG00000022345	ENSTDLP00000046482.1*	ND
Triplophysa rosa	col1a1a	LG11	LOC130561512	XP_057201888.1	1448	51	UPI002216DC4C	ENSTRSG00005002330	ENSTRSP00005008877.1	A0A9W7TR82
	col1a1b	LG18	LOC130568995	XP_057214290.1	1452	51	UPI002545FA85	ENSTRSG00005001816	ENSTRSP00005003105.1	ND
	col1a2	16	LOC130567013	XP_057210867.1	1352	50	UPI002545DEC5	ENSTRSG00005008258	ENSTRSP00005021811.1*	ND
Onychostoma macrolepis	col1a1a	3	LOC131536845	XP_058625976.1	1447	51	UPI0018468B53	ND	ND	A0A7J6DBF9
	col1a1b	12	LOC131551090	XP_058649725.1	1451	51	UPI001794F62F	ND	ND	A0A7J6CKA1
	col1a2	19	LOC131525893	KAF4100397.1	1352	50	UPI0017D4D93D	ND	ND	A0A7J6BZI0
Carassius auratus	col1a1a1	3	LOC113044312	XP_026059983.1	1448	51	UPI000E3F921C	ENSCARG00000001155	ENSCARP00000004144.1	A0A6P6JJT0
	col1a1a2	28	LOC113046705	XP_026063409.1	1448	51	UPI000E401F3F	ENSCARG00000023493	ENSCARP00000059044.1	A0A6P6JU61
	col1a1b1	37	LOC113056433	XP_026079004.1	1450	51	UPI000E421C56	ENSCARG00000009049	ENSCARP00000040111.1	A0A6P6L374

	col1a1b2	12	LOC113112073	XP_026133239.1	1443	51	UPI000E3FF0F1	ENSCARG00000002562	ENSCARP00000005684.1	A0A6P6QI39
	col1a2a	44	LOC113062051	XP_026087352.1	1352	50	UPI000E4066BB	ENSCARG00000029332	ENSCARP00000071562.1*	A0A6P6LU57
	col1a2b	19	LOC113119369	XP_026144583.1	1352	50	UPI000E406F88	ENSCARG00000051222	ENSCARP00000102398.1*	A0A6P6RHE3
Carassius carassius	col1a1a1	9	LOC132148948	XP_059413741.1	1448	51	UPI002869788E	ENSYGUG00000035218	ENSYGUP00000 075888.1	ND
	col1a1a2	1	LOC132142072	XP_059407697.1	1448	51	UPI0028684C87	ENSYGUG00000037020	ENSYGUP00000079422.1	ND
	col1a1b1	39	LOC132121195	XP_059386367.1	1451	51	UPI002868B830	ENSYGUG00000012193	ENSYGUP00000022218.1	ND
	col1a1b2	40	ND	ND	1441	51	ND	ENSYGUG00000015506	ENSYGUP00000035257.1	ND
	col1a2a	8	LOC132145424	XP_059412438.1	1352	50	UPI002868FDB6	ENSYGUG00000048108	ENSYGUP00000101314.1*	ND
	col1a2b	25	LOC132104277	XP_059365615.1	1352	50	UPI002868733A	ENSYGUG00000029706	ENSYGUP00000059321.1*	ND
Carassius gibelio	col1a1a1	A3	LOC127946536	XP_052399132.1	1448	51	UPI000E3F921C	ENSCGBG00020051913	ENSCGBP00020070321.1	ND
	col1a1a2	B3	LOC127951946	XP_052406069.1	1448	51	UPI002278D0F9	ENSCGBG00020041188	ENSCGBP00020050108.1*	ND
	col1a1b1	A12	LOC128025089	XP_052467049.1	1450	51	UPI00227750B2	ENSCGBG00020052277	ENSCGBP00020082421.1	ND
	col1a1b2	B12	LOC127968702	XP_052426001.1	1443	51	UPI0022785AFE	ENSCGBG00020002413	ENSCGBP00020005793.1	ND
	col1a2a	B19	LOC127979389	XP_052440772.1	1352	50	UPI0022779B7C	ENSCGBG00020003295	ENSCGBP00020023831.1*	ND
	col1a2b	A19	LOC127935260	XP_052388916.1*	1270*	46*	UPI0022792CAB*	ENSCGBG00020071545	ENSCGBP00020113557.1*	ND
Cyprinus carpio carpio	col1a1a1	A3	LOC109053879	XP_042584098.1	1448	51	UPI001C5A2622	ENSCCRG00000009806	ENSCCRP00000017971.2	A0 A9Q9WIM3
	col1a1a2	B3	LOC109092016	XP_042576683.1	1448	51	UPI001C55AECF	ENSCCRG00000023476	ENSCCRP00000045612.2*	A0 A9Q9ZWQ3
	col1a1b1	A12	LOC109099110	XP_042623041.1	1453	51	UPI001337ED1C	ENSCCRG00000061457	ENSCCRP00000128983.1	A0A9J7ZCX2
	col1a1b2	B12	ND	ND	1444	51	UPI0021EC136B	ENSCCRG00000018053	ENSCCRP00000158048.1	A0A9J8BNI6
	col1a2a	B19	LOC109063376	XP_042601316.1	1352	50	UPI001C59A0FB	ENSCCRG00025015796	ENSCCRP00025042931.1	A0A9Q9XHP7
	col1a2b	A19	LOC109112311	XP_042632906.1	1352	50	UPI001C5A13D7	ENSCCRG00000013700	ENSCCRP00000036271.2	A0A8C1BQ90
Sinocyclocheilus anshuiensis	col1a1a	Unplaced	LOC107667754	XP_016314956.1	1448	51	UPI0007B99FC3	ENSSANG00000049412	ENSSANP00000101572.1*	A0A671 SYE2*
	col1a1a2	Unplaced	LOC107662397	XP_016307878.1*	1327*	41*	UPI001350F090*	ENSSANG00000010623	ENSSANP00000021443.1*	A0A671 LLU9*
	col1a1b1	Unplaced	LOC107693674	XP_016348621.1	1452	51	UPI0007BA91C6	ENSSANG00000013091	ENSSANP00000031442.1*	A0A671 MLG1*
	col1a1b2	Unplaced	LOC107689922	XP_016343587.1	1451	51	UPI0007B811BF	ENSSANG00000012272	ENSSANP00000024763.1*	A0A671 LW44*
	col1a2a	Unplaced	LOC107667770	XP_016314980.1	1352	50	UPI0007B96509	ENSSANG00000050293	ENSSANP00000102755.1*	A0A671SZL3*
	col1a2b	Unplaced	LOC107669816	XP_016317714.1	1352	50	UPI0007B847C1	ENSSANG00000039661	ENSSANP00000081499.1*	A0A671 RDG4*
Sinocyclocheilus grahami	col1a1a1	Unplaced	LOC107585052	XP_016127375.1	1447	51	UPI0007AC8E48	ENSSGRG00000035415	ENSSGRP00000099127.1*	A0A672SCJ7*
	col1a1a2	Unplaced	LOC107582926	XP_016124437.1*	1391*	50*	UPI0007AD61A4*	ENSSGRG00000035415	ENSSGRP00000098954.1*	A0A672 SAV3*

(continued)

Table 8.3 (continued)

NCBI gene	Gene	Chromosome	NCBI gene	NCBI protein	aa	exons	Uniparc	Ensembl gene	Ensembl protein	Uniprot
	col1a1b1	Unplaced	LOC107561997	XP_016102039.1*	1388*	49*	UPI0007ACB0B4*	ENSSGRG00000035415	ENSSGRP00000057351.1*	A0A672P1A8*
	col1a1b2	Unplaced	LOC107571757	XP_016113605.1*	1381*	49*	UPI0007AC5D86*	ENSSGRG00000035415	ENSSGRP00000005029.1*	A0A672 K7U9*
	col1a2a	Unplaced	LOC107591591	XP_016136266.1	1352	50	UPI0007AD482F	ENSSGRG00000035415	ENSSGRP00000106373.1*	A0A672 SXB0*
	col1a2b	Unplaced	LOC107593526	XP_016138752.1	1352	50	UPI0007ACD62F	ENSSGRG00000035415	ENSSGRP00000070494.1*	A0A672Q2 M7*
Sinocyclocheilus rhinocerous	col1a1a1	Unplaced	LOC107717204	XP_016379370.1*	1273*	42*	UPI0007B7D9B3*	ENSSRHG00000027141	ENSSRHP00000056814.1*	A0A673K2 N7*
	col1a1a2	Unplaced	LOC107740634	XP_016408602.1	1447	51	UPI0007BA2B24	ENSSRHG00000002682	ENSSRHP00000017352.1*	A0A673 GVM3*
	col1a1b1	Unplaced	LOC107708989	XP_016368866.1	1452	51	UPI0007B9C273	ENSSRHG00000006848	ENSSRHP00000013133.1*	A0A673GJD7*
	col1a1b2	Unplaced	LOC107720033	XP_016383096.1	1451	51	UPI0007B8A2E9	ENSSRHG00000024349	ENSSRHP00000060721.1*	A0A673K8J4*
	col1a2a	Unplaced	LOC107731699	XP_016398335.1	1352	50	UPI0007B8AC76	ENSSRHG00000029663	ND	ND
	col1a2b	Unplaced	LOC107722140	XP_016385859.1	1352	50	UPI0007B95BB8	ENSSRHG00000012097	ENSSRHP00000024288.1*	A0A673HF93*
Cirrhinus molitorella	col1a1a	1	Q8A67_000444	KAK2916070.1	1447	51	UPI0029440E60	ND	ND	A0AA88QJW0
	col1a1b	12	QQF64_004159	KAL1263804.1	1454	51	UPI0034D883CB	ND	ND	ND
	col1a2	15	Q8A67_015219	KAK2886991.1	1329*	49*	UPI0029454A17	ND	ND	A0AA88PKU9
Labeo rohita	col1a1a	3	LOC127161719	XP_050960393.1	1448	51	UPI0021E20B9E	ENSLROG00010012532	ENSLROP00010024405.1	ND
	col1a1b	12	LOC127173943	XP_050980005.1	1443	51	UPI0021E31863	ENSLROG00010005949	ENSLROP00010015828.1*	ND
	col1a2	19	LOC127181701	XP_050992542.1	1352	50	UPI0021E1CE5D	ENSLROG00010005023	ENSLROP00010007844.1*	ND
Puntigrus tetrazona	col1a1a	3	LOC122341499	XP_043090837.1	1446	51	UPI001C89D43B	ENSPRZG00000011377	ENSPRZP00000018945.1	ND
	col1a1b	Unplaced	LOC122333356	XP_043086861.1	1445	51	UPI001C89AA67	ENSPRZG00000003874	ENSPRZP00000005033.1*	ND
	col1a2	19	LOC122323755	XP_043073788.1	1352	50	UPI001C8A904F	ENSPRZG00000027710	ENSPRZP00000058897.1*	ND
Danio rerio	col1a1a	3	LOC337158	AAH63249	1447	51	UPI00002262A0	ENSDARG00000012405	ENSDARP00000003177.6	Q6P4U1
	col1a1b	12	LOC325675	NP_958886.1	1449	51	UPI00001CDFC4	ENSDARG00000035809	ENSDARP00000003038.8	Q6PEI9
	col1a2	19	LOC336471	NP_892013.2	1352	50	UPI0000378110	ENSDARG00000020007	ENSDARP00000118091.1	Q6IQX2
Danio aesculapii	col1a1a	3	LOC130221061	XP_056309348.1	1447	51	UPI0024BFD827	ENSDAEG00005023374	ENSDAEP00005050470.1	ND
	col1a1b	12	LOC130238159	XP_056325037.1	1449	51	UPI0024C04CBC	ENSDAEG00005018431	ENSDAEP00005039477.1	ND
	col1a2	19	LOC130246548	XP_056335515.1	1352	50	UPI0024C0C1DE	ENSDAEG00005017225	ENSDAEP00005036901.1*	ND
Pseudorasbora parva	col1a1a	2	LOC137038038	XP_067268579.1	1448	51	UPI00351F6C81	ENSPPBG00000017023	ENSPPBP00000023605.1	ND
	col1a1b	21	LOC137055458	XP_067285329.1	1446	51	UPI00351DB7FD	ENSPPBG00000019985	ENSPPBP00000029161.1	ND
	col1a2	19	LOC137047662	XP_067281577.1	1270*	48*	UPI00351F13BC	ENSPPBG00000026678	ENSPPBP00000042949.1*	ND
Phoxinus phoxinus	col1a1a	2	R3I93_002131	KAK7175136.1	1448	51	UPI0030ADA40A	ENSXXPG00000011332	ENSXXPP00000011986.1	A0 AAN9DKU8
	col1a1b	21	R3I93_017591	KAK7134226.1	1450	51	UPI00309FB1B1	ENSXXPG00000036097	ENSXXPP00000059234.1	

										A0 AAN9GX61
	col1a2	19	ND	ND	1270*	46*	ND	ENSXXPG00000031195	ENSXXPP00000047533.1*	ND
Pimephales promelas	col1a1a	Unplaced	LOC120489494	XP_039542288.1	1448	51	UPI0019556102	ND	ND	ND
	col1a1b	Unplaced	LOC120469020	XP_039513638.1	1451	31*	UPI0019555731	ND	ND	ND
	col1a2	Unplaced	LOC120485187	XP_039536717.1	1352	50	UPI001955BDB0	ND	ND	ND
Rhinichthys klamathensis	col1a1a	Unplaced	LOC130070332	XP_056090680.1	1448	51	UPI0024B53220	ND	ND	ND
	col1a1b	Unplaced	LOC130080626	XP_056102041.1	1451	51	UPI0024B60813	ND	ND	ND
	col1a2	Unplaced	LOC130084324	XP_056105662.1	1352	50	UPI0024B5B125	ND	ND	ND
Chanodichthys erythropterus	col1a1a	3	LOC137017409	XP_067237731.1	1448	51	UPI00351F4BB2	ENSCEYG00000010353	ENSCEYP00000019470.1	ND
	col1a1b	19	LOC137007409	XP_067224952.1	1450	51	UPI00351DEF2A	ENSCEYG00000010461	ENSCEYP00000024623.1	ND
	col1a2	15	LOC137037497	XP_067267624.1*	1352	49	UPI00351E6EF8*	ENSCEYG00000002044	ENSCEYP00000018674.1	ND
Ctenopharyngodon idella	col1a1a	3	LOC127509112	XP_051743565.1	1448	51	UPI0001DE156A	ND	ND	E2GK07
	col1a1b	12	LOC127523768	XP_051770805.1	1445	51	UPI00223196CE	ND	ND	ND
	col1a2	19	LOC127501255	XP_051729126.1	1352	49	UPI00222E712E	ND	ND	ND
Culter alburnus	col1a1a	3	ABG768_019221	KAK9977407.1	1448	51	UPI0032E54808	ENSCSLG00000021788	ENSCSLP00000054447.1*	A0AAW2 AXE1
	col1a1b	12	ABG768_004170	KAK9965056.1	1450	51	UPI0032E76519	ENSCSLG00000009831	ENSCSLP00000023690.1*	A0AAW1 ZUA5
	col1a2	19	ABG768_011801	KAK9957561.1	1352	49	UPI0032E650F3	ENSCSLG00000002425	ENSCSLP00000010256.1*	A0AAW1 Z8C5
Hypophthalmichthys molitrix	col1a1a	2	ND	AIL02135.1	1448	51	UPI0004EFA120	ENSHMSG00005006977	ENSHMSP00005021226.1*	A0A077B3P8
	col1a1b	Unplaced	ND	ND	1335*	48*	ND	ENSHMSG00005025009	ENSHMSP00005065054.1*	ND
	col1a2	Unplaced	ND	AUF74474.1	1352	49	UPI000CA0B034	ENSHMSG00005026321	ENSHMSP00005070098.1*	A0A2H4ZEX8
Megalobrama amblycephala	col1a1a	LG1	LOC125272910	XP_048054052.1	1448	51	UPI00159CC52C	ENSMAYG00005022637	ENSMAYP00005057696.1	A0A7D4Y5M8
	col1a1b	LG20	LOC125255084	XP_048026002.1	1450	51	UPI002014709D	ENSMAYG00005017325	ENSMAYP00005044054.1	A0A7D4 XVD3
	col1a2	LG9	LOC125274981	XP_048057531.1	1352	50	UPI00159CE25E	ENSMAYG00005012704	ENSMAYP00005033805.1	A0A7D4 XUU0
Esox lucius	col1a1a	11	LOC105005744	XP_034151134.1	1450	47	UPI00147733FB	ENSELUG00000014218	ENSELUP00000014166.1	A0A3P8YDI4
	col1a1b	5	LOC105021174	XP_010886968.4	1453	49	UPI0014769F14	ENSELUG00000015762	ENSELUP00000036350.3	A0A3P9A5A0
	col1a2	10	LOC105022787	XP_010889810.1	1352	50	UPI0005766662	ENSELUG00000020073	ENSELUP00000080175.2	A0A6Q2ZR41
Coregonus clupeaformis	col1a1a1	35	LOC121550483	XP_041718685.2	1449	47	UPI001E1C797F	ENSCCFG00005005198	ENSCCFP00005011384.1	ND
	col1a1a2	Unplaced	LOC121569686	XP_041736766.1	1449	47	UPI001BE01717	ENSCCFG00005002208	ENSCCFP00005007022.1	ND
	col1a1b1	1	LOC121577829	XP_041747688.1	1455	49	UPI001BE09FE2	ENSCCFG00005000797	ENSCCFP00005001665.1	ND
	col1a1b2	31	LOC121547583	XP_041714861.1	1456	49	UPI001BE09657	ENSCCFG00005012297	ENSCCFP00005027090.1	ND
	col1a2a	8	LOC121572093	XP_041739932.1	1352	49	UPI001BDF8934	ENSCCFG00005035125	ENSCCFP00005079596.1*	ND

(continued)

Table 8.3 (continued)

NCBI gene	Gene	Chromosome	NCBI gene	NCBI protein	aa	exons	Uniparc	Ensembl gene	Ensembl protein	Uniprot
	col1a2b	17	LOC121585876	XP_041758112.1	1352	49	UPI001BE09D90	ENSCCFG00005020990	ENSCCFP00005043205.1*	ND
Salmo salar	col1a1a1	ssa03	LOC106600852	XP_014048044.1	1449	47	UPI0006B6E1C6	ENSSSAG00000003820	ENSSSAP00000007857.1	A0A1S3R8F9
	col1a1 a2	ssa06	LOC100286406	XP_014059932.1	1449	47	UPI0006B716C4	ENSSSAG00075008806	ENSSSAP00075011913.1	A0A1S3S6G4
	col1a1b1	ssa01	LOC106610502	XP_014065361.1	1460	49	UPI0006B7DEE3	ENSSSAG00000005527	ENSSSAP00000011150.1	A0A1S3SM10
	col1a1b2	ssa28	LOC106589632	XP_014035319.1	1455	49	UPI0006B8483B	ENSSSAG00000069947	ENSSSAP00000127030.1	A0A1S3Q7E3
	col1a2a	ssa14	LOC106570460	XP_013998298.1	1352	49	UPI0006B7B1DA	ENSSSAG00000046988	ENSSSAP00000046469.2	A0A1S3M4X1
	col1a2b	ssa27	LOC106588939	XP_014033986.1	1352	49	UPI0006B6FB37	ENSSSAG00000067809	ENSSSAP00000085198.2	A0A1S3Q269
Salmo trutta	col1a1a1	32	LOC115170991	XP_029583278.1	1449	47	UPI0011302838	ENSSTUG00000043409	ENSSTUP00000096570.1	A0A674DL09
	col1a1a2	1	LOC115196785	XP_029613558.1	1449	47	UPI0011316276	ENSSTUG00000013912	ENSSTUP00000032406.1*	A0A673 YCX6*
	col1a1b1	18	LOC115152868	XP_029553618.1	1460	49	UPI001130DA67	ENSSTUG00000042980	ENSSTUP00000095570.1	A0A674DI56
	col1a1b2	10	LOC115201445	XP_029620930.1	1453	49	UPI001131AD81	ENSSTUG00000007914	ENSSTUP00000017601.1*	A0A673XB29
	col1a2a	36	LOC115175524	XP_029590672.1	1352	49	UPI00113129FA	ENSSTUG00000043917	ENSSTUP00000097749.1	A0A674 DQD8*
	col1a2b	34	LOC115174015	XP_029588473.1	1352	49	UPI001131B5A5	ENSSTUG00000038533	ENSSTUP00000088845.1*	A0A674 CYQ9*
Salvelinus fontinalis	col1a1a1	2	LOC129828559	XP_055745565.1	1449	47	UPI0024861E90	ENSFAAG00000018809	ENSFAAP00000040967.1	ND
	col1a1a2	34	LOC129832892	XP_055752969.1	1449	47	UPI0024854BF5	ENSFAAG00000021251	ENSFAAP00000047282.1	ND
	col1a1b1	1	LOC129863615	XP_055791704.1	1460	49	UPI0024854EAA	ENSFAAG00000005634	ENSFAAP00000010328.1	ND
	col1a1b2	30	LOC129828865	XP_055746100.1	1460	49	UPI0024850CF1	ENSFAAG00000046336	ENSFAAP00000110850.1	ND
	col1a2b	22	LOC129820186	XP_055733143.1	1352	49	UPI002484FFB4	ENSFAAG00000013844	ENSFAAP00000029989.1*	ND
Salvelinus namaycush	col1a1a1	3	LOC120034849	XP_038837430.1	1449	47	UPI001901C39F	ENSSNMG00000047178	ENSSNMP00000069407.1	A0A8U0Q6A1
	col1a1a2	41	LOC120034486	XP_038836996.1	1449	47	UPI001901742E	ENSSNMG00000032650	ENSSNMP00000046547.1	A0A8U0Q3B1
	col1a1b1	4	LOC120046838	XP_038848313.1	1460	49	UPI0019020705	ENSSNMG00000037712	ENSSNMP00000054571.1	A0A8U0UD30
	col1a1b2	35	LOC120029978	XP_038831219.1	1460	49	UPI0018FFF27F	ENSSNMG00000029392	ENSSNMP00000041517.1	A0A8U0TVY6
	col1a2a	32	LOC120026738	XP_038827385.1	1352	49	UPI0019008301	ENSSNMG00000033830	ENSSNMP00000048404.1*	A0A8U0PLR6
	col1a2b	27	LOC120022140	XP_038821923.1	1352	49	UPI00190073D1	ENSSNMG00000037589	ENSSNMP00000054406.1*	A0A8U0PB59
Oncorhynchus clarkii lewisi	col1a1a1	12	LOC139421079	XP_071027704.1	1449	47	ND	ND	ND	ND
	col1a1a2	13	LOC139364584	XP_070957549.1	1449	47	ND	ND	ND	ND
	col1a1b1	23	LOC139381400	XP_070975824.1	1460	49	ND	ND	ND	ND
	col1a1b2	20	LOC139376794	XP_070981001.1	1458	49	ND	ND	ND	ND
	col1a2a	23	LOC139392337	XP_070996353.1	1352	49	ND	ND	ND	ND
	col1a2b	18	LOC139373782	XP_070970875.1	1352	49	ND	ND	ND	ND
Oncorhynchus gorbuscha	col1a1a1	LG16	LOC124000418	XP_046162722.1	1449	47	UPI001EAF03DE	ENSOGOG00000043572	ENSOGOP00000092505.1	ND

	col1a1a2	LG26	LOC124016310	XP_046187816.1	1449	47	UPI001EAF0C86	ENSOGOG00000020104	ENSOGOP00000037569.1	ND
	col1a1b1	LG06	LOC124038600	XP_046210609.1	1459	49	UPI001EAEA267	ENSOGOG00000012674	ENSOGOP00000021225.1*	ND
	col1a1b2	LG07	LOC124040237	XP_046213173.1*	1460	49	UPI001EAE9711*	ENSOGOG00000047576	ENSOGOP00000101119.1	ND
	col1a2a	LG09	LOC124043250	XP_046217582.1	1352	49	UPI001EAEBA11	ENSOGOG00000036079	ENSOGOP00000075744.1	ND
	col1a2b	LG24	LOC124013049	XP_046183126.1	1352	49	UPI001EAECC54	ENSOGOG00000044715	ENSOGOP00000097135.1*	ND
Oncorhynchus keta	col1a1a1	5	LOC118384419	XP_035626829.2	1449	47	UPI00227D3B6C	ENSOKEG00005046925	ENSOKEP00005085367.1*	ND
	col1a1a2	32	LOC118365182	XP_035603055.2	1449	47	UPI00227ADD74	ENSOKEG00005009353	ENSOKEP00005015848.1	ND
	col1a1b1	2	LOC118373431	XP_052327313.1	1460	49	UPI00227BB3E9	ENSOKEG00005000493	ENSOKEP00005000801.1	ND
	col1a1b2	24	LOC118402876	XP_035657123.1	1460	49	UPI0015FC8898	ENSOKEG00005011005	ENSOKEP00005018742.1	ND
	col1a2a	19	LOC118370848	XP_052325920.1	1352	49	UPI00227A9402	ENSOKEG00005015513	ENSOKEP00005026653.1*	ND
	col1a2b	20	LOC118399622	XP_052328572.1	1352	49	UPI0015FC2034	ENSOKEG00005032696	ENSOKEP00005052837.1	ND
Oncorhynchus kisutch	col1a1a1	LG6	LOC109893232	XP_031683073.1	1449	47	UPI000D09B3B6	ENSOKIG00005002517	ENSOKIP00005005393.1	A0A8C7CN52
	col1a1a2	LG10	LOC109898318	XP_031689743.1	1449	47	UPI0012DCFA44	ENSOKIG00005007949	ENSOKIP00005018080.1	A0A8C7DC43
	col1a1b1	LG11	LOC109879686	XP_020327438.1	1456	49	UPI00099F40E8	ENSOKIG00005048480	ENSOKIP00005111215.1	A0A8C7L7M1
	col1a1b2	LG25	LOC109870104	XP_020316038.1	1460	49	UPI00099FD420	ENSOKIG00005007321	ENSOKIP00005016403.1	A0A8C7D5I0
	col1a2a	LG14	LOC109903380	XP_020355639.1	1352	49	UPI0009A061F6	ENSOKIG00005016895	ENSOKIP00005039909.1*	A0 A8C7GAI0*
	col1a2b	LG17	LOC109907202	XP_031651017.1	1352	49	UPI00099F5D10	ENSOKIG00005034501	ENSOKIP00005081243.1*	A0A8C7J2 W7*
Oncorhynchus masou masou	col1a1a1	15	LOC135555820	XP_064844650.1	1449	47	UPI0031840BF2	ND	ND	ND
	col1a1a2	14	LOC135553953	XP_064841987.1	1449	47	UPI003183A004	ND	ND	ND
	col1a1b1	24	LOC135511916	XP_064789507.1	1460	49	UPI003183E2DD	ND	ND	ND
	col1a1b2	31	LOC135523593	XP_064806444.1	1461	49	UPI003184189F	ND	ND	ND
	col1a2a	12	LOC135549652	XP_064835896.1	1270*	47*	UPI0031836A7F	ND	ND	ND
	col1a2b	13	LOC135551802	XP_064839098.1	1270*	47*	UPI0031835290	ND	ND	ND
Oncorhynchus mykiss	col1a1a1	12	LOC100135778	NP_001117649.1	1449	47	UPI00000FBCA3	ENSOMYG00000020647	ENSOMYP00000045908.2	Q910C0
	col1a1a2	13	LOC110486153	XP_036796700.1	1449	47	UPI0018785DE6	ENSOMYG00000041314	ENSOMYP00000045972.2	A0A8C7R225
	col1a1b1	23	LOC110503033	XP_036816680.1	1460	49	UPI0018788109	ENSOMYG00000032355	ENSOMYP00000069811.2	A0A8C7W553
	col1a1b2	20	LOC100135810	NP_001117678.1	1458	49	UPI00000FC387	ENSOMYG00000040194	ENSOMYP00000086840.2	Q910B9
	col1a2a	32	LOC100135811	NP_001117679.1	1356	50	UPI0000126D3A	ENSOMYG00000013646	ENSOMYP00000029222.2	O93484
	col1a2b	18	LOC110496633	XP_036809307.1	1352	49	UPI000B4FA3DE	ENSOMYG00000009373	ENSOMYP00000020143.2	A0A8C7PBP3
Oncorhynchus nerka	col1a1a1	LG26	LOC115110151	XP_029491415.2	1449	47	UPI0031B8A1CB	ENSONEG00010012823	ENSONEP00010027929.1*	ND
	col1a1a2	LG21	LOC115104043	XP_029481061.2	1449	47	UPI0031B8AC32	ENSONEG00010038801	ENSONEP00010094661.1*	ND
	col1a1b1	Unplaced	LOC115118054	XP_029502449.1	1460	49	UPI001131B4F7	ENSONEG00010000751	ENSONEP00010001428.1	ND
	col1a1b2	LG16	LOC115144285	XP_029541053.1	1459	49	UPI0011327DAD	ENSONEG00010013141	ENSONEP00010028889.1	ND
	col1a2a	LG7	LOC115131316	XP_064876380.1*	1352	49	UPI0011323C5D*	ENSONEG00010015338	ENSONEP00010033892.1	ND

(continued)

Table 8.3 (continued)

NCBI gene	Gene	Chromosome	NCBI gene	NCBI protein	aa	exons	Uniparc	Ensembl gene	Ensembl protein	Uniprot
	col1a2b	LG4	LOC115119901	XP_064874050.1	1352	49	UPI00112FEA30	ENSONEG00010014988	ENSONEP00010033048.1	ND
Oncorhynchus tshawytscha	col1a1a1	LG09	LOC112258533	XP_024288724.1	1449	47	UPI000D09B3B6	ENSOTSG00005010996	ENSOTSP00005023514.1	A0A8C8ESB4
	col1a1a2	LG27	LOC112225909	XP_024245808.1	1449	47	UPI000D0A1A93	ENSOTSG00005038360	ENSOTSP00005081566.2*	A0A8C8IUS4*
	col1a1b1	LG01	LOC112259622	XP_042183044.1	1460	49	UPI001C3C9E89	ENSOTSG00005072515	ENSOTSP00005121680.1	A0AAZ3PZP2
	col1a1b2	LG25	LOC112224508	XP_024243872.1	1460	49	UPI000D0A4A1F	ENSOTSG00005008078	ENSOTSP00005016281.2*	A0 A8C8CZU3*
	col1a2a	LG31	LOC112229484	XP_042166179.1	1352	49	UPI001C3C60FC	ENSOTSG00005014170	ENSOTSP00005132562.1	A0AAZ3QTB1
	col1a2b	LG13	LOC112264350	XP_024296660.1	1352	49	UPI000D09ECF4	ENSOTSG00005003152	ENSOTSP00005126234.1	A0AAZ3QE32
Hypomesus transpacificus	col1a1a	17	LOC124479588	XP_046894343.1	1449	50	UPI001F087472	ENSHTAG00005009211	ENSHTAP00005023167.1	ND
	col1a1b	13	LOC124476042	XP_046888974.1	1446	51	UPI001F07EA85	ENSHTAG00005021278	ENSHTAP00005054978.1	ND
	col1a2	4	LOC124466841	XP_046874696.1	1354	50	UPI001F0790AF	ENSHTAG00005022716	ENSHTAP00005057771.1	ND
Osmerus eperlanus	col1a1a	2	LOC134015407	XP_062310897.1	1449	50	UPI002E0DD0DE	ND	ND	ND
	col1a1b	12	LOC134031054	XP_062330512.1	1445	51	UPI002E167026	ND	ND	ND
	col1a2	20	LOC134040679	XP_062342687.1	1356	50	UPI002E11FDD0	ND	ND	ND
Osmerus mordax	col1a1a	3	LOC136941013	XP_067089483.1	1449	50	UPI00350FADB6	ND	ND	ND
	col1a1b	10	LOC136950867	XP_067101471.1	1445	51	UPI00350EC921	ND	ND	ND
	col1a2	18	LOC136962268	XP_067112086.1	1357	48	UPI00350ED1CB	ND	ND	ND
Lampris incognitus	col1a1a	LG10	LOC130132780	XP_056157805.1	1449	49	UPI0024B4A402	ENSLING00000013938	ENSLINP00000027681.1	ND
	col1a1b	LG18	LOC130127269	XP_056152827.1	1448	51	UPI0024B56D08	ENSLING00000022616	ENSLINP00000041447.1	ND
	col1a2	LG19	LOC130128368	XP_056153964.1*	1357	51	UPI0024B59C1C*	ENSLING00000020171	ENSLINP00000037462.1	ND
Gadus morhua	col1a1a	2	LOC115531299	XP_030196341.1	1449	49	UPI0011B61265	ENSGMOG00000008472	ENSGMOP00000063843.1*	A0 A8C5CNW5*
	col1a1b	18	LOC115530900	XP_030195580.1	1453	51	UPI0011B46927	ENSGMOG00000015290	ENSGMOP00000016465.2	A0A8C4ZK0 6*
	col1a2	22	LOC115535609	XP_030202823.1	1350	50	UPI0011B37903	ENSGMOG00000013455	ENSGMOP00000014420.3*	A0A8C4ZJ33*
Gadus macrocephalus	col1a1a	2	LOC132475343	XP_059932470.1	1449	49	UPI0028CB2BC7	ND	ND	ND
	col1a1b	18	LOC132446847	XP_059893361.1	1453	51	UPI0028CB9D35	ND	ND	ND
	col1a2	22	LOC132451726	XP_059900329.1	1275*	48*	UPI0028CB8C50	ND	ND	ND
Gadus chalcogrammus	col1a1a	2	LOC130404761	XP_056465631.1	1449	49	UPI0024C2123D	ND	ND	ND
	col1a1b	18	LOC130371906	XP_056433752.1	1453	51	UPI0024C4B0B5	ND	ND	ND
	col1a2	22	LOC130375331	XP_056438136.1	1356	49	UPI0024C4CAE0	ND	ND	ND
Myripristis murdjan	col1a1a	8	LOC115363492	XP_029913605.1	1445	49	UPI001175FDFC	ENSMMDG00005000 844	ENSMMDP000050 01588.1*	ND
	col1a1b	19	LOC115377705	XP_029933507.1	1446	51	UPI001175FADB	ENSMMDG0000502092 7	ENSMMDP0000504 5639.1*	ND

	col1a2	11	LOC115368033	XP_029919850.1	1353	50	UPI001175D6AA	ENSMMDG00005013559	ENSMMDP00005028324.1*	ND
Anabas testudineus	col1a1a	8	LOC113153278	XP_026202622.1	1446	49	UPI000E45F05C	ENSATEG00000011778	ENSATEP00000066349.2	A0A7N6FJN9
	col1a1b	19	LOC113156291	XP_026207127.1	1445	51	UPI000E45DFA4	ENSATEG00000005367	ENSATEP00000060907.2	A0A7N6BCJ3
	col1a2	11	LOC113153813	XP_026203436.1	1352	50	UPI000E455AEC	ENSATEG00000018607	ENSATEP00000077983.1*	ND
Betta splendens	col1a1a	8	LOC114860628	XP_029015179.1	1447	49	UPI0010F932B0	ENSBSLG00000012242	ENSBSLP00000034171.1	A0A6P7NBJ9
	col1a1b	19	LOC114845965	XP_028990436.1	1450	51	UPI0010F49C1D	ENSBSLG00000005774	ENSBSLP00000019249.1	A0A6P7L713
	col1a2	11	LOC114865721	XP_055369069.1*	1352	50	UPI0010F871D3	ENSBSLG00000005986	ENSBSLP00000063963.1	A0A6P7NY32
Channa argus	col1a1a	15	LOC137099540	XP_067332616.1	1447	49	UPI0014174DFC	ENSAHCG00000002421	ENSAHCP00000002472.1	A0A6G1QBU3
	col1a1b	20	LOC137105249	XP_067343254.1	1446	51	UPI002946AF33	ENSAHCG00000018253	ENSAHCP00000050953.1	ND
	col1a2	1	LOC137127128	KAK2921202.1	1352	50	UPI002944A233	ENSAHCG00000008899	ENSAHCP00000023350.1	ND
Channa maculata	col1a1a	2	ND	ND	1447	49	ND	ENSMHCG00000014047	ENSMHCP00000030590.1	ND
	col1a1b	20	ND	ND	1444	51	ND	ENSMHCG00000002505	ENSMHCP00000015420.1	ND
	col1a2	1	ND	ND	1352	50	ND	ENSMHCG00000013529	ENSMHCP00000031895.1	ND
Channa striata	col1a1a	LG13	Q5P01_016730	KAK2832841.1	1447	49	UPI0029480A7C	ENSNWLG00000007799	ENSNWLP00000019198.1*	A0AA88MC27
	col1a1b	LG19	Q5P01_023592	KAK2820633.1	1435*	52*	UPI0029480AB	ENSNWLG00000009618	ENSNWLP00000023647.1*	A0AA88J6H5
	col1a2	LG14	Q5P01_018178	KAK2830247.1	1358	49	UPI002944E990	ENSNWLG00000013680	ENSNWLP00000033472.1*	A0AA88M6R2
Mastacembelus armatus	col1a1a	8	LOC113140882	XP_026180754.1	1447	49	UPI000E45C9FD	ENSMAMG00015013683	ENSMAMP00015042199.1	ND
	col1a1b	19	LOC113135359	XP_026171118.1	1452	51	UPI000E45F0CC	ENSMAMG00015006215	ENSMAMP00015014673.1	ND
	col1a2	11	LOC113126189	XP_026155865.1	1355	50	UPI000E46455E	ENSMAMG00015011142	ENSMAMP00015031600.1	ND
Monopterus albus	col1a1a	Unplaced	LOC109951101	XP_020441049.1	1447	49	UPI0009B2F7B3	ENSMALG00000018340	ENSMALP00000027253.1*	A0A3Q3KE32*
	col1a1b	Unplaced	LOC109974887	XP_020480987.1	1450	51	UPI0009B4CB8B	ENSMALG00000011026	ENSMALP00000015761.1*	A0A3Q3QMD5*
	col1a2	Unplaced	LOC109968174	XP_020469902.1	1345	49	UPI0009B3CA83	ENSMALG00000020065	ENSMALP00000029093.1*	A0A3Q3KIB6*
Thalassophryne amazonica	col1a1a	16	LOC117527226	XP_034045336.1	1444	49	UPI0014723376	ENSTAMG00005014973	ENSTAMP00005037686.1*	ND
	col1a1b	18	LOC117530443	XP_034049214.1	1452	51	UPI0014721415	ENSTAMG00005013010	ENSTAMP00005026894.1*	ND
	col1a2	20	LOC117501433	XP_034016214.1	1354	51	UPI001471A398	ENSTAMG00005004430	ND	ND
Lates calcarifer	col1a1a	LG11	LOC108898561	XP_018553992.1	1447	49	UPI0008737804	ENSLCAG00010006888	ENSLCAP00010014889.2	A0AAJ7QEE6
	col1a1b	LG23	LOC108876589	XP_018521723.1	1449	51	UPI000873DD47	ENSLCAG00010014814	ENSLCAP00010062184.1	A0AAJ7PF75
	col1a2	LG3	LOC108876841	XP_050921738.1	1350	50	UPI0021D7B990	ENSLCAG00010018745	ENSLCAP00010040654.2	A0AAJ8AXW6
Toxotes jaculatrix	col1a1a	18	LOC121198022	XP_040917763.1	1447	48	UPI001B3A9050	ENSTJCG00005024137	ENSTJCP00005057671.1	ND
	col1a1b	21	LOC121201026	XP_040922565.1	1447	51	UPI001B3B0652	ENSTJCG00005019835	ENSTJCP00005048003.1	ND
	col1a2	13	LOC121191649	XP_040908848.1	1352	50	UPI001B3AA1F8	ENSTJCG00005006683	ENSTJCP00005015422.1	ND

(continued)

Table 8.3 (continued)

NCBI gene	Gene	Chromosome	NCBI gene	NCBI protein	aa	exons	Uniparc	Ensembl gene	Ensembl protein	Uniprot
Seriola aureovittata	col1a1a	17	LOC130185876	XP_056258579.1	1447	49	UPI0024BEBDD6	ENSUEOG00000020812	ENSUEOP00000061570.1	ND
	col1a1b	21	LOC130162103	XP_056221624.1	1450	51	UPI0024BE90C9	ENSUEOG00000012870	ENSUEOP00000037391.1	ND
	col1a2	16	LOC130184150	XP_056255936.1*	1352	50	UPI0024 BDC7D6*	ENSUEOG00000014073	ENSUEOP00000040845.1	ND
Seriola dumerili	col1a1a	Unplaced	LOC111223728	XP_022603698.1	1447	49	UPI000BBE77B9	ENSSDUG00000006335	ENSSDUP00000008629.1*	A0A3B4 TQU3*
	col1a1b	Unplaced	LOC111219718	XP_022598172.1	1450	51	UPI000BBEE2EA	ENSSDUG00000012841	ENSSDUP00000017680.1*	A0A3B4 UIY5*
	col1a2	Unplaced	LOC111223033	XP_022602721.1	1352	50	UPI000BBE17C7	ENSSDUG00000006239	ENSSDUP00000008659.1*	A0A3B4TQX3
Seriola lalandi dorsalis	col1a1a	Unplaced	LOC111662465	XP_023272098.1	1447	49	UPI000C6FA551	ENSSLDG00000012463	ENSSLDP00000015781.1*	A0A3B4 XHU9*
	col1a1b	Unplaced	LOC111656147	XP_023263510.1	1450	51	UPI000C6FC104	ENSSLDG00000008127	ENSSLDP00000013726.1*	A0A3B4 XCD2*
	col1a2	Unplaced	LOC111656921	XP_023264598.1	1352	50	UPI000C6F93E8	ENSSLDG00000012616	ENSSLDP00000015983.1*	A0A3B4 XXH2*
Trachinotus anak	col1a1a	Unplaced	LOC139618651	XP_071330872.1	1447	49	ND	ND	ND	ND
	col1a1b	Unplaced	LOC139609208	XP_071319595.1	1450	51	ND	ND	ND	ND
	col1a2	Unplaced	LOC139633971	XP_071356254.1*	1269*	47*	ND	ND	ND	ND
Trachinotus ovatus	col1a1a	15	ND	ND	1447	49	ND	ENSTOVG00005019807	ENSTOVP00005052076.2	ND
	col1a1b	21	ND	ND	1441*	51	ND	ENSTOVG00005019040	ENSTOVP00005050586.2	ND
	col1a2	19	ND	ATI25517.1	1352	50	ND	ENSTOVG00005018181	ENSTOVP00005044973.2	ND
Echeneis naucrates	col1a1a	8	LOC115048217	XP_029365427.1	1447	49	UPI0011141F49	ENSENLG00000001956	ENSENLP00000023971.1	A0A665UX31
	col1a1b	19	LOC115059530	XP_029382891.1	1448	51	UPI001113CB3B	ENSENLG00000015172	ENSENLP00000038570.1*	A0A665W40 7*
	col1a2	11	LOC115050611	XP_029369389.1	1352	50	UPI001113853A	ENSENLG00000008664	ENSENLP00000026470.1*	A0A665V4 C6*
Xiphias gladius	col1a1a	LG3	LOC120807554	XP_040015681.1	1447	49	UPI001A987433	ENSXGLG00000015994	ENSXGLP00000051437.1	ND
	col1a1b	LG9	LOC120794174	XP_039990881.1	1449	51	UPI001A9A244C	ENSXGLG00000010378	ENSXGLP00000030515.1	ND
	col1a2	LG24	LOC120786179	XP_039977337.1	1352	50	UPI001A986594	ENSXGLG00000005940	ENSXGLP00000016973.1	ND
Cynoglossus semilaevis	col1a1a	17	LOC103393603	XP_008328838.1	1446	49	UPI0004977E18	ENSCSEG00000012919	ENSCSEP00000020695.1*	A0A3 P8W833*
	col1a1b	8	LOC103381814	XP_008312583.1	1447	51	UPI00049732AF	ENSCSEG00000014131	ENSCSEP00000022472.1*	A0A3 P8W7H0*
	col1a2	18	LOC103394149	XP_008329548.1	1352	50	UPI0004958640	ENSCSEG00000012737	ENSCSEP00000019953.1*	A0A3P8VXI7*
Paralichthys olivaceus	col1a1a	5	LOC109641372	XP_019961379.1	1447	49	UPI000049B90F	ENSPOLG00010013781	ENSPOLP00010048932.1	Q5NT96
	col1a1b	21	LOC109640645	XP_019960310.1	1447	51	UPI00097DC873	ENSPOLG00010018473	ENSPOLP00010055413.1	ND

	col1a2	20	LOC109631736	XP_069372114.1	1351	50	UPI00375104AC	ENSPOLG00010016226	ENSPOLP00010044784.1	ND
Hippoglossus hippoglossus	col1a1a	8	LOC117765964	XP_034448504.1	1447	49	UPI00148E52C3	ENSHHIG00000013963	ENSHHIP00000051154.1*	ND
	col1a1b	19	LOC117753054	XP_034426781.1*	1450	51	UPI00148C01FC	ENSHHIG00000020002	ENSHHIP00000055513.1	ND
	col1a2	11	LOC117770892	XP_034456600.1	1352	50	UPI00148C8AA9	ENSHHIG00000006740	ENSHHIP00000020331.1	ND
Hippoglossus stenolepis	col1a1a	16	LOC118122763	XP_035035491.1	1447	49	UPI00148E52C3	ENSHSTG00010014080	ENSHSTP00010036914.1	ND
	col1a1b	21	LOC118100704	XP_035001935.1	1450	51	UPI00159C3B54	ENSHSTG00010009995	ENSHSTP00010028705.1	ND
	col1a2	17	LOC118124348	XP_035037911.2	1352	50	UPI001FAF7FC1	ENSHSTG00010015384	ENSHSTP00010040923.1	ND
Limanda limanda	col1a1a	17	LOC133022782	XP_060945663.1	1447	49	UPI0029C94F5C	ENSNDLG00000005379	ENSNDLP00000015136.1	ND
	col1a1b	21	LOC133027497	XP_060950492.1	1449	51	UPI0029C6BBB2	ENSNDLG00000024221	ENSNDLP00000052452.1	ND
	col1a2	19	LOC133025721	XP_060948440.1	1352	50	UPI0029C94BF0	ENSNDLG00000003658	ENSNDLP00000009400.1	ND
Platichthys flesus	col1a1a	16	LOC133970833	XP_062263880.1	1447	49	UPI001A86957B	ENSAYYG00005036085	ENSAYYP00005060376.1	ND
	col1a1b	20	LOC133975939	XP_062269958.1	1450	51	UPI002DBF01E4	ENSAYYG00005028572	ENSAYYP00005042860.1	ND
	col1a2	17	LOC133972299	XP_062265648.1*	1352	50	UPI002DBF5316*	ENSAYYG00005020930	ENSAYYP00005016597.1	ND
Platichthys stellatus	col1a1a	5	ND	UUW46935.1	1447	49	UPI0021F9294E	ENSSPLG00000018579	ENSSPLP00000039985.1	ND
	col1a1b	7	ND	ND	1450	51	ND	ENSSPLG00000021618	ENSSPLP00000043466.1	ND
	col1a2	15	ND	ND	1352	50	ND	ENSSPLG00000013690	ENSSPLP00000033196.1	ND
Pleuronectes platessa	col1a1a	16	LOC128458624	XP_053299497.1	1447	49	UPI00232A36A4	ENSOESG00000018527	ENSOESP00000033495.1	ND
	col1a1b	21	LOC128426720	XP_053269725.1	1450	51	UPI00232A495D	ENSOESG00000025801	ENSOESP00000042621.1	ND
	col1a2	18	LOC128461659	XP_053302663.1	1352	50	UPI00232A42AE	ENSOESG00000011105	ENSOESP00000022417.1	ND
Scophthalmus maximus	col1a1a	17	LOC118288607	XP_035470772.1	1447	49	UPI000F3058F1	ENSSMAG00025009565	ENSSMAP00025032193.1	A0A8D3B554
	col1a1b	16	LOC118287288	XP_047183839.1	1451	51	UPI000F346B33	ENSSMAG00025018438	ENSSMAP00025059618.1	ND
	col1a2	22	LOC118291958	XP_035476418.2	1352	50	UPI001FA90A68	ENSSMAG00025006002	ENSSMAP00025019241.1	ND
Solea senegalensis	col1a1a	LG19	LOC122784805	XP_043906095.1	1446	49	UPI001C422C6A	ENSSSGG00010019341	ENSSSGP00010052456.1	A0AAV6RJR8
	col1a1b	LG18	LOC122759117	XP_043869618.1	1449	51	UPI001C429D65	ENSSSGG00010009805	ENSSSGP00010024791.1	A0 AAV6RRH8
	col1a2	LG20	LOC122786914	XP_043909384.1	1353	50	UPI001C421BC2	ENSSSGG00010007347	ENSSSGP00010016982.1	A0AAV6R7S9
Solea solea	col1a1a	16	LOC131476157	XP_058510671.1	1447	49	UPI00272DA3E6	ENSXQHG00000020001	ENSXQHP00000045907.1	ND
	col1a1b	21	LOC131448244	XP_058476506.1	1449	51	UPI00272D87F0	ENSXQHG00000020458	ENSXQHP00000046383.1	ND
	col1a2	20	LOC131447168	XP_058474693.1*	1353	50	UPI00272BEA50*	ENSXQHG00000012667	ENSXQHP00000031237.1	ND
Pempheris klunzingeri	col1a1a	17	LOC139217052	XP_070704422.1	1447	49	ND	ND	ND	ND
	col1a1b	20	LOC139220349	XP_070708531.1	1447	51	ND	ND	ND	ND
	col1a2	16	LOC139215046	XP_070701969.1	1351	50	ND	ND	ND	ND
Chaetodon trifascialis	col1a1a	15	LOC139343150	XP_070836733.1	1447	49	ND	ND	ND	ND
	col1a1b	21	LOC139349377	XP_070846231.1	1450	51	ND	ND	ND	ND
	col1a2	17	LOC139346007	XP_070841017.1	1351	50	ND	ND	ND	ND
Chelmon rostratus	col1a1a	17	LOC121620257	XP_041812173.1	1447	49	UPI001BECE210	ENSCRTG00005011504	ENSCRTP00005034384.1	ND
	col1a1b	21	LOC121624434	XP_041817996.1	1450	51	UPI001BECFD9D	ENSCRTG00005013794	ENSCRTP00005035117.1	ND

(continued)

Table 8.3 (continued)

NCBI gene	Gene	Chromosome	NCBI gene	NCBI protein	aa	exons	Uniparc	Ensembl gene	Ensembl protein	Uniprot
	col1a2	16	LOC121619337	XP_041810901.1	1352	50	UPI001BE7BE27	ENSCRTG00005009801	ENSCRTP00005024418.1	ND
Micropterus dolomieu	col1a1a	LG02	LOC123983899	XP_045926300.1	1447	48	UPI001E8CDCD0	ENSMDLG00000016775	ENSMDLP00000046463.1*	ND
	col1a1b	LG14	LOC123982638	XP_045924259.1	1449	51	UPI001E8CD8D6	ENSMDLG00000005686	ENSMDLP00000012148.1	ND
	col1a2	LG09	LOC123976120	XP_045913930.1	1352	50	UPI001E8DEE52	ENSMDLG00000023284	ENSMDLP00000061219.1	ND
Micropterus salmoides	col1a1a	15	LOC119882269	XP_038548195.1	1447	48	UPI0018EBD334	ENSMSDG00010019125	ENSMSDP00010058285.1	ND
	col1a1b	21	LOC119916117	XP_038591933.1	1449	51	UPI0018EC9DC8	ENSMSDG00010018904	ENSMSDP00010052081.1	ND
	col1a2	14	LOC119906985	XP_038580425.1	1352	50	UPI0018ED7FEB	ENSMSDG00010014820	ENSMSDP00010040642.1	ND
Enoplosus armatus	col1a1a	17	LOC139300261	XP_070779392.1	1447	49	ND	ND	ND	ND
	col1a1b	20	LOC139303222	XP_070783063.1	1450	51	ND	ND	ND	ND
	col1a2	16	LOC139299445	XP_070778449.1	1355	50	ND	ND	ND	ND
Siniperca chuatsi	col1a1a	LG21	LOC122868580	XP_044036604.1	1446	49	UPI001CE15EB9	ENSSCTG00010015724	ENSSCTP00010059985.1*	ND
	col1a1b	LG20	LOC122867154	XP_044033481.1	1450	51	UPI001CE08D33	ENSSCTG00010017909	ENSSCTP00010057828.1*	ND
	col1a2	LG17	LOC122864951	XP_044028706.1	1351	50	UPI001CE1104E	ENSSCTG00010009373	ENSSCTP00010030178.1	ND
Dicentrarchus labrax	col1a1a	Unplaced	LOC127356970	XP_051245032.1	1447	49	UPI001FC3A21C	ENSDLAG00005017636	ENSDLAP00005054292.2	A0A8C4NXC0
	col1a1b	Unplaced	LOC127373888	XP_051274718.1	1450	51	UPI0021F65008	ENSDLAG00005001815	ENSDLAP00005010972.2*	A0A8C4 DY99*
	col1a2	Unplaced	LOC127357364	XP_051245476.1	1351	50	UPI0021F68009	ENSDLAG00005015255	ENSDLAP00005036943.2*	A0A8C4H4 U5*
Morone saxatilis	col1a1a	Unplaced	LOC118323043	XP_035510856.1	1458*	50*	UPI0015E1F8CA	ND	ND	ND
	col1a1b	Unplaced	LOC118334950	XP_035526952.1	1450	51	UPI0015E1C4F5	ND	ND	ND
	col1a2	Unplaced	LOC118342857	XP_035537418.1	1352	50	UPI0015E1F600	ND	ND	ND
Larimichthys crocea	col1a1a	XII	LOC104932212	XP_019117249.1	1447	49	UPI00054C0E21	ENSLCRG00035005568	ENSLCRP00035024841.1	A0A6G0HX74
	col1a1b	XVI	LOC104933943	XP_010747796.3	1451	51	UPI00054B83F6	ENSLCRG00035019998	ENSLCRP00035047713.1	A0A6G0HXS0
	col1a2	XIII	LOC104927463	XP_027142672.1	1351	50	UPI00054B778D	ENSLCRG00035018038	ENSLCRP00035045259.1	ND
Scatophagus argus	col1a1a	16	LOC124073187	XP_046271183.1	1447	49	UPI001ED84E42	ENSSAGG00005010963	ENSSAGP00005024914.1	ND
	col1a1b	21	LOC124052640	XP_046233080.1	1450	51	UPI001ED844F7	ENSSAGG00005015268	ENSSAGP00005034935.1*	ND
	col1a2	17	LOC124074134	XP_046272687.1	1352	50	UPI001ED7ECCD	ENSSAGG00005013568	ENSSAGP00005030886.1	ND
Cheilinus undulatus	col1a1a	21	LOC121529426	XP_041673197.1	1447	49	UPI001BD6CA92	ENSCUDG00000022601	ENSCUDP00000060707.1	ND
	col1a1b	20	LOC121528110	XP_041671208.1	1446	51	UPI001BD2B6EA	ENSCUDG00000017169	ENSCUDP00000045240.1	ND
	col1a2	16	LOC121523413	XP_041664241.1	1352	50	UPI001BD6C70A	ENSCUDG00000014213	ENSCUDP00000034897.1*	ND
Labrus bergylta	col1a1a	16	LOC110000581	XP_065820317.1	1447	49	UPI003313BE77	ENSLBEG00000012092	ENSLBEP00000015541.1*	A0A3Q3F6P3*
	col1a1b	21	LOC109981840	XP_020486465.2	1454	51	UPI0033138BDD	ENSLBEG00000027905	ENSLBEP00000037437.1*	A0A3Q3 GSY4*
	col1a2	19	LOC109982443	XP_065803887.1	1352	50	UPI0009B4406D	ENSLBEG00000023364	ENSLBEP00000030985.1*	A0A3Q3 NAA8*

Labrus mixtus	col1a1a	20	LOC132954267	XP_060882779.1	1447	49	UPI0029C0519F	ENSHRMG00000027860	ENSHRMP00000051249.1	ND
	col1a1b	21	LOC132955913	XP_060884988.1*	1373*	48*	UPI0029C0ED8F*	ENSHRMG00000019814	ENSHRMP00000022428.1*	ND
	col1a2	16	LOC132991442	XP_060916199.1	1352	50	UPI0029C0B561	ENSHRMG00000019266	ENSHRMP00000022512.1	ND
Notolabrus celidotus	col1a1a	20	LOC117831987	XP_034566810.1	1447	49	UPI0014900BA8	ENSNCEG00005017240	ENSNCEP00005043653.1*	ND
	col1a1b	18	LOC117830252	XP_034564180.1	1451	51	UPI00149066C4	ENSNCEG00005010703	ENSNCEP00005029274.1*	ND
	col1a2	16	LOC117827549	XP_034560025.1	1352	50	UPI00148FD9C2	ENSNCEG00005007790	ND	ND
Semicossyphus pulcher	col1a1a	Unplaced	LOC138602735	XP_069530396.1	1447	49	UPI0037E96011	ND	ND	ND
	col1a1b	Unplaced	LOC138593767	XP_069520958.1	1448	51	UPI0037E918E8	ND	ND	ND
	col1a2	Unplaced	LOC138609549	XP_069538278.1	1352	50	UPI0037E7BE4C	ND	ND	ND
Antennarius striatus	col1a1a	16	LOC137609135	XP_068192029.1	1447	49	UPI0035B099F8	ND	ND	ND
	col1a1b	21	LOC137588134	XP_068161088.1	1447	51	UPI0035AEF527	ND	ND	ND
	col1a2	14	LOC137607605	XP_068189594.1	1351	50	UPI0035B181DC	ND	ND	ND
Brachionichthys hirsutus	col1a1a	16	LOC137905598	XP_068605941.1	1447	49	UPI003604D71E	ND	ND	ND
	col1a1b	21	LOC137910012	XP_068610539.1	1448	51	UPI0036050299	ND	ND	ND
	col1a2	13	LOC137903531	XP_068603783.1	1351	50	UPI0036044909	ND	ND	ND
Anoplopoma fimbria	col1a1a	11	LOC129097917	XP_054462789.1	1447	49	UPI0023EA8796	ND	ND	ND
	col1a1b	5	LOC129091393	XP_054454966.1	1450	51	UPI0023EBE487	ND	ND	ND
	col1a2	10	LOC129096791	XP_054461410.1	1351	50	UPI0023EB766B	ND	ND	ND
Clinocottus analis	col1a1a	Unplaced	LOC137786647	XP_068434761.1	1447	49	UPI0035BEE25C	ND	ND	ND
	col1a1b	Unplaced	LOC137789208	XP_068438627.1	1451	51	UPI0035C01714	ND	ND	ND
	col1a2	Unplaced	LOC137784539	XP_068431403.1*	1269*	47*	UPI0035C0AB0C	ND	ND	ND
Cyclopterus lumpus	col1a1a	8	LOC117735334	XP_034395777.1	1446	49	UPI001486973C	ENSCLMG00020018948	ENSCLMP00020051732.1	A0A8C2ZG29*
	col1a1b	19	LOC117749049	XP_034415142.1	1451	51	UPI0014865D64	ENSCLMG00020002597	ENSCLMP00020006995.1	A0A8C2XG12*
	col1a2	11	LOC117739063	XP_034401231.1	1352	50	UPI001485DBA6	ENSCLMG00020003359	ENSCLMP00020009540.1	A0A8C2ZMT0*
Gasterosteus aculeatus aculeatus	col1a1a	11	LOC120827772	XP_040046813.1	1447	49	UPI0000E3CB66	ENSGACG00000005143	ENSGACP00000006903.2	G3NNJ9
	col1a1b	5	LOC120819786	XP_040033420.1	1445	51	UPI001A9A1070	ENSGACG00000003361	ENSGACP00000004433.2	G3NGH7
	col1a2	10	LOC120826558	XP_040044885.1	1353	50	UPI001A97EC41	ENSGACG00000006511	ENSGACP00000008672.2	G3NTK4
Pungitius pungitius	col1a1a	12	LOC119220242	XP_037331988.2	1447	49	UPI002E14CD65	ENSPPNG00035029156	ENSPPNP00035045191.1	ND
	col1a1b	21	LOC119212855	XP_037319573.2	1442	51	UPI002E156964	ENSPPNG00035017233	ENSPPNP00035011377.1	ND
	col1a2	17	LOC119219285	XP_062414433.1	1352	50	UPI002E137B0E	ENSPPNG00035023618	ENSPPNP00035031162.1	ND
Pseudoliparis swirei	col1a1a	23	LOC130188200	XP_056262384.1	1448	49	UPI0024BE06DB	ENSRIWG00000005580	ENSRIWP00000021240.1	ND
	col1a1b	13	LOC130203252	XP_056285209.1	1455	51	UPI0024BE3438	ENSRIWG00000003345	ENSRIWP00000007166.1	ND
	col1a2	1	LOC130197371	XP_056276001.1*	1351	50	UPI0024BEAA14*	ENSRIWG00000002333	ENSRIWP00000005031.1	ND

(continued)

Table 8.3 (continued)

NCBI gene	Gene	Chromosome	NCBI gene	NCBI protein	aa	exons	Uniparc	Ensembl gene	Ensembl protein	Uniprot
Anarrhichthys ocellatus	col1a1a	Unplaced	LOC116399239	XP_031732101.1	1447	49	UPI0012EE66EE	ND	ND	ND
	col1a1b	Unplaced	LOC116381468	XP_031699507.1	1454	51	UPI0012EDC5B6	ND	ND	ND
	col1a2	Unplaced	LOC116387113	XP_031709775.1	1352	50	UPI0012ED34FF	ND	ND	ND
Cebidichthys violaceus	col1a1a	Unplaced	LOC137888223	XP_068587672.1	1447	49	UPI0035CC9E15	ND	ND	ND
	col1a2	Unplaced	LOC137872695	XP_068567169.1	1351	50	UPI0035CA3913	ND	ND	ND
Gymnodraco acuticeps	col1a1a	Unplaced	LOC117534517	XP_034054619.1	1447	49	UPI00147128D7	ENSGAUG00000023360	ENSGAUP00000060121.1*	A0A6P8SUL8
	col1a1b	Unplaced	LOC117547364	XP_034073947.1	1447	51	UPI0014721052	ENSGAUG00000025409	ENSGAUP00000058390.1*	A0A6P8VA93
	col1a2	Unplaced	LOC117562871	XP_034096707.1	1352	50	UPI0014714D90	ENSGAUG00000020457	ENSGAUP00000045517.1*	A0A6P8X9E2
Cottoperca gobio	col1a1a	8	LOC115012626	XP_029294185.1	1447	49	UPI00110E66E0	ENSCGOG00000017849	ENSCGOP00000043849.1*	A0A6J2Q9U5
	col1a1b	19	LOC115024512	XP_029312017.1	1449	51	UPI00110E658A	ENSCGOG00000009581	ENSCGOP00000029503.1*	A0A6J2RRG5
	col1a2	11	LOC115015492	XP_029298708.1	1353	50	UPI00110E2011	ENSCGOG00000014901	ENSCGOP00000032760.1	A0A6J2QJX6
Chaenocephalus aceratus	col1a1a	Unplaced	KUCAC02_ 026728	KAI4796819.1	1447	49	ND	ND	ND	ND
	col1a1b	19	KUCAC02_ 027384	KAI4807582.1	1450	51	UPI0021982C63	ND	ND	ND
	col1a2	11	KUCAC02_ 010902	KAI4817511.1	1164*	39*	UPI00218DF296	ND	ND	ND
Pseudochaenichthys georgianus	col1a1a	Unplaced	LOC117444715	XP_033936030.1	1447	49	UPI00146D1E8E	ENSPGEG00000004549	ENSPGEP00000065055.1*	ND
	col1a1b	19	LOC117464657	XP_033963088.1	1450	51	UPI00146C2713	ENSPGEG00000016672	ENSPGEP00000045851.2	ND
	col1a2	11	LOC117455300	XP_033950637.1	1352	50	UPI00146C7EDE	ENSPGEG00000024269	ENSPGEP00000060870.1	ND
Eleginops maclovinus	col1a1a	16	LOC134877997	XP_063759805.1	1447	49	UPI00307FD941	ENSEMLG00005002817	ENSEMLP00005011469.1*	ND
	col1a1b	10	LOC134870754	XP_063749211.1	1446	51	UPI0030803F7D	ENSEMLG00005000399	ENSEMLP00005001443.1*	ND
	col1a2	13	LOC134874809	XP_063755091.1	1352	50	UPI003080870B	ENSEMLG00005017420	ENSEMLP00005048535.1*	ND
Notothenia coriiceps	col1a1a	Unplaced	LOC104944135	XP_010767913.1	1317*	47*	UPI00054FD27E	ND	ND	A0 A6I9MX97*
	col1a1b	Unplaced	LOC104945045	XP_010768975.1	1448	51	UPI00054F7560	ND	ND	A0A6I9N4J7
	col1a2	Unplaced	LOC104948445	XP_010772950.1	1352	50	UPI00055362AC	ND	ND	A0A6I9NEZ5
Notothenia rossii	col1a1a	8	ND	ND	1444	49	ND	ENSNROG00005012398	ENSNROP00005022078.1	ND
	col1a1b	11	ND	ND	1448	51	ND	ENSNROG00005016264	ENSNROP00005026371.1	ND
	col1a2	10	ND	ND	1322*	49*	ND	ENSNROG00005030090	ENSNROP00005047495.1*	ND
Trematomus bernacchii	col1a1a	Unplaced	LOC117499360	XP_034007595.1	1389*	48*	UPI00146ADA60*	ENSTBNG00005003318	ENSTBNP00005005902.1*	ND
	col1a1b	Unplaced	LOC117473789	XP_033975555.1	1447	51	UPI00146D6AEE	ENSTBNG00005010587	ENSTBNP00005024355.1*	ND
	col1a2	Unplaced	LOC117477721	XP_033980337.1	1352	50	UPI00146F6EA2	ENSTBNG00005021538	ENSTBNP00005049430.1*	ND
Etheostoma cragini	col1a1a	15	LOC117958710	XP_034751171.1	1447	49	UPI00155E56EB	ND	ND	ND
	col1a1b	21	LOC117937362	XP_034717160.1	1455	51	UPI00155E6E5B	ND	ND	ND

	col1a2	14	LOC117956680	XP_034747766.1	1352	50	UPI00155E462E	ND	ND	ND
Etheostoma spectabile	col1a1a	15	LOC116703513	XP_032394176.1	1447	49	UPI0013AE96CD	ND	ND	ND
	col1a1b	21	LOC116671215	XP_032358182.1	1455	51	UPI0013AEA57E	ND	ND	ND
	col1a2	14	LOC116701464	XP_032391070.1	1352	50	UPI0013AE8B76	ND	ND	ND
Sander lucioperca	col1a1a	2	LOC116045007	XP_031148362.1	1447	49	UPI00125E4342	ENSSLUG00000024030	ENSSLUP00000056681.2	ND
	col1a1b	22	LOC116061156	XP_035853346.1	1451	51	UPI001653E92D	ENSSLUG00000000479	ENSSLUP00000001160.2	ND
	col1a2	16	LOC116066927	XP_031179004.1	1352	50	UPI00125D4FA8	ENSSLUG00000014735	ENSSLUP00000037655.2	ND
Sander vitreus	col1a1a	15	ND	ND	1447	49	ND	ENSSVIG00005016474	ENSSVIP00005050436.1	ND
	col1a1b	21	ND	ND	1453	51	ND	ENSSVIG00005009124	ENSSVIP00005050436.1	ND
	col1a2	14	ND	ND	1352	50	ND	ENSSVIG00005012060	ENSSVIP00005033112.1	ND
Perca flavescens	col1a1a	15	LOC114569989	XP_028455973.1	1447	49	UPI00106E471E	ND	ND	A0A484CIB2
	col1a1b	21	LOC114548727	XP_028424561.1	1450	51	UPI00106ED070	ND	ND	ND
	col1a2	14	LOC114567629	XP_028452516.1	1352	50	UPI00106E2B07	ND	ND	ND
Perca fluviatilis	col1a1a	15	LOC120574258	XP_039680441.1	1447	49	UPI00144B9BB7	ENSPFLG00020018179	ENSPFLP00020044673.1	A0A6A5EZU1
	col1a1b	21	LOC120551189	XP_039644344.1	1451	51	UPI00144C1D9F	ENSPFLG00020013767	ENSPFLP00020031751.1	A0A6A5ECB7
	col1a2	13	LOC120571573	XP_039676540.1	1352	50	UPI0019645089	ENSPFLG00020024341	ENSPFLP00020059341.1*	ND
Sebastes umbrosus	col1a1a	14	LOC119502225	XP_037648990.1	1447	49	UPI00189DECCF	ENSSUBG00005011847	ENSSUBP00005031838.1	ND
	col1a1b	20	LOC119479164	XP_037610383.1	1453	51	UPI0018A03D58	ENSSUBG00005006670	ENSSUBP00005045291.1	ND
	col1a2	15	LOC119502985	XP_037650307.1	1352	50	UPI00189FB902	ENSSUBG00005006129	ENSSUBP00005047852.1*	ND
Sebastes schlegelii	col1a1a	14	ND	ND	1447	49	ND	ENSSSHG00005002167	ENSSSHP00005006945.1	ND
	col1a1b	18	ND	ND	1453	51	ND	ENSSSHG00005008276	ENSSSHP00005022044.1	ND
	col1a2	16	ND	ND	1352	50	ND	ENSSSHG00005003946	ENSSSHP00005010154.1	ND
Epinephelus fuscoguttatus	col1a1a	LG19	LOC125879272	XP_049416991.1	1447	49	UPI0020D0B06C	ENSEFSG00000021053	ENSEFSP00000065442.1	ND
	col1a1b	LG20	LOC125880323	XP_049418603.1	1451	51	UPI0020CFEB8F	ENSEFSG00000003183	ENSEFSP00000021735.1	ND
	col1a2	LG17	LOC125904181	XP_049457373.1*	1352	50	UPI0020D12747*	ENSEFSG00000006908	ENSEFSP00000019819.1	ND
Epinephelus lanceolatus	col1a1a	16	LOC117269031	XP_033501735.1	1447	49	UPI0014489117	ENSELGG00000003680	ENSELGP00000016613.1	ND
	col1a1b	21	LOC117246975	XP_033466929.1	1452	51	UPI001447B5BA	ENSELGG00000011357	ENSELGP00000038803.1	ND
	col1a2	12	LOC117263211	XP_033492304.1	1352	50	UPI001447136D	ENSELGG00000012300	ENSELGP00000027055.1	ND
Epinephelus moara	col1a1a	18	LOC126405546	XP_049925274.1	1447	49	UPI00214EF8CB	ENSEMOG00000017407	ENSEMOP00000043538.1	ND
	col1a1b	13	LOC126399391	XP_049915244.1	1452	51	UPI00214F0BFC	ENSEMOG00000004840	ENSEMOP00000017949.1	ND
	col1a2	22	LOC126384103	XP_049890956.1	1356	49	UPI00214F6073	ENSEMOG00000007315	ENSEMOP00000016568.1	ND
Plectropomus leopardus	col1a1a	17	LOC121956874	XP_042361230.1	1447	49	UPI001C4B6735	ENSPLRG00005019250	ENSPLRP00005053020.1	ND
	col1a1b	19	LOC121958474	XP_042363352.1	1452	51	UPI001C4ADCB4	ENSPLRG00015018568	ENSPLRP00015054183.1*	ND
	col1a2	18	LOC121957433	XP_042361939.1	1352	50	UPI001C4D0872	ENSPLRG00015015818	ENSPLRP00015049313.1	ND
Centropristis striata	col1a1a	13	LOC131983285	XP_059203972.1	1447	49	UPI0027DEE15E	ENSPUXG00000003263	ENSPUXP00000022923.1	ND
	col1a1b	21	LOC131959857	XP_059181061.1	1452	50	UPI0027E0D1EC	ENSPUXG00000003647	ENSPUXP00000014205.1	ND
	col1a2	14	LOC131984747	XP_059205669.1	1352	50	UPI0027DF5031	ENSPUXG00000028248	ENSPUXP00000056837.1*	ND

(continued)

Table 8.3 (continued)

NCBI gene	Gene	Chromosome	NCBI gene	NCBI protein	aa	exons	Uniparc	Ensembl gene	Ensembl protein	Uniprot
Acanthopagrus latus	col1a1a	23	LOC119014162	XP_036945008.1	1445	49	UPI00187BE94A	ENSALTG00010015300	ENSALTP00010042524.1	ND
	col1a1b	20	LOC119010173	XP_036938011.1	1448	51	UPI00187BE062	ENSALTG00010002039	ENSALTP00010003659.1	ND
	col1a2	3	LOC119015386	XP_036947233.1	1352	50	UPI00187C4154	ENSALTG00010019175	ENSALTP00010052131.1	ND
Sparus aurata	col1a1a	23	LOC115575148	XP_030262871.1	1445	49	UPI0011C1C59D	ENSSAUG00010019190	ENSSAUP00010064601.1	A0A671YMN4
	col1a1b	20	LOC115570781	XP_030255366.1	1448	51	UPI0011C13196	ENSSAUG00015011773	ENSSAUP00015030171.1	A0A671 YFV1*
	col1a2	3	LOC115578305	XP_030267061.1	1352	50	UPI0011C0D3A5	ENSSAUG00010001273	ENSSAUP00010004019.1	A0A671TPX8
Takifugu flavidus	col1a1a	18	LOC130515205	XP_056871322.1	1448	49	UPI0025442A80	ENSTFLG00005022470	ENSTFLP00005047332.1	ND
	col1a1b	1	LOC130539601	XP_056914071.1	1448	51	UPI00254467DD	ENSTFLG00005018632	ENSTFLP00005041028.1	ND
	col1a2	21	LOC130517856	XP_056876016.1	1355	50	UPI00254454CC	ENSTFLG00005018801	ENSTFLP00005039676.1*	ND
Takifugu rubripes	col1a1a	5	LOC101076114	XP_003964321.2	1448	49	UPI001145F7BB	ENSTRUG00000013913	ENSTRUP00000058951.1	A0A674MC31
	col1a1b	1	LOC101063107	XP_003961059.2	1448	51	UPI0011456110	ENSTRUG00000007520	ENSTRUP00000018631.3	H2T1U3
	col1a2	12	LOC101070597	XP_011607785.1	1349	50	UPI0005D28092	ENSTRUG00000015407	ENSTRUP00000039390.3	H2UR17*
Takifugu bimaculatus	col1a1a	16	ND	ND	1448	49	UPI00116E6039	ENSTBMG00005011992	ENSTBMP00005024602.1	A0A4Z2C1B8
	col1a1b	1	ND	ND	1345*	48*	ND	ENSTBMG00005012490	ENSTBMP00005019893.1	ND
	col1a2	20	ND	ND	1349	50	ND	ENSTBMG00005024816	ENSTBMP00005037521.1	ND
Tetraodon nigroviridis	col1a1a	Unplaced	ND	CAG02093.1*	1459*	49	UPI00017B46AB	ENSTNIG00000011168	ENSTNIP00000002716.1	H3C397
	col1a1b	2	ND	CAG06575.1*	1455	51	UPI00017B3A84	ENSTNIG00000014440	ENSTNIP00000017459.1	H3DAB8
	col1a2	21	ND	CAG11117.1*	1352	50	UPI00017B0A04	ENSTNIG00000017765	ENSTNIP00000002806.1	H3C3I7
Mugilogobius chulae	col1a1a	2	AAG832_ 002644	KAL3562258.1	1446	48	UPI00312FF106	ND	ND	A0AAW0 PU51
	col1a1b	7	AAG832_ 010541	KAL3556445.1	1345*	48*	UPI0031311E83	ND	ND	A0AAW0 PBA4*
	col1a2	14	AAG832_ 019628	KAL3549243.1	1876*	49*	UPI00312DFA30	ND	ND	A0AAW0 NJQ0*
Boleophthalmus pectinirostris	col1a1a	Unplaced	LOC110162385	XP_055008232.1	1446	48	UPI000A1C5CCC	ND	ND	ND
	col1a1b	Unplaced	LOC110154336	XP_055006378.1	1445	50	UPI0024315785	ND	ND	ND
	col1a2	Unplaced	LOC110174114	XP_055013243.1	1350	51	UPI0024302D64	ND	ND	ND
Periophthalmus magnuspinnatus	col1a1a	8	LOC117375240	XP_033827407.1	1446	48	UPI00145B88AE	ENSPMGG00005001624	ENSPMGP00005003301.1	ND
	col1a1b	19	LOC117387832	XP_033841296.1	1445	50	UPI00145A8D8A	ENSPMGG00005019715	ENSPMGP00005044588.1	ND
	col1a2	11	LOC117393397	XP_055081275.1	1352	50	UPI0024372DBC	ENSPMGG00005000063	ENSPMGP00005000049.1	ND
Sphaeramia orbicularis	col1a1a	Unplaced	LOC115415944	XP_029985486.1	1447	49	UPI0011811F96	ENSSORG00005017731	ENSSORP00005037786.1*	ND
	col1a1b	19	LOC115410411	XP_029977918.1	1448	51	UPI00117F091F	ENSSORG00005005390	ENSSORP00005024101.1*	ND
	col1a2	11	LOC115428582	XP_030003567.1	1351	50	UPI0011811B49	ENSSORG00005011598	ENSSORP00005029627.1*	ND

Melanotaenia boesemani	col1a1a	2	LOC121648284	XP_041854212.1*	1464*	50*	UPI001C048FC9	ENSMBOG00000001547	ENSMBOP00000003016.1*	ND
	col1a1b	21	LOC121632354	XP_041829721.1	1445	51	UPI001C042E0B	ENSMBOG00000015110	ENSMBOP00000035935.1	ND
	col1a2	17	LOC121656843	XP_041867948.1	1352	50	UPI001C041128	ENSMBOG00000009511	ENSMBOP00000020741.1*	ND
Oryzias javanicus	col1a1a	8	OJAV_ G00073390	RVE68979.1	1445	49	UPI000FDFF226	ENSOJAG00000007607	ND	A0A3S2Q3J5
	col1a1b	19	OJAV_ G00192700	RVE59944.1*	1305*	37*	UPI000 FDFCBFE*	ENSOJAG00000023691	ENSOJAP00000049051.1*	A0A3S2 MIK4*
	col1a2	11	OJAV_ G00112400	RVE66956.1*	1290*	45*	UPI0023FF7B94*	ENSOJAG00000009445	ENSOJAP00000021826.1*	ND
Oryzias latipes	col1a1a	8	LOC100144368	NP_001116390.2	1446	49	UPI0000EA1020	ENSORLG00000017013	ENSORLP00000021285.1	H2MRA6
	col1a1b	19	LOC101162163	XP_004080643.2	1446	51	UPI0005CC19E1	ENSORLG00000011273	ENSORLP00000014129.2*	H2M6N2*
	col1a2	11	LOC105355091	XP_023815576.1	1356	50	UPI0009D9BF7E	ENSORLG00000004319	ENSORLP00000042954.1	A0A3B3IGL8
Oryzias melastigma	col1a1a	Unplaced	LOC112138211	XP_024116552.1	1446	49	UPI000CF8136E	ENSOMEG00005000194	ENSOMEP00005001268.1	A0A834BUP3
	col1a1b	LG19	LOC112154218	XP_024140767.1	1446	51	UPI000CF7B3F5	ENSOMEG00005003428	ENSOMEP00005015095.1	ND
	col1a2	LG11	LOC112162784	XP_024154448.1	1356	50	UPI000CF7E425	ENSOMEG00005017166	ENSOMEP00005040607.1	ND
Oryzias sinensis	col1a1a	Unplaced	ND	ND	1446	49	UPI0016365565	ENSOSIG00000005123	ENSOSIP00000007814.1	A0A8C7X540
	col1a1b	Unplaced	ND	ND	1319*	37*	UPI001636C6A2	ENSOSIG00000019541	ENSOSIP00000040066.1	A0A8C7ZDC1
	col1a2	Unplaced	ND	ND	1356	49	UPI00162F52A8	ENSOSIG00000020542	ENSOSIP00000046005.1	A0A8C7ZNY6
Oryzias curvinotus	col1a1a	17	ND	ND	1479*	48*	ND	ENSOCVG00000016404	ENSOCVG00000016404	ND
	col1a1b	19	ND	ND	1446	51	ND	ENSOCVG00000013832	ENSOCVP00000033257.1	ND
	col1a1c		ND	ND	1356	50	ND	ENSOCVG00000006486	ENSOCVP00000029609.1	ND
Cololabis saira	col1a1a	21	LOC133422776	XP_061568850.1	1445	49	UPI002AD510E9	ENSLTZG00000023037	ENSLTZP00000047185.1	ND
	col1a1b	19	LOC133419443	XP_061564615.1	1446	51	UPI002AD49796	ENSLTZG00000007131	ENSLTZP00000003937.1*	ND
	col1a2	15	LOC133461598	XP_061598661.1	1361	49	UPI002AD5940C	ENSLTZG00000034463	ENSLTZP00000059092.1*	ND
Austrofundulus limnaeus	col1a1a	Unplaced	LOC106536191	XP_013888853.1	1447	49	UPI0006B3997F	ND	ND	A0A2I4D9D6
	col1a1b	Unplaced	LOC106535181	XP_013887577.1	1453	51	UPI0006B3AFA0	ND	ND	A0A2I4D5P7
	col1a2	Unplaced	LOC106519536	XP_013866709.1	1352	50	UPI0006B365D7	ND	ND	A0A2I4BG36
Nothobranchius furzeri	col1a1a	18	LOC107391613	XP_015824467.1	1447	49	UPI00077D1560	ENSNFUG00015006971	ND	ND
	col1a1b	16	LOC107387680	XP_015818276.1	1453	51	UPI00077D195F	ENSNFUG00015024901	ENSNFUP00015054324.1	A0A8C6W0P7
	col1a2	11	LOC107383302	XP_070397560.1	1352	50	UPI00077D3D1D	ENSNFUG00015024212	ENSNFUP00015051781.1	A0A1 A8UMG6
Nematolebias whitei	col1a1a	15	LOC119421425	XP_037544711.1	1447	49	UPI0018999FFE	ND	ND	ND
	col1a1b	17	LOC119424068	XP_037547466.1	1454	51	UPI001896F257	ND	ND	ND
	col1a2	19	LOC119425425	XP_037548878.1	1353	50	UPI00189A240B	ND	ND	ND
Kryptolebias marmoratus	col1a1a	LG12	LOC108233705	XP_017267795.1	1447	49	UPI0007F89ECA	ENSKMAG00000009262	ENSKMAP00000012327.1	A0A3Q3FGS3
	col1a1b	LG17	LOC108248231	XP_017292349.1	1452	51	UPI0007F8E62E	ENSKMAG00010004468	ENSKMAP00010012472.1	A0A3Q3A2 N5*
	col1a2	LG5	LOC108240667	XP_017279795.1	1352	50	UPI0007F8E75C	ENSKMAG00010014921	ENSKMAP00010033490.1*	

(continued)

Table 8.3 (continued)

NCBI gene	Gene	Chromosome	NCBI gene	NCBI protein	aa	exons	Uniparc	Ensembl gene	Ensembl protein	Uniprot
										A0A3Q2 ZBK2*
Kryptolebias hermaphroditus	col1a1a	5	ND	ND	1447	49	ND	ENSKHEG00000000573	ENSKHEP00000001664.1	ND
	col1a1b	17	ND	ND	1452	51	ND	ENSKHEG00000002124	ENSKHEP00000003559.1	ND
	col1a2	5	ND	ND	1352	50	ND	ENSKHEG00000007070	ENSKHEP00000015889.1	ND
Cyprinodon tularosa	col1a1a	Unplaced	LOC119792170	XP_038154601.1	1445	49	UPI0018E241AB	ND	ND	ND
	col1a1b	Unplaced	LOC119791079	XP_038152857.1	1447	51	UPI0018E1F1B2	ND	ND	ND
	col1a2	Unplaced	LOC119785888	XP_038144702.1	1349	50	UPI0018E24850	ND	ND	ND
Cyprinodon variegatus	col1a1a	Unplaced	LOC107097738	XP_015250501.1	1445	49	UPI000742BF89	ENSCVAG00000010540	ENSCVAP00000008692.1	A0A3Q2CSU4
	col1a1b	Unplaced	LOC107102379	XP_015257153.1	1447	51	UPI000742C857	ENSCVAG00000012073	ENSCVAP00000026422.1*	A0A3Q2E2 Q0*
	col1a2	Unplaced	LOC107085874	XP_015231985.1	1349	50	UPI000742865B	ENSCVAG00000015066	ENSCVAP00000012535.1*	A0A3Q2 FX51*
Cyprinodon diabolis	col1a1a	16	ND	ND	1445	49	ND	ENSHOPG00000010233	ENSHOPP00000027936.1	ND
	col1a1b	10	ND	ND	1447	51	ND	ENSHOPG00000012439	ENSHOPP00000026029.1	ND
	col1a2	13	ND	ND	1349	50	ND	ENSHOPG00000013915	ENSHOPP00000030306.1	ND
Fundulus heteroclitus	col1a1a	16	LOC105925362	XP_012716616.2	1445	49	UPI00165B2C0B	ENSFHEG00000009359	ENSFHEP00000008090.2	A0A3Q2 P8B7*
	col1a1b	10	LOC105932502	XP_021176749.2	1446	51	UPI00165C02AA	ENSFHEG00000020368	ENSFHEP00000044531.1	A0A3Q2 PWY4*
	col1a2	13	LOC105917032	XP_012707154.2	1352	49	UPI00165CD176	ENSFHEG00000012802	ENSFHEP00000053088.1	A0A3Q2 QLQ8*
Girardinichthys multiradiatus	col1a1a1	X	LOC124862227	XP_047211997.1	1442	49	UPI001FAE471D	ENSGMLG00005005245	ENSGMLP00005009238.1*	ND
	col1a1a2	Y	LOC124864205	XP_047214839.1	1442	49	UPI001FABCA8E	ENSGMLG00005020598	ENSGMLP00005036461.1*	ND
	col1a1b	10	LOC124874764	XP_047232232.1	1446	51	UPI001FACA87B	ENSGMLG00005004444	ENSGMLP00005007824.1*	ND
	col1a2	13	LOC124879892	XP_047240669.1	1352	50	UPI001FADCFAC	ENSGMLG00005013689	ENSGMLP00005024144.1*	ND
Gambusia affinis	col1a1a	LG19	LOC122821247	XP_043954940.1	1445	49	UPI001CDB91E0	ENSGAFG00005013942	ENSGAFP00005047087.1	ND
	col1a1b	LG20	LOC122823113	XP_043958322.1	1446	51	UPI001CDB9BBF	ENSGAFG00005021053	ENSGAFP00005059970.1	ND
	col1a2	LG14	LOC122843269	XP_043993847.1	1351	50	UPI001CDCA457	ENSGAFG00005012713	ENSGAFP00005035626.1	ND
Poecilia formosa	col1a1a	16	LOC103147239	XP_007565541.1	1445	49	UPI0004447156	ENSPFOG00000017127	ENSPFOP00000017292.3	A0A0 87YGY9*
	col1a1b	10	LOC103151433	XP_007571557.1	1447	51	UPI0004439CD6	ENSPFOG00000015217	ENSPFOP00000015896.3	A0A0 87YCZ3*
	col1a2	13	LOC103152520	XP_007573076.1	1351	50	UPI000443FAE4	ENSPFOG00000012624	ENSPFOP00000013100.3*	A0A087Y4Z7*
Poecilia latipinna	col1a1a	Unplaced	LOC106941728	XP_014880315.1	1445	49	UPI00072EDB02	ENSPLAG00000009482	ENSPLAP00000007111.1*	A0A3B3 U341*

	col1a1b	10	LOC106946740	XP_014886816.1	1447	51	UPI00072DC6A4	ENSPLAG00000021746	ENSPLAP00000017031.1*	A0A3B3 UWI7*
	col1a2	13	LOC106952337	XP_014895871.1	1351	50	UPI00072EAF20	ENSPLAG00000009652	ENSPLAP00000007252.1*	A0A3B3U3 S4*
Poecilia mexicana	col1a1a	Unplaced	LOC106906588	XP_014827396.1	1445	49	UPI0004447156	ENSPMEG00000012374	ENSPMEP00000027662.1*	A0A3B3YK0 5*
	col1a1b	Unplaced	LOC106932422	XP_014866612.1	1447	51	UPI00072E1906	ENSPMEG00000023263	ENSPMEP00000019904.1*	A0A3B3 XXT0*
	col1a2	Unplaced	LOC106923972	XP_014852900.1	1351	50	UPI00072E46DB	ENSPMEG00000000390	ENSPMEP00000030684.1*	A0A3B3 YTU0*
qPoecilia reticulata	col1a1a	LG8	LOC103468138	XP_008413225.1	1445	49	UPI0004A3FDD9	ENSPREG00000006150	ENSPREP00000009056.1*	A0A3 P9NHJ7*
	col1a1b	LG19	LOC103481655	XP_008435501.1	1447	51	UPI0004A40D6B	ENSPREG00000010085	ENSPREP00000014997.1*	A0A3P9NZJ7*
	col1a2	LG11	ND	ND	1282*	47*	UPI000F361493*	ENSPREG00000002508	ENSPREP00000003635.1*	A0A3P9N261*
Poeciliopsis prolifica	col1a1a	Unplaced	LOC129354930	XP_054880454.1	1445	49	UPI002413F185	ND	ND	ND
	col1a1b	Unplaced	LOC129367965	XP_054898665.1	1446	51	UPI0024134972	ND	ND	ND
	col1a2	Unplaced	LOC129378566	XP_054914852.1	1355	50	UPI0024142463	ND	ND	ND
Xiphophorus couchianus	col1a1a	7	LOC114147802	XP_027878237.1	1445	49	UPI001015E6C2	ND	ND	ND
	col1a1b	10	LOC114151867	XP_027885139.1	1446	51	UPI001016AB93	ND	ND	ND
	col1a2	13	LOC114155637	XP_027891419.1	1351	50	UPI0010171B6B	ND	ND	ND
Xiphophorus hellerii	col1a1a	16	LOC116735097	XP_032442604.1	1445	49	UPI0013B476EF	ND	ND	ND
	col1a1b	10	LOC116727223	XP_032430391.1	1446	51	UPI0013B477B9	ND	ND	ND
	col1a2	13	LOC116730825	XP_032436216.1	1351	50	UPI0013B37F95	ND	ND	ND
Xiphophorus maculatus	col1a1a	16	LOC102232000	XP_023204469.1	1445	49	UPI000C6D7D90	ENSXMAG00000016028	ENSXMAP00000026914.1*	A0A3B5Q9I1*
	col1a1b	10	LOC102229978	XP_023196785.1	1446	51	UPI000293BA40	ENSXMAG00000017341	ENSXMAP00000017599.2*	M4AT04*
	col1a2	13	LOC102219489	XP_023200744.1	1351	50	UPI000C6E2D3A	ENSXMAG00000000698	ENSXMAP00000000701.2	M3ZEQ9*
Xiphophorus birchmanni	col1a1a	16	ND	ND	1445	49	ND	ENSJTHG00000013374	ENSJTHP00000035529.1	ND
	col1a1b	10	ND	ND	1446	51	ND	ENSJTHG00000021681	ENSJTHP00000053739.1	ND
	col1a2	13	ND	ND	1266*	45*	ND	ENSJTHG00000006462	ENSJTHP00000014776.1*	ND
Salarias fasciatus	col1a1a	4	LOC115386887	XP_029945204.1	1325*	48*	UPI0011765ABD	ENSSFAG00005018859	ENSSFAP00005037832.1*	A0A672I9J7*
	col1a1b	8	LOC115393090	XP_029953772.1	1452	51	UPI00117667F4	ENSSFAG00005000472	ENSSFAP00005006585.1*	A0A672FPE9*
	col1a2	22	LOC115380945	XP_029938120.1	1352	50	UPI00117665E6	ENSSFAG00005018265	ENSSFAP00005037812.1*	A0A672I9I0*
Gouania willdenowi	col1a1a	8	LOC114468996	XP_028311994.1	1446	49	UPI0010566202	ENSGWIG00000001631	ENSGWIP00000040994.1	A0A8C5H709
	col1a1b	19	LOC114481048	XP_028331171.1	1450	51	UPI0010566822	ENSGWIG00000018954	ENSGWIP00000038375.1*	A0A8C5H3 92*
	col1a2	11	LOC114471747	XP_028316412.1	1351	50	UPI001054FD33	ENSGWIG00000024361	ENSGWIP00000050009.1*	A0 A8C5HTT7*
Astatotilapia calliptera	col1a1a	4	LOC113021346	XP_026021750.1	1447	49	UPI000E4008AA	ENSACLG00000006444	ENSACLP00000041904.1	A0AAX7SCI9

(continued)

Table 8.3 (continued)

NCBI gene	Gene	Chromosome	NCBI gene	NCBI protein	aa	exons	Uniparc	Ensembl gene	Ensembl protein	Uniprot
	col1a1b	8	LOC113028167	XP_026033997.1	1444	51	UPI000E41CD4C	ENSACLG00000026412	ENSACLP00000056178.1	ND
	col1a2	22	LOC113014520	XP_026011889.1	1353	50	UPI000E41A6BE	ENSACLG00000023609	ENSACLP00000045219.1	A0AAX7SL03
Haplochromis burtoni	col1a1a	Unplaced	LOC102311964	XP_005946024.1	1447	49	UPI0003BDC440	ENSHBUG00000020378	ENSHBUP00000033453.1	A0A3Q3CY74
	col1a1b	Unplaced	LOC102304089	XP_005913627.1	1444	51	UPI0003BD4ECA	ENSHBUG00000007919	ENSHBUP00000006378.1	A0A3Q2VF15
	col1a2	Unplaced	LOC102292988	XP_005928634.1	1353	50	UPI0003BCD153	ENSHBUG00000019130	ENSHBUP00000017074.1*	A0A3Q2 W722*
Maylandia zebra	col1a1a	LG4	LOC101476419	XP_004572575.1	1447	49	UPI0003297509	ENSMZEG00005018722	ENSMZEP00005025081.1*	A0A3 P9CTF5*
	col1a1b	LG8	LOC101475669	XP_004556770.2	1444	51	UPI000D2FA4B4	ENSMZEG00005005660	ENSMZEP00005007428.1*	A0A3 P9BC04*
	col1a2	LG22	LOC101486099	XP_004547977.1	1353	50	UPI000329C6A8	ENSMZEG00005008145	ENSMZEP00005010948.1*	A0A3 P9BM02*
Pundamilia nyererei	col1a1a	Unplaced	LOC102204473	XP_005750908.1	1447	49	UPI0003AF5B8D	ENSPNYG00000016239	ENSPNYP00000021607.1*	A0A9Y3RW60
	col1a1b	Unplaced	LOC102204830	XP_005725945.1	1444	51	UPI0003AFB5B0	ENSPNYG00000003644	ENSPNYP00000005020.1*	A0A9Y3QVF3
	col1a2	Unplaced	LOC102199993	XP_005728787.1	1353	50	UPI0003AFB71C	ENSPNYG00000019493	ENSPNYP00000025895.1*	A0A9Y3VGZ0
Neolamprologus brichardi	col1a1a	Unplaced	LOC102798806	XP_006808934.1	1447	49	UPI0003EBF829	ENSNBRG00000024309	ENSNBRP00000032022.1*	A0A3Q4IH39*
	col1a1b	Unplaced	LOC102789211	XP_006790993.1	1444	51	UPI0003EC5AD6	ENSNBRG00000006458	ENSNBRP00000008564.1*	A0A3Q4 MFA6*
	col1a2	Unplaced	LOC102779727	XP_006787734.1	1353	50	UPI0003EC5AB4	ENSNBRG00000006663	ENSNBRP00000009615.1*	A0A3Q4 GJP4*
Oreochromis niloticus	col1a1a	LG4	LOC100702721	NP_001266373.1	1447	49	UPI00022B0047	ENSONIG00000006366	ENSONIP00000008023.2	G9M6I5
	col1a1b	LG8	LOC100694532	NP_001269818.1	1444	51	UPI00023B37F0	ENSONIG00000007944	ENSONIP00000010008.2	G9M6I7
	col1a2	LG22	LOC100710094	NP_001269826.1	1350	50	UPI00023B37EF	ENSONIG00000010792	ENSONIP00000013570.2	G9M6I6
Oreochromis aureus	col1a1a	4	LOC116321934	XP_031597733.2	1447	49	UPI001952BCCE	ENSOABG00000001674	ENSOABP00000009095.2	A0A668S3R5
	col1a1b	8	LOC116336325	XP_031616039.1	1444	51	UPI0012BD5280	ENSOABG00000002023	ENSOABP00000005752.2	A0A668RN02
	col1a2	22	LOC116319458	XP_031594655.2	1350	50	UPI001953D9E2	ENSOABG00000019042	ENSOABP00000045892.2	A0A668V4N4
Pelmatolapia mariae	col1a1a	LG4	LOC134625861	XP_063327246.1	1447	49	UPI002FE67877	ND	ND	ND
	col1a1b	LG8	LOC134633703	XP_063338753.1	1443	51	UPI002FE51B0F	ND	ND	ND
	col1a2	LG22	LOC135933129	XP_065327128.1	1356	50	UPI003211CBA6	ND	ND	ND
Simochromis diagramma	col1a1a	Unplaced	LOC120736860	XP_039893135.1	1447	49	UPI001A7EC5FB	ND	ND	ND
	col1a1b	Unplaced	LOC120738257	XP_039895398.1	1444	51	UPI001A7EF23A	ND	ND	ND
	col1a2	Unplaced	LOC120726706	XP_039876436.1	1353	50	UPI001A7E37E9	ND	ND	ND
Amphilophus citrinellus	col1a1a	LG4	ND	ND	1440	49	ND	ENSACIG00005010563	ENSACIP00005039844.1	ND
	col1a1b	LG8	ND	ND	1447	51	ND	ENSACIG00005006085	ENSACIP00005015399.1	ND
	col1a2	LG22	ND	ND	1352	50	ND	ENSACIG00005013162	ENSACIP00005036645.1	ND
Archocentrus centrarchus	col1a1a	8	LOC115785259	XP_030592656.1	1446	49	UPI0011E9CA10	ENSACEG00005001193	ENSACEP00005012095.1*	ND

	col1a1b	19	LOC115797863	XP_030610255.1	1447	51	UPI0011EA14BE	ENSACEG00005019538	ENSACEP00005049670.1	ND
	col1a2	11	LOC115787824	XP_030596418.1	1352	50	UPI0011E9DF00	ENSACEG00005001673	ENSACEP00005007412.1*	ND
Mugil cephalus	col1a1a	20	LOC124998126	XP_047428285.1	1446	48	UPI001FB56F30	ENSMGCG00000020674	ENSMGCP00000061499.1	ND
	col1a1b	16	LOC125023025	XP_047466046.1*	1300*	46*	UPI001FB634BE*	ENSMGCG00000026713	ENSMGCP00000066255.1*	ND
	col1a2	14	LOC125020056	XP_047461248.1	1352	50	UPI001FB782DA	ENSMGCG00000003564	ENSMGCP00000012334.1	ND
Parambassis ranga	col1a1a	8	LOC114439468	XP_028267212.1	1447	49	UPI0010425D31	ENSPRNG00000017208	ENSPRNP00000044143.1*	A0A6P7IU28
	col1a1b	6	LOC114437151	XP_028263435.1	1450	50	UPI001041E35C	ENSPRNG00000004002	ENSPRNP00000014600.1	A0A6P7I5U3
	col1a2	2	LOC114450388	XP_028284246.1	1351	50	UPI0010427857	ENSPRNG00000005467	ENSPRNP00000011840.1*	A0A6P7K9S7
Brachyistius frenatus	col1a1a	Unplaced	LOC138627671	XP_069566265.1	1447	49	UPI0037E9379E	ND	ND	ND
	col1a1b	Unplaced	LOC138633602	XP_069576048.1	1448	51	UPI0037E8052F	ND	ND	ND
	col1a2	Unplaced	LOC138618627	XP_069550595.1	1352	50	UPI0037E905C9	ND	ND	ND
Embiotoca jacksoni	col1a1a	Unplaced	LOC138215466	XP_069033742.1	1447	49	UPI003703AEF4	ND	ND	ND
	col1a1b	Unplaced	LOC138204762	XP_069016328.1	1448	51	UPI0037041350	ND	ND	ND
	col1a2	Unplaced	LOC138214545	XP_069032353.1	1352	50	UPI003704CE2A	ND	ND	ND
Acanthochromis polyacanthus	col1a1a	2	LOC110957415	XP_022059103.1	1447	49	UPI000B8F03DC	ENSAPOG00000009538	ENSAPOP00000025957.1*	A0A3Q1 GEL1*
	col1a1b	19	LOC110949374	XP_022047059.2	1443	51	UPI002233F939	ENSAPOG00000008178	ENSAPOP00000006217.1*	A0A3Q1 ERK6*
	col1a2	11	LOC110964782	XP_051811715.1	1352	50	UPI0022343298	ENSAPOG00000003588	ENSAPOP00000002123.1*	A0A3Q1 EEQ2*
Amphiprion ocellaris	col1a1a	19	LOC111588590	XP_023154805.1	1447	49	UPI000C30589A	ENSAOCG00000000904	ENSAOCP00000056624.1	A0 AAQ5YUL2
	col1a1b	18	LOC111579186	XP_023142131.2	1446	51	UPI002410D463	ENSAOCG00000014737	ENSAOCP00000010420.2*	A0A3Q1 B8X8*
	col1a2	15	LOC111573347	XP_054873633.1*	1352	50	UPI0032D05AAA	ENSAOCG00000015957	ENSAOCP00000027131.2	A0A3Q1D832
Amphiprion percula	col1a1a	14	ND	ND	1447	49	ND	ENSAPEG00000007891	ENSAPEP00000025462.1	ND
	col1a1b	19	ND	ND	1443	51	ND	ENSAPEG00000004385	ENSAPEP00000010700.1	ND
	col1a2	15	ND	ND	1352	50	ND	ENSAPEG00000005351	ENSAPEP00000013470.1	ND
Amphiprion clarkii	col1a1a	17	ND	ND	1447	49	ND	ENSRAIG00000000813	ENSRAIP00000029590.1	ND
	col1a1b	20	ND	ND	1452	51	ND	ENSRAIG00000006559	ENSRAIP00000015862.1	ND
	col1a2	14	ND	ND	1352	50	ND	ENSRAIG00000014162	ENSRAIP00000033059.1	ND
Stegastes partitus	col1a1a	Unplaced	LOC103367367	XP_008293593.1	1447	49	UPI000497D739	ENSSPAG00000009486	ENSSPAP00000012784.1*	A0A9Y4NCV4
	col1a1b	Unplaced	LOC103372024	XP_008299760.1	1447	51	UPI00049507A5	ENSSPAG00000015210	ENSSPAP00000020253.1*	A0A9Y4NM32
	col1a2	Unplaced	LOC103362790	XP_008287499.1	1353	50	UPI000497C9C2	ENSSPAG00000003453	ENSSPAP00000004559.1*	A0A9Y4K3N2
Scomber japonicus	col1a1a	18	LOC128378365	XP_053193834.1	1447	49	UPI002305A795	ND	ND	ND
	col1a1b	20	LOC128380888	XP_053196740.1	1450	51	UPI0023052F6F	ND	ND	ND
	col1a2	15	LOC128374490	XP_053190734.1	1353	50	UPI002306B831	ND	ND	ND

(continued)

Table 8.3 (continued)

NCBI gene	Gene	Chromosome	NCBI gene	NCBI protein	aa	exons	Uniparc	Ensembl gene	Ensembl protein	Uniprot
Scomber scombrus	col1a1a	18	LOC133999778	XP_062295068.1	1447	49	UPI002DD7BC14	ENSSXIG00000008800	ENSSXIP00000026347.1	ND
	col1a1b	21	LOC134003045	XP_062298136.1	1441	51	UPI002DD8AC97	ENSSXIG00000011140	ENSSXIP00000028016.1*	ND
	col1a2	16	LOC133996656	XP_062292246.1	1269*	45*	UPI002DDAA22C	ENSSXIG00000020486	ENSSXIP00000050962.1*	ND
Thunnus albacares	col1a1a	17	LOC122966220	XP_044186193.1	1447	49	UPI001CF606E6	ENSTABG00005016521	ENSTABP00005050457.1	ND
	col1a1b	20	LOC122971298	XP_044193815.1	1452	51	UPI001CF64E31	ENSTABG00005011070	ENSTABP00005027707.1	ND
	col1a2	19	LOC122969546	XP_044191233.1	1352	50	UPI001CF6340D	ENSTABG00005010744	ENSTABP00005027399.1	ND
Thunnus maccoyii	col1a1a	18	LOC121884875	XP_042249887.1	1447	49	UPI001C4C0336	ENSTMYG00005010998	ENSTMYP00005032615.1	ND
	col1a1b	20	LOC121887005	XP_042253453.1	1452	51	UPI001C4C1E1D	ENSTMYG00005016196	ENSTMYP00005041692.1	ND
	col1a2	15	LOC121912953	XP_042291469.1	1352	50	UPI001C4C7F2E	ENSTMYG00005016374	ENSTMYP00005042941.1	ND
Thunnus thynnus	col1a1a	17	LOC137201095	XP_067472083.1	1447	49	UPI003528405E	ENSTTHG00005015728	ENSTTHP00005024026.1	ND
	col1a1b	20	LOC137171575	XP_067431685.1	1452	51	UPI00352921EB	ENSTTHG00005001697	ENSTTHP00005002481.1	ND
	col1a2	15	LOC137197797	XP_067467313.1	1352	50	UPI003527EE0B	ENSTTHG00005019552	ENSTTHP00005028405.1	ND
Synchiropus splendidus	col1a1a	19	LOC128751151	XP_053707967.1	1447	49	UPI00237DED03	ENSSSDG00005006133	ENSSSDP00005010814.1	ND
	col1a1b	1	LOC128771830	XP_053743599.1	1449	51	UPI00237D6644	ENSSSDG00005018044	ENSSSDP00005029219.1	ND
	col1a2	4	LOC128758017	XP_053719734.1	1349	50	UPI00237DCDED	ENSSSDG00005002909	ENSSSDP00005004332.1	ND
Hippocampus comes	col1a1a	Unplaced	LOC109524772	XP_019740420.1	1448	49	UPI00094E4C41	ENSHCOG00000000752	ENSHCOP00000007782.1*	A0A3Q2 XU67*
	col1a1b	Unplaced	LOC109513273	XP_019721149.1	1447	51	UPI00094E7976	ENSHCOG00000018831	ENSHCOP00000027145.1*	A0A3Q2 Z8D3*
	col1a2	Unplaced	LOC109527674	XP_019745274.1	1349	50	UPI00094EA638	ENSHCOG00000009651	ENSHCOP00000022137.1*	A0A3Q2 YTP9*
Hippocampus zosterae	col1a1a	17	LOC127589407	XP_051904497.1	1448	49	UPI00223CF449	ENSHZOG00010017531	ENSHZOP00010037566.1	ND
	col1a1b	7	LOC127603963	XP_051926669.1	1447	51	UPI00223E7A93	ENSHZOG00010007482	ENSHZOP00010016561.1	ND
	col1a2	7	LOC127604631	XP_051927778.1	1356	49	UPI00223E6867	ENSHZOG00010016055	ENSHZOP00010032159.1*	ND
Hippocampus abdominalis	col1a1a	16	ND	ND	1267*	38*	ND	ENSHABG00000013051	ENSHABP00000033255.1*	ND
	col1a1b	21	ND	ND	1449	51	ND	ENSHABG00000017777	ENSHABP00000043101.1	ND
	col1a2	20	ND	ND	1349	50	ND	ENSHABG00000021887	ENSHABP00000051499.1	ND
Corythoichthys intestinalis	col1a1a	16	LOC130904692	XP_057673614.1	1448	49	UPI0025A5D784	ENSYQEG00000016514	ENSYQEP00000039756.1*	ND
	col1a1b	21	LOC130909916	XP_057682846.1	1449	51	UPI0025A5F610	ENSYQEG00000017925	ENSYQEP00000035003.1	ND
	col1a2	22	LOC130910855	XP_057684445.1	1358	49	UPI0025A5109D	ENSYQEG00000025876	ENSYQEP00000044901.1	ND
Dunckerocampus dactyliophorus	col1a1a	18	LOC129170629	XP_054614332.1	1448	49	UPI002405BC76	ENSDDAG00005025074	ENSDDAP00005049325.1	ND
	col1a1b	15	LOC129194798	XP_054656165.1	1447	51	UPI0024072CE1	ENSDDAG00005002759	ENSDDAP00005001475.1*	ND
	col1a2	20	LOC129173622	XP_054620653.1	1356	49	UPI002405BA72	ENSDDAG00005003990	ENSDDAP00005004119.1*	ND
Doryrhamphus excisus	col1a1a	15	LOC131103507	XP_057905816.1	1448	49	UPI0025ADBD24	ENSRDMG00005006033	ENSRDMP00005014939.1	ND

	col1a1b	7	LOC131132471	XP_057934126.1	1446	51	UPI0025AE8940	ENSRDMG00005016729	ENSRDMP00005029278.1	ND
	col1a2	14	LOC131102213	XP_057903724.1	1351	50	UPI0025AE69AD	ENSRDMG00005015392	ENSRDMP00005026149.1*	ND
Entelurus aequoreus	col1a1a	LG25	LOC133642355	XP_061892482.1	1447	49	UPI002B1D2213	ENSEAEG00010017898	ENSEAEP00010031146.1	ND
	col1a1b	LG08	LOC133656298	XP_061913295.1	1448	51	UPI002B1E2AAA	ENSEAEG00010006146	ENSEAEP00010014410.1	ND
	col1a2	LG20	LOC133636159	XP_061885874.1	1356	49	UPI002B1E5F59	ENSEAEG00010026356	ENSEAEP00010043361.1*	ND
Nerophis lumbriciformis	col1a1a1	Unplaced	LOC133596661	XP_061805487.1	1448	49	UPI0025A5D784	ENSIQYG00000018601	ENSIQYP00000022559.1*	ND
	col1a1a2	LG14	LOC133615184	XP_061829556.1	1447	49	UPI002ADFC0EE	ENSIQYG00000048380	ENSIQYP00000065129.1	ND
	col1a1b1	Unplaced	LOC133598481	XP_061807157.1	1441	50	UPI002ADF0DC9	ENSIQYG00000010553	ND	ND
	col1a1b2	LG09	LOC133610258	XP_061822374.1	1448	51	UPI002ADF6C5B	ENSIQYG00000002569	ENSIQYP00000007405.1	ND
	col1a2	Unplaced	LOC133599427	XP_061808200.1	1358	49	UPI0025A5109D	ND	ND	ND
	col1a2	LG22	LOC133620894	XP_061838509.1	1356	49	UPI002AE051A1	ENSIQYG00000002434	ENSIQYP00000004410.1	ND
Nerophis ophidion	col1a1a	LG23	LOC133541086	XP_061740157.1	1447	49	UPI002AE07FC3	ENSNOPG00005027705	ENSNOPP00005048842.1	ND
	col1a1b	LG08	LOC133557280	XP_061763619.1	1448	51	UPI002ADF5CF8	ENSNOPG00005025327	ENSNOPP00005045575.1	ND
	col1a2	LG21	LOC133539961	XP_061738350.1	1355	50	UPI002ADF6D7C	ENSNOPG00005011665	ENSNOPP00005024169.1	ND
Phyllopteryx taeniolatus	col1a1a	16	LOC133465493	XP_061604358.1	1448	49	UPI002AD436E7	ENSPTLG00010018145	ENSPTLP00010044901.1	ND
	col1a1b	19	LOC133469711	XP_061613102.1	1449	51	UPI002AD47580	ENSPTLG00010024013	ENSPTLP00010051220.1	ND
	col1a2	21	LOC133470636	XP_061615181.1	1357	49	UPI002AD4A8E6	ENSPTLG00010007317	ENSPTLP00010015763.1	ND
Phycodurus eques	col1a1a	16	LOC133414741	XP_061556322.1	1448	49	UPI002ACECB9A	ENSPEQG00000014696	ENSPEQP00000041700.1	ND
	col1a1b	19	LOC133417863	XP_061562034.1	1447	51	UPI002ACEA3FD	ENSPEQG00000014866	ENSPEQP00000037862.1	ND
	col1a2	20	LOC133395449	XP_061520325.1	1357	49	UPI002ACD5460	ENSPEQG00000019045	ENSPEQP00000047750.1	ND
Syngnathoides biaculeatus	col1a1a	16	LOC133514372	XP_061702005.1	1448	49	UPI002ADE7E8D	ENSSBIG00005016686	ENSSBIP00005035746.1	ND
	col1a1b	22	LOC133495094	XP_061665288.1	1449	51	UPI002ADE5582	ENSSBIG00005011679	ENSSBIP00005024498.1	ND
	col1a2	1	LOC133497385	XP_061669344.1	1352	50	UPI002 ADD5AAA	ENSSBIG00005018581	ENSSBIP00005036104.1	ND
Syngnathus acus	col1a1a	8	LOC119125226	XP_037111387.1	1448	49	UPI001885F518	ENSSACG00010023203	ENSSACP00010033223.1	ND
	col1a1b	19	LOC119138516	XP_037134528.1	1447	51	UPI0018860B11	ENSSACG00010018448	ENSSACP00010027024.1	ND
	col1a2	11	LOC119130718	XP_037120492.1	1349	50	UPI001885B44C	ENSSACG00010013360	ENSSACP00010019430.1*	ND
Syngnathus scovelli	col1a1a	16	LOC125983028	XP_049599816.1	1446	49	UPI0021106D69	ENSSICG00000022655	ENSSICP00000034531.1	ND
	col1a1b	21	LOC125991452	XP_049615322.1	1449	51	UPI00210F2C1B	ENSSICG00000024293	ENSSICP00000036886.1	ND
	col1a2	19	LOC125986950	XP_049606868.1	1348	50	UPI002110AEFD	ENSSICG00000018165	ENSSICP00000027004.1	ND
Syngnathus typhle	col1a1a	LG17	LOC133170124	XP_061158782.1	1447	49	UPI002A69F282	ENSSTYG00005006114	ENSSTYP00005021189.1	ND
	col1a1b	LG18	LOC133171300	XP_061160527.1	1447	51	UPI002A6A714E	ENSSTYG00005011717	ENSSTYP00005020557.1	ND
	col1a2	LG20	LOC133144298	XP_061122849.1	1348	50	UPI002A6A1A10	ENSSTYG00005017122	ENSSTYP00005034791.1	ND

*Entries marked with an asterisk are fragments or partial sequences other fish show two genes in their genomes, *col2a1a* and *col2a1b*, which are differentially expressed [32]. Similarly, there are duplicates for the minor fibrillar collagens V, XI, and XXVII. Two taxonomic groups have an even more expanded family of fibrillar collagen genes: the Salmonidae family, which includes salmons and trouts, and the subfamily Cyprininae, which includes the common carp and members of the *Carassius* and *Sinocyclocheilus* genera. Salmonidae and Cyprininae are tetraploids and have four sets of chromosomes as a result of additional WGD events that occurred around 50–80 million years in the Salmoniformes and 5.6–11.3 million years in the Cyprinoformes [24, 55, 95] (Fig. 8.1). Accordingly, these two groups initially doubled their numbers of fibrillar collagen genes from the already extended number of genes in teleosts. Many of these duplicate genes have remained to this day, and for example cyprininae and salmonids have six genes for type I collagen (two *col1a1a* coding for α1(I) chains, two *col1a1b* for α3(I) chains and two *col1a2* for α2(I) chains), three or four genes for type II collagen, at least nine genes for type V collagen, and so on (Tables 8.2 and 8.3). Not all gene duplicates resulting from previous WGD events have survived. Most fish genomes only contain a single *col24a1* gene for type XXIV collagen or one *col1a2* gene for the α2(I) chain (Tables 8.2 and 8.3).

Polypteridae (bichirs and reedfish), holostei (bowfins and gars), and chondrostei (sturgeons and paddlefish) are ancient taxonomic groups that form the basal ray-finned fish lineage that diverged from the most numerous taxonomic group, the teleosts, before the WGD event of 320–350 million years ago (Fig. 8.1). Both holostei and polypteridae show the same numbers of fibrillar collagen genes as mammals (Table 8.2), including a gene coding for type III collagen that has been lost in the teleost group. Chondrostei, however, have duplicated numbers of genes for most fibrillar types, which probably result from other independent WGD events [126].

Genome duplication events create functionally redundant gene copies that can evolve into new functions and help the organism adapt to environmental changes. In the case of fibrillar collagens, the existence of multiple genes for each collagen type will probably require some control of their expression to minimise the risk of formation of unwanted trimer combinations. There are a few studies investigating the expression of duplicated fibrillar collagens in zebrafish and medaka [27, 32, 33, 40, 53, 63, 103] and analysing the impact of mutations on the different redundant genes for a given type [54]. However, our current understanding of the expression and biological function of all the different fibrillar collagen genes in most other fish and the occurrence across tissues of the different homotrimeric or heterotrimeric combinations is very limited.

Structures of Fibrillar Collagen Genes in Bony Fish

The gene structures of bony fish fibrillar collagens follow the same trends as mammalian and other vertebrates. The gene regions coding for the major triple-helical domain (~1014 amino acids) are made from multiple short exons with lengths of 54 bp, 45 bp, and multiples or combinations of these values (Fig. 8.2). These characteristic "collagen exons" correspond to whole numbers of Gly-X-Y triplets (5, 6, 12 triplets, 15, 18, 36 amino acids), and they start with a glycine codon and end with the codon for the third residue of the last Gly-X-Y triplet. This exon organisation has been well documented and has been related to the origin of the first ancestral fibrillar collagen genes [37–39, 43]. Small differences between the number and lengths of the collagen exons can be useful for the identification of the different collagen types on genomic sequences. The exon organisation of the C-terminal propeptide (ColFi domain, discussed below) is also highly conserved in all fibrillar fish collagens and is very similar to that of human and mammalian collagens.

Domain Architectures of the Procollagen Chains

Fibrillar collagen chains are synthesised as multidomain precursor proteins, procollagens, that need to be processed to reach the mature state. Processing of mammalian fibrillar collagens is well understood, and it has been reviewed many times [6, 39, 77, 105, 109]. It is assumed that the same occurs with the fibrillar collagens of other organisms, including bony fish [43]. Briefly, each fibrillar procollagen α chain contains one long triple-helical domain (abbreviated as COL) of approximately 1000 amino acids, with the characteristic Gly-X-Y repeating sequence. This COL domain is flanked at the N- and C-termini by non-collagenous domains (Fig. 8.3). Traditional convention is to name the individual domains on each α chain sequentially, starting from the C-terminus (NC1), continuing with two collagen domains COL1 (long, major helix) and COL2 (short, minor helix), and ending at the N-terminal domain (NC2) if present (Fig. 8.3). All collagen α chains are synthesised first as procollagens, and different procollagen types experience different levels of processing to reach their mature form. In all fibrillar collagens, the NC1 domain has the same type of structure, the fibrillar collagen C-terminal domain (abbreviated as ColFi) [6, 13, 39], which mediates the initial steps of trimerisation forming a trimeric C-propeptide structure. At the N-terminal end, the combination of NC2 domains and the minor collagen helices COL2 from the three α chains forms the N-propeptide. Once the three procollagen α chains have trimerised, specialised proteases remove the C-propeptide and in some collagen types the N-propeptide by cleaving all three α chains at specific points. Propeptide removal leaves the mature collagen form ready to assemble into collagen fibres [6, 39, 77, 105, 109]. Short non-collagenous telopeptides are left at the end of the α chains where the propeptides have been removed (Fig. 8.3). In the case of fish type I collagens, the mature α chains (COL1 plus telopeptides) are predicted to have around 1055 amino acids.

Based on their sequence similarity, the fibrillar collagen genes can be divided into three subfamilies, or clades, named A, B, and C [9, 39, 148]. Proteins coded by genes from different clades show differences in their domain organisation, and fish fibrillar

Fig. 8.2 Exon organisation of fibrillar fish collagens transcripts. (**a**) Amino acid sequence of the precursor α1(I) collagen chain from *Oreochromis niloticus* (Nile tilapia). Exons are represented as alternate segments in black and blue. Residues overlapping a splice site are shown in red. The predominant translated exon length in the Gly-X-Y repeating sequence of the collagen domain is 18 amino acids or 6 Gly-X-Y triplets, corresponding to 54 bp. The other common exon lengths are 15 amino acids (5 Gly-X-Y triplets, 45 bp) and 36 amino acids (12 Gly-X-Y triplets, 108 bp). (**b**) Exon-intron structure of transcripts for the *col1a1a* and *col1a1b* genes of *Oreochromis niloticus* (Nile tilapia) (red), and for the human *COL1A1* gene (gold). Exons are drawn at scale but introns are not. Regions corresponding to the minor and major triple helices are indicated by the green boxes. The transcripts for *O. niloticus col1a1a* and *col1a1b* contain 49 and 51 exons, respectively. The transcript for *H. sapiens COL1A1* contains 51 exons. The exon lengths in the *col1a1b* and *COL1A1* transcripts are identical over their major collagen domains. The blue boxes indicate regions where two pairs of exons in the *col1a1b* and *COL1A1* transcripts (54–54 bp, 108–54 bp) correspond to fused exons in the *col1a1a* transcript (108 bp and 162 bp, respectively)

collagens are consistent with this classification. Thus, in bony fish clade A includes the major fibrillar collagen genes *col1a1a* and *col1a1b*, *col1a2*, *col2a1a*, and *col2a1b*, the minor fibrillar collagen genes *col5a2a* and *col5a2b*, and the additional duplicates of these genes in the tetraploid groups cyprininae and salmonids. The few instances of *col3a1* genes in polypteridae, holostei, and chondrostei also belong to clade A. Procollagen proteins coded by clade A genes usually have an N-terminal von Willebrand factor type C domain (VWC) followed by a short triple-helical domain (minor helix) (Fig. 8.4a). Together these form the N-propeptide which is proteolytically removed in the processing of the procollagen proteins to mature collagens. Exceptions are the *col1a2* genes, for which the N-propeptide only contains the minor triple-helical domain, and in some fish the *col5a2b* genes code for slightly longer procollagen proteins, with two VWC domains in tandem at the

A

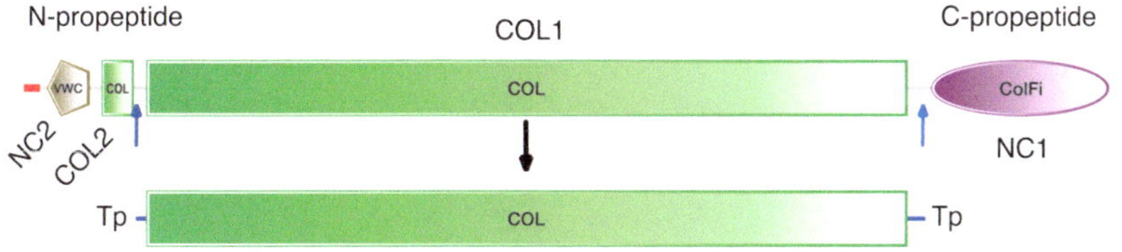

B

MFSFVDIRLGLLLAAVVLAVRGQGEDDISFGSCTLEGQNYNDKDVWKPEPCQICVCDSGTVMCDEVICEDTSDCANPEI
PDGECCPICPDEEVTAAPEPPYVDTVGPAGPRGDPGPVGPPGRDGTPGENGLPGPPGPPGTSGLGGGSFLSQMAYGSEK
SSGPPQPGPPGPMGPRGLPGPAGSPGPQGFTGPPGEPGEPGAPGPMGSRGPPGPPGKNGEDGEAGKAGRPGERGAAGPQ
GARGFPGTPGLPGIKGHRGFPGLDGAKGDAGIAGPKGEAGSPGEAGTPGAMGARGLPGERGRPGPPGPSGARGNDGNTG
AAGPPGPTGPAGPPGFPGGAGAKGETGAAGGRGSEGPQGPRGEPGNPGPAGAAGPAGGPGSDGTPGGKGSPGAAGIAGA
PGFPGARGPPGPAGPNGAPGPKGNNGDPGLQGPKGDAGPKGEPGPVGPQGLAGNAGEEGKRGARGEPGAAGPVGPPGAR
GAPGNRGLPGSDGAPGPLGAPGERGANGPMGAQGASGESGRPGEPGLPGSKGMTGAPGSPGPDGKAGPSGAPGQDGRPG
PPGPAGSRGAPGVMGFPGPKGTDGESGKPGERGLAGPTGPLGAPGKDGDIGSPGAPGPAGPAGEKGEQGSAGPSGFQGL
PGAQGATGETGKPGEQGLPGEAGPSGPSGPRGDRGFPGERGAPGPGGPTGPRGAPGSPGNDGAKGEPGAAGANGGVGPA
GMQGMPGERGAGGMPGARGDRGDGGPKGPDGAPGKDGVRGMTGPIGPPGPSGAPGDKGEPGAVGPPGLTGPRGGPGERG
EVGAPGPAGFAGPPGGNGQPGAKGETGDSGPKGDAGAPGPAGPVGASGPQGPAGATGPKGARGGAGPPGATGFPGAAGR
VGPPGPSGASGPPGPTGPAGKEGQRGARGEKGPAGRPGEAGAAGPPGPSGERGSPGPDGPLGLAGAPGPPGVIGSRGIV
GLPGQRGERGFGGLAGPSGEPGKQGPSGPVGERGPPGPMGPPGLSGAPGEAGREGNPGHDGVSGRDGPPGPKGDRGESG
NSGPPGAPGAPGAPGPIGPSGKTGDRGEAGPAGPAGPSGPAGARGALGPAGPRGDRGEAGEAGERGMKGHRGLSGMPGV
PGPPGTSGEQGPAGATGPAGPRGPAGSNGSPGKDGMNGVPGPVGPPGPRGRSGEMGPAGAPGLPGPPGPPGSAGPGEPF
LPMPQQEKGPDPLRGGYRADDASVRDRDSEVDTSLKSLSQKIENIRSPEGTQANPARMCRDLRMCHPEWKSGSYWVDPN
QGSPLDAIKVFCNMETGETCVSPYHSDIPRKNWYTSKNIREKKHVWFGESMPNGFQFQYGSDGSDPEDVNIQLTFMRLM
SNQGSQNITYHCKNSIAYMDETSGNLKKALLLQGSNDIEIRAEGNSRFTYRVSEDGCTSHTGTWGKTVIDYKTTKTSRL
PIIDIAPMDIGAPDQEFGVEVGPVCFL

Fig. 8.3 Modular structure of the α3(I) chain from zebrafish. (**a**) Domain architecture, based on the amino acid sequence (**b**) coded by the *col1a1b* gene (NP_958886.1, Q6PEI9). Domains are numbered starting from the C-terminal domain as NC1, COL1, COL2, and NC2. NC1 has the characteristic structure of the C-terminal domains of fibrillar collagens (ColFi, purple). There are two triple-helical domains (COL, green). The major collagen domain (COL1) is 1014 amino acids long, and the shorter collagen domain, or minor helix (COL2), is 39 amino acids long (these numbers change by three amino acids if additional triplets GGS or GPG are included). The N-terminal domain is a von Willebrand factor type C domain (VWC, gold), and the chain ends with a 22-amino-acid-long signal peptide (red). The blue arrows indicate the predicted sites of proteolytic cleavage for the removal of the N- and C-propeptides. When the N-propeptide is removed in fibrillar collagens, the site of cleavage usually occurs between the COL1 and COL2 domains (major and minor helices). The mature α3(I) chain consists of the COL1 domain flanked at both ends by short non-collagenous sequences known as telopeptides (Tp, indicated in blue in the amino acid sequence). The length of the entire α3(I) procollagen chain is 1449 amino acids, while the mature α3(I) chain is 1055 amino acids long

N-terminus followed by the minor triple-helical domain (Fig. 8.4a). In mammals, clade A fibrillar collagens are either homotrimers (types II and III, less often type I) or heterotrimers (type I) [39, 128]. The multiplicity of duplicate chains in fish makes it difficult to predict if any clade A fish collagen will form always homotrimers, always heterotrimers, or will change chain composition depending on tissue localisation.

In bony fish, clade B includes the minor fibrillar collagen genes *col5a1*, *col5a3a* and *col5a3b*, *col11a1a* and *col11a1b*, *col11a2*, and any additional duplicates of these genes. Often, there is a high degree of sequence similarity between type V and

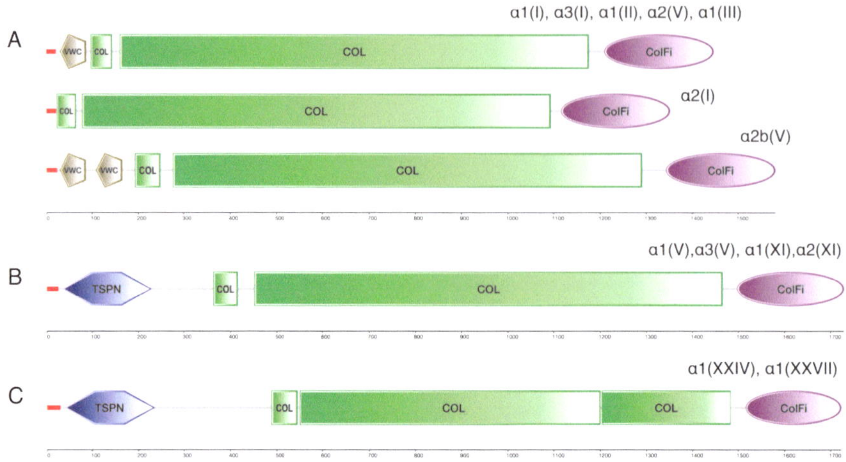

Fig. 8.4 Domain architectures of the three clades of fibrillar collagens. (**a**) Clade A collagens have at least three different architectures differing in the number of von Willebrand factor type C domains (VWC) in their N-propeptides. (**b**) All collagens from clade B have the same architecture, with a thrombospondin N-terminal domain (TSPN) in the N-propeptide. (**c**) Collagens from clade C have also a TSPN domain but no clear minor collagen helix as those from clades A and B. The major collagen triple-helical domain of clade C collagens shows two short interruptions

type XI chains, and it is not uncommon in sequence databases to find a type V gene annotated as type XI and vice versa. Procollagen proteins coded by clade B genes have an N-terminal Thrombospondin domain (TSPN) followed by the minor triple-helical domain (Fig. 8.4b). Mammalian collagen types V and XI are unusual in many respects, and it is largely unknown if the equivalent fish collagen types V and XI show the same peculiar characteristics. In mammals, collagens type V and XI form usually heterotrimers with different chain compositions, although an α1(V) homotrimer has also been described [97, 99, 128]. The different chain combinations (isoforms) are expressed in different tissues and probably have different biological roles [99]. A third chain for mammalian type XI collagen, α3(XI), is in fact coded by the mammalian *col2a1* gene (which is sometimes referred to as *col11a3* gene), but differs in posttranslational processing [128, 150]. Types V and XI can form a heterotypic heterotrimer α1(XI)α1(V)α3(XI) made of chains from both types [86, 99]. To make the picture more complex, additional type V and XI chains can be produced as alternatively spliced variants from the same genes in different tissues or developmental stages [41]. An alternative splicing site in the ColFi region of the mammalian *col5a1* gene is conserved in the *col5a1* zebrafish gene, and both isoforms are detected in zebrafish embryos and adults [69]. The sequences for alternative splicing of *col5a1* are conserved in many fish, and sequences for both isoforms have been predicted in each case, suggesting that splicing may also occur in other fish α1(V) chains.

The third clade C includes the *col24a1*, *col27a1a*, and *col27a1b* genes, plus any additional duplicates. Members of this clade also contain an N-terminal TSPN domain, followed by a relatively long sequence of undefined structure, and a single triple-helical domain with less than 1000 amino acids and with a couple of short interruptions of the Gly-X-Y repetitive sequence (Fig. 8.4c). The lengths of the interrupted triple-helical domains on clade C proteins are shorter than those uninterrupted domains on members of clades A and B. In mammals, both clade C collagens XXIV and XXVII form homotrimers, and for most fish this would be expected for type XXIV as there only seems to be one *col24a1* gene. Conversely, most fish have two type XXVII genes in their genomes, and in zebrafish, both of them are expressed, at least during embryonic and larval stages [27]. It is unclear if the two type XXVII chains remain segregated to form respective homotrimers, or if they come together to form heterotrimers, or some situation in between.

It is expected that the processing of N-propeptides during the synthesis of fibrillar collagens of bony fish will be similar to what has been observed for mammalian collagens. This processing differs depending on the clade of the chain. For α chains of clade A, the N-propeptide formed by the VWC domain(s) and the COL2 minor helix is removed by endopeptidases, and the cleaving site is located between the minor and major triple helices. The mature form consists of three α chains forming a single triple-helical domain, with the short telopeptide extensions. For α chains of clades B and C, the N-propeptide is partially retained. This means that processing of procollagens from types V or XI will end up with some retained N-terminal propeptides in their mature forms. When these collagens are incorporated into fibres, the retained N-propeptides are thought to project outwards from the surface of the fibres, preventing incorporation of further collagen molecules and helping to regulate the diameter of the fibres [39, 90, 97].

Three additional fibrillar collagen genes have been identified in zebrafish, named *si:dkey-21p1.3*, *si:dkey-61 l1.4*, and *si:ch211-196i2.1*. Not much is known about the function of these genes other than being predicted as having an extracellular matrix role, but orthologs have been identified in the genomes of other bony fish. However, the quality of annotation of all these genes is highly variable. The predicted protein sequences for genes *si:dkey-61l1.4* and *si:ch211-196i2.1* in zebrafish have 1813 amino acids each, with a TSPN domain at the N-terminus and a ColFi domain at the C-terminus, a minor helix domain COL2 with 51 amino acids, and a major, uninterrupted triple-helical domain COL1 with 1014 amino acids. The domain architectures of these two predicted proteins would classify the *si:dkey-61 l1.4* and *si:ch211-196i2.1* genes as clade B members. Orthologs in other fish genomes with predicted complete sequences show the same domain architecture. The predicted protein sequence for gene *si:dkey-21p1.3* is clearly incomplete, starting with a short collagen sequence of 19 amino acids, followed by a major uninterrupted triple-helical domain of 1011 amino acids and the ColFi domain at the C-terminus. Orthologs in other fish genomes are highly variable in length, and at present, it is unclear to which clade this gene belongs.

Procollagen C-Propeptide Trimerisation and Chain Selection

All fibrillar collagens have the same C-terminal non-collagenous domain (ColFi) as their C-propeptide. The ColFi domain is the nucleation point for the folding of the procollagen chains, drives the assembly of three procollagen α chains to form trimers, and promotes chain registration, which is then followed by a zipper-like folding of the collagen triple-helical domain from the C-terminus towards the N-terminus [12]. For heterotrimer fibrillar collagens such as type I, the ColFi domain has a key role in selecting the right assembly of α chains; in the mammalian case, two α1(I) chains and one α2(I) chain. The trimerisation process needs to ensure the correct combination of chains avoiding the formation of incorrect assemblies. The ColFi domain is approximately 250 residues long and its sequence is highly conserved. It has, however, a variable chain recognition sequence (CRS) that is important for chain recognition. The crystal structure of a ColFi trimer was first determined in 2012 [13], and additional structures of homo- and heterotrimeric trimers were obtained a few years later [137]. These structures guided studies of mutation of key residues that confirmed their role in trimerisation [137]. Several highly conserved residues at specific positions of the ColFi domain form interchain disulphide bonds that stabilise the C-propeptide trimeric structure [13, 137]. It was thought that only the assemblies that could form stabilising interchain disulphide bonds are observed. For instance, the mammalian α2(I) chain has one critical cysteine residue changed to a serine, and for this reason, trimer combinations with two or more copies of the α2 (I) chain have never been reported.

In bony fish, type I collagen heterotrimers can have two different combinations, α1(I)α2(I)α3(I) or $\alpha 1(\mathrm{I})_2\alpha 2(\mathrm{I})$, depending on the tissue (see section "Multiple Fibrillar Collagen Genes in Bony Fish Genomes"). Figure 8.5 shows a structural model for the α1(I)α2(I)α3(I) heterotrimer where two disulphide bonds are predicted, one between the α1(I) and the α2(I) chains, and another between the α1(I) and the α3 (I) chains. The last disulphide between the α2(I) and α3(I) chains cannot form as both cysteine residues have been changed to serine. In a similar situation to mammalian collagens, these substitutions would prevent the formation of heterotrimer combinations with two α2(I) chains or with two α3(I) chains, although the pattern of cysteines alone would not explain why a heterotrimer is preferred to an α1 (I) homotrimer.

A recent study, however, has reported that the region of the triple-helical domain near the ColFi domain is also important in committing the correct assembly to irreversible folding, while the incorrect combinations dissociate and are available to search for the correct binding partners [15, 158]. It is likely that a similar two-step mechanism will be at play for the correct chain selection in heterotrimeric fibrillar fish collagens.

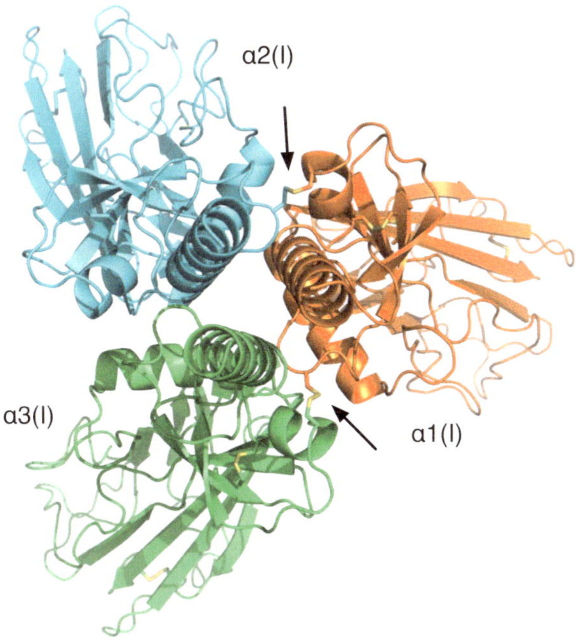


```
α1   DLEVDTTLKSLSQQIESIISPDGTKKNPARTCRDLKMCHPDWKSGEYWIDPDQGCNQDAI
α3   DSEVDTSLKSLSQKIENIRSPEGTQANPARMCRDLRMCHPEWKSGSYWVDPNQGSPLDAI
α2   DYEVDATIKSLNTQIENLLSPEGSKKNPARTCRDIRLSHPEWSSGFYWIDPNQGCTMDAI
```

Fig. 8.5 Structural model of the C-propeptide for zebrafish type I procollagen with heterotrimer α1(I)α2(I)α3(I) chain composition. The alignment shows the first 60 residues of each chain. Boxed residues show the sites where cysteine residues have been replaced by serines. Due to the serine substitutions, only two intermolecular disulphide bonds can be formed, between the α1(I) and the α3(I) chains and between the α2(I) and the α1(I) chains, indicated with arrows

Amino Acid Sequence Preferences in Fish Collagens and Their Relation to Thermal Stability

Fish inhabit a broad range of thermal environments, from the freezing waters of the Antarctic to tropical seas. Therefore, their fibrillar collagens must be stable and functional at the temperatures of the waters in which they live in. Thus, fish collagens exhibit a wide range of different denaturation temperatures. Fibrillar collagens from cold-water fish such as those living in the Arctic and Antarctic show very low thermal stability, with denaturation temperatures around 6 °C [18, 129]. In contrast, fibrillar collagens of warm-water or tropical fish have higher thermal stability, similar to that of mammalian collagens, which typically denature in the range of 30–40 °C [18, 19]. The wide variation in the thermal stability of fish collagens has made them a natural tool for the investigation of collagen stability. The study of the amino acid composition of fibrillar fish collagens (for example from their skins) has helped to understand how the amino acid sequence influences the thermal stability of the collagen triple helix. With the availability of many fish genomes, it is now straightforward to calculate amino acid sequence preferences for fish fibrillar collagens from the protein sequences obtained from the genomic data (Table 8.4).

A key insight from comparative studies between fish and mammalian collagens was establishing the role of imino acids, in particular hydroxyproline, in the stabilisation of the collagen triple helix. Early researchers on collagen stability (see section "Early Years" and references therein) established that fish collagens, especially those from cold-adapted species, have lower levels of hydroxyproline (Table 8.4). They deduced that hydroxyproline content must be an important factor in the thermal stability of fibrillar collagens. Nowadays, we know that a high imino acid content stabilises the collagen triple helix via stereochemical restrictions imposed by the imino acid rings, which favour the polyproline II conformation adopted by the three polypeptide chains forming the collagen triple-helical structure (see [5] for a review). Hydroxylation of prolines in the Y position of the Gly-X-Y repetitive sequence increases the stability even more, through a combination of stereoelectronic effects and hydrogen bonding to water [5, 7, 138].

Through evolution, cold-water fish have adapted to live in cold temperatures sacrificing the molecular stability of their fibrillar collagens in exchange for functional adaptation. Research from the 2000s suggests that fibrillar collagens should not be "too stable" for the particular thermal conditions in which the animals containing these collagens live [91, 118]. The collagen triple helices must be able to denature and refold locally, giving resilience and flexibility to the collagens that form the basis of connective tissues such as skin or bone. This can be achieved if the melting temperature of these triple helices is slightly below the body temperature [91].

Thus, fibrillar fish collagens offer a unique window into how collagen thermal stability depends largely (but not uniquely) on the amino acid sequences of their

Table 8.4 Amino acid residue preferences (%) in the X and Y positions of the Gly-X-Y sequences of COL domains from type I collagens of mammals and of different fish groups

Amino acids	Mammals	Warm water	Carps	Salmonids	Ice fish
P(O)	34/32	32/29	30/29	27/25	26/20
ED	16/6	16/6	16/6	16/6	16/6
KR	6/19	6/20	6/19	6/19	6/19
VILMF	15/9	14/7	15/8	14/7	15/8
A	15/16	18/18	17/17	15/16	15/16
TS	7/10	6/12	8/13	9/13	11/18
QN	5/7	5/7	5/7	7/9	5/9
G	1/1	2/1	2/1	5/4	5/4
HYCW	1/0	1/0	1/0	2/1	1/1

Standard one-letter abbreviations for all amino acids. O indicates 4-hydroxyproline, which only occurs at the Y position. C (Cys) and W (Trp) are almost completely absent from the COL domains of fibrillar collagens triple-helical domains. The analysis of collagen sequences in these fish helps to understand the evolution of proteins under environmental pressure and provides a blueprint for biomedical applications requiring tuneable collagen properties.

Biomedical Applications of Fibrillar Collagens from Bony Fish

Fibrillar collagen is used for a wide variety of biomedical applications due to its ability to polymerise into supramolecular structures which accurately resemble the extracellular matrix (ECM) of mammalian tissues. Collagen facilitates key cellular behaviours such as differentiation, proliferation, and migration [77, 125, 144], and it can be prepared in a variety of physical forms depending on the intended application. Collagen can be prepared as a viscoelastic material such as hydrogels, emulsions, or injectable liquids, and in solid forms such as films, 3D scaffolds, and solid implants [22, 29, 44, 93, 123, 139, 142], making it a truly versatile graft material for biomedical therapies.

Biomedical products that incorporate collagen or are entirely collagen-based almost exclusively use fibrillar type I collagen, as it is the most abundant protein in animals and large quantities can be extracted from animal tissues [6, 128]. Collagen is typically sourced from waste products of livestock farming such as hooves, bones, skins, tendons, and cartilage from bovine and porcine origin [65, 115]. The main limitation to the widespread adoption of collagen derived from these sources into biomedical products is the regulation of bovine and porcine products due to the risk of zoonotic disease transmission, in particular bovine spongiform encephalopathy and foot and mouth disease [157]. Bovine and porcine products are also unfeasible to a significant portion of the global population due to widely held Hindu, Jewish, and Muslim beliefs [56].

For these reasons, alternative collagen sources such as marine organisms present a promising resource with potentially fewer limitations to widespread adoption into biomedical products. In a similar manner to the sourcing of collagen from the waste products of abattoirs, fibrillar collagens are extracted in large quantities from the waste products of fisheries, such as fish scales, bones, organs, and heads [111]. These fibrillar collagens are typically referred to as type I collagen for both vertebrate and invertebrate marine organisms, despite their biochemical differences and the decreasing homology between the collagen genes of organisms that diverged early on in the phylogenetic tree, e.g., between marine invertebrates and mammals.

A large number of research articles have been published to date which investigate the application of marine collagens as biomaterials. These collagens come from a variety of marine sources including bony fish, cartilaginous fish, and a huge range of invertebrates such as sea cucumbers, sea urchins, sponges, sea anemones, cephalopods, and jellyfish [25, 30, 45, 46, 48, 116]. Given the lack of conservation of collagen genes between such a broad spectrum of organisms, it is difficult to compare these studies since they investigate fundamentally different proteins—possibly not even fibrillar collagens in all cases. Additionally, many studies focus on the application of marine collagen peptides where the extracted collagens are hydrolysed into small fragments which are no longer fibrillar or triple-helical. Collagen peptides are typically used in the cosmetic and nutraceutical industries as supplements [8, 10, 26, 92, 147]. However, there is a substantial number of research articles which investigate the pharmacological applications of collagen peptides as anti-hypertensive, anti-oxidant, and anti-inflammatory small molecules. These studies fall outside of the scope of fibrillar collagens and are comprehensively reviewed elsewhere [20, 87, 130].

The primary biomedical application of fibrillar collagens is in the field of tissue engineering and regenerative medicine (TERM). TERM attempts to recreate functional tissues and organs as a panacea for injured or diseased human tissue. Focussing on the application of fibrillar collagens from bony fish, current studies centre on the regeneration of skin (commonly referred to as wound healing), bone, articular cartilage, and ligament tissues. Recent TERM publications that incorporate fibrillar collagens from bony fish are listed in Table 8.5.

Wound Healing

Wound healing is a popular area of TERM research which frequently incorporates collagen as a 3D scaffold (sponge) to accelerate the closure of skin lesions. The ability of collagen to facilitate cell proliferation and migration is the key driver in accelerating wound closure; however, there is also a significant amount of literature highlighting the haemostatic properties of collagen which is also a key attribute in facilitating the early stages of wound healing.

Freeze-dried sponges made from fish skin-derived collagen have been shown to have a haemostatic effect using in vitro and in vivo models. Li et al. demonstrated that sponges made from acid-solubilised and pepsin-solubilised collagen from bighead bream scales were significantly more effective at inducing blood clotting in vitro than synthetic controls (gauze) or gelatine sponges [94]. This finding is interesting as it would suggest that the haemostatic effect of collagen is dependent on its natural triple-helical conformation, since the denatured collagen (gelatine) sponge was entirely ineffective. The similar efficacy of acid-solubilised and pepsin-solubilised collagens would also suggest that the N- and C-terminus collagen telopeptides are not involved in the haemostatic mechanism of action. Sun et al. verified this haemostatic effect in vivo with the use of freeze-dried tilapia skin collagen sponges to accelerate blood clotting in liver and artery haemorrhage models in rats [145]. The tilapia collagen sponges were equally effective to mammalian type I collagen scaffolds in both models, except when cross-linked with the carbodiimide EDC/NHS which significantly accelerated the clotting ability of the sponge for arterial blood haemorrhage. The authors comment that this effect is due to the higher absorption capacity of the cross-linked sponge which accelerated clot formation and

Table 8.5 Publications investigating the use of bony fish fibrillar collagens for TERM research

Application	Source	Final material	Form	Biological evaluation	Reference
Wound healing	Barramundi	Collagen from skin was freeze-dried	Sponge	In vitro human fibroblast and keratinocyte proliferation, in vivo wound healing and angiogenesis	[96]
	Bighead bream	Collagen from scales was freeze-dried	Sponge	Viability assay on mouse embryonic fibroblasts and coagulation	[94]
	Black pomfret	Collagen from fins and tails was mixed with pectin and freeze-dried	Sponge	Haemolysis, coagulation, metabolic activity of human keratinocytes, scratch test and in vivo angiogenesis	[134]
	Blue grenadier	Collagen from skin was pepsin-digested and methacrylated	Hydrogel	In vitro fibroblast viability and metabolic activity assays and 3D bioprinting	[98]
	Tilapia and cod	Collagen from tilapia skin was methacrylated and blended with cod-skin collagen	Hydrogel	In vitro fibroblast and keratinocyte viability assays and 3D bioprinting	[89]
	Tilapia	Collagen from skin was formed as a hydrogel or freeze-dried	Hydrogel/sponge	Adhesion, proliferation, and migration of human fibroblasts in vitro and in vivo wound healing of sponge	[73]
	Tilapia	Collagen from skin was freeze-dried and cross-linked with EDC/NHS	Sponge	In vivo biocompatibility, in vitro fibroblast growth rate, and in vivo haemostasis	[145]
	Tilapia	Collagen from skin was freeze-dried and the outside coated with chitosan	Bi-layered sponge	In vivo wound healing, histological analysis, inflammatory cytokine levels in skin, and in vitro fibroblast proliferation	[146]
	Tilapia	Collagen from skin was formed as a hydrogel	Hydrogel	In vivo wound healing, histological analysis, and gene expression	[36]

(continued)

Table 8.5 (continued)

Application	Source	Final material	Form	Biological evaluation	Reference
	Tilapia	Collagen from skin was formed as a hydrogel	Hydrogel	Viability assay on mouse embryonic fibroblasts, in vivo wound healing, and histological analysis	[52]
	Tilapia	Collagen from skin was freeze-dried	Sponge	In vivo wound healing, histological analysis, and gene expression	[23]
Bone engineering	Goldfish	Scales were decellularised and functionalised with osteogenic exosomes	Sponge	Viability assay on bone marrow-derived stem cells (BMSCs), in vivo biocompatibility, in vitro osteogenesis, and calvarial defect repair in vivo	[156]
	Olive flounder	Collagen from skins was functionalised to polycaprolactone and hydroxyapatite scaffolds	Sponge	Viability and alkaline phosphatase assays on pre-osteoblasts in vitro and calvarial defect repair in vivo	[82]
	Salmon	Collagen from skin was mixed with hydroxyapatite precursors and allowed to form mineralised fibrils before freeze-drying	Sponge	Proliferation and alkaline phosphatase assays on mesenchymal stem cell (MSC)-derived osteoblasts in vitro	[71]
	Silver carp	Collagen from skin was formed as a hydrogel and combined with a commercial bone substitute (Geistlich Bio-Oss)	Hydrogel/sponge	Calvarial defect repair in vivo	[61]
	Tilapia	Collagen from scales was formed as a hydrogel before dehydrating in ethanol	Membrane	Alkaline phosphatase assay and gene expression on MSC-derived osteoblasts in vitro	[104]

(continued)

Table 8.5 (continued)

Application	Source	Final material	Form	Biological evaluation	Reference
	Tilapia	Collagen from skin was cross-linked with EDC/NHS, then blended with hydroxyapatite and freeze-dried	Sponge	Viability, migration, toxicity, and osteogenic differentiation assays for BMSCs in vitro and calvarial defect repair in vivo	[160]
Cartilage engineering	Salmon	Collagen from skin was pepsin-digested, freeze-dried, cross-linked with EDC, and freeze-dried again	Sponge	Knee joint defect repair in vivo, histological analysis, and gene expression	[81]
	Tilapia	Commercial collagen solution from scales	Coating	MSC chondrogenesis assay and gene expression in vitro	[72]
Ligament engineering	Salmon	Collagen from skin was formed as a hydrogel and cross-linked with EDC	Hydrogel	Ligament fibroblast proliferation and differentiation in vitro	[113]
	Salmon	Collagen from skin was freeze-dried, cross-linked with EDC, and freeze-dried again	Sponge	Ligament fibroblast proliferation and differentiation in vitro	[112]

reduced the total blood loss. In a later study, however, Sun et al. observed that non-cross-linked tilapia collagen sponges showed significantly accelerated wound closure compared to EDC/NHS cross-linked sponges in a full-thickness skin wound in vivo model, suggesting that a cross-linked scaffold is beneficial for initial haemostasis but not for longer term wound closure—possibly due to the cross-linked material being more difficult to degrade during the tissue remodelling process [146].

Several studies evidence the ability of fish skin collagen materials to promote the proliferation and migration of the principal skin cells—fibroblasts and keratinocytes. Huang et al. showed that tilapia skin collagen-coated tissue culture (TC) wells were comparable to type I collagen coatings from bovine, porcine, and rodent sources for supporting the adhesion, proliferation, and migration of human fibroblasts in vitro [73]. They also demonstrated the ability of freeze-dried tilapia collagen sponges to assist in wound closure in a full-thickness skin wound model in mice. The rate of wound closure was comparable to bovine and porcine type I collagen sponge controls; however, the tilapia collagen sponge only showed a slight improvement to the non-collagen control sponge (gauze) at one time point.

Elbialy et al. demonstrated that tilapia skin collagen hydrogels could accelerate wound closure in vivo using a full-thickness skin wound model [36]. This could suggest that collagen accessibility to cells during tissue remodelling is improved when provided as a soft hydrogel as opposed to a scaffold; however, only an untreated wound control was included here and an equivalent scaffold control would be required to verify this hypothesis.

Bone Engineering

The ECM of bone tissue is primarily composed of type I collagen, making fibrillar collagens from bony fish viable substrates for TERM approaches. Bony fish collagens are frequently employed in tandem with the inorganic mineral hydroxyapatite, which is also a major component of bone tissue, to create composite materials which can assist in mesenchymal stem cell (MSC) osteogenic differentiation and in the deposition of de novo bone tissue by osteoblasts.

Hoyer et al. demonstrated that freeze-dried salmon skin collagen scaffolds incorporating hydroxyapatite promoted the proliferation of MSC-derived osteoblasts as well as increasing the activity of the bone mineralisation enzyme alkaline phosphatase (ALP) in vitro [71]. One limitation of this study to consider is that the control used was simply MSC-derived osteoblasts cultured in 2D on polystyrene TC plates, which does not allow the comparison of the salmon skin collagen to a mammalian type I collagen equivalent and does not account for the potential differences in cell phenotype and activity seen between 2D and 3D culture methods. Matsumoto et al. took a similar approach investigating the potential for tilapia scale collagen membranes to stimulate the osteogenic differentiation of MSCs towards osteoblasts [104]. They observed increased ALP activity and the upregulation of osteogenic genes for MSC-derived osteoblasts cultured on tilapia collagen-coated TC plates compared to porcine type I collagen coatings. Interestingly, they also report a significant increase in ALP activity when cultured on fibrillated tilapia collagen compared to non-fibrillated tilapia skin collagen coatings, suggesting that the supramolecular structure of collagen plays a role in MSC osteogenesis.

Several studies investigated the potential for bony fish collagens to assist in bone repair in vivo through the use of a calvaria defect model. These studies all demonstrated favourable results; however, only one study to date has included an appropriate mammalian type I collagen control to compare to [160]. Wang et al. showed how decellularised goldfish scales scaffolds could be combined with osteogenic exosomes to enable de novo bone formation in vivo, albeit when compared to an untreated control group [156]. It is also worth noting that this study employed a decellularised ECM material which would likely contain a combination of fibrous proteins in addition to collagen, such as elastin, keratin, and fibronectin. Hadzik et al. showed similarly positive results when using a silver carp skin collagen hydrogel in combination with a commercial bone substitute material to repair bone tissue in a rat calvaria defect model [61]. They demonstrated de novo bone formation to a comparable level to that of a type I collagen membrane used in conjunction with the same bone substitute material; however, it is difficult to determine what extent of bone formation was due to the silver carp skin hydrogel as opposed to the bone substitute material which was included in both treatment groups. Kim et al. demonstrated how a synthetic polymer such as polycaprolactone (PCL) could be used as a 3D-printable scaffold material which could subsequently be functionalised with collagen to improve the cytocompatibility of the printed constructs [82]. They combined 3D-printed PCL scaffolds with hydroxyapatite before functionalising the printed constructs with olive flounder skin atelocollagen. This material was shown to accelerate the deposition of de novo bone tissue in a mouse calvaria defect model when compared to a non-functionalised PCL control, demonstrating the potential for 3D-printable materials to be functionalised with bioactive molecules post fabrication. Yao et al. showed comparable de novo bone formation in a rat calvarial defect model between tilapia skin collagen scaffolds and bovine type I collagen scaffolds, suggesting that mammalian type I collagen is not essential for stimulating bone regeneration in mammals [160]. They also demonstrated the relationship between increasing hydroxyapatite content incorporated into the bone graft material and the increase in de novo bone formation in vivo.

Cartilage Engineering

Articular cartilage is an attractive target for TERM therapies due to its poor intrinsic regenerative capacity. This tissue is rich in type II collagen and has already been repaired in a clinical setting using a collagen-assisted regenerative medicine approach called matrix-assisted chondrocyte implantation (MACI) [3]. Porcine and bovine type I collagen are currently used in MACI therapies which means there is a clear clinical impact which can be achieved if an alternative collagen source can be proven to have superior chondrogenic potential.

Hsu et al. showed the ability of tilapia scale collagen-coated TC wells to support MSCs during chondrogenic differentiation [72]. Quantification of glycosaminoglycan production and chondrogenic marker gene expression showed mixed results with no clear trends other than a significant upregulation in expression of the proteoglycan aggrecan for both tilapia scale collagen and porcine type I collagen-coated TC wells when compared to uncoated control wells. Kawaguchi et al. used a freeze-dried salmon skin atelocollagen sponge to repair articular cartilage in a femoral trochlea defect model in rabbits [81]. Gross observation and histological evaluation scores showed that the salmon skin atelocollagen sponge produced more de novo cartilage than the untreated control in addition to producing a higher quality of hyaline cartilage in terms of type II collagen and proteoglycan content. These values were not significantly different to porcine type I collagen sponge controls. However, it was observed that the salmon skin atelocollagen sponge produced higher amounts of type II collagen mRNA transcripts than the porcine equivalent which could suggest that a significant difference may have been observed at a later time point.

Ligament Engineering

Ligament tissue has a very high proportion of type I collagen (80% of the dry mass) which means that any fish collagens which closely resemble mammalian type I collagen would be ideally suited as grafts which accurately mimic native ligament ECM [74]. A key consideration with ligament TERM strategies is that they must be able to support the alignment of large collagen fibres, whether initially or after cellular remodelling of the implanted graft, in order to address the mechanical forces which ligaments naturally withstand.

Nagai et al. are the only authors, to our knowledge, who have investigated ligament repair strategies using collagen derived from bony fish. In two separate studies, they investigated the use of salmon skin atelocollagen freeze dried sponges and hydrogels to facilitate the in vitro expansion of human periodontal ligament cells (HPDLs) [112, 113]. In the former study, salmon skin atelocollagen scaffolds exhibited a similar potential for supporting the proliferation of HPDLs to bovine type I atelocollagen sponges, in addition to producing a higher level of ALP activity than the bovine control sponge. These findings were also observed using salmon skin atelocollagen hydrogels in the latter study when compared to porcine type I atelocollagen hydrogels. Gene expression analysis of HPDLs cultured on the hydrogels also validated this finding by showing an upregulation in the expression of ALP mRNA. Though these findings are encouraging, studies which include in vivo models of ligament repair using bony fish collagen materials are required to properly investigate their potential for ligament TERM therapies.

Concluding Remarks

Collagens sourced from marine organisms are emerging as valuable biomaterials with promising potential in many biotechnological and biomedical applications, particularly in the field of tissue regeneration. Fibrillar collagens from bony fish present economic advantages, as they can be extracted from skin, scales, bones, and other byproducts of the fish industry. They have attracted growing interest from both academic and industrial sectors. While type I collagen is the most studied due to its abundance and ease of extraction, other fibrillar collagens remain largely unexplored and their biological roles are expected to be similar to those of their mammalian collagen counterparts. Early research efforts concentrated on the biochemical characterisation of type I collagen from fish skins, bones, or swim bladders, but attention has gradually shifted toward understanding the function and regulation of the multiple fibrillar collagen genes present in bony fish, using model organisms such as zebrafish. This shift reflects the complexity introduced by whole genome duplication events in the evolutionary history of teleost fishes, which have resulted in multiple gene copies and diverse expression patterns for most fibrillar collagens. Such multiplicity complicates the functional interpretation of individual collagens and their trimeric assemblies, although ongoing studies are working to resolve these intricacies. One of the unique features of bony fish collagens is their high degree of amino acid sequence variability, which contributes to a broad spectrum of thermal stabilities. This diversity reflects the evolutionary adaptation of bony fish to a wide range of aquatic environments and may offer tailored functional properties for specific biotechnological applications. While significant genomic data are available for various species, there remains a notable gap in our understanding at the biochemical and functional levels. As research progresses, expanding our knowledge of fish fibrillar collagens beyond type I collagen will be essential. Unlocking their full potential could lead to novel applications spanning from nutraceutical and cosmetic formulations to advanced scaffolds for tissue engineering and regenerative therapies.

References

1. Amores A, Force A, Yan Y-L, Joly L, Amemiya C, Fritz A, Ho RK, Langeland J, Prince V, Wang Y-L, Westerfield M, Ekker M, Postlethwait JH (1998) Zebrafish *hox* clusters and vertebrate genome evolution. Science 282(5394):1711–1714. https://doi.org/10.1126/science.282.5394.1711
2. An B, Lin Y-S, Brodsky B (2016) Collagen interactions: drug design and delivery. Adv Drug Deliv Rev 97:69–84. https://doi.org/10.1016/j.addr.2015.11.013
3. Basad E, Wissing FR, Fehrenbach P, Rickert M, Steinmeyer J, Ishaque B (2015) Matrix-induced autologous chondrocyte implantation (MACI) in the knee: clinical outcomes and challenges. Knee Surg Sports Traumatol Arthrosc 23(12):3729–3735. https://doi.org/10.1007/s00167-014-3295-8
4. Bear RS (1952) The structure of collagen fibrils. Adv Protein Chem 7:69–160. https://doi.org/10.1016/s0065-3233(08)60018-2
5. Bella J (2016) Collagen structure: new tricks from a very old dog. Biochem J 473(8):1001–1025. https://doi.org/10.1042/BJ20151169
6. Bella J, Hulmes DJ (2017) Fibrillar Collagens. Subcell Biochem 82:457–490. https://doi.org/10.1007/978-3-319-49674-0_14
7. Bella J, Brodsky B, Berman HM (1995) Hydration structure of a collagen peptide. Structure 3(9):893–906. https://doi.org/10.1016/S0969-2126(01)00224-6
8. Bhagwat PK, Dandge PB (2016) Isolation, characterization and valorizable applications of fish scale collagen in food and agriculture industries. Biocatal Agric Biotechnol 7:234–240. https://doi.org/10.1016/j.bcab.2016.06.010
9. Boot-Handford RP, Tuckwell DS, Plumb DA, Rock CF, Poulsom R (2003) A novel and highly conserved collagen (proα1(XXVII)) with a unique expression pattern and unusual molecular characteristics establishes a new clade within the vertebrate fibrillar collagen family. J Biol Chem 278(33):31067–31077. https://doi.org/10.1074/jbc.M212889200
10. Boran G, Regenstein JM (2010) Fish gelatin. Adv Food Nutr Res 60(1):119–143. https://doi.org/10.1016/S1043-4526(10)60005-8
11. Boraschi-Diaz I, Wang J, Mort JS, Komarova SV (2017) Collagen type I as a ligand for receptor-mediated signaling. Front Phys 5:12. https://doi.org/10.3389/fphy.2017.00012

12. Boudko SP, Engel J, Bachinger HP (2012) The crucial role of trimerization domains in collagen folding. Int J Biochem Cell Biol 44(1):21–32. https://doi.org/10.1016/j.biocel.2011.09.009
13. Bourhis JM, Mariano N, Zhao Y, Harlos K, Exposito JY, Jones EY, Moali C, Aghajari N, Hulmes DJ (2012) Structural basis of fibrillar collagen trimerization and related genetic disorders. Nat Struct Mol Biol 19(10):1031–1036. https://doi.org/10.1038/nsmb.2389
14. Bradford YM, Van Slyke CE, Ruzicka L, Singer A, Eagle A, Fashena D, Howe DG, Frazer K, Martin R, Paddock H, Pich C, Ramachandran S, Westerfield M (2022) Zebrafish information network, the knowledgebase for *Danio rerio* research. Genetics 220(4). https://doi.org/10.1093/genetics/iyac016
15. Brodsky B, Persikov AV (2024) A two-step dance commits collagen to folding. Proc Natl Acad Sci USA 121(52):e2422338121. https://doi.org/10.1073/pnas.2422338121
16. Brodsky B, Ramshaw JAM (2017) Bioengineered Collagens. Subcell Biochem 82:601–629. https://doi.org/10.1007/978-3-319-49674-0_18
17. Browne S, Zeugolis DI, Pandit A (2013) Collagen: finding a solution for the source. Tissue Eng Part A 19(13–14):1491–1494. https://doi.org/10.1089/ten.TEA.2012.0721
18. Burjanadze TV (1979) Hydroxyproline content and location in relation to collagen thermal stability. Biopolymers 18(4):931–938. https://doi.org/10.1002/bip.1979.360180413
19. Burjanadze TV (2000) New analysis of the phylogenetic change of collagen thermostability. Biopolymers 53(6):523–528. https://doi.org/10.1002/(SICI)1097-0282(200005)53:6<523::AID-BIP8>3.0.CO;2-7
20. Cadar E, Pesterau A-M, Sirbu R, Negreanu-Pirjol BS, Tomescu CL (2023) Jellyfishes—significant marine resources with potential in the wound-healing process: a review. Mar Drugs 21(4):201. https://doi.org/10.3390/md21040201
21. Cao L, Zhang Z, Yuan D, Yu M, Min J (2024) Tissue engineering applications of recombinant human collagen: a review of recent progress. Front Bioeng Biotechnol 12:1358246. https://doi.org/10.3389/fbioe.2024.1358246
22. Chattopadhyay S, Raines RT (2014) Collagen-based biomaterials for wound healing. Biopolymers 101(8):821–833. https://doi.org/10.1002/bip.22486
23. Chen J, Gao K, Liu S, Wang S, Elango J, Bao B, Dong J, Liu N, Wu W (2019a) Fish collagen surgical compress repairing characteristics on wound healing process in vivo. Mar Drugs 17(1):33. https://doi.org/10.3390/md17010033
24. Chen Z, Omori Y, Koren S, Shirokiya T, Kuroda T, Miyamoto A, Wada H, Fujiyama A, Toyoda A, Zhang S, Wolfsberg TG, Kawakami K, Phillippy AM, Program NCS, Mullikin JC, Burgess SM (2019b) De novo assembly of the goldfish (*Carassius auratus*) genome and the evolution of genes after whole-genome duplication. Sci Adv 5(6):eaav0547. https://doi.org/10.1126/sciadv.aav0547
25. Cheung RCF, Ng TB, Wong JH (2015) Marine peptides: bioactivities and applications. Mar Drugs 13(7):4006–4043. https://doi.org/10.3390/md13074006
26. Choi BD, Choi YJ (2012) Nutraceutical functionalities of polysaccharides from marine invertebrates. Adv Food Nutr Res 65:11–30. https://doi.org/10.1016/B978-0-12-416003-3.00002-0
27. Christiansen HE, Lang MR, Pace JM, Parichy DM (2009) Critical early roles for *col27a1a* and *col27a1b* in zebrafish notochord morphogenesis, vertebral mineralization and post-embryonic axial growth. PLoS One 4(12):e8481. https://doi.org/10.1371/journal.pone.0008481
28. Cohen C (1955) Optical rotation and helical polypeptide chain configuration in collagen and gelatin. J Biophys Biochem Cytol 1(3):203–214. https://doi.org/10.1083/jcb.1.3.203
29. Coppola D, Oliviero M, Vitale GA, Lauritano C, D'Ambra I, Iannace S, De Pascale D (2020) Marine collagen from alternative and sustainable sources: extraction, processing and applications. Mar Drugs 18(4):214. https://doi.org/10.3390/md18040214
30. De Domenico S, De Rinaldis G, Paulmery M, Piraino S, Leone A (2019) Barrel jellyfish (*Rhizostoma pulmo*) as source of antioxidant peptides. Mar Drugs 17(2):134. https://doi.org/10.3390/md17020134

31. Driever W, Fishman MC (1996) The zebrafish: heritable disorders in transparent embryos. J Clin Invest 97(8):1788–1794. https://doi.org/10.1172/JCI118608
32. Durán I, Marí-Beffa M, Santamaría JA, Becerra J, Santos-Ruiz L (2011) Actinotrichia collagens and their role in fin formation. Dev Biol 354(1):160–172. https://doi.org/10.1016/j.ydbio.2011.03.014
33. Durán I, Csukasi F, Taylor SP, Krakow D, Becerra J, Bombarely A, Marí-Beffa M (2015) Collagen duplicate genes of bone and cartilage participate during regeneration of zebrafish fin skeleton. Gene Expr Patterns 19(1–2):60–69. https://doi.org/10.1016/j.gep.2015.07.004
34. Eastoe JE (1957) The amino acid composition of fish collagen and gelatin. Biochem J 65(2): 363–368. https://doi.org/10.1042/bj0650363
35. Elango J, Hou C, Bao B, Wang S, Maté Sánchez De Val JE, Wenhui W (2022) The molecular interaction of collagen with cell receptors for biological function. Polymers 14(5):876. https://doi.org/10.3390/polym14050876
36. Elbialy ZI, Atiba A, Abdelnaby A, Al-Hawary II, Elsheshtawy A, El-Serehy HA, Abdel-Daim MM, Fadl SE, Assar DH (2020) Collagen extract obtained from Nile tilapia (*Oreochromis niloticus* L.) skin accelerates wound healing in rat model via up regulating VEGF, bFGF, and α-SMA genes expression. BMC Vet Res 16(1):352. https://doi.org/10.1186/s12917-020-02566-2
37. Exposito JY, Garrone R (1990) Characterization of a fibrillar collagen gene in sponges reveals the early evolutionary appearance of two collagen gene families. Proc Natl Acad Sci USA 87(17):6669–6673. https://doi.org/10.1073/pnas.87.17.6669
38. Exposito JY, Cluzel C, Garrone R, Lethias C (2002) Evolution of collagens. Anat Rec 268(3): 302–316. https://doi.org/10.1002/ar.10162
39. Exposito JY, Valcourt U, Cluzel C, Lethias C (2010) The fibrillar collagen family. Int J Mol Sci 11(2):407–426. https://doi.org/10.3390/ijms11020407
40. Fang M, Adams JS, Mcmahan BL, Brown RJ, Oxford JT (2010) The expression patterns of minor fibrillar collagens during development in zebrafish. Gene Expr Patterns 10(7–8): 315–322. https://doi.org/10.1016/j.gep.2010.07.002
41. Fang M, Jacob R, McDougal O, Oxford JT (2012) Minor fibrillar collagens, variable regions alternative splicing, intrinsic disorder, and tyrosine sulfation. Protein Cell 3(6):419–433. https://doi.org/10.1007/s13238-012-2917-5
42. Farndale RW, Slatter DA, Siljander PR, Jarvis GE (2007) Platelet receptor recognition and cross-talk in collagen-induced activation of platelets. J Thromb Haemost 5(Suppl 1):220–229. https://doi.org/10.1111/j.1538-7836.2007.02521.x
43. Fassini D, Wilkie IC, Pozzolini M, Ferrario C, Sugni M, Rocha MS, Giovine M, Bonasoro F, Silva TH, Reis RL (2021) Diverse and productive source of biopolymer inspiration: marine collagens. Biomacromolecules 22(5):1815–1834. https://doi.org/10.1021/acs.biomac.1c00013
44. Felician FF, Xia C, Qi W, Xu H (2018) Collagen from marine biological sources and medical applications. Chem Biodivers 15(5):1700557. https://doi.org/10.1002/cbdv.201700557
45. Felician FF, Yu RH, Li MZ, Li CJ, Chen HQ, Jiang Y, Tang T, Qi WY, Xu HM (2019) The wound healing potential of collagen peptides derived from the jellyfish *Rhopilema esculentum*. Chin J Traumatol 22(1):12–20. https://doi.org/10.1016/j.cjtee.2018.10.004
46. Ferrario C, Rusconi F, Pulaj A, Macchi R, Landini P, Paroni M, Colombo G, Martinello T, Melotti L, Gomiero C, Candia Carnevali MD, Bonasoro F, Patruno M, Sugni M (2020) From food waste to innovative biomaterial: sea urchin-derived collagen for applications in skin regenerative medicine. Mar Drugs 18(8):414. https://doi.org/10.3390/MD18080414
47. Fisher S, Jagadeeswaran P, Halpern ME (2003) Radiographic analysis of zebrafish skeletal defects. Dev Biol 264(1):64–76. https://doi.org/10.1016/s0012-1606(03)00399-3
48. Frazão B, Antunes A (2016) Jellyfish bioactive compounds: methods for wet-lab work. Mar Drugs 14(4):75. https://doi.org/10.3390/md14040075
49. Gallop PM (1955) Studies on a parent gelatin from ichthyocol. Arch Biochem Biophys 54(2): 501–512. https://doi.org/10.1016/0003-9861(55)90062-x

50. Gallop PM, Seifter S, Meilman E (1957) The solubility and properties of a purified ichthyocol in salt solutions of neutral pH. J Biophys Biochem Cytol 3(4):545–557. https://doi.org/10.1083/jcb.3.4.545
51. Garnero P (2015) The role of collagen organization on the properties of bone. Calcif Tissue Int 97(3):229–240. https://doi.org/10.1007/s00223-015-9996-2
52. Ge B, Wang H, Li J, Liu H, Yin Y, Zhang N, Qin S (2020) Comprehensive assessment of Nile tilapia skin (*Oreochromis niloticus*) collagen hydrogels for wound dressings. Mar Drugs 18(4):178. https://doi.org/10.3390/md18040178
53. Gistelinck C, Gioia R, Gagliardi A, Tonelli F, Marchese L, Bianchi L, Landi C, Bini L, Huysseune A, Witten PE, Staes A, Gevaert K, De Rocker N, Menten B, Malfait F, Leikin S, Carra S, Tenni R, Rossi A, De Paepe A, Coucke P, Willaert A, Forlino A (2016) Zebrafish collagen type I: molecular and biochemical characterization of the major structural protein in bone and skin. Sci Rep 6:21540. https://doi.org/10.1038/srep21540
54. Gistelinck C, Kwon RY, Malfait F, Symoens S, Harris MP, Henke K, Hawkins MB, Fisher S, Sips P, Guillemyn B, Bek JW, Vermassen P, De Saffel H, Witten PE, Weis M, De Paepe A, Eyre DR, Willaert A, Coucke PJ (2018) Zebrafish type I collagen mutants faithfully recapitulate human type I collagenopathies. Proc Natl Acad Sci USA 115(34):E8037–E8046. https://doi.org/10.1073/pnas.1722200115
55. Glasauer SMK, Neuhauss SCF (2014) Whole-genome duplication in teleost fishes and its evolutionary consequences. Mol Gen Genomics 289(6):1045–1060. https://doi.org/10.1007/s00438-014-0889-2
56. Gómez-Guillén MC, Giménez B, López-Caballero ME, Montero MP (2011) Functional and bioactive properties of collagen and gelatin from alternative sources: a review. Food Hydrocoll 25(8):1813–1827. https://doi.org/10.1016/j.foodhyd.2011.02.007
57. Gordon MK, Hahn RA (2010) Collagens. Cell Tissue Res 339(1):247–257. https://doi.org/10.1007/s00441-009-0844-4
58. Gross J, Highberger JH, Schmitt FO (1954) Collagen structures considered as states of aggregation of a kinetic unit. The tropocollagen particle. Proc Natl Acad Sci USA 40(8):679–688. https://doi.org/10.1073/pnas.40.8.679
59. Gross J, Highberger JH, Schmitt FO (1955) Extraction of collagen from connective tissue by neutral salt solutions. Proc Natl Acad Sci USA 41(1):1–7. https://doi.org/10.1073/pnas.41.1.1
60. Gustavson KH (1955) The function of hydroxyproline in collagens. Nature 175(4445):70–74. https://doi.org/10.1038/175070a0
61. Hadzik J, Kubasiewicz-Ross P, Kunert-Keil C, Jurczyszyn K, Nawrot-Hadzik I, Dominiak M, Gedrange T (2016) A silver carp skin derived collagen in bone defect treatment—a histological study in a rat model. Ann Anat 208:123–128. https://doi.org/10.1016/j.aanat.2016.07.009
62. Hamaia S, Farndale RW (2014) Integrin recognition motifs in the human collagens. Adv Exp Med Biol 819:127–142. https://doi.org/10.1007/978-94-017-9153-3_9
63. Hardy MJ, Reeck JC, Fang M, Adams JS, Oxford JT (2020) *Col11a1* expression is required for zebrafish development. J Dev Biol 8(3):16. https://doi.org/10.3390/jdb8030016
64. Harrington WF, von Hippel PH (1961) Formation and stabilization of the collagen-fold. Arch Biochem Biophys 92:100–113. https://doi.org/10.1016/0003-9861(61)90224-7
65. Hashim P, Mohd Ridzwan MS, Bakar J, Mat Hashim D (2015) Collagen in food and beverage industries. Int Food Res J 22(1):1–8. Available at www.ifrj.upm.edu.my/22(01)2015/(1).pdf
66. Heino J (2014) Cellular signaling by collagen-binding integrins. Adv Exp Med Biol 819:143–155. https://doi.org/10.1007/978-94-017-9153-3_10
67. Henriksen K, Karsdal MA (2016) Chapter 1 - Type I collagen. In: Karsdal MA (ed) Biochemistry of collagens, laminins and elastin, 2nd edn. Academic Press, pp 1–12. https://doi.org/10.1016/B978-0-12-817068-7.00001-X
68. Highberger JH, Gross J, Schmitt FO (1951) The interaction of mucoprotein with soluble collagen; an electron microscope study. Proc Natl Acad Sci USA 37(5):286–291. https://doi.org/10.1073/pnas.37.5.286

69. Hoffman GG, Branam AM, Huang G, Pelegri F, Cole WG, Wenstrup RM, Greenspan DS (2010) Characterization of the six zebrafish clade B fibrillar procollagen genes, with evidence for evolutionarily conserved alternative splicing within the pro-α1(V) C-propeptide. Matrix Biol 29(4):261–275. https://doi.org/10.1016/j.matbio.2010.01.006
70. Howe K, Clark MD, Torroja CF, Torrance J, Berthelot C, Muffato M, Collins JE, Humphray S, McLaren K, Matthews L, McLaren S, Sealy I, Caccamo M, Churcher C, Scott C, Barrett JC, Koch R, Rauch GJ, White S, Chow W, Kilian B, Quintais LT, Guerra-Assunção JA, Zhou Y, Gu Y, Yen J, Vogel JH, Eyre T, Redmond S, Banerjee R, Chi J, Fu B, Langley E, Maguire SF, Laird GK, Lloyd D, Kenyon E, Donaldson S, Sehra H, Almeida-King J, Loveland J, Trevanion S, Jones M, Quail M, Willey D, Hunt A, Burton J, Sims S, McLay K, Plumb B, Davis J, Clee C, Oliver K, Clark R, Riddle C, Elliot D, Threadgold G, Harden G, Ware D, Begum S, Mortimore B, Kerry G, Heath P, Phillimore B, Tracey A, Corby N, Dunn M, Johnson C, Wood J, Clark S, Pelan S, Griffiths G, Smith M, Glithero R, Howden P, Barker N, Lloyd C, Stevens C, Harley J, Holt K, Panagiotidis G, Lovell J, Beasley H, Henderson C, Gordon D, Auger K, Wright D, Collins J, Raisen C, Dyer L, Leung K, Robertson L, Ambridge K, Leongamornlert D, McGuire S, Gilderthorp R, Griffiths C, Manthravadi D, Nichol S, Barker G, Whitehead S, Kay M, Brown J, Murnane C, Gray E, Humphries M, Sycamore N, Barker D, Saunders D, Wallis J, Babbage A, Hammond S, Mashreghi-Mohammadi M, Barr L, Martin S, Wray P, Ellington A, Matthews N, Ellwood M, Woodmansey R, Clark G, Cooper J, Tromans A, Grafham D, Skuce C, Pandian R, Andrews R, Harrison E, Kimberley A, Garnett J, Fosker N, Hall R, Garner P, Kelly D, Bird C, Palmer S, Gehring I, Berger A, Dooley CM, Ersan-Ürün Z, Eser C, Geiger H, Geisler M, Karotki L, Kirn A, Konantz J, Konantz M, Oberländer M, Rudolph-Geiger S, Teucke M, Lanz C, Raddatz G, Osoegawa K, Zhu B, Rapp A, Widaa S, Langford C, Yang F, Schuster SC, Carter NP, Harrow J, Ning Z, Herrero J, Searle SMJ, Enright A, Geisler R, Plasterk RHA, Lee C, Westerfield M, de Jong PJ, Zon LI, Postlethwait JH, Nüsslein-Volhard C, Hubbard TJP, Crollius HR, Rogers J, Stemple DL (2013) The zebrafish reference genome sequence and its relationship to the human genome. Nature 496(7446):498–503. https://doi.org/10.1038/nature12111
71. Hoyer B, Bernhardt A, Heinemann S, Stachel I, Meyer M, Gelinsky M (2012) Biomimetically mineralized salmon collagen scaffolds for application in bone tissue engineering. Biomacromolecules 13(4):1059–1066. https://doi.org/10.1021/bm201776r
72. Hsu H-H, Uemura T, Yamaguchi I, Ikoma T, Tanaka J (2016) Chondrogenic differentiation of human mesenchymal stem cells on fish scale collagen. J Biosci Bioeng 122(2):219–225. https://doi.org/10.1016/j.jbiosc.2016.01.001
73. Huang J-Y, Wong T-Y, Tu T-Y, Tang M-J, Lin H-H, Hsueh Y-Y (2024) Assessment of tilapia skin collagen for biomedical research applications in comparison with mammalian collagen. Molecules 29(2):402. https://doi.org/10.3390/molecules29020402
74. Hudson DM, Archer M, Rai J, Weis M, Fernandes RJ, Eyre DR (2021) Age-related type I collagen modifications reveal tissue-defining differences between ligament and tendon. Matrix Biol Plus 12:100070. https://doi.org/10.1016/j.mbplus.2021.100070
75. Jimenez SA, Bashey RI, Benditt M, Yankowski R (1977) Identification of collagen α_1(I) trimer in embryonic chick tendons and calvaria. Biochem Biophys Res Commun 78(4):1354–1361. https://doi.org/10.1016/0006-291x(77)91441-3
76. Kadler KE (2017) Fell Muir lecture: collagen fibril formation *in vitro* and *in vivo*. Int J Exp Pathol 98(1):4–16. https://doi.org/10.1111/iep.12224
77. Kadler KE, Baldock C, Bella J, Boot-Handford RP (2007) Collagens at a glance. J Cell Sci 120 (Pt 12):1955–1958. https://doi.org/10.1242/jcs.03453
78. Kadler KE, Hill A, Canty-Laird EG (2008) Collagen fibrillogenesis: fibronectin, integrins, and minor collagens as organizers and nucleators. Curr Opin Cell Biol 20(5):495–501. https://doi.org/10.1016/j.ceb.2008.06.008
79. Kalamajski S, Oldberg Å (2010) The role of small leucine-rich proteoglycans in collagen fibrillogenesis. Matrix Biol 29(4):248–253. https://doi.org/10.1016/j.matbio.2010.01.001
80. Karsdal MA, Henriksen K, Nielsen MJ, Byrjalsen I, Leeming DJ, Gardner S, Goodman Z, Patel K, Krag A, Christiansen C, Schuppan D (2016) Fibrogenesis assessed by serological type

III collagen formation identifies patients with progressive liver fibrosis and responders to a potential antifibrotic therapy. Am J Physiol Gastrointest Liver Physiol 311(6):G1009–G1017. https://doi.org/10.1152/ajpgi.00283.2016
81. Kawaguchi Y, Kondo E, Kitamura N, Arakaki K, Tanaka Y, Munekata M, Nagai N, Yasuda K (2011) In vivo effects of isolated implantation of salmon-derived crosslinked atelocollagen sponge into an osteochondral defect. J Mater Sci Mater Med 22(2):397–404. https://doi.org/10.1007/s10856-010-4215-1
82. Kim S-C, Heo S-Y, Oh G-W, Yi M, Jung W-K (2022) A 3D-printed polycaprolactone/marine collagen scaffold reinforced with carbonated hydroxyapatite from fish bones for bone regeneration. Mar Drugs 20(6):344. https://doi.org/10.3390/md20060344
83. Kimura S (1992) Wide distribution of the skin type I collagen α3 chain in bony fish. Comp Biochem Physiol B 102(2):255–260. https://doi.org/10.1016/0305-0491(92)90119-c
84. Kimura S, Ohno Y (1987) Fish type I collagen: tissue-specific existence of two molecular forms, (α1)2α2 and α1α2α3, in Alaska pollack. Comp Biochem Physiol B 88(2):409–413. https://doi.org/10.1016/0305-0491(87)90320-8
85. Kimura S, Ohno Y, Miyauchi Y, Uchida N (1987) Fish skin type I collagen: wide distribution of an α3 subunit in teleosts. Comp Biochem Physiol B 88(1):27–34. https://doi.org/10.1016/0305-0491(87)90074-5
86. Kleman JP, Hartmann DJ, Ramirez F, Van Der Rest M (1992) The human rhabdomyosarcoma cell line A204 lays down a highly insoluble matrix composed mainly of α1 type-XI and α2 type-V collagen chains. Eur J Biochem 210(1):329–335. https://doi.org/10.1111/j.1432-1033.1992.tb17425.x
87. Laasri I, Bakkali M, Mejias L, Laglaoui A (2023) Marine collagen: unveiling the blue resource-extraction techniques and multifaceted applications. Int J Biol Macromol 253(6): 127253. https://doi.org/10.1016/j.ijbiomac.2023.127253
88. Lamandé SR, Bateman JF (2020) Genetic disorders of the extracellular matrix. Anat Rec (Hoboken) 303(6):1527–1542. https://doi.org/10.1002/ar.24086
89. Lee H, Chun W, Kim G (2023) Three-dimensional artificial skin construct bioprinted with a marine-based biocomposite. Biomacromolecules 24(6):2864–2878. https://doi.org/10.1021/acs.biomac.3c00253
90. Leeming DJ, Karsdal MA (2024) Chapter 5 - Type V collagen. In: Karsdal MA (ed) Biochemistry of collagens, laminins and elastin, 3rd edn. Academic Press, pp 55–60. https://doi.org/10.1016/B978-0-443-15617-5.00036-6
91. Leikina E, Mertts MV, Kuznetsova N, Leikin S (2002) Type I collagen is thermally unstable at body temperature. Proc Natl Acad Sci USA 99(3):1314–1318. https://doi.org/10.1073/pnas.032307099
92. Leone A, Lecci RM, Durante M, Meli F, Piraino S (2015) The bright side of gelatinous blooms: nutraceutical value and antioxidant properties of three Mediterranean jellyfish (Scyphozoa). Mar Drugs 13(8):4654–4681. https://doi.org/10.3390/md13084654
93. León-López A, Morales-Peñaloza A, Martínez-Juárez VM, Vargas-Torres A, Zeugolis DI, Aguirre-Álvarez G (2019) Hydrolyzed collagen-sources and applications. Molecules 24(22): 4031. https://doi.org/10.3390/molecules24224031
94. Li G, Guo T, Chen H, Huang Z, Chen Q, Wu C, Wang Y (2025) Extraction, characterization, and hemostatic effect of collagen from the scales of *Megalonibea fusca*. J Food Sci 90(1): e17644. https://doi.org/10.1111/1750-3841.17644
95. Lien S, Koop BF, Sandve SR, Miller JR, Kent MP, Nome T, Hvidsten TR, Leong JS, Minkley DR, Zimin A, Grammes F, Grove H, Gjuvsland A, Walenz B, Hermansen RA, von Schalburg K, Rondeau EB, Di Genova A, Samy JKA, Vik JO, Vigeland MD, Caler L, Grimholt U, Jentoft S, Våge DI, de Jong P, Moen T, Baranski M, Palti Y, Smith DR, Yorke JA, Nederbragt AJ, Tooming-Klunderud A, Jakobsen KS, Jiang X, Fan D, Hu Y, Liberles DA, Vidal R, Iturra P, Jones SJM, Jonassen I, Maass A, Omholt SW, Davidson WS (2016) The Atlantic salmon genome provides insights into rediploidization. Nature 533(7602):200–205. https://doi.org/10.1038/nature17164

96. Liu S, Wen F, Muthukumaran P, Rakshit M, Lau C-S, Yu N, Suryani L, Dong Y, Teoh SH (2021) Self-assembled nanofibrous marine collagen matrix accelerates healing of full-thickness wounds. ACS Appl Bio Mater 4(9):7044–7058. https://doi.org/10.1021/acsabm.1c00685
97. Luo YY, Szlarski PM, Nissen NI, Jensen C, Karsdal MA (2024) Chapter 11 - Type XI collagen. In: Biochemistry of collagens, laminins and elastin, 3rd edn. MA Karsdal. Academic Press, Editor, pp 105–112. https://doi.org/10.1016/B978-0-443-15617-5.00033-0
98. Maher M, Glattauer V, Onofrillo C, Duchi S, Yue Z, Hughes TC, Ramshaw JAM, Wallace GG (2022) Suitability of marine- and porcine-derived collagen type I hydrogels for bioprinting and tissue engineering scaffolds. Mar Drugs 20(6):366. https://doi.org/10.3390/md20060366
99. Mak KM, Png CYM, Lee DJ (2016) Type V collagen in health, disease, and fibrosis. Anat Rec (Hoboken) 299(5):613–629. https://doi.org/10.1002/ar.23330
100. Makareeva E, Han S, Vera JC, Sackett DL, Holmbeck K, Phillips CL, Visse R, Nagase H, Leikin S (2010) Carcinomas contain a matrix metalloproteinase-resistant isoform of type I collagen exerting selective support to invasion. Cancer Res 70(11):4366–4374. https://doi.org/10.1158/0008-5472.CAN-09-4057
101. Matsui R, Yamauchi H, Suganuma H, Ishida M, Kimura S (1989) Subunit composition of type I collagen from fish in salmonoidei. Nippon Suisan Gakkaishi 55(8):1421–1426. https://doi.org/10.2331/suisan.55.1421
102. Matsui R, Ishida M, Kimura S (1991) Characterization of an α3 chain from the skin type I collagen of chum salmon (*Oncoorhynchus keta*). Comp Biochem Physiol B 99(1):171–174. https://doi.org/10.1016/0305-0491(91)90025-9
103. Matsumoto T, Deguchi T, Kawasaki T, Yuba S, Sato J (2012) Molecular cloning and expression of the *col2a1a* and *col2a1b* genes in the medaka, *Oryzias latipes*. Gene Expr Patterns 12(1–2):46–52. https://doi.org/10.1016/j.gep.2011.11.002
104. Matsumoto R, Uemura T, Xu Z, Yamaguchi I, Ikoma T, Tanaka J (2015) Rapid oriented fibril formation of fish scale collagen facilitates early osteoblastic differentiation of human mesenchymal stem cells. J Biomed Mater Res A 103(8):2531–2539. https://doi.org/10.1002/jbm.a.35387
105. Mienaltowski MJ, Birk DE (2014) Structure, physiology, and biochemistry of collagens. Adv Exp Med Biol 802:5–29. https://doi.org/10.1007/978-94-007-7893-1_2
106. Miyauchi Y, Kimura S (1990) Characterization of an α3 chain from carp skin type I collagen. Nippon Suisan Gakkaishi 56(9):1509–1514. https://doi.org/10.2331/suisan.56.1509
107. Morvan Dubois G, Haftek Z, Crozet C, Garrone R, Le Guellec D (2002) Structure and spatio temporal expression of the full length DNA complementary to RNA coding for alpha2 type I collagen of zebrafish. Gene 294(1–2):55–65. https://doi.org/10.1016/s0378-1119(02)00770-9
108. Morvan-Dubois G, Le Guellec D, Garrone R, Zylberberg L, Bonnaud L (2003) Phylogenetic analysis of vertebrate fibrillar collagen locates the position of zebrafish α3(I) and suggests an evolutionary link between collagen α chains and hox clusters. J Mol Evol 57(5):501–514. https://doi.org/10.1007/s00239-003-2502-x
109. Myllyharju J, Kivirikko KI (2001) Collagens and collagen-related diseases. Ann Med 33(1): 7–21. https://doi.org/10.3109/07853890109002055
110. Myllyharju J, Kivirikko KI (2004) Collagens, modifying enzymes and their mutations in humans, flies and worms. Trends Genet 20(1):33–43. https://doi.org/10.1016/j.tig.2003.11.004
111. Nagai T, Suzuki N (2000) Isolation of collagen from fish waste material—skin, bone and fins. Food Chem 68(3):277–281. https://doi.org/10.1016/S0308-8146(99)00188-0
112. Nagai N, Yunoki S, Suzuki T, Sakata M, Tajima K, Munekata M (2004) Application of cross-linked salmon atelocollagen to the scaffold of human periodontal ligament cells. J Biosci Bioeng 97(6):389–394. https://doi.org/10.1016/S1389-1723(04)70224-8
113. Nagai N, Mori K, Satoh Y, Takahashi N, Yunoki S, Tajima K, Munekata M (2007) *In vitro* growth and differentiated activities of human periodontal ligament fibroblasts cultured on

salmon collagen gel. J Biomed Mater Res A 82A(2):395–402. https://doi.org/10.1002/jbm.a.31110
114. Nedeva IR, Vitale M, Elson A, Hoyland JA, Bella J (2021) Role of OSCAR signaling in osteoclastogenesis and bone disease. Front Cell Dev Biol 9:641162. https://doi.org/10.3389/fcell.2021.641162
115. Noorzai S, Verbeek CJR, Lay MC, Swan J (2020) Collagen extraction from various waste bovine hide sources. Waste Biomass Valoriz 11:5687–5698. https://doi.org/10.1007/s12649-019-00843-2
116. Ohnishi A, Osaki T, Matahira Y, Tsuka T, Imagawa T, Okamoto Y, Minami S (2013) Evaluation of the chondroprotective effects of glucosamine and fish collagen peptide on a rabbit ACLT model using serum biomarkers. J Vet Med Sci 75(4):421–429. https://doi.org/10.1292/jvms.12-0240
117. Padhi BK, Joly L, Tellis P, Smith A, Nanjappa P, Chevrette M, Ekker M, Akimenko M-A (2004) Screen for genes differentially expressed during regeneration of the zebrafish caudal fin. Dev Dyn 231(3):527–541. https://doi.org/10.1002/dvdy.20153
118. Persikov AV, Brodsky B (2002) Unstable molecules form stable tissues. Proc Natl Acad Sci USA 99(3):1101–1103. https://doi.org/10.1073/pnas.042707899
119. Piez KA (1964) Nonidentity of the three α chains in codfish skin collagen. J Biol Chem 239: 4315–4316. Available at www.jbc.org/article/S0021-9258(18)91176-2/pdf
120. Piez KA (1965) Characterization of a collagen from codfish skin containing three chromatographically different α chains. Biochemistry 4(12):2590–2596. https://doi.org/10.1021/bi00888a007
121. Piez KA, Gross J (1960) The amino acid composition of some fish collagens: the relation between composition and structure. J Biol Chem 235: 995–998. Available at www.jbc.org/article/S0021-9258(18)69466-9/pdf
122. Rahman MA (2019) Collagen of extracellular matrix from marine invertebrates and its medical applications. Mar Drugs 17(2):118. https://doi.org/10.3390/md17020118
123. Ramshaw JAM (2016) Biomedical applications of collagens. J Biomed Mater Res B Appl Biomater 104(4):665–675. https://doi.org/10.1002/jbm.b.33541
124. Ramshaw JAM, Werkmeister JA, Bremner HA (1988) Characterization of type I collagen from the skin of blue grenadier (*Macruronus novaezelandiae*). Arch Biochem Biophys 267(2): 497–502. https://doi.org/10.1016/0003-9861(88)90056-2
125. Ramshaw JAM, Werkmeister JA, Glattauer V (1996) Collagen-based biomaterials. Biotechnol Genet Eng Rev 13(1):335–382. https://doi.org/10.1080/02648725.1996.10647934
126. Redmond AK, Casey D, Gundappa MK, Macqueen DJ, McLysaght A (2023) Independent rediploidization masks shared whole genome duplication in the sturgeon-paddlefish ancestor. Nat Commun 14(1):2879. https://doi.org/10.1038/s41467-023-38714-z
127. Reinhardt M, Drechsler M, Paululat A (2023) *Drosophila* collagens in specialised extracellular matrices. Biol Chem 404(5):535–550. https://doi.org/10.1515/hsz-2022-0297
128. Ricard-Blum S (2011) The collagen family. Cold Spring Harb Perspect Biol 3(1):a004978. https://doi.org/10.1101/cshperspect.a004978
129. Rigby BJ (1968) Amino-acid composition and thermal stability of the skin collagen of the Antarctic ice-fish. Nature 219(5150):166–167. https://doi.org/10.1038/219166a0
130. Rigogliuso S, Campora S, Notarbartolo M, Ghersi G (2023) Recovery of bioactive compounds from marine organisms: focus on the future perspectives for pharmacological, biomedical and regenerative medicine applications of marine collagen. Molecules 28(3):1152. https://doi.org/10.3390/molecules28031152
131. Sainio A, Järvelainen H (2020) Extracellular matrix-cell interactions: focus on therapeutic applications. Cell Signal 66:109487. https://doi.org/10.1016/j.cellsig.2019.109487
132. Saito M, Kunisaki N, Hirono I, Aoki T, Ishida M, Urano N, Kimura S (1998) Partial characterization of cDNA clones encoding the three distinct proα chains of type I collagen from rainbow trout. Fish Sci 64(5):780–786. https://doi.org/10.2331/fishsci.64.780

133. Saito M, Takenouchi Y, Kunisaki N, Kimura S (2001) Complete primary structure of rainbow trout type I collagen consisting of α1(I)α2(I)α3(I) heterotrimers. Eur J Biochem 268(10): 2817–2827. https://doi.org/10.1046/j.1432-1327.2001.02160.x
134. Sathyaraj WV, Pravin YR, Prabakaran L, Gokulnath A, Bhoopathy J, Rajendran S (2024) Therapeutic potency of marine collagen/pectin scaffolds—fabrication, characterization and evaluation. Eur J Pharmacol 984:177066. https://doi.org/10.1016/j.ejphar.2024.177066
135. Schier AF (2013) Genomics: zebrafish earns its stripes. Nature 496(7446):443–444. https://doi.org/10.1038/nature12094
136. Senadheera TRL, Dave D, Shahidi F (2020) Sea cucumber derived type I collagen: a comprehensive review. Mar Drugs 18(9):471. https://doi.org/10.3390/md18090471
137. Sharma U, Carrique L, Vadon-Le Goff S, Mariano N, Georges RN, Delolme F, Koivunen P, Myllyharju J, Moali C, Aghajari N, Hulmes DJ (2017) Structural basis of homo- and heterotrimerization of collagen I. Nat Commun 8:14671. https://doi.org/10.1038/ncomms14671
138. Shoulders MD, Raines RT (2009) Collagen structure and stability. Annu Rev Biochem 78: 929–958. https://doi.org/10.1146/annurev.biochem.77.032207.120833
139. Silva TH, Alves A, Ferreira BM, Oliveira JM, Reys LL, Ferreira RJF, Sousa RA, Silva SS, Mano JF, Reis RL (2012) Materials of marine origin: a review on polymers and ceramics of biomedical interest. Int Mater Rev 57(5):276–307. https://doi.org/10.1179/1743280412Y.0000000002
140. Silver FH, Freeman JW, Seehra GP (2003) Collagen self-assembly and the development of tendon mechanical properties. J Biomech 36(10):1529–1553. https://doi.org/10.1016/s0021-9290(03)00135-0
141. Singh D, Rai V, Agrawal DK (2023) Regulation of collagen I and collagen III in tissue injury and regeneration. Cardiol Cardiovasc Med 7(1):5–16. https://doi.org/10.26502/fccm.92920302
142. Sionkowska A, Skrzyński S, Śmiechowski K, Kołodziejczak A (2017) The review of versatile application of collagen. Polym Adv Technol 28(1):4–9. https://doi.org/10.1002/pat.3842
143. Smith IP, Domingos M, Richardson SM, Bella J (2023) Characterization of the biophysical properties and cell adhesion interactions of marine invertebrate collagen from *Rhizostoma pulmo*. Mar Drugs 21(2):59. https://doi.org/10.3390/md21020059
144. Sorushanova A, Delgado LM, Wu Z, Shologu N, Kshirsagar A, Raghunath R, Mullen AM, Bayon Y, Pandit A, Raghunath M, Zeugolis DI (2019) The collagen suprafamily: from biosynthesis to advanced biomaterial development. Adv Mater 31(1):e1801651. https://doi.org/10.1002/adma.201801651
145. Sun L, Li B, Song W, Zhang K, Fan Y, Hou H (2020) Comprehensive assessment of Nile tilapia skin collagen sponges as hemostatic dressings. Mater Sci Eng C Mater Biol Appl 109: 110532. https://doi.org/10.1016/j.msec.2019.110532
146. Sun L, Li L, Wang Y, Li M, Xu S, Zhang C (2022) A collagen-based bi-layered composite dressing for accelerated wound healing. J Tissue Viability 31(1):180–189. https://doi.org/10.1016/j.jtv.2021.09.003
147. Swatschek D, Schatton W, Kellermann J, Müller WE, Kreuter J (2002) Marine sponge collagen: isolation, characterization and effects on the skin parameters surface—pH, moisture and sebum. Eur J Pharm Biopharm 53(1):107–113. https://doi.org/10.1016/S0939-6411(01)00192-8
148. Takahara K, Hoffman GG, Greenspan DS (1995) Complete structural organization of the human α1(V) collagen gene (*COL5A1*): divergence from the conserved organization of other characterized fibrillar collagen genes. Genomics 29(3):588–597. https://doi.org/10.1006/geno.1995.9961
149. Teuscher AC, Jongsma E, Davis MN, Statzer C, Gebauer JM, Naba A, Ewald CY (2019) The *in-silico* characterization of the *Caenorhabditis elegans* matrisome and proposal of a novel collagen classification. Matrix Biol Plus 1:100001. https://doi.org/10.1016/j.mbplus.2018.11.001

150. Thom JR, Morris NP (1991). Biosynthesis and proteolytic processing of type XI collagen in embryonic chick sterna. J Biol Chem 266(11): 7262–7269. Available at www.jbc.org/article/S0021-9258(20)89639-2/pdf
151. Ton C, Hwang DM, Dempsey AA, Tang HC, Yoon J, Lim M, Mably JD, Fishman MC, Liew CC (2000) Identification, characterization, and mapping of expressed sequence tags from an embryonic zebrafish heart cDNA library. Genome Res 10(12):1915–1927. https://doi.org/10.1101/gr.10.12.1915
152. Uitto J (1979) Collagen polymorphism: isolation and partial characterization of α1(I)-trimer molecules in normal human skin. Arch Biochem Biophys 192(2):371–379. https://doi.org/10.1016/0003-9861(79)90105-x
153. Van Eeden FJ, Granato M, Schach U, Brand M, Furutani-Seiki M, Haffter P, Hammerschmidt M, Heisenberg CP, Jiang YJ, Kane DA, Kelsh RN, Mullins MC, Odenthal J, Warga RM, Nüsslein-Volhard C (1996) Genetic analysis of fin formation in the zebrafish, *Danio rerio*. Development 123:255–262. https://doi.org/10.1242/dev.123.1.255
154. Veldman MB, Lin S (2008) Zebrafish as a developmental model organism for pediatric research. Pediatr Res 64(5):470–476. https://doi.org/10.1203/PDR.0b013e318186e609
155. Von Hippel PH, Harrington WF (1959) Enzymic studies of the gelatin to collagen-fold transition. Biochim Biophys Acta 36:427–447. https://doi.org/10.1016/0006-3002(59)90185-4
156. Wang Y, Kong B, Chen X, Liu R, Zhao Y, Gu Z, Jiang Q (2022) BMSC exosome-enriched acellular fish scale scaffolds promote bone regeneration. J Nanobiotechnol 20(1):444. https://doi.org/10.1186/s12951-022-01646-9
157. Webb M, Gibson J, Strutt A (2018) The impact of diseases on international beef trade: market switching and persistent effects. Food Policy 75:93–108. https://doi.org/10.1016/j.foodpol.2018.01.006
158. Yammine KM, Li RC, Borgula IM, Mirda Abularach S, DiChiara AS, Raines RT, Shoulders MD (2024) An outcome-defining role for the triple-helical domain in regulating collagen-I assembly. Proc Natl Acad Sci USA 121(46):e2412948121. https://doi.org/10.1073/pnas.2412948121
159. Yan Y-L, Hatta K, Riggleman B, Postlethwait JH (1995) Expression of a type II collagen gene in the zebrafish embryonic axis. Dev Dyn 203(3):363–376. https://doi.org/10.1002/aja.1002030308
160. Yao S, Shang Y, Ren B, Deng S, Wang Z, Peng Y, Huang Z, Ma S, Peng C, Hou S (2022) A novel natural-derived tilapia skin collagen mineralized with hydroxyapatite as a potential bone-grafting scaffold. J Biomater Appl 37(2):219–237. https://doi.org/10.1177/08853282221086246
161. Young EG, Lorimer JW (1960) The acid-soluble collagen of cod skin. Arch Biochem Biophys 88:373–381. https://doi.org/10.1016/0003-9861(60)90253-8
162. Young EG, Lorimer JW (1961) A comparison of the acidsoluble collagens from the skin and swim bladder of the cod. Arch Biochem Biophys 92:183–190. https://doi.org/10.1016/0003-9861(61)90234-x
163. Zeltz C, Gullberg D (2016) The integrin-collagen connection—a glue for tissue repair? J Cell Sci 129(4):653–664. https://doi.org/10.1242/jcs.180992
164. Zylberberg L, Bonaventure J, Cohen-Solal L, Hartmann DJ, Bereiter-Hahn J (1992) Organization and characterization of fibrillar collagens in fish scales in situ and in vitro. J Cell Sci 103(1):273–285. https://doi.org/10.1242/jcs.103.1.273

Chapter 9
Comparative Study of the Positions of Glycosylated Residues in the Chains of Fibril-Forming Collagens and the Role that These Residues Play in Axial and Lateral Assembly

Gillian E. Norris and David A. D. Parry

Abstract In this review, we have collected and analysed collagen glycosylation data in an effort to better understand its possible function in fibrillogenesis. Towards that end, we have identified glycosylation sites in many different fibril-forming collagen molecules (Types I, II, III, V, XI and XXIV) and, more importantly, the glycan itself as well as the percentage of collagen molecules in the sample that contain it. As the numbers of quantitative analyses of different collagen types from different species and tissues taken under different physiological conditions grows, a more accurate picture is emerging of the role that collagen *O*-glycosylation may play in fibrillogenesis. It is becoming increasingly clear that glycosylation does play an important part in the initial stages of the axial assembly of collagen molecules as well as in the lateral control of fibril size. Much remains unknown of the precise mechanisms involved, however, and further detailed analyses will be required before a complete picture of the various in vivo roles played by the glycans emerges.

Keywords Glycosylation · Fibril-forming collagens · Post-translational modifications · Propeptides

Abbreviations

COPII	Coat protein complex II
DOM1	Triple-helical collagenous rod region of Type V chains
DOM2	Short triple-helical region (95 residues) in the N-propeptides of Type V chains
HK	Hydroxylysine

G. E. Norris (✉) · D. A. D. Parry
School of Natural Sciences, Massey University, Palmerston North, New Zealand
e-mail: G.Norris@massey.ac.nz

D. A. D. Parry, P. K. Luther (eds.), *Fibrous Proteins: Structures and Mechanisms*, Subcellular Biochemistry 113, https://doi.org/10.1007/978-3-032-05273-5_9

HKG	Galactosylated-hydroxylysine
HKGG	Glucosylgalactosylated-hydroxylysine
NMR	Nuclear magnetic resonance
PARP	Proline/arginine-rich protein that forms part of the N-propeptides of Type V chains
PTM	Post-translational modification
rER	Rough endoplasmic reticulum
VAR	Variable sequence region

Introduction

The presence of carbohydrate molecules in collagens was first reported by Grassman and Schleich [22] and Beek [4], who used a colorimetric assay to show that hydrolysates of collagen isolated from skin, tendon and bone contained carbohydrate. Furthermore, the carbohydrate was shown to be composed of galactose and glucose in roughly equal amounts. Later, Spiro and colleagues showed that the bound carbohydrates were glucosylgalactose and galactose *O*-linked to hydroxylysine residues [59, 60]. These early analyses were mainly carried out using traditional biochemical techniques to first isolate collagen and then fractionate it from the other components of the extracellular matrix before assaying for carbohydrate content and using Amino Acid Analysis in an attempt to identify possible modified residues. Since then, rapid technological advances in mass spectrometry have resulted in more accurate analyses of glycosylation sites in collagen types I, II, III, IV, V and VI [2, 31, 43, 52, 53, 57, 65, 66, 68, 69, 81, 83]. However, the exact function of these collagen-linked glycans has remained unclear.

Collagen Biosynthesis

In order to probe the possible functions of collagen glycosylation, particularly of the fibrillar collagens, it is necessary to first understand the process of collagen biosynthesis, specifically where and when the post-translational modifications (PTMs) occur. Several excellent reviews have covered this topic in detail [1, 20, 44, 48, 73]. As this chapter is primarily concerned with the fibril-forming collagens, namely types I (Col1), II (Col2), III (Col3), V (Col5), XI (Col11) and XXIV (Col24), a brief summation of Col1 biosynthesis follows. The mRNA, encoding three α-chain COL genes, specific for a type of collagen is transcribed to produce three pre-procollagen molecules consisting of a signal peptide, a non-collagenous N-terminal domain (N-propeptide and N-telopeptide), a left-handed helical domain and a C-terminal noncollagenous domain (C-telopeptide and C-propeptide). Pro-collagen, formed by removal of the signal peptide, then enters the lumen of the rough endoplasmic reticulum (rER), where hydroxylation of selected proline and lysine residues occurs. Glycosylation of hydroxylysines in the Y position of Gly-X-Y triplets to form 1-*O*-β-D-galactosylhydroxylysine (Gal-Hyl or HKG) or the longer but less abundant 2-*O*-α-D-glucopyranosyl-*O*-β-D-galactopyranosylhydroxylysine (Glc-Gal-Hyl or HKGG) then occurs, a process that is necessary for the formation of the correct tertiary and quaternary collagen structures [80]. In order for these processes to occur, the pro-α-chains need to remain untangled long enough for modification to occur, a process that is mediated by several chaperones [35]. The correct three pro-α-chains must then align precisely to form a right-handed triple helix. To facilitate this, interchain disulphide bonds form between the highly homologous 250 residue C-terminal propeptide domains, each of which contains eight conserved cysteine residues, and a 23 amino acid sequence that is responsible for type-specific pro-α-chain association [37]. Next, the approximately 1000 residues with a G-X-Y triplet substructure that comprise the helical domains of the procollagen chains begin to coil around one another in the direction of the N-terminus to form a right-handed supercoil. Each helix is twisted by approximately 30° at every turn and staggered by one residue relative to each other.

It has been shown that once the triple helix has formed, Hyl can no longer be glycosylated [30, 61]. These modifications have been reported as being specific for collagen-type [31], and vary between tissues [80] and the physiological state of the tissue [20]. The triple helix is then transferred to the Golgi before being exported to the extracellular space, a procedure that requires it to interact with a COPII-containing multi-protein complex, an interaction that may be dependent on the collagen di-saccharide Glc(α1–2)Gal(β1-*O*). After secretion into the extracellular space, the N and C-terminal propeptide domains are cleaved to form tropocollagen, which consists of a large central triple-helical domain flanked by N- and C-terminal non-collagenous domains termed telopeptides. For Col1, Col2 and Col3, the N-terminal and C-terminal propeptides domains are completely removed, while Col5 and Col11 may retain a partially processed N-terminal propeptide domain. It has been suggested that the N-propeptides of Col5 and Col11 may have a role in controlling fibril assembly [49]. At this stage, specific lysines or hydroxylysines are further enzymatically deamidated to allow the spontaneous formation of crosslinks between tropocollagen molecules to promote the development of stable fibrils of various sizes in different tissues. In Col1α1, these are $K9^N$ and $K16^C$ in the telopeptide regions, (numbered from the N-terminus of each telopeptide domain) and K87 and K930 in the triple-helical region (numbered from the N-terminus of the triple helical domain); in Col1α2, they are $K5^N$ and $K16^C$ in the telopeptide region and K87 and K933 in the helical region (Fig. 9.1)

Once propeptide processing is complete, the tropocollagen molecules self-assemble longitudinally in a quarter-staggered manner to produce fibrils with a 67 nm D-periodic banding pattern [58]. At this point, many other molecules can interact with collagen fibrils to regulate their assembly and size. These include the glycoprotein tenascin-X [40], other collagens such as collagen type IX (Col9) and the proteoglycans perlecan [36], decorin, biglycan [50] and lumican [10].

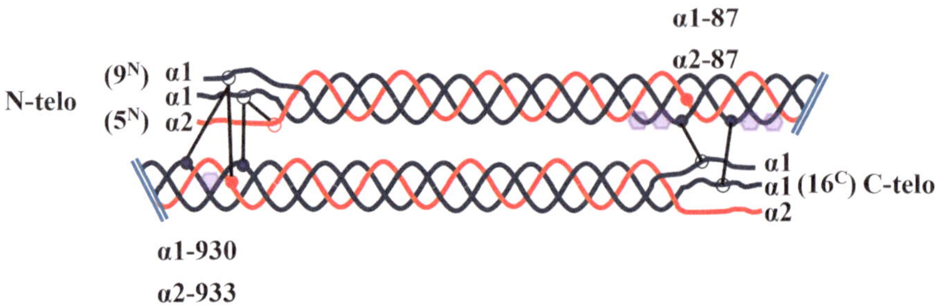

Fig. 9.1 Enzymatic collagen crosslinking in type 1 collagen. The dark blue lines represent the α1 chains, the red lines the α2 chains, and the black lines indicate intra- and inter-molecular crosslinking. The numbers are the residue numbers of the lysines involved (open circles and numbers in brackets, telopeptide; filled circles helical), and the purple semi-transparent hexagons represent the saccharide moieties attached to these lysines

Glycosylation of the Fibrous Collagens (Table 9.1)

Type I Collagen

The positions of the glycosylated residues in the rod domain of the Type I collagen chains, as determined from the skins of sheep, goat, deer and cow, are numbered relative to the beginning of the rod domain (length 1014 residues). The data are taken directly from Visser et al. [75].

Type II Collagen

The numbering of residues for the amino acid sequence of the bovine cartilage collagen chain [57, 65] commenced at the N-terminal residue of the N-propeptide. The triplet region defining the rod domain is easily identified and stretches from residues 201–1214. Thus, by subtracting 200 from the stated positions of the glycosylated residues, the positions can be directly compared with those determined for Col1. Whilst the results are not identical, there is a very high degree of correspondence between the two sets of data.

Table 9.1 Glycosylated lysine residues (Y) in collagens 1, 2, 3, 5 and 11

Position	Type I	Type II	Type III	Position	Type V/XI
27				24	Y/Y
72				69	Y/Y
87	Y	Y	Y	84	Y/Y
99	Y	Y			
108		Y			
174	Y	Y			
219	Y	Y		216	Y/Y
240				237	Y/Y
249				246	Y/Y
252	Y	Y		249	Y/Y
255				252	Y/Y
264	Y	Y		261	Y/Y
270	Y	Y		267	Y/Y
285	Y				
290	Y				
291				288	Y/Y
309				306	Y/Y
327	Y	Y	Y	324	Y/Y
342		Y		339	Y/Y
360	Y				
378				375	Y/Y
408	Y	Y	Y		
420	Y	Y			
434	Y				
453	Y				
479	Y				
483				480	Y/Y
486			Y		
519				516	Y/Y
531		Y			
564	Y	Y	Y	561	
567	Y		Y		
603	Y	Y	Y	600	Y/Y
612				609	Y/Y
615				612	Y/Y
618				615	Y/Y
648	Y	Y			
654	Y				
657	Y	Y		654	Y/Y
684	Y	Y		681	Y/Y
729		Y		726	Y/Y
750				747	Y/Y
756	Y	Y	Y	753	Y/Y
780				777	Y/Y
806	Y				
852				849	Y/Y
855	Y	Y		852	Y/Y
884	Y				
918	Y	Y			
924				921	Y/Y
927				924	Y/Y
933	Y				
945				942	Y/Y
960				957	Y/Y

(continued)

Table 9.1 (continued)

Numbering: Residue 1 is the beginning of the triplet region that defines the rod domain and aligned to that used for Col1 (1–1014) in Visser et al. [75]
Collagen I: Data from Merle-Pham et al. [43], Terajima et al. [70], Visser et al. [75] (Residues 1–1014)
Collagen II: Data from Taga et al. [65], Song and Mechref [57], Zhang et al. [83] (Residues 210–1214)
Collagen III: Data from Taga et al. [66], Merl-Pham et al. [43] (Residues 177–1190)
Collagen V: Data from Merle-Pham et al. [43], Yang et al. [81]
Collagen XI: No mass spectrometry data, but lysines in the correct position
Shading represents the the residue boundaries of clumps of glycosylated lysines in multiples of 234/6 residues. The clumps 96–114 (light grey), 135–153 (pale pink), 174–192 (pale blue), 213–231 (light green) were observed by Visser et al. [75], whereas 18–38 (deep pink) and 57–75 (sky blue) were not. These observations showed that the true axial D-period was 234 residues rather than 234/6 residues

Type III Collagen

Taga et al. [66] identified two glycosylated residues—residues 96 and 336—in their Col3 peptides from patients with *Osteogenesis imperfecta*. By studying these sequences and determining the beginning of the rod domain by its homology to the Col1 and Col2 sequences, it can be seen that residues 96 and 336 line up with residues 87 and 327 in the rod domain of Col1. Similarly, Merl-Pham et al. [43] analysed human lung tissue and identified six glycosylated residues in Col3—584, 662, 740, 743, 779 and 932. Although the triplet region in human lung Col3 is five triplets longer than that of Col1 and Col2 (see also mice skin, which is six triplets longer), the region corresponding to the rod domain of Col1 is nonetheless identifiable (residues 177–1190 for human lung). By subtracting 176 from the stated positions of the glycosylated residues in the rod domain of Col3, their positions were identified as 408, 486, 564, 567, 603 and 756. Five of these are in the same position in Col1. The results are again very convincing—there is a high degree of correspondence between the positions of the glycosylated residues in Col1, Col2 and Col3.

Type V Collagen

Col5 is different from Col1, Col2 and Col3 in that the N-terminal propeptide domains of the α1 and α3 isoforms are not completely post-translationally processed. The N-propeptides are composed of a proline/arginine-rich protein (PARP) domain, a variable domain (VAR), a short 95 residue triple-helical domain (DOM2) and a flexible or hinge domain. The triple-helical domain (DOM1) links the N- and C-propeptides.

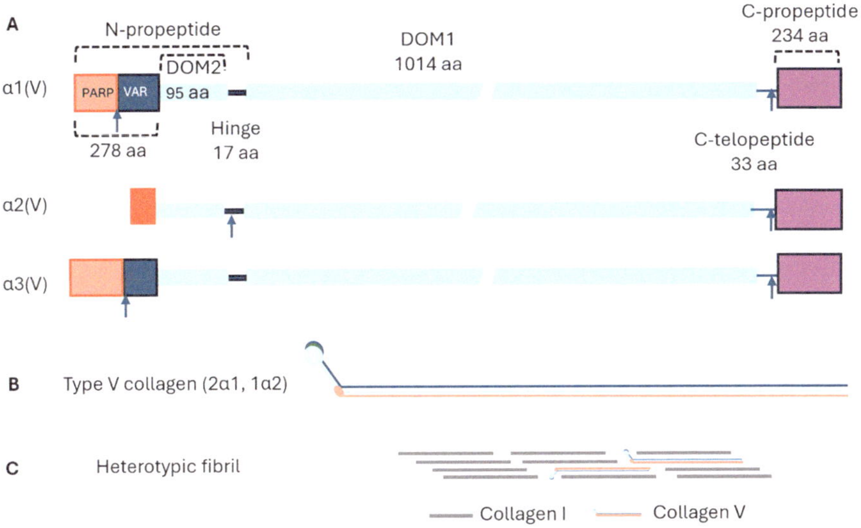

Fig. 9.2 (**a**) α-chain domain structures, with the number of amino acids in each for the α1(V) chain. The N- and C- proteinase cleavage sites are shown by an arrow. The major collagenous domain (DOM1) is composed of 338 (α1V), 339 (α2V) or 337 (α3V) (Col1) uninterrupted (Gly-X-Y) repeats. The DOM2 domain is the N-terminal collagenous domain separated by a non-collagenous hinge region. (**b**) The type V collagen molecule comprises two α1 chains and one α2 chain. (**c**) The collagen fibrils contain both collagen 1 (grey) and collagen V molecules (orange) axially staggered by D or a multiple

Partial processing removes the PARP region whilst retaining the variable domain, DOM2 and the hinge [23, 39] (Fig. 9.2a). Collagen V is a trimer most commonly containing 2 Colα1(V) and 1 Colα2(V) chains (Fig. 9.2b). It combines with collagen I molecules to form a heterotypic collagen fibril. However, because its N-terminal domain cannot be integrated into the staggered packing of the helical domains, fibril packing is affected. The hinge allows the DOM2 and VAR domains to project toward the fibril surface in the gap region of the assembled fibril (Fig. 9.2c) [55].

Many glycosylation sites in the rod domain have been identified in bovine placenta and human embryonic kidney Col5 [81], and, interestingly, the site numbers differ from those recorded in Col1, Col2 and Col3 by three residues. The question arises as to whether the 1014 residue rod domain in Col5 might start three residues earlier, thereby bringing the positions of the glycosylated residues into register. This would be on the basis of: (1) the rod domain terminating after $(G\text{-}P\text{-}P)_3$ as in Col1, rather than after the "additional" triplet G-E-V that follows $(G\text{-}P\text{-}P)_3$ in Col5; (2) the close sequence homology between Col1 and Col5 around residue 87 that are nominally three residues apart (AGLPGEKGHRG) and (AGLPGMKGHRG); (3) the fact that 14 glycosylation sites would then line up with those established for Col1, 2 and 3. However, if this was the case, the rod domain would begin with a triplet lacking the characteristic G. On balance, the evidence cannot be ignored that the three-residue difference may be a genuine one and that it has structural/functional importance. Nonetheless, the glycosylation sites for Col5, while not identical to those Col1, Col2 and Col3 because of the three-residue shift, are, at the very least, closely related.

Type XI Collagen

Col11 is a heterotrimeric molecule comprised of three isoforms Col11α1, Col11α2 and Col11α3 [46]. While the α1 and α2 chains are unique, the α3 procollagen chain of Col11 and the α1 chain of Col2 procollagen are encoded by the same gene, but Col11α3 is more highly glycosylated [18]. Col11 is found in thin cartilage fibrils composed of four micro-fibrils, two Col2 and two Col11, surrounded by ten Col2 micro-fibrils [25, 71] and is thought to be responsible for regulating their diameters [8]. It is broadly distributed in tendons, trabecular bone, testis, trachea, skeletal muscle, placenta, ovarian, lung and the neoepithelium of the brain. The observed rod domain of this collagen is of standard length (1014 res). Interestingly, it is very similar to Type V collagen in that it too has a G-E-V triplet after the "normal" end of the rod domain characterised by several G-P-P motifs. However, comparing directly just the glycosylation sites in Col5 with the lysines in Col11, the number of alignments between them is 34/34, a spectacular 100% overlap. While no quantitative analyses of these glycosylation sites have been reported, a very early analysis showed that the proportion of carbohydrate in the Col11α1, α2 and α3 chains was similar to that of the Col5α1 and α2 chains and more than in the Col2α1 chain [15]. Another similarity to Col5 is that the N-terminal propeptide of the Col11α1 isoform is not completely post-translationally processed and retains the VAR domain [51]. A lack of experimental evidence from intact Col11 has given rise to some controversy over the N-terminal processing of the Col11α2 isoform. Some reports indicate it is not completely processed, while others indicate that it is [55]. More recent work showed that Col11α3 (Col2α1) is most likely to retain a 69 amino acid N-terminal domain [42]. As Col11 has been shown to form striated fibrils it is likely that the N-terminal extension of Col11α1 protrudes from the fibril towards the surface in the gap region like that of Col5 (Fig. 9.1c).

There is no question now that Col1, Col2 and Col3 should be grouped together based on their biochemical properties. Col5 and Col11, and probably Col24, also form a group. Apart from the differences in post-translational trimming of their N-terminal domains (at least one isoform of each type has an N-terminal extension) and slight differences in the length of their triple-helical domain, Col5 and Col11 are more heavily glycosylated than Col1, Col2 and Col3 (Table 9.1). Although there is no direct evidence, sequence similarities to Col5 and Col11 suggest Col24 may also be similarly glycosylated [34] and may very well be incompletely post-translationally processed at its N-terminus.

Further, there is now clear evidence that Col1, Col2, Col3, Col5 and Col11 have major similarities to one another as regards the axial positioning of their glycosylation sites.

Type XXIV Collagen

Col24 is a minor collagen that is similar to collagen Col5. The triplet region in the amino acid sequence of the collagen Type Col24 c (human α isoform X3) is 8 triplets shorter than the Col1 chain. It extends for 708 residues (236 triplets) from the N-terminal end before the triplet region is interrupted by a four-residue insertion (S-T-V-L). This is followed by 279 residues (93 triplets) before the C-propeptide. The positions of lysine residues before the break have been compared with the glycosylation sites in the rod domains of Types I, II and III collagen and with the glycosylation sites in Types V collagen and the putative lysine sites in XI collagen. The residue numbering was increased by 24 residues (8 triplets) accordingly, this being the observed shortness of the Type XXIV rod domain compared to that observed in Types I, II and III chains. The conclusion is that 8 sites would line up with Types I, II and III, and that 12 sites would line up with Types V and XI. It is difficult to say with any certainty that this observation is statistically significant, but the observation remains of interest. Col24 is produced in the skull and appendicular skeleton in the early development of the embryo, then later may form D-periodic fibrils.

O-Glycosylation and Fibrillogenesis

Many have questioned the role of *O*-glycosylation in fibrillogenesis. It is well known that it is essential for the correct assembly of tropocollagen and fibril formation, and that disruption of lysine hydroxylation and the subsequent addition of either a mono- or di-saccharide results in many severe or lethal diseases [3, 14, 19, 21, 45, 52, 74]. Exactly how these documented changes in collagen glycosylation affect its structure has, however, not been elucidated due to the technological difficulties in imaging intact collagen fibres at sufficient resolution to see the molecular interactions involving these glycans, which are bulky and very mobile. Several researchers have reported that collagen *O*-linked glycosylation appears to be correlated with fibril diameter, with an increase in collagen *O*-glycosylation resulting in smaller diameter fibrils and vice versa [62, 69, 70, 72, 76]. However, it is known that different tissues comprising mainly collagen type I (Col1), which is glycosylated to the same extent, contain fibrils with very different diameters. For example, corneal collagen fibrils are uniformly small being 17 nm for bony fish and 24–25 nm for cartilaginous fish, amphibians, birds, reptiles and mammals when measured using conventional electron microscopy preparation techniques [13].

Control of Fibril Size

The control of collagen fibril size in the cornea has been attributed in part to Col5, which is present as a minor component, and which acts as a template in fibrillogenesis [7]. This was verified in both the cornea and dermis by targeted deletion of the *Col5a1* gene in mice (embryonic lethal), and by fibril assembly assays which showed that reduction of Col5 in relation to Col1 resulted in larger diameter fibrils [6, 54, 77–79]. Yet Col5 is more highly glycosylated than Col1, so based on the assumption that an increase in glycosylation results in larger diameter fibrils, its inclusion in the collagen alloy should result in larger diameter fibrils. Therefore, its effect in decreasing fibril size is more likely to be due to the structural effect of the extended N-terminus suggested for Col11 [25]. Col5 has also been shown to regulate fibrillogenesis in tendons and ligaments which contain mainly Col2 [11, 12, 63]. Others have found it in cartilage and other collagen type II-containing tissues, where it may play a similar role [9, 41].

Col11 is implicated in the regulation of fibril diameter in tendons and cartilage, which are predominantly made up of Col2, and which contain both thick and thin fibrils [17, 32], as do other tissues, particularly during the early stages of development [38, 82]. In these tissues, both very large (>500 nm diameter), medium-sized and small diameter fibrils are observed, yet the same Col2 molecules with the same or similar PTMs are involved. Again, the difference seems to be attributable to the ratio of Col11 to Col2, with higher concentrations of Col11 resulting in thinner very homologous fibrils. In vitro experiments showed that Col11 can also affect fibril assembly of Col1. As the Col11 concentration is increased relative to Col1, fibrils with variable but smaller diameters were produced [24]. In contrast, in vivo experiments with mice showed that the elimination of production of the α1 isoform of Col11, resulted in tendons containing fibrils not only with smaller diameters, but also with disrupted fibril alignment [64]. These results are not incompatible as Sun et al. demonstrated that reduced Col11 production was sufficient for the regulation of fibrillogenesis seen in the in vitro studies.

The question is, do these patterns of glycosylation play a role in fibrillogenesis and/or the specification of fibril size, or is fibril size controlled entirely by the constituent collagen molecules as discussed above. It is hard to believe that the cell would waste precious resources on a post-translational modification that did not play a role in the structure and properties of the parent molecule. Furthermore, the fact that further glycosylation cannot occur once the triple helix has been formed would suggest that the different post-translational modifications of collagen are not random. Recently, Visser et al. [75] showed that the positions of the glycosylated resides along the length of the collagen molecule display a D-period, implying that they may play a role in facilitating a coarse alignment of the collagen molecules prior to the detailed alignment provided by the interactions of oppositely charged residues and also between apolar residues in the amino acid sequence [29]. The degree of similarity in the positions of the glycosylated hydroxylysines of collagens 1, 2 and 3 is striking. Yet these collagens, with a similar glycosylation pattern, participate in

Table 9.2 Relative abundances (rounded up %) of lysine (K) and its modified forms hydroxylysine (HK), galactosylated-HK (HKG) and glucosylgalactosyl-HK (HKGG) of Col1α1 isolated from bovine cartilage (columns 2–5) and mouse skin and tendon (columns 6–9). ND not determined

	Bovine cartilage collagen [70]				Mouse skin (mouse tendon) collagen [31]			
Residue	K	HK	HKG	HKGG	K	HK	HKG	HKGG
87 α1	2	4	8	86	16 (−)	2 (97)	7 (3)	75(1)
87 α2	8	92	ND	ND	6 (ND)	94 (100)	ND (ND)	ND (ND)
99 α1	69	20	7	4	77 (77)	20 (19)	3 (2)	ND (ND)
174 α1	53	41	3	2	66 (60)	32 (39)	2 (1)	ND (ND)
174 α2	36	6	38	20	41 (15)	5 (82)	51 (3)	4 (ND)
219 α1	85	15	ND	ND	86 (73)	14 (27)	ND (ND)	ND (ND)
219 α2	34	62	1	3	55 (52)	34 (30)	1 (17)	10 (1)
564 α1	67	24	4	5	77 (79)	20 (20)	2 (1)	1 (ND)
933 α2					93 (1)	ND (94)	7 95)	ND (ND)

the formation of fibrils with very different diameters. As with all glycosylated proteins, however, there are several glycoforms for each molecule and the occupancies of the sites may vary, i.e., hydroxylysines on each collagen molecule may not be consistently glycosylated, and a single site may bear either a Gal or a GalGlu [66, 83]. Is it possible that the pattern of glycoforms at each position or the occupancy of each glycan, or both, contribute to the control of fibril size? Obtaining sufficient quantitative data to analyse the abundance of each modification is hampered by the fact that only limited data are suitable for analysis [28, 31, 57, 70, 83] (Tables 9.2 and 9.3). From these limited reports, there are, nonetheless, a few differences worth noting.

K87, a residue involved in the formation of crosslinks for both Col1α1 and Col1α2, is found mainly as HKGG in Col1 isolated from many different tissues and species. The abundances for this modification shown in Table 9.2 are similar to those reported for K87 from mouse skin (100%), bone (78%), aorta (100%) and cornea (93%) [28]. An exception is tendon, where K87 from Col1α1 was present as HK only. Taga and Visser, on the other hand, both found K87 present as both HKG and HKGG in roughly equal proportions in Col1α1 from bovine cartilage [65] and in several mammalian skins [75]. In Col1α2, however, K87 is predominantly seen only as the hydroxylated species (HK) in skin, cartilage and tendon from many different species.

The remaining few residues in Table 9.2 show that the modifications and their abundances are, unexpectedly, remarkably similar between tissues, but differ between isoforms, in contrast to other reports that state there are differences in the glycosylation patterns of collagens extracted from different tissues, species and types. The most significant differences seen in this analysis are as follows: K174, where the α2 isoform is 40–50% glycosylated, while the α1 is either unmodified or hydroxylated, depending on the source tissue; K684 was seen as HKG and HKGG in Col1α1 but not in Col1α2 by Taga [65], although this was a qualitative analysis rather than a quantitative one.

Table 9.3 Relative abundances (rounded up %) of lysine (K) and its modified forms—hydroxylysine (HK), galactosylated-HK (HKG) and glucosylgalactosyl-HK (HKGG) of collagen IIα1 isolated from bovine cartilage. ND not determined

	Zhang et al. [83]				Song and Mechref [57]				Taga et al. [66]			
Residue	K	HK	HKG	HKGG	K	HK	HKG	HKGG	K	HK	HKG	HKGG
87	0	71	14	14	ND	82	2	16			Y	Y
99	1.5	50	31	16	ND	21	42	37			Y	
108	2.5	75	12	10	ND	86	6	8			Y	
174	0	24	43	32	ND	19	35	46			Y	–
219	10	47	15	28	ND	18	10	72			Y	Y
252	39	26	17	19	ND	89	3	7			Y	Y
264	9	61	16	15	39	21	22	18				
270	43	42	10	6	11	74	11	4				
327	44	15	22	17	54	20	13	12			Y	Y
342	30	0	28	41	9	24	31	36			Y	Y
408	53	12	16	13	6	31	24	39			Y	–
420	39	35	7	20	4	54	15	27			Y	Y
531	6	2	18	74	ND	8	15	77			Y	Y
564	33	35	15	19	2	28	38	32			Y	Y
573	87	13	0	0	87	ND	ND	ND				
603	8	44	15	34	2	45	6	46			Y	Y
648	16	40	32	10	13	62	17	8				
657	14	71	9	7	11	8	49	32				
684	9	4	11	75	0	2	2	96			Y	
729	0	0	0	100	ND	88	7	5				
756	57	37	5	1	27	67	5	1			Y	
855	35	46	12	7	36	50	9	5				
918	25	71	2	2	ND	97	1	3				

Columns 2–5, data from Zhang et al. [83]; Columns 6–9, data from Song and Mechref [57]; Columns 10–13 collagen IIα1, data from Taga et al. [66], identification only. *ND* not determined.

Qualitative analyses of Col1 lysine PTMs by Merl-Pham (human lung), Taga (skin fibroblasts) and Visser (cow, sheep, goat and deer skins) showed there was a degree of congruence of modification between tissues [43, 65, 66, 75]. As Visser found more PTMs, modifications identified by Merl-Pham and Taga are compared to these, and a lack of modification in these latter two reports is assumed to be a result of technical issues. Differences are seen for: K219, where both Visser and Merl-Pham did not detect either HKG or HKGG in contrast to Taga who detected both modifications in Col1α2; K270 where Visser detected only HK, yet Taga found HKG and Merle-Pham found both HKG and HKGG in Col1α1; Taga [66] reported that K648 was detected as both HKG and HKGG in Col1α2 only, whereas Visser detected only very low amounts. K657 was detected as HKG in Col1α2 by Taga, whereas Visser found both HKGG and HKG in Col1α1, but only GHK in Col1α2. Two residues not identified by Visser were K342, which Merl-Pham detected as HKG and HKGG in Col1α1, and K531, which Taga detected as HKG and HKGG in Col1α1 but only as HKG in Col1α2. Whether these differences are significant or not cannot be determined until more data become available.

A note of caution must be exercised in analysing such data until many more quantitative analyses are carried out on collagens isolated from different tissues and from different species under a range of physiological conditions. The well-documented technical issues associated with collagen extraction and glycan analyses by mass spectrometry should also be considered [83], and differences in abundances should be looked at as trends rather than absolutes.

A comparison of the results shown in Table 9.3 for Col2a1 shows the modifications in terms of the type of glycosylation are similar with a few exceptions. These are lysines 99, 108, 174, 408 and 684 that Taga found to be mono-galactosylated, whereas both Zhang et al. [83] and Song and Mechref [57] detected them as both HKG and HKGG.

In summary, the fibres and fibrils found in many different tissues, tendon, cartilage and bone display a wide range of diameters, with fibrils ranging from 20 to 500 nm in diameter and fibres from 1 to 100 μm. They are, however, all made from collagens I, II and III, which have very similar sequences and, as has been shown here, very similar glycosylation patterns within isoforms. Thin fibrils have been shown to be more highly glycosylated and contain more Col5/11 fibrils, which are also more highly glycosylated than Col1, 2 and 3, and the α1 isoform of Col1 also seems to have a different glycosylation pattern to that of Col1α2.

Structural analysis of collagen model peptides that form triple helices has shown that extensive hydrogen-bonded water networks occupy the space between them connected through hydroxyproline anchors (Fig. 9.3) [5, 47]. The distances between these model triple helices are similar to those between collagen molecules in tendons and other tissues that have been measured using fibre diffraction, and it is well known that hydroxylation of proline plays an essential role in maintaining this cushion.

So, what does hydroxylation of lysine and its subsequent glycosylation achieve? In contrast to the hydroxylation of proline, which is very consistent, hydroxylation of lysine is reported to vary between 15% and 90% depending on collagen types, tissue

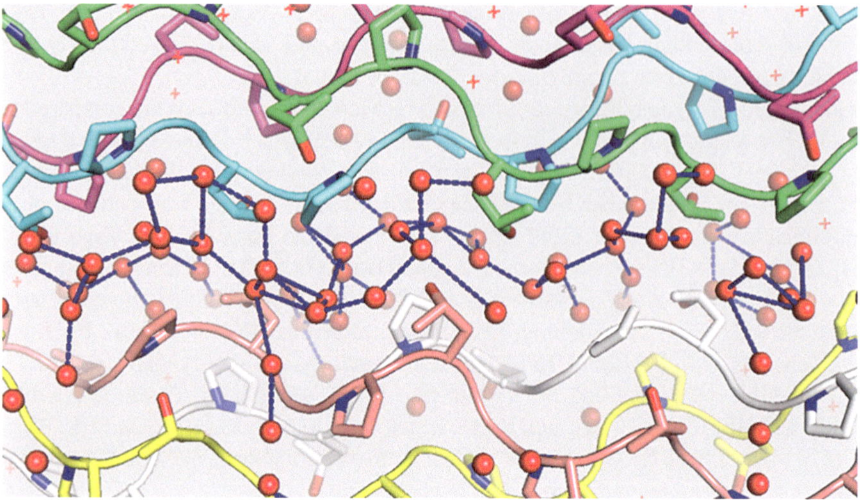

Fig. 9.3 Ordered water network between neighbouring triple helices in the crystal structure of the $(POG)_9$ peptide (PDB 3B0S) [47]. Figure generated using PyMOL (The PyMOL Molecular Graphics System, Version 1.8, Schrödinger, LLC). Top helix magenta, cyan and green ribbons, bottom helix white, pink and yellow ribbons. Waters are represented by spheres, and hydrogen bonds between them are dark blue dashed lines

and the physiological condition of the tissue [80]. As most of these lysines or hydroxylysines lie in regions of the triple helix that are relatively bereft of imino residues, there will be small local changes in the structure of the triple helix. The helical twist will relax from approximately 102.6° to 108° with a concomitant increase in the height of a turn from 2.84 A to 2.9 A [5], allowing for the introduction of a bulky flexible group on the lysine sidechain. Although both modifications are not going to remove the charge on the sidechains, model structures of collagen-like polypeptides show there is little direct hydrogen-bonding, charge-charge or hydrophobic contact between sidechains of adjacent triple helices [5]. It is possible that the introduction of a bulky flexible group may disrupt the space between adjacent triple helices in these regions, assuming they are found on the outside. Modelling experiments using collagen-like peptides showed that galactosylation of hydroxylysine rendered the triple helix less stable than glucosylgalactosylation due to its greater conformational mobility [67]. The lack of hydroxyproline groups in these regions means there will be a lack of anchor points for the network of water molecules between triple helices. These will, however, be replaced by the hydroxyl groups of the saccharides attached to the hydoxylysines, and so provide minimum interruption to the cushion. However, unlike water, the saccharides have additional properties. They have a polar face and a hydrophobic face (Fig. 9.4). The polar face prefers to be oriented towards the solvent between strands, while the hydrophobic face interacts with hydrophobic surfaces of the triple helices, reducing the surface available for hydrophobic interactions between neighbouring collagen chains. Both these interactions contribute to the stabilisation of the quaternary structure because they reduce

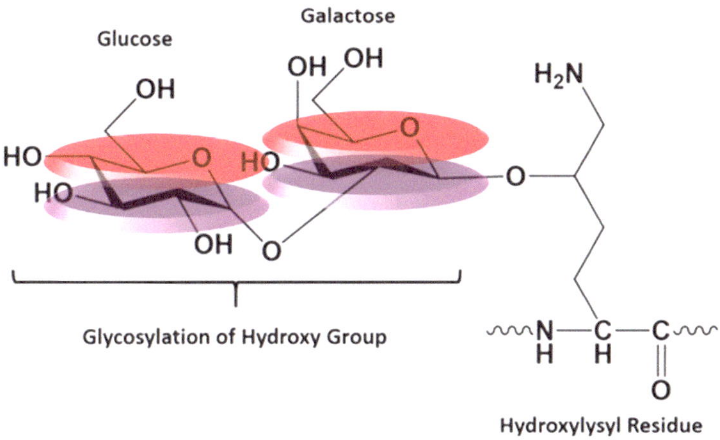

Fig. 9.4 Structure of glucosylgalactosyl hydroxylysine showing the hydrophilic (red) and hydrophobic (purple) faces

disruption to the surrounding water caused by the presence of the collagen molecules themselves [67] and inhibit the lateral growth of the collagen fibre.

It appears, therefore, that circumstantial evidence strongly implicates collagens 5, 11 and possibly 24, as well as *O*-glycosylation, in the control of fibril size. For collagens 5 and 11, which have been shown to nucleate fibres composed of mainly Col1 and Col2, respectively, a common denominator is the lack of complete processing of the N-terminal propeptides, which protrude from the gap region of the resulting composite microfibrils, thereby disrupting their packing. The current model for the Col11/Col2 alloy involves a core of 2 Col11 and 2 Col 2 chains surrounded by 10 Col2 chains. Modelling shows that the non-helical termini of the 2 Col11 chains protrude through the gap regions in the microfibril and have sufficient length to wrap around the microfibril, preventing further lateral growth [25]. A similar process presumably occurs for the Col5/Col1 alloys, where Col5 has been shown to nucleate the growth of Col1 microfibrils. However, both these collagen types are also heavily glycosylated, in fact more so than collagens 1, 2 and 3. As it is known that the presence of glycans on protein molecules reduce their propensity to aggregate [56], this may be their function in the nucleation of Col1, 2 and 3 fibrils, where it is important to limit the number of Col5 or Col11 molecules in the centre of the nascent microfibril.

Table 9.1 confirms that for all the collagens listed, glycosylation of lysines falls into the pattern of bands both shown and predicted by Visser et al. to follow the D period of the collagen triple helix. It is therefore likely that they too may be positioned in or near the gap regions of the microfibrils. The diameter of a tropocollagen molecule is between 16 and 20 Å, and the space between them is >20 Å. Crystal and NMR structures of model short triple helices containing G-X-Y triplets, where X is lysine, arginine, aspartic or glutamic acid, show the presence of interhelical hydrogen bonds between the charged residues at the interfaces of the

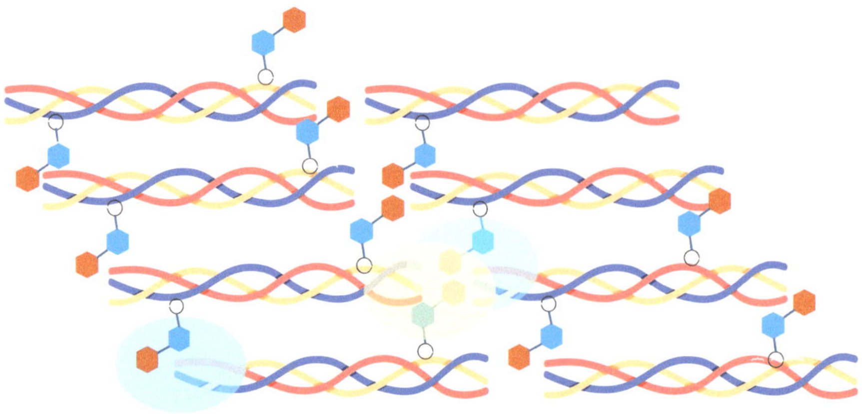

Fig. 9.5 Schematic diagram of a hypothetical section of a Col1 or Col2 microfibril showing the possible positions of disaccharides attached to specific lysine or hydroxylysine residues. The oval shapes (coloured according to the chain they are linked to) show how the flexible disaccharide moieties might interact to roughly align the collagen molecules and induce fibrillogenesis. Or alternatively, how they might mask the binding sites of molecules that regulate fibril size, such as lysyl oxidase, decorin or even other collagen types

triple helices [16]. It is therefore not unlikely that there could be interaction between the glycans on adjacent chains, which help in correctly aligning the tropocollagen molecules for subsequent crosslinking. However, such a hypothesis does not account for the observation that more highly glycosylated collagen molecules produce thinner fibrils. Indeed, model collagen-like peptides that are not glycosylated are still able to form triple helices and microfibrils [26, 27]. An alternative hypothesis is that these glycans, especially the disaccharide, could mask the residues near the N- and C-telopeptide lysines that are required for recognition by lysyl oxidase, preventing subsequent crosslinking, thus limiting the stable lateral growth of the fibril. They could also mask binding sites of molecules known to be involved in regulation of fibril diameter and length, such as decorin, which has been shown to bind to Col1α1 near the C-terminus [33] and collagen type IX [25] (Fig. 9.5). As lysine hydroxylation is variable and seems to differ between types of collagens and collagens of different species, tissues and even the physiological state of the tissue, this would explain why collagens of the same type with similar compositions form fibrils of different sizes.

Summary

In this review, we have attempted to show why and how collagen glycosylation may affect fibrillogenesis. Unfortunately, the technology to enable visualisation of collagen fibres in tissue samples at high enough resolution to see the precise interactions between the different molecules, especially those that are glycosylated, is not yet available. Therefore, much of our current knowledge about the structure-function relationships has come from collagen model peptides, all of which contain no glycosylated lysine or hydroxylysine residues. Technological advances in mass spectrometry have, however, allowed the identification of glycosylation sites in many different collagens and, more importantly, have allowed the identification of the glycan itself and the percentage of collagen molecules in the sample that contain it. As the number of such quantitative analyses of different collagen types from different species and tissues taken under different physiological conditions grows, a more accurate picture of the role of collagen *O*-glycosylation plays in fibrillogenesis will emerge. These may change the conclusions made here, but until then, we will have to rely on trying to make sense of the reports available as we try to understand the complexities of collagen fibrillogenesis.

References

1. Amirrah IN, Lokanathan Y, Zulkiflee I, Wee M, Motta A, Fauzi MB (2022) A comprehensive review on collagen type I development of biomaterials for tissue engineering: from biosynthesis to bioscaffold. Biomedicine 10
2. Basak T, Vega-Montoto L, Zimmerman LJ, Tabb DL, Hudson BG, Vanacore RM (2016) Comprehensive characterization of glycosylation and hydroxylation of basement membrane collagen IV by high-resolution mass spectrometry. J Proteome Res 15:245–258
3. Batge B, Winter C, Notbohm H, Acil Y, Brinckmann J, Muller PK (1997) Glycosylation of human bone collagen I in relation to lysylhydroxylation and fibril diameter. J Biochem 122: 109–115
4. Beek JJ (1941) The carbohydrate content of collagen. J Res Natl Inst Stand 27:507–517
5. Bella J (2016) Collagen structure: new tricks from a very old dog. Biochem J 473:1001–1025
6. Birk DE (2001) Type V collagen: heterotypic type I/V collagen interactions in the regulation of fibril assembly. Micron 32:223–237
7. Birk DE, Fitch JM, Babiarz JP, Doane KJ, Linsenmayer TF (1990) Collagen fibrillogenesis in vitro: interaction of types I and V collagen regulates fibril diameter. J Cell Sci 95(Pt 4): 649–657
8. Blaschke UK, Eikenberry EF, Hulmes DJ, Galla HJ, Bruckner P (2000) Collagen XI nucleates self-assembly and limits lateral growth of cartilage fibrils. J Biol Chem 275:10370–10378
9. Bos KJ, Holmes DF, Kadler KE, Mcleod D, Morris NP, Bishop PN (2001) Axial structure of the heterotypic collagen fibrils of vitreous humour and cartilage. J Mol Biol 306:1011–1022
10. Chakravarti S, Magnuson T, Lass JH, Jepsen KJ, Lamantia C, Carroll H (1998) Lumican regulates collagen fibril assembly: skin fragility and corneal opacity in the absence of lumican. J Cell Biol 141:1277–1286
11. Connizzo BK, Freedman BR, Fried JH, Sun M, Birk DE, Soslowsky LJ (2015) Regulatory role of collagen V in establishing mechanical properties of tendons and ligaments is tissue dependent. J Orthop Res 33:882–888
12. Connizzo BK, Adams SM, Adams TH, Birk DE, Soslowsky LJ (2016) Collagen V expression is crucial in regional development of the supraspinatus tendon. J Orthop Res 34:2154–2161
13. Craig AS, Parry DAD (1981) Collagen fibrils of the vertebrate corneal stroma. J Ultrastruct Res 74:232–239
14. Ewans LJ, Colley A, Gaston-Massuet C, Gualtieri A, Cowley MJ, Mccabe MJ, Anand D, Lachke SA, Scietti L, Forneris F, Zhu Y, Ying K, Walsh C, Kirk EP, Miller D, Giunta C,

Sillence D, Dinger M, Buckley M, Roscioli T (2019) Pathogenic variants in Plod3 result in a Stickler syndrome-like connective tissue disorder with vascular complications. J Med Genet 56: 629–638

15. Eyre D, Wu J-J (1987) Type Xi Or 1α2α3α Collagen. In: Maynes RAB, R.E. (eds) Structure and function of collagen types. Academic Press, Orlando
16. Fallas JA, Dong J, Tao YJ, Hartgerink JD (2012) Structural insights into charge pair interactions in triple helical collagen-like proteins. J Biol Chem 287:8039–8047
17. Fernandes RJ, Weis M, Scott MA, Seegmiller RE, Eyre DR (2007) Collagen XI chain misassembly in cartilage of the chondrodysplasia (cho) mouse. Matrix Biol 26:597–603
18. Furuto DK, Miller EJ (1983) Different levels of glycosylation contribute to the heterogeneity of alpha 1(II) collagen chains derived from a transplantable rat chondrosarcoma. Arch Biochem Biophys 226:604–611
19. Geister KA, Lopez-Jimenez AJ, Houghtaling S, Ho TH, Vanacore R, Beier DR (2019) Loss of function of Colgalt1 disrupts collagen post-translational modification and causes musculoskeletal defects. Dis Model Mech 12
20. Gelse K, Poschl E, Aigner T (2003) Collagens--structure, function, and biosynthesis. Adv Drug Deliv Rev 55:1531–1546
21. Gistelinck C, Weis M, Rai J, Schwarze U, Niyazov D, Song KM, Byers PH, Eyre DR (2021) Abnormal bone collagen cross-linking in osteogenesis Imperfecta/Bruck syndrome caused by compound heterozygous PLOD2 mutations. JBMR Plus 5:e10454
22. Grassmann W, Schleich H (1935) Uber den Kohlenhydratgehalt des Kollagens II. Biochem Z 277:320–328
23. Gregory KR, Oxford JT, Chen Y, Gambee JE, Gygi SP, Aebersold R, Neame PJ, Mechling DE, Bachinger HP, Morris NP (2000) Structural organization of distinct domains within the non-collagenous N-terminal region of collagen type XI. J Biol Chem 275:11498–11506
24. Hansen U, Bruckner P (2003) Macromolecular specificity of collagen fibrillogenesis: fibrils of collagens I and XI contain a heterotypic alloyed core and a collagen I sheath. J Biol Chem 278: 37352–37359
25. Holmes DF, Kadler KE (2006) The 10+4 microfibril structure of thin cartilage fibrils. Proc Natl Acad Sci USA 103:17249–17254
26. Holmes DF, Capaldi MJ, Chapman JA (1986) Reconstitution of collagen fibrils in vitro; the assembly process depends on the initiating procedure. Int J Biol Macromol 8:161–166
27. Holmes DF, Yinhui L, Starborg T, Kadler KE (2018) Collagen fibril assembly and function. Curr Top Dev Biol 130:107–142
28. Hudson DM, Weis M, Rai J, Joeng KS, Dimori M, Lee BH, Morello R, Eyre DR (2017) P3h3-null and Sc65-null mice phenocopy the collagen lysine under-hydroxylation and cross-linking abnormality of Ehlers-Danlos syndrome type VIA. J Biol Chem 292:3877–3887
29. Hulmes DJS, Miller A, Parry DAD, Piez KA, Woodhead-Galloway J (1973) Analysis of the primary structure of collagen for the origins of molecular packing. J Mol Biol 79:137–148
30. Ishikawa Y, Bachinger HP (2013) A molecular ensemble in the rER for procollagen maturation. Biochim Biophys Acta 1833:2479–2491
31. Ishikawa Y, Taga Y, Zientek K, Mizuno N, Salo AM, Semenova O, Tufa SF, Keene DR, Holden P, Mizuno K, Gould DB, Myllyharju J, Bachinger HP (2021) Type I and type V procollagen triple helix uses different subsets of the molecular ensemble for lysine posttranslational modifications in the rER. J Biol Chem 296:100453
32. Keene DR, Oxford JT, Morris NP (1995) Ultrastructural localization of collagen types II, IXX, and XI in the growth plate of human rib and fetal bovine epiphyseal cartilage: type XI collagen is restricted to thin fibrils. J Histochem Cytochem 43:967–979
33. Keene DR, San Antonio JD, Mayne R, Mcquillan DJ, Sarris G, Santoro SA, Iozzo RV (2000) Decorin binds near the C terminus of type I collagen. J Biol Chem 275:21801–21804
34. Koch M, Laub F, Zhou P, Hahn RA, Tanaka S, Burgeson RE, Gerecke DR, Ramirez F, Gordon MK (2003) Collagen XXIV, a vertebrate fibrillar collagen with structural features of

invertebrate collagens: selective expression in developing cornea and bone. J Biol Chem 278: 43236–43244
35. Koide TNK (2005) Collagen biosynthesis. In: Brinckmann J, Notbohm H, Muller PK (eds) Collagen: primer in structure, processing and assembly. Springer, Berlin Heidelberg
36. Kvist AJ, Johnson AE, Morgelin M, Gustafsson E, Bengtsson E, Lindblom K, Aszodi A, Fassler R, Sasaki T, Timpl R, Aspberg A (2006) Chondroitin sulfate perlecan enhances collagen fibril formation. Implications for perlecan chondrodysplasias. J Biol Chem 281:33127–33139
37. Lees JF, Tasab M, Bulleid NJ (1997) Identification of the molecular recognition sequence which determines the type-specific assembly of procollagen. EMBO J 16:908–916
38. Lincoln J, Florer JB, Deutsch GH, Wenstrup RJ, Yutzey KE (2006) ColVa1 and ColXIa1 are required for myocardial morphogenesis and heart valve development. Dev Dyn 235:3295–3305
39. Linsenmayer TF, Gibney E, Igoe F, Gordon MK, Fitch JM, Fessler LI, Birk DE (1993) Type V collagen: molecular structure and fibrillar organization of the chicken alpha 1(V) NH2-terminal domain, a putative regulator of corneal fibrillogenesis. J Cell Biol 121:1181–1189
40. Mao JR, Taylor G, Dean WB, Wagner DR, Afzal V, Lotz JC, Rubin EM, Bristow J (2002) Tenascin-X deficiency mimics Ehlers-Danlos syndrome in mice through alteration of collagen deposition. Nat Genet 30:421–425
41. Mayne R, Brewton RG, Mayne PM, Baker JR (1993) Isolation and characterization of the chains of type V/type XI collagen present in bovine vitreous. J Biol Chem 268:9381–9386
42. McAlinden A, Traeger G, Hansen U, Weis MA, Ravindran S, Wirthlin L, Eyre DR, Fernandes RJ (2014) Molecular properties and fibril ultrastructure of types II and XI collagens in cartilage of mice expressing exclusively the alpha1(IIA) collagen isoform. Matrix Biol 34:105–113
43. Merl-Pham J, Basak T, Knuppel L, Ramanujam D, Athanason M, Behr J, Engelhardt S, Eickelberg O, Hauck SM, Vanacore R, Staab-Weijnitz CA (2019) Quantitative proteomic profiling of extracellular matrix and site-specific collagen post-translational modifications in an in vitro model of lung fibrosis. Matrix Biol Plus 1:100005
44. Mienaltowski MJ, Birk DE (2014) Structure, physiology, and biochemistry of collagens. Adv Exp Med Biol 802:5–29
45. Miyatake S, Schneeberger S, Koyama N, Yokochi K, Ohmura K, Shiina M, Mori H, Koshimizu E, Imagawa E, Uchiyama Y, Mitsuhashi S, Frith MC, Fujita A, Satoh M, Taguri M, Tomono Y, Takahashi K, Doi H, Takeuchi H, Nakashima M, Mizuguchi T, Takata A, Miyake N, Saitsu H, Tanaka F, Ogata K, Hennet T, Matsumoto N (2018) Biallelic COLGALT1 variants are associated with cerebral small vessel disease. Ann Neurol 84:843–853
46. Morris NP, Bachinger HP (1987) Type XI collagen is a heterotrimer with the composition (1 alpha, 2 alpha, 3 alpha) retaining non-triple-helical domains. J Biol Chem 262:11345–11350
47. Okuyama K, Miyama K, Mizuno K, Bachinger HP (2012) Crystal structure of (Gly-Pro-Hyp) (9): implications for the collagen molecular model. Biopolymers 97:607–616
48. Onursal C, Dick E, Angelidis I, Schiller HB, Staab-Weijnitz CA (2021) Collagen biosynthesis, processing, and maturation in lung ageing. Front Med (Lausanne) 8:593874
49. Revell CK, Jensen OE, Shearer T, Lu Y, Holmes DF, Kadler KE (2021) Collagen fibril assembly: new approaches to unanswered questions. Matrix Biol Plus 12:100079
50. Robinson KA, Sun M, Barnum CE, Weiss SN, Huegel J, Shetye SS, Lin L, Saez D, Adams SM, Iozzo RV, Soslowsky LJ, Birk DE (2017) Decorin and biglycan are necessary for maintaining collagen fibril structure, fiber realignment, and mechanical properties of mature tendons. Matrix Biol 64:81–93
51. Rousseau JC, Farjanel J, Boutillon MM, Hartmann DJ, van der Rest M, Moradi-Ameli M (1996) Processing of type XI collagen. Determination of the matrix forms of the alpha1 (XI) chain. J Biol Chem 271:23743–23748
52. Sarohi V, Basak T (2023) Perturbed post-translational modification (PTM) network atlas of collagen I during stent-induced neointima formation. J Proteome 276:104842
53. Sarohi V, Srivastava S, Basak T (2022) Comprehensive mapping and dynamics of site-specific prolyl-hydroxylation, Lysyl-hydroxylation and Lysyl O-glycosylation of collagens deposited in ECM during zebrafish heart regeneration. Front Mol Biosci 9:892763

54. Segev F, Heon E, Cole WG, Wenstrup RJ, Young F, Slomovic AR, Rootman DS, Whitaker-Menezes D, Chervoneva I, Birk DE (2006) Structural abnormalities of the cornea and lid resulting from collagen V mutations. Invest Ophthalmol Vis Sci 47:565–573
55. Smith SM, Birk DE (2012) Focus on molecules: collagens V and XI. Exp Eye Res 98:105–106
56. Sola RJ, Al-Azzam W, Griebenow K (2006) Engineering of protein thermodynamic, kinetic and colloidal stability: chemical glycosylation with monofunctionally activated Glycans. Biotechnol Bioeng 94:1072–1079
57. Song E, Mechref Y (2013) LC-MS/MS identification of the O-glycosylation and hydroxylation of amino acid residues of collagen alpha-1 (II) chain from bovine cartilage. J Proteome Res 12: 3599–3609
58. Sorushanova A, Delgado LM, Wu Z, Shologu N, Kshirsagar A, Raghunath R, Mullen AM, Bayon Y, Pandit A, Raghunath M, Zeugolis DI (2019) The collagen suprafamily: from biosynthesis to advanced biomaterial development. Adv Mater 31:e1801651
59. Spiro RG (1969) Characterization and quantitative determination of the hydroxylysine-linked carbohydrate units of several collagens. J Biol Chem 244:602–612
60. Spiro RG, Fukushi S (1969) The lens capsule. Studies on the carbohydrate units. J Biol Chem 244:2049–2058
61. Sricholpech M, Perdivara I, Nagaoka H, Yokoyama M, Tomer KB, Yamauchi M (2011) Lysyl hydroxylase 3 glucosylates galactosylhydroxylysine residues in type I collagen in osteoblast culture. J Biol Chem 286:8846–8856
62. Sricholpech M, Perdivara I, Yokoyama M, Nagaoka H, Terajima M, Tomer KB, Yamauchi M (2012) Lysyl hydroxylase 3-mediated glucosylation in Type I Collagen. J Biol Chem 287: 22998–23009
63. Sun M, Connizzo BK, Adams SM, Freedman BR, Wenstrup RJ, Soslowsky LJ, Birk DE (2015) Targeted deletion of collagen V in tendons and ligaments results in a classic Ehlers-Danlos syndrome joint phenotype. Am J Pathol 185:1436–1447
64. Sun M, Luo EY, Adams SM, Adams T, Ye Y, Shetye SS, Soslowsky LJ, Birk DE (2020) Collagen XI regulates the acquisition of collagen fibril structure, organization and functional properties in tendon. Matrix Biol 94:77–94
65. Taga Y, Kusubata M, Ogawa-Goto K, Hattori S (2012) Development of a novel method for analyzing collagen O-glycosylations by hydrazide chemistry. Mol Cell Proteomics 11: M111.010397
66. Taga Y, Kusubata M, Ogawa-Goto K, Hattori S (2013) Site-specific quantitative analysis of overglycosylation of collagen in osteogenesis imperfecta using hydrazide chemistry and SILAC. J Proteome Res 12:2225–2232
67. Tang M, Wang X, Gandhi NS, Foley BL, Burrage K, Woods RJ, Gu Y (2020) Effect of hydroxylysine-O-glycosylation on the structure of type I collagen molecule: a computational study. Glycobiology 30:830–843
68. Terajima M, Taga Y, Chen Y, Cabral WA, Hou-Fu G, Srisawasdi S, Nagasawa M, Sumida N, Hattori S, Kurie JM, Marini JC, Yamauchi M (2016) Cyclophilin-B modulates collagen cross-linking by differentially affecting lysine hydroxylation in the helical and telopeptidyl domains of tendon type I collagen. J Biol Chem 291:9501–9512
69. Terajima M, Taga Y, Sricholpech M, Kayashima Y, Sumida N, Maeda N, Hattori S, Yamauchi M (2019) Role of glycosyltransferase 25 domain 1 in type I collagen glycosylation and molecular phenotypes. Biochemistry 58:5040–5051
70. Terajima M, Taga Y, Nakamura T, Guo HF, Kayashima Y, Maeda-Smithies N, Parag-Sharma-K, Kim JS, Amelio AL, Mizuno K, Kurie JM, Yamauchi M (2022) Lysyl hydroxylase 2 mediated collagen post-translational modifications and functional outcomes. Sci Rep 12: 14256
71. Tong D, Lonnblom E, Yau ACY, Nandakumar KS, Liang B, Ge C, Viljanen J, Li L, Balan M, Klareskog L, Chagin AS, Gjertsson I, Kihlberg J, Zhao M, Holmdahl R (2018) A shared epitope of collagen type XI and type II is recognized by pathogenic antibodies in mice and humans with arthritis. Front Immunol 9:451

72. Tsuneizumi K, Kasamatsu A, Saito T, Fukushima R, Taga Y, Mizuno K, Sunohara M, Uzawa K, Yamauchi M (2024) Generation of bone-specific lysyl hydroxylase 2 knockout mice and their phenotypes. Biochem Biophys Rep 39:101790
73. Tvaroska I (2024) Glycosylation modulates the structure and functions of collagen: a review. Molecules 29
74. Vahidnezhad H, Youssefian L, Saeidian AH, Touati A, Pajouhanfar S, Baghdadi T, Shadmehri AA, Giunta C, Kraenzlin M, Syx D, Malfait F, Has C, Lwin SM, Karamzadeh R, Liu L, Guy A, Hamid M, Kariminejad A, Zeinali S, McGrath JA, Uitto J (2019) Mutations in PLOD3, encoding lysyl hydroxylase 3, cause a complex connective tissue disorder including recessive dystrophic epidermolysis bullosa-like blistering phenotype with abnormal anchoring fibrils and type VII collagen deficiency. Matrix Biol 81:91–106
75. Visser DR, Loo TS, Norris GE, Parry DAD (2023) Potential implications of the glycosylation patterns in collagen alpha1(I) and alpha2(I) chains for fibril assembly and growth. J Struct Biol 215:107938
76. Wang S, He L, Xiao F, Gao M, Wei H, Yang J, Shu Y, Zhang F, Ye X, Li P, Hao X, Zhou X, Wei H (2023) Upregulation of GLT25D1 in hepatic stellate cells promotes liver fibrosis via the TGF-beta1/SMAD3 pathway in vivo and in vitro. J Clin Transl Hepatol 11:1–14
77. Wenstrup RJ, Florer JB, Brunskill EW, Bell SM, Chervoneva I, Birk DE (2004a) Type V collagen controls the initiation of collagen fibril assembly. J Biol Chem 279:53331–53337
78. Wenstrup RJ, Florer JB, Cole WG, Willing MC, Birk DE (2004b) Reduced type I collagen utilization: a pathogenic mechanism in COL5A1 haplo-insufficient Ehlers-Danlos syndrome. J Cell Biochem 92:113–124
79. Wenstrup RJ, Florer JB, Davidson JM, Phillips CL, Pfeiffer BJ, Menezes DW, Chervoneva I, Birk DE (2006) Murine model of the Ehlers-Danlos syndrome. col5a1 haploinsufficiency disrupts collagen fibril assembly at multiple stages. J Biol Chem 281:12888–12895
80. Yamauchi M, Sricholpech M (2012) Lysine post-translational modifications of collagen. Essays Biochem 52:113–133
81. Yang C, Park AC, Davis NA, Russell JD, Kim B, Brand DD, Lawrence MJ, Ge Y, Westphall MS, Coon JJ, Greenspan DS (2012) Comprehensive mass spectrometric mapping of the hydroxylated amino acid residues of the alpha1(V) collagen chain. J Biol Chem 287:40598–40610
82. Yoshioka H, Iyama K, Inoguchi K, Khaleduzzaman M, Ninomiya Y, Ramirez F (1995) Developmental pattern of expression of the mouse alpha 1 (XI) collagen gene (Col11a1). Dev Dyn 204:41–47
83. Zhang Y, Yu CY, Song E, Li SC, Mechref Y, Tang H, Liu X (2015) Identification of glycopeptides with multiple hydroxylysine O-glycosylation sites by tandem mass spectrometry. J Proteome Res 14:5099–5108

Chapter 10
Collagen Mechanics

Phoebe Szarek and Jeffrey W. Ruberti

Abstract The collagen superfamily has evolved over nearly a billion years to produce a set of at least 28 proteins that are present in all vertebrates. Of these, the fibrillar collagens (Types I–III, V, XI, XIV, and XVII) comprise a diverse, fibrous structural polymer system that extensively invests and mechanically supports connective tissue. The chemistry, structure, and mechanics of single collagen molecules will be reviewed to provide insight into why collagen's triple helical motif is such an effective, efficient, and versatile structural building block. Review of the integration of molecular collagen into each successive length scale: fibrils, fibers, and tissues will show how the mechanical signature of each level in the hierarchy reflects the mechanical behavior of the smaller length scales from which it is composed. The remarkable structural and mechanical versatility of multiple whole connective tissues with completely different structural roles will demonstrate how each of them relies on collagen to perform their diverse mechanical functions. Finally, multiple themes that address more transcendent phenomena such as fibril diameter modulation, crosslinking, fatigue, and collagen mechanochemistry will be examined to provide a broader view of the field and open new directions for research.

Keywords Collagen · Mechanics · Fibrils · Fibers · Connective tissue

Abbreviations

ACL	Anterior cruciate ligament
AFM	Atomic force microscopy
AGE	Advanced glycation end
BAPN	β-aminoproprionitrile
BMP-1	Bone morphogenic protein
C-	C terminal

P. Szarek · J. W. Ruberti (✉)
Department of Bioengineering, Northeastern University, Boston, MA, USA
e-mail: p.szarek@northeastern.edu; j.ruberti@northeastern.edu

D. A. D. Parry, P. K. Luther (eds.), *Fibrous Proteins: Structures and Mechanisms*, Subcellular Biochemistry 113, https://doi.org/10.1007/978-3-032-05273-5_10

Col(X)A(Y)	Collagen (type) alpha (chain number)
DDR	Discoidin domain receptor
DEJ	Dermal-epidermal junction
DNA	Deoxyribonucleic acid
ECM	Extracellular matrix
FACIT	Fibril-associated collagen with interrupted triple helices
FCD	Fixed charge density
Fe	Iron
GAG	Glycosaminoglycan
Gly	Glycine
GPa	Gigapascals
Hyp	Hydroxyproline
IOP	Intraocular pressure
LOX	Lysyl oxidase
MACIT	Membrane-associated collagen with interrupted triple helices
MCL	Medial collateral ligament
MEM	Microelectromechanical
mM	millimolar
mmHg	Millimeters of mercury
MPa	Megapascals
MYA	Million years ago
N-	N terminal
nm	Nanometer
OI	Osteogenesis imperfecta
PG	Proteoglycan
pN	picoNewtons
PNAS	Proceedings of the National Academy of Science
Pro	Proline
PTM	Post-translational modification
QLV	Quasi-linear viscoelasticity
RNA	Ribonucleic acid
SEM	Scanning electron microscopy
TEM	Transmission electron microscopy
WLC	Worm-like chain
Xaa	Amino acid in the X position of Gly-X-Y
Yaa	Amino acid in the Y position of Gly-X-Y
μm	Micrometer

Introduction

The Extracellular Matrix

The hallmark of multicellularism is the presence of an intercellular adhesive material that is referred to as the extracellular matrix or ECM. The ECM binds what would otherwise be wayward, isolated cells into a unified, purposeful organism. Plants, animals, and even bacteria can produce ECM that provides adhesive strength, organization, and security to their cellular colonies [50]. In animals, the variation in the structure of ECM proteins can be traced to the extensive recombination of modular domains through exon/domain shuffling [63, 117]. Comparative analyses among species have clearly demonstrated that all bilateral taxa share a common set of ECM proteins with few exceptions, generally due to gene loss. With few exceptions, in animals, the most structurally important ECM molecule is collagen.

The Collagens: Structure, Function, and Evolution

Structure Collagen is defined as repeats of the triplet gly-x-y, where x and y are often proline and hydroxyproline, but they can be other amino acids. The triplets repeat linearly along the axis of a left-handed helical α-chain (minor helix) that is further coiled around two additional collagen α-chains (which may or may not be identical [165]) to form the canonical right-handed major triple helix as described by Ramachandran and Kartha [207] (Fig. 10.1). Ramachandran and Reddi [208] suggest that any other sequence of amino acids would perturb the triple helical structure (see also [20] for the implications of substitutions/mutations in diseases such as osteogenesis imperfecta). The quintessential structural motif of collagen is offset, repeating triplets tightly coiled about one another in a pattern that buries the glycine at the center of the major helix, but exposes the other two amino (imino, in the cases of proline and hydroxyproline) acids at the surface of the molecule. Figure 10.2 shows multiple models of the main triple helical repeat of collagen, depicting the offset chains and externally facing X and Y motifs [19, 174]. Thus, as one can imagine, the surface amino (imino) acids are subject to both co- and post-translational modification (PTM) (i.e., hydroxylation, glycosylation, and phosphorylation) that add a layer of informational complexity and chemical diversity to the overall structure (see [115] for a well-thought out review of collagen diversity, synthesis and assembly). Even genetically identical collagens (e.g., type I collagen, comprising two α1(I) alpha chains and one 1α(II) alpha chain) can display tissue-specific variations in their PTM decoration [200], suggesting there are functional differences for a particular collagen type within the same animal (e.g., type I collagen in tendon vs. skin).

The vertebrate collagen superfamily comprises over 50 molecules [218] that can be referred to as collagen or collagen-like due to the presence of the triple helical

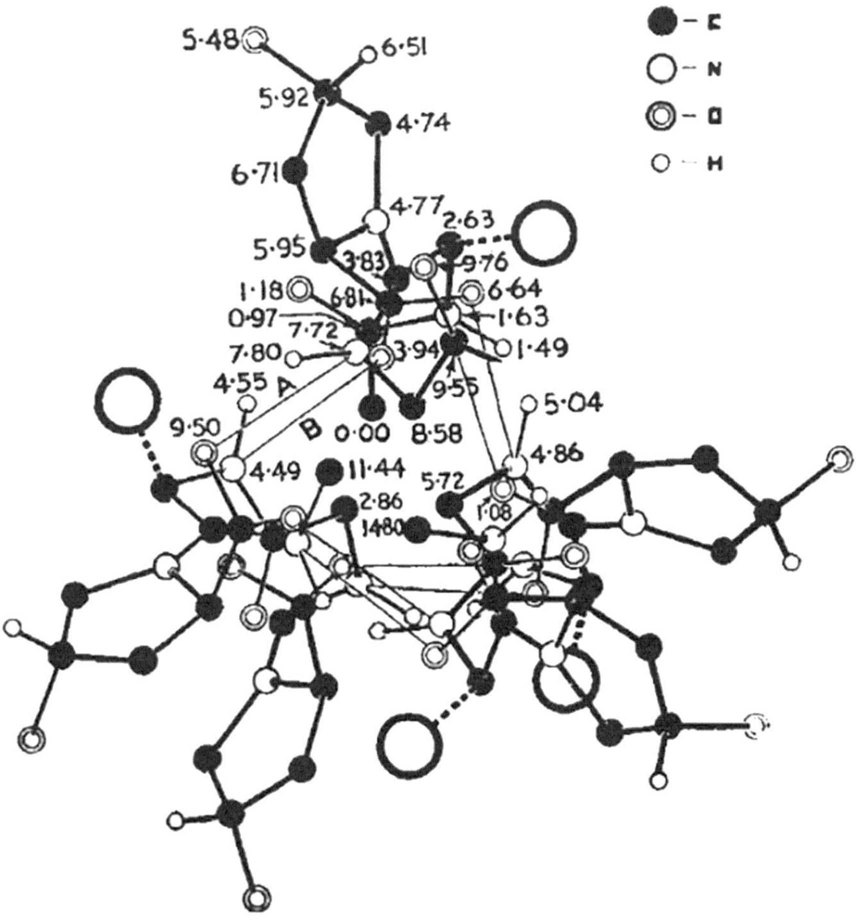

Fig. 10.1 Proposed structure of collagen (projection down the axis of the triple helix) as shown by Ramachandran and Kartha in their 1955 Nature paper. (Reproduced from Ramachandran and Kartha [207] with permission)

motifs. The distinction between what is truly collagen and what is merely collagen-adjacent is an academic pursuit we will avoid for the purposes of this review. Within the collagen superfamily, there are at least 28 types that are common to both humans and vertebrates (referred to by their roman numbers I–XXVIII) comprising combinations of 46 distinct α-chains. The collagens can be subdivided into: fibrillar collagens—Types I, II, III, V, XI, XXIV, and XXVII; network-forming collagens—Types IV, VIII, and X; fibril-associated collagens with interrupted triple helices (FACITs)—Types IX, XII, XIV, XVI, XX, XIX, XXI, and XXII; membrane-associated collagens (MACITs)—Types XIII, XV, XVII, XVIII, XXIII, and XXV; and other unique collagens—Type VII, VI, XXVI, and XXVIII. See

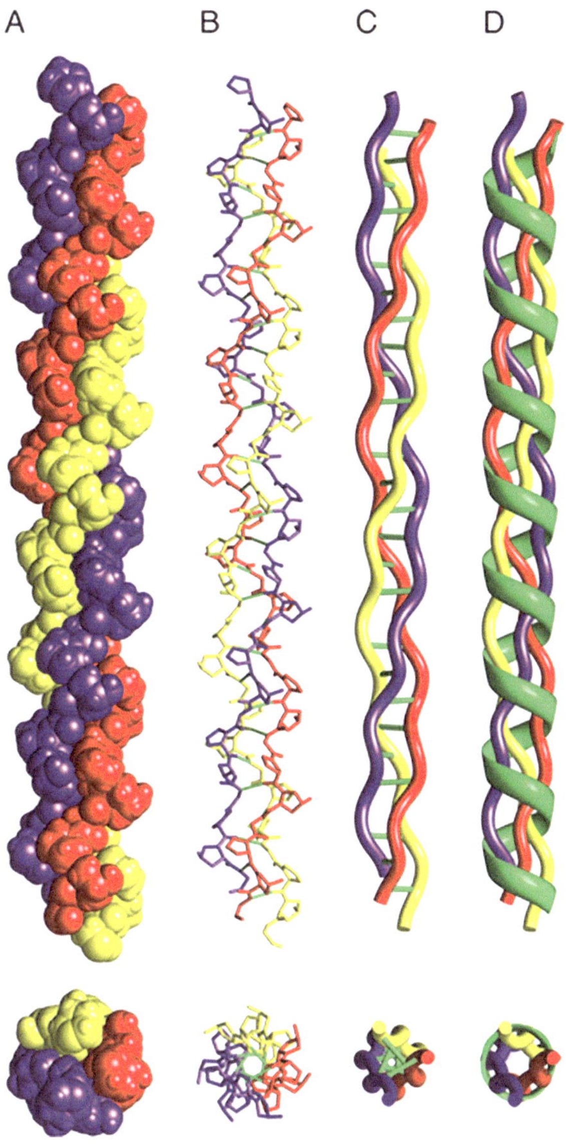

Fig. 10.2 Collagen triple helical structure representations. The three alpha chains are shown in purple (trailing), red (middle), and green (leading strand). The strands are named for their vertical position in the direction from the N- (top) to the C-terminus. (**a**) Space-filling diagram. (**b**) Stick diagram. (**c**) Ribbon diagram showing hydrogen bond topology. (**d**) Ribbon diagram including the left-handed superhelix (green). (Reproduced from Bella [19] with permission)

Mienaltowski et al. [165] for a comprehensive review of the collagen types and functions.

Function As extracellular matrix proteins, the collagens were originally thought to perform purely structural roles such as imparting mechanical strength to tissues (e.g., tendon), protecting and supporting sheets of epithelium (e.g., barrier basement membranes), or binding cells to matrix (i.e., desmosomes and participation in focal adhesions). However, as one might expect, collagen has evolved over time to perform other roles including signaling to resident matrix cells (e.g., DDR1/2 receptors [48]), to transient cell populations (e.g., immune cells), and to platelets [187]. Fragments of collagen can act as signaling molecules as well: XVIII-endostatin [190]. As our knowledge of the collagens has grown substantially, more functional diversity has been uncovered. It is certainly not surprising that Nature would, over millions of years of evolution, imbue these highly stable, ubiquitous molecules with multiple layers of functionality in addition to their structural role.

Dysfunction *Cancer*: Collagen can also exhibit pathological behavior. Collagen-DDR1 signaling has been shown to play an important role in cancer progression [246] and in the establishment of a resistant cancer ECM niche. There is ostensibly a rich, complex, and underexplored relationship between cancer and collagen. The discovery that tumor cells synthesize non-canonical homotrimeric type I collagen [172] has altered the perception of the tumor ECM, changing it from a passive matrix in which the tumor resides to a collaborative niche that enhances the survival of the cancer. Recent work by Chen et al. [45] confirms that cancer fibroblasts in both mice and humans produce oncogenic type I collagen homotrimer that is resistant to enzymatic degradation and which may be considered "*an overarching feature of human and mouse pancreatic cancer cells.*" *Mutations*: There are more than 1000 mutations in collagen genes (see Myllyharju and Kivirikko [176] for an excellent, comprehensive early review). These sequence disruptions cause a range of disorders from spontaneous early lethality to diseases such as Alport syndrome (Col4X mutations), osteogenesis imperfecta (Col 1A1/2 mutation), and Ehlers-Danlos (Col5A1/A2). *Fibrosis*. "Too much of a good thing" can be a significant problem as well. The excess deposition of ostensibly normal collagen contributes to approximately 45% of all deaths in the developed world [107, 284]. Fibrosis is generally considered the final common endpoint of a number of chronic inflammatory diseases [284] and is characterized principally by the over-deposition of insoluble collagen fibrils (typically comprising types I and III). Excess collagen can lead to the disruption of tissue function and structure. Common sites of fibrotic diseases are the liver, lung, heart, kidney, and skin [8]. Within fibrotic lesions, fibroblasts transition to more muscular cells or myofibroblasts [94] suggesting that mechanics may play an important role in the abnormal over-deposition of collagen (vide infra for discussion of collagen mechanochemistry).

Evolution A hallmark of all metazoan life is the presence of collagen [70]. Thus, the evolution of collagen can be considered an essential event underpinning the emergence of multicellular animals. Of all the collagens, only a few have been conserved from the first glass sponges to humans: Type IV collagen and the fibrillar collagens [69]. *Type IV Collagen*. Type IV collagen is considered the "ancient" primordial collagen [76], and is part of the common basement membrane toolkit [279] comprising auxiliary nidogen, laminin, a laminin-binding glycoprotein, and perlecan [289]. Type IV collagen arrived over 500 MYA and is present in all subsequent animal phyla (Fig. 10.3). It has persisted throughout Chordata evolution and can be

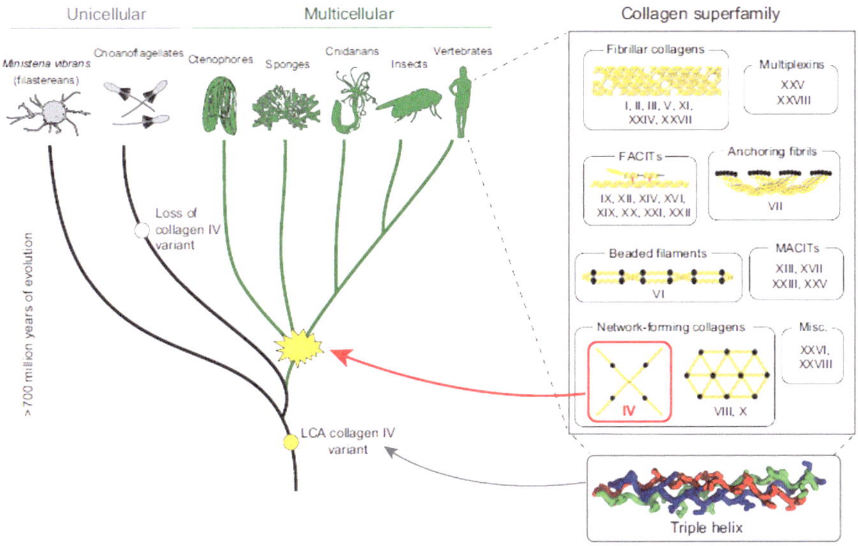

Fig. 10.3 Evolutionary tree of the collagen superfamily. Network-forming collagen IV is the likely original "collagen" with its last common ancestor appearing more than 700 MYA. The triple helix is the common "thread" that persists across all animal taxa. (Reproduced from Fidler et al. [76] with permission)

found in all extant vertebrates in the form of a network typically secreted beneath epi- or endothelial cell layers. Type IV basement membrane networks can provide both function and structure to tissues. Functional basement membranes include the glomerular basement membrane in the kidney where its primary role is as a dynamic barrier to control the filtration rate of fluid entering the glomerular capsule (for an excellent review of the functional role of collagens in the kidney, see De Gregorio et al. [54]). Structural basement membranes are ubiquitous in metazoans and can be found lining the basal side of epithelium or endothelium in vertebrates. Figure 10.4 shows an excellent high-resolution QFDE image of the basement membrane under the lens capsule epithelium in the eye [15].

Fibrillar Collagens

The principal structural elements in animals are crosslinked collagen fibrils. Fibrils are supramolecular linear arrays formed through the fibrillogenesis (entropically driven assembly) of collagen molecules from a subset of the collagens (types I, II, III, V, XI, XXIV, and XXVII). Fibrillar collagen molecules are derived from a set of genes that are highly conserved at the sequence level from sponges to humans and throughout the animal kingdom [69]. The genes coding for the fibril forming collagens have a homologous organization with the triple helical coding exons likely

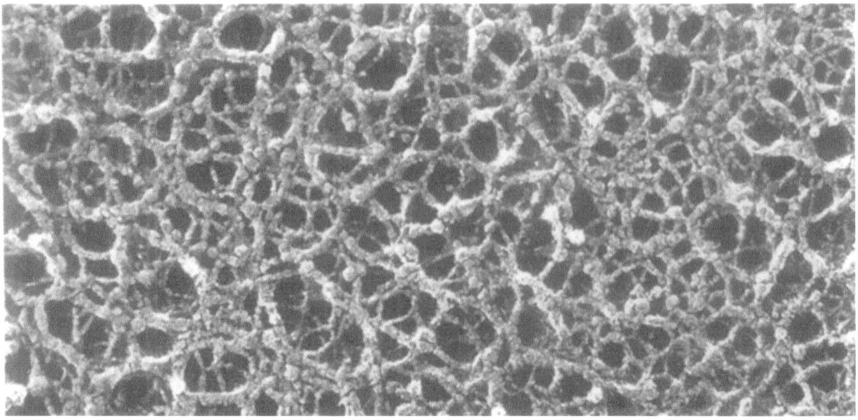

Fig. 10.4 Quick Freeze Deep Etch image of the type IV collagen basement membrane of the lens in the human eye. The tissue was extracted with trypsin, hyaluronidase, and heparinase prior to imaging to improve the clarity of the collagen network. (Reproduced from Barnard et al. [15] with permission)

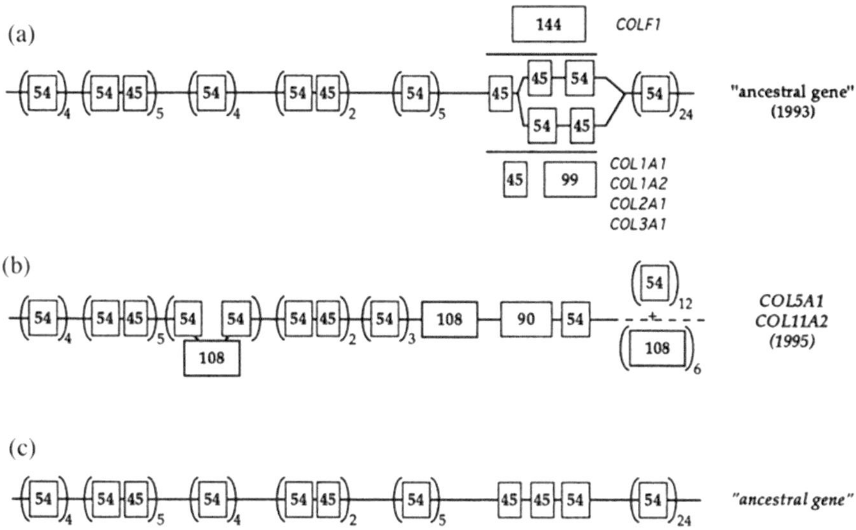

Fig. 10.5 Convergence on the putative fibrillar collagen ancestral gene comprising. (**a**) Region coding for the triple helical domain of a deduced ancestral fibrillar collagen gene. The proposed structure aligns the gene coding for a freshwater sponge (COLF1) with human types I–III fibrillar genes [68] (**b**) Organization of Col5A1 and Col11A2 genes as depicted by Takahara et al. [261]. (**c**) Ancestral gene as proposed by Exposito et al. [69]. Note the prominent exon domains of 54 and 54-9 base pairs and their duplications. (Reproduced from Exposito et al. [69] with permission)

derived from an ancestral unit comprising 54 base pairs encoding six Gly-Xaa-Yaa repeating "mers" [67]. Figure 10.5 shows the convergence in the literature on the putative ancestral gene for fibrillar collagen as proposed by Exposito et al. [69]. A

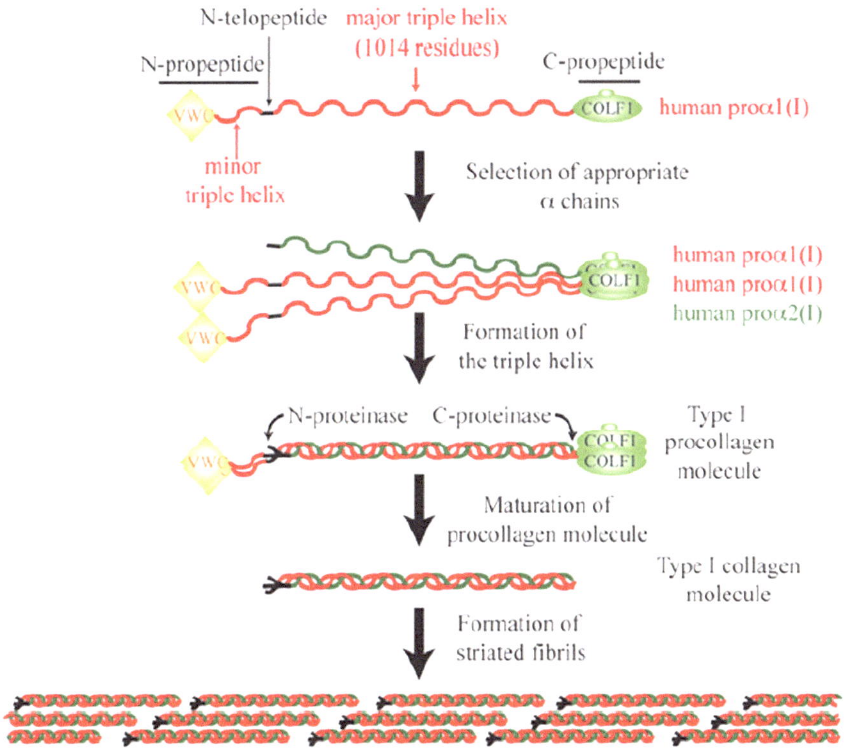

Fig. 10.6 The general assembly process for the fibrillar collagens using type I human collagen as the model. The three alpha chains are synthesized and conjoined at the C-terminal from where they "zip" into the final triple helical structure. The N- and C- terminal domains prevent premature assembly in the cell and are cleaved by their pro-peptidases generally post-secretion into the extracellular space. The model is general for all seven fibrillar collagen types with differences in the N-propeptide (I, II, III vs. V, XI, XXIV, XXVII), glycine substitutions (XXIV and XXVII), and Gly-Xaa-Yaa-Zaa imperfections (XXIV and XXVII). (Reproduced from Exposito et al. [71] with permission)

prototypical fibrillar procollagen molecule is an uninterrupted major triple helical domain comprising a perfect series of Gly-Xaa-Yaa triplets flanked by C- and N-propeptides which are not triple helical. Upon post-translational cleavage of the terminal propeptides, the resulting tropocollagen (telocollagen) molecule is approximately 1.5 nm in diameter and 300 nm in length with two short non-helical telopeptides remaining at the cleaved ends [33]. Figure 10.6 illustrates the general process by which fibril-forming collagens are produced [71]. Once assembled, collagen fibrils display a characteristic D-periodic banding pattern that repeats approximately 67 nm [17, 230]. The phylum Porifera are the most primitive multicellular animals. At the cusp of metazoan inception, sponges developed a material that would hold their cellular colonies together in rudimentary but functional organization. From this moment forward, cells and matrix began a billion year

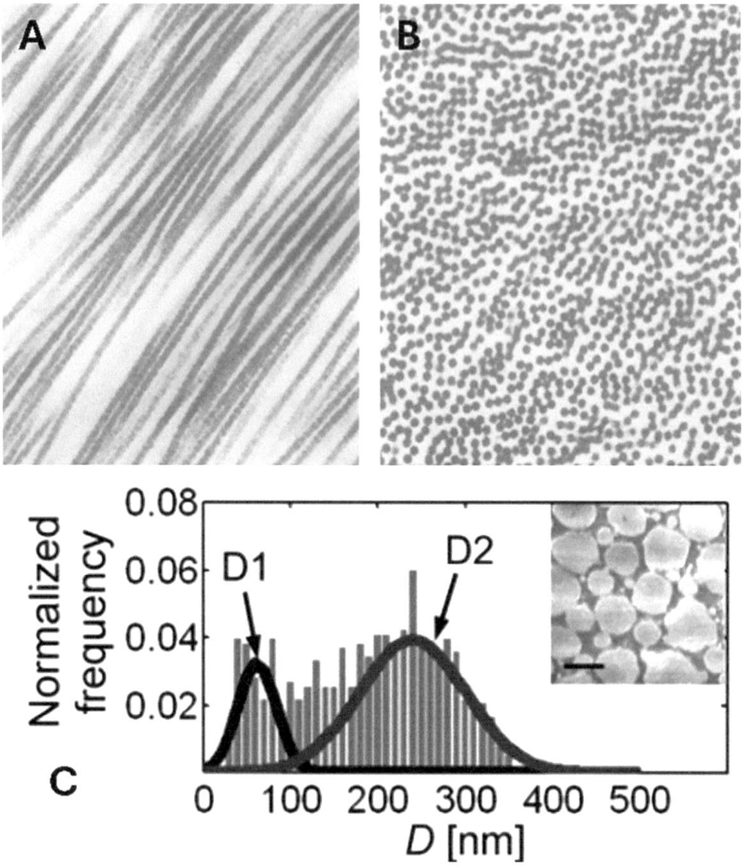

Fig. 10.7 (**a**) TEM image in longitudinal section of a cortical bundle of sponge (*Chondrosia renformis Nardo*) collagen fibrils showing periodic banding and long-range organization. (**b**) TEM of a bundle in cross-section showing the highly uniform diameter distribution of the fibrils. From [89] with permission. (**c**) Typical biomodal fibril diameter distribution in a modern mammal (mouse). (From Goh et al. [96] with permission)

cooperative dance that led to the plethora of mechanically sophisticated extant body forms. Sponges express multiple types of collagen including small (~20–25 nm) fibrils that display banding patterns similar to fibrillar collagens found in vertebrates [89]. Figure 10.7 shows fibrillar collagen from the marine sponge *Chondrosia renformis Nardo* with clear periodic banding (Fig. 10.7a) and a highly monodisperse diameter distribution (Fig. 10.7b) [90]. In Fig. 10.7c for comparison, we see the typical arrangements of predominantly type I fibrillar collagen in a mammal (adult mouse) with its generally bimodal diameter distribution of type I collagen fibrils. The distinct "bimodal" fibril diameter distribution in mammalian tendons is thought to be optimized to increase the resilience and rupture strength of the tendon (vide infra for more discussion in Theme I) [96, 192].

The Mechanics of Collagenous Tissue

Origins of Collagenous Tissue Mechanical Strength

Macroscopic load-bearing fibrous tissues comprise multiple levels of structural hierarchy [125]. Each level of structure will affect the aggregate mechanics of the tissue. While, holistically, the most important mechanical behavior is the response of the entire structure to applied loads, we will take a short reductionist detour to understand the fundamental mechanical building blocks that contribute to collagen's mechanical signature. Doing so enables researchers to tie genetic defects, which affect proteins at the amino acid level, to the large-scale behavior of the resulting collagenous tissue (e.g., Col1a1/2mutations in osteogenesis imperfecta).

Collagen Molecular Structure and Mechanics

Molecular Structure Figure 10.8 shows a rotary shadowed image of the prototypical type I collagen molecule in both pro- and telo-collagen form Fleischmajer et al. [79]. Their high aspect ratio and thin cross-section are highly suggestive of their role as a tensile load-bearing element. The triple helical main axis, a structure unique to collagen (with a few notable exceptions—e.g., triplex DNA/RNA [83]), places three polypeptide chains in staggered, close alignment along the collagen molecule's major axis (vide supra—Fig. 10.2). The tight helix is derivative of the Gly-X-Y repeat which assures that the smallest amino acid, glycine, is at the core of the molecule. The staggering of the three alpha chains serves to create the space for each successive glycine in the core, ensuring that the molecule has a small diameter (~1.3–1.5 nm—[33, 46, 47, 207, 219]) and that the radial density of collagen molecules is high.

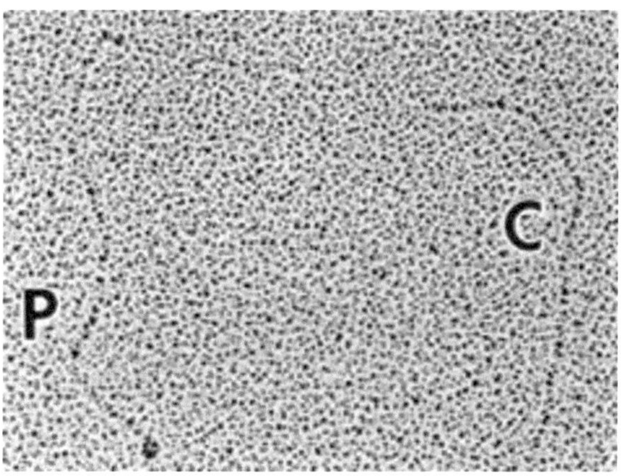

Fig. 10.8 Rotary Shadowed TEM images of type I pro-collagen (P) and telocollagen (C) from chick tendon. Note the short globular C-propeptide and the linear N-propeptide on the pro-collagen molecule (P). (Reproduced from Fleischmajer et al. [79] with permission)

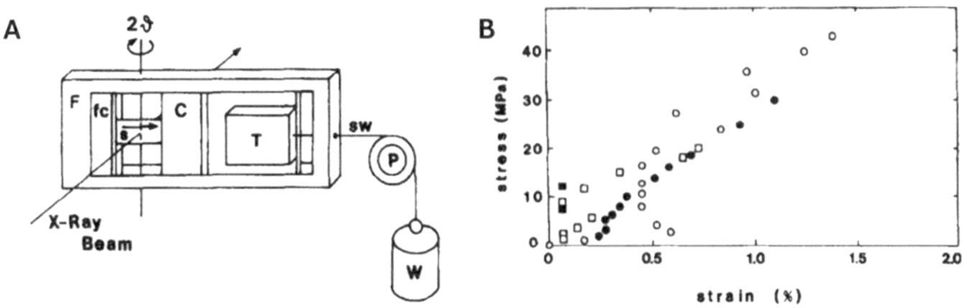

Fig. 10.9 (**a**) Mechanical testing apparatus of Sasaki et al. [228] with X-ray diffraction readout of molecular strain. (**b**) Stress/strain relationship for molecular collagen in bovine Achilles' tendon (shapes represent four different experiments). (Reproduced from [228]) with permission)

Molecular Mechanics The mechanics of the type I collagen triple helix have been examined both experimentally [227, 228, 254, 255] and theoretically [41–43]. Early attempts by Cowan et al. to observe the mechanical response of hierarchical structures to applied force suffered from a general lack of information about the organization of the tissue at multiple length scales [46, 47]. However, in 1996, Naoki Sasaki and Singo Odajima, armed with more complete information about tendon structure, reported measurements of the modulus of the type I collagen triple helix (bovine Achilles' tendon) taken directly from X-ray diffraction data obtained during mechanical loading [228]. Figure 10.9 shows their simple, but very effective, apparatus and the stress/strain data they obtained from multiple measurements of collagen molecular extension. Using a simple regression analysis, they obtained a molecular Young's modulus of 2.9 GPa. Because their system operates in static equilibrium, we expect their modulus measurements to provide a lower value than those taken under dynamic conditions (e.g., Brillouin light scattering), which have been reported to be as high as 9 GPa [102]. In 2002, Sun et al. [254] directly measured the flexibility (persistence length) of type I procollagen using an optical trap. The data also provided information about the stiffness of the collagen triple helix at different levels of molecular extension. Figure 10.10 shows the force extension curves for both type I and type II collagen obtained over the course of two different investigations [254, 255]. Both collagen monomers were found to be flexible (14.5 nm and 11.2 nm estimated persistence lengths for type I and type II collagen monomers, respectively) and very stiff in enthalpic extension. Note the very low resistance to load and high extensibility until about 3 pN. This is the entropic phase of the extension curve. Once the molecule is loaded in enthalpic extension, it rapidly stiffens, approaching a very high spring constant as it is extended to and past its contour length (~300 nm for telocollagen). Type I collagen had an estimated modulus ranging from 350 MPa to 12.2 GPa depending on your choice of molecular radius (0.28 nm or 0.68 nm). The high flexibility (short persistence length) found by this method was inconsistent with previous measurements taken using more indirect methods (hydrodynamic ~170 nm [183] and electron microscopic measurement—57 nm [110]).

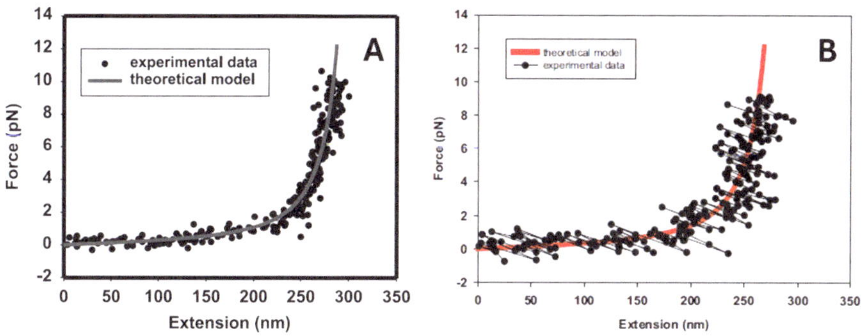

Fig. 10.10 Direct mechanical data obtained by stretching single collagen molecules in an optical trap. (**a**) Force extension curve for human type I procollagen (**b**) Force extension curve for recombinant type II procollagen. The two molecules exhibited high flexibility and stiff resistance in energetic extension. (Reproduced from Sun et al. [254, 255] with permission)

The elegance of directly measuring the extension curve led to a general acceptance of the optical trap data and the shorter persistence length. However, most studies of molecular collagen, including those of Sun et al., assume that it is a homogeneous, deformable rod with one mode of deformation, which it is not. The inconsistencies associated with persistence length measurement variability led Nancy Forde and colleagues to consider the mechanics and stability of the collagen triple helix in greater detail [130]. They implicate several factors that could confound measurements of persistence length including: (1) complex chemical composition of collagen (hetero vs. homotrimer); (2) the ionic strength of the solution; (3) the pH of the solution; (4) the short contour length of collagen (in extension measurements); (5) inappropriate use of the worm-like chain (WLC) model; (6) alternate modes of deformation (i.e. stretch and twist); and (7) sequence dependent behavior. Before one can integrate the molecular mechanical behavior of the collagen triple helix into larger-scale structures, it is necessary to more fully understand how it behaves as a single molecule under the many different conditions in which it finds itself. In general, we can conclude that the triple helix has a high tensile modulus, low extensibility, and a fairly constrained molecular conformation (resistance to bending). Certainly, in soft tissues, we can expect that the triple helical stiffness will be modulated by integration into the macroscopic structures, generally leading to a softer aggregate modulus. Bone is an exception to this rule and possesses a very high modulus relative to soft tissue ~20GPa [215]. We will address the structure/mechanics of bone's mineral/collagen composite behavior in a later section. In any case, we look forward to more experimental studies of collagen's molecular mechanics to inform our understanding of collagenous tissue's macroscopic mechanical behavior. Before moving on to the next level of integration, it is important to mention that there has been a substantial effort to model the mechanics of collagen at the molecular level. For computational atomistic assessments of collagen, one should start with the work of Markus Buehler who has put the collagen helix through a battery of perturbations at very high molecular resolution [41].

Collagen Fibril Structure, Chemistry, and Mechanics

Structure The next level of hierarchy and what is considered the fundamental, continuous [109, 258] load-bearing structure in collagenous tissue is the collagen fibril [230]. Among the most exciting discoveries in connective tissue biology was the realization that collagen molecules could be reprecipitated to form fibrils that had the same ordered morphology as native fibrils [98]. Figure 10.11 shows a prototypical type I collagen fibril via rotary shadowing (intact *A* and swollen/assembling *B*) [79]. The intact collagen fibril possesses the hallmark ~64–67 nm D-periodic striations [17, 230] that are found to be consistent across all the fibrillar collagens. Unlike the D-periodic banding pattern, the fibril diameter is not consistent. The

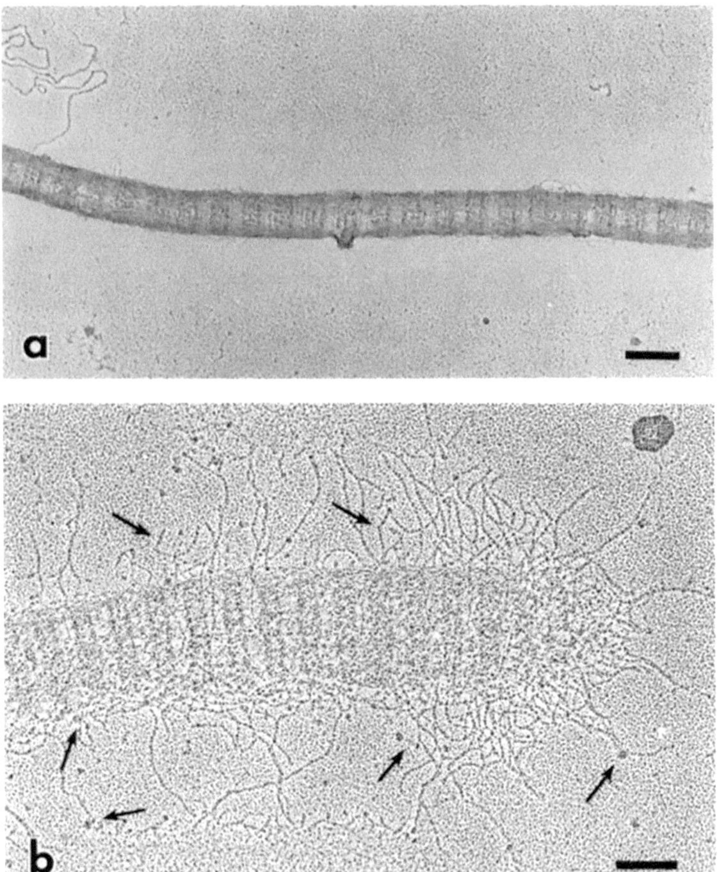

Fig. 10.11 Rotary-shadowed TEM images of type I collagen fibrils from chicken tendon. (**a**) Intact fibril. (**b**) Swollen, assembling fibril showing interaction with molecular collagen. Globular and flexible pro-domains indicated by arrows. Bar = 100 nm. (Reproduced from Fleischmajer et al. [79] with permission)

average diameter and the diameter dispersity of collagen fibrils vary widely across animal taxa (vide supra Fig. 10.7 and [192]) and within the same animal (e.g., cornea and tendon). In the human cornea, the heterotypic collagen fibrils (20% type V, 80% type I) are ~31 + −1.0 nm [53] in diameter while human adult tendons comprising primarily type I collagen possess bimodal distributions of fibrils that range in size from 50 to 400 nm [59]. Fibril diameters and their dispersity also change with the age of the tissue [192]. The young, swollen fibril that is under construction in Fig. 10.11b clearly shows the interaction of the fibril surface with molecular collagen from which it is ostensibly being assembled during tissue development (Tissue: chick tendon). Understanding the mechanical performance of these molecular aggregates is critical to understanding the integrated performance of tissues that generally comprise arrays of fibrils working together to resist and transmit mechanical forces (i.e., tendon, ligament, anulus fibrosus, cartilage, skin, bone, the tough ocular tunic, and other specialized collagenous connective tissues such as the enthesis and meniscus). The packing of collagen molecules into linear, semi-crystalline molecular aggregates (fibrils) suggests that the purpose of fibrillar collagen is to place as many well-supported peptide bonds as possible into the path of tensile loads directed along the axis of the aligned, peptide chains. X-ray data suggest that collagen fibrils are mixed crystals of quasihexagonally packed fibril-forming monomers that are tilted somewhat away (~15° [166, 191]) from the molecular axis [116]. More recent and detailed crystallographic data suggest that within the fibrils writ large, five 1-D, staggered, "pleated" molecules are packed quasihexagonally into supertwisted, discontinuous right-handed microfibrils that interdigitate with their neighbors [191]. Figure 10.12 shows a model of the packing of collagen molecules into a putative microfibril and the probable juxtaposition of three microfibrils as they would be arranged within a fibril [191]. The high concentration of fibrillar collagens and their organization into linear arrays in many load-bearing tissues provides strong evidence that the primary function of fibrillar collagens is to efficiently resist/transmit forces that would otherwise disperse large structures. Indeed, the singular advantage of metazoans over protozoans is the maintenance of functional, long-range organization that increases the probability of survival in a mechanically disruptive environment. While the collagen fibrils are clearly organized to resist/transmit tensile forces along their principal structural axis, their detailed mechanical signature and general stability are highly dependent on co-translational, post-translational, and post-depositional modifications that substantially modify the structure of molecular collagen (reviewed comprehensively in Myllyharju and Kivirikko [177]). Of these modifications, proline hydroxylation, lysyl hydroxylation, and oxidative deamination of lysyl and hydroxylysyl residues are primarily responsible for stabilizing the triple helix and strengthening fibrillar aggregates.

Chemistry: Proline Hydroxylation Proline is the only proteinogenic amino acid that has a secondary amine. In human tendon, there are about 125 prolines per 1000 residues [60]. Due to the presence of the imino nitrogen atom which is devoid of a proton, proline cannot participate substantially in inter- or intrachain hydrogen bonding [103]. Stabilization of collagen is achieved through the modification of a

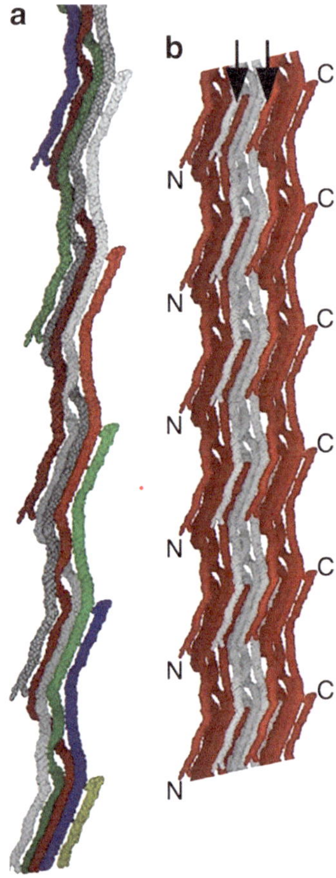

Fig. 10.12 Probable molecular packing of collagen into a putative microfibril (**a**) and the supramolecular packing of microfibrils as they would reside within a collagen fibril (**b**). (Reproduced from Orgel et al. [191] with permission)

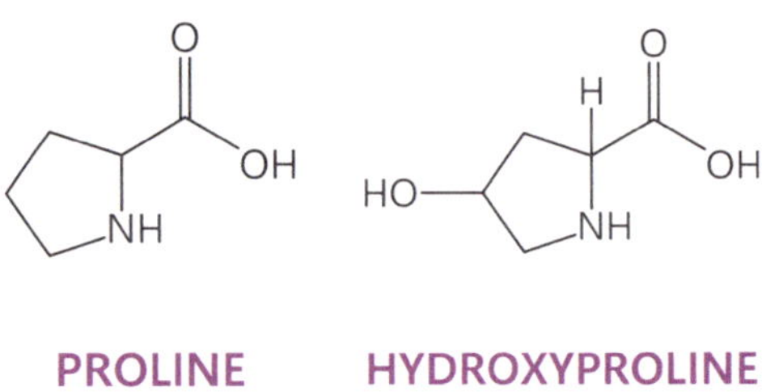

Fig. 10.13 Structures of the two imino acids, proline and hydroxypoline which, together, make up about 21% of the residues in fibrillar collagens. The consequence of nitrogen's embedding in the pyrrolidine ring is the loss of an available proton to participate in stabilizing hydrogen bonding. Thus, the hydroxylation of proline is necessary to stabilize the triple helix

substantial number of the prolines by hydroxylation [24] (e.g., 92 of 217 in human tendon [60]) by two families of enzymes: prolyl 4- and prolyl 3-hydroxylases (see Rappu et al. [210] for an excellent review of collagen hydroxylation). Figure 10.13

shows the structures of the imino acids proline and hydroxyproline with their characteristic pyrrolidine rings. Because hydroxyproline is generally unique to collagens with only small amounts identified in other proteins (e.g., elastin and acetylcholine esterase), it is often used to measure collagen concentration in tissue samples and in cell culture [24]. The most abundant modified proline is 4-hydroxyproline (4-Hyp) which, in vertebrates, is processed by a family of three isoenzymes (prolyl-4-hydroxylases—C-P4Hs) that mediate the reaction [133, 134, 175]. Prolyl 4-hydroxylation is dependent on 2-oxoglutarate, Fe^{2+}, molecular oxygen, and ascorbic acid [177]. The co-translational modification of proline in the Y position (X-*Y*-Gly) by prolyl 4-hydroxylase is required for the folding of newly synthesized procollagen [210]. The melting temperature, t_m, of the collagenous triple helix is proportional to the content of 4-Hyp indicating its importance in stabilizing the molecule [224]. The mechanism of stability enhancement is not fully understood, but strong evidence suggests that it is due at least in part to stereoelectronic effects associated with proline ring puckering [239]. In addition, 4-Hyp-mediated hydrogen bonding via water bridges between the collagen chains may also enhance stability [19, 131] but the net effect of these interactions in natural collagen may be quite low [238, 239]. Another form of hydroxylated proline (3-hydroxyproline; 3-Hyp) is present in the fibrillar collagens, but in much lower abundance: 1–2 residues per chain in types I and II collagens. The function of 3-Hyp is not known, but there is some evidence that 3-Hyp may be important in supporting the D-periodic stagger of fibrillar collagens through interchain hydrogen bonding [277]. The most recognized pathological condition secondary to hydroxyproline deficiency is scurvy (see Alberts et al. [5] for a general review of the role of vitamin C in scorbutic pathologies). The principal cause of scurvy is insufficient intake of the essential vitamin, ascorbic acid, which prevents the oxidation of Fe^{2+} to the inactive form Fe^{3+}. Without active Fe^{2+}, the hydroxylases are inactive, reducing collagen molecular stability and impairing inter- and intramolecular crosslinking.

Chemistry: Lysyl Oxidase (LOX) Reconstituted collagen gels are mechanically weak because they do not possess extensive native collagen crosslinking and because they are not organized into linear arrays of fibrils. The linear tensile modulus of a 2 mg/ml reconstituted collagen gel is approximately 25 kPa at high extensional strain rates [222] while the human patellar tendon at 200 mg/ml has a linear tensile modulus in the range of 660 MPa [120]—a five order of magnitude difference with only two orders of magnitude more collagen content. While the organization and alignment of the collagen fibrils is critical to the increase in modulus, the intermolecular collagen crosslinking [36, 52] is a major factor as well. The process of strengthening the fibrils through crosslinking (both inter and intramolecular [35, 123]) is post-depositional and involves, as a first step, the interaction of telocollagen with the enzyme lysyl oxidase (LOX) [122, 199, 244, 249, 268] in the extracellular space. The short non-triple helical regions at the C- and N-terminals of telocollagen possess LOX modifiable substrates (lysine and hydroxylysine) and are, thus, critical players in this process. It should be noted that atelocollagen (collagen extracted from animal tissue using pepsin at low pH which cleaves the telopeptides) is an important medical implant material due to its biocompatibility (for a comprehensive review of collagen-based biomaterials, see Meyer [164]). However, to retain atelocollagen for appreciable lengths of time in vivo, it must be artificially crosslinked (e.g., with glutaraldehyde) and is considered a transient scaffold that is reabsorbed through metabolism/degradation [57]. In contrast, native, crosslinked collagen in the core fibrils of human tendons is essentially inert for the life of the person [106]. To understand why this is the case, it is important to examine how individual collagen monomers are chemically bound together to produce a strong fibril. In 1933, it was observed fortuitously that feeding growing rats sweet peas would cause a variety of connective tissue defects including spinal curvature and costochondral junction enlargement [93]. The cause of the defects was traced to the compound β-(a-glutamyl)aminoproprionitrile, with the effect specifically isolated to β-aminoproprionitrile (BAPN). In growing animals, BAPN prevents the initiation of crosslinking of collagen by irreversibly binding LOX [262]. Animals with BAPN-induced lathyrism produce normal amounts of thermally stable collagen, but the collagen does not form stable fibrils [264].

In normal tissues, once secreted into the extracellular space, interstitial collagen molecules are processed by specialized groups of extracellular enzymes ([202]. First the collagen propeptidases (e.g., BMP-1) remove the C- and N-terminal propeptides which permits collagen fibril self-assembly (for a review of collagen pro-peptide removal, see Prockop et al. [203]). The time course of this conversion is not known, but is likely to be soon after deposition. Silverman et al. [245] tracked the conversion of punctate collagen aggregates, exported to the cell surface by human corneal fibroblasts in culture, in real time. Following their arrival at the cell surface, the collagen aggregates almost immediately began forming fibrils at an average rate of ~45 nm/min with some fibrils reaching ~5 microns in length in 55 minutes. LOX modifies the assembling collagen molecules to facilitate the subsequent complex process of intramolecular and intermolecular covalent crosslink formation [35, 74, 88, 122, 123, 199, 244, 268]. LOX appears to be more effective on collagen molecules already assembled into fibrils than it is on molecules in solution [202]. The detailed complex chemistry of collagen crosslinking is well beyond the scope of this review but has been well-documented by Seigel, Eyre, and others [74, 88, 244, 268]. Figure 10.14 shows the general process of post-depositional modification of collagen that is essential to produce strong fibrils.

Briefly, multiple types of aldehyde-derived crosslinks are formed in a sequence of steps that begin with the modification of telopeptidyl lysine to allysine, hydroxylysine, or hydroxyallysine. LOX-mediated oxidative deamination is the common initiating step which is why BAPN is so effective at preventing crosslinking. LOX operates *only* on substrates residing in the non-helical C- or N-telopeptides ([249]; [199]). The resulting aldehydes then spontaneously "condense" with neighboring amino groups or other peptidyl aldehydes to form covalent crosslinks. The crosslinks start as "immature" divalent connections but then progress to mature, pyrrole containing trivalent crosslinks in mature tissue. The particular type of immature cross-link depends on the rate of hydroxylation of the telopeptide lysine and of the corresponding helical crosslinking site. Different tissues will favor

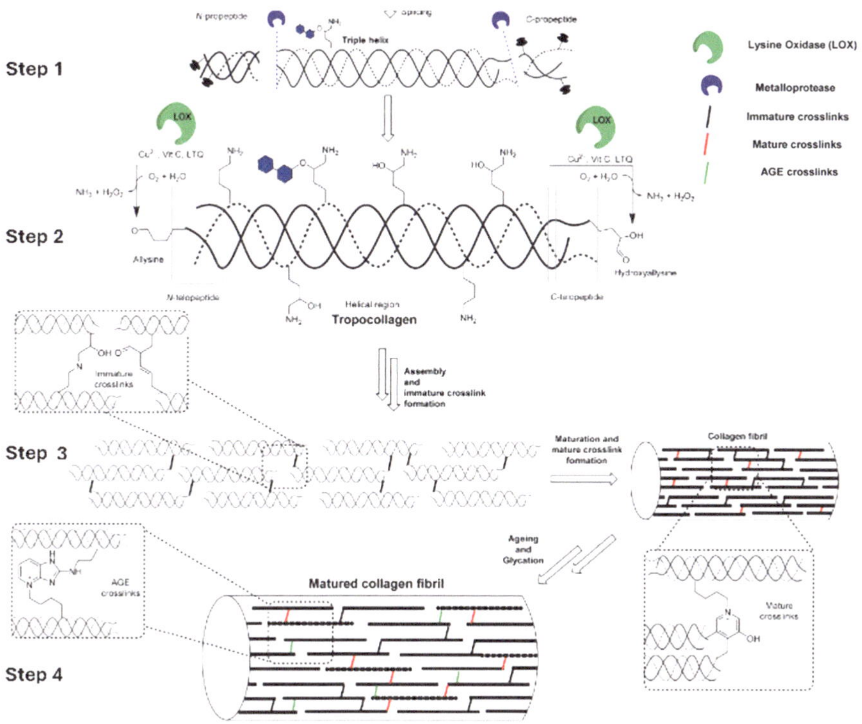

Fig. 10.14 Extracellular biochemical transformation of molecular collagen into mature, mechanically strong fibrillar collagen. Step 1: Procollagen is transformed into telocollagen (tropocollagen) via propeptidase cleavage of the C- and N-terminal propeptides; Step 2: Lysyl oxidase transforms telopeptidyl lysine and hydroxylysine into allysine and hydroxyallysine; Step 3: Immature crosslinks form between allysine and hydroxyallysine moieties and neighboring amino groups or peptidyl aldehydes; Step 4: Divalent crosslinks mature to trivalent crosslinks with time; Step 5: AGE non-enzymatic crosslinks form with age and glycation. (Reproduced from Gaar et al. [88] with permission)

different immature crosslinks with load-bearing tissues possessing more hydroxylysine-derived cross-linking [285, 286]. In older tissue, a form of non-enzymatic crosslinking begins to appear in the fibrillar collagens: advanced glycation end products or AGE products [91]. These are produced via Maillard reactions between sugars and lysine (also arginine) and are accelerated by diabetes and radiation [171, 253]. The effect of AGEs on fibril mechanics is deleterious to tissues in general, reducing their toughness and strength [247]. In summary, the effect of LOX on collagen fibrils is to strengthen them by connecting the assembled molecules together into a single, tensile structural unit. The crosslinking process is relatively slow, multistep, variable, and complex. If the reader is interested, there are many good resources [74, 88] that delve into the large number of reactions and crosslinks that are possible within collagen fibrils that all begin with LOX-mediated aldehyde formation in the terminal telopeptides.

Fibril Mechanics As the principal continuous structural unit integrated into load-bearing collagenous tissues, the collagen fibril is expected to dominate their response to applied loads. As a general rule, the fibril's role is to resist and/or transmit *tensile* mechanical loads along its major axis (even in cartilage, a compressive load-bearing material, the type II collagen fibrils are in a state of continuous tension [155]). While collagen fibrils cover an extensive range of diameters across time, tissues, and taxa, we are interested in their inherent (normalized), isolated mechanical properties (e.g., tensile Young's modulus). Fortunately, over the last decade, there has been a substantial effort to extract, preserve, and test the mechanics of individual collagen fibrils directly. A recent, comprehensive review of collagen fibril mechanics has been published by Andriotis et al. [7]. While Andriotis et al. consider fibrils in a number of different physicochemical states, we do not view the testing of dry fibrils to be particularly relevant for our purposes as we are interested in the contribution of fibril mechanics to the mechanical behavior of functioning, living tissues. Likewise, we are not interested in the mechanics of reconstituted or synthetically crosslinked fibrils except for comparison to native fibrils nor do we review the results from non-physiological modes of loading (e.g., nanoindentation) which has been performed extensively on collagen fibrils.

As one can imagine, it is quite difficult to test the mechanics of a single isolated collagen fibril in tension. First, fibrils are difficult to extract from tissue while confirming that they are both intact and free of other confounding materials (e.g., extensive decoration with proteoglycans or bound to other collagen fibrils). Second, once extracted, they require fixation to a system that can control displacements on the order of microns. Third, once secured into a displacement control system, the forces applied to the fibrils, in the pico-to-micronewton range, must be detectable. Finally, to make the measurements widely useful, they need to be normalized to the fibril dimensions. For tensile loads, the relevant dimension is fibril diameter, which must be measured with good accuracy (i.e., to within 10 nanometers or so). Consider that most fibrils are in the 100–200 nm diameter range. An error of 10% in the diameter would lead to a ~20% error in the modulus without any other measurement errors (displacement or force readouts). Further, one typically has to make separate measurements of the fibril diameter using an imaging method that works well below the diffraction limit such as AFM, TEM, or SEM (see Siadat et al. [242] for a non-destructive method to measure collagen fibril diameter optically with DIC in situ). As of the writing of the Andriotis review in 2023, only 94 fibrils had been tested in tension by atomic force microscopy (AFM) or by microelectromechanical (MEM) devices since the first attempts by van der Rijt et al. [272] and Graham et al. [97] (see Figs. 10.15 and 10.16). Of the fibrils tested, only 66 of those were tested under hydrated conditions. In PBS, van der Rijt et al. found that bovine Achilles' tendon fibrils had a tensile Young's modulus in the range of 200–500 MPa and could be strained up to 4%. The mechanical response is non-linear and exponential in nature, suggesting a possible entropic straightening, viscoelastic, or strain-stiffening effect. Eppell et al. (sea cucumber fibril) used an improved system that can apply forces in the μN range [64] and give visual access to the fibrils as they were strained. Their estimate of low-strain modulus was in the range of van der Rijt et al.

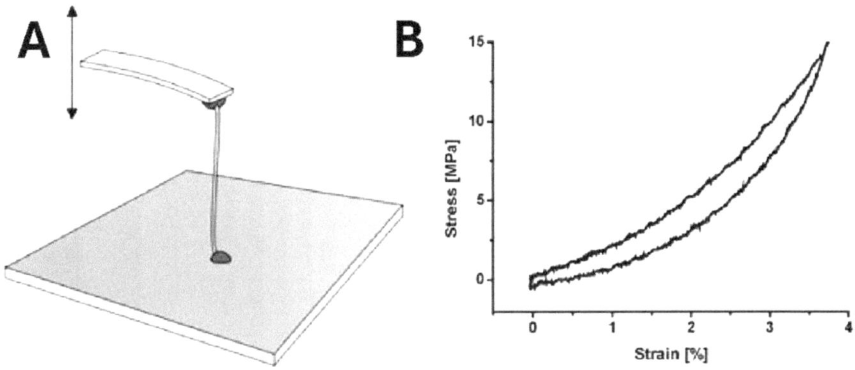

Fig. 10.15 Typical experimental approach to measuring the mechanical properties of isolated single bovine Achilles's tendon collagen fibrils. (**a**) AFM fixed by glue drop to a collagen fibril that is fixed to a glass plate. (**b**) Representative stress-strain curve for single collagen fibril obtained in PBS. (Reproduced from van der Rijt et al. [272] with permission)

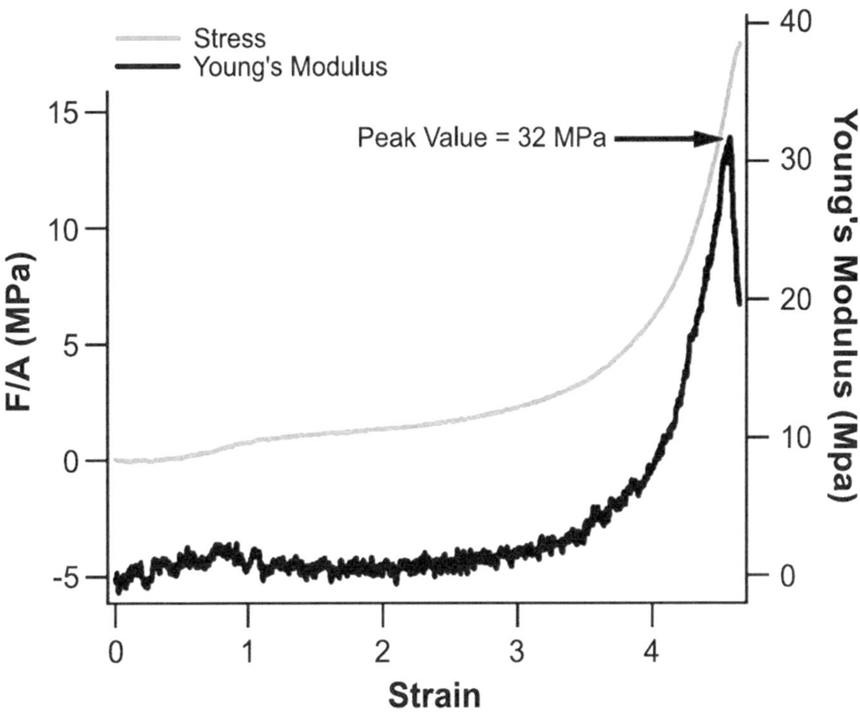

Fig. 10.16 Stress/strain and Young's modulus/strain curves for human type I collagen synthesized by dermal fibroblasts. Note the long toe region where it requires very little stress to produce large changes in strain. (From Graham et al. [97] with permission)

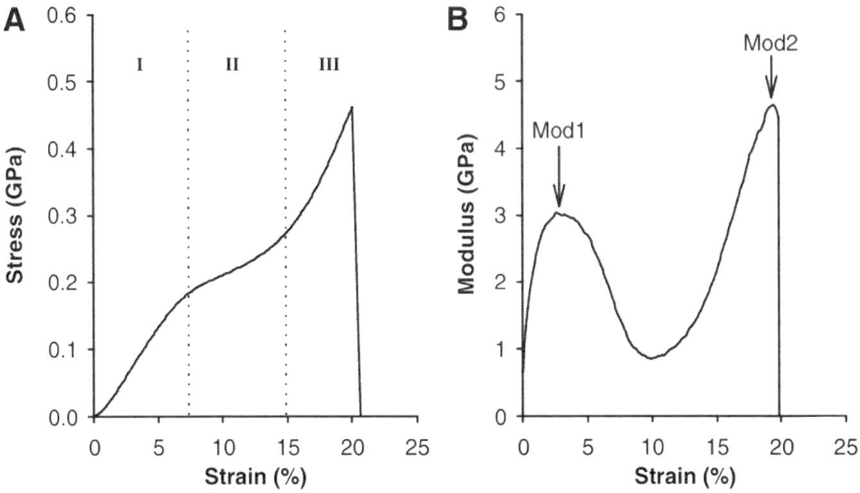

Fig. 10.17 Mechanical behavior of human patellar tendon collagen fibrils. (**a**) Three-phase stress/strain mechanical signature showing the entire mechanical loading curve from initial engagement to failure. (**b**) Derivative of (**a**) showing two definitive moduli for collagen fibrils. (Reproduced from Svensson et al. [257] with permission)

(500 MPa), and they observed a "toe" region similar to that found by Graham et al. [97] and similar to that produced by whole tendon extensional strain experiments [84]. Because of the well-organized, uniform structure of fibrils, there is little reason to expect an entropic "toe" unless the tilt angle of the molecules is rotated from ~15° toward 0°. Attempts to fit the general exponential shape of fibril extension curves with models such as the WLC or Langevin function found that the underlying assumptions were incompatible with fibril physical parameters (e.g., too short a predicted persistence length) [97, 99]. In Svensson et al. [257] published a relatively definitive investigation using AFM to load human patellar tendon and rat tail tendon fibrils to failure [257]. Thus, they were able to observe the full mechanical curve from initial loading to rupture. Figure 10.17 shows the resulting three-phase plots. They obtain two moduli: (1) a low strain modulus: 3.5 GPa and (2) a high strain modulus: 4.3 GPa. They also provide data comparing human patellar tendon fibrils, which have a substantial number of mature trivalent crosslinks, to rat tail tendon fibrils which possess only divalent immature crosslinks. The mechanical curves had very different shapes. Human fibrils failed at high strains with increasing modulus while rat tail fibrils plateaued prior to failure at a lower modulus and at lower stress. The implication being that mature crosslinks imbue better high-load mechanical performance to the human collagen fibrils, as might be expected (at the cost of embrittlement). The mechanisms that govern the tensile mechanical behavior of the fibrils have been reviewed extensively [7]. Briefly, in phase I (refer to Fig. 10.17 for plot), the collagen molecules in the fibril, which are at an angle to the fibril axis, begin to rotate into alignment, gaining order. At about 2% strain, the molecules begin to uncoil as the fibril enters a linear region in the stress/strain curve. At the end

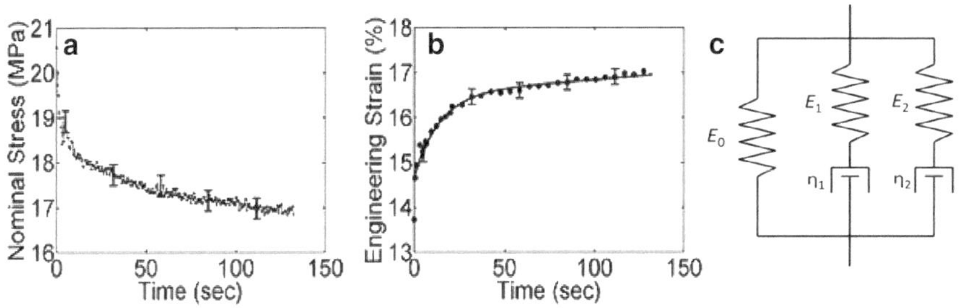

Fig. 10.18 Collagen fibril viscoelasticity. (**a**) Stress-relaxation response of a single fibril. (**b**) Creep response of a single fibril. (**c**) Best fit model to the fibril response was Maxwell-Weichert model with two separate time constants. Reproduced from Shen et al. [234] with permission

of phase I, the molecules begin sliding past one another. In Phase II, beginning at 10% strain, the modulus decreases as the molecules with lower crosslinking density or immature crosslinks deform to failure. In Phase III, the modulus increases as the more fully crosslinked collagen molecules' peptide backbones begin to stretch.

Fibril Viscoelasticity Viscous effects are a substantial component in the mechanical response of collagenous tissues, particularly at high-strain rates [87, 104]. The origin of viscous behavior likely begins at the molecular and fibril level. In 2011, Shen et al. [234] tested collagen (type I: sea cucumber, *Cucumaria frondosa*) in an MEM device that was capable of performing tensile creep and stress relaxation tests on single fibrils. They found that the fibril's temporal response was best described with a Maxwell-Weichert model (as opposed to the Fung QLV model [87]). Figure 10.18 shows a representative stress-relaxation and creep test with the model used to extract the moduli and time constants. The two time constants were: ~7 s and ~102 s. The short relaxation time was longer than that of bulk tissue, and the long relaxation time was an order of magnitude shorter than the response of bulk tissue (e.g. goat MCL [1]; τ_s:2s; τ_l:2222s; improved QLV model). While it is difficult to explain the short relaxation time difference between fibrils and bulk tissue, the long relaxation time difference suggests that tissue level viscous responses are dominated not by collagen fibrils but by other non-collagenous materials such as proteoglycans.

Summary The fundamental mechanical properties of fibrillar collagen at multiple scales: the molecular level and the fibril level have been reviewed. The tight triple helix, the intramolecular stabilization (via post-translational modification of proline to hydroxyproline), the molecular packing, and the interfibrillar enzymatic crosslinking all cooperate to place a high density of protein bonds into a thin cross-section. The result is a material with a very high maximum tensile modulus at each scale: single molecule ~4.7 GPa; single fibril: ~2.19 GPa (data averaged from table 1 in Sherman et al. [237] and includes multiple methods of measurement and tissue conditions—wet/dry). In addition, there is intrinsic viscoelasticity at the fibril level, likely due to an inherent disorder-to-order transition in the fibril and or the movement of water driven by mechanical strain in the fibril. The relatively short timescale of the viscosity of the fibril suggests that it will not significantly influence the bulk viscosity of macroscopic load-bearing tissues. The fundamental theoretical mechanisms that create composite fibril strength and toughness have been examined by Markus Buehler in his 2006 PNAS paper [42]. He suggests that collagen possesses long tropocollagen molecules along which intermolecular crosslinks are distributed to maximize both strength and energy dissipation by "shear-like" mechanisms. The result is "tough" collagen, naturally.

Integrated Collagenous Mechanical Structures

Now that an appreciation for the fundamental structural materials from which nearly all load-bearing vertebrate tissues are built has been developed, we are nearly in position to examine the macroscopic mechanical response of the tissues into which they are integrated. However, it is important to understand that the coupling of the collagen elements to the tissue mechanical structure is not direct. X-ray investigations of whole tissue loading have shown that the strain experienced by fibrils within the tissues, even in tendon, where the fibrils are most aligned with the load direction, is always "considerably smaller" than the strain on the whole tissue [85, 204, 227]. Figure 10.19 clearly shows the decoupling of the collagen fibrils and molecules from the applied load in bovine Achilles' tendon. Thus, the precise tissue architecture constitutes yet a third layer of complexity that must be considered. Figure 10.20 shows the classical hierarchical arrangement of collagen in a tendon where arrays of fibrils (e.g., fibers) are further organized into the fascicles of tendon (the same figure applies generally to ligament as well) [125].

Fig. 10.19 Decoupling of the strain at each level of hierarchy in a tendon. Single molecule (closed circle); Fibril (open circle); Whole tendon (square). (From Sasaki and Odajima [227] with permission)

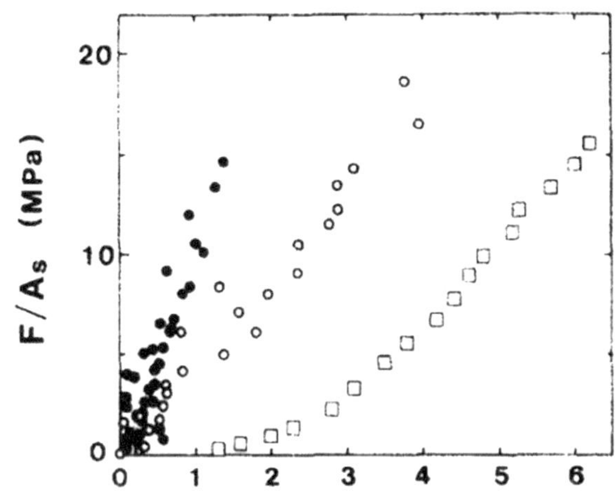

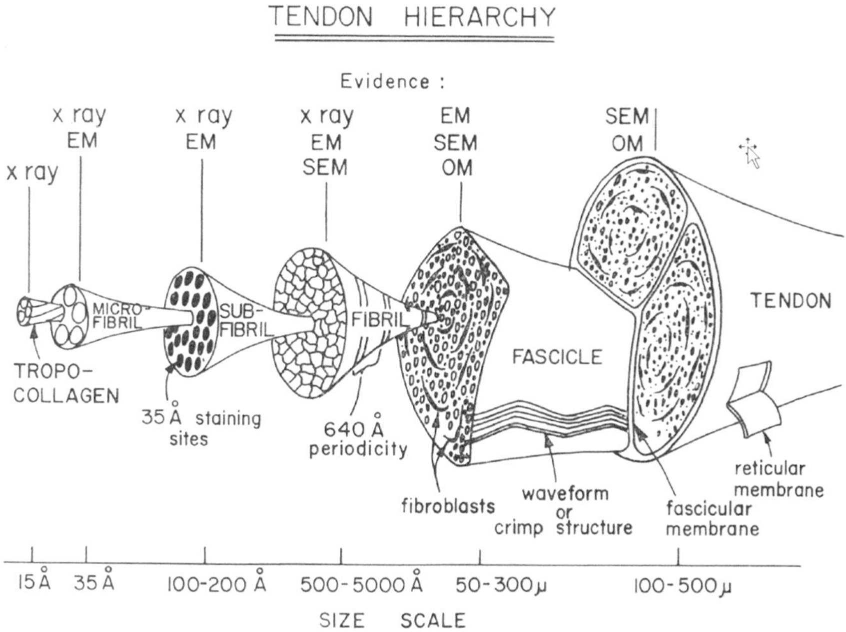

Fig. 10.20 Classical representation of collagen structural hierarchy in tendon. (Reproduced from Kastelic et al. [125] with permission)

Mechanics of Fibrillar Arrays

Before we examine the macroscopic mechanical properties of a number of different important connective tissues, there is one more level of organization that requires attention: Discrete arrays of fibrils. Collagen in its fibrillar form has been used to produce mechanically important industrial and military materials such as bows, bowstrings, cordage, and lashings for millennia. The Mongolian composite bow, a weapon that was remarkably advanced technologically (Fig. 10.21), used sinew (typically arrays of collagen fibrils stripped from bovine tendons) on its tensile backing to take advantage of collagen's strength in tension, horn on its compressive "belly," bone as a stiffener, and animal glue to fuse the components into a single unit [25]. The results of the construction produced a shorter, more energy efficient and more powerful bow than the contemporary European long (self) bow [212]. The leather industry has motivated research on the properties of collagenous materials, their strength, and the chemistry of preserving them (tanning). In Germany, a Tanner's school in 1889 was formed to ensure standardization of collagenous tissue tanning. It was soon followed by the formation of the German Research Institute for the Leather Industry in 1897. These were the predecessors of the FILK Freiberg Institute which holds the Freiberg Collagen Symposium quadrennially [118]. The *Journal of the International Society of Leather Trades' Chemists* has produced some

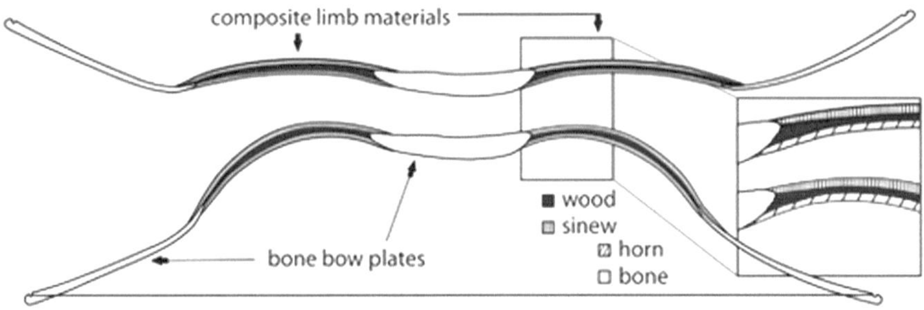

Fig. 10.21 The Mongolian composite bow utilized sinew on its back (tensile side), horn on its belly (compressive side), and bone as a stiffener. It also used collagen-based glues to fuse the entire composite structure. It thus took advantage of two collagen-rich materials and placed them under mechanical conditions for which they were appropriate. (Reproduced from Reisinger [212] with permission)

excellent investigations. In 1943, R.G. Mitton extensively tested the mechanical properties of vegetable-tanned, steer hide collagen fibers from which leathers are typically produced [169]. His insights, though derived from testing fibers in non-physiological conditions, were remarkably sophisticated and informed by a deep, up-to-date knowledge of collagen's chemical composition and structure as determined by both X-ray and electron microscopy investigations. What he found was that not all fibers from the same tissue are the same, that fiber mechanics depends on from where they were taken (suggesting anisotropy and heterogeneity of the source material), and that they had mechanical "memory" due to their stress history. He also found that fibers exhibit time-dependent or "viscous" behavior, that bulk material, comprising a multitude of fibers, deforms more than its constituent fibers, and that failure of whole leather fabric is due to fracture of the fibers and not sliding of fibers past one another. These observations are remarkably foundational, and many have been supported by subsequent investigations on fibers tested more recently and under physiological conditions [85, 227].

In 1997, Misof et al. [168] attempted to determine the mechanism behind the low force extension heel region seen in tendon fiber extension. They isolated individual fibers from rat tail tendons and subjected them to mechanical perturbation while observing them with X-ray diffraction to track molecular organization. The X-ray data were able to provide an explanation for the low force extension of the fibrils in the "heel" region of the stress-strain curve where light microscopy showed no discernible structural changes. Figure 10.22 summarizes a typical rat-tail tendon fiber stress-strain curve overlayed with explanations for each phase of the curve presented in Fratzl et al. [85]. The figure characterizes the integrated mechanical stress/strain signature of molecules, fibrils, and fibers into one unified stress response to tensile load. The "toe" portion of the curve is related to the removal of macroscopic crimp in the fiber and is where all the fibrils are brought into general alignment. It is a low-force "entropic" region and is easily observed by polarization microscopy. In the heel phase, the fibrils begin taking on more load, and their

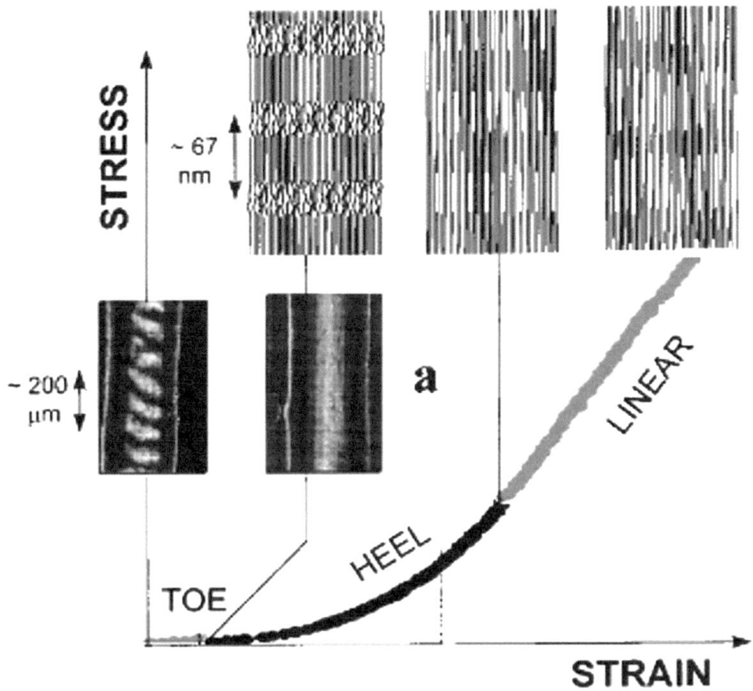

Fig. 10.22 Stress-strain curve for excised rat-tail tendon fibers with overlayed structural explanation for the toe, heal, and linear phases of the curve. Toe: macroscopic kink removed; Heel: straightening molecular kinks in the gap region of the fibrils; Linear: gliding of molecules. (From Fratzl et al. [85] with permission)

constituent molecules increase in order. The order effect is only detectable by X-ray, and Fratzl et al. [85] attribute it to straightening molecular kinks in the fibril gaps. Finally, the linear region is where the molecules themselves are loaded and sliding relative to one another in the fiber.

Fibrillar Array Viscoelasticity Fibrillar arrays, like fibrils, also exhibit time-dependent mechanical behavior. Fratzl's group's combination of sophisticated fiber mechanical testing and X-ray observation provides quantitative information on the viscous response of fibers [204]. While mechanically extending rat-tail fibers, they observed strain-rate dependence of the ratio of fibril elongation to whole tendon elongation. To explain the data, they proposed a simple "two-in-series" Kelvin model (Fig. 10.23) and were able to capture, to some extent, the viscoelastic behavior of the tendon fiber and its constituent fibrils. Note that Figure 10.23a depicts the tendon fiber as a structure comprising discontinuous collagen fibrils in series and in parallel. That depiction remains controversial [258]. In 2014, Szczesny and Elliott [259] combined mechanical tensile loading, microscopic experimental techniques, and a shear lag model to explicitly test the hypothesis that shear between

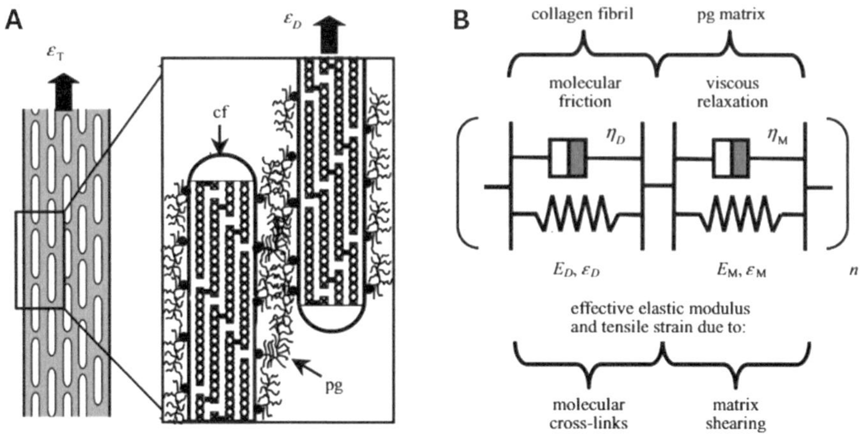

Fig. 10.23 Model of the origin of fiber viscoelasticity. (**a**) Schematic representation of collagen fibrils (cf) embedded in a proteoglycan milieu (pg) within a fiber that is under mechanical strain (ε_T). The effective strain on the fibrils is ε_D. (**b**) In series double Kelvin model proposed to capture the viscoelastic behavior of the collagen fibril/pg matrix unit. η_D—viscosity of the fibrils; η_M—viscosity of the matrix; E_D—effective elastic modulus of the fibrils; E_M—effective elastic modulus of the matrix. (From Puxkandl et al. [204] with permission)

discontinuous fibrils is the fundamental mode of load transfer between fibers and fibrils in rat tail tendon. Their data and model strongly support the idea of load transfer and suggest that fibrils or arrays of fibrils may be discontinuous in load-bearing tissues comprising aligned collagen.

Mechanics of Whole Tissues

It is quite clear that each level of tissue hierarchy further diminishes the direct influence of collagen's molecular mechanical signature on the materials' mechanical performance. This is, of course, necessary. Connective tissues have substantial variability in their mechanical requirements. None of these requirements can be met with a simple linear array of collagen molecules. As has been already shown, even in tendon, where there is maximum co-alignment of the collagen with loading axis, the mechanical performance of the whole tissue is substantially different from that of molecular collagen. Nonetheless, collagen is the principal load-bearing material, which can be readily appreciated in the anatomy. The form-follows function axiom in physiology certainly applies to load-bearing collagenous tissue. Figure 10.24 shows, schematically, the general orientation of collagen fibrils within a number of geometrically and mechanically distinct tissues that possess high concentrations of fibrillar collagen. In tissues that are subjected to unidirectional forces (e.g., ligament or tendon), fibrillar arrays are collected into highly aligned, crimped fascicles [125, 211]. In tissues that resist pressure (e.g., cornea and sclera), fibrillar arrays are organized into flattened lamellae that alternate in direction,

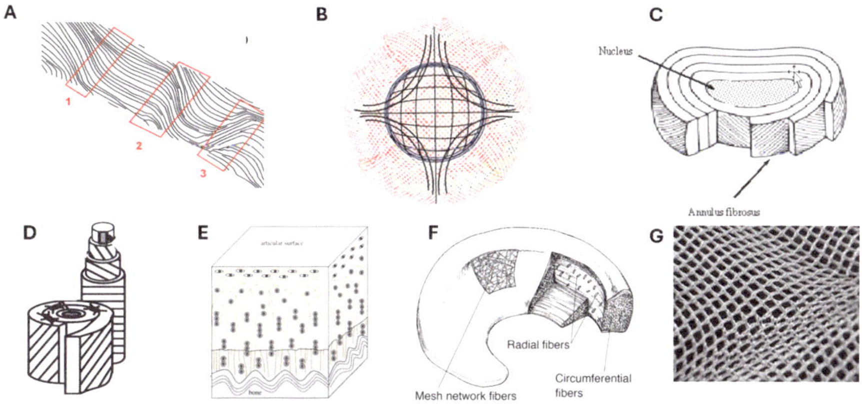

Fig. 10.24 Schematic views of collagen fibril orientation in different tissues. (**a**) Tendon (red boxes denote regions of crimp) [211]; (**b**) cornea [163]; (**c**) annulus fibrosus [198]; (**d**) secondary bone [216]; (**e**) cartilage [151]; (**f**) meniscus [153]; and (**g**) skin [217]. (Reproduced with permission)

forming a spherical shell [34, 108, 162, 163]. In the annulus fibrosus, which is subject to static tension and transient torsional loads, the fibrils are arranged in lamellae as well, but their orientation is optimized to resist torsional stresses produced in the spinal column [108]. In 3-D tissues such as bone and cartilage, fibrils are arranged into complex structures [151, 216]. In cartilage, the type II fibril network is arranged in vertical arcades to resist the internal swelling pressure created by the high fixed-charge density of the embedded proteoglycans [155]. In addition, there is a specialized superficial zone comprising arrays of fibrils oriented tangential to the frictional forces produced by the relative motion of the tibial plateau and femoral condyles. In primary bone, arrays of fibrils are mineralized and, in the case of secondary cortical bone, the tissue is revised (removed and replaced) continuously by Haversian systems comprising osteoclasts (catabolic function) and osteoblasts (anabolic function) [216]. Collagen fibrils also invest other tissues in arrays arranged to support their mechanical function (i.e., longitudinal and circumferential arrays in artery or radial and circumferential arrays in meniscus [213]). Even where the organization is presumed to be “random” (e.g., skin), there is still some level of organization among the fibrils comprising the tissue that serves to support the mechanical performance of the tissue [217]. We will briefly examine the mechanical properties of some of these dense collagenous tissues.

Tensile Load-Bearing Tissues

Collagen molecules, fibrils, and fibers are designed to carry tensile loads. In many tissues, collagen is “placed” directly in the path of maximum mechanical tensile force. This is excellent for the tissue’s ability to withstand the forces that should dissipate it, which in some soft tissues are quite substantial. It is also odd, because typically, materials do not accumulate in the path of load, they eschew it. The placing of collagen in the path of a disrupting force however is fundamental to metazoan life. The mechanisms that enable that particular process are interesting to consider. They will be touched on later.

Tendons Of all the load-bearing tissues, tendons most directly reflect their constituent collagen organization. Tendons constitute the critical mechanical link that enables efficient transfer of muscle-generated forces to their targets, typically bones [23]. Tendons must be stiff to ensure rapid actuation but also mechanically strong because, in general, the musculoskeletal system is built to mechanically disadvantage limb movement. Tendon attachment points are typically very close to joint rotation centers which increases limb range of motion at the expense of high force production at the extremities. Structurally, the arrangement of collagen directly reflects these mechanical requirements [23]. Within the tendon, the collagen molecules aggregate into fibrils, and the fibrils are collected into fibers and fascicles that are all colinear with the loading axis (vide supra Fig. 10.20). The structural format and mechanical properties are thus anisotropic, reflecting the function. The collagen fibrils are embedded in a hygroscopic soft matrix that typically comprises proteoglycans [125]. The primary PGs (decorin and biglycan) in tendinous tissues vary temporally depending on the phase of tissue development (see Siadat et al. for a review of tendon ECM assembly [243]). In the adult tendon, PG content is diminished relative to development (~1% of the dry weight of tensile tendons), leaving collagen as the major component of the material with PGs providing spacing, lubricity, and possibly load transference capacity [259]. An important characteristic of fibril arrays is the formation of "crimp" which causes the whole fiber to change orientation away from the loading axis in a periodic fashion (vide supra Fig. 10.24a). The highly ordered organization of collagen molecules, fibrils, and fibers enables tendons to effectively resist uniaxial mechanical loads along the fiber direction, while maintaining relatively low levels of deformation (strain). Like other soft tissues, water is the major component ~70% of the hydrated weight [189, 232, 269]. Collagen accounts for about 60–80% of the dry weight with the remainder primarily comprising proteoglycans and other glycoproteins [125]. Type I collagen is ~60% of the total collagen with types III, IV, V, and VI accounting for the remaining 40% [135]. Type III collagen, a homotrimer, is often associated with type I collagen and is thought to be a fibril nucleator during tendon fibrillogenesis, wound healing, and scar tissue formation [30, 143, 220]. Within tendons, type V collagen assists in fibrillogenesis and possibly in controlling fibril diameter dispersity [30, 31]. Tendons are heterogeneous and exhibit regional variation in collagen distribution, with high-load regions often more enriched with type I collagen [40]. Some regions and fibrils contain a higher percentage of type III collagen and are thus more compliant (as in other type III-rich tissue, e.g., arteries [11, 144]).

Tendon composition and structure vary considerably within the animals, with gender and age (see Hefferan et al. [105] for a survey of tendon variation). Collagen organization and response to loading differ between tendons that serve postural or dynamic functions. Postural tendons primarily stabilize body positions with minimal displacement, while energy-storing tendons facilitate dynamic movements, the transfer of mechanical energy during activities such as locomotion, and the ability to sustain high cycle counts. Dynamic tendons typically experience larger mechanical displacements (~10%) relative to postural tendons (2–3%) [16, 29, 149]. The stiffness matches the mechanical requirements: stabilizing postural tendons are stiffer while energy-storing tendons are more compliant [28, 266]. However, similarities in isolated tissue mechanics independent of function have been reported [128, 148, 201]. Energy-storing tendons have lower total collagen content but higher levels of type III collagen [265, 267]. Damage mechanisms also differ between tendon types: postural tendons tend to exhibit fibrillar-level damage under cyclic loading, whereas energy-storing tendons are more prone to fiber-level kinking with minimal fibrillar damage [197, 273]. In dynamic tendons, interfibrillar sliding is thought to act as a protective mechanism, preventing excessive damage to individual collagen fibrils. In contrast, the increased stiffness of postural tendons is associated with elevated levels of collagen crosslinking, which enhances stiffness but limits interfibrillar sliding and may result in increased fibrillar damage during cyclic loading [266].

Uniaxial tension testing provides valuable insights into the structural behavior and mechanical function of tendons as it mimics in vivo loading. The load-displacement curve typically displays an initial low-stiffness region, referred to as the "toe/heel region" followed by a transition into a region with higher stiffness as the collagen fibers are progressively stretched, referred to as the "linear region." Figure 10.25 shows the now familiar-looking stress-strain curve which clearly reflects contributions from the mechanical behavior of collagen fibers, fibrils, and monomers which have already been considered. As observed in fiber mechanics, the unloaded crimped or wavy arrangement of collagen fibers within tendon straightens as the tendon is loaded (Fig. 10.26) [100, 274, 278]. The strength of alignment and realignment of collagen fibers during tensile testing may produce additional variability and nonlinearity in the bulk tissue response [139]. The stiffness of the linear region of human tendons generally ranges from 50 to 500 MPa, with variations that depend on the specific function and aging processes, as well as the mechanical testing method, i.e., in vivo versus ex vivo measurements and tissue subunit/scale [22, 120, 148, 251]. Tendons also exhibit viscoelastic behaviors, where the rate of strain impacts the observed behavior, e.g., stiffness [112, 120, 170, 231]. The viscous response, as seen with collagen fibers [204], is mainly due to water and proteoglycan contributions and their interactions with the collagenous matrix although collagen fibrils themselves have viscoelastic properties [234].

Ligaments Ligaments are also axially loaded, collagenous tissues [82] that, despite their similarities to tendons, differ in structure, composition, and mechanical behavior [158, 193, 226, 251]. Ligaments are highly hydrated [82], and of the solid content, the majority is collagen type I, with the remainder comprising elastin, collagens III and V, glycoproteins (e.g., laminin, actin), and other matrix proteins. Compared to tendons, ligament collagen fibrils are more glycosylated (~80% vs. ~25%), more extensively crosslinked (~90% vs. ~60%), and less densely

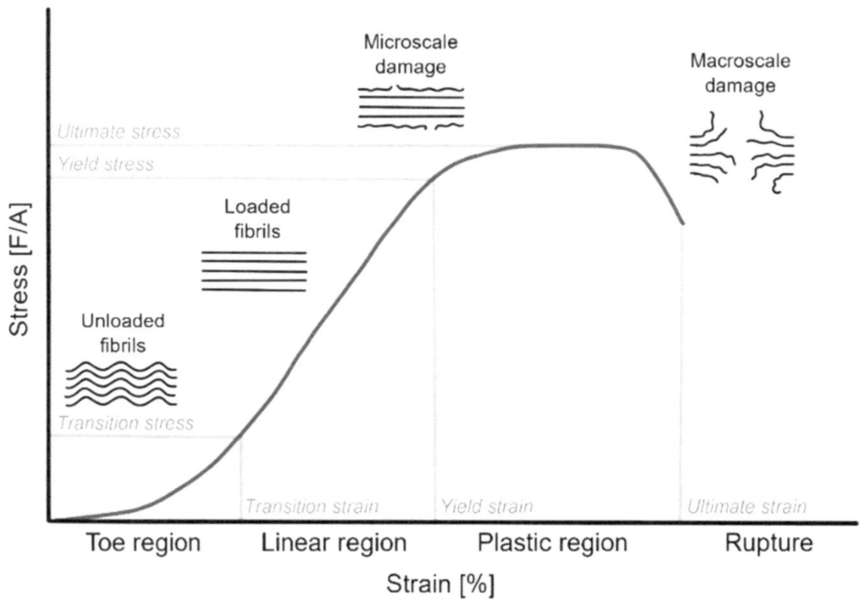

Fig. 10.25 The characteristic stress-strain curve of tendon reveals a behavior typical of collagenous materials. Unloaded, crimped fibrils become aligned and loaded under applied strain that causes microscale damage to individual fibrils or molecules past a yield threshold. Damage accumulates and ultimately causes tissue rupture. This two-region profile is shared with other collagenous tissues with aligned collagen fibrils like ligaments, skin (along fibril direction), and cornea (in lamellae). (Image courtesy of Dr. Phoebe Szarek, 2025)

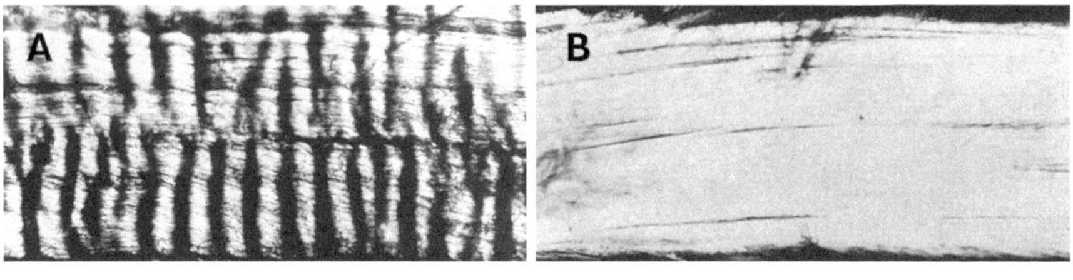

Fig. 10.26 Polarized light microscopy reveals (**a**) wavy, banded collagen within rabbit tendon sections that are removed under low load producing a (**b**) bright, aligned signal. (Reproduced from Viidik and Elkholm [274] with permission)

packed. They also have considerably more elastin (up to 70% in some specialized ligaments), which is the most important differentiator in their mechanical signature [158]. Ligaments exhibit their own version of the characteristic stress-strain curve with which we are now familiar. Initially, collagen fibers uncrimp under load, with no observable fibril deformation strain, as confirmed by macroscopic and small-angle scattering analyses [124]. At higher strains than tendon, the stress-strain response becomes linear, and the fibers begin taking on substantial load. Because

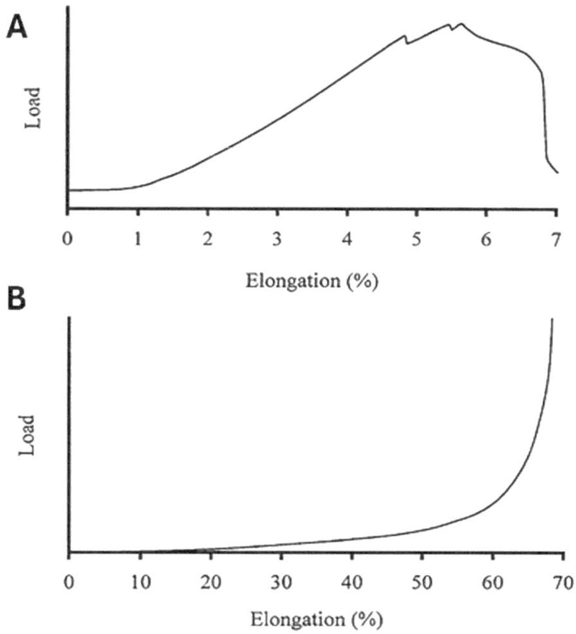

Fig. 10.27 Mechanical performance difference between the AC (**a**) and the LF (**b**) ligaments. Note the difference in the x-axes. (Reproduced from Martin et al. [158] with permission)

they generally join two stiff members (bone-to-bone connections), ligaments typically have an extended toe/heel region and a lower modulus [158]. Like tendons, ligaments exhibit a tremendous range of behavior both within the same animal and across species. They provide structural stability to synovial joints (e.g., the collateral and cruciate ligaments of the knee [221, 223, 248, 276]and to less mobile joints (e.g., the longitudinal, interspinous, and ligamentum flavum ligaments of the vertebral column). The anterior cruciate ligament (ACL) is important for stabilizing the knee. Its tensile strength is highest along the anatomical axis, where collagen fiber alignment enhances load bearing [281]. The ligamentum flavum (LF) connects adjacent vertebra in the human spine and must maintain function over a large range of relative vertebral angles. It has a very low elastic modulus (100 MPa at low elongations [179]). Figure 10.27 schematically compares the mechanical performance of the ACL and the LF ligament in uniaxial tension. The data show that in ligaments, it is clear that the constituent collagen fibers are not the main drivers of the mechanical response (at least not until high levels of strain). Instead, the tissue is a complex mixture of components combined with variable structural organization that modify the whole tissue response to suit the specific application.

Skin The skin is the largest organ in the human body and must withstand various types of mechanical loading, including uniaxial and biaxial tension, compression, and shear, while maintaining its reference geometry after deformation. The skin's structure varies across different regions, particularly in terms of thickness. It comprises multiple layers, each with specialized compositions and organization of collagen. The three primary layers are the epidermis, dermis, and hypodermis (subcutaneous fat), listed from the outermost to innermost. Each layer can be further subdivided into functional sublayers with distinct compositions.

The epidermis is predominantly composed of keratinocytes (95% of total cells). The basement membrane, or dermal-epidermal junction (DEJ), lies between the epidermis and dermis and contributes to the mechanical integrity of the skin. The human basement membrane contains topographical undulations (rete ridges) that influence the mechanical and biochemical environment, as well as the transfer of load between layers [205, 235]. It comprises laminins, collagen IV, proteoglycans, and nidogen, with the collagen forming a cross-linked meshwork that contributes to tissue stiffness.

The dermis contains collagen type I and III fibers that provide mechanical strength via interwoven structures [217]. Types I and III collagen predominate in the dermis, along with elastin, proteoglycans, and glycoproteins, that affect tissue mechanics at small strains and under conditions where osmotic or viscous effects would be in play [78, 160, 236]. Finally, the hypodermis is primarily adipose tissue with functions including temperature regulation, fluid homeostasis, and interfacing with deeper tissues [27, 145, 280].

Skin is often modeled as a perfectly elastic material due to its ability to return to its original conformation after the removal of a loading force [196]. In vivo, skin may be deformed up to 30% during motions like elbow deflection without experiencing plastic deformation. Like other collagenous tissues, skin exhibits a similar mechanical response with the same toe, heel, and linear phases as tendon and ligament. Initially, it shows low stiffness under low strain, transitioning to high stiffness at higher strains [121]. In the linear region, collagen fibers align and are recruited, and under further deformation, the tissue eventually softens due to the gradual failure of collagen fibers. Skin's Young's modulus is low and ranges from 0.05 to 20 MPa, depending on testing methods such as suction, indentation, or sliding in vivo [56, 119, 173, 290].

Skin's mechanical properties are variable and sensitive to the testing conditions. Unlike other collagenous tissues, which are typically internal and consistently well hydrated, skin's water content significantly influences its mechanical behavior [271]. Aging reduces skin stiffness (>50% between ages 25 and 60), in vivo loading, and failure strain [39, 147]. Photoaging also causes a loss of elasticity, paired with observable damage and disorganization in the dermal collagen fiber bundle structures [185, 188]. Collagen accounts for 70–80% of the dry weight of skin [270] and is not isotropically distributed. Langer lines (Fig. 10.28) define the lines of maximum skin tension over the body that align with the collagen fibers within the tissue and the axis of greatest retraction when a sample is excised from the body. In vitro uniaxial extension tests of human skin strips parallel and perpendicular to the Langer lines show that skin is stiffer parallel to the principal direction of collagen fibers and that tensile specimen and fiber alignment correlate with specimen stiffness (Fig. 10.29) [184]. Additionally, aging and repeated loading can also increase the anisotropy of the tissue, akin to a preconditioning protocol, with evidence that the remodeling occurs in the top-most layers of the dermis first.

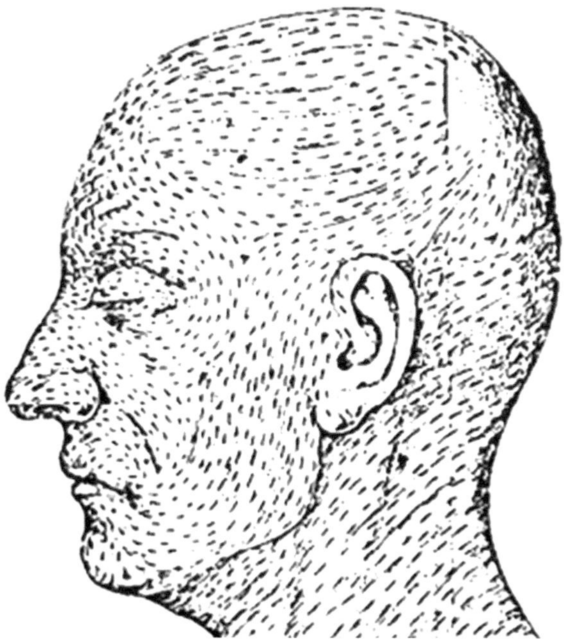

Fig. 10.28 Langer lines on the head and neck of an adult. (Reproduced from Langer (1861))

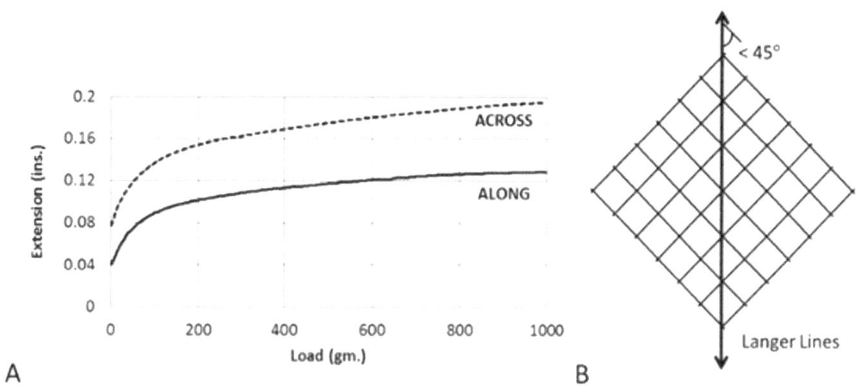

Fig. 10.29 Mechanical performance of skin along and across Langer's lines (direction of collagen fibers in the skin) exposing the mechanical influence of collagen in the tissue. (Reproduced from Ni Annaidh and Destrade [184] with permission)

Cornea The tough ocular tunic comprises the cornea and sclera. Together, they maintain the optical functions of the eye by supporting the delicate sensory structures, by resisting intraocular pressure, by sustaining transient mechanical deformations, and by refracting light onto the retina. The cornea is a transparent and highly organized multilayered structure comprising the stroma (~90% of total thickness) and its supporting membranes. The stroma is bound on the exterior surface with an

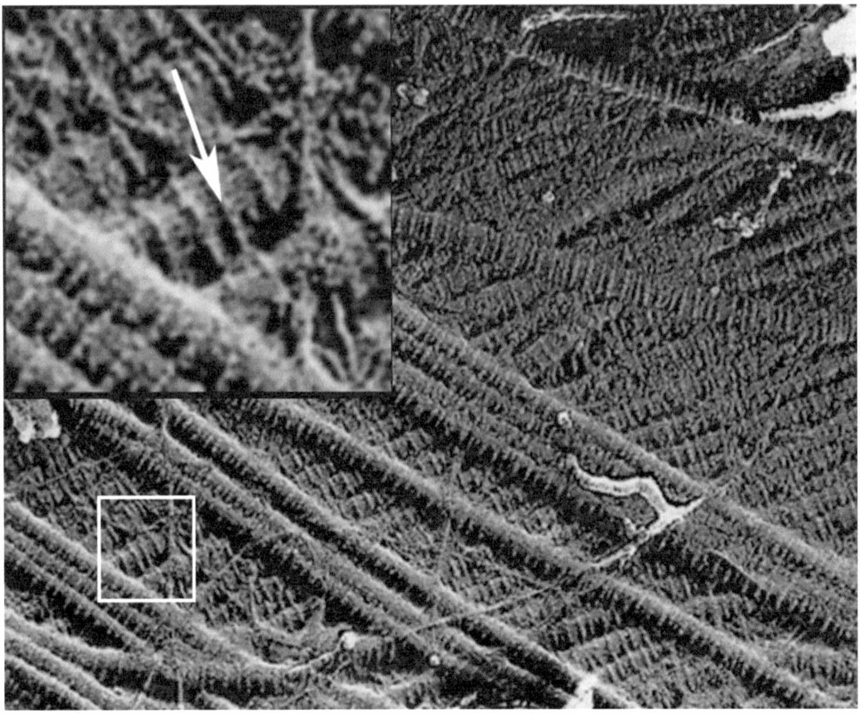

Fig. 10.30 Quick-freeze deep-etch TEM image of bovine corneal stromal collagen. The fibrils are uniform in diameter (~36 nm). Their spacing and diameter is controlled by proteoglycans that associate with the collagen fibrils (inset white arrow) at precise intervals. (Image courtesy Dr. Jeffrey Ruberti)

epithelium (barrier) and on the interior surface with an endothelium (pump). Optical transparency is maintained through the cornea's tightly regulated collagen organization (diameter and spacing) [21] and the relative state of dehydration produced by the membranes. The corneal stroma is approximately 78% water and 15% collagen, with the remaining materials primarily proteoglycans: lumican and decorin [159]. Collagen fibrils within the stroma are uniformly thin and uniformly spaced (Fig. 10.30; [14, 111, 162]). Fibrils organize to form lamellae that run parallel to the corneal surface, with alternating in-plane orientations, ~40% of which run orthogonal to one another and aligned with the four rectus muscles [161], and regionally variable interlacing of lamellae. There are approximately 250 lamellae across the ~550 micron thickness of the human stroma (Fig. 10.31) [65, 206, 209]. Type I collagen, co-assembled with type V (80%/20%), forms the heterotypic stromal fibrils, while other collagen types (e.g., III, IV, VI, and XII) are expressed and modulate the organization and material properties, including under pathological conditions such as keratoconus and conjunctivitis [2, 181, 256]. Normal intraocular pressure (IOP) is typically approximated to be ~15 mmHg, although values up to

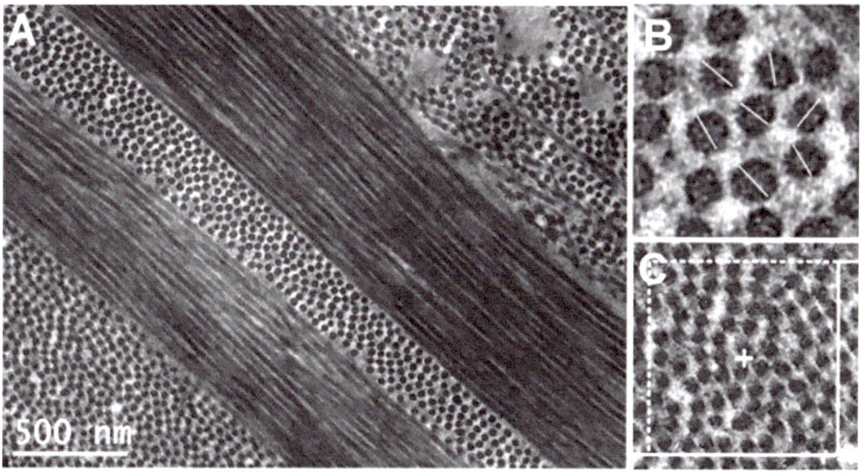

Fig. 10.31 TEM micrograph of the corneal cross-section showing the lamellar architecture of the collagen (**a**) in the stroma. Note the uniform fibril diameter (**b**) and the packing of the fibrils (**c**). (Reproduced from Bergmanson et al. [26] with permission)

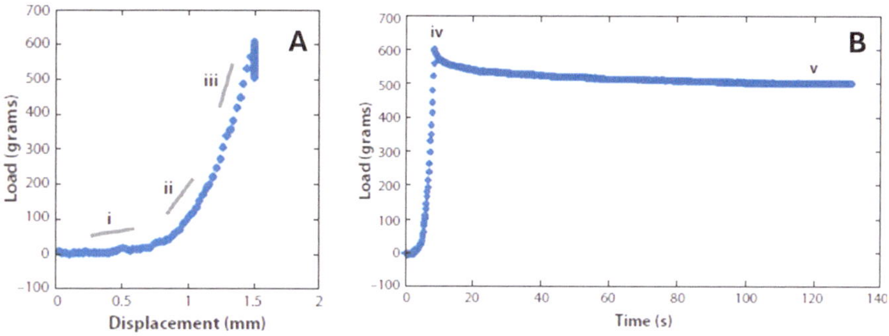

Fig. 10.32 Human corneal stromal mechanical response to tensile loading under inflation conditions. (**a**) Elastic response. (**b**) Viscoelastic stress relaxation. (Reproduced from Dupps and Wilson [58] with permission)

21 mmHg have also been reported). Ocular blood flow induces cyclic IOP fluctuations of approximately 2 mmHg during each pulse cycle [62, 126]. Additionally, IOP varies diurnally, with fluctuations of ~5 mmHg and peak pressures occurring in the early morning; both baseline IOP and its fluctuations are elevated in patients with glaucoma [12, 51]. IOP generates circumferential stresses in the cornea. The cornea behaves as a linear viscoelastic material, with its viscous response governed by proteoglycan-water interactions and its elastic response primarily driven by collagen (Fig. 10.32; [58]). Corneal stiffness is often assessed via inflation or bulge testing, which reveal a low-strain modulus of ~0.2 MPa and a high-strain modulus of ~5 MPa. The transition between these regimes occurs at 12–20 mmHg, aligning

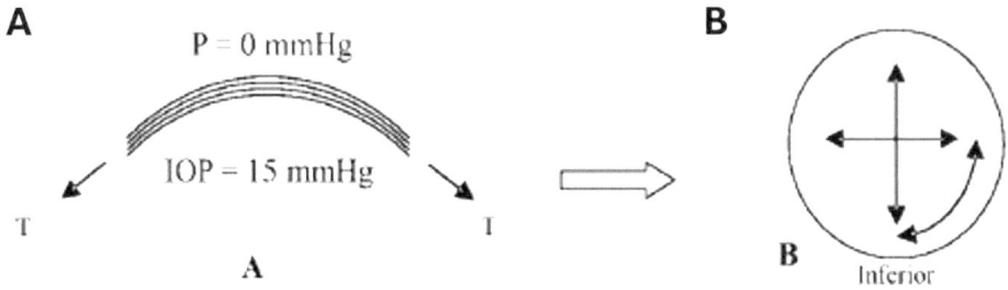

Fig. 10.33 Corneal loading in tension (**a**) and the regional preferential direction of the collagen fibrils in the plane of the stroma (**b**). The max Young's modulus is in the peripheral ring (13.0 MPa) and along the vertical and horizontal lines (8.6 MPa). (Reproduced with from Ethier et al. [66] with permission)

with physiological IOP levels [38]. Uniaxial testing of corneal strips shows that the in-plane modulus is maximum along the lines of collagen fibril alignment in the stroma ([66]; Fig. 10.33). Corneal stiffness is influenced by systemic factors such as disease (e.g., diabetes), and notably, aging causes increase in stiffness by a factor of 2 between the ages of 20 and 100 years old [32, 138].

Summary: Tensile load-bearing tissue For all of these soft-tensile load-bearing tissues, a recurring pattern is emerging in their mechanical signature. There is always an exponential elastic curve that echoes the fundamental tensile mechanical properties of the collagen fibrils that find themselves in the path of force within the tissue. The moduli for each tissue vary substantially, typically because of the inclusion of other materials (e.g., elastin in ligament) or because the tissue architecture changes the number of collagen fibrils that are in the path of the force, but the shape of the load-extension/stress-strain curves is relatively constant. We can surmise that at high strains, the elastic curve is likely reflecting the number of collagen fibrils that are in tension in the particular tissue. It is also interesting to note that the collagen fibrils in the two planar tissues display preferred orientation (along Langer's lines in skin and along the path of the rectus muscle load lines in cornea). In addition to the elastic response, all of these tissues display viscous behavior. This is in part due to collagen's intrinsic viscosity, but other components in the tissues contribute substantially to the viscous response.

Compressive Load-Bearing Tissues

Collagen is also present in tissues that are not loaded in tension. Bone and cartilage are heavily invested with fibrillar collagen (type I in bone and type II in cartilage). In the case of bone, the collagen is deposited and then mineralized to support compression. In cartilage, the collagen is actually in tension in the tissue which constrains the swelling pressure of the proteoglycans and thus creates a compressively robust material.

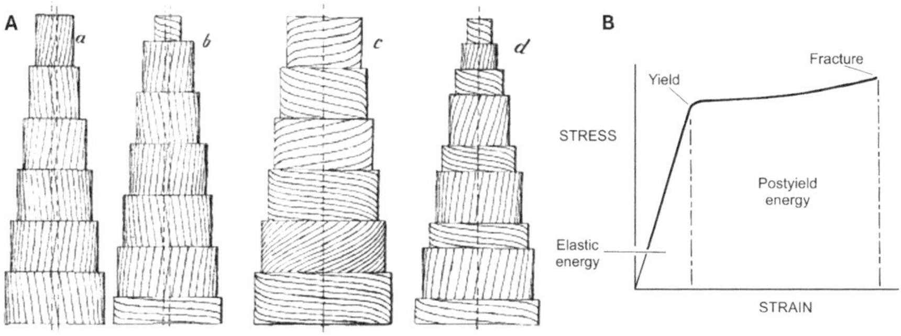

Fig. 10.34 The structure of collagen lamellae in secondary bone osteons. Note the organization of the collagen would tend to favor tensile loading of (**a**) and (**b**), compressive loading of (**c**), and mixed loading of (**d**). (**e**) Bone mechanical signature in tension. Note the distinctly different nature of the stress-strain relationship for bone. (**a**–**d**) Reproduced from (**a**) [92] and (**b**) [49] with permission

Bone Bone is a unique collagenous material as the collagen that defines its internal structure is mineralized rather than hydrated. Bone behaves more like a fiber-reinforced ceramic than it does a collagenous soft tissue. It is 60% inorganic, 30% organic, and 10% water, drastically differing from the 70–80% hydration of other collagenous tissues [37]. The hierarchical structure and organization of bone have been studied at many scales [214, 216]. The 30% organic fraction is broken further into 15–20% collagen and 10–15% non-collagenous proteins with a small percentage of cells including osteoblasts which form bone, osteocytes which maintain bone, and osteoclasts which degrade bone. Collagen molecules assemble into fibrils that are further organized into lamellae that form the structure of osteons or trabeculae. The lamellar structure of collagen fibrils has been described as "twisted plywood" and "orthogonal plywood" which highlights the transitions of local collagen alignment (Fig. 10.34) [92, 95]. Collagen plays an important role in initiating and guiding mineral formation. The post-translational modifications on the bone collagen molecules are understood to induce their association with hydroxyapatite [72, 275]. Collagen types I, III, V, and XI are found within bone with a transition from type V to XI with age [186]. Type XI is more often found in cartilage (associated with type II), but both types V and XI assist in regulating collagen fibril diameter in bone when associated with type I. In mature bovine tissue, type I collagen is the predominant collagen in the fibrils with approximately 30 type I molecules for every type V or XI molecule [283].

Mechanically, bone does not have a toe region in its stress-strain curve and cannot undergo large tensile strains unlike other collagenous tissues. Under bending, plexiform (unorganized) and osteonal bone both exhibit a linear stiffness of ~20 GPa with a maximum strength of ~220 MPa at a failure strain of 0.02% [156]. Molecular modeling of bone mechanics shows that the mineral content holds up to four times the stress of the collagen content, explaining bone's distinct behavior [180]. The ultimate strength and toughness of bone are attributed to collagenous structures [49] while the stiffness and yield strength are believed to be due to the mineral content. The collagen fibers in bone provide its intrinsic and extrinsic toughening [140, 182]. Intrinsic toughening works by impeding crack propagation through a material plasticity effect at the crack tip. Extrinsic toughening is produced by the varying directionality of bone's structure at the macroscale which is produced by the collagen orientation, which, in turn, drives the mineral organization. Collagen organization also contributes to bone mechanics, manifesting in both anisotropic material properties and sample-to-sample variability. Bone is strong along the direction of the osteons which aligns with the principal direction of the collagen fiber distribution [9, 10]. The strength of collagen alignment also affects the bending strength and stiffness of bone but not the flexural fatigue life, potentially related to an advantage in the dispersion of fibril alignment and recruitment under repeated loading [157].

Bone mineralization and the AGE non-enzymatic crosslinking of collagen both increase with age leading to stiffer and less tough bone [49, 150]. Genetic disorders of collagen (OI—type I collagen defect) can reduce the strength and toughness of bone possibly through a sharp reduction in the toughness and strength of the collagen [167]. Unloading (e.g., space travel, rehabilitation, bedrest) of bone decreases tissue mass and stiffness by altering the regular turnover cycle [77, 250, 252]. While morphological changes are measurable, material property changes dominate, e.g., one week of spaceflight during the growth phase of rats resulted in a 19% decrease in bending stiffness of the tibia compared to the non-flight control with insignificant changes in bone cross-sectional morphology [233]. Loss of material strength may be due to an altered crosslinking profile of collagen due to a disruption in the maintenance of the osteoblast population with less altered osteoclast activity [233].

Cartilage Unlike all the other tissues considered so far, the predominant collagen in cartilage is type II, the properties of which are far less defined than for type I. Cartilage is also the only soft tissue discussed that is loaded in compression under normal physiological conditions (the annulus fibrosus is another example of compressive load-bearing collagenous tissue). Within cartilage, thin type II collagen fibrils form a network that entangles and constrains hyaluronan-bound aggrecan molecules which harbor a very large fixed-charge density (FCD) associated with their GAG chains: ~300 mM [142]. The general cellular and fibrillar structures of cartilage are shown schematically in Fig. 10.35 ([152] #365). The collagen organization is characterized by vertical fibrils emanating from the osteochondral interface that progress in arcades or "leaves" that roll over to horizontal as they approach the cartilage surface. The effect of this structure is to create a tough mechanical trap that constrains the large aggrecan molecules. The resulting high fixed charge density is translated to an osmotic pressure that is balanced by tension in the collagen network. Thus, the collagen in cartilage is in a state of mechanical tension under normal static conditions [127, 155]. This, of course, is sensible as collagen is clearly designed to carry tensile loads. Typically, cartilage is mechanically tested in confined compression which can only inform us indirectly that collagen is constraining a pressure of about 1.7 MPa [154, 155]. However, noting that cartilage mechanics is likely

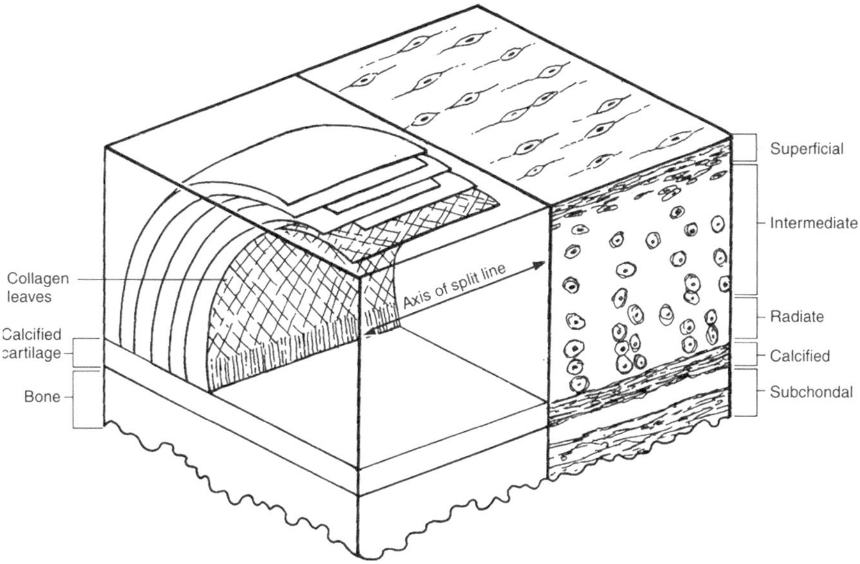

Fig. 10.35 The structure of cartilage. (Reproduced from Mansaur [152] with permission)

governed by the mechanics of type II collagen network, Kempson et al. [127] measured the mechanical properties of collagen in cartilage in tension. Figure 10.36 shows stress versus extension curves for four levels of cartilage (surface layer to the deep layer). The surface layer, against which the tibial plateau would translate (these are condylar samples), has directional, dense fibrils and thus has the highest modulus. The curves once again reflect the classic mechanical signature of collagen arrays that we have seen repeatedly with a toe, heel, and linear region. Unlike collagen arrays, there does not appear to be a non-linear rise in the modulus at higher extensions in this data. The collagen network in cartilage is critical to the health of the tissue, and the loss of the collagen network integrity is, thus far, irreversible and portends an inexorable progression to osteoarthritis.

Summary: Mechanics of whole tissue Examination of mechanically important macroscale tissues produced through integration of the fundamental collagenous structural units (molecules and fibrils) shows remarkable similarities in their tensile mechanical signatures. While the magnitude of the tensile modulus and the overall curve shapes are altered, they are clearly derivative of collagen fibril array mechanics and universally, with the exception of bone, exhibit a toe, heal, and generally linear region. These three phases are undoubtedly related to the initial low force uncrimping and alignment of the constituent fibers, followed by increasing load stiffening where the fibrils in the fibers gain order internally and begin to extend under load, and then fibril array gliding as the load is increased. The contribution of collagen fibrils to these tissues is critical to their performance. The *form* of the collagen always *follows* the mechanical *function* of the tissue. While it is easy to

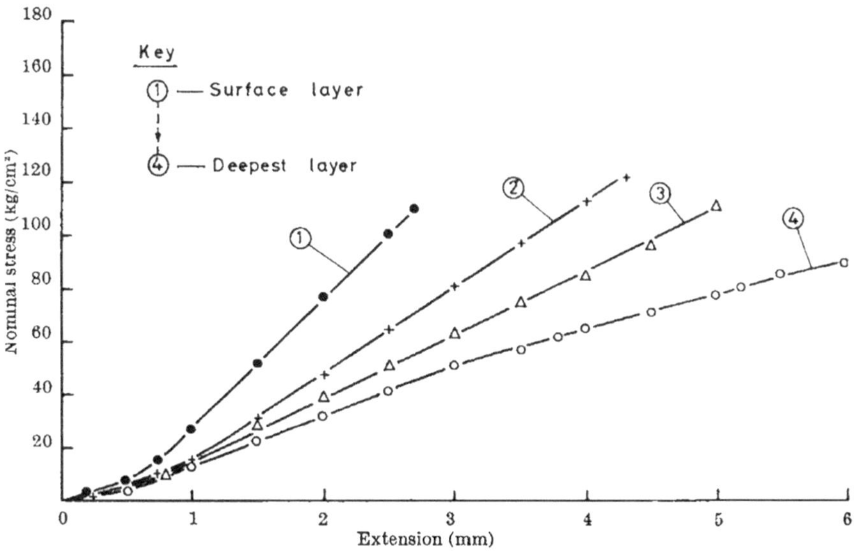

Fig. 10.36 Tensile mechanical behavior of human femoral condyle articular cartilage at multiple levels from the surface (1) to the deep layer adjacent to the bone (4). (Kempson et al. [127] with permission)

appreciate this in tensile load-bearing tissue, it is not as clear in bone or cartilage, which carry compressive loads. Even in these tissues, Nature managed to place its premier load-bearing polymer, which excels principally in resisting tension, into positions where its mechanical performance is used to great advantage. In bone, where one is hard pressed to detect collagen's mechanical trace in the stress-strain curve, the tensile properties of collagen appear to be critical to its ultimate strength and toughness.

While the tensile Young's modulus of collagenous materials at multiple scales does not convey the whole story of tissue mechanics (Young's modulus conveys information about a single location on the stress/strain curve or an average value over a region), they are relatively useful guideposts. [237] compiled tensile moduli for a number of tissue fibers and whole tissues (skin, tendon, cornea, and mitral valve). The average values were 721 MPa for fibers and 425 MPa for whole tissues [237]. We see in the data that the trend that each level of hierarchy has a reduced stiffness holds across all scales and is characteristic of collagenous tissues.

Themes in Collagenous Tissues

Although collagenous tissues are distinct from one another, they share a dependence on the biomechanical and biochemical properties of the fibrillar collagens, their primary structural and functional constituent. Identifying transcendent themes that apply to collagenous load-bearing tissues may reveal fundamental processes and mechanisms that inform the interpretation of the data with which we are regularly presented.

Theme I: Fibril Size and Mechanics

In a foundational study, Parry et al. [192, 193] conducted inter- and intra-tissue comparisons of fibril diameters across tissues, age, and species and investigated whether diameter distribution predicted tissue-level mechanics [192]. They reported a general increase in mass-average fibril diameter from birth to maturity (except in the cornea), a positive correlation between mass average fibril diameter and ultimate tensile strength, and a bimodal diameter distribution was found in tissues subjected to long-term stress. The authors hypothesized that larger fibrils enhance tensile strength through increased intrafibrillar crosslinking, whereas smaller fibrils promote elasticity and creep resistance due to greater surface area and interaction with neighboring fibrils and matrix components. This is supported by observations in tendinopathy, where a shift toward smaller fibrils correlates with decreased stiffness and a transition from a bimodal to unimodal diameter distribution [18]. Similarly, skin aging is associated with reduced fibril diameter and tissue softening. However, average fibril diameter may be less influential than total collagen volume fraction. For instance, in the cornea, there is a 30% increase in fibril diameter toward the periphery, which should increase the stiffness of the tissue. However, the fibril diameter does not overcome the loss of stiffness secondary to a 15% reduction in fiber volume fraction. The result is a net peripheral stiffness (Fig. 10.37) [26]. Other perspectives have focused on the interaction of collagen fibrils with the interfibrillar

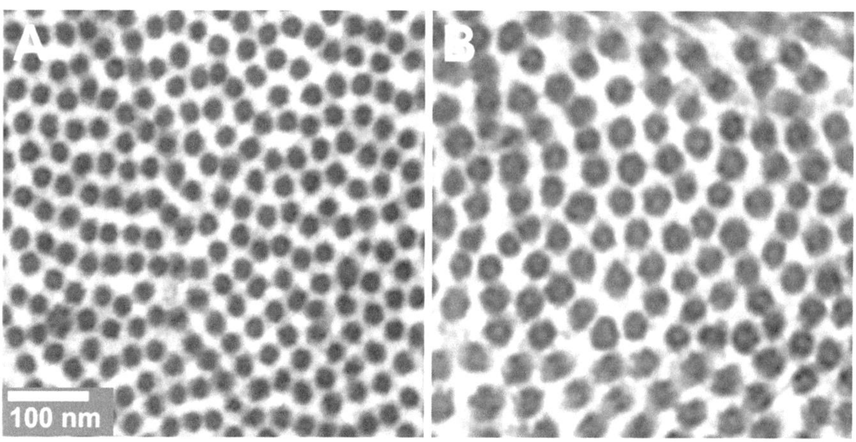

Fig. 10.37 Fibril diameters in the cornea. (**a**) Central cornea. (**b**) Peripheral cornea. (Reproduced from Bergamson et al. [26] with permission)

proteoglycan matrix generating a greater frictional shear stress with smaller but more abundant fibrils (e.g., more surface area) [96]. There is also the possibility that biomodal distributions of fibril diameters just increase the packing density of the fibrils, placing more collagen in the path of the load and increasing the surface area for interactions between the fibrils themselves and the other matrix components. Regardless of the mechanism, it appears that bimodal fibril distributions are more effective than single-mode distributions, particularly where loads are expected to be high.

Theme II: Crosslinking and Mechanics

Crosslinking within and between fibrils is critical for establishing a material capable of bearing mechanical loads within an organism (vide supra for a more comprehensive discussion of crosslinking chemistry) [263]. Mechanical loading regulates crosslinking by signaling to mechanosensitive cells to produce crosslinking enzymes [194]. Thus, tissues which have higher mechanical loading may be stiffer through the mechanobiological action of cells producing more crosslinks.

As touched on briefly, type I collagen can be modified heavily co- and post-translationally [132]. Crosslinking profiles vary across collagenous tissues due to variable LOX modification of telopeptide lysines and hydroxylysines. Hydroxylysine aldehydes mature to form pyridinoline (divalent) and pyrrole (trivalent) crosslinks. In ligaments, stable hydroxylysylpyridinoline crosslinks predominate, while tendons exhibit a mix of lysyl and hydroxylysyl pyridinoline crosslinks, which in turn support their capacity for remodeling and repair [114]. Lysyl forms are only significantly present in dentin and bone, yet in human bone, the ratio of hydroxylysyl to lysyl pyridinoline crosslinks is still only approximately 3.5 to 1. Both hydroxylysine-derived crosslink types are essentially absent in skin, membranes, cornea, and rat tail tendon in line with the lower relative stiffness in these tissues [73]. In these tissue, lysine aldehydes undergo aldol and Schiff base reaction to produce divalent and tetravalent crosslinks [74]. The population of immature crosslink species also declines with age with an increase of stable, mature crosslinks, explaining increases in tissue stiffness often seen with growth and aging [75]. Pyrrole crosslinks are localized to mineralized tissues such as bone, where mature bone contains roughly equal proportions of pyrrolic and pyridinoline crosslinks [101]. Unlike pyridinoline crosslinks, these crosslinks are hypothesized to influence mechanical strength, potentially through effects on trabecular architecture or mineralization spaces [13, 136, 137]. The aromatic structure of pyrrole may contribute to crosslink stability. However, due to bone turnover, even these crosslinks are not permanent.

There has been very little investigation of the role mechanics may play in directing the activity of crosslinking chemistry. Thus far, there is only anecdotal evidence that mechanical stimulation can influence the distribution of crosslinking. How it may do so is an interesting question to explore.

Theme III: Fatigue and Mechanics

Many load-bearing tissues are subjected to millions of cyclic mechanical loads over long time periods (>50 years). The duration and frequency of loading often play a more critical role in altering tissue mechanical properties than the load magnitude alone, as observed in discrete loading events. Repeated or prolonged loading produces cyclic fatigue and creep fatigue, corresponding to cyclic and sustained loading, respectively [128]. Recent investigations have demonstrated that molecular-level collagen damage accumulates with repeated loading of tendons. This damage has been quantified using collagen hybridizing peptides [292, 293] and observed through imaging techniques that reveal changes in collagen crimp patterns and native collagen organization following mechanical loading [6, 86]. These structural alterations are associated with progressive changes in the mechanical behavior of the tissue over time. Specifically, changes are observed in fatigue life (cycles-to-failure), tissue material stiffness changes during individual loading cycles, and energy dissipation per cycle based on cyclic loading profiles (Fig. 10.38) [128, 282]. However, some studies report that despite detectable changes in collagen organization due to fatigue loading, local tissue stiffness (as assessed by AFM) remains unaffected [260].

Collectively, these findings suggest that cyclic loading induces structural changes in collagenous tissues at the molecular level. In vivo, load-bearing tissues such as tendons *must* possess recovery mechanisms beyond those captured in simplified in vitro models because they are able to sustain function over a lifetime that can easily exceed 100 million cycles. Appropriately modified in vitro experiments can be leveraged to simulate in vivo-like conditions to identify the key mechanisms that contribute to tissue longevity. Predictive models that integrate in vitro experimental data with known in vivo tissue lifespans can provide insights into rates of damage accumulation, repair, and adaptation under steady-state mechanical loading [128, 229]. Given the progressive accumulation of damage at the molecular scale, dynamically tracking the behavior of the fundamental structural units (e.g., collagen) may offer insights about collagen's mechanochemical response to fatigue loading. Understanding how mechanics may influence the interaction of molecular collagen with fibrils under mechanical stress [240] may enhance tissue repair and prevent fatigue-related pathologies such as tendinopathy and other overuse or degenerative conditions.

Theme IV: Collagen Mechanochemistry

The principal role of the fibrillar collagens, particularly collagen type I, is to resist and efficiently transmit the mechanical forces that would otherwise dissipate the structure of the animals in which they reside. It is very clear from its primary, secondary, tertiary, and quaternary structure and the integration of those structures

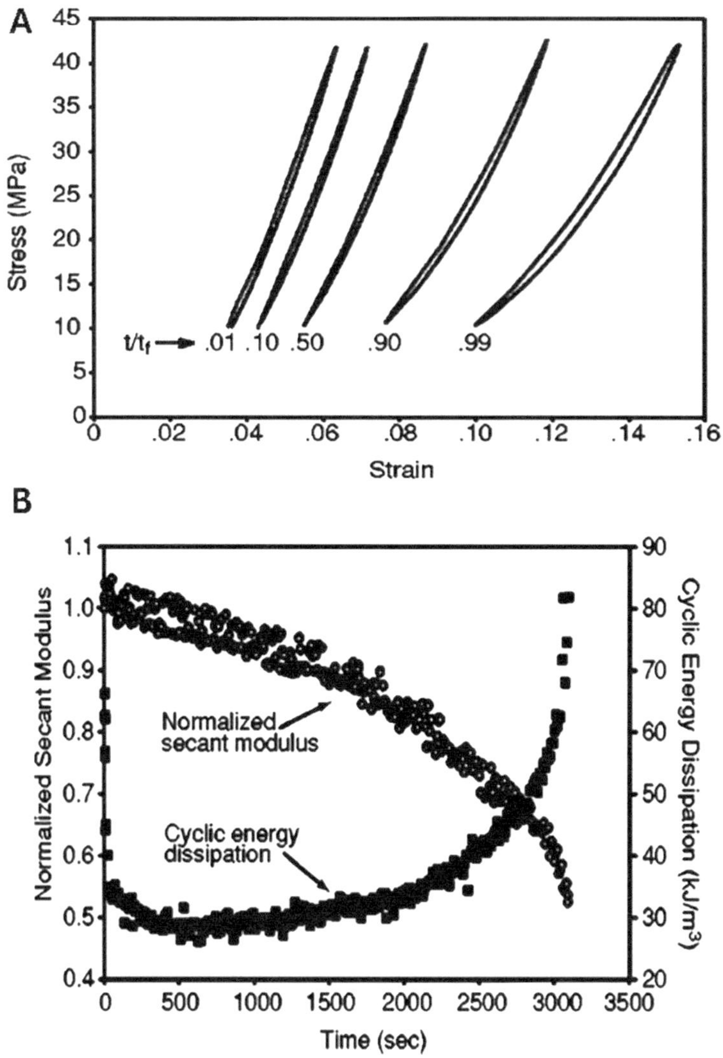

Fig. 10.38 Cyclic testing of human Achilles tendons exhibit a change in mechanics over cycles of loading. (**a**) With each stress-driven cycle, the strain increases and illustrates creep paired with (**b**) a decrease in secant modulus (i.e., tissue stiffness) and increase in energy dissipation. (Reproduced from Wren et al. [282] with permission)

into fibrillar arrays that tensile loads are the mechanical "signal" to which fibrillar collagens are exquisitely tuned. We have already seen that even the type II collagen network in cartilage is actually under tension [155] and that reducing the tensile strength of the collagen in bone (in a rodent model of OI) leads to poor mechanical performance [167]. In another important compressive load-bearing tissue, the outer annulus fibrosus, the type I collagen is arranged such that its nominal loading is

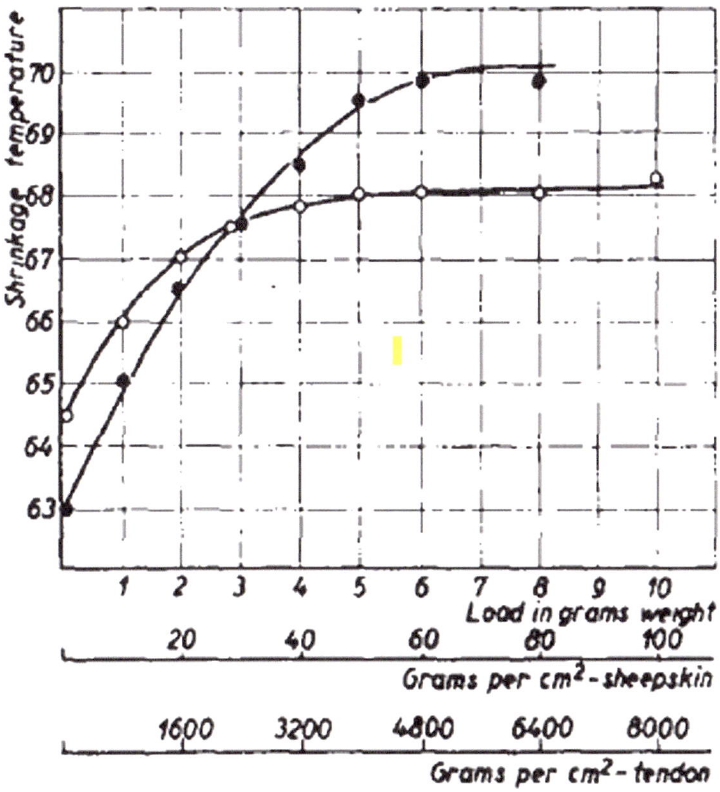

Fig. 10.39 Shrinkage temperature of skin and tendon as a function of tensile load. (Reproduced from Lennox [141] with permission)

tensile (resisting torsion and hoop stress) [146]. Indeed, it seems as if tension is the signal and that fibrillar collagens tend to "accumulate" where that signal is strongest, in the path of mechanical force, precisely where they are needed. Mechanical signals in metazoans are robust, long-range, and predictable. It is highly improbable that the collagen molecule *itself* does not interpret the mechanical environment and respond in a way that is beneficial to the cells from which it has been secreted.

Strain-modulation of collagen stability Some of the first evidence that collagen might be mechanoresponsive was provided by Lennox [141] who found that mechanical tension increased the shrinkage temperature of sheep skin and tendon collagen (Fig. 10.39). In 1977, Huang and Yannas showed that tensile loading of reconstituted bovine tendon collagen tapes was not enzymatically degraded by bacterial collagenase as quickly as unloaded tapes [113]. In 1996, Nabeshima et al. ran a well-controlled investigation demonstrating that rabbit patellar tendons under tension were less susceptible to collagenase than unloaded tendon [178]. The Nabeshima paper, while well-run and profound, surprisingly did not garner much attention at the time of its publication.

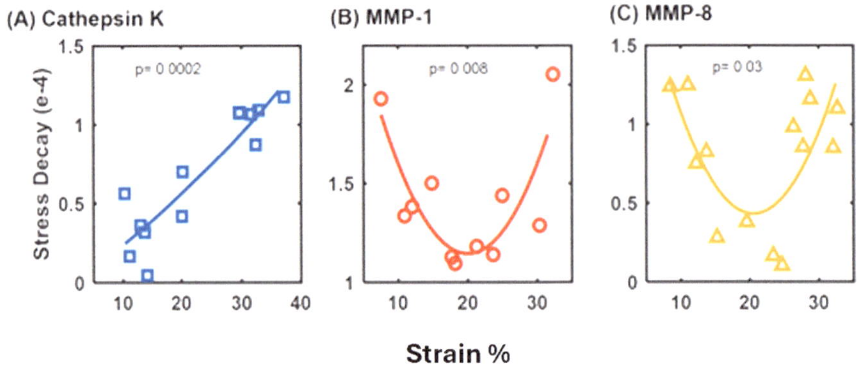

Fig. 10.40 Effect of mechanical tension on the degradation rate (measured as stress decay rate) of porcine pericardium in biaxial tension. (**a**) Cathepsin K; (**b**) MMP-1; (**c**) MMP-8. (Reproduced from Yeganegi et al. [287] with permission)

In 2005, Ruberti and Hallab demonstrated that unloaded collagen fibrils in uniaxially loaded strips of bovine cornea could be preferentially degraded by bacterial collagenase in the direction orthogonal to an applied tensile load [225]. This line of mechanochemical investigation was extended to single molecules [3, 4, 44, 129], single fibrils [81], collagen gels [80], and whole tissues [291]. While there were some contradictory results at the molecular scale, ultimately the field has converged on consensus that at the tissue level, nominal physiological strain slows the rate of degradation by both bacterial collagenase and MMPs (while Cathepsin K appears to degrade tissue faster under tensile load) (Fig. 10.40; [287]). The implications of the strain-modulating enzyme predilection for collagen cannot be overstated. If collagen is more resistant to removal when under load (in systems with MMP constitutive expression), it would thus persist in the path of force right where it is needed. This is remarkable. Conversely, unloaded collagen would be removed preferentially. The removal of unloaded collagen is a potential problem. In osteoarthritis, for instance, the progression of the disease could be quite simply due to enhanced degradation of the type II network after the escape of the aggrecan molecules which had been keeping the collagen in tension and protected from MMP enzymes. There are numerous other implications of this mechanochemical behavior of collagen and its complementary enzymes (e.g., mechanically protected tumor ECM) which have been reviewed in [241]).

Mechanically driven assembly of collagen While strain-stabilization has the potential to explain loaded collagen's persistence, it does not explain how collagen is deposited in the path of force in the first place. In 2016, Paten et al. showed that extensional strain rates in the neighborhood of 0.5 Hz can catalyze the conversion of soluble type I collagen molecules to insoluble fibrils [195] (Fig. 10.41). At the time, Paten et al. suggested that this mechanism could be used by cells to initiate the deposition of collagenous structures in the path of force and that larger scale movements (e.g., from developing muscles or animal motion) could drive the cell-

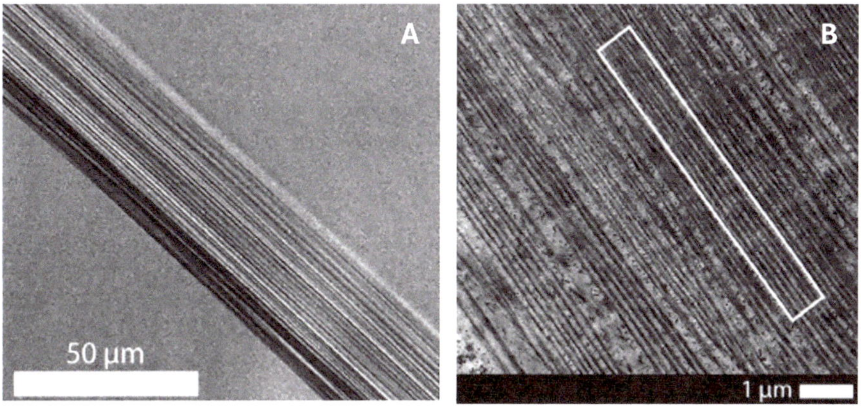

Fig. 10.41 Collagen fiber pulled directly from a solution of type I collagen monomers (extensional strain rate: 0.3 Hz; concentration of collagen: 15 mg/ml). (**a**) DIC optical image of fiber. (**b**) TEM of fibrils in the fiber, all aligned in the path of force. (Reproduced from Paten et al. [195] with permission)

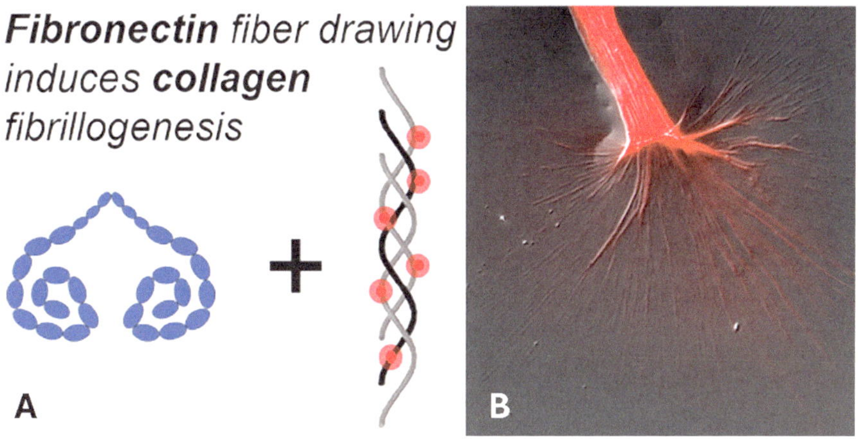

Fig. 10.42 Fibronectin, which readily flow-aggregates, binds to and enhances the flow-induced crystallization of collagen (red label). Fibronectin lowers the physicochemical requirements (extensional strain rate and collagen concentration) to convert soluble monomers to insoluble fibrils ostensibly bringing them within reach of the resident fibroblasts and myofibroblasts. (Reproduced from Paten et al. [195] with permission)

produced segments to continuity, completing structure via the self-healing equation (see fig. 6, fig. S8b, and eq. 1 in Paten et al. [195]). In 2019, Paten et al. extended the work to show that the ubiquitous molecule, fibronectin (another mechanosensitive assembling molecular system [61] associated both with fibrosis and with collagen fibrillogenesis), reduced the necessary strain rates and collagen concentrations for flow-induced crystallization to values well within the physiological range (Fig. 10.42).

In a recent paper, Silverman et al. showed that primary human fibroblasts do indeed pull cooperatively to produce collagen and fibronectin filaments [245]. The data suggest that connective tissue fibrils are not necessarily "deposited" by cells. Rather they are precipitated by tension created by cooperative cell contractions. The contractions would be in line with the cell's orientation, which, in most developing tissues, coincides with the intended tissue architecture. The corneal fibroblasts in the Silverman paper often pulled in two directions at about 90 degrees to one another, suggesting that they would create orthogonal structures.

While there are many implications of the idea that the mechanical tension can directly drive structure formation (see Siadat and Ruberti [241]), the most important is likely how it relates to a particularly difficult pathology: fibrosis. Fibrosis and organ scarring are leading causes of death, with data suggesting they are responsible for up to 45% of deaths in the developed world. Mechanochemistry of collagen could provide a "common thread" that leads to pathological collagen accumulation driven by myofibroblast contractions [55, 288].

Conclusion

Collagen is the most abundant protein in vertebrates, and the family of collagens has evolved over nearly a billion years from the primordial Gly-X-Y ancestral motif into multiple classes of complex molecules coded by 50 or more genes. The utility of the triple helix as a stable structural element has led to its conservation, its extensive modification, and its utilization in virtually all animals in a multiplicity of tissues performing a multiplicity of functions. Of the 28 types of collagen found in vertebrates, the fibrillar collagens (Types I, II, III, V, XI, XXIV, and XXVII) are responsible for transmitting and resisting the mechanical forces incipient on the tissues in which they reside. From the delicate rectus muscle insertions in the ocular globe to the compact primary bone in a femur, fibrillar collagens *enable* vertebrates to deftly negotiate the mechanical world. We have seen that the long, tight triple helical structure of fibrillar collagen produces a mechanical signature that is reflected across scales and in the mechanics of the tissue within which it is integrated. Collagen's versatile chemistry permits modification of the mechanical properties of tissues during the translational, post-translational, and post-depositional phases of its synthesis and secretion. At the tissue level, collagen is integrated into the structures to produce the desired mechanical performance through variation of collagen fibrillar alignment, concentration, and interaction with other ECM components (e.g., elastin). While the passive mechanics of fibrillar collagen are remarkable, it is intriguing to consider that collagen is more than just a material manipulated by cells. Recent evidence suggests that type I collagen is, at a fundamental level, mechanoresponsive. A rather large body of work has now established that collagen's enzyme susceptibility can be altered by mechanical tension, implying that collagen retention in the path of force is favored at the molecular/fibril level. More recent investigations suggest that collagen's assembly may also be influenced and directed by mechanical forces. The implications of collagen possessing inherent mechanosensitivity are far-reaching, but also quite logical. It only makes sense that Nature's premier structural polymer would be mechanosensitive and capable of providing an independent "material assist" to the cells that are responsible for building and maintaining the matrix. The prospect of collagen exhibiting fundamental mechanochemical behavior is likely to drive new and exciting investigations and further expand our appreciation of this already remarkable structural material.

References

1. Abramowitch SD, Woo SLY (2004) An improved method to analyze the stress relaxation of ligaments following a finite ramp time based on the quasi-linear viscoelastic theory. J Biomech Eng 126(1):92–97. https://doi.org/10.1115/1.1645528
2. Abu el-Asrar AM, Geboes K, Al-Kharashi SA, Tabbara KF, Missotten L (1998) Collagen content and types in trachomatous conjunctivitis. Eye (Lond) 12(Pt 4):735–739. https://doi.org/10.1038/eye.1998.179
3. Adhikari AS, Chai J, Dunn AR (2011) Mechanical load induces a 100-fold increase in the rate of collagen proteolysis by MMP-1. J Am Chem Soc 133(6):1686–1689. https://doi.org/10.1021/ja109972p
4. Adhikari AS, Glassey E, Dunn AR (2012) Conformational dynamics accompanying the proteolytic degradation of trimeric collagen I by collagenases. J Am Chem Soc 134(32): 13259–13265. https://doi.org/10.1021/ja212170b
5. Alberts A, Moldoveanu E-T, Niculescu A-G, Grumezescu AM (2025) Vitamin C: a comprehensive review of its role in health, disease prevention, and therapeutic potential. Molecules 30(3):748. https://doi.org/10.3390/molecules30030748
6. Andarawis-Puri N, Flatow EL (2011) Tendon fatigue in response to mechanical loading. J Musculoskelet Neuronal Interact 11(2):106–114
7. Andriotis OG, Nalbach M, Thurner PJ (2023) Mechanics of isolated individual collagen fibrils. Acta Biomater 163:35–49. https://doi.org/10.1016/j.actbio.2022.12.008
8. Antar SA, Ashour NA, Marawan ME, Al-Karmalawy AA (2023) Fibrosis: types, effects, markers, mechanisms for disease progression, and its relation with oxidative stress, immunity, and inflammation. Int J Mol Sci 24(4):4004. https://doi.org/10.3390/ijms24044004
9. Ascenzi A, Bonucci E (1967) The tensile properties of single osteons. Anat Rec 158(4): 375–386. https://doi.org/10.1002/ar.1091580403
10. Ascenzi A, Bonucci E (1968) The compressive properties of single osteons. Anat Rec 161(3): 377–391. https://doi.org/10.1002/ar.1091610309
11. Asgari M, Latifi N, Heris HK, Vali H, Mongeau L (2017) In vitro fibrillogenesis of tropocollagen type III in collagen type I affects its relative fibrillar topology and mechanics. Sci Rep 7(1):1392. https://doi.org/10.1038/s41598-017-01476-y
12. Asrani SG, McGlumphy EJ, Al-Aswad LA, Chaya CJ, Lin S, Musch DC, Pitha I, Robin AL, Wirostko B, Johnson TV (2024) The relationship between intraocular pressure and glaucoma: an evolving concept. Prog Retin Eye Res 103:101303. https://doi.org/10.1016/j.preteyeres.2024.101303
13. Banse X, Devogelaer JP, Lafosse A, Sims TJ, Grynpas M, Bailey AJ (2002) Cross-link profile of bone collagen correlates with structural organization of trabeculae. Bone 31(1):70–76. https://doi.org/10.1016/s8756-3282(02)00800-1
14. Bard JB, Bansal MK (1987) The morphogenesis of the chick primary corneal stroma. I. New observations on collagen organization in vivo help explain stromal deposition and growth. Development 100(1):135–145. https://doi.org/10.1242/dev.100.1.135

15. Barnard K, Burgess SA, Carter DA, Woolley DM (1992) Three-dimensional structure of type IV collagen in the mammalian lens capsule. J Struct Biol 108(1):6–13. https://doi.org/10.1016/1047-8477(92)90002-r
16. Batson EL, Paramour RJ, Smith TJ, Birch HL, Patterson-Kane JC, Goodship AE (2003) Are the material properties and matrix composition of equine flexor and extensor tendons determined by their functions? Equine Vet J 35(3):314–318. https://doi.org/10.2746/042516403776148327
17. Bear RS (1944) X-ray diffraction studies on protein fibers. I The large fiber-axis period of collagen. J Am Chem Soc 66(8):1297–1305. https://doi.org/10.1021/ja01236a027
18. Beisbayeva Z, Zhanbassynova A, Kulzhanova G, Mukasheva F, Erisken C (2021) Change in collagen fibril diameter distribution of bovine anterior cruciate ligament upon injury can be mimicked in a nanostructured scaffold. Molecules 26(5). https://doi.org/10.3390/molecules26051204
19. Bella J (2016) Collagen structure: new tricks from a very old dog. Biochem J 473(8):1001–1025. https://doi.org/10.1042/BJ20151169
20. Bella J, Eaton M, Brodsky B, Berman HM (1994) Crystal and molecular structure of a collagen-like peptide at 1.9 A resolution. Science 266(5182):75–81. https://doi.org/10.1126/science.7695699
21. Benedek GB (1971) Theory of transparency of the eye. Appl Opt 10(3):459–473. https://doi.org/10.1364/AO.10.000459
22. Benedict JV, Walker LB, Harris EH (1968) Stress-strain characteristics and tensile strength of unembalmed human tendon. J Biomech 1(1):53–63. https://doi.org/10.1016/0021-9290(68)90038-9
23. Benjamin M, Kaiser E, Milz S (2008) Structure-function relationships in tendons: a review. J Anat 212(3):211–228. https://doi.org/10.1111/j.1469-7580.2008.00864.x
24. Berg RA (1973) Determination of 3- and 4-Hydroxyproline. In: Methods in enzymology, vol 82. Academic Press
25. Bergman CA, McEwan E (1997) Sinew-reinforced and composite bows. In: Bergman CA (ed) Projectile technology. Springer, New York, pp 143–160
26. Bergmanson JPG, Burns AR, Walker MK (2024) Central versus peripheral thickness in the human cornea explained. Cont Lens Anterior Eye 47(3):102165. https://doi.org/10.1016/j.clae.2024.102165
27. Biggs LC, Kim CS, Miroshnikova YA, Wickstrom SA (2020) Mechanical forces in the skin: roles in tissue architecture, stability, and function. J Invest Dermatol 140(2):284–290. https://doi.org/10.1016/j.jid.2019.06.137
28. Birch HL (2007) Tendon matrix composition and turnover in relation to functional requirements. Int J Exp Pathol 88(4):241–248. https://doi.org/10.1111/j.1365-2613.2007.00552.x
29. Birch HL, Worboys S, Eissa S, Jackson B, Strassburg S, Clegg PD (2008) Matrix metabolism rate differs in functionally distinct tendons. Matrix Biol 27(3):182–189. https://doi.org/10.1016/j.matbio.2007.10.004
30. Birk DE, Mayne R (1997) Localization of collagen types I, III and V during tendon development. Changes in collagen types I and III are correlated with changes in fibril diameter. Eur J Cell Biol 72(4):352–361
31. Birk DE, Fitch JM, Babiarz JP, Doane KJ, Linsenmayer TF (1990) Collagen fibrillogenesis in vitro: interaction of types I and V collagen regulates fibril diameter. J Cell Sci 95(Pt 4):649–657
32. Blackburn BJ, Jenkins MW, Rollins AM, Dupps WJ (2019) A review of structural and biomechanical changes in the cornea in aging, disease, and photochemical crosslinking. Front Bioeng Biotechnol 7:66. https://doi.org/10.3389/fbioe.2019.00066
33. Boedtker H, Doty P (1956) The native and denatured states of soluble collagen. J Am Chem Soc 78(17):4267–4280. https://doi.org/10.1021/ja01598a024

34. Boote C, Sigal IA, Grytz R, Hua Y, Nguyen TD, Girard MJA (2020) Scleral structure and biomechanics. Prog Retin Eye Res 74:100773. https://doi.org/10.1016/j.preteyeres.2019.100773
35. Bornstein P, Piez KA (1964) A biochemical study of human skin collagen and the relation between intra- and intermolecular cross-linking. J Clin Invest 43(9):1813–1823. https://doi.org/10.1172/JCI105055
36. Bornstein P, Martin GR, Piez KA (1964) Intermolecular cross-linking of collagen and the identification of a new beta-component. Science 144(3623):1220–1221. https://doi.org/10.1126/science.144.3623.1220
37. Boskey AL Robey PG (2013) The composition of bone. In: CJ Rosen (ed) Primer on the metabolic bone diseases and disorders of mineral metabolism
38. Boyce BL, Grazier JM, Jones RE, Nguyen TD (2008) Full-field deformation of bovine cornea under constrained inflation conditions. Biomaterials 29(28):3896–3904. https://doi.org/10.1016/j.biomaterials.2008.06.011
39. Boyer G, Pailler Mattei C, Molimard J, Pericoi M, Laquieze S, Zahouani H (2012) Non contact method for in vivo assessment of skin mechanical properties for assessing effect of ageing. Med Eng Phys 34(2):172–178. https://doi.org/10.1016/j.medengphy.2011.07.007
40. Buckley MR, Evans EB, Matuszewski PE, Chen YL, Satchel LN, Elliott DM, Soslowsky LJ, Dodge GR (2013) Distributions of types I, II and III collagen by region in the human supraspinatus tendon. Connect Tissue Res 54(6):374–379. https://doi.org/10.3109/03008207.2013.847096
41. Buehler MJ (2006a) Atomistic and continuum modeling of mechanical properties of collagen: elasticity, fracture, and self-assembly. J Mater Res 21(8):1947–1961. https://doi.org/10.1557/jmr.2006.0236
42. Buehler MJ (2006b) Nature designs tough collagen: explaining the nanostructure of collagen fibrils. Proc Natl Acad Sci 103(33):12285–12290. https://doi.org/10.1073/pnas.0603216103
43. Buehler MJ, Wong SY (2007) Entropic elasticity controls nanomechanics of single tropocollagen molecules. Biophys J 93(1):37–43. https://doi.org/10.1529/biophysj.106.102616
44. Camp RJ, Liles M, Beale J, Saeidi N, Flynn BP, Moore E, Murthy SK, Ruberti JW (2011) Molecular mechanochemistry: low force switch slows enzymatic cleavage of human type I collagen monomer. J Am Chem Soc 133(11):4073–4078. https://doi.org/10.1021/ja110098b
45. Chen Y, Yang S, Tavormina J, Tampe D, Zeisberg M, Wang H, Mahadevan KK, Wu C-J, Sugimoto H, Chang C-C, Jenq RR, McAndrews KM, Kalluri R (2022) Oncogenic collagen I homotrimers from cancer cells bind to α3β1 integrin and impact tumor microbiome and immunity to promote pancreatic cancer. Cancer Cell 40(8):818–834.e819. https://doi.org/10.1016/j.ccell.2022.06.011
46. Cowan PM, McGavin S, North ACT (1955a) The polypeptide chain configuration of collagen. Nature 176(4492):1062–1064. https://doi.org/10.1038/1761062a0
47. Cowan PM, North ACT, Randall JT (1955b) X-ray diffraction studies of collagen fibers. Symp Soc Exp Biol 9:115–126
48. Curat CA, Eck M, Dervillez X, Vogel WF (2001) Mapping of epitopes in discoidin domain receptor 1 critical for collagen binding. J Biol Chem 276(49):45952–45958. https://doi.org/10.1074/jbc.m104360200
49. Currey JD (2003) Role of collagen and other organics in the mechanical properties of bone. Osteoporos Int 14:29–36. https://doi.org/10.1007/s00198-003-1470-8
50. Davey ME, O'Toole GA (2000) Microbial biofilms: from ecology to molecular genetics. Microbiol Mol Biol Rev 64(4):847–867. https://doi.org/10.1128/MMBR.64.4.847-867.2000
51. David R, Zangwill L, Briscoe D, Dagan M, Yagev R, Yassur Y (1992) Diurnal intraocular pressure variations: an analysis of 690 diurnal curves. Br J Ophthalmol 76(5):280–283. https://doi.org/10.1136/bjo.76.5.280
52. Davison PF (1989) The contribution of labile crosslinks to the tensile behavior of tendons. Connect Tissue Res 18(4):293–305. https://doi.org/10.3109/03008208909019078

53. Daxer A, Misof K, Grabner B, Ettl A, Fratzl P (1998) Collagen fibrils in the human corneal stroma: structure and aging. Invest Ophthalmol Vis Sci 39(3):644–648
54. De Gregorio V, Barua M, Lennon R (2025) Collagen formation, function and role in kidney disease. Nat Rev Nephrol 21(3):200–215. https://doi.org/10.1038/s41581-024-00902-5
55. Desmouliere A, Darby IA, Gabbiani G (2003) Normal and pathologic soft tissue remodeling: role of the myofibroblast, with special emphasis on liver and kidney fibrosis. Lab Investig 83(12):1689–1707
56. Diridollou S, Patat F, Gens F, Vaillant L, Black D, Lagarde JM, Gall Y, Berson M (2000) In vivo model of the mechanical properties of the human skin under suction. Skin Res Technol 6(4):214–221. https://doi.org/10.1034/j.1600-0846.2000.006004214.x
57. Dong C, Lv Y (2016) Application of collagen scaffold in tissue engineering: recent advances and new perspectives. Polymers 8(2):42. https://doi.org/10.3390/polym8020042
58. Dupps WJ Jr, Wilson SE (2006) Biomechanics and wound healing in the cornea. Exp Eye Res 83(4):709–720. https://doi.org/10.1016/j.exer.2006.03.015
59. Dyer RF, Enna CD (1976) Ultrastructural features of adult human tendon. Cell Tissue Res 168(2):247–259. https://doi.org/10.1007/BF00215881
60. Eastoe JE (1955) The amino acid composition of mammalian collagen and gelatin. Biochem J 61(4):589–600. https://doi.org/10.1042/bj0610589
61. Ejim OS, Blunn GW, Brown RA (1993) Production of artificial-orientated mats and strands from plasma fibronectin: a morphological study. Biomaterials 14(10):743–748. https://doi.org/10.1016/0142-9612(93)90038-4
62. Elsheikh A, McMonnies CW, Whitford C, Boneham GC (2015) In vivo study of corneal responses to increased intraocular pressure loading. Eye Vis (Lond) 2:20. https://doi.org/10.1186/s40662-015-0029-z
63. Engel J (1996) Domain organizations of modular extracellular matrix proteins and their evolution. Matrix Biol 15(5):295–299. https://doi.org/10.1016/s0945-053x(96)90130-4
64. Eppell SJ, Smith BN, Kahn H, Ballarini R (2006) Nano measurements with micro-devices: mechanical properties of hydrated collagen fibrils. J R Soc Interface 3(6):117–121. https://doi.org/10.1098/rsif.2005.0100
65. Espana EM, Birk DE (2020) Composition, structure and function of the corneal stroma. Exp Eye Res 198:108137. https://doi.org/10.1016/j.exer.2020.108137
66. Ethier CR, Johnson M, Ruberti J (2004) Ocular biomechanics and biotransport. Annu Rev Biomed Eng 6:249–273. https://doi.org/10.1146/annurev.bioeng.6.040803.140055
67. Exposito JY, Garrone R (1990) Characterization of a fibrillar collagen gene in sponges reveals the early evolutionary appearance of two collagen gene families. Proc Natl Acad Sci USA 87(17):6669–6673. https://doi.org/10.1073/pnas.87.17.6669
68. Exposito JY, van der Rest M, Garrone R (1993) The complete intron/exon structure of Ephydatia mulleri fibrillar collagen gene suggests a mechanism for the evolution of an ancestral gene module. J Mol Evol 37(3):254–259. https://doi.org/10.1007/BF00175502
69. Exposito J, Cluzel C, Lethias C, Garrone R (2000) Tracing the evolution of vertebrate fibrillar collagens from an ancestral alpha chain. Matrix Biol 19(3):275–279. https://doi.org/10.1016/s0945-053x(00)00067-6
70. Exposito JY, Cluzel C, Garrone R, Lethias C (2002) Evolution of collagens. Anat Rec 268(3):302–316. https://doi.org/10.1002/ar.10162
71. Exposito J-Y, Valcourt U, Cluzel C, Lethias C (2010) The fibrillar collagen family. Int J Mol Sci 11(2):407–426. https://doi.org/10.3390/ijms11020407
72. Eyre DR, Weis MA (2013) Bone collagen: new clues to its mineralization mechanism from recessive osteogenesis imperfecta. Calcif Tissue Int 93(4):338–347. https://doi.org/10.1007/s00223-013-9723-9
73. Eyre DR, Koob TJ, Van Ness KP (1984a) Quantitation of hydroxypyridinium crosslinks in collagen by high-performance liquid chromatography. Anal Biochem 137(2):380–388. https://doi.org/10.1016/0003-2697(84)90101-5

74. Eyre DR, Paz MA, Gallop PM (1984b) Cross-linking in collagen and elastin. Annu Rev Biochem 53:717–748. https://doi.org/10.1146/annurev.bi.53.070184.003441
75. Eyre DR, Dickson IR, Van Ness K (1988) Collagen cross-linking in human bone and articular cartilage. Age-related changes in the content of mature hydroxypyridinium residues. Biochem J 252(2):495–500. https://doi.org/10.1042/bj2520495
76. Fidler AL, Boudko SP, Rokas A, Hudson BG (2018) The triple helix of collagens—an ancient protein structure that enabled animal multicellularity and tissue evolution. J Cell Sci 131(7): jcs203950. https://doi.org/10.1242/jcs.203950
77. Firschein HE, Shill JP (1966) The determination of total hydroxyproline in urine and bone extracts. Anal Biochem 14(2):296–304. https://doi.org/10.1016/0003-2697(66)90140-0
78. Fleischmajer R, MacDonald ED, Perlish JS, Burgeson RE, Fisher LW (1990) Dermal collagen fibrils are hybrids of type I and type III collagen molecules. J Struct Biol 105(1–3):162–169. https://doi.org/10.1016/1047-8477(90)90110-x
79. Fleischmajer R, Perlish JS, Faraggiana T (1991) Rotary shadowing of collagen monomers, oligomers, and fibrils during tendon fibrillogenesis. J Histochem Cytochem 39(1):51–58. https://doi.org/10.1177/39.1.1983873
80. Flynn BP, Bhole AP, Saeidi N, Liles M, Dimarzio CA, Ruberti JW (2010) Mechanical strain stabilizes reconstituted collagen fibrils against enzymatic degradation by mammalian collagenase matrix metalloproteinase 8 (MMP-8). PLoS One 5(8):e12337. https://doi.org/10.1371/journal.pone.0012337
81. Flynn BP, Tilburey GE, Ruberti JW (2013) Highly sensitive single-fibril erosion assay demonstrates mechanochemical switch in native collagen fibrils. Biomech Model Mechanobiol 12(2):291–300. https://doi.org/10.1007/s10237-012-0399-2
82. Frank CB (2004) Ligament structure, physiology and function. J Musculoskelet Neuronal Interact 4(2):199–201
83. Frank-Kamenetskii MD, Mirkin SM (1995) Triplex DNA structures. Annu Rev Biochem 64: 65–95. https://doi.org/10.1146/annurev.bi.64.070195.000433
84. Fratzl P (2008) Collagen: structure and mechanics, an introduction. In: Fratzl P (ed) Collagen: structure and mechanics. Springer
85. Fratzl P, Misof K, Zizak I, Rapp G, Amenitsch H, Bernstorff S (1998) Fibrillar structure and mechanical properties of collagen. J Struct Biol 122(1–2):119–122. https://doi.org/10.1006/jsbi.1998.3966
86. Freedman BR, Zuskov A, Sarver JJ, Buckley MR, Soslowsky LJ (2015) Evaluating changes in tendon crimp with fatigue loading as an ex vivo structural assessment of tendon damage. J Orthop Res 33(6):904–910. https://doi.org/10.1002/jor.22875
87. Fung YC (1967) Elasticity of soft tissues in simple elongation. Am J Phys 213(6):1532–1544. https://doi.org/10.1152/ajplegacy.1967.213.6.1532
88. Gaar J, Naffa R, Brimble M (2020) Enzymatic and non-enzymatic crosslinks found in collagen and elastin and their chemical synthesis. Organ Chem Front 7(18):2789–2814. https://doi.org/10.1039/d0qo00624f
89. Garrone R (1978) Phylogenesis of connective tissue: morphological aspects and biosynthesis of sponge intercellular matrix. Frontiers of Matrix Biology. Karger
90. Garrone R, Huc A, Junqua S (1975) Fine structure and physiocochemical studies on the collagen of the marine sponge Chondrosia reniformis nardo. J Ultrastruct Res 52(2):261–275. https://doi.org/10.1016/s0022-5320(75)80117-1
91. Gautieri A, Passini FS, Silvan U, Guizar-Sicairos M, Carimati G, Volpi P, Moretti M, Schoenhuber H, Redaelli A, Berli M, Snedeker JG (2017) Advanced glycation end-products: mechanics of aged collagen from molecule to tissue. Matrix Biol 59:95–108. https://doi.org/10.1016/j.matbio.2016.09.001
92. Gebhardt W (1905) Über funktionell wichtige Anordnungsweisen der feineren und gröberen Bauelemente des Wirbeltierknochens. Archiv für Entwicklungsmechanik der Organismen 20(2):187–322. https://doi.org/10.1007/bf02162810

93. Geiger BJ, Steenbock H, Parsons HT (1933) Lathyrism in the rat. J Nutr 6(5):427–442. https://doi.org/10.1093/jn/6.5.427
94. Gibb AA, Lazaropoulos MP, Elrod JW (2020) Myofibroblasts and fibrosis. Circ Res 127(3): 427–447. https://doi.org/10.1161/circresaha.120.316958
95. Giraud-Guille MM (1988) Twisted plywood architecture of collagen fibrils in human compact bone osteons. Calcif Tissue Int 42(3):167–180. https://doi.org/10.1007/BF02556330
96. Goh KL, Holmes DF, Lu Y, Purslow PP, Kadler KE, Bechet D, Wess TJ (2012) Bimodal collagen fibril diameter distributions direct age-related variations in tendon resilience and resistance to rupture. J Appl Physiol 113(6):878–888. https://doi.org/10.1152/japplphysiol.00258.2012
97. Graham JS, Vomund AN, Phillips CL, Grandbois M (2004) Structural changes in human type I collagen fibrils investigated by force spectroscopy. Exp Cell Res 299(2):335–342. https://doi.org/10.1016/j.yexcr.2004.05.022
98. Gross J (1956) The behavior of collagen units as a model in morphogenesis. J Biophys Biochem Cytol 2(4 Suppl):261–274. https://doi.org/10.1083/jcb.2.4.261
99. Gutsmann T, Fantner GE, Kindt JH, Venturoni M, Danielsen S, Hansma PK (2004) Force spectroscopy of collagen fibers to investigate their mechanical properties and structural organization. Biophys J 86(5):3186–3193. https://doi.org/10.1016/S0006-3495(04)74366-0
100. Hansen KA, Weiss JA, Barton JK (2002) Recruitment of tendon crimp with applied tensile strain. J Biomech Eng 124(1):72–77. https://doi.org/10.1115/1.1427698
101. Hanson DA, Eyre DR (1996) Molecular site specificity of pyridinoline and pyrrole cross-links in type I collagen of human bone. J Biol Chem 271(43):26508–26516. https://doi.org/10.1074/jbc.271.43.26508
102. Harley R, James D, Miller A, White JW (1977) Phonons and the elastic moduli of collagen and muscle. Nature 267(5608):285–287. https://doi.org/10.1038/267285a0
103. Harrington WF, Von Hippel PH (1962) The structure of collagen and gelatin. Adv Protein Chem 16:1–138
104. Haut RC, Little RW (1972) A constitutive equation for collagen fibers. J Biomech 5(5): 423–430. https://doi.org/10.1016/0021-9290(72)90001-2
105. Hefferan SA, Blaker CL, Ashton DM, Little CB, Clarke EC (2025) Structural variations of tendons: a systematic search and narrative review of histological differences between tendons, tendon regions, sex, and age. J Orthop Res 43(5):994–1011. https://doi.org/10.1002/jor.26060
106. Heinemeier KM, Schjerling P, Heinemeier J, Magnusson SP, Kjaer M (2013) Lack of tissue renewal in human adult Achilles tendon is revealed by nuclear bomb ¹⁴ C. FASEB J 27(5):2074–2079. https://doi.org/10.1096/fj.12-225599
107. Henderson NC, Rieder F, Wynn TA (2020) Fibrosis: from mechanisms to medicines. Nature 587(7835):555–566. https://doi.org/10.1038/s41586-020-2938-9
108. Hickey DS, Hukins DW (1980) X-ray diffraction studies of the arrangement of collagenous fibres in human fetal intervertebral disc. J Anat 131(Pt 1):81–90
109. Hijazi KM, Singfield KL, Veres SP (2019) Ultrastructural response of tendon to excessive level or duration of tensile load supports that collagen fibrils are mechanically continuous. J Mech Behav Biomed Mater 97:30–40. https://doi.org/10.1016/j.jmbbm.2019.05.002
110. Hofmann H, Voss T, Kuhn K, Engel J (1984) Localization of flexible sites in thread-like molecules from electron micrographs. Comparison of interstitial, basement membrane and intima collagens. J Mol Biol 172(3):325–343. https://doi.org/10.1016/s0022-2836(84)80029-7
111. Holmes DF, Gilpin CJ, Baldock C, Ziese U, Koster AJ, Kadler KE (2001) Corneal collagen fibril structure in three dimensions: structural insights into fibril assembly, mechanical properties, and tissue organization. Proc Natl Acad Sci USA 98(13):7307–7312. https://doi.org/10.1073/pnas.111150598
112. Hooley CJ, McCrum NG, Cohen RE (1980) The viscoelastic deformation of tendon. J Biomech 13(6):521–528. https://doi.org/10.1016/0021-9290(80)90345-0

113. Huang C, Yannas IV (1977) Mechanochemical studies of enzymatic degradation of insoluble collagen fibers. J Biomed Mater Res 11(1):137–154. https://doi.org/10.1002/jbm.820110113
114. Hudson DM, Archer M, Rai J, Weis M, Fernandes RJ, Eyre DR (2021) Age-related type I collagen modifications reveal tissue-defining differences between ligament and tendon. Matrix Biol Plus 12:100070. https://doi.org/10.1016/j.mbplus.2021.100070
115. Hulmes DJS (2008) Collagen diversity, synthesis and assembly. In: Fratzl P (ed) Collagen. Springer, London, pp 15–47
116. Hulmes DJS, Miller A (1979) Quasi-hexagonal molecular packing in collagen fibrils. Nature 282(5741):878–880. https://doi.org/10.1038/282878a0
117. Hynes RO (2012) The evolution of metazoan extracellular matrix. J Cell Biol 196(6):671–679. https://doi.org/10.1083/jcb.201109041
118. Institute FF (2025) FILK Freiberg Institute. https://www.filkfreiberg.de/
119. Jayawardana K, Ovenden NC, Cottenden A (2017) Quantifying the frictional forces between skin and nonwoven fabrics. Front Physiol 8:107. https://doi.org/10.3389/fphys.2017.00107
120. Johnson GA, Tramaglini DM, Levine RE, Ohno K, Choi NY, Woo SL (1994) Tensile and viscoelastic properties of human patellar tendon. J Orthop Res 12(6):796–803. https://doi.org/10.1002/jor.1100120607
121. Joodaki H, Panzer MB (2018) Skin mechanical properties and modeling: a review. Proc Inst Mech Eng H 232(4):323–343. https://doi.org/10.1177/0954411918759801
122. Kagan HM, Li W (2003) Lysyl oxidase: properties, specificity, and biological roles inside and outside of the cell. J Cell Biochem 88(4):660–672. https://doi.org/10.1002/jcb.10413
123. Kang AH, Gross J (1970) Relationship between the <i>Intra</i> and <i>inter</i> molecular Cross-links of Collagen. Proc Natl Acad Sci 67(3):1307–1314. https://doi.org/10.1073/pnas.67.3.1307
124. Karunaratne A, Li S, Bull AMJ (2018) Nano-scale mechanisms explain the stiffening and strengthening of ligament tissue with increasing strain rate. Sci Rep 8(1):3707. https://doi.org/10.1038/s41598-018-21786-z
125. Kastelic J, Galeski A, Baer E (1978) The multicomposite structure of tendon. Connect Tissue Res 6(1):11–23. https://doi.org/10.3109/03008207809152283
126. Kaufmann C, Bachmann LM, Robert YC, Thiel MA (2006) Ocular pulse amplitude in healthy subjects as measured by dynamic contour tonometry. Arch Ophthalmol 124(8):1104–1108. https://doi.org/10.1001/archopht.124.8.1104
127. Kempson GE, Freeman MAR, Swanson SAV (1968) Tensile properties of articular cartilage. Nature 220(5172):1127–1128. https://doi.org/10.1038/2201127b0
128. Ker RF, Wang XT, Pike AV (2000) Fatigue quality of mammalian tendons. J Exp Biol 203 (Pt 8):1317–1327. https://doi.org/10.1242/jeb.203.8.1317
129. Kirkness MWH, Forde NR (2018) Single-molecule assay for proteolytic susceptibility: force-induced collagen destabilization. Biophys J 114(3):570–576. https://doi.org/10.1016/j.bpj.2017.12.006
130. Kirkness MW, Lehmann K, Forde NR (2019) Mechanics and structural stability of the collagen triple helix. Curr Opin Chem Biol 53:98–105. https://doi.org/10.1016/j.cbpa.2019.08.001
131. Kivirikko KI, Myllyharju J (1998) Prolyl 4-hydroxylases and their protein disulfide isomerase subunit. Matrix Biol 16(7):357–368. https://doi.org/10.1016/s0945-053x(98)90009-9
132. Kivirikko KI, Myllyla R (1982) Posttranslational enzymes in the biosynthesis of collagen: intracellular enzymes. Methods Enzymol 82 Pt A:245–304. https://doi.org/10.1016/0076-6879(82)82067-3
133. Kivirikko KI, Pihlajaniemi T (1998) Collagen hydroxylases and the protein disulfide isomerase subunit of prolyl 4-hydroxylases. Adv Enzymol Relat Areas Mol Biol 72:325–398. https://doi.org/10.1002/9780470123188.ch9
134. Kivirikko KI, Helaakoski T, Tasanen K, Vuori K, Myllyla R, Parkkonen T, Pihlajaniemi T (1990) Molecular biology of prolyl 4-hydroxylase. Ann N Y Acad Sci 580:132–142. https://doi.org/10.1111/j.1749-6632.1990.tb17925.x

135. Kjaer M (2004) Role of extracellular matrix in adaptation of tendon and skeletal muscle to mechanical loading. Physiol Rev 84(2):649–698. https://doi.org/10.1152/physrev.00031.2003
136. Knott L, Bailey AJ (1998) Collagen cross-links in mineralizing tissues: a review of their chemistry, function, and clinical relevance. Bone 22(3):181–187. https://doi.org/10.1016/s8756-3282(97)00279-2
137. Knott L, Whitehead CC, Fleming RH, Bailey AJ (1995) Biochemical changes in the collagenous matrix of osteoporotic avian bone. Biochem J 310(Pt 3):1045–1051. https://doi.org/10.1042/bj3101045
138. Knox Cartwright NE, Tyrer JR, Marshall J (2011) Age-related differences in the elasticity of the human cornea. Invest Ophthalmol Vis Sci 52(7):4324–4329. https://doi.org/10.1167/iovs.09-4798
139. Lake SP, Miller KS, Elliott DM, Soslowsky LJ (2009) Effect of fiber distribution and realignment on the nonlinear and inhomogeneous mechanical properties of human supraspinatus tendon under longitudinal tensile loading. J Orthop Res 27(12):1596–1602. https://doi.org/10.1002/jor.20938
140. Launey ME, Buehler MJ, Ritchie RO (2010) On the mechanistic origins of toughness in bone. Annu Rev Mater Res 40:25–53
141. Lennox FG (1949) Shrinkage of collagen. Biochim Biophys Acta 3:170–187
142. Lesperance LM, Gray ML, Burstein D (1992) Determination of fixed charge density in cartilage using nuclear magnetic resonance. J Orthop Res 10(1):1–13. https://doi.org/10.1002/jor.1100100102
143. Liu SH, Yang RS, al-Shaikh R, Lane JM (1995) Collagen in tendon, ligament, and bone healing. A current review. Clin Orthop Relat Res 318:265–278
144. Liu X, Wu H, Byrne M, Krane S, Jaenisch R (1997) Type III collagen is crucial for collagen I fibrillogenesis and for normal cardiovascular development. Proc Natl Acad Sci USA 94(5):1852–1856. https://doi.org/10.1073/pnas.94.5.1852
145. Liu M, Lu F, Feng J (2024) Aging and homeostasis of the hypodermis in the age-related deterioration of skin function. Cell Death Dis 15(6):443. https://doi.org/10.1038/s41419-024-06818-z
146. Lotz JC, Hsieh AH (2014) The effects of mechanical forces on nucleus pulposus and annulus fibrosus cells. In: Springer, Vienna, pp 109–124. https://doi.org/10.1007/978-3-7091-1535-0_7
147. Lynch B, Pageon H, Le Blay H, Brizion S, Bastien P, Bornschlogl T, Domanov Y (2022) A mechanistic view on the aging human skin through ex vivo layer-by-layer analysis of mechanics and microstructure of facial and mammary dermis. Sci Rep 12(1):849. https://doi.org/10.1038/s41598-022-04767-1
148. Maganaris CN (2002) Tensile properties of in vivo human tendinous tissue. J Biomech 35(8):1019–1027. https://doi.org/10.1016/s0021-9290(02)00047-7
149. Maganaris CN, Paul JP (1999) In vivo human tendon mechanical properties. J Physiol 521 Pt 1 (Pt 1):307–313. https://doi.org/10.1111/j.1469-7793.1999.00307.x
150. Manilay Z, Novitskaya E, Sadovnikov E, McKittrick J (2013) A comparative study of young and mature bovine cortical bone. Acta Biomater 9(2):5280–5288. https://doi.org/10.1016/j.actbio.2012.08.040
151. Mansfield JC, Mandalia V, Toms A, Winlove CP, Brasselet S (2019) Collagen reorganization in cartilage under strain probed by polarization sensitive second harmonic generation microscopy. J R Soc Interface 16(150):20180611. https://doi.org/10.1098/rsif.2018.0611
152. Mansour JM (2003) Biomechanics of cartilage. Kinesiol Mech Pathomech Hum Mov 2:66–75
153. Markes AR, Hodax JD, Ma CB (2020) Meniscus form and function. Clin Sports Med 39(1):1–12. https://doi.org/10.1016/j.csm.2019.08.007
154. Maroudas A (1975) Biophysical chemistry of cartilaginous tissues with special reference to solute and fluid transport. Biorheology 12(3–4):233–248. https://doi.org/10.3233/bir-1975-123-416

155. Maroudas A (1976) Balance between swelling pressure and collagen tension in normal and degenerate cartilage. Nature 260(5554):808–809. https://doi.org/10.1038/260808a0
156. Martin RB, Boardman DL (1993) The effects of collagen fiber orientation, porosity, density, and mineralization on bovine cortical bone bending properties. J Biomech 26(9):1047–1054. https://doi.org/10.1016/s0021-9290(05)80004-1
157. Martin RB, Lau ST, Mathews PV, Gibson VA, Stover SM (1996) Collagen fiber organization is related to mechanical properties and remodeling in equine bone. A comparison of two methods. J Biomech 29(12):1515–1521
158. Martin RB, Burr DB, Sharkey NA, Fyhrie DP (2015) Mechanical properties of ligament and tendon. In: Springer, New York, pp 175–225. https://doi.org/10.1007/978-1-4939-3002-9_4
159. Maurice DM (1984) Chapter 1. The cornea and sclera. In: Davson H (ed) The eye, 3rd edn. Academic Press, Orlando, Florida, pp 1–158
160. Mecham RP (2001) Overview of extracellular matrix. Curr Protoc Cell Biol Chapter 10:Unit 10 11. https://doi.org/10.1002/0471143030.cb1001s00
161. Meek KM (2009) Corneal collagen—its role in maintaining corneal shape and transparency. Biophys Rev 1(2):83–93. https://doi.org/10.1007/s12551-009-0011-x
162. Meek KM, Leonard DW (1993) Ultrastructure of the corneal stroma: a comparative study. Biophys J 64(1):273–280. https://doi.org/10.1016/S0006-3495(93)81364-X
163. Meek KM, Knupp C, Lewis PN, Morgan SR, Hayes S (2025) Structural control of corneal transparency, refractive power and dynamics. Eye (Lond) 39(4):644–650. https://doi.org/10.1038/s41433-024-02969-7
164. Meyer M (2019) Processing of collagen based biomaterials and the resulting materials properties. Biomed Eng Online 18(1). https://doi.org/10.1186/s12938-019-0647-0
165. Mienaltowski MJ, Gonzales NL, Beall JM, Pechanec MY (2021) Basic structure, physiology, and biochemistry of connective tissues and extracellular matrix collagens. In: Springer International Publishing, pp 5–43. https://doi.org/10.1007/978-3-030-80614-9_2
166. Miller A, Wray JS (1971) Molecular packing in collagen. Nature 230(5294):437–439. https://doi.org/10.1038/230437a0
167. Misof K, Landis WJ, Klaushofer K, Fratzl P (1997a) Collagen from the osteogenesis imperfecta mouse model (oim) shows reduced resistance against tensile stress. J Clin Invest 100(1):40–45. https://doi.org/10.1172/jci119519
168. Misof K, Rapp G, Fratzl P (1997b) A new molecular model for collagen elasticity based on synchrotron X-ray scattering evidence. Biophys J 72(3):1376–1381. https://doi.org/10.1016/S0006-3495(97)78783-6
169. Mitton RG (1945) Mechanical properties of leather fibers. J Int Soc Leath Trades' Chem 29(1): 169–194
170. Mlyniec A, Dabrowska S, Heljak M, Weglarz WP, Wojcik K, Ekiert-Radecka M, Obuchowicz R, Swieszkowski W (2021) The dispersion of viscoelastic properties of fascicle bundles within the tendon results from the presence of interfascicular matrix and flow of body fluids. Mater Sci Eng C Mater Biol Appl 130:112435. https://doi.org/10.1016/j.msec.2021.112435
171. Monnier VM, Kohn RR, Cerami A (1984) Accelerated age-related browning of human collagen in diabetes mellitus. Proc Natl Acad Sci 81(2):583–587. https://doi.org/10.1073/pnas.81.2.583
172. Moro L, Smith BD (1977) Identification of collagen alpha1(I) trimer and normal type I collagen in a polyoma virus-induced mouse tumor. Arch Biochem Biophys 182(1):33–41. https://doi.org/10.1016/0003-9861(77)90280-6
173. Mostafavi Yazdi SJ, Baqersad J (2022) Mechanical modeling and characterization of human skin: a review. J Biomech 130:110864. https://doi.org/10.1016/j.jbiomech.2021.110864
174. Motooka D, Kawahara K, Nakamura S, Doi M, Nishi Y, Nishiuchi Y, Kang YK, Nakazawa T, Uchiyama S, Yoshida T, Ohkubo T, Kobayashi Y (2012) The triple helical structure and

stability of collagen model peptide with 4(S)-hydroxyprolyl-Pro-Gly units. Biopolymers 98(2):111–121. https://doi.org/10.1002/bip.21730

175. Myllyharju J (2003) Prolyl 4-hydroxylases, the key enzymes of collagen biosynthesis. Matrix Biol 22(1):15–24. https://doi.org/10.1016/s0945-053x(03)00006-4
176. Myllyharju J, Kivirikko KI (2001) Collagens and collagen-related diseases. Ann Med 33(1): 7–21. https://doi.org/10.3109/07853890109002055
177. Myllyharju J, Kivirikko KI (2004) Collagens, modifying enzymes and their mutations in humans, flies and worms. Trends Genet 20(1):33–43. https://doi.org/10.1016/j.tig.2003.11.004
178. Nabeshima Y, Grood ES, Sakurai A, Herman JH (1996) Uniaxial tension inhibits tendon collagen degradation by collagenase in vitro. J Orthop Res 14(1):123–130. https://doi.org/10.1002/jor.1100140120
179. Nachemson AL, Evans JH (1968) Some mechanical properties of the third human lumbar interlaminar ligament (ligamentum flavum). J Biomech 1(3):211–220. https://doi.org/10.1016/0021-9290(68)90006-7
180. Nair AK, Gautieri A, Chang SW, Buehler MJ (2013) Molecular mechanics of mineralized collagen fibrils in bone. Nat Commun 4:1724. https://doi.org/10.1038/ncomms2720
181. Nakayasu K, Tanaka M, Konomi H, Hayashi T (1986) Distribution of types I, II, III, IV and V collagen in normal and keratoconus corneas. Ophthalmic Res 18(1):1–10. https://doi.org/10.1159/000265406
182. Nalla RK, Kruzic JJ, Ritchie RO (2004) On the origin of the toughness of mineralized tissue: microcracking or crack bridging? Bone 34(5):790–798. https://doi.org/10.1016/j.bone.2004.02.001
183. Nestler FH, Hvidt S, Ferry JD, Veis A (1983) Flexibility of collagen determined from dilute solution viscoelastic measurements. Biopolymers 22(7):1747–1758. https://doi.org/10.1002/bip.360220710
184. Ni Annaidh A, Bruyere K, Destrade M, Gilchrist MD, Maurini C, Ottenio M, Saccomandi G (2012) Automated estimation of collagen fibre dispersion in the dermis and its contribution to the anisotropic behaviour of skin. Ann Biomed Eng 40(8):1666–1678. https://doi.org/10.1007/s10439-012-0542-3
185. Nishimori Y, Edwards C, Pearse A, Matsumoto K, Kawai M, Marks R (2001) Degenerative alterations of dermal collagen fiber bundles in photodamaged human skin and UV-irradiated hairless mouse skin: possible effect on decreasing skin mechanical properties and appearance of wrinkles. J Invest Dermatol 117(6):1458–1463. https://doi.org/10.1038/jid.2001.2
186. Niyibizi C, Eyre DR (1989) Identification of the cartilage alpha 1(XI) chain in type V collagen from bovine bone. FEBS Lett 242(2):314–318. https://doi.org/10.1016/0014-5793(89)80492-2
187. Nuyttens BP, Thijs T, Deckmyn H, Broos K (2011) Platelet adhesion to collagen. Thromb Res 127(Suppl 2):S26–S29. https://doi.org/10.1016/S0049-3848(10)70151-1
188. Oba A, Edwards C (2006) Relationships between changes in mechanical properties of the skin, wrinkling, and destruction of dermal collagen fiber bundles caused by photoaging. Skin Res Technol 12(4):283–288. https://doi.org/10.1111/j.0909-752X.2006.00154.x
189. O'Brien M (1997) Structure and metabolism of tendons. Scand J Med Sci Sports 7(2):55–61. https://doi.org/10.1111/j.1600-0838.1997.tb00119.x
190. O'Reilly MS, Boehm T, Shing Y, Fukai N, Vasios G, Lane WS, Flynn E, Birkhead JR, Olsen BR, Folkman J (1997) Endostatin: an endogenous inhibitor of angiogenesis and tumor growth. Cell 88(2):277–285. https://doi.org/10.1016/s0092-8674(00)81848-6
191. Orgel JPRO, Irving TC, Miller A, Wess TJ (2006) Microfibrillar structure of type I collagen <i>in situ</i>. Proc Natl Acad Sci 103(24):9001–9005. https://doi.org/10.1073/pnas.0502718103
192. Parry DA, Barnes GR, Craig AS (1978a) A comparison of the size distribution of collagen fibrils in connective tissues as a function of age and a possible relation between fibril size

distribution and mechanical properties. Proc R Soc Lond B Biol Sci 203(1152):305–321. https://doi.org/10.1098/rspb.1978.0107
193. Parry DA, Craig AS, Barnes GR (1978b) Tendon and ligament from the horse: an ultrastructural study of collagen fibrils and elastic fibres as a function of age. Proc R Soc Lond B Biol Sci 203(1152):293–303. https://doi.org/10.1098/rspb.1978.0106
194. Passini FS, Jaeger PK, Saab AS, Hanlon S, Chittim NA, Arlt MJ, Ferrari KD, Haenni D, Caprara S, Bollhalder M, Niederost B, Horvath AN, Gotschi T, Ma S, Passini-Tall B, Fucentese SF, Blache U, Silvan U, Weber B, Silbernagel KG, Snedeker JG (2021) Shear-stress sensing by PIEZO1 regulates tendon stiffness in rodents and influences jumping performance in humans. Nat Biomed Eng 5(12):1457–1471. https://doi.org/10.1038/s41551-021-00716-x
195. Paten JA, Siadat SM, Susilo ME, Ismail EN, Stoner JL, Rothstein JP, Ruberti JW (2016) Flow-induced crystallization of collagen: a potentially critical mechanism in early tissue formation. ACS Nano 10(5):5027–5040. https://doi.org/10.1021/acsnano.5b07756
196. Pawlaczyk M, Lelonkiewicz M, Wieczorowski M (2013) Age-dependent biomechanical properties of the skin. Postepy Dermatol Alergol 30(5):302–306. https://doi.org/10.5114/pdia.2013.38359
197. Pedaprolu K, Szczesny SE (2023) Mouse Achilles tendons exhibit collagen disorganization but minimal collagen denaturation during cyclic loading to failure. J Biomech 151:111545. https://doi.org/10.1016/j.jbiomech.2023.111545
198. Peng XQ, Guo Z, Moran W (2006) An anisotropic hyperelastic constitutive model with fiber-matrix shear interaction for the human annulus fibrosus. J Appl Mech 73(5):815–824
199. Pinnell SR, Martin GR (1968) The cross-linking of collagen and elastin: enzymatic conversion of lysine in peptide linkage to alpha-aminoadipic-delta-semialdehyde (allysine) by an extract from bone. Proc Natl Acad Sci 61(2):708–716. https://doi.org/10.1073/pnas.61.2.708
200. Pokidysheva E, Zientek KD, Ishikawa Y, Mizuno K, Vranka JA, Montgomery NT, Keene DR, Kawaguchi T, Okuyama K, Bachinger HP (2013) Posttranslational modifications in type I collagen from different tissues extracted from wild type and prolyl 3-hydroxylase 1 null mice. J Biol Chem 288(34):24742–24752. https://doi.org/10.1074/jbc.M113.464156
201. Pollock CM, Shadwick RE (1994) Relationship between body mass and biomechanical properties of limb tendons in adult mammals. Am J Phys 266(3 Pt 2):R1016–R1021. https://doi.org/10.1152/ajpregu.1994.266.3.R1016
202. Prockop DJ, Tuderman L (1982) Posttranslational enzymes in the biosynthesis of collagen: extracellular enzymes. In: Cunningham LW, Frederickson DW (eds) Methods in enzymology: structural and contractile proteins part A: extracellular matrix, vol 82. Academic Press, pp 3–913
203. Prockop DJ, Sieron AL, Li S-W (1998) Procollagen N-proteinase and procollagen C-proteinase. Two unusual metalloproteinases that are essential for procollagen processing probably have important roles in development and cell signaling. Matrix Biol 16(7):399–408. https://doi.org/10.1016/s0945-053x(98)90013-0
204. Puxkandl R, Zizak I, Paris O, Keckes J, Tesch W, Bernstorff S, Purslow P, Fratzl P (2002) Viscoelastic properties of collagen: synchrotron radiation investigations and structural model. Philos Trans R Soc Lond B Biol Sci 357(1418):191–197. https://doi.org/10.1098/rstb.2001.1033
205. Qiao H, Gao C, Lu C, Liu H, Zhang Y, Jin A, Dai Q, Yang S, Zhang B, Liu Y (2024) A novel method for fabricating the undulating structures at dermal-epidermal junction by composite molding process. J Funct Biomater 15(4). https://doi.org/10.3390/jfb15040102
206. Radner W, Zehetmayer M, Aufreiter R, Mallinger R (1998) Interlacing and cross-angle distribution of collagen lamellae in the human cornea. Cornea 17(5):537–543. https://doi.org/10.1097/00003226-199809000-00012
207. Ramachandran GN, Kartha G (1955) Structure of collagen. Nature 176(4482):593–595. https://doi.org/10.1038/176593a0
208. Ramachandran GN, Reddi AH (eds) (1976) Biochemistry of collagen. Plenum, New York

209. Raoux C, Chessel A, Mahou P, Latour G, Schanne-Klein MC (2023) Unveiling the lamellar structure of the human cornea over its full thickness using polarization-resolved SHG microscopy. Light Sci Appl 12(1):190. https://doi.org/10.1038/s41377-023-01224-0
210. Rappu P, Salo AM, Myllyharju J, Heino J (2019) Role of prolyl hydroxylation in the molecular interactions of collagens. Essays Biochem 63(3):325–335. https://doi.org/10.1042/ebc20180053
211. Reiser K, Stoller P, Knoesen A (2017) Three-dimensional geometry of collagenous tissues by second harmonic polarimetry. Sci Rep 7(1):2642. https://doi.org/10.1038/s41598-017-02326-7
212. Reisinger MA (2017) New evidence about composite bows and their arrows in inner Asia. Silk Road J 8:42–62
213. Renstrom P, Johnson RJ (1990) Anatomy and biomechanics of the menisci. Clin Sports Med 9(3):523–538
214. Reznikov N, Bilton M, Lari L, Stevens MM, Kroger R (2018) Fractal-like hierarchical organization of bone begins at the nanoscale. Science 360(6388). https://doi.org/10.1126/science.aao2189
215. Rho JY, Ashman RB, Turner CH (1993) Young's modulus of trabecular and cortical bone material: ultrasonic and microtensile measurements. J Biomech 26(2):111–119. https://doi.org/10.1016/0021-9290(93)90042-d
216. Rho JY, Kuhn-Spearing L, Zioupos P (1998) Mechanical properties and the hierarchical structure of bone. Med Eng Phys 20(2):92–102. https://doi.org/10.1016/s1350-4533(98)00007-1
217. Ribeiro JF, dos Anjos EH, Mello ML, de Campos Vidal B (2013) Skin collagen fiber molecular order: a pattern of distributional fiber orientation as assessed by optical anisotropy and image analysis. PLoS One 8(1):e54724. https://doi.org/10.1371/journal.pone.0054724
218. Ricard-Blum S, Ruggiero F, Van Der Rest M (2005) The collagen superfamily. In: Springer, Berlin Heidelberg, pp 35–84. https://doi.org/10.1007/b103819
219. Rich A, Crick FHC (1955) The structure of collagen. Nature 176(4489):915–916. https://doi.org/10.1038/176915a0
220. Riley GP, Harrall RL, Constant CR, Chard MD, Cawston TE, Hazleman BL (1994) Tendon degeneration and chronic shoulder pain: changes in the collagen composition of the human rotator cuff tendons in rotator cuff tendinitis. Ann Rheum Dis 53(6):359–366. https://doi.org/10.1136/ard.53.6.359
221. Ristaniemi A, Stenroth L, Mikkonen S, Korhonen RK (2018) Comparison of elastic, viscoelastic and failure tensile material properties of knee ligaments and patellar tendon. J Biomech 79:31–38. https://doi.org/10.1016/j.jbiomech.2018.07.031
222. Roeder BA, Kokini K, Sturgis JE, Robinson JP, Voytik-Harbin SL (2002) Tensile mechanical properties of three-dimensional type I collagen extracellular matrices with varied microstructure. J Biomech Eng 124(2):214–222. https://doi.org/10.1115/1.1449904
223. Roldan E, Reeves ND, Cooper G, Andrews K (2017) In vivo mechanical behaviour of the anterior cruciate ligament: a study of six daily and high impact activities. Gait Posture 58:201–207. https://doi.org/10.1016/j.gaitpost.2017.07.123
224. Rosenbloom J, Harsch M, Jimenez S (1973) Hydroxyproline content determines the denaturation temperature of chick tendon collagen. Arch Biochem Biophys 158(2):478–484. https://doi.org/10.1016/0003-9861(73)90539-0
225. Ruberti JW, Hallab NJ (2005) Strain-controlled enzymatic cleavage of collagen in loaded matrix. Biochem Biophys Res Commun 336(2):483–489. https://doi.org/10.1016/j.bbrc.2005.08.128
226. Rumian AP, Wallace AL, Birch HL (2007) Tendons and ligaments are anatomically distinct but overlap in molecular and morphological features--a comparative study in an ovine model. J Orthop Res 25(4):458–464. https://doi.org/10.1002/jor.20218

227. Sasaki N, Odajima S (1996a) Elongation mechanism of collagen fibrils and force-strain relations of tendon at each level of structural hierarchy. J Biomech 29(9):1131–1136. https://doi.org/10.1016/0021-9290(96)00024-3
228. Sasaki N, Odajima S (1996b) Stress-strain curve and Young's modulus of a collagen molecule as determined by the X-ray diffraction technique. J Biomech 29(5):655–658. https://doi.org/10.1016/0021-9290(95)00110-7
229. Schechtman H, Bader DL (1997) In vitro fatigue of human tendons. J Biomech 30(8):829–835. https://doi.org/10.1016/s0021-9290(97)00033-x
230. Schmitt FO, Hall CO, Jakus MA (1942) Electron microscope investigations of the structure of collagen. J Cell Comp Physiol 20(1):11–33
231. Screen HRC (2008) Investigating load relaxation mechanics in tendon. J Mech Behav Biomed Mater 1(1):51–58. https://doi.org/10.1016/j.jmbbm.2007.03.002
232. Sharma P, Maffulli N (2006) Biology of tendon injury: healing, modeling and remodeling. J Musculoskelet Neuronal Interact 6(2):181–190
233. Shaw SR, Vailas AC, Grindeland RE, Zernicke RF (1988) Effects of a 1-wk spaceflight on morphological and mechanical properties of growing bone. Am J Phys 254(1 Pt 2):R78–R83. https://doi.org/10.1152/ajpregu.1988.254.1.R78
234. Shen ZL, Kahn H, Ballarini R, Eppell SJ (2011) Viscoelastic properties of isolated collagen fibrils. Biophys J 100(12):3008–3015. https://doi.org/10.1016/j.bpj.2011.04.052
235. Shen Z, Cao Y, Li M, Yan Y, Cheng R, Zhao Y, Shao Q, Wang J, Sang S (2021) Construction of tissue-engineered skin with rete ridges using co-network hydrogels of gelatin methacrylated and poly(ethylene glycol) diacrylate. Mater Sci Eng C Mater Biol Appl 129:112360. https://doi.org/10.1016/j.msec.2021.112360
236. Shergold OA, Fleck NA, Radford D (2006) The uniaxial stress versus strain response of pig skin and silicone rubber at low and high strain rates. Int J Impact Eng 32(9):1384–1402. https://doi.org/10.1016/j.ijimpeng.2004.11.010
237. Sherman VR, Yang W, Meyers MA (2015) The materials science of collagen. J Mech Behav Biomed Mater 52:22–50. https://doi.org/10.1016/j.jmbbm.2015.05.023
238. Shoulders MD, Raines RT (2009a) Collagen structure and stability. Annu Rev Biochem 78(1):929–958. https://doi.org/10.1146/annurev.biochem.77.032207.120833
239. Shoulders MD, Raines RT (2009b) Modulating collagen triple-helix stability with 4-Chloro, 4-Fluoro, and 4-Methylprolines. In: Springer, New York, pp 251–252. https://doi.org/10.1007/978-0-387-73657-0_115
240. Siadat SM (2020) On the mechanobiology of collagen growth and remodelling. Ph.D., Northeastern University, Boston
241. Siadat SM, Ruberti JW (2023) Mechanochemistry of collagen. Acta Biomater 163:50–62. https://doi.org/10.1016/j.actbio.2023.01.025
242. Siadat SM, Silverman AA, DiMarzio CA, Ruberti JW (2021a) Measuring collagen fibril diameter with differential interference contrast microscopy. J Struct Biol 213(1):107697. https://doi.org/10.1016/j.jsb.2021.107697
243. Siadat SM, Zamboulis DE, Thorpe CT, Ruberti JW, Connizzo BK (2021b) Tendon extracellular matrix assembly, maintenance and dysregulation throughout life. Adv Exp Med Biol 1348:45–103. https://doi.org/10.1007/978-3-030-80614-9_3
244. Siegel RC (1979) Lysyl oxidase. Int Rev Connect Tissue Res 8:73–118. https://doi.org/10.1016/b978-0-12-363708-6.50009-6
245. Silverman AA, Olszewski JD, Siadat SM, Ruberti JW (2024) Tension in the ranks: cooperative cell contractions drive force-dependent collagen assembly in human fibroblast culture. Matter 7(4):1533–1557. https://doi.org/10.1016/j.matt.2024.01.023
246. Sirvent A, Espie K, Papadopoulou E, Naim D, Roche S (2022) New functions of DDR1 collagen receptor in tumor dormancy, immune exclusion and therapeutic resistance. Front Oncol 12. https://doi.org/10.3389/fonc.2022.956926

247. Sloseris D, Forde NR (2025) AGEing of collagen: the effects of glycation on collagen's stability, mechanics and assembly. Matrix Biol 135:153–160. https://doi.org/10.1016/j.matbio.2024.12.007
248. Smeets K, Slane J, Scheys L, Claes S, Bellemans J (2017) Mechanical analysis of extra-articular knee ligaments. Part One: Native knee ligaments. Knee 24(5):949–956. https://doi.org/10.1016/j.knee.2017.07.013
249. Smith-Mungo LI, Kagan HM (1998) Lysyl oxidase: properties, regulation and multiple functions in biology. Matrix Biol 16(7):387–398. https://doi.org/10.1016/s0945-053x(98)90012-9
250. Spengler DM, Morey ER, Carter DR, Turner RT, Baylink DJ (1983) Effects of spaceflight on structural and material strength of growing bone. Proc Soc Exp Biol Med 174(2):224–228. https://doi.org/10.3181/00379727-174-41729
251. Staubli HU, Schatzmann L, Brunner P, Rincon L, Nolte LP (1999) Mechanical tensile properties of the quadriceps tendon and patellar ligament in young adults. Am J Sports Med 27(1):27–34. https://doi.org/10.1177/03635465990270011301
252. Stupakov GP, Volozhin AI, Korzhen'iants VA, Korolev VV, Iagodovskii VS (1979) Change in the properties of the femoral bones of rats due to exarticulation of the crus and hypokinesia. Kosm Biol Aviakosm Med 13(1):35–40
253. Suji G, Sivakami S (2004) Glucose, glycation and aging. Biogerontology 5(6):365–373. https://doi.org/10.1007/s10522-004-3189-0
254. Sun YL, Luo ZP, Fertala A, An KN (2002) Direct quantification of the flexibility of type I collagen monomer. Biochem Biophys Res Commun 295(2):382–386. https://doi.org/10.1016/s0006-291x(02)00685-x
255. Sun YL, Luo ZP, Fertala A, An KN (2004) Stretching type II collagen with optical tweezers. J Biomech 37(11):1665–1669. https://doi.org/10.1016/j.jbiomech.2004.02.028
256. Sun M, Zafrullah N, Devaux F, Hemmavanh C, Adams S, Ziebarth NM, Koch M, Birk DE, Espana EM (2020) Collagen XII is a regulator of corneal stroma structure and function. Invest Ophthalmol Vis Sci 61(5):61. https://doi.org/10.1167/iovs.61.5.61
257. Svensson RB, Mulder H, Kovanen V, Magnusson SP (2013) Fracture mechanics of collagen fibrils: influence of natural cross-links. Biophys J 104(11):2476–2484. https://doi.org/10.1016/j.bpj.2013.04.033
258. Svensson RB, Herchenhan A, Starborg T, Larsen M, Kadler KE, Qvortrup K, Magnusson SP (2017) Evidence of structurally continuous collagen fibrils in tendons. Acta Biomater 50:293–301. https://doi.org/10.1016/j.actbio.2017.01.006
259. Szczesny SE, Elliott DM (2014) Interfibrillar shear stress is the loading mechanism of collagen fibrils in tendon. Acta Biomater 10(6):2582–2590. https://doi.org/10.1016/j.actbio.2014.01.032
260. Szczesny SE, Aeppli C, David A, Mauck RL (2018) Fatigue loading of tendon results in collagen kinking and denaturation but does not change local tissue mechanics. J Biomech 71:251–256. https://doi.org/10.1016/j.jbiomech.2018.02.014
261. Takahara K, Hoffman GG, Greenspan DS (1995) Complete structural organization of the human alpha 1 (V) collagen gene (COL5A1): divergence from the conserved organization of other characterized fibrillar collagen genes. Genomics 29(3):588–597. https://doi.org/10.1006/geno.1995.9961
262. Tang SS, Trackman PC, Kagan HM (1983) Reaction of aortic lysyl oxidase with beta-aminopropionitrile. J Biol Chem 258(7):4331–4338
263. Tanzer ML (1973) Cross-linking of collagen. Science 180(4086):561–566. https://doi.org/10.1126/science.180.4086.561
264. Tanzer ML, Gross J (1964) Collagen metabolism in the normal and lathyritic chick. J Exp Med 119(2):275–289
265. Thorpe CT, Stark RJ, Goodship AE, Birch HL (2010) Mechanical properties of the equine superficial digital flexor tendon relate to specific collagen cross-link levels. Equine Vet J Suppl 38:538–543. https://doi.org/10.1111/j.2042-3306.2010.00175.x

266. Thorpe CT, Udeze CP, Birch HL, Clegg PD, Screen HR (2012) Specialization of tendon mechanical properties results from interfascicular differences. J R Soc Interface 9(76): 3108–3117. https://doi.org/10.1098/rsif.2012.0362
267. Thorpe CT, Birch HL, Clegg PD, Screen HRC (2015) Chapter 1—Tendon physiology and mechanical behavior: Structure–function relationships. In: Gomes ME, Reis R Rodrigues MT (ed) Tendon regeneration. Academics, Boston, pp. 3–39
268. Traub W, Piez KA (1971) The chemistry and structure of collagen. Adv Protein Chem 25:243–352. https://doi.org/10.1016/s0065-3233(08)60281-8
269. Tresoldi I, Oliva F, Benvenuto M, Fantini M, Masuelli L, Bei R, Modesti A (2013) Tendon's ultrastructure. Muscles Ligaments Tendons J 3(1):2–6. https://doi.org/10.11138/mltj/2013.3.1.002
270. Tzaphlidou M (2004) The role of collagen and elastin in aged skin: an image processing approach. Micron 35(3):173–177. https://doi.org/10.1016/j.micron.2003.11.003
271. Uchegbulam I, Danby SG, Lewis R, Carre MJ, Maiti R (2022) Effect of seasonal change on the biomechanical and physical properties of the human skin. J Mech Behav Biomed Mater 127: 105058. https://doi.org/10.1016/j.jmbbm.2021.105058
272. van der Rijt JA, van der Werf KO, Bennink ML, Dijkstra PJ, Feijen J (2006) Micromechanical testing of individual collagen fibrils. Macromol Biosci 6(9):697–702. https://doi.org/10.1002/mabi.200600063
273. Veres SP, Lee JM (2012) Designed to fail: a novel mode of collagen fibril disruption and its relevance to tissue toughness. Biophys J 102(12):2876–2884. https://doi.org/10.1016/j.bpj.2012.05.022
274. Viidik A, Ekholm R (1968) Light and electron microscopic studies of collagen fibers under strain. Zeitschrift fur Anatomie und Entwicklungsgeschichte 127:154–164
275. Wang Y, Azais T, Robin M, Vallee A, Catania C, Legriel P, Pehau-Arnaudet G, Babonneau F, Giraud-Guille MM, Nassif N (2012) The predominant role of collagen in the nucleation, growth, structure and orientation of bone apatite. Nat Mater 11(8):724–733. https://doi.org/10.1038/nmat3362
276. Wang JH, Kato Y, Ingham SJ, Maeyama A, Linde-Rosen M, Smolinski P, Fu FH, Harner C (2014) Effects of knee flexion angle and loading conditions on the end-to-end distance of the posterior cruciate ligament: a comparison of the roles of the anterolateral and posteromedial bundles. Am J Sports Med 42(12):2972–2978. https://doi.org/10.1177/0363546514552182
277. Weis MA, Hudson DM, Kim L, Scott M, Wu J-J, Eyre DR (2010) Location of 3-hydroxyproline residues in collagen types I, II, III, and V/XI implies a role in fibril supramolecular assembly. J Biol Chem 285(4):2580–2590. https://doi.org/10.1074/jbc.m109.068726
278. Whittaker P, Canham PB (1991) Demonstration of quantitative fabric analysis of tendon collagen using two-dimensional polarized light microscopy. Matrix 11(1):56–62. https://doi.org/10.1016/s0934-8832(11)80227-1
279. Whittaker CA, Bergeron KF, Whittle J, Brandhorst BP, Burke RD, Hynes RO (2006) The echinoderm adhesome. Dev Biol 300(1):252–266. https://doi.org/10.1016/j.ydbio.2006.07.044
280. Wong R, Geyer S, Weninger W, Guimberteau JC, Wong JK (2016) The dynamic anatomy and patterning of skin. Exp Dermatol 25(2):92–98. https://doi.org/10.1111/exd.12832
281. Woo SL, Hollis JM, Adams DJ, Lyon RM, Takai S (1991) Tensile properties of the human femur-anterior cruciate ligament-tibia complex. The effects of specimen age and orientation. Am J Sports Med 19(3):217–225. https://doi.org/10.1177/036354659101900303
282. Wren TA, Lindsey DP, Beaupre GS, Carter DR (2003) Effects of creep and cyclic loading on the mechanical properties and failure of human Achilles tendons. Ann Biomed Eng 31(6): 710–717. https://doi.org/10.1114/1.1569267
283. Wu JJ, Weis MA, Kim LS, Carter BG, Eyre DR (2009) Differences in chain usage and cross-linking specificities of cartilage type V/XI collagen isoforms with age and tissue. J Biol Chem 284(9):5539–5545. https://doi.org/10.1074/jbc.M806369200

284. Wynn TA, Ramalingam TR (2012) Mechanisms of fibrosis: therapeutic translation for fibrotic disease. Nat Med 18(7):1028–1040. https://doi.org/10.1038/nm.2807
285. Yamauchi M, Shiiba M (2008) Lysine hydroxylation and cross-linking of collagen. In: Methods in molecular biology. Humana Press, pp 95–108. https://doi.org/10.1007/978-1-60327-084-7_7
286. Yamauchi M, Sricholpech M (2012) Lysine post-translational modifications of collagen. Essays Biochem 52:113–133. https://doi.org/10.1042/bse0520113
287. Yeganegi A, Whitehead K, de Castro Bras LE, Richardson WJ (2023) Mechanical strain modulates extracellular matrix degradation and byproducts in an isoform-specific manner. Biochim Biophys Acta Gen Subj 1867(3):130286. https://doi.org/10.1016/j.bbagen.2022.130286
288. Younesi FS, Miller AE, Barker TH, Rossi FMV, Hinz B (2024) Fibroblast and myofibroblast activation in normal tissue repair and fibrosis. Nat Rev Mol Cell Biol 25(8):617–638. https://doi.org/10.1038/s41580-024-00716-0
289. Yurchenco PD (2011) Basement membranes: cell scaffoldings and signaling platforms. Cold Spring Harb Perspect Biol 3(2):a004911. https://doi.org/10.1101/cshperspect.a004911
290. Zahouani H, Pailler-Mattei C, Sohm B, Vargiolu R, Cenizo V, Debret R (2009) Characterization of the mechanical properties of a dermal equivalent compared with human skin in vivo by indentation and static friction tests. Skin Res Technol 15(1):68–76. https://doi.org/10.1111/j.1600-0846.2008.00329.x
291. Zareian R, Church KP, Saeidi N, Flynn BP, Beale JW, Ruberti JW (2010) Probing collagen/enzyme mechanochemistry in native tissue with dynamic, enzyme-induced creep. Langmuir 26(12):9917–9926. https://doi.org/10.1021/la100384e
292. Zitnay JL, Jung GS, Lin AH, Qin Z, Li Y, Yu SM, Buehler MJ, Weiss JA (2020) Accumulation of collagen molecular unfolding is the mechanism of cyclic fatigue damage and failure in collagenous tissues. Sci Adv 6(35):eaba2795. https://doi.org/10.1126/sciadv.aba2795
293. Zitnay JL, Lin AH, Weiss JA (2021) Tendons exhibit greater resistance to tissue and molecular-level damage with increasing strain rate during cyclic fatigue. Acta Biomater 134:435–442. https://doi.org/10.1016/j.actbio.2021.07.045

Chapter 11
Collagen in Fibrotic Diseases

Andrzej Fertala

Abstract This chapter explores the role of fibrillar collagens, mainly collagen I, in developing fibrotic disorders associated with acute or chronic injuries. While collagen molecules' fundamental structure, composition, and intracellular biosynthesis steps remain similar in healthy and scar tissues, their extracellular architecture and physical properties significantly differ. These differences arise from the excessive production of collagen I and auxiliary proteins associated with collagen I folding and posttranslational modifications. As a result, the overaccumulation of collagen I-based fibrotic deposits creates a rigid mechanical environment that, through mechanotransduction, amplifies pro-fibrotic signaling in resident fibroblasts.

In reviewing the literature, this chapter highlights key players that create, transmit, and sustain these signals, thereby perpetuating fibrosis. Given the growing recognition of mechanotransduction as a valid therapeutic target to limit fibrosis, this chapter also discusses strategies to inhibit different elements of this process. A significant challenge with these strategies is that both balanced and excessive scarring rely on the exact underlying mechanisms of scar tissue formation. Consequently, conventional anti-fibrotic agents may inadvertently impair the essential scarring needed to preserve tissue integrity after injury. Therefore, mechanotherapeutics that reduce collagen accumulation-driven scar stiffness represent a novel approach for developing more targeted anti-fibrotic therapies.

Keywords Collagen · Collagen fibril · Fibrosis · Scar · Mechanotransduction · Procollagen · Myofibroblasts · Extracellular matrix · Matrix stiffness

---

A. Fertala (✉)
Department of Orthopaedic Surgery, Sidney Kimmel Medical College, Thomas Jefferson University, Curtis Building, Philadelphia, PA, USA
e-mail: andrzej.fertala@jefferson.edu

D. A. D. Parry, P. K. Luther (eds.), *Fibrous Proteins: Structures and Mechanisms*, Subcellular Biochemistry 113, https://doi.org/10.1007/978-3-032-05273-5_11

Introduction

A balance between anabolic and catabolic processes maintains the homeostasis of cellular and extracellular components across various tissues and organs. This homeostasis is disrupted when injuries occur, triggering a wound-healing response essential for preserving vital tissue functions.

While some extant anamniotes can regenerate damaged tissues and organs to their pre-injury state, this ability is largely absent in most amniotes, including humans. Most vertebrates can restore specific tissues, such as the epidermis, cochlear hair cells, intestinal epithelium, blood, and olfactory neurons. However, the regeneration of complex organs composed of multiple integrated tissues generally does not occur in amniotes. Instead, wound healing typically forms scar neotissue, which maintains the morphological integrity of the injury site but lacks the original tissue's critical biological, structural, and mechanical properties. Consequently, our tissue repair system, which relies on scar formation, represents a compromise between the inability to regenerate damaged tissues and the necessity of sustaining tissue function. Although this compromise is effective in addressing acute or systemic injuries, inadequate regulation of scar formation can lead to life-threatening fibrotic disorders [3].

Collagen-Based Scar: A Universal Material for Supporting Injured Sites

Acute and chronic injuries trigger inflammation, orchestrating the initial stages of scar formation. While a detailed discussion of inflammation's role in wound healing is beyond the scope of this chapter, a few critical players are worth mentioning. For instance, mast cells and macrophages contribute to tissue healing by producing essential growth factors, including transforming growth factor-β1 (TGF-β1), platelet-derived growth factor (PDGF), vascular endothelial growth factor (VEGF), interleukin (IL)-6, IL-13, and others [15, 125, 196]. These factors stimulate sedentary resident and migratory fibroblasts, promoting their proliferation and motility. Moreover, they activate fibroblasts to produce key scar tissue components, primarily collagenous proteins. In many cases, activated fibroblasts differentiate into myofibroblasts, characterized by increased contractility and an upregulated production of scar components (Fig. 11.1) [58, 172].

Regardless of the injury type or the anatomical location of the wounded tissue, the primary component of scar tissue is a dense, collagen-based fibrillar matrix. For example, assays measuring collagen-specific hydroxyproline have shown that collagen constitutes about 80% of the dry scar mass isolated from arthrofibrotic knee capsules [159]. Similarly, scars formed in the skin, liver, kidney, heart, and many other organs exhibit high collagen content [95, 119, 176, 198]. Still, other

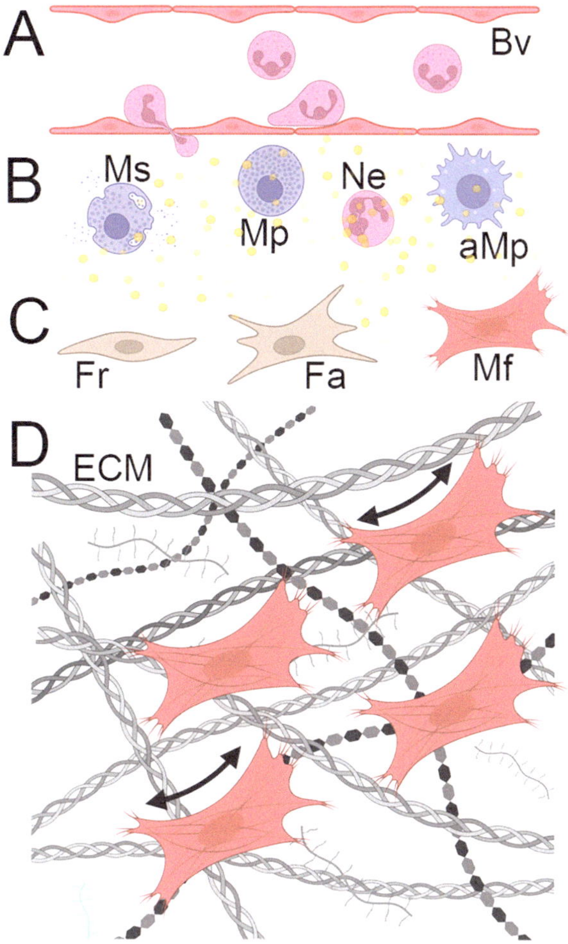

Fig. 11.1 Inflammation and fibroblast activation processes that drive balanced wound healing and fibrotic scarring. Blood-derived (**a**) and tissue-resident (**b**) inflammatory cells migrate to injury sites. Growth factors (**b**) secreted by these cells act on resting fibroblasts (**c**, Fr) and activate them (Fa). Subsequently, the activated fibroblasts differentiate into myofibroblasts (Mf). Myofibroblasts contract the wound site (**d**, arrows) by pulling the elements of the extracellular matrix (ECM). Symbols: Bv blood vessel, Ms mast cells, Mp macrophages, aMp activated macrophages, Ne neutrophils. (Created with BioRender.com)

macromolecules, including fibronectin (FN) and proteoglycans (PGs), also contribute to the scar mass, albeit to a lesser extent [185].

Interestingly, collagen I-based scars also form in tissues where this collagen type and its fibroblast producers are not prevalent under healthy conditions. For instance, in healthy cartilage, a crucial scaffold comprising collagen II, collagen IX, collagen XI, and proteoglycans produced by chondrocytes resists compression forces in articular joints. However, cartilage defects heal through the formation of fibrocartilage, where collagen I is the dominant component [5]. Similarly, while healthy kidneys contain significant amounts of collagen IV and cellular components such as podocytes, pericytes, smooth muscle cells, and fibroblasts, injury-induced scar deposits primarily consist of collagen I [16].

These examples illustrate that, despite variations in the cellular and molecular composition of different healthy tissues, nature has developed a universal repair mechanism reliant on collagen I-based scar formation. Although these scars lack the biological and mechanical properties of uninjured native tissues, they provide a supportive patch that restores the matrix void at injury sites. While this repair mechanism does not restore native tissue structure, its near-universal applicability negates the need for diverse, tissue-specific repair processes. Moreover, the relatively rapid formation of collagen I-based scars enables subsequent, often long-term, tissue remodeling, during which scar tissue evolves into a more functional structure.

A significant downside of scar formation is its potential to trigger fibrosis, a process characterized by excessive scar-tissue production.

Fibrotic Scarring

The severity of fibrotic diseases ranges from mild to fatal. For example, posttraumatic fibrosis of musculoskeletal tissues, including joint capsules, ligaments, tendons, and muscles, can cause pain and restrict the range of motion. Excessive scarring of peripheral nerves hampers their regenerative capacity and induces pain by forming fibrotic adhesions that impair nerve gliding during body movement [186].

Fibrotic scarring is also a significant issue in ocular tissues. Corneal scarring, for instance, affects transparency and leads to vision blurring. Similarly, subretinal fibrosis, fibrosis associated with glaucoma surgery, and fibrotic scarring following cataract surgery can pose serious risks, potentially resulting in blindness [105]. In another example, surgeries in the abdomen frequently lead to extensive fibrotic adhesions, which may cause more severe complications than the original medical condition requiring surgical intervention [18].

While localized excessive scarring imposes significant medical burdens on patients and their families, systemic fibrosis affecting internal organs is often life-threatening. In idiopathic pulmonary fibrosis (IPF), for example, collagen-rich fibrotic deposits accumulate around the alveoli, severely impairing gas exchange and causing respiratory distress. Many IPF patients require supplemental oxygen, and a significant number ultimately require lung transplants. Unfortunately, even with modern treatment options, IPF has a poor prognosis, with a life expectancy of approximately 2–5 years [100]. Similarly, fibrotic scarring in the kidneys, liver, and heart leads to life-threatening complications. As in IPF, treatment options to halt or reverse fibrosis in these organs are limited, and patient prognoses remain poor [101].

Fibrotic diseases may vary in anatomy and cellular and molecular composition of fibrotic scars, yet their hallmark across different organs is the accumulation of collagen-rich scar deposits. These scars disrupt native tissue function, ultimately leading to organ failure (Fig. 11.2).

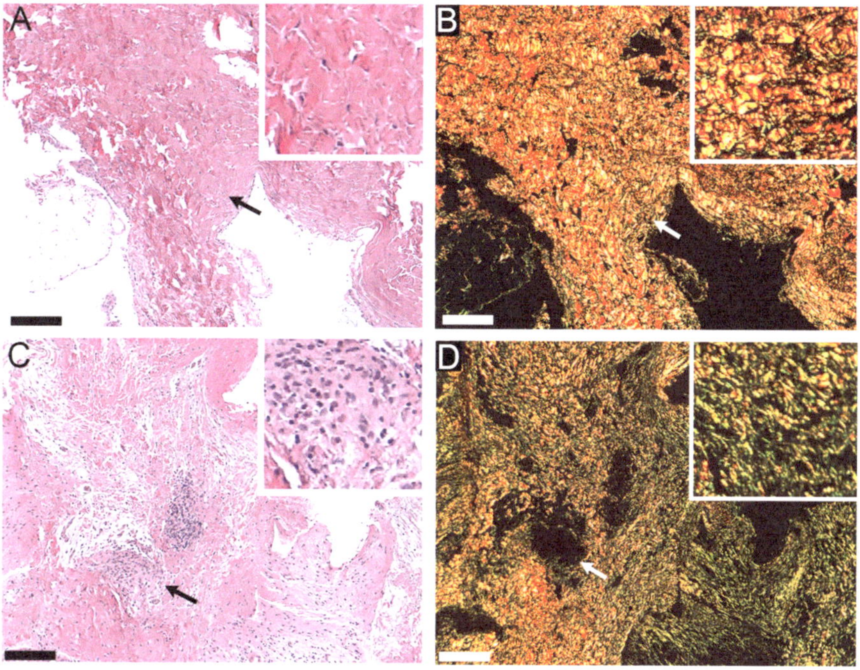

Fig. 11.2 Histology of posterior joint capsules from a healthy (**a**, **b**) and fibrotic (**c**, **d**) rabbit's knees. Hematoxylin and eosin staining patterns indicate sparse cellularity in the healthy (**a**) posterior capsule. In contrast, similar staining shows an invasion of the inflammatory cells into the injured capsule (**c**). Inserts in panels A and C show more detailed views of the areas indicated by the arrows. Picrosirius red staining of the collagenous matrix (**b**, **d**) observed in a polarized-light microscope reveals increased green birefringence in the injured capsule (**d**, insert) in comparison to the healthy counterpart (**c**, insert). As described by Steplewski et al., this observation suggests changes in the organization and composition of scar neotissue formed in response to knee injury [159]. Scale bars = 500 μm

Balanced Versus Excessive Scar Formation

As noted above, scar formation is a natural response to injury, enabling damaged tissues to preserve vital functions. In a balanced wound-healing response, subsequent processes remodel the rudimentary scar patch formed during the initial rapid scarring stage. Scar remodeling involves matrix metalloproteinases (MMPs), which degrade scar components, allowing tissue rearrangement into more functional structures. Fibroblasts, through their matrix-specific receptors, sense mechanical cues from scar tissue and facilitate remodeling. Once remodeling is complete, activated fibroblasts revert to quiescence or undergo apoptosis, ultimately ceasing the scar formation process [27]. In contrast, excessive production of collagen-rich matrices and dysregulated remodeling define fibrotic scarring. These pathological processes accumulate scar tissue, causing the failure of affected tissues and organs.

The complexity and redundancy of these mechanisms make developing anti-fibrotic therapies highly challenging. Because regular and fibrotic healing rely on similar scar-forming processes, identifying valid anti-fibrotic targets that block pathological fibrosis without impairing proper tissue healing remains difficult.

Stiff Matrix: A Valid Target for Limiting Fibrosis

Recent studies strongly suggest that fibrotic, rigid scar tissue plays a pro-fibrotic role. Their findings have demonstrated that the composition and stiffness of fibrotic deposits provide potent signals that drive fibroblasts into a pro-fibrotic state [68, 152]. Consequently, targeting the stiff extracellular matrix (ECM) provides a new avenue for developing effective anti-fibrotic therapies [175].

The following sections focus on the collagen I-based component of scar tissue, outlining key steps in forming the stiff ECM and discussing its pro-fibrotic functions.

Intracellular Procollagen I Production Steps

Hydroxylation of Proline and Lysine Residues

Procollagen I, the precursor of collagen I, is a heterotrimer composed of two pro-α1 (I) chains and one pro-α2(I) chain. Like other fibrillar collagens, procollagen I chains contain a large central collagenous domain formed by uninterrupted Gly-X-Y triplet repeats, where proline frequently occupies the -X- and -Y- positions. Short telopeptides and relatively large propeptides flank this central domain (Fig. 11.3).

During biosynthesis, nascent procollagen chains undergo various posttranslational modifications (Fig. 11.4). Proline and lysine residues at the -Y- positions of Gly-X-Y triplets are hydroxylated by specific enzymes. Prolyl-4-hydroxylase (P4H) catalyzes proline hydroxylation, while the lysyl hydroxylase family of enzymes (LH1–3) hydroxylates lysine residues. While LH1 catalyzes the hydroxylation of lysine residues in the triple-helical domain, LH2 primarily catalyzes the hydroxylation of selected lysine residues within collagen telopeptides [171].

These enzymes require nascent procollagen chains as substrates and use Fe^{2+}, 2-oxoglutarate, molecular O_2, and ascorbate as cofactors [134]. P4H consists of two catalytic α subunits and two non-catalytic β subunits. The β subunit also functions as protein disulfide isomerase (PDI) and a chaperone in procollagen synthesis [133]. Hydroxylation of proline residues stabilizes the triple-helical structure of collagen molecules by forming hydrogen bonds that, along with water molecules, bind individual chains, keeping them together at physiological temperatures [14].

Prolyl-3-hydroxylase 1 (P3H1) catalyzes the formation of 3-hydroxyproline residues, targeting prolyl residues at the -X- position [40, 142]. While collagen IV

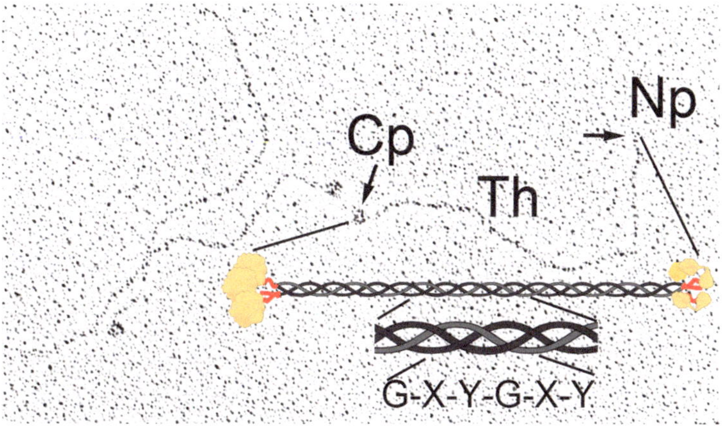

Fig. 11.3 An electron microscopy rotary shadowing image of procollagen I molecules produced by cultured fibroblasts. The three procollagen molecules visible in the image are 300 nm long. The C-terminal (Cp), N-terminal (Np) propeptides, and triple-helical domain (Th) are indicated. A procollagen model depicts crucial domains and highlights G-X-Y triplets where proline residues often occupy X and Y positions. Created, in part, with BioRender.com

and collagen V contain relatively high amounts of 3-hydroxyproline, in procollagen I, only a single proline residue at position 986 of the α1(I) chain is hydroxylated (position numbering considers the first glycine of the triple helix as "1") [72, 113, 142].

P3H1 forms a complex with cartilage-associated protein (CRTAP) and cyclophilin B (CyPB), an immunophilin family member. Another immunophilin, FKPB65, facilitates lysine hydroxylation in the telopeptide region of procollagen I by promoting LH2 dimerization [47, 171]. Hydroxylation of these residues is essential for intermolecular cross-linking of fibril-incorporated collagen I molecules.

Collagen Glycosylation

Beyond its role in fibril-stabilizing cross-links, hydroxylation of lysine residues within the triple-helical domain enables O-glycosylation of collagen chains (Fig. 11.4). Glycosylation facilitates procollagen secretion and assembly collagen molecules into mature fibrils [57].

The most abundant O-glycan in collagen is a disaccharide of glucose and galactose linked via hydroxylysine residues, Glc(α1-2)Gal(β1-O). This simple glycosylation pattern is highly conserved among animals expressing collagenous proteins [57].

COLGALT1 and COLGALT2 catalyze the β1-O glycosylation, while the glucosyltransferase responsible for α1-2 glycosylation remains less defined. However, research presented by Heikkinen et al. suggested that LH3 acts as a collagen glucosyltransferase [55].

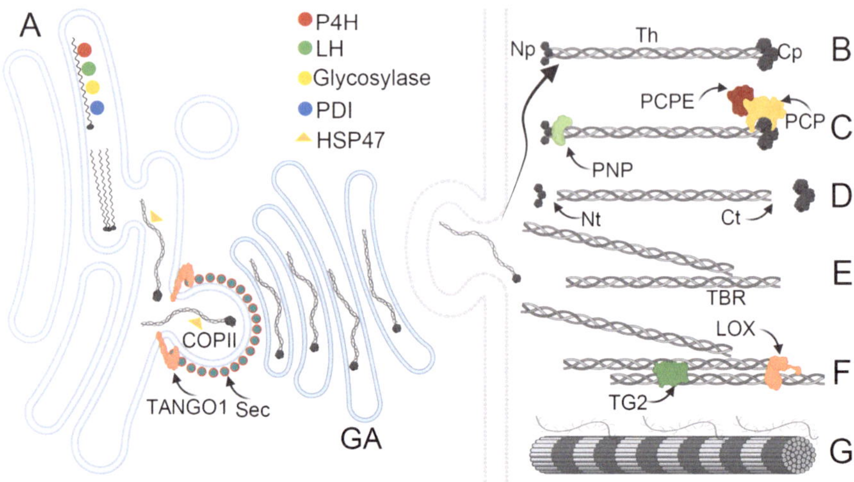

Fig. 11.4 A schematic depicting crucial intracellular (**a**) and extracellular (**b–g**) steps of procollagen production and collagen assembly into fibrils. (**a**) Procollagen chains synthesized in the endoplasmic reticulum (ER) undergo complex posttranslational modifications. Subsequently, they fold into triple-helical molecules and get secreted via the Golgi apparatus (GA). Coat protein complex II (COPII) vesicles that include Sec proteins, and the transport and Golgi organization 1 (TANGO1) protein family participate in the secretion process. (**b–d**) In the extracellular space, the procollagen propeptides are cleaved off by procollagen C proteinase (PCP) and procollagen N proteinase (PNP). PCP enhancer (PCPE) participates in the C propeptide cleavage. Cleavage of the propeptides exposes the N-terminal (Nt) and C-terminal (Ct) collagen telopeptides. (**e**) The C-terminal telopeptides of one collagen molecule bind to an interacting partner's telopeptide binding region (TBR), initiating the formation of collagen fibrils. F: Collagen fibrils undergo cross-linking in which lysyl oxidases (LOX) and transglutaminase 2 (TG2) play a crucial role. G: Mature collagen fibrils interact with collagenous and non-collagenous molecules to form complex biological alloys. Symbols: ER endoplasmic reticulum, GA Golgi apparatus, COPII coat protein complex II, Sec Sec proteins, TANGO1 transport and Golgi organization 1, P4H prolyl-4-hydroxylase, LH lysyl hydroxylase, PDI protein disulfide isomerase, HSP47 heat shock protein 47, Th collagen triple helix, Np, Cp N-terminal and C-terminal procollagen propeptides, PNP, PCP procollagen N and C proteinases, PCPE procollagen C proteinase enhancer, Nt, Ct collagen N terminal and C terminal telopeptides, TBR telopeptide-binding region, LOX lysyl oxidase, TG2 transglutaminase 2. (Created with BioRender.com)

Folding of Collagen Chains into the Triple-Helical Conformation

Appropriately modified procollagen chains assemble into a triple-helical structure in a zipper-like manner [31]. Site-specific interactions within the C-propeptide region initiate this assembly (Fig. 11.4) [86, 108].

Chaperone proteins transiently bind procollagen chains to stabilize them and prevent non-specific aggregation. These include heat shock protein 47 (HSP47), heat shock 70 kDa-related luminal binding protein (BiP), and PDI. Moreover, FKBP65 and CyPB also participate in collagen triple-helix formation [13, 60, 118, 157, 169, 184].

Secretion of Procollagen into Extracellular Space

Procollagen secretion into the extracellular space is challenging due to the elongated structure of this relatively large protein (Fig. 11.3). Coat protein complex II (COPII) that includes Sec proteins, and the transport and Golgi organization 1 (TANGO1) protein family facilitate the secretion process (Fig. 11.4) [6].

Extracellular Procollagen Processing and Fibril Assembly

Enzymatic Cleavage of Procollagen Propeptides

Cleavage of the N- and C-terminal propeptides is essential for fibril formation [69, 70]. A disintegrin and metalloprotease with thrombospondin motifs (ADAMTS)-2, known as procollagen N proteinase (PNP), removes the N propeptide [22]. Procollagen C proteinase (PCP), initially identified as bone morphogenetic protein-1 (BMP1), cleaves the C-terminal propeptide (Fig. 11.4) [74, 92].

Collagen Fibril Formation

Collagen fibrillogenesis is crucial for forming stable scaffolds of connective tissues and organs. During this process, thermolabile collagen molecules gain thermostability upon incorporation into fibrils [87, 109].

The precise fibril formation mechanism remains unclear. Studies in vitro suggest entropy-driven aggregation of individual collagen molecules in the extracellular space, but precise mechanisms governing their D-periodic staggering within formed fibrils remain unidentified.

A challenge in understanding fibrillogenesis stems partly from suboptimal models used to study this process. For instance, tissue-derived collagen material utilized in these models to reconstruct the fibrils often contains a mix of intact monomers, telopeptide-deficient molecules, and cross-linked oligomers. Although these molecules form D-staggered fibrils, their assembly characteristics and morphology are not native-like (Fig. 11.5).

A more relevant model utilizes a de novo fibril formation approach. Here, intact procollagen is processed by enzymes that remove propeptides while preserving telopeptides [21, 110]. This processing yields well-defined, not cross-linked, collagen monomers assembled into native-like fibrils (Fig. 11.5).

The significance of telopeptides is evident in the low critical concentration of about 0.42 μg/ml required for the polymerization of intact collagen I at 37 °C [69]. In contrast, collagen with enzymatically degraded telopeptides requires much higher concentrations to aggregate into fibrils [45].

Furthermore, intact telopeptides are crucial in forming fibrils with proper morphology. Numerous studies have demonstrated that intact collagen I molecules form

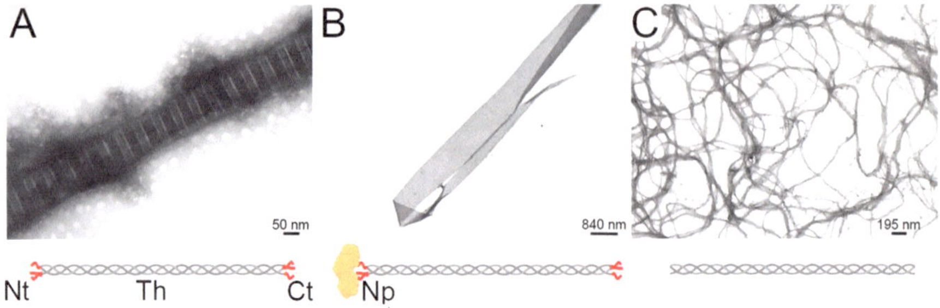

Fig. 11.5 Transmission electron microscopy analysis of assemblies formed by various collagen variants. (**a**) A collagen fibril formed by aggregation of collagen molecules with intact telopeptides. (**b**) pN-sheets formed by assembly of pN-collagen molecules in which procollagen C proteinase cleaved off the C propeptides. (**c**) Fibrillar structures formed by aggregation of collagen molecules in which propeptides and telopeptides were digested by pepsin. Symbols: Th collagen triple helix, Nt, Ct collagen N-terminal and C-terminal telopeptides, Np N-terminal propeptides. (Created, in part, with BioRender.com)

well-structured fibrils characterized by regular banding patterns observed using electron microscopy. In contrast, collagen I molecules in which telopeptides have been degraded with pepsin form aberrant fibrillar aggregates lacking distinct banding patterns (Fig. 11.5). Another example of a crucial role of telopeptides in proper fibrillogenesis is the formation of pN sheets from pN-collagen molecules generated by enzymatic cleavage of the C-terminal propeptide from procollagen I (Fig. 11.5) [59]. Although pN collagen molecules include intact telopeptide regions, the steric hindrance of their N telopeptides caused by bulky N propeptides prevents the formation of proper fibrils.

Early studies suggested collagen telopeptides drive fibrillogenesis via site-specific interactions [45]. Later, researchers identified a hydrophobic interaction between the α2(I) C-terminal telopeptide and a telopeptide-binding region (TBR) in the triple-helical domain of an interacting partner, crucial for the D-periodic arrangement of collagen molecules within the fibrils (Fig. 11.4) [36, 131, 132]. These interactions also facilitate the formation of fibril-stabilizing covalent cross-links between collagen molecules that form the fibrils.

Collagen Cross-Linking

Following collagen assembly into fibrils, non-covalent forces hold collagen molecules together in a transient aggregate state. The subsequent formation of covalent cross-links between collagen molecules stabilizes the fibrils, enhancing their mechanical strength.

The first step in forming collagen cross-links is the hydroxylation of selected lysine residues in the telopeptides and triple-helical domains by LH variants (Figs. 11.4 and 11.6).

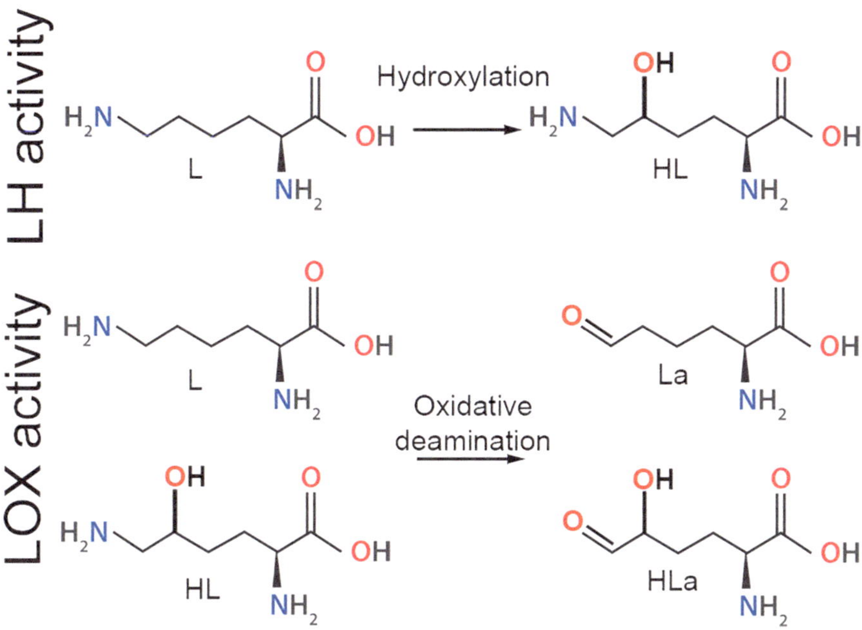

Fig. 11.6 A diagram depicting the enzymatic reactions catalyzed by lysyl hydroxylase (LH) and lysyl oxidase (LOX). Symbols: L lysine, HL hydroxylysine, La lysine aldehyde, HLa hydroxylysine aldehyde. (Created with BioRender.com)

Subsequently, lysyl oxidase (LOX) and LOX-like (LOXL) enzymes, including LOXL1, LOXL2, LOXL3, and LOXL4, participate in the cross-linking process (Fig. 11.6). These enzymes contain a highly conserved C-terminal catalytic region with a copper-binding site, while their more variable N-terminal regions determine tissue specificity [24]. LOX enzymes catalyze the oxidative deamination of lysine or hydroxylysine residues within the N- and C-terminal telopeptides, leading to the aldehydes allysine and hydroxyallysine formation (Fig. 11.6). These aldehydes then condense with lysine or hydroxylysine residues in adjacent collagen chains (Fig. 11.7).

If LH does not modify either interacting amino acid, the reaction yields lysine aldehyde with lysine, forming dehydro-lysino-norleucine (deH-LNL). If one or more reacting molecules are LH-modified, the resulting product contains one or two hydroxyl groups. Consequently, lysine aldehyde reacting with hydroxylysine forms dehydro-hydroxylysino-norleucine (deH-HLNL). When both interacting molecules are LH-hydroxylated, i.e., hydroxylysine aldehyde and hydroxylysine, the product is dehydro-dihydroxylysino-norleucine (deH-DHLNL) (Fig. 11.7).

Divalent cross-links such as deH-HLNL and deH-DHLNL are considered immature and may undergo further reactions to form mature trivalent cross-links. These maturation reactions produce pyrroles or pyridinolines through various lysine forms. Pyrroles form with lysine aldehydes: deH-DHLNL with a lysine aldehyde forms

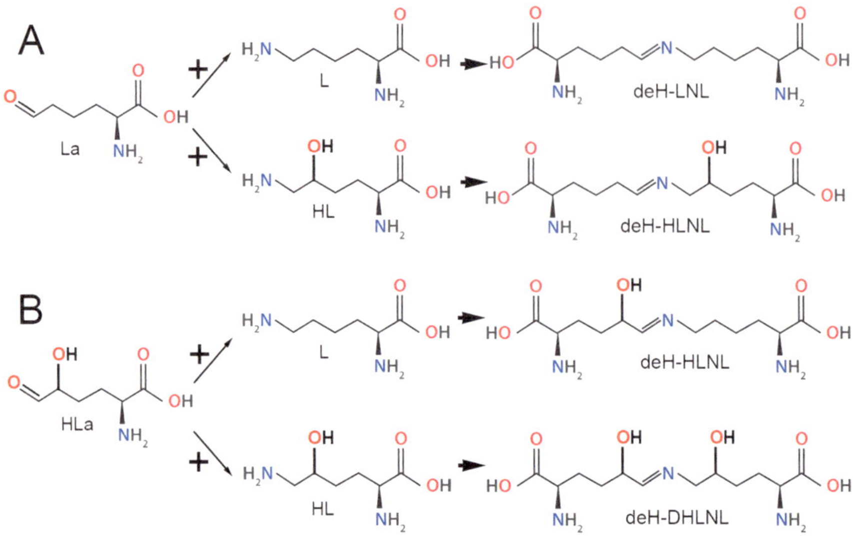

Fig. 11.7 The formation of immature collagen cross-links. Lysine (**a**) or hydroxylysine (**b**) aldehydes react with unmodified lysine or hydroxylysine residues to form immature divalent cross-links. Symbols: La lysine aldehyde, Hla hydroxylysine aldehyde, deH-LNL dehydro-lysino-norleucine, deH-HLNL dehydro-hydroxylysino-norleucine, deH-DHLNL dehydro-dihydroxylysino-norleucine. (Created with BioRender.com)

pyrrole (Prl), whereas deH-HLNL with a lysine aldehyde generates deoxy-pyrrole (DPrl). Likewise, deH-DHLNL reacting with a hydroxylysine aldehyde forms pyridinoline (Pyr), while deH-HLNL with a hydroxylysine aldehyde produces deoxypyridinoline (DPyr) (Fig. 11.8) [34, 77, 151, 189].

In addition to LOX-mediated cross-links, transglutaminase 2 (TG2) contributes to collagen fibril cross-linking by catalyzing covalent bonds between lysine and glutamine residues (Fig. 11.4) [23].

Alterations in Procollagen Modifications in Fibrosis

As noted above, fibrotic scarring leads to excessive ECM deposition, with collagenous proteins as the most abundant scar component. This increased collagen production necessitates the upregulation of genes encoding specific procollagen chains and auxiliary proteins essential for their modifications and transport (Fig. 11.4).

A central regulator of this upregulation process is transforming growth factor-beta 1 (TGF-β1), which acts through heteromeric complexes of transmembrane serine/threonine kinase receptors: type I (TGFβRI) and type II (TGFβRII). TGF-β1 binding to TGFβRII recruits and phosphorylates TGFβRI, transducing signals to intracellular SMAD2 and SMAD3. Once phosphorylated, these proteins heterodimerize with

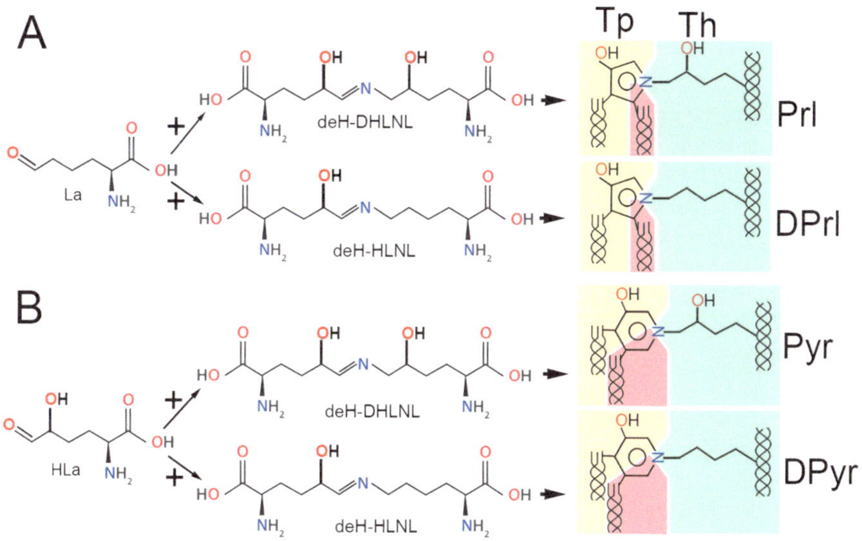

Fig. 11.8 Maturation of the divalent cross-links into trivalent cross-link variants. (**a**) Pyrroles form with lysine aldehydes (La): dehydro-dihydroxylysino-norleucine (deH-DHLNL) with a lysine aldehyde forms pyrrole (Prl), whereas dehydro-hydroxylysino-norleucine (deH-HLNL) with a lysine aldehyde (La) generates deoxy-pyrrole (DPrl). (**b**) Likewise, deH-DHLNL reacting with a hydroxylysine aldehyde (HLa) forms pyridinoline (Pyr), while deH-HLNL with a hydroxylysine aldehyde produces deoxypyridinoline (DPyr). Telopeptide (Tp) and collagen triple helix (Th) cross-linking sites are indicated. (Created with BioRender.com)

SMAD4 and translocate into the nucleus to regulate the transcription of target genes (Fig. 11.9).

Although SMAD2 and SMAD3 have overlapping regulatory functions, SMAD3 plays a particularly critical role in fibrotic processes. Studies have shown that SMAD3 binds to the CAGA box in many collagen promoters, activating them [20, 30, 182].

TGF-β also activates SMAD-independent pathways implicated in fibrosis, including mitogen-activated protein kinase (MAPK) pathways such as p38 and Jun N-terminal kinase (JNK), as well as Rho-like GTPase and phosphatidylinositol-3-kinase (PI3K/Akt) pathways [114, 193]. Additionally, other cytokines, particularly interleukin-4 (IL-4) and interleukin-13 (IL-13), promote collagen I expression [38, 66]. IL-13 further enhances macrophage TGF-β expression, amplifying the fibrotic response [39].

These collagen production pathways also upregulate key auxiliary proteins involved in posttranslational modifications, folding, and transport of collagen molecules. Rabbit-based studies have demonstrated significant upregulation of P4H, lysyl hydroxylase 1 (LH1), BiP, HSP47, and LOX in arthrofibrosis and peripheral nerve scarring [144, 159]. Furthermore, a TGF-β1-mediated increase in TANGO1 expression has been observed in fibrotic disorders, such as liver fibrosis and scleroderma [6].

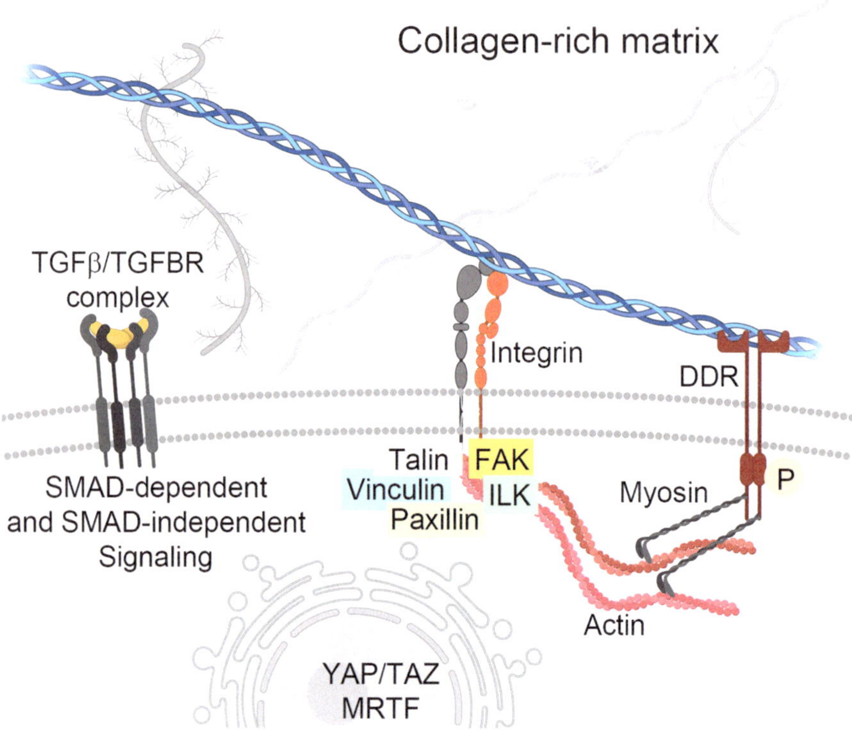

Fig. 11.9 A drawing demonstrating crucial extracellular matrix and cellular elements involved in mechanotransduction. Symbols: TGFβ/TGFBR transforming growth factor beta/transforming growth factor beta receptor, DDR discoidin domain receptor, FAK focal adhesion kinase, ILK integrin-linked kinase, YAP/TAZ yes-associated protein/transcriptional co-activator with PDZ-binding motif, MRTF myocardin-related transcription factor. (Created with BioRender.com)

Additionally, the expression of PNP, PCP, and PCP enhancer-1 (PCPE-1) is upregulated during the fibrotic processes (Fig. 11.4). Elevated production of these proteins facilitates the efficient processing of excess procollagen molecules, enabling their incorporation into fibrotic scar tissue [62, 117, 122].

Beyond increased collagen deposition in fibrosis, the cross-linking of collagen molecules within fibrils further stabilizes them, making them more resistant to proteolytic degradation. Cross-linked fibrils also stiffen fibrotic tissues, altering their mechanical properties and promoting pro-fibrotic mechanotransduction [29].

Notably, increased formation of pyridinoline, a trivalent cross-link, has been observed in many fibrotic disorders, including liver fibrosis and experimental arthrofibrosis (Fig. 11.8) [99, 161, 195]. Moreover, catalyzed by LH2, telopeptide hydroxylysine residues participate in pyridinoline formation. Upregulation of LH2 has been demonstrated in fibroblasts from patients with scleroderma and mice with liver fibrosis, underscoring this enzyme's role in fibrotic progression [178, 192].

Similarly, increased expression of LOX enzymes involved in collagen cross-linking has been observed in fibrotic disorders across multiple tissues and organs (for review, see [130]). Further, studies of fibrotic tissues have revealed changes in the relative abundance of various cross-link types, with deH-HLNL, deH-DHLNL, Pyr, and DPyr increasing in many fibrotic conditions (Figs. 11.7 and 11.8). These changes alter ECM structure, biomechanical properties, and susceptibility to MMP-dependent degradation [65, 80, 127, 179].

The composition of collagen cross-links varies by tissue. In skeletal tissues such as cartilage and bone, deH-HLNL, deH-DHLNL, Pyr, and DPyr predominate, with Pyr being particularly abundant in mineralized tissues and tendons. In contrast, lysine aldehyde-derived cross-links are more prevalent in the cornea and skin. Proper collagen cross-link composition is essential for normal tissue function, and restoring its balance is crucial for effective anti-fibrotic therapies [177, 180, 189].

Collagen Fibril Organization

Although collagen molecules can spontaneously self-assemble into fibrils, the distinct, tissue-specific 3D fibrillar architectures suggest that additional mechanisms contribute to their organization. One such mechanism involves fibroblastic cells applying small traction forces on the surrounding stroma. These forces deform ECM, break weak bonds, and promote the assembly of new fibrils. These processes, called tissue homeostasis, maintain ECM dynamic equilibrium, whose one consequence is tension homeostasis. The existence of this baseline tension is illustrated by an experiment where fibroblasts are seeded onto stress-free ECM. Soon after seeding, cell traction forces generate an internal stress of about 4 kPa, comparable to that measured at focal adhesions [106]. This stress is then maintained at various conditions by actively rearranging the cytoskeleton of the seeded cells and adjusting cell-cell and cell-ECM interactions [162].

Moreover, research indicates that FN plays a crucial role in matrix assembly, including the organization of collagen fibrils. First, soluble FN monomers must assemble into FN fibrils. However, they cannot do so without first being stretched by cells via FN-specific $\alpha5\beta1$ integrin. This stretching exposes cryptic sites on FN monomers, allowing them to interact and drive FN fibrillogenesis. Subsequently, FN fibrils serve as guiding paths for collagen fibril deposition, which, once cross-linked, define tissue stiffness [71, 88].

Another mechanism for organizing collagen fibrils has been described in studies of chick embryo tendons [17]. Researchers observed the formation of cell membrane protrusions, termed fibripositors, in which collagen fibrils are pre-positioned along the tendon axis. This actomyosin-driven process involves collagen-specific $\alpha2\beta1$ integrins, which pull and orient the fibrils to establish the tendon-specific matrix architecture.

However, the role of cells in collagen fibril organization is not always clear. For instance, Trelstad proposed that the chick cornea's lamellar structure assembly is purely physical. Supporting this hypothesis, he noted that the direction of lamellar rotation through the corneal depth is identical in both eyes. The absence of bilateral symmetry suggests that corneal fibroblasts may not actively participate in organizing collagen fibril-based lamellae [173]. Further supporting this conclusion, molecular crowding experiments successfully recreated the corneal lamellar structure without cellular components [146]. Additionally, the highly organized fibrillar structure of tendons, which contain only sparsely distributed tenocytes, suggests a limited role for these cells in tendon matrix organization.

Regardless of the specific mechanisms involved, proper collagen fibril organization, established during embryonic development, is maintained postnatally through a balanced process of matrix degradation and biosynthesis under physiological conditions. This equilibrium is disrupted during fibrosis, where excessive collagen-rich scar formation creates an abnormal mechanical environment that perpetuates scarring and ultimately leads to tissue dysfunction.

Pro-fibrotic Cellular Mediators of Mechanotransduction

Mechanical signals generated by the stiff ECM enhance the production of macromolecules that contribute to fibrotic scar formation. This mechanotransduction process is primarily mediated by integrins, which promote intracellular signaling by assembling focal adhesion complexes composed of talin, vinculin, focal adhesion kinase (FAK), and paxillin (Fig. 11.9) [29].

Discoidin domain receptors (DDRs) also play a role in pro-fibrotic mechanotransduction [49, 158]. Evidence suggests that DDR1 influences cell adhesion and movement through collagen-based matrices by interacting with non-muscle myosin IIA [61]. This enzyme, a hexamer with ATPase activity, links to the actin cytoskeleton, generating forces that shape cellular structure and promote motility (Fig. 11.9) [33, 46].

TGFβRI and TGFβRII are critical regulators of fibrosis, mediating ECM-cell interactions by binding to TGF-β ligands and activating SMAD-dependent signaling pathways (Fig. 11.9). Studies indicate that under low shear stress, TGF-β1-induced SMAD2 signaling is enhanced [1]. This mechanism promotes endothelial-to-mesenchymal transition, intensifying fibrotic responses in various tissues [104]. Additional mechanotransduction pathways regulate mechanosensitive ion channels, which respond to forces transmitted through extracellular and intracellular components [48, 190].

The nucleus also functions as a mechanosensor [76]. In one example, lamin-A, a key nuclear structural protein, exhibits increased expression in response to ECM stiffness caused by collagen deposition and cross-linking [165]. Yes-associated protein (YAP) and transcriptional co-activator with PDZ-binding motif (TAZ) also contribute to this process. Under low mechanical stress, YAP and TAZ remain in the cytoplasm, where they are targeted for degradation by proteasomes. Conversely, they evade degradation and translocate to the nucleus under high mechanical stress, promoting pro-fibrotic fibroblast functions such as proliferation, differentiation, apoptosis inhibition, and ECM production [50, 82, 97, 166]. In addition to YAP/TAZ, myocardin-related transcription factor (MRTF) also modulates fibroblast activity in response to ECM stiffness (Fig. 11.9) [153, 181].

Because stiffness-dependent mediators are pivotal in promoting pro-fibrotic cell activity, they became attractive anti-fibrotic targets. However, efforts to inhibit pro-fibrotic intracellular mediators driving fibrotic healing have had limited success [84]. Significant barriers to translating these discoveries into clinical therapies include the broad activity of potential inhibitors and their adverse side effects [185].

Experimental Mechanotherapeutic Aiming at Matrix Stiffness

Most mechanotherapeutic interventions have targeted downstream elements affected by ECM stiffness. Targeting upstream mechanisms, such as matrix-stiffening processes like collagen fibrillogenesis, may offer a novel approach to reducing fibrosis (Fig. 11.9).

Blocking Procollagen Processing

As previously outlined, the enzymatic removal of procollagen N-terminal and C-terminal propeptides is essential for proper collagen fibril assembly (Fig. 11.4). Retaining both propeptides prevents stable fibril formation while removing only one results in abnormal structures, such as tapes or sheets (Fig. 11.5) [59].

Thus, PNP and PCP have been recognized as potential anti-fibrotic targets. The rationale for this concept is that inhibiting these enzymes would prevent propeptide cleavage, not allowing fibril formation and, as a result, reducing excessive scarring (Fig. 11.4). Acidic dipeptide hydroxamate inhibitors targeting PCP have succeeded in vitro [124]. Although inhibitors for PNP have also been identified, their lack of specificity limits their effectiveness in selectively targeting N-propeptide cleavage [10].

Knockout studies in mice have revealed compensatory enzymes capable of processing procollagen propeptides following PCP or PNP gene deletion, leading to a decline in interest in these targets [93, 164]. Consequently, researchers have considered attenuating PCP activity by targeting PCPE-1 (Fig. 11.4), which is upregulated in several fibrotic conditions [83]. However, no targeted inhibitors for PCPE-1 have been developed, and concerns remain about potential side effects due to the role of the enhancer in angiogenesis, cell proliferation, and RNA stabilization [83].

Inhibiting Collagen Fibrillogenesis

An innovative strategy to limit collagen I-rich fibrotic buildup involves preventing collagen molecules from self-assembling into fibrils [21]. The author's group has developed a monoclonal antibody targeting the C-terminal telopeptide of collagen I, effectively blocking fibril formation (Fig. 11.4) [35, 36]. This anti-collagen antibody (ACA) has demonstrated anti-fibrotic effects in animal models of various fibrotic disorders [21, 37, 160, 161]. The ACA-based strategy thus presents a promising approach to controlling fibrosis by targeting the early stages of fibril formation, a critical step in scar development across tissues.

Other fibrillogenesis inhibitors, such as (±)-α-lipoic acid, trigonelline hydrochloride, oleuropein, capsaicin, soluble DDR, and fibromodulin, have shown in vitro efficacy by directly binding to collagen molecules [12, 41, 42, 129, 139, 140]. However, their anti-fibrotic potential in animal models remains unexplored and warrants further investigation.

Knuppel et al. also proposed that the anti-fibrotic mechanisms of nintedanib and pirfenidone, used to treat IPF, may include direct inhibition of collagen fibrillogenesis [78]. These FDA-approved drugs could serve as effective mechanotherapeutics if their inhibitory effects on collagen fibrillogenesis are confirmed.

Interfering with Collagen Cross-Linking

Scar tissue remodeling involves enzymatic degradation by MMPs and the synthesis of new ECM components. Ideally, this process would replace rigid scars with structures that better integrate with surrounding tissues and restore their function. However, remodeling mature, collagen-based, fibrotic matrices is challenging due to their inherent resistance to proteolysis and the dense covalent cross-links that reinforce them.

Moreover, cross-linking stiffens the ECM, promoting pro-fibrotic signaling [80]. Rigid matrices induce macrophages and fibroblasts to adopt pro-fibrotic phenotypes characterized by excessive ECM production and accumulation [155, 163]. Thus, targeting collagen cross-linking represents a promising strategy to soften scar tissue and reduce fibrosis.

Targeting LOX Activity

β-Aminopropionitrile (BAPN), which chelates copper ions essential for LOX activity, has been shown to reduce fibrosis in animal models [2, 111, 143]. Further research has identified LOXL2 as a key driver of fibrosis in various disorders, making it a prime therapeutic target [32].

A monoclonal antibody targeting LOXL2 significantly reduced fibrotic scarring in preclinical models [9]. Subsequently, Gilead Sciences developed simtuzumab, a humanized IgG_4 version of this antibody, for clinical trials to treat fibrosis. However, the trials were discontinued due to simtuzumab's lack of efficacy [11, 53, 54, 136]. This failure may have resulted from compensatory activities by other cross-linking enzymes, including additional LOX family members and TGs, which simtuzumab did not inhibit [135].

Inhibiting TG2

As noted earlier, TG2 catalyzes cross-links between lysine and glutamine residues in the collagen matrix, significantly contributing to fibrosis. Supporting this notion, TG2-deficient mice exhibited reduced bleomycin-induced pulmonary fibrosis due to decreased collagen accumulation [123]. Irreversible TG2 inhibitors have also been effective in reducing fibrosis in models of kidney fibrosis, nephrosclerosis, myocardial infarction, and peritoneal fibrosis [8, 67, 81, 187].

TG2 stiffens the ECM and enhances TGF-β1's pro-fibrotic effects by multimerizing latent TGF-β-binding protein 1 (LTBP1) [174]. While various competitive, reversible, and irreversible TG2 inhibitors have been identified, concerns remain regarding their specificity and potential side effects [52, 73]. For example, some inhibitors have been associated with hyperproliferation, parakeratosis, and inhibition of coagulation factor XIIIa [43, 52].

Researchers have also developed therapeutic anti-TG2 antibodies with well-defined binding epitopes to improve specificity. These antibodies have shown promise in reducing ECM accumulation in cell-based fibrosis models [102], suggesting the potential for safe and effective TG2-targeted therapies.

Focusing on Mediators of TGF-β1 Activity to Mitigate Fibrosis

Targeting Integrins

TGF-β homodimers are synthesized as pro-TGF-β precursors linked to latency-associated protein (LAP) (Fig. 11.10) [4, 85]. After intracellular cleavage by furins, mature TGF-β remains non-covalently bound to LAP, forming the small latent complex (SLC), which must be released to activate TGF-β receptors [28].

Activating TGF-β1 involves integrins that release latent forms of this factor from ECM depots and various mediators that modulate the mechanisms of its actions. Consequently, integrins containing the αv subunit, paired with β1, β3, β5, β6, or β8, play crucial roles in organ fibrosis, primarily by activating latent TGF-β family members [138].

In most cells, the SLC is secreted as part of the large latent complex (LLC) through its association with latent TGF-β-binding proteins (LTBPs) [26, 147]. Once

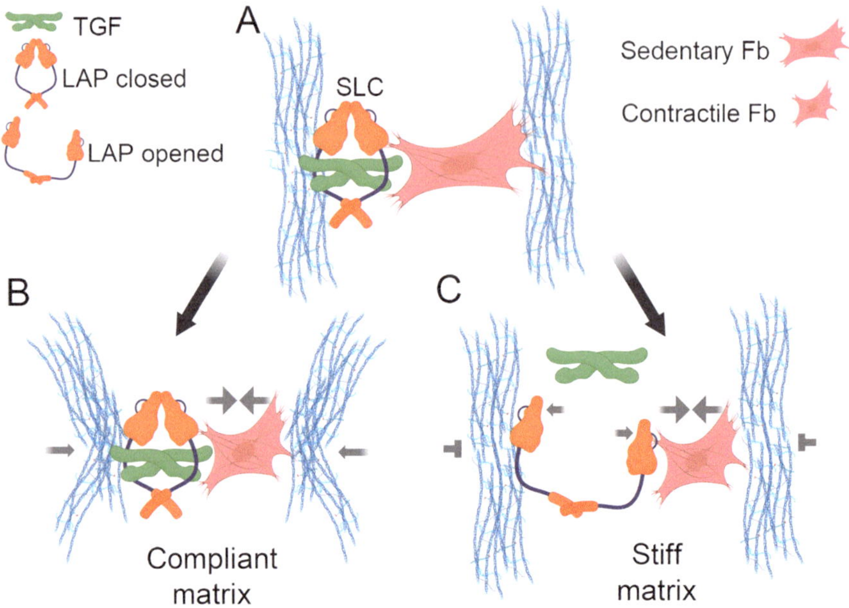

Fig. 11.10 Panel (**a**) presents potential mechanisms for transforming growth factor (TGF) release from the small latent complex (SLC), driven by forces generated by fibroblasts (Fb). Compared to a compliant matrix (**b**), a stiff matrix (**c**) enhances TGF release from its complex with latency-associated protein (LAP), facilitated by Fb-generated forces (illustrated by arrows). (Created with BioRender.com)

in the extracellular environment, LLC binds to ECM components, serving as a reservoir for latent TGF-β1. Consequently, TGF-β1 must be liberated from the ECM to exert biological activity (Fig. 11.10) [150, 167, 170].

TGF-β1 activation occurs via two primary pathways. The first involves enzymatic cleavage of the LLC by proprotein convertases (PCPs) and various MMPs [25, 44, 103]. Integrins containing the αv subunit facilitate this process by bringing LLC into proximity with proteolytic enzymes, thereby promoting TGF-β1 release. Additionally, these integrins optimize TGF-β signaling by positioning LLC near TGF-β1 receptors [149, 188].

The second pathway enables TGF-β release from LLC without enzymatic cleavage [115, 148]. In this mechanism, αv integrins bind to the LAP component of the LLC, which is covalently linked to the ECM (Fig. 11.10). In stiff ECM environments, cellular traction forces exerted through αv integrins deform LAP, liberating active TGF-β. Conversely, in a more compliant ECM, the lack of mechanical resistance prevents LAP deformation and subsequent active TGF-β release, reducing pro-fibrotic signaling modulated by this growth factor (Fig. 11.10).

Consequently, discovery of αv integrins' role in activating TGF-β1 has opened new avenues for therapeutic intervention. Studies involving CWHM12, a synthetic pan-inhibitor targeting all αv integrins, demonstrated reduced fibrosis in mouse models of liver, lung, heart, and skeletal muscle [56]. MK-0429, an αv integrin inhibitor, also effectively mitigated kidney fibrosis in rat models [197].

Mediators of TGF-β1 Activity as Potential Targets to Limit Fibrosis

Selective integrin inhibitors have also been explored. For instance, compound 8 (c8), a small molecule inhibitor of αvβ1 integrin, demonstrated efficacy in treating renal, hepatic, and pulmonary fibrosis in mice [19, 141]. Furthermore, cilengitide, a cyclic RGD peptide, exhibited anti-fibrotic effects by blocking αvβ3 and αvβ5 integrins in a mouse model of scleroderma. Similarly, an antibody targeting αvβ5 integrin reduced the pro-fibrotic activity of fibroblasts derived from localized scleroderma patients [7].

The αvβ6 integrin has also been a focal point for anti-fibrotic therapies. Patsenker et al. demonstrated that the αvβ6 antagonist EMD527040 attenuated biliary fibrosis in mice by decreasing TGF-β1 activation, suggesting its broader utility in combating biliary fibrosis [126].

In murine models, neutralizing antibodies against αvβ6 integrin reduced renal fibrosis in Alport syndrome mice [51]. This observation led to developing the therapeutic antibody STX-100 (BG00011), which underwent Phase 2 clinical trials to assess its efficacy in reducing kidney transplant fibrosis. However, these trials were halted due to undisclosed safety concerns [145].

STX-100 was also tested in clinical trials for IPF. Unfortunately, these studies revealed that patients receiving the antibody experienced worsening fibrosis compared to the placebo group, leading to the trial's termination [137].

Other integrin subtypes may contribute to fibrotic disease mechanisms and are thus considered potential anti-fibrotic targets (Fig. 11.9) [191]. Among integrins that bind to the RGD motif, including αIIbβ3, α5β1, and α8β1, the α8β1 integrin stands out as a particularly promising target. Blocking this integrin with neutralizing antibodies reduced liver and lung fibrosis in mouse models [120, 121, 191].

Although none of these integrin inhibitors have yet advanced to clinical trials, integrins remain a compelling therapeutic focus for mitigating TGF-β-mediated fibrotic processes [156].

Direct targeting of TGF-β1 has proven to be a practical anti-fibrotic approach; however, direct inhibition of TGF-β1 has been associated with significant adverse effects [128, 194]. Consequently, researchers have shifted their focus to indirectly suppressing TGF-β1 signaling by aiming at mediators of its activity. One such mediator, connective tissue growth factor (CTGF), has emerged as a promising candidate, with studies indicating that its inhibition reduces fibrotic responses in various diseases [63, 79]. Nevertheless, clinical trials in patients with IPF treated with the anti-CTGF antibody pamrevlumab failed to demonstrate a significant difference in forced vital capacity (FVC) between the antibody-treated and placebo groups, leading to the discontinuation of these trials [64]. These disappointing clinical results highlight the need for further research to identify more effective anti-fibrotic targets.

Blocking DDRs to Limit Fibrosis

DDR1 and DDR2 are collagen-specific receptors classified under receptor tyrosine kinases (RTKs) (Fig. 11.9) [89, 90]. Under normal physiological conditions, DDRs play essential roles in embryonic development, growth, wound repair, and tissue maintenance, and they are expressed across various tissues. DDR1 is found on epithelial cells in both normal and fibrotic tissues of the skin, liver, lung, and kidneys and is activated predominantly by collagen types I, II, III, IV, and V. DDR2 is primarily expressed in mesenchymal cells. It is activated by collagen types I, II, III, V, and X [91, 154, 183].

Evidence suggests that DDR expression is elevated in fibrotic diseases, including IPF [112, 168]. Moreover, deleting DDR1 reduced adipose tissue fibrosis in a mouse model of cardiometabolic disease, while DDR2 knockout diminished renal interstitial fibrosis [94, 96]. These findings highlight DDRs as contributors to pro-fibrotic processes and promising targets for anti-fibrotic therapies.

CQ-061, a DDR1 inhibitor, significantly decreased collagen and fibronectin accumulation in TGF-β1-stimulated human lung fibroblast cultures [168]. Additionally, broad-spectrum kinase inhibitors like imatinib and dasatinib were studied as prototypes for developing more selective DDR1 inhibitors [75, 98, 116].

Despite advancements in DDR inhibitor development using general RTK blockers, researchers have encountered challenges due to the conserved ATP-binding sites across all RTKs. These sites complicate efforts to achieve DDR-specific inhibition, suggesting the need to explore type IV kinase inhibitors that compete with substrates rather than ATP [107].

Concluding Remarks

Decades of research have established the amino acid sequence, posttranslational modifications, structure, and physicochemical properties of individual collagen molecules. However, the mechanisms underlying the assembly of complex homotypic and heterotypic collagenous matrices remain less understood. Even greater gaps exist in our knowledge of how these matrices influence the behavior of resident cells.

One approach to addressing these gaps is to study pathological processes characterized by aberrant collagenous matrix formation. A prime example is the investigation of heritable diseases caused by mutations in fibrillar collagens. By comparing mutated and normal collagen molecules, such studies have been instrumental in elucidating the structure and function of these proteins.

This chapter explores fibrotic scarring in the context of collagen-rich matrices and their role in fibrosis. Specifically, it highlights fibrosis-associated matrix stiffening as a key driver of disease progression. Reducing matrix stiffness has emerged as a promising strategy to mitigate fibrosis. Potential approaches include directly inhibiting various steps of collagen synthesis and fibrillogenesis, as well as blocking pro-fibrotic signaling from the stiffened matrix to inflammatory and fibroblastic cells. Addressing the challenge of reducing collagenous matrix stiffness in fibrosis not only holds therapeutic potential but also offers valuable insights into matrix assembly, matrix-cell interactions, and matrix-dependent cellular functions.

References

1. Abuammah A, Maimari N, Towhidi L, Frueh J, Chooi KY, Warboys C, Krams R (2018) New developments in mechanotransduction: cross talk of the Wnt, TGF-β and Notch signalling pathways in reaction to shear stress. Curr Opin Biomed Eng 5:96–104
2. Aciksari K, Yanar HT, Hepgul G, Ozucelik DN, Yanar F, Agcaoglu O, Eser M, Tanriverdi G, Topacoglu H, Ayvaci BM, Dogan H, Gunay K, Ertekin C, Celikmen F (2013) The effect of Beta-Aminopropionitrile and prednisolone on the prevention of fibrosis in alkali esophageal burns: an experimental study. Gastroenterol Res Pract 2013:574260
3. Alibardi L (2022) Regeneration in anamniotes was replaced by regengrow and scarring in amniotes after land colonization and the evolution of terrestrial biological cycles. Dev Dyn 251(9):1404–1413
4. Annes JP, Munger JS, Rifkin DB (2003) Making sense of latent TGFβ activation. J Cell Sci 116(2):217–224
5. Armiento AR, Alini M, Stoddart MJ (2019) Articular fibrocartilage-Why does hyaline cartilage fail to repair? Adv Drug Deliv Rev 146:289–305
6. Artlett CM, Connolly LM (2023) TANGO1 dances to export of procollagen from the endoplasmic reticulum. Fibrosis (Hong Kong, China) 1(2):10008
7. Asano Y, Ihn H, Jinnin M, Mimura Y, Tamaki K (2006) Involvement of αvβ5 integrin in the establishment of autocrine TGF-β signaling in dermal fibroblasts derived from localized scleroderma. J Invest Dermatol 126(8):1761–1769
8. Badarau E, Wang Z, Rathbone DL, Costanzi A, Thibault T, Murdoch CE, El Alaoui S, Bartkeviciute M, Griffin M (2015) Development of potent and selective tissue transglutaminase inhibitors: their effect on TG2 function and application in pathological conditions. Chem Biol 22(10):1347–1361
9. Barry-Hamilton V, Spangler R, Marshall D, McCauley S, Rodriguez HM, Oyasu M, Mikels A, Vaysberg M, Ghermazien H, Wai C, Garcia CA, Velayo AC, Jorgensen B, Biermann D, Tsai D, Green J, Zaffryar-Eilot S, Holzer A, Ogg S, Thai D, Neufeld G, Van Vlasselaer P, Smith V (2010) Allosteric inhibition of lysyl oxidase-like-2 impedes the development of a pathologic microenvironment. Nat Med 16(9):1009–1017
10. Bekhouche M, Colige A (2015) The procollagen N-proteinases ADAMTS2, 3 and 14 in pathophysiology. Matrix Biol 44-46:46–53
11. Benson AB III, Wainberg ZA, Hecht JR, Vyushkov D, Dong H, Bendell J, Kudrik F (2017) A phase II randomized, double-blind, placebo-controlled study of Simtuzumab or placebo in combination with gemcitabine for the first-line treatment of pancreatic adenocarcinoma. Oncologist 22(3):241–e215
12. Bharathy H, Fathima NN (2017) Exploiting oleuropein for inhibiting collagen fibril formation. Int J Biol Macromol 101:179–186
13. Bottomley MJ, Batten MR, Lumb RA, Bulleid NJ (2001) Quality control in the endoplasmic reticulum: PDI mediates the ER retention of unassembled procollagen C-propeptides. Curr Biol 11(14):1114–1118
14. Brodsky B (1999) Hydrogen bonding in the triple-helix. Springer, Proceedings of the Indian academy of sciences-chemical sciences

15. Budi EH, Schaub JR, Decaris M, Turner S, Derynck R (2021) TGF-β as a driver of fibrosis: physiological roles and therapeutic opportunities. J Pathol 254(4):358–373
16. Bülow RD, Boor P (2019) Extracellular matrix in kidney fibrosis: more than just a scaffold. J Histochem Cytochem 67(9):643–661
17. Canty EG, Lu Y, Meadows RS, Shaw MK, Holmes DF, Kadler KE (2004) Coalignment of plasma membrane channels and protrusions (fibripositors) specifies the parallelism of tendon. J Cell Biol 165(4):553–563
18. Carmichael SP II, Shin J, Vaughan JW, Chandra PK, Holcomb JB, Atala AJ (2022) Regenerative medicine therapies for prevention of abdominal adhesions: a scoping review. J Surg Res 275:252–264
19. Chang Y, Lau WL, Jo H, Tsujino K, Gewin L, Reed NI, Atakilit A, Nunes ACF, DeGrado WF, Sheppard D (2017) Pharmacologic blockade of αvβ1 integrin ameliorates renal failure and fibrosis in vivo. J Am Soc Nephrol 28(7):1998–2005
20. Chen S-J, Yuan W, Mori Y, Levenson A, Varga J, Trojanowska M (1999) Stimulation of type I collagen transcription in human skin fibroblasts by TGF-β: involvement of Smad 3. J Invest Dermatol 112(1):49–57
21. Chung HJ, Steplewski A, Chung KY, Uitto J, Fertala A (2008) Collagen fibril formation. A new target to limit fibrosis. J Biol Chem 283(38):25879–25886
22. Colige A, Li SW, Sieron AL, Nusgens BV, Prockop DJ, Lapiere CM (1997) cDNA cloning and expression of bovine procollagen I N-proteinase: a new member of the superfamily of zinc-metalloproteinases with binding sites for cells and other matrix components. Proc Natl Acad Sci USA 94(6):2374–2379
23. Collighan R, Griffin M (2009) Transglutaminase 2 cross-linking of matrix proteins: biological significance and medical applications. Amino Acids 36:659–670
24. Csiszar K (2001) Lysyl oxidases: a novel multifunctional amine oxidase family. Prog Nucleic Acid Res Mol Biol 70:1–32
25. D'Angelo M, Billings PC, Pacifici M, Leboy PS, Kirsch T (2001) Authentic matrix vesicles contain active metalloproteases (MMP): a role for matrix vesicle-associated MMP-13 in activation of transforming growth factor-β. J Biol Chem 276(14):11347–11353
26. Dallas S, Park-Snyder S, Miyazono K, Twardzik D, Mundy G, Bonewald L (1994) Characterization and autoregulation of latent transforming growth factor beta (TGF beta) complexes in osteoblast-like cell lines. Production of a latent complex lacking the latent TGF beta-binding protein. J Biol Chem 269(9):6815–6821
27. Darby IA, Zakuan N, Billet F, Desmouliere A (2016) The myofibroblast, a key cell in normal and pathological tissue repair. Cell Mol Life Sci 73(6):1145–1157
28. Dubois CM, Laprise M-H, Blanchette F, Gentry LE, Leduc R (1995) Processing of transforming growth factor β1 precursor by human furin convertase. J Biol Chem 270(18): 10618–10624
29. Duscher D, Maan ZN, Wong VW, Rennert RC, Januszyk M, Rodrigues M, Hu M, Whitmore AJ, Whittam AJ, Longaker MT, Gurtner GC (2014) Mechanotransduction and fibrosis. J Biomech 47(9):1997–2005
30. Ellis LR, Warner DR, Greene RM, Pisano MM (2003) Interaction of Smads with collagen types I, III, and V. Biochem Biophys Res Commun 310(4):1117–1123
31. Engel J, Prockop DJ (1991) The zipper-like folding of collagen triple helices and the effects of mutations that disrupt the zipper. Annu Rev Biophys Biophys Chem 20:137–152
32. Erasmus M, Samodien E, Lecour S, Cour M, Lorenzo O, Dludla P, Pheiffer C, Johnson R (2020) Linking LOXL2 to cardiac interstitial fibrosis. Int J Mol Sci 21. https://doi.org/10.3390/ijms21165913
33. Even-Ram S, Doyle AD, Conti MA, Matsumoto K, Adelstein RS, Yamada KM (2007) Myosin IIA regulates cell motility and actomyosin–microtubule crosstalk. Nat Cell Biol 9(3):299–309
34. Eyre DR, Weis MA, Wu JJ (2008) Advances in collagen cross-link analysis. Methods 45(1): 65–74

35. Fertala J, Steplewski A, Kostas J, Beredjiklian P, Williams G, Arnold W, Abboud J, Bhardwaj A, Hou C, Fertala A (2013) Engineering and characterization of the chimeric antibody that targets the C-terminal telopeptide of the alpha2 chain of human collagen I: a next step in the quest to reduce localized fibrosis. Connect Tissue Res 54(3):187–196
36. Fertala J, Kostas J, Hou C, Steplewski A, Beredjiklian P, Abboud J, Arnold WV, Williams G, Fertala A (2014) Testing the anti-fibrotic potential of the single-chain Fv antibody against the alpha2 C-terminal telopeptide of collagen I. Connect Tissue Res 55(2):115–122
37. Fertala J, Romero F, Summer R, Fertala A (2017) Target-specific delivery of an antibody that blocks the formation of collagen deposits in skin and lung. Monoclonal Antibodies Immunodiagn Immunother 36(5):199–207
38. Fertin C, Nicolas J, Gillery P, Kalis B, Banchereau J, Maquart F (1991) Interleukin-4 stimulates collagen synthesis by normal and scleroderma fibroblasts in dermal equivalents. Cell Mol Biol 37(8):823–829
39. Fichtner-Feigl S, Strober W, Kawakami K, Puri RK, Kitani A (2006) IL-13 signaling through the IL-13α2 receptor is involved in induction of TGF-β1 production and fibrosis. Nat Med 12(1):99–106
40. Fietzek PP, Rexrodt FW, Wendt P, Stark M, Kuhn K (1972) The covalent structure of collagen. Amino-acid sequence of peptide 1-CB6-C2. Eur J Biochem 30(1):163–168
41. Flynn LA, Blissett AR, Calomeni EP, Agarwal G (2010) Inhibition of collagen fibrillogenesis by cells expressing soluble extracellular domains of DDR1 and DDR2. J Mol Biol 395(3): 533–543
42. Font B, Eichenberger D, Goldschmidt D, Boutillon MM, Hulmes DJ (1998) Structural requirements for fibromodulin binding to collagen and the control of type I collagen fibrillogenesis–critical roles for disulphide bonding and the C-terminal region. Eur J Biochem 254(3):580–587
43. Freund KF, Doshi KP, Gaul SL, Claremon DA, Remy DC, Baldwin JJ, Pitzenberger SM, Stern AM (1994) Transglutaminase inhibition by 2-[(2-oxopropyl) thio] imidazolium derivatives: mechanism of factor XIIIa inactivation. Biochemistry 33(33):10109–10119
44. Ge G, Greenspan DS (2006) BMP1 controls TGFβ1 activation via cleavage of latent TGF-β-binding protein. J Cell Biol 175(1):111–120
45. Gelman RA, Poppke DC, Piez KA (1979) Collagen fibril formation in vitro. The role of the nonhelical terminal regions. J Biol Chem 254(22):11741–11745
46. Giannone G, Dubin-Thaler BJ, Rossier O, Cai Y, Chaga O, Jiang G, Beaver W, Döbereiner H-G, Freund Y, Borisy G (2007) Lamellipodial actin mechanically links myosin activity with adhesion-site formation. Cell 128(3):561–575
47. Gjaltema RA, van der Stoel MM, Boersema M, Bank RA (2016) Disentangling mechanisms involved in collagen pyridinoline cross-linking: the immunophilin FKBP65 is critical for dimerization of lysyl hydroxylase 2. Proc Natl Acad Sci 113(26):7142–7147
48. Goswami C, Kuhn J, Heppenstall PA, Hucho T (2010) Importance of non-selective cation channel TRPV4 interaction with cytoskeleton and their reciprocal regulations in cultured cells. PLoS One 5(7):e11654
49. Guerrot D, Kerroch M, Placier S, Vandermeersch S, Trivin C, Mael-Ainin M, Chatziantoniou C, Dussaule J-C (2011) Discoidin domain receptor 1 is a major mediator of inflammation and fibrosis in obstructive nephropathy. Am J Pathol 179(1):83–91
50. Haak AJ, Kostallari E, Sicard D, Ligresti G, Choi KM, Caporarello N, Jones DL, Tan Q, Meridew J, Diaz Espinosa AM, Aravamudhan A, Maiers JL, Britt RD Jr, Roden AC, Pabelick CM, Prakash YS, Nouraie SM, Li X, Zhang Y, Kass DJ, Lagares D, Tager AM, Varelas X, Shah VH, Tschumperlin DJ (2019) Selective YAP/TAZ inhibition in fibroblasts via dopamine receptor D1 agonism reverses fibrosis. Sci Transl Med 11(516):eaau6296
51. Hahm K, Lukashev ME, Luo Y, Yang WJ, Dolinski BM, Weinreb PH, Simon KJ, Wang LC, Leone DR, Lobb RR (2007) αvβ6 integrin regulates renal fibrosis and inflammation in Alport mouse. Am J Pathol 170(1):110–125

52. Harrison CA, Layton CM, Hau Z, Bullock AJ, Johnson TS, MacNeil S (2007) Transglutaminase inhibitors induce hyperproliferation and parakeratosis in tissue-engineered skin. Br J Dermatol 156(2):247–257
53. Harrison SA, Abdelmalek MF, Caldwell S, Shiffman ML, Diehl AM, Ghalib R, Lawitz EJ, Rockey DC, Schall RA, Jia C, McColgan BJ, McHutchison JG, Subramanian GM, Myers RP, Younossi Z, Ratziu V, Muir AJ, Afdhal NH, Goodman Z, Bosch J, Sanyal AJ, Gs US, Investigators G-U (2018) Simtuzumab is ineffective for patients with bridging fibrosis or compensated cirrhosis caused by nonalcoholic Steatohepatitis. Gastroenterology 155(4): 1140–1153
54. Hecht JR, Benson AB III, Vyushkov D, Yang Y, Bendell J, Verma U (2017) A phase II, randomized, double-blind, placebo-controlled study of Simtuzumab in combination with FOLFIRI for the second-line treatment of metastatic KRAS mutant colorectal adenocarcinoma. Oncologist 22(3):243–e223
55. Heikkinen J, Risteli M, Wang C, Latvala J, Rossi M, Valtavaara M, Myllylä R (2000) Lysyl hydroxylase 3 is a multifunctional protein possessing collagen glucosyltransferase activity. J Biol Chem 275(46):36158–36163
56. Henderson NC, Arnold TD, Katamura Y, Giacomini MM, Rodriguez JD, McCarty JH, Pellicoro A, Raschperger E, Betsholtz C, Ruminski PG, Griggs DW, Prinsen MJ, Maher JJ, Iredale JP, Lacy-Hulbert A, Adams RH, Sheppard D (2013) Targeting of αv integrin identifies a core molecular pathway that regulates fibrosis in several organs. Nat Med 19(12):1617–1624
57. Hennet T (2019) Collagen glycosylation. Curr Opin Struct Biol 56:131–138
58. Hinz B (2016) Myofibroblasts. Exp Eye Res 142:56–70
59. Holmes DF, Mould AP, Chapman JA (1991) Morphology of sheet-like assemblies of pN-collagen, pC-collagen and procollagen studied by scanning transmission electron microscopy mass measurements. J Mol Biol 220(1):111–123
60. Hu G, Gura T, Sabsay B, Sauk J, Dixit SN, Veis A (1995) Endoplasmic reticulum protein Hsp47 binds specifically to the N- terminal globular domain of the amino-propeptide of the procollagen I alpha 1 (I)-chain. J Cell Biochem 59(3):350–367
61. Huang Y, Arora P, McCulloch CA, Vogel WF (2009) The collagen receptor DDR1 regulates cell spreading and motility by associating with myosin IIA. J Cell Sci 122(Pt 10):1637–1646
62. Hultstrom M, Leh S, Skogstrand T, Iversen BM (2008) Upregulation of tissue inhibitor of metalloproteases-1 (TIMP-1) and procollagen-N-peptidase in hypertension-induced renal damage. Nephrol Dial Transplant 23(3):896–903
63. Ihn H (2002) Pathogenesis of fibrosis: role of TGF-beta and CTGF. Curr Opin Rheumatol 14(6):681–685
64. Isshiki T, Naiel S, Vierhout M, Otsubo K, Ali P, Tsubouchi K, Yazdanshenas P, Kumaran V, Dvorkin-Gheva A, Kolb MR (2024) Therapeutic strategies to target connective tissue growth factor in fibrotic lung diseases. Pharmacol Ther 253:108578
65. Ištok R, Bely M, Stančíková M, Rovenský J (2001) Evidence for increased pyridinoline concentration in fibrotic tissues in diffuse systemic sclerosis. Clin Exp Dermatol 26(6): 545–547
66. Jinnin M, Ihn H, Yamane K, Tamaki K (2004) Interleukin-13 stimulates the transcription of the human α2 (I) collagen gene in human dermal fibroblasts. J Biol Chem 279(40): 41783–41791
67. Johnson TS, Fisher M, Haylor JL, Hau Z, Skill NJ, Jones R, Saint R, Coutts I, Vickers ME, El Nahas AM (2007) Transglutaminase inhibition reduces fibrosis and preserves function in experimental chronic kidney disease. J Am Soc Nephrol 18(12):3078–3088
68. Junker JP, Kratz C, Tollback A, Kratz G (2008) Mechanical tension stimulates the transdifferentiation of fibroblasts into myofibroblasts in human burn scars. Burns 34(7): 942–946
69. Kadler KE, Hojima Y, Prockop DJ (1987) Assembly of collagen fibrils de novo by cleavage of the type I pC-collagen with procollagen C-proteinase. Assay of critical concentration

demonstrates that collagen self-assembly is a classical example of an entropy-driven process. J Biol Chem 262(32):15696–15701
70. Kadler KE, Hojima Y, Prockop DJ (1990) Collagen fibrils in vitro grow from pointed tips in the C- to N- terminal direction. Biochem J 268(2):339–343
71. Kadler KE, Hill A, Canty-Laird EG (2008) Collagen fibrillogenesis: fibronectin, integrins, and minor collagens as organizers and nucleators. Curr Opin Cell Biol 20(5):495–501
72. Kefalides NA (1973) Structure and biosynthesis of basement membranes. Int Rev Connect Tissue Res 6:63–104
73. Keillor JW, Apperley KYP, Akbar A (2015) Inhibitors of tissue transglutaminase. Trends Pharmacol Sci 36(1):32–40
74. Kessler E, Takahara K, Biniaminov L, Brusel M, Greenspan DS (1996) Bone morphogenetic protein-1: the type I procollagen C-proteinase. Science 271(5247):360–362
75. Kim H-G, Tan L, Weisberg EL, Liu F, Canning P, Choi HG, Ezell SA, Wu H, Zhao Z, Wang J (2013) Discovery of a potent and selective DDR1 receptor tyrosine kinase inhibitor. ACS Chem Biol 8(10):2145–2150
76. Kirby TJ, Lammerding J (2018) Emerging views of the nucleus as a cellular mechanosensor. Nat Cell Biol 20(4):373–381
77. Knott L, Bailey AJ (1998) Collagen cross-links in mineralizing tissues: a review of their chemistry, function, and clinical relevance. Bone 22(3):181–187
78. Knuppel L, Ishikawa Y, Aichler M, Heinzelmann K, Hatz R, Behr J, Walch A, Bachinger HP, Eickelberg O, Staab-Weijnitz CA (2017) A novel antifibrotic mechanism of nintedanib and pirfenidone. inhibition of collagen fibril assembly. Am J Respir Cell Mol Biol 57(1):77–90
79. Kok HM, Falke LL, Goldschmeding R, Nguyen TQ (2014) Targeting CTGF, EGF and PDGF pathways to prevent progression of kidney disease. Nat Rev Nephrol 10(12):700–711
80. Kong W, Lyu C, Liao H, Du Y (2021) Collagen crosslinking: effect on structure, mechanics and fibrosis progression. Biomed Mater 16(6):062005
81. Kunoki S, Tatsukawa H, Sakai Y, Kinashi H, Kariya T, Suzuki Y, Mizuno M, Yamaguchi M, Sasakura H, Ikeno M (2023) Inhibition of transglutaminase 2 reduces peritoneal injury in a chlorhexidine-induced peritoneal fibrosis model. Lab Investig 103:100050
82. Lagares D, Santos A, Grasberger PE, Liu F, Probst CK, Rahimi RA, Sakai N, Kuehl T, Ryan J, Bhola P, Montero J, Kapoor M, Baron M, Varelas X, Tschumperlin DJ, Letai A, Tager AM (2017) Targeted apoptosis of myofibroblasts with the BH3 mimetic ABT-263 reverses established fibrosis. Sci Transl Med 9(420):eaal3765
83. Lagoutte P, Bettler E, Vadon-Le Goff S, Moali C (2021) Procollagen C-proteinase enhancer-1 (PCPE-1), a potential biomarker and therapeutic target for fibrosis. Matrix Biol Plus 11: 100062
84. Lampi MC, Reinhart-King CA (2018) Targeting extracellular matrix stiffness to attenuate disease: from molecular mechanisms to clinical trials. Sci Transl Med 10(422):eaao0475
85. Lawrence DA, Pircher R, Kryceve-Martinerie C, Jullien P (1984) Normal embryo fibroblasts release transforming growth factors in a latent form. J Cell Physiol 121(1):184–188
86. Lees JF, Tasab M, Bulleid NJ (1997) Identification of the molecular recognition sequence which determines the type-specific assembly of procollagen. EMBO J 16(5):908–916
87. Leikina E, Mertts MV, Kuznetsova N, Leikin S (2002) Type I collagen is thermally unstable at body temperature. Proc Natl Acad Sci USA 99(3):1314–1318
88. Leiss M, Beckmann K, Girós A, Costell M, Fässler R (2008) The role of integrin binding sites in fibronectin matrix assembly in vivo. Curr Opin Cell Biol 20(5):502–507
89. Leitinger B (2011) Transmembrane collagen receptors. Annu Rev Cell Dev Biol 27:265–290
90. Leitinger B (2014) Chapter 2: Discoidin domain receptor functions in physiological and pathological conditions. In: Jeon KW (ed) International review of cell and molecular biology, vol 310. Academic, pp 39–87
91. Leitinger B, Kwan APL (2006) The discoidin domain receptor DDR2 is a receptor for type X collagen. Matrix Biol 25(6):355–364

92. Li SW, Sieron AL, Fertala A, Hojima Y, Arnold WV, Prockop DJ (1996) The C-proteinase that processes procollagens to fibrillar collagens is identical to the protein previously identified as bone morphogenic protein-1. Proc Natl Acad Sci USA 93(10):5127–5130
93. Li SW, Arita M, Fertala A, Bao Y, Kopen GC, Langsjo TK, Hyttinen MM, Helminen HJ, Prockop DJ (2001) Transgenic mice with inactive alleles for procollagen N-proteinase (ADAMTS-2) develop fragile skin and male sterility. Biochem J 355(Pt 2):271–278
94. Li X, Bu X, Yan F, Wang F, Wei D, Yuan J, Zheng W, Su J, Yuan J (2019) Deletion of discoidin domain receptor 2 attenuates renal interstitial fibrosis in a murine unilateral ureteral obstruction model. Ren Fail 41(1):481–488
95. Li L, Fu H, Liu Y (2022) The fibrogenic niche in kidney fibrosis: components and mechanisms. Nat Rev Nephrol 18(9):545–557
96. Lino M, Ngai D, Liu A, Mohabeer A, Harper C, Caruso L-L, Schroer SA, Fu F, McKee T, Giacca A, Woo M, Bendeck MP (2020) Discoidin domain receptor 1-deletion ameliorates fibrosis and promotes adipose tissue beiging, brown fat activity, and increased metabolic rate in a mouse model of cardiometabolic disease. Mol Metab 39:101006
97. Liu F, Lagares D, Choi KM, Stopfer L, Marinkovic A, Vrbanac V, Probst CK, Hiemer SE, Sisson TH, Horowitz JC, Rosas IO, Fredenburgh LE, Feghali-Bostwick C, Varelas X, Tager AM, Tschumperlin DJ (2015) Mechanosignaling through YAP and TAZ drives fibroblast activation and fibrosis. Am J Physiol Lung Cell Mol Physiol 308(4):L344–L357
98. Liu L, Hussain M, Luo J, Duan A, Chen C, Tu Z, Zhang J (2017) Synthesis and biological evaluation of novel dasatinib analogues as potent DDR 1 and DDR 2 kinase inhibitors. Chem Biol Drug Des 89(3):420–427
99. Lloyd SM, He Y (2024) Exploring extracellular matrix crosslinking as a therapeutic approach to fibrosis. Cells 13(5):438
100. Luppi F, Kalluri M, Faverio P, Kreuter M, Ferrara G (2021) Idiopathic pulmonary fibrosis beyond the lung: understanding disease mechanisms to improve diagnosis and management. Respir Res 22(1):109
101. Lurje I, Gaisa NT, Weiskirchen R, Tacke F (2023) Mechanisms of organ fibrosis: emerging concepts and implications for novel treatment strategies. Mol Asp Med 92:101191
102. Maamra M, Benayad OM, Matthews D, Kettleborough C, Atkinson J, Cain K, Bon H, Brand H, Parkinson M, Watson PF (2022) Transglutaminase 2: development of therapeutic antibodies reveals four inhibitory epitopes and confirms extracellular function in fibrotic remodelling. Br J Pharmacol 179(11):2697–2712
103. Maeda S, Dean D, Gomez R, Schwartz Z, Boyan B (2002) The first stage of transforming growth factor β1 activation is release of the large latent complex from the extracellular matrix of growth plate chondrocytes by matrix vesicle stromelysin-1 (MMP-3). Calcif Tissue Int 70: 54–65
104. Mahler GJ, Frendl CM, Cao Q, Butcher JT (2014) Effects of shear stress pattern and magnitude on mesenchymal transformation and invasion of aortic valve endothelial cells. Biotechnol Bioeng 111(11):2326–2337
105. Mallone F, Costi R, Marenco M, Plateroti R, Minni A, Attanasio G, Artico M, Lambiase A (2021) Understanding drivers of ocular fibrosis: current and future therapeutic perspectives. Int J Mol Sci 22(21):11748
106. Marenzana M, Wilson-Jones N, Mudera V, Brown RA (2006) The origins and regulation of tissue tension: identification of collagen tension-fixation process in vitro. Exp Cell Res 312(4): 423–433
107. Martinez R, Defnet A, Shapiro P (2020) Avoiding or co-opting ATP inhibition: overview of type III, IV, V, and VI kinase inhibitors. In: Shapiro P (ed) Next generation kinase inhibitors: moving beyond the ATP binding/catalytic sites. Springer, Cham, pp 29–59
108. McLaughlin SH, Bulleid NJ (1998) Molecular recognition in procollagen chain assembly. Matrix Biol 16(7):369–377
109. Miles CA, Ghelashvili M (1999) Polymer-in-a-box mechanism for the thermal stabilization of collagen molecules in fibers. Biophys J 76(6):3243–3252

110. Miyahara M, Njieha FK, Prockop DJ (1982) Formation of collagen fibrils in vitro by cleavage of procollagen with procollagen proteinases. J Biol Chem 257(14):8442–8448
111. Mohseni R, Arab Sadeghabadi Z, Goodarzi MT, Karimi J (2019) Co-administration of resveratrol and beta-aminopropionitrile attenuates liver fibrosis development via targeting lysyl oxidase in CCl4-induced liver fibrosis in rats. Immunopharmacol Immunotoxicol 41(6):644–651
112. Moll S, Desmoulière A, Moeller MJ, Pache JC, Badi L, Arcadu F, Richter H, Satz A, Uhles S, Cavalli A, Drawnel F, Scapozza L, Prunotto M (2019) DDR1 role in fibrosis and its pharmacological targeting. Biochim Biophys Acta, Mol Cell Res 1866(11):118474
113. Morello R, Bertin TK, Chen Y, Hicks J, Tonachini L, Monticone M, Castagnola P, Rauch F, Glorieux FH, Vranka J, Bachinger HP, Pace JM, Schwarze U, Byers PH, Weis M, Fernandes RJ, Eyre DR, Yao Z, Boyce BF, Lee B (2006) CRTAP is required for prolyl 3- hydroxylation and mutations cause recessive osteogenesis imperfecta. Cell 127(2):291–304
114. Mu Y, Gudey SK, Landström M (2012) Non-Smad signaling pathways. Cell Tissue Res 347(1):11–20
115. Munger JS, Harpel JG, Giancotti FG, Rifkin DB (1998) Interactions between growth factors and integrins: latent forms of transforming growth factor-β are ligands for the integrin αvβ1. Mol Biol Cell 9(9):2627–2638
116. Murray CW, Berdini V, Buck IM, Carr ME, Cleasby A, Coyle JE, Curry JE, Day JE, Day PJ, Hearn K (2015) Fragment-based discovery of potent and selective DDR1/2 inhibitors. ACS Med Chem Lett 6(7):798–803
117. N'Diaye E-N, Cook R, Wang H, Wu P, LaCanna R, Wu C, Ye Z, Seshasayee D, Hazen M, Lin W (2021) Extracellular BMP1 is the major proteinase for COOH-terminal proteolysis of type I procollagen in lung fibroblasts. Am J Phys Cell Phys 320(2):C162–C174
118. Natsume T, Koide T, Yokota S, Hirayoshi K, Nagata K (1994) Interactions between collagen-binding stress protein HSP47 and collagen. Analysis of kinetic parameters by surface plasmon resonance biosensor. J Biol Chem 269(49):31224–31228
119. Neff LS, Bradshaw AD (2021) Cross your heart? Collagen cross-links in cardiac health and disease. Cell Signal 79:109889
120. Nishimichi N, Kanno K, Sentani K, Yasui W, Yokosaki Y (2013) A possible therapeutic agent for pulmonary fibrosis: antibody against integrin α8β1. Eur Respir J 42(Suppl 57):P4885
121. Nishimichi N, Tsujino K, Kanno K, Sentani K, Kobayashi T, Chayama K, Sheppard D, Yokosaki Y (2021) Induced hepatic stellate cell integrin, α8β1, enhances cellular contractility and TGFβ activity in liver fibrosis. J Pathol 253(4):366–373
122. Ogata I, Auster AS, Matsui A, Greenwel P, Geerts A, D'Amico T, Fujiwara K, Kessler E, Rojkind M (1997) Up-regulation of type I procollagen C-proteinase enhancer protein messenger RNA in rats with CCl4-induced liver fibrosis. Hepatology 26(3):611–617
123. Oh K, Park H-B, Byoun O-J, Shin D-M, Jeong EM, Kim YW, Kim YS, Melino G, Kim I-G, Lee D-S (2011) Epithelial transglutaminase 2 is needed for T cell interleukin-17 production and subsequent pulmonary inflammation and fibrosis in bleomycin-treated mice. J Exp Med 208(8):1707–1719
124. Ovens A, Joule JA, Kadler KE (2000) Design and synthesis of acidic dipeptide hydroxamate inhibitors of procollagen C-proteinase. J Pept Sci 6(9):489–495
125. Paolini C, Agarbati S, Benfaremo D, Mozzicafreddo M, Svegliati S, Moroncini G (2022) PDGF/PDGFR: a possible molecular target in scleroderma fibrosis. Int J Mol Sci 23(7):3904
126. Patsenker E, Popov Y, Stickel F, Jonczyk A, Goodman SL, Schuppan D (2008) Inhibition of integrin αvβ6 on cholangiocytes blocks transforming growth factor-β activation and retards biliary fibrosis progression. Gastroenterology 135(2):660–670
127. Pehrsson M, Mortensen JH, Manon-Jensen T, Bay-Jensen A-C, Karsdal MA, Davies MJ (2021) Enzymatic cross-linking of collagens in organ fibrosis–resolution and assessment. Expert Rev Mol Diagn 21(10):1049–1064
128. Peng D, Fu M, Wang M, Wei Y, Wei X (2022) Targeting TGF-β signal transduction for fibrosis and cancer therapy. Mol Cancer 21(1):104

129. Perumal S, Dubey K, Badhwar R, George KJ, Sharma RK, Bagler G, Madhan B, Kar K (2015) Capsaicin inhibits collagen fibril formation and increases the stability of collagen fibers. Eur Biophys J 44(1–2):69–76
130. Poe A, Martinez Yus M, Wang H, Santhanam L (2023) Lysyl oxidase like-2 in fibrosis and cardiovascular disease. Am J Phys Cell Phys 325(3):C694–C707
131. Prockop DJ, Fertala A (1998a) The collagen fibril: the almost crystalline structure. J Struct Biol 122(1–2):111–118
132. Prockop DJ, Fertala A (1998b) Inhibition of the self-assembly of collagen I into fibrils with synthetic peptides. Demonstration that assembly is driven by specific binding sites on the monomers. J Biol Chem 273(25):15598–15604
133. Prockop DJ, Kivirikko KI (1995) Collagens: molecular biology, diseases, and potentials for therapy. Annu Rev Biochem 64:403–434
134. Prockop DJ, Berg RA, Kivirikko KI, Uitto J (1976) Intracellular steps in the biosynthesis of collagen. In: Ramachandran GN, Reddi AH (eds) Biochemistry of collagen. Plenum, New York, pp 163–237
135. Puente A, Fortea JI, Cabezas J, Arias Loste MT, Iruzubieta P, Llerena S, Huelin P, Fabrega E, Crespo J (2019) LOXL2-a new target in Antifibrogenic therapy? Int J Mol Sci 20(7):1634
136. Raghu G, Brown KK, Collard HR, Cottin V, Gibson KF, Kaner RJ, Lederer DJ, Martinez FJ, Noble PW, Song JW, Wells AU, Whelan TP, Wuyts W, Moreau E, Patterson SD, Smith V, Bayly S, Chien JW, Gong Q, Zhang JJ, O'Riordan TG (2017) Efficacy of simtuzumab versus placebo in patients with idiopathic pulmonary fibrosis: a randomised, double-blind, controlled, phase 2 trial. Lancet Respir Med 5(1):22–32
137. Raghu G, Mouded M, Chambers DC, Martinez FJ, Richeldi L, Lancaster LH, Hamblin MJ, Gibson KF, Rosas IO, Prasse A, Zhao G, Serenko M, Novikov N, McCurley A, Bansal P, Stebbins C, Arefayene M, Ibebunjo S, Violette SM, Gallagher D, Behr J (2022) A phase IIb randomized clinical study of an anti-alpha(v)beta(6) monoclonal antibody in idiopathic pulmonary fibrosis. Am J Respir Crit Care Med 206(9):1128–1139
138. Rahman SR, Roper JA, Grove JI, Aithal GP, Pun KT, Bennett AJ (2022) Integrins as a drug target in liver fibrosis. Liver Int 42(3):507–521
139. Rasheeda K, Fathima NN (2018) Trigonelline hydrochloride: a promising inhibitor for type I collagen fibrillation. Colloids Surf B Biointerfaces 170:273–279
140. Rasheeda K, Muvva C, Fathima NN (2019) Governing the inhibition of reconstituted collagen type I assemblies mediated through noncovalent forces of (+/−)-alpha Lipoic acid. Langmuir 35(4):980–989
141. Reed NI, Jo H, Chen C, Tsujino K, Arnold TD, DeGrado WF, Sheppard D (2015) The αvβ1 integrin plays a critical in vivo role in tissue fibrosis. Sci Transl Med 7(288):288ra279
142. Rexrodt FW, Hopper KE, Fietzek PP, Kuhn K (1973) The covalent structure of collagen. 1. The chymotrypsin, trypsin, and thermolysin-derived peptides of alpha1-CB7 from calf-skin collagen. Eur J Biochem 38(2):384–395
143. Riley DJ, Kerr JS, Berg RA, Ianni BD, Pietra GG, Edelman NH, Prockop DJ (1982) beta-Aminopropionitrile prevents bleomycin-induced pulmonary fibrosis in the hamster. Am Rev Respir Dis 125(1):67–73
144. Rivlin M, Miller A, Tulipan J, Beredjiklian PK, Wang ML, Fertala J, Steplewski A, Kostas J, Fertala A (2017) Patterns of production of collagen-rich deposits in peripheral nerves in response to injury: a pilot study in a rabbit model. Brain Behav 7(7):e00659
145. Ruiz-Ortega M, Lamas S, Ortiz A (2022) Antifibrotic agents for the management of CKD: a review. Am J Kidney Dis 80(2):251–263
146. Saeidi N, Karmelek KP, Paten JA, Zareian R, DiMasi E, Ruberti JW (2012) Molecular crowding of collagen: a pathway to produce highly-organized collagenous structures. Biomaterials 33(30):7366–7374
147. Saharinen J, Keski-Oja J (2000) Specific sequence motif of 8-Cys repeats of TGF-β binding proteins, LTBPs, creates a hydrophobic interaction surface for binding of small latent TGF-β. Mol Biol Cell 11(8):2691–2704

148. Sarrazy V, Koehler A, Chow ML, Zimina E, Li CX, Kato H, Caldarone CA, Hinz B (2014) Integrins αvβ5 and αvβ3 promote latent TGF-β1 activation by human cardiac fibroblast contraction. Cardiovasc Rse 102(3):407–417
149. Scaffidi AK, Petrovic N, Moodley YP, Fogel-Petrovic M, Kroeger KM, Seeber RM, Eidne KA, Thompson PJ, Knight DA (2004) αvβ3 integrin interacts with the transforming growth factor β (TGFβ) type II receptor to potentiate the proliferative effects of TGFβ1 in living human lung fibroblasts. J Biol Chem 279(36):37726–37733
150. Schoppet M, Chavakis T, Al-Fakhri N, Kanse SM, Preissner KT (2002) Molecular interactions and functional interference between vitronectin and transforming growth factor-β. Lab Investig 82(1):37–46
151. Shetty SS, Sharma M, Kabekkodu SP, Kumar NA, Satyamoorthy K, Radhakrishnan R (2021) Understanding the molecular mechanism associated with reversal of oral submucous fibrosis targeting hydroxylysine aldehyde-derived collagen cross-links. J Carcinog 20:9
152. Shi Y, Dong Y, Duan Y, Jiang X, Chen C, Deng L (2013) Substrate stiffness influences TGF-beta1-induced differentiation of bronchial fibroblasts into myofibroblasts in airway remodeling. Mol Med Rep 7(2):419–424
153. Shiwen X, Stratton R, Nikitorowicz-Buniak J, Ahmed-Abdi B, Ponticos M, Denton C, Abraham D, Takahashi A, Suki B, Layne MD, Lafyatis R, Smith BD (2015) A role of Myocardin related transcription factor-a (MRTF-A) in scleroderma related fibrosis. PLoS One 10(5):e0126015
154. Shrivastava A, Radziejewski C, Campbell E, Kovac L, McGlynn M, Ryan TE, Davis S, Goldfarb MP, Glass DJ, Lemke G, Yancopoulos GD (1997) An orphan receptor tyrosine kinase family whose members serve as nonintegrin collagen receptors. Mol Cell 1(1):25–34
155. Simoes FC, Cahill TJ, Kenyon A, Gavriouchkina D, Vieira JM, Sun X, Pezzolla D, Ravaud C, Masmanian E, Weinberger M, Mayes S, Lemieux ME, Barnette DN, Gunadasa-Rohling M, Williams RM, Greaves DR, Trinh LA, Fraser SE, Dallas SL, Choudhury RP, Sauka-Spengler-T, Riley PR (2020) Macrophages directly contribute collagen to scar formation during zebrafish heart regeneration and mouse heart repair. Nat Commun 11(1):600
156. Slack R, Macdonald S, Roper J, Jenkins R, Hatley R (2022) Emerging therapeutic opportunities for integrin inhibitors. Nat Rev Drug Discov 21(1):60–78
157. Smith T, Ferreira LR, Hebert C, Norris K, Sauk JJ (1995) Hsp47 and cyclophilin B traverse the endoplasmic reticulum with procollagen into pre-Golgi intermediate vesicles. A role for Hsp47 and cyclophilin B in the export of procollagen from the endoplasmic reticulum. J Biol Chem 270(31):18323–18328
158. Song S, Shackel NA, Wang XM, Ajami K, McCaughan GW, Gorrell MD (2011) Discoidin domain receptor 1: isoform expression and potential functions in cirrhotic human liver. Am J Pathol 178(3):1134–1144
159. Steplewski A, Fertala J, Beredjiklian PK, Abboud JA, Wang ML, Namdari S, Barlow J, Rivlin M, Arnold WV, Kostas J, Hou C, Fertala A (2016) Auxiliary proteins that facilitate formation of collagen-rich deposits in the posterior knee capsule in a rabbit-based joint contracture model. J Orthop Res 34(3):489–501
160. Steplewski A, Fertala J, Beredjiklian PK, Abboud JA, Wang MLY, Namdari S, Barlow J, Rivlin M, Arnold WV, Kostas J, Hou C, Fertala A (2017) Blocking collagen fibril formation in injured knees reduces flexion contracture in a rabbit model. J Orthop Res 35(5):1038–1046
161. Steplewski A, Fertala J, Tomlinson RE, Wang ML, Donahue A, Arnold WV, Rivlin M, Beredjiklian PK, Abboud JA, Namdari S, Fertala A (2021) Mechanisms of reducing joint stiffness by blocking collagen fibrillogenesis in a rabbit model of posttraumatic arthrofibrosis. PLoS One 16(9):e0257147
162. Sun B (2021) The mechanics of fibrillar collagen extracellular matrix. Cell Rep Phys Sci 2(8): 100515
163. Sutherland TE, Dyer DP, Allen JE (2023) The extracellular matrix and the immune system: a mutually dependent relationship. Science 379(6633):eabp8964

164. Suzuki N, Labosky PA, Furuta Y, Hargett L, Dunn R, Fogo AB, Takahara K, Peters DM, Greenspan DS, Hogan BL (1996) Failure of ventral body wall closure in mouse embryos lacking a procollagen C-proteinase encoded by Bmp1, a mammalian gene related to Drosophila tolloid. Development 122(11):3587–3595
165. Swift J, Ivanovska IL, Buxboim A, Harada T, Dingal PC, Pinter J, Pajerowski JD, Spinler KR, Shin JW, Tewari M, Rehfeldt F, Speicher DW, Discher DE (2013) Nuclear lamin-A scales with tissue stiffness and enhances matrix-directed differentiation. Science 341(6149):1240104
166. Szeto SG, Narimatsu M, Lu M, He X, Sidiqi AM, Tolosa MF, Chan L, De Freitas K, Bialik JF, Majumder S, Boo S, Hinz B, Dan Q, Advani A, John R, Wrana JL, Kapus A, Yuen DA (2016) YAP/TAZ are Mechanoregulators of TGF-beta-Smad signaling and renal Fibrogenesis. J Am Soc Nephrol 27(10):3117–3128
167. Taipale J, Miyazono K, Heldin C-H, Keski-Oja J (1994) Latent transforming growth factor-beta 1 associates to fibroblast extracellular matrix via latent TGF-beta binding protein. J Cell Biol 124(1):171–181
168. Tao J, Zhang M, Wen Z, Wang B, Zhang L, Ou Y, Tang X, Yu X, Jiang Q (2018) Inhibition of EP300 and DDR1 synergistically alleviates pulmonary fibrosis in vitro and in vivo. Biomed Pharmacother 106:1727–1733
169. Tasab M, Batten MR, Bulleid NJ (2000) Hsp47: a molecular chaperone that interacts with and stabilizes correctly-folded procollagen. EMBO J 19(10):2204–2211
170. Ten Dijke P, Arthur HM (2007) Extracellular control of TGFβ signalling in vascular development and disease. Nat Rev Mol Cell Biol 8(11):857–869
171. Terajima M, Taga Y, Nakamura T (2022) Lysyl hydroxylase 2 mediated collagen post-translational modifications and functional outcomes. Sci Rep 12:14256
172. Tomasek JJ, Gabbiani G, Hinz B, Chaponnier C, Brown RA (2002) Myofibroblasts and mechano-regulation of connective tissue remodelling. Nat Rev Mol Cell Biol 3(5):349–363
173. Trelstad RL (1982) The bilaterally asymmetrical architecture of the submammalian corneal stroma resembles a cholesteric liquid crystal. Dev Biol 92(1):133–134
174. Troilo H, Steer R, Collins RF, Kielty CM, Baldock C (2016) Independent multimerization of Latent TGFβ Binding Protein-1 stabilized by cross-linking and enhanced by heparan sulfate. Sci Rep 6(1):34347
175. Tschumperlin DJ, Lagares D (2020) Mechano-therapeutics: targeting mechanical signaling in fibrosis and tumor stroma. Pharmacol Ther 212:107575
176. Tsukada S, Parsons CJ, Rippe RA (2006) Mechanisms of liver fibrosis. Clin Chim Acta 364(1):33–60
177. Vallet SD, Ricard-Blum S (2019) Lysyl oxidases: from enzyme activity to extracellular matrix cross-links. Essays Biochem 63(3):349–364
178. Van der Slot AJ, Zuurmond A-M, Bardoel AF, Wijmenga C, Pruijs HE, Sillence DO, Brinckmann J, Abraham DJ, Black CM, Verzijl N (2003) Identification of PLOD2 as telopeptide lysyl hydroxylase, an important enzyme in fibrosis. J Biol Chem 278(42): 40967–40972
179. Van Der Slot AJ, Zuurmond A-M, Van Den Bogaerdt AJ, Ulrich MM, Middelkoop E, Boers W, Ronday HK, DeGroot J, Huizinga TW, Bank RA (2004) Increased formation of pyridinoline cross-links due to higher telopeptide lysyl hydroxylase levels is a general fibrotic phenomenon. Matrix Biol 23(4):251–257
180. van Huizen NA, Ijzermans JN, Burgers PC, Luider TM (2020) Collagen analysis with mass spectrometry. Mass Spectrom Rev 39(4):309–335
181. Velasquez LS, Sutherland LB, Liu Z, Grinnell F, Kamm KE, Schneider JW, Olson EN, Small EM (2013) Activation of MRTF-A-dependent gene expression with a small molecule promotes myofibroblast differentiation and wound healing. Proc Natl Acad Sci USA 110(42): 16850–16855
182. Verrecchia F, Chu M-L, Mauviel A (2001) Identification of novel TGF-β/Smad gene targets in dermal fibroblasts using a combined cDNA microarray/promoter transactivation approach. J Biol Chem 276(20):17058–17062

183. Vogel W, Gish GD, Alves F, Pawson T (1997) The discoidin domain receptor tyrosine kinases are activated by collagen. Mol Cell 1(1):13–23
184. Walmsley AR, Batten MR, Lad U, Bulleid NJ (1999) Intracellular retention of procollagen within the endoplasmic reticulum is mediated by prolyl 4-hydroxylase. J Biol Chem 274(21): 14884–14892
185. Walraven M, Hinz B (2018) Therapeutic approaches to control tissue repair and fibrosis: extracellular matrix as a game changer. Matrix Biol 71-72:205–224
186. Wang ML, Rivlin M, Graham JG, Beredjiklian PK (2018a) Peripheral nerve injury, scarring, and recovery. Connect Tissue Res 60:1–7
187. Wang Z, Stuckey DJ, Murdoch CE, Camelliti P, Lip GY, Griffin M (2018b) Cardiac fibrosis can be attenuated by blocking the activity of transglutaminase 2 using a selective small-molecule inhibitor. Cell Death Dis 9(6):613
188. Wipff P-J, Hinz B (2008) Integrins and the activation of latent transforming growth factor β1–an intimate relationship. Eur J Cell Biol 87(8–9):601–615
189. Yamauchi M, Sricholpech M (2012) Lysine post-translational modifications of collagen. Essays Biochem 52:113–133
190. Yang S, Plotnikov SV (2021) Mechanosensitive regulation of fibrosis. Cells 10(5):994
191. Yokosaki Y, Nishimichi N (2021) New therapeutic targets for hepatic fibrosis in the integrin family, α8β1 and α11β1, induced specifically on activated stellate cells. Int J Mol Sci 22(23): 12794
192. Yue W, Zhang H, Gao Y, Ding J, Xue R, Dong C, Liu F, Yang L, Yang L, Li L (2024) Procollagen-lysine 2-oxoglutarate 5-dioxygenase 2 promotes collagen cross-linking and ECM stiffening to induce liver fibrosis. Biochim Biophys Acta (BBA) – Mol Basis Dis 1870(5): 167205
193. Zhang EY (2009) Non-Smad pathways in TGF-β signaling. Cell Res 19(1):128–139
194. Zhang M, Zhang YY, Chen Y, Wang J, Wang Q, Lu H (2021) TGF-β signaling and resistance to cancer therapy. Front Cell Dev Biol 9:3310
195. Zhao W, Yang A, Chen W, Wang P, Liu T, Cong M, Xu A, Yan X, Jia J, You H (2018) Inhibition of lysyl oxidase-like 1 (LOXL1) expression arrests liver fibrosis progression in cirrhosis by reducing elastin crosslinking. Biochim Biophys Acta (BBA) – Mol Basis Dis 1864(4):1129–1137
196. Zhao M, Wang L, Wang M, Zhou S, Lu Y, Cui H, Racanelli AC, Zhang L, Ye T, Ding B (2022) Targeting fibrosis: mechanisms and clinical trials. Signal Transduct Target Ther 7(1): 206
197. Zhou X, Zhang J, Haimbach R, Zhu W, Mayer-Ezell R, Garcia-Calvo M, Smith E, Price O, Kan Y, Zycband E, Zhu Y, Hoek M, Cox JM, Ma L, Kelley DE, Pinto S (2017) An integrin antagonist (MK-0429) decreases proteinuria and renal fibrosis in the ZSF1 rat diabetic nephropathy model. Pharmacol Res Perspect 5(5):e00354
198. Zhou B, Tu T, Gao Z, Wu X, Wang W, Liu W (2021) Impaired collagen fibril assembly in keloids with enhanced expression of lumican and collagen V. Arch Biochem Biophys 697: 108676

Chapter 12
The Impact of Ageing on Fibrillar Collagens

Elizabeth G. Canty-Laird, Himadri S. Gupta, and Helen L. Birch

Abstract The world population is ageing rapidly. The over-60s now outnumber the under- 5s, and 1 in 6 people will be over 60 by 2030 (WHO). Collagen is a key structural component of many tissues and organs, and although a fraction of the collagenous component of tissues is remarkably long-lived, it progressively accumulates damage over a lifetime. The capacity for new collagen synthesis and post-translational modification is altered and dysregulated during ageing. The mature crosslinks that stabilise collagenous tissues can remain stable or increase with age, whereas age-related glycation end-products can increase and affect tissue biomechanics. At the fibrillar nanoscale, changes associated with ageing and disease influence fibril deformation and stress transfer in a tissue-specific manner. Age-related loss of collagen can be caused by proteolytic degradation, but normal collagen turnover is also affected by ageing and its dysregulation is detrimental to tissue homeostasis. Age-related accumulation of senescent cells may contribute to the aberrant turnover of collagen during ageing. Finally, collagen itself may hold the key to counteracting some of the detrimental effects of ageing, with ingested hydrolysed collagen peptides demonstrating beneficial effects on skin and the musculoskeletal system.

Keywords Ageing · Collagen · Crosslinking · Nanostructure · Nanomechanics · Degradation · Turnover · Senescence · Supplements

---

E. G. Canty-Laird (✉)
Department of Musculoskeletal and Ageing Science, Institute of Life Course and Medical Sciences, William Henry Duncan Building, Liverpool, UK
e-mail: elizabeth.laird@liverpool.ac.uk

H. S. Gupta
School of Engineering and Materials Science, Queen Mary University of London, London, UK
e-mail: h.gupta@qmul.ac.uk

H. L. Birch
University College London, Stanmore Campus, Brockley Hill, Stanmore, UK
e-mail: h.birch@ucl.ac.uk

D. A. D. Parry, P. K. Luther (eds.), *Fibrous Proteins: Structures and Mechanisms*, Subcellular Biochemistry 113, https://doi.org/10.1007/978-3-032-05273-5_12

Introduction

The collagens are a diverse family of proteins with ancient origins, and the most abundant protein in all multicellular animals from simple sponges to complex mammalian species [44]. Collagen is best recognised as a structural protein in animals, providing support to cells and mechanical resilience to tissues. More recent work, however, has highlighted additional roles for collagen in cell signalling and cell behaviour. Given the broad function of the collagens, alterations to their structure, from primary structure to the supramolecular assemblies they form, have widespread implications for mammalian health and disease. The characteristics of collagen allow a high degree of variation between genetically identical proteins, in addition to simple up- or down-regulation of collagen synthesis. For example, following translation into polypeptide chains, a large amount of enzyme-mediated post-translational modification occurs providing opportunity for chemical variety. Second, the relatively long half-life of some collagens provides a chance for adventitious chemical modifications resulting in further variation in collagen structure. In this chapter, the changes to collagen that occur during the ageing process and the consequence of these changes will be discussed.

The Collagen Family

The collagen family has at least 28 members, characterised by common features relating to primary and quaternary structure, post-translational modifications and extracellular function [136]. The defining features of a collagen are an extracellular protein with a structural role in tissues with a triple-helical structure formed from three polypeptide chains. The primary structure of a repeating glycine-containing triplet of amino acids is an essential feature, as is the predominance of the secondary amino acids, proline and hydroxyproline. Another feature of the collagens is extensive post-translational modification, including covalent crosslinking between collagen molecules. For ease of understanding, the collagens have been grouped according to the supramolecular assemblies they form (see [77] for a review of collagen types). The most abundant of the collagens are fibril-forming collagens and include collagen types I, II, III, V, XI, XXIV and XXVII. The long uninterrupted triple-helical domains of these collagen types result in rod-like molecules allowing assembly into fibrils and further hierarchical packing into fibres. A group of collagens known as fibril-associated collagens with interrupted triple-helices (FACIT), as the name implies, do not themselves form fibrils, but associate themselves with the fibrillar collagens. Network-forming collagens interact through non-collagenous domains at the carboxy-terminus forming sheet-like networks found in membranes, such as the basement membrane. Beaded filament-forming collagens have large globular domains at the carboxy- and amino-terminus that form a bead-like structure when collagens self-assemble end-to-end, the relatively short triple- helical part of the molecule forming the filament between the beads. Type VII collagen forms anchoring filaments that attach the basement membrane to the underlying dermis. Although collagens, by definition, are extracellular proteins, a group of four collagens have a transmembrane domain and a short amino-terminal domain that resides within the cell cytosol, although the majority of the molecule has triple-helical extracellular domains. The final group of collagens are endostatin-producing collagens as they can be cleaved to produce the potent anti-angiogenic factor, endostatin. The fibril-forming collagens, as fibrous proteins, are the focus of this chapter.

Alterations in the Collagen Content of Tissues with Age

With ageing, the fibrillar collagen content of tissues and organs can either decrease due to reduced production and increased degradation or increase due to focal overproduction (Fig. 12.1) that may occur due to an altered mechanical cellular environment.

Skin collagen content decreases by approximately 1% per year during adulthood [146]. Skin is particularly affected by photo-ageing caused by UV damage to collagen itself and to resident cells, which makes sun-exposed skin particularly prone to age-related loss of collagen. The collagen content of bone [10] and either bone mass or bone mineral density as a proxy measure of collagen content are also known to decrease with age [29, 111, 118], which manifests as a reduced bone mineral density leading to osteopenia or osteoporosis [145].

Conversely, collagen content increases with age in lung [21], liver [99], heart ventricles [14, 35, 36, 42, 67, 68, 157], vasculature [60] and intramuscular connective tissue (epimysium, perimysium and endomysium) of skeletal muscle [47].

Collagen Structure

The Collagen Molecule

Collagen molecules are formed from three polypeptide chains wrapped around each other to form a right-handed superhelix, first described by Ramachandran and Kartha [133]. In the fibril-forming collagens, these chains are approximately 1000 amino acids long and composed of a long uninterrupted triple-helical domain flanked by non-helical regions at the N- and C-terminals (telopeptides) and a pro-peptide globular domain at each end, which is later removed in processing. The sequence of amino acids in the polypeptide chains (protein primary structure) is key to the triple-helix formation and is characterised by a repeating triplet of amino acids with glycine every third residue. The triplet, represented by Gly-Xaa-Yaa, has an abundance of the secondary amino acid proline, some of which are hydroxylated in a post-translational modification to form hydroxyproline. Other amino acids occurring

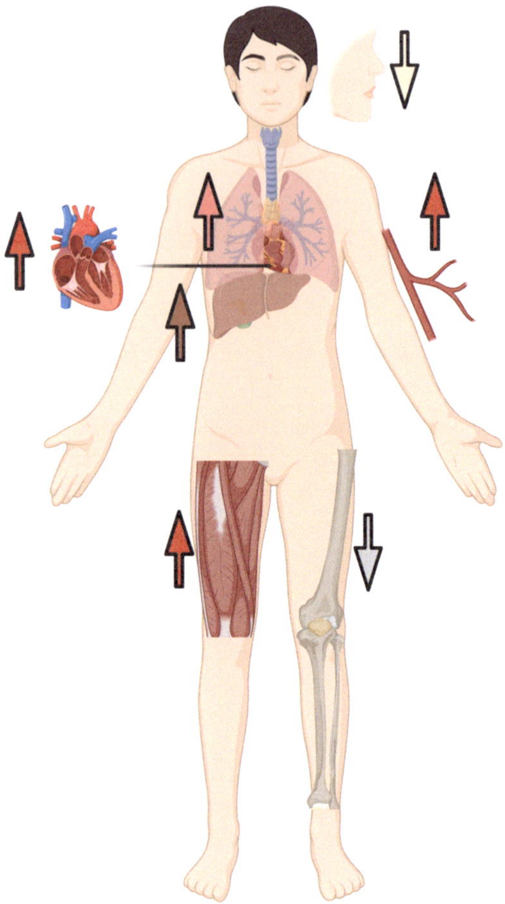

Fig. 12.1 Age-related alterations to collagen content in tissues and organs. The directionality of the change in collagen content is indicated by arrows and depicted in relation to the human body. Collagen content decreases in skin and bone but increases in heart, lung, liver, muscle and the vasculature with age. (Figure created in Biorender)

frequently in the triple-helical region tend to be smaller, such as alanine and serine or have charged side chains such as arginine and glutamic acid. Lysine residues are common, often being subjected to hydroxylation and are important due to their role in crosslink formation between collagen molecules. The polypeptide chains differ between collagen types and, in some cases, within the collagen molecule. For example, type I collagen is a heterotrimer composed of two alpha I chains (α1(I)) and one alpha 2 chain (α2(I)), while type III collagen is a homotrimer formed from three identical polypeptide chains (α1(III)). These different polypeptide chains are distinct gene products, resulting in 11 genes for the 7 members of the fibril-forming collagen group.

Collagen Biosynthesis and Post-translational Modifications

Proline Hydroxylation

Following transcription and translation of mRNA into polypeptide chains, a series of enzyme-driven post-transitional modifications occur, initially within the cell, prior to folding of the triple helix. Proline residues, predominantly in the Y position of the triplet, undergo hydroxylation mediated by prolyl 4-hydroxylase to produce 4-hydroxyproline and, to a lesser extent, prolyl 3-hydroxylase to produce 3-hydroxyproline. Most of the proline residues in the Y position are hydroxylated in α1(I) and α2(I) chains to form 4-hydroxyproline. Proline hydroxylation is important as it increases the thermal stability of the collagen triple helix [63]. The degree of proline hydroxylation in the Y position is considered relatively stable, although differences are known to occur between species and have been observed between tissue types. For example, in cold water fish, proline hydroxylation levels are lower, resulting in a denaturation temperature about 8 °C lower than in warm-blooded mammals [63, 89]. In human skin, 41% of proline residues in α1(I) are hydroxylated while in bone 46% are hydroxylated [115]. The higher hydroxylation for bone is reflected in a 1.9 °C higher denaturation temperature for the collagen molecule in bone. Variation seems to occur primarily in the helical portion of the α1(I) chain close to the N-terminus, where in chick bone 49% of the proline resides theoretically susceptible to hydroxylation were found to be hydroxylated but only 35% in tendon [13]. In rat tissue, proline hydroxylation in the same site was found to be 80% for available proline residues in skin compared to 50% in tail tendon [19].

The presence of 3-hydroxyproline was first reported in bovine tendon collagen [116] and later suggested to be the only hydroxylation of proline in the Xaa position when found in proline position 983 in the chick α1(I) chain [49]. Further studies have revealed partial modification to 3-hydroxyproline at other X positions of proline in type I, II and V collagens in human and bovine species but no 3-hydroxyproline in type III collagen [175]. There are three different forms of the prolyl 3-hydroxylase enzyme, and studies have suggested that hydroxylation at proline 986 is under different control to those at other sites [48]. The distance between sites of proline hydroxylation and 3-hydroxyproline was noted to be regular and close to the D-period, and therefore, it was suggested that 3-hydroxyproline may play a role in fibril formation [175]. The formation of 3-hydroxyproline in type I collagen differs between tissues, with rat tail tendon showing an enriched region at the C-terminus which is absent in skin and bone [46].

Lysine Hydroxylation

In addition to proline hydroxylation, lysine residues in the nascent polypeptide chains are converted to 5-hydroxylysine [127, 128]. The degree of lysine hydroxylation is variable between collagen types, tissue types and region of the polypeptide chains. Type V collagen is highly hydroxylated while type I collagen has low to moderate levels of lysine hydroxylation [72]. Levels of hydroxylation are higher in chick tendon than in bone, and levels in bone higher than in skin [13]. In human type I collagen, the α1(I) chain contains 38 lysine residues while the α2(I) chain contains 31 (UniProt sequence accession numbers P02452 and P08123) although the percentage of lysine residues hydroxylated is higher in the α1(2) chain than in the α1 (I) chain, as measured in chick tissues [13]. The extent of hydroxylation was found to vary between the helical and non-helical regions of the collagen molecule, suggesting these sites are under separate enzymatic control [13]. In fact, in addition to lysyl hydroxylase 1 (LH1), two further forms; lysyl hydroxylase 2 (LH2) and lysyl hydroxylase 3 (LH3), have been identified [165, 166], with LH2 having two alternatively spliced variants, LH2a and LH2b [186]. The three different isoforms of lysyl hydroxylase are coded for by PLOD1, PLOD2 and PLOD3 genes, respectively. LH1 hydroxylates lysine residues in the helical region of the polypeptide chain in the Yaa position [184] while LH2b hydroxylates lysine residues in the telopeptide regions [163]. LH3 is multifunctional with lysine hydroxylation and glycosylation activity [151], glycosylation being the next step in post-translational modification. Initially, a galactose sugar is added to some of the hydroxylysine residues in the helical domain of the polypeptide chain, catalysed by hydroxylysyl galactosyltransferase (GT). Some of these glycosylation sites are further modified by the addition of glucose to form glucosyl-galactosyl-hydroxylysine catalysed by galactosylhydroxylysyl glucosyltransferase (GGT). Evidence suggests that the main function of LH3 is as a GGT [151]. The ratio of hydroxylysine residues modified by addition of galactose followed by glucose and those modified by addition of galactose alone differs in different tissues with collagen in skin having a higher ratio than in tendon and cartilage and bone having the lowest ratio [143]. Further information on collagen glycosylation can be found in Chap. 10 [114].

Crosslinking

The modified polypeptide chains fold to form the triple helix inside the cell before secretion into the extracellular matrix. Following cleavage of N- and C-terminal pro-peptides, the newly synthesised collagen molecules assemble in a quarter-staggered arrangement to form fibrils. The strength of the fibrils is greatly enhanced by the formation of covalent crosslinks between the collagen molecules at specific sites.

The first step in crosslink formation (Fig. 12.2) is the action of the amine oxidase enzyme lysyl oxidase (LOX) which converts lysine and hydroxylysine residues in the telopeptide region of the collagen molecules to an aldehyde. LOX, like LH, is a family of proteins with five different members, namely, LOX and four LOX-like (LOXL) proteins (LOXL1–4) [106]. Differing expression of these isoforms may play a role in controlling the type of crosslink formed. Hydroxylysine aldehydes in the telopeptide undergo a further spontaneous reaction with a hydroxylysine or lysine residue in the helical region of neighbouring molecules to form the immature bivalent crosslinks dehydro-dihydroxylysinonorleucine (deH-DHLNL) and

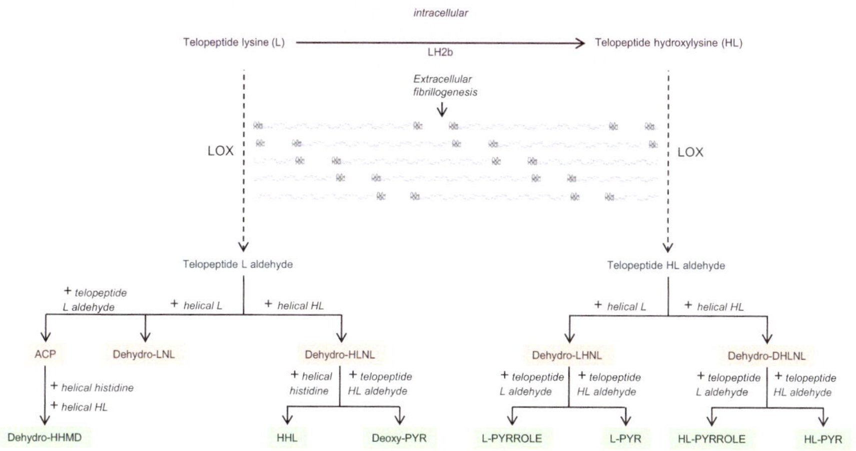

Fig. 12.2 Pathways for the formation of lysyl oxidase-mediated crosslinks between neighbouring collagen molecules

dehydro-lysinohydroxynorleucine (deH-LHNL), respectively. However, if a lysine aldehyde is involved in the reaction with a helical hydroxylysine residue, the crosslink dehydro-hydroxylysinonorleucine (deH-HLNL) results. The bivalent immature crosslinks undergo further reaction with a hydroxylysine or lysine aldehyde in the telopeptide of a third collagen molecule to produce a series of pyridinoline and pyrrole mature crosslinks (Fig. 12.2). The type of crosslink formed depends on the degree of lysine hydroxylation in the helical part of the molecule and the non-helical telopeptide region (for a review of collagen crosslinking, see [9, 185]). Extensive hydroxylation of lysine residues throughout the collagen molecule results in the formation of hydroxylysyl-pyridinoline (HL-Pyr). In tissues where less hydroxylation of the helical or telopeptide lysine residues occurs, the mature crosslinks lysyl-pyridinoline (L-Pyr) and pyrrole form, respectively. A further mature crosslink, histidino-hydroxylysinonorleucine (HHL), found predominantly in skin but also present in some tendons, can form from the interaction between an immature deH-HLNL crosslink and a helical histidine residue. In tendon and cartilage, HL-PYR is the dominant mature crosslink. In mineralising tissues, a relatively low level of lysyl hydroxylation results in low levels of HL-PYR, and the occurrence of the largely bone-specific L-Pyr and pyrrolic crosslinks [9].

Age-Related Alterations in the Synthesis and Post-translational Modification of Fibrillar Collagens

Direct measurements of collagen synthesis with age can be challenging, especially in human tissue. Measures that can be employed include monitoring mRNA levels by RT-qPCR as a proxy for protein synthesis, immunoassays to detect the procollagen propeptides that are present on procollagen and removed prior to fibrillogenesis, measuring incorporation of radiolabelled proline, or determining the amount of radiolabelled hydroxyproline.

Age-Related Decline in Gene Expression

Type I and type III collagen mRNA levels generally decrease with age. In human dermal fibroblasts, COL1A1 gene expression decreased with UVB irradiation, serial passage and in fibroblasts from older donors, although though COL1A2 increased in fibroblasts from elderly donors [81]. COL1A1 gene expression has been shown to decrease with age in human skin as assessed by Northern blotting and in-situ hybridisation [27] as has 'type I procollagen mRNA' by qPCR [131]. In mouse lung fibroblasts, Col1a1, Col1a2 and Col3a1 decreased with age although fibroblasts in older mice were less resistant to pro-fibrotic induction [123]. In the heart, Col1a1 and Col3a1 mRNAs were found to be decreased in the left ventricular wall of aged rats by Northern blot analysis [5, 14, 157] although some reports indicate that Col1 and Col3a1 mRNAs may be unchanged with age in the heart [67].

Age-Related Changes in Collagen Protein Synthesis

In accordance with decreased type I and type III procollagen gene expression with age, there appears to be a decrease in new collagen protein synthesis with age. Decreased procollagen production has been reported in human skin and in cultured dermal fibroblasts using immunoassays for the propeptides [167, 168], an effect mediated by age-related suppression of the TGF-b/Smad/CTGF axis [131]. In rat skin, heart, lung and muscle, there was a decreased collagen fractional synthesis rate with age, though with a temporary increase at 15 months in heart and muscle that was attributed to new collagen synthesis derived entirely from recycling of newly synthesised collagen in these tissues [100]. Conversely, in liver, the collagen fractional synthesis rate dipped at 15 months in rats but recovered by 24 months; however, the fractional collagen deposition rate (distinct from the synthesis rate) decreased with age [99] (Fig. 12.3). A more recent study in mice using pulsed SILAC labelling and proteomics demonstrated reduced collagen incorporation into tissues in aged skin, bone and cartilage [6]. In human muscle, a higher collagen fractional synthesis rate was detected in elderly male subjects but without a corresponding increase in muscle collagen concentration leading the authors to conclude that collagen breakdown was also elevated in the elderly [7].

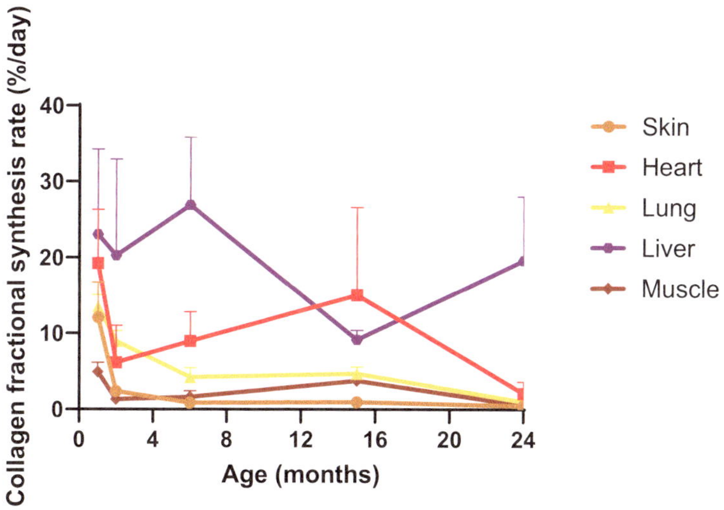

Fig. 12.3 Collagen fractional synthesis rates at various ages in rat tissues. Data presented in tables in [99, 100] were plotted. SDs were calculated from both SEM values and the number of samples and are shown as positive error bars. $N = 6$ except for 2-month-old samples, where $n = 4$ and, 1-month-old samples, where $n = 8$, for all tissues except liver

Age-Related Changes to Alpha Chain Ratios in Type I Collagen

Although most type I collagen molecules are heterotrimers comprising two alpha-1 (I) chains and an alpha-2(I) chain, an abnormal form of type I collagen, comprising three alpha-1(I) chains has been reported (Fig. 12.4). The homotrimer was first reported in chick embryo chondrocytes grown in culture [97] where type II collagen was initially synthesised but over time synthesis switched to heterotrimeric and homotrimeric type I collagen. The same authors later observed that the change in collagen synthesis was associated with factors leading to cellular senescence [98]. The homotrimeric form has been genetically or biochemically associated with age-related musculoskeletal [125, 182, 188] and cardiovascular diseases [20, 83, 94] as well as with fibrosis [179] and cancer [190, 191].

Although a mouse model of osteogenesis imperfecta ('oim') comprises solely homotrimeric type I collagen, by comparison to a Col1a2 null, it has been shown that the homotrimer itself does not directly cause bone fragility [83]. However, the mechanical strength of myocardium, aorta and tendon is altered in the oim model [26, 104, 122, 170, 174], and Col1a2 null mice show cardiovascular abnormalities including dilated left ventricle, cardiac hypertrophy and reactive cardiac fibrosis [73, 148]. In humans, complete absence of the alpha-2(I) chain causes joint hypermobility and cardiac valve defects [61, 94]. Altered mechanical properties of tissues comprising or containing homotrimeric type I collagen may be due to less

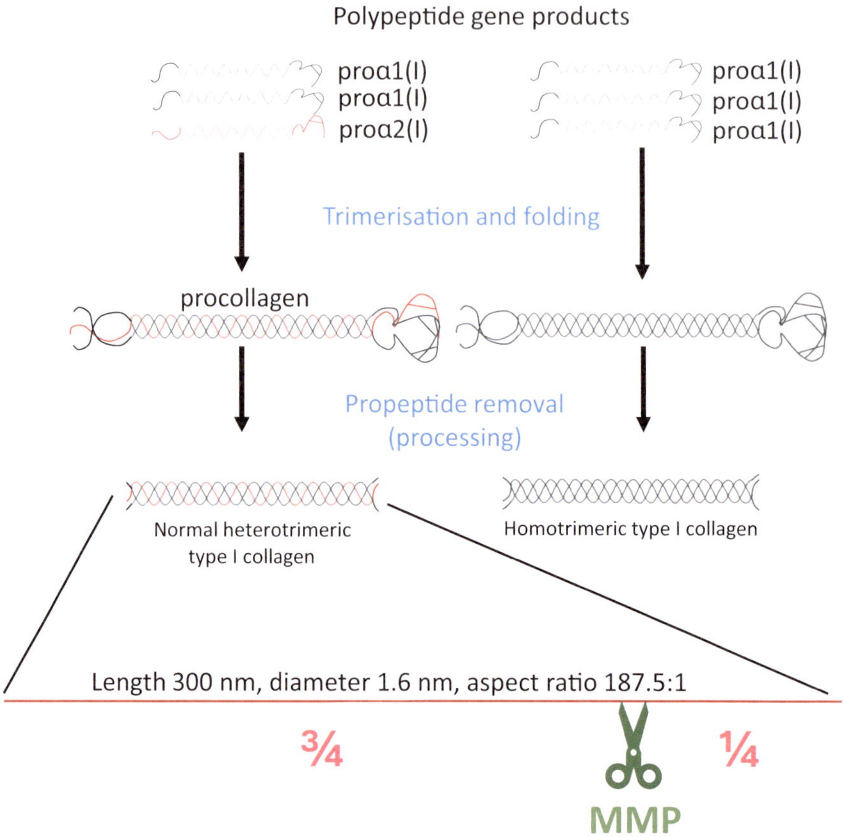

Fig. 12.4 MMP cleavage of the collagen triple helix produces ¾ and ¼ length fragments. Normal type I procollagen is a $(\alpha 1)_2(\alpha 2)_1$ heterotrimer of the gene products of the COL1A1 and COL1A2 genes. The $(\alpha 1)_3$ homotrimer is derived from the COL1A1 gene alone. The proα chains are co-translationally translocated into the endoplasmic reticulum where post-translational modification and trimerisation occur. Subsequent proteolytic removal of the N- and C-terminal propeptides results in 'tropocollagen' molecules 300 nm in length, which form the building blocks of collagen fibrils. The full-length collagen triple helix is thinner than can be depicted with an aspect ratio of 187.5:1. Homotrimeric type I collagen is resistant to MMP-mediated degradation

ordered packing of collagen in the fibril, as demonstrated by a lower enthalpy of denaturation [11], or due to the homotrimeric collagen molecule being more resistant to degradation by collagenase. The action of collagenase requires local unwinding of the triple helix at the cleavage site [28], and this is less efficient in homotrimeric collagen, such that type I collagen homotrimer is resistant to MMP-mediated degradation [65].

Age-Related Changes to Alpha Chain Hydroxylation and Glycosylation

The overall levels of 4-hydroxyproline are relatively stable through development and ageing although there appears to be a decrease in proline hydroxylation close to the N-terminus in α1(I) chains of chick from the embryo to 18 months of age [13]. Conversely, 3-hydroxyproline levels were found to increase from 0.5 months to 3 months after birth and then remain constant in rat tail tendon, while there were no age-related changes to skin and bone collagen [152]. An increase in 3-hydroxyproline was also seen in bovine Achilles tendon from foetal to adult age. The site of increased proline hydroxylation occurred not at the known site of proline 986 but at another previously identified modification site at proline 707 and newly identified sites, proline 716 and proline 719 [152].

Although lysine hydroxylation shows considerable variation, age-related changes have not been clearly defined. Lysine hydroxylation was found to decrease rapidly in chicks from 1 day old to 3 weeks in skin, bone and tendon [13] although in bone, and to some extent tendon, hydroxylation in the N-terminal telopeptide remained at the high levels seen in the embryo. There appeared to be no change in lysine hydroxylation between 16 weeks and 18 months of age. The expression of LH2 increased sixfold as bone marrow stromal cells underwent differentiation in vitro, and this was associated with increased lysine hydroxylation in the telopeptides of type I collagen [163].

Glycosylation of hydroxylysine residues is thought to play a role in controlling the morphology of collagen fibrils with higher levels of glycosylation being associated with smaller diameter fibrils. An increase in glycosylation has been shown in several age-related diseases including osteoarthritis and osteoporosis [162].

Age-Related Changes to LOX-Mediated Crosslinks

The levels of bivalent crosslinks decline with increasing chronological age reaching a steady state at approximately 4–5 years in bovine Achilles tendon and 17–20 years in human Achilles tendon, in parallel with an increase in the trifunctional crosslinks [8]. It was this observation that paved the way to understanding that mature crosslinks form from further reactions between immature crosslinks and neighbouring collagen molecules. At maturity, the levels of immature crosslinks are related to the turnover rate of the tissue; hence, in tissues such as cartilage and tendon, immature crosslinks virtually disappear while in bone low levels remain [45]. Levels of the mature pyridinoline crosslinks increased in human cancellous and cortical bone and cartilage, reaching a maximum at 10–15 years and then remained relatively stable throughout adult life to 80 years of age [45]. Similarly, in human iliac crest bone samples, levels of L-PYR, HL-PYR and pyrrole appear to remain relatively stable during ageing from 18 to 96 years [10]. In mouse tail tendon, levels of pyridinoline increased from 8 to 100 weeks. However, in human Achilles tendon, total pyridinoline crosslink levels dropped significantly from a young group of individuals (14–21 years) to an old group (65–81 years), and this was due to a decrease in the major crosslink HL-PYR, as the minor crosslink L-PYR showed a significant increasing with increasing age. A similar, but smaller, trend was seen in the tibialis anterior tendon [91]. In equine tendon, levels of L-PYR also showed a significant increase in two tendon types from horses between four and 30 years of age [158].

Alterations in Cellular Composition with Age

Alterations to the cellular composition of tissues with age could affect whether fibrillar collagen synthesis is increased or decreased with age. For example, in liver, single-cell transcriptome analysis demonstrated that hepatic stellate cells (HSCs) acquired an activated phenotype with ageing [87]. HSC activation is characterised by fibrillar collagen synthesis and liver fibrosis [160]. Similarly, in the ageing heart, fibroblast populations were altered with a subset adopting an osteogenic transcriptome [169]. Conversely, in skin, CD26/CD90 double-positive fibroblasts with high COL1A1 expression do not change abundance with age [181].

Spontaneous Chemical Modifications to Fibrillar Collagen In Situ with Age

Proteins in the body are susceptible to spontaneous non-enzymatic chemical modifications, and these can be reversible or irreversible. In general, non-enzymatic post-translational modifications are considered to be detrimental; however, there is an increasing awareness that some reversible modifications play an important role in a range of cell processes, adding an additional layer of complexity to control of cell behaviour under normal physiological conditions [66]. As the fibrous collagen framework is turned over relatively slowly in most tissues, irreversible modifications accumulate with increasing age, and these are likely to be detrimental. Irreversible modifications include racemisation, oxidation, deamidation, carbonylation, glycation and carbamylation. Racemisation is the simplest of these modifications and results from a re-arrangement of functional groups to form stereoisomers. The rate of racemisation varies between different amino acids and is most rapid in aspartic acid, hence its use as a marker for collagen age. The other chemical modifications involve the removal or addition of chemical groups following attack on the protein by reactive species. Most relevant to collagen are those modifications that affect amino acids with a high representation and importance in collagen, such as lysine. Lysine residues are susceptible to glycation by the addition of sugars, and these modifications are the most well-studied due to their prominence in diabetic patients, although the impact of their formation is still relatively poorly understood.

Collagen Glycation and Formation of Advanced Glycation End-Products

Advanced glycation end-products (AGE) form through a series of reactions initiated by a reactive carbonyl group on a reducing sugar attacking the nucleophilic amino group on lysine and arginine amino acid side-chains in proteins. This process is known as the Maillard reaction or commonly referred to as 'browning'. A series of re-arrangements occurs from a Schiff base to a more stable Amadori product, and then over time, a series of re-arrangements which may culminate in a further reaction with an amino group on a side-chain of a second arginine or lysine amino acid resulting in a crosslink. In theory, this series of reactions can result in numerous different types of AGEs, some within a protein and some that crosslink between proteins (Fig. 12.5). AGEs are often fluorescent, and it has been well documented that collagen fluorescence increases with increasing age due to accumulation of these compounds [91].

There are a number of factors that determine the type of AGE formed in vivo including the source of reactive carbonyl group, amino acid residues involved and the metabolic status of the individual. Hexose sugars, such as glucose, galactose and fructose can take part in the initial reaction or a pentose sugar such as ribose and arabinose. As glucose is most abundant in healthy individuals, being present at levels between 4.4 and 6.1 mM, glucose is most likely to be involved. Other sugars such as fructose are more reactive in the glycation reaction, but they are only present at

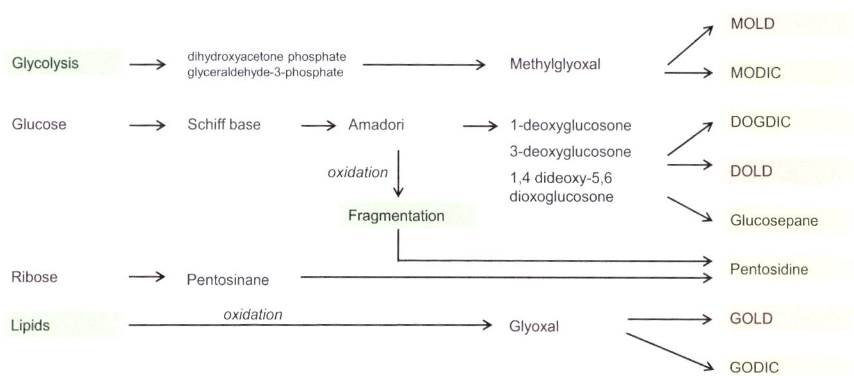

Fig. 12.5 Pathways leading to the formation of AGE (advanced glycation end-product) crosslinks following the Maillard reaction. Lysine-arginine crosslinks are shown in yellow and lysine-lysine crosslinks shown in orange. (Adapted from [16])

micro-molar levels [141]. AGE crosslinks can also be formed from other dicarbonyl species, namely glyoxal, methylglyoxal, 1-deoxyglucosone and 3-deoxyglucosone. Glycolysis can be a source of methylglyoxal through dephosphorylation of dihydroxyacetone phosphate and glyceraldehyde-3-phosphate under conditions of high glucose flux and oxidative stress [124]. Glyoxal can be produced by oxidation degradation of polyunsaturated fatty acids. Glyoxal, methylglyoxal and 3-deoxyglucosone react with two lysine residues on neighbouring molecules to form the imidazolium crosslinks, GOLD, MOLD and DOLD, respectively. Glyoxal, methylglyoxal and 3-deoxyglucosone can also react with a lysine and arginine residue to form the AGE crosslinks MODIC, GODIC and DOGDIC. It should be noted that glyoxal and methylglyoxal are effectively detoxified by mammalian glyoxalase enzyme systems in healthy individuals, so levels are low and subsequently, levels of GOLD, GODIC, MOLD and MODIC are not likely to accumulate in vivo. In fact, DOGDIC, MODIC and GODIC crosslinks have all been detected in collagen from human skin and glomerular basement membrane samples although levels are low [144].

Pentosidine is a lysine-arginine crosslink resulting from reaction with pentose sugars or Amadori breakdown products, although pentose sugars are thought to be the predominant pathway, through the intermediate crosslink pentosinane [15]. Pentosidine is the most widely measured and reported AGE crosslink in the literature due to its acid stability and inherent fluorescence. Pentosidine shows a strong correlation with increasing age in tendon [32, 158]; however, absolute levels in tissues are relatively low. In skin samples, pentosidine levels of between 6 and 10 mmoles/mole collagen for healthy and diabetic individuals, respectively, have been measured [107]. In tendon tissue, measured levels range between 5 mmoles/mole collagen for human posterior tibialis and flexor digitorum longus tendons [31], 14 mmoles/mole collagen for equine tendons [158] and 11–73 mmoles/mole collagen for old and young human patellar tendon [32]. This is likely due to low plasma levels of pentose sugars of about 44 μM in human plasma.

Glucosepane, an AGE crosslink formed from glucose through 1,4-dideoxy-5,6-dioxoglucosone with starting reactants of glucose, lysine and arginine, is considered to be of prime physiological significance [15]. Indeed, glucosepane has been measured at considerably higher levels than other AGE crosslinks in skin samples from diabetic (5000 pmoles/mg collagen) and non-diabetic people (2000 pmoles/mg collagen) [108, 144]. The levels measured approach those for the lysyl oxidase-mediated collagen crosslinks, suggesting that multiple crosslinks may form for each collagen molecule. Furthermore, glucosepane levels show a significant increase with ageing in skin [144] and tendon [91] suggesting that glucosepane may play a part in age-related decline in collagen function. Using a combination of computational and experimental techniques, the presence of glucosepane has been shown to be associated with reduced density of collagen packing and increased porosity to water molecules [91].

Protein Carbamylation

Carbamylation is a non-enzymatic post-translational modification of proteins that occurs when the amino group of a lysine residue binds to isocyanic acid forming homocitrulline. Although any of the lysine residues may be susceptible to carbamylation, four preferential sites in the α1(I) chain and one in the α2(I) chain have been identified, some of which overlap with glycation sites [59]. Additionally, hydroxylysine can also undergo carbamylation to form hydroxyhomocitrulline [153]. Breakdown of urea produces isocyanate; thus, conditions which cause an increase in blood urea levels such as chronic kidney disease are associated with higher levels of homocitrulline. Homocitrulline is mostly associated with collagen and has been shown to accumulate in dermis, bone and tendon with increasing age in murine species. In longer-lived species, homocitrulline reaches higher levels and has been shown to accumulate in bovine and human dermis [59]. The relative importance of carbamylation compared to glycation was investigated by comparing the absolute levels of homocitrulline and the glycoxidation product carboxymethyl-lysine (CML). It was found that in human skin collagen samples over 70 years of age, homocitrulline levels were 5.15 mmol/mole lysine compared to 1.96 mmoles/mole of collagen for CML [59], although it should be noted that CML is not the major AGE in human collagen samples, being some four times lower than glucosepane levels [109].

Nanostructure and Nanomechanics

Nanostructure and Mechanics of Ageing Collagen

The collagenous component of the ECM exhibits a hierarchical structure, ranging from the molecular to the macroscopic level. At the nanoscale, tropocollagen triple-helical molecules undergo guided self-assembly to form microfibrils which form fibrils ~20–200 nm in diameter [53]. These fibrils can form uniaxial fibre bundles (in tendons and ligaments), cylindrical calcified lamellae (in bone), planar isotropic patterns (e.g. in skin) or other geometries, such as the arcing structure in cartilage. Ageing will induce biochemical alterations at the fibrillar- and intra-fibrillar level; however, the impact of these changes on function (e.g. biomechanics) will be different across tissues due to their distinct architectures. In this section, we review the effect of age on collagenous tissue structure and function at the nanometre (fibrillar) level, across a range of tissues. It will be seen that some changes are necessarily convolved with other age-related disorders, such as diabetes.

Fibrillar collagens exhibit a characteristic axial periodicity in their electron density with a period of D~65–67 nm, due to the inhomogeneity in charge distributions along the tropocollagen molecule. The D-period serves as a marker for fibril strain, and intrafibrillar organisation, which can be accessed by nanoscale imaging

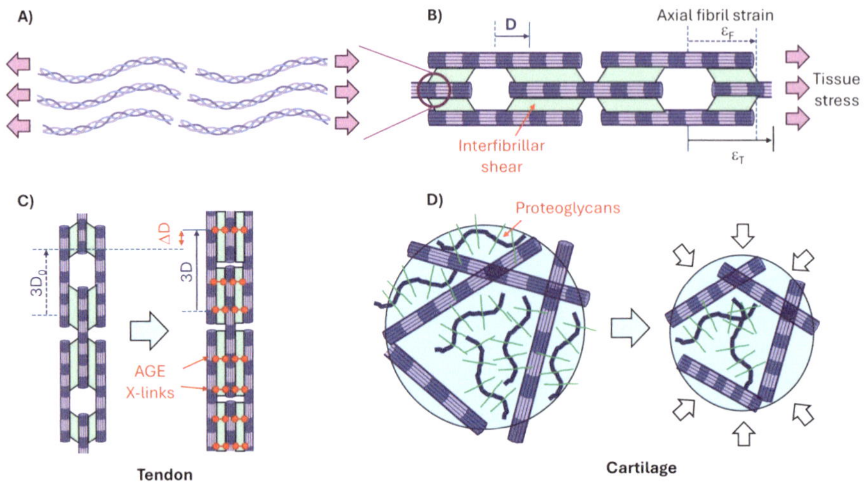

Fig. 12.6 (**a**) Molecular and (**b**) fibrillar collagen ultrastructure showing D-period and interfibrillar shear. (**c**) Increased D-period in tendon (AGE X-links) and (**d**) reduced D-period in cartilage (PG depletion). (Figure created in Biorender)

and structural techniques like scanning probe microscopy imaging and microfocus small-angle X-ray scattering [3, 70, 74], and is shown schematically in Fig. 12.6a, b.

Techniques like atomic force microscopy and small-angle X-ray scattering with in situ mechanics have been used to characterise the fibrillar-level mechanics of collagenous tissues. A common theme in healthy tissue (neither specifically aged nor diseased) is the presence of a hierarchical deformation pattern, with shear between stiff elements at the fibrillar and molecular- level, corresponding to the interfibrillar matrix and between the triple helices, respectively [54, 62, 139]. Crosslinking makes fibrils stiffer but can also increase connections between fibrils; the first option would reduce individual fibril strain and increase strain in the interfibrillar matrix, while the second would increase fibril strain due to greater stress transfer at the interfibrillar level.

Organs with a wider distribution of fibre orientations across the tissue will also exhibit differing fibrillar-level strain in response to external strain, but the specifics are not fully clear: on the one hand, since fewer fibrils will be aligned along any given load direction, many fibrils will remain unstrained, while others will bear the load disproportionately. However, network effects (entangled fibres) in such systems may also recruit off-axis fibres into the loading direction. In such cases, age-related changes, such as inter- and intrafibrillar glycation-induced crosslinks, will affect both the individual fibril deformation and the stress transfer in fibril arrays and will be expected to be distinct in tissues with differing architectures. These differences are highlighted in the review of collagen behaviour in various tissues in the following tissues.

Tendon

Geometrically, tendons exhibit a relatively simple nanoscale fibril architecture: all aligned along the tendon long axis and principal tensile loading direction. Age-related changes occur in and within fibrils, primarily driven by non-enzymatic glycation-linked crosslinking (advanced glycation end products or AGEs). By scanning electron microscopy investigation of tendons which had been synthetically crosslinked with ribose to induce AGE formation and then mechanically deformed to failure, Lee and Veres found that the fibrillar-level plasticity and damage mechanisms were inhibited relative to controls. They speculated that inflammatory cellular response to damaged collagen thus might be suppressed in diabetic and aged tissue [82]. The compressive modulus of aged supraspinatus tendons in mice was larger at the nanoscale than in mature tendons, and aged tissue also showed greater energy dissipation and permeability; the latter was linked to potential alterations and organisation of the collagen fibrils in aged tissue [30]. These results are consistent with earlier work where tail tendons from mice with diabetes and increased AGE concentration showed increased fibrillar D-period as well as higher stiffness as measured by atomic-force microscopy-based nanoindentation [58]. Ligaments of diabetic patients exhibit fewer collagen fibres. As reviewed by Adamska et al. [2] and reported by Li et al. [85], AGE-related crosslinking in such tendons reduced inter-fibre sliding, but as a compensation, intra-fibre stretching (which is proportional to fibril stretching) increased relative to control; the authors speculated this may increase the risk of overload at the fibrillar level during daily use.

Cartilage

Articular cartilage at the ends of long bones is essential for frictionless pain-free joint articulation. In contrast to tendon, the role of collagen fibrils in articular cartilage ECM is to act as a fibrillar meshwork constraining the swelling pressure of hydrated glycosaminoglycans and proteoglycans, which give the compressive resistance necessary for cartilage to function. The fibrils are thus in a constant state of tensile pre-strain (which will be visible via changes in D-period), dependent on the internal swelling pressure from the non-collagenous matrix, the intrinsic stiffness of the collagen fibrils and other factors like hydration and ionic balance. Ageing reduces the proteoglycan content and hydration; thus, aged cartilage may be expected to have less pre-strained, potentially more disordered fibrils and less compressive resistance. Consistent with this view, researchers used synchrotron X-ray scattering with in situ mechanical testing [75] and showed that synthetic modification of cartilage in ways that resemble ageing (removal of chondroitin sulphate) led to a lowered fibril pre-strain as well as reduced elastic modulus; notably, the level of pre-strain reduction was like physiological compression of healthy cartilage, suggesting aged tissue may exist in a pre-compressed state, predisposing it to injury. Thus, other factors being equal, lowered pre-strain in collagen fibrils in cartilage may serve as a marker for sites with less compressive resistance and a greater chance of tissue damage. Inflammation of cartilage (as in chronic inflammation of aged tissue) alters how fibrils under cyclic loading undergo stress relaxation, by reducing the ability of the fibrils to spread out reversibly to accommodate the load [76]. Atomic force microscopy-based nanoindentation of aged human menisci revealed a broadened and bimodal distribution of elastic moduli at the fibrillar level, with an overall shift to higher values compared to normal tissue, which was attributed to age-related stiffening of the extracellular matrix (ECM) [80]. Nanoindentation studies of collagen fibrils from cartilage-bone slices in human joints showed that collagen fibrils in aged articular cartilage exhibited a greater variation in indentation modulus; osteoarthritic tissue exhibited collagen fibrillar stiffening [176].

Bone

In ageing human bone, fibrillar plasticity was found to be suppressed, which was linked to increased nonenzymatic crosslinking (e.g. via AGEs) [189]. In patients with age-related hip fractures, earlier mineral/fibril decoupling was proposed as a possible reason for higher fragility [93]. Age can also affect collagen nanostructure in fracture healing: immobilisation led to a well-structured and organised collagen fibril network in aged mice, while a more disordered fibrillar structure was found in young mice [90]. Interestingly, suppression of collagen crosslinking enzymes in young male mice exhibiting osteocytic-intrinsic repression of TGF-β signalling led to lowered collagen fibrillar-level extensibility, which matched the behaviour in aged male controls [142]. Bone from diabetic mice also had limited fibrillar deformability, associated with reductions in macroscopic mechanical strength [1]. Commonalities between changes in bone nanomechanics in diabetes and age-related degeneration were also identified as a reduction of fibrillar plasticity and increased fibrillar stiffness, due to alterations in the ECM like AGE-accumulation [138]. Collagen crosslink ratios (mature to reducible crosslinks) increased with age post-ovariectomy, and these changes correlated with decline in bone strength [177]. Collagen fibrils from bone and tendon of diabetic mice had higher average D-period as well as a wider distribution of values compared to controls [64]. Giannini et al. observed a tendency to greater meridional order in collagen fibrils from coxarthritis-affected patients with increasing age, which they linked to increased AGE-derived crosslinking and fibril stiffening [56].

Skin

Increased collagen fibril fragmentation, linked to MMP production, is a hallmark of ageing in skin collagen [130]. As discussed in [50], the integrin-mediated mechanical equilibrium between collagen fibrils (which provide tension) and fibroblasts is necessary for regular levels of collagen and protease expression. In ageing skin, it was proposed [50] that the loss of fibroblast attachment to integrins led to fragmented collagen fibrils, fibroblast collapse, causing reduction of collagen production and relative increase in collagen-degrading enzymes. In skin sections across a range of ages, ultrastructural collagen fibrillar parameters like interfibrillar packing, regions of organised and visible D-periods and others were shown to vary and act as markers to discriminate between regions of the dermis as well as for chronological ageing [70]. Investigations of the transverse modulus of collagen fibrils from human dermis showed a significant age-related decrease by ~50%, which was attributed to age-related differences in water retention and fibril density [3]. Studies of changes in collagen often measure a different factor (e.g., photoaging) along with chronological age as a covariate, sometimes showing no significant effect on its own. For example, collagen precursor levels (types I and III) are both reduced with photodamage but not with chronological age [156]. However, a later study used immunohistochemical imaging to show collagen content and fibre architecture in the skin exhibited signs of reduction and disorganisation for subjects over 50 years of age [43].

Arteries and Cardiac Tissues

Pulse-wave velocity and nanoindentation measurements showed that the collagen-rich adventitia of arteries exhibited fibrillar stiffening in aged patients [24]. Alteration of fibrillar gap/overlap ratio and reduction of D-period with increased glycation-induced crosslinking in the bovine pericardium [57]. The col. I/III ratio increases with age, along with increased fibril diameter [36]. Type I collagen increases with age in the myocardial collagen network: existing type I collagen fibrils increase in diameter, while more are newly synthesised (possibly linked to a concomitant decrease in myocytes, and an inhibition of collagen degradation). Since the network consists of a mixture of type I and III, with type III believed to be more deformable, this process also reduces the ventricular elasticity [35].

Variability of Fibril Ultrastructure Across Tissues in Ageing

From the above, ageing affects the nanostructural organisation of collagen fibrils in a range of ways, but some common themes can be identified. Fibrillar stiffening (likely driven by AGE-related crosslinking) is a common feature across tissues. However, differences between relatively stable tissues (as in the cardiovascular system and cartilage) and more rapidly remodelling tissues like skin can be inferred; more actively remodelling tissues show greater collagen fragmentation possibly due to MMP-induced degradation, while more stable tissues like the aorta can show increase in fibril diameter and stiffening, which will be deleterious to healthy functioning. The effect of age-related crosslinking on the fibrillar D-period and intrafibrillar architecture is not always consistent, and a schematic is shown in Fig. 12.6c, d—some studies above in pericardium show a reduced D-period which might be explained by intrafibrillar stiffening by crosslinks, reducing the amount by which the fibrils can get pre-strained. However, in tendons, a higher D-period is found in aged tissue, which could be linked to an increase in intrafibrillar order by crosslinking. Thus, the knowledge of the local microenvironment is necessary to interpret properly the biomechanical implications of such age-related alterations.

Alterations to Collagen Degradation with Age

With ageing there is an increase in collagen fibre fragmentation particularly in skin [150]. Collagen fibres can be degraded by matrix metalloproteinases (MMPs) and cathepsins, and fragmented fibres may also be internalised for lysosomal degradation [4]. Furthermore, there is evidence for internal degradation of a fraction of newly synthesised procollagens [52, 135].

Proteolytic Degradation of Collagen

Fibrillar collagens can be degraded by MMP-1, −2, −8 and −13 as well as membrane-bound MT1-MMP and MT3-MMP. Types I and III are primarily degraded by MMP-1 and type II by MMP-13, whilst MMPs −3 and −10 only degrade type III collagen [149]. These MMPs cleave tropocollagen into ¾ and ¼ length fragments (Fig. 12.4). Once unwound and denatured, fibrillar collagens can be degraded by the gelatinases MMP-2 and -9. Cathepsin K has a particularly important role in collagen degradation in bone, where it is secreted into bone resorption pits by osteoclasts, whereas cathepsins B and L participate in intracellular collagen degradation via the lysosomal route.

In skin, age-related collagen degradation is mediated by MMPs that are induced by UV-radiation [55]. Indeed, expression of catalytically active MMP-1 in in vitro skin models results in fragmented collagen fibrils [183]. Positive feedback loops appear to further drive the ageing process as fragmented collagen can itself induce MMP-1 expression and activation [51]. UV-radiation can also cause indirect damage via the formation of free radicals, as well as by directly damaging collagen [173]. Even in sun-protected skin, MMP-1, −2 and −9 are present at higher levels in the elderly [167]. Furthermore, using serial passage in culture as a model for dermal fibroblast ageing showed that MMP-1 (collagenase) and MMP-3 (stromelysin) mRNA and protein were increased in later passage cultures [18, 102, 103, 137], demonstrating that age-related loss of collagen in skin is both an intrinsic and an extrinsically driven process.

Although collagen content increases with age in the heart and vasculature, there are known age-related changes in MMPs and in their natural inhibitors, the TIMPs (tissue inhibitors of matrix metalloproteinases). MMP-2 activity increases with age in the left ventricles of sheep and in the human aorta [68, 101]. Age-related increases in the protein levels of MMP-3, −8, −9, −12, MT1-MMP and decreases in TIMP-3 and -4 have also been demonstrated in the left ventricle in mice [88], and similarly in rat tenocytes where the activity of MMP-2 and -9 increases with age [187]. In humans, circulating MMP-2, MMP-7, TIMP-1 and TIMP-2 were found to increase with age and MMP-9 to decrease [67].

Aberrant Collagen Turnover in Ageing and Disease

The importance of continued collagen turnover in healthy ageing is demonstrated by the pro-senescent effect of collagenase-resistant type I collagen on vascular smooth muscle cells and its associated premature bone loss and hypertension in mice [164]. Aberrant production of collagenase-resistant type I collagen homotrimer (Fig. 12.4) has the potential to affect tissue mechanics and cellular phenotype. In humans, a common COL1A1 polymorphism (rs1800012) can increase COL1A1 transcription, resulting in type I collagen homotrimer production [95]. The SNP is associated with age-related diseases [22, 147, 182, 188], and it is possible that the homotrimeric form could accumulate over a lifetime in genetically predisposed individuals to alter the course of extracellular and cellular ageing.

In the liver, ageing enhances the fibrotic response in the CCL_4 model in mice, via reduced induction of MMP-13 and a corresponding reduction in the ability of liver homogenates to degrade exogenous soluble type I collagen [37]. Similarly, in lung, the fibrotic response is enhanced in aged mice but via reduced cell-mediated collagen update by MRC2 (the protein mannose receptor, C-type 2/Endo180/uPARAP) and degradation [126]. Age-associated changes to collagen can themselves affect collagen turnover, with age-related glycation end product (AGE) formation reducing collagen fibre degradation by MMPs -1,-8 and -13 [120] and by cathepsin K [119].

The circadian clock is an important regulator of collagen homeostasis with diurnal fluctuations in a soluble collagen pool in tendon likely providing a protective mechanism against tissue damage [25]. Circadian clocks are themselves dysregulated with age which can influence fibrillar collagen secretion and degradation [41]. Indeed rhythmicity in the soluble fraction of collagen is lost with ageing, which may reflect reduced synthesis or secretion, and/or slower and less effective collagen degradation [69].

The Role of Cellular Senescence and the Senescent-Associated Secretory Phenotype (SASP) in Collagen Ageing

Localised over-production of collagen in ageing can lead to tissue and organ fibrosis, for which effective treatments are still limited.

Senescence is a cellular phenotype characterised by cell-cycle arrest, which is one of the key hallmarks of ageing [92]. Senescence is important in normal development and physiology and acts as a tumour suppressor by preventing the proliferation of damaged cells [33]. However, accumulation of senescent cells with age is linked to numerous age-related diseases, which has led to the threshold theory of senescent cell accumulation [23]. The phenotype of senescent cells includes the 'senescence-associated secretory phenotype' (SASP) characterised by the production and secretion of cytokines, chemokines, bioactive lipids, proteases and nucleotides. Extracellular matrix remodelling proteases are a key component of the SASP with several MMPs being upregulated at senescence.

Senescent cells may drive fibrosis by changing the phenotype of neighbouring cells (Fig. 12.7). Senescent cells have been implicated in fibrotic lung disease, such as idiopathic lung fibrosis (IPF) by secretion of SASP factors such as TGFβ-1, IL-6 and MM12 [140]. Indeed, SASP components, such as the cytokine TGFβ-1, can produce myofibroblasts via epithelial to mesenchymal transition (EMT), macrophage-myofibroblast transition (MMT) or endothelial to mesenchymal transition (EndoMT) to drive fibrosis [171], though acute senescence may promote resolution of fibrosis [134].

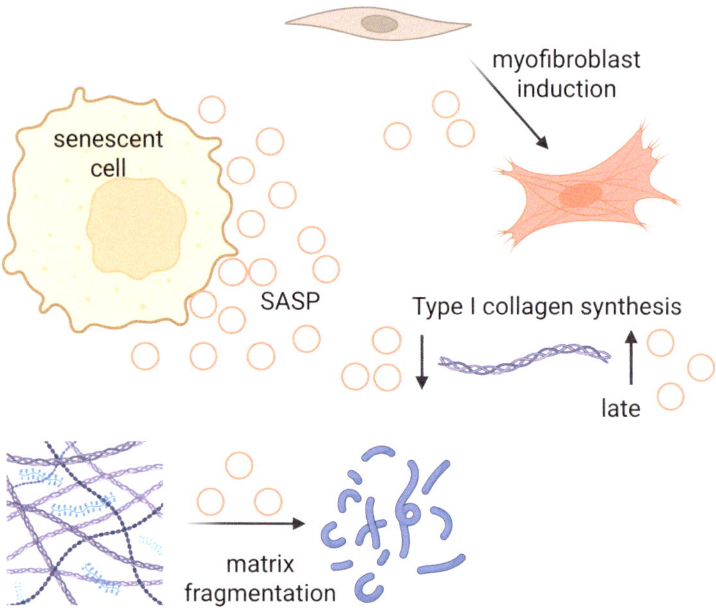

Fig. 12.7 Illustration of the potential effects of the senescent-associated secretory phenotype (SASP) on the collagen content of tissues. The SASP derived from senescent cells can drive myofibroblast activation leading to over-production of collagen; it can degrade collagen fibres leading to matrix fragmentation and senescent cells may themselves down-regulate or over-produce fibrillar collagen. (Figure created in Biorender)

The expression and production of collagens can also be altered once a cell becomes senescent (Fig. 12.7). The expression of type I and III collagen mRNAs is decreased in senescent cells [81, 96]. There is evidence that this is reflected in a reduction in type I collagen protein production; for example, the amount of hydroxyproline in media (as a proxy for collagen protein synthesis) is decreased with serial passage in dermal fibroblasts [79], replicative senescence reduced the amount of type I procollagen protein in media [132] and supernatants of mitomycin-treated skin equivalents contained less type-1-procollagen aminoterminal propeptide [39]. However, a meta-analysis of 20 studies of the composition of the SASP protein secretome, which included 70 datasets derived from senescent induction by replicative stress, hydrogen peroxide treatment, irradiation and drug treatment, found that collagen alpha-1(I) chain production as well as that of MMP-2 and metalloproteinase inhibitor-1 (TIMP-1) were common to all stressors and specific to late stage senescence [117]. For replicative stress, other ECM proteins were induced including collagen alpha-2(I) and type III, V and VI collagen chains. Hence, fibrillar collagen production may either decrease or increase in senescent cells, and the stage of senescence may influence the biosynthetic response.

As well as driving myofibroblast activation, the SASP has the potential to remodel the extracellular matrix environment by degrading the native stroma (Fig. 12.7), by producing fragmented collagens that may themselves affect cellular responses and potentially by fabricating a new ECM. Whether the type I collagen protein produced by late-stage senescent cells is assembled into new fibrils or represents a rapidly turned-over fraction is currently unknown. Conflicting results between studies could reflect differences in the amount of new procollagen that is processed and assembled into fibrils, versus that present in the cell culture medium.

The Emerging Potential for Collagen Supplements to Halt or Reverse Age-Induced Changes in Collagen-Rich Tissues

There has been recent interest in the potential for collagen-derived nutritional supplements to counteract the signs and effects of ageing. This has led to an explosion of commercial suppliers, celebrity endorsements and a market size of several billion US dollars in 2025.

Human studies do indeed support the claims of the beneficial effects of collagen supplements for skin and joint health. A meta-analysis of 19 studies with a total of 1125 participants concluded that that ingestion of hydrolysed collagen (typically 3–6 kDa (kilodaltons) consisting of 10–20 linked Gly-Xaa-Yaa triplets [84]) improves skin elasticity and hydration whilst reducing wrinkles [34], whilst a systematic review and meta-analysis of 14 studies with a collective cohort of 967 participants, concluded that oral supplementation with hydrolysed collagen improved skin moisture levels and elasticity [38]. A further systematic review and meta-analysis of 26 randomised controlled trials involving 1721 participants concluded that hydrolysed collagen supplementation improved skin hydration and elasticity [129]. All of the studies considered by de Miranda et al. [34] and by Dewi et al. [38] were also included in the study by Pu et al. [129]. Of the 27 studies, most were commercially funded, but 22 were double-blind. Fourteen of the studies considered a hydrolysed collagen supplement alone, whereas 13 only considered the supplement in combination with at least one other ingredient, often antioxidants or vitamins. Most studies used objective measures of skin parameters, rather than relying on subjective reporting. Pu et al. [129] identified some potential concerns about bias for deviations from the intended intervention, missing outcome data or selection of the reported result for some studies. Similarly, Dewi et al. [38] identified an unclear risk of bias for selection, blinding and outcome detection in around half of the 14 studies, whilst De Miranda et al. [34] identified some risk of bias in blinding and incomplete data but no publication bias.

In terms of joint health, a systematic review of 15 randomised controlled trials concluded that there was strong evidence that combination of collagen peptide supplementation and exercise improved joint functionality and reduced joint pain [78]. There was evidence of improved body composition and strength, particularly in the elderly, and there was a modest effect on muscle recovery. A further systematic review and meta-analysis of 19 studies concluded that a combination of collagen peptide supplementation and long-term physical training improved fat-free mass, tendon morphology, muscle mass, maximal strength and recovery [17].

Animal studies support beneficial effects of hydrolysed collagen supplements on the musculoskeletal system including anti-oxidant activity and reduced muscle fatigue in mice [40], improved bone fracture healing in rats [161] and increased muscle failure strain in chickens [172]. There is also evidence of beneficial effects on the immune system [112], cognitive function [121], mental health [105, 113] and on metabolic health [159, 180] in mice. Studies in rats and worms even indicate a potential beneficial effect on lifespan [86, 110].

Whilst ingestion of hydrolysed collagen di- or tri-peptides may increase the availability of the amino acids glycine and proline in the circulation to support new collagen synthesis, cell culture studies support a direct effect of collagen peptides on cells. Media supplementation can increase osteoblast and tenocyte differentiation, collagen protein production, tenocyte migration and hippocampal cell proliferation [71, 105, 113, 155, 161]. As well as direct effects on cells, potential mechanisms of action include direct inhibition of angiotensin converting enzyme (ACE) [154] or induction of regulatory T-cells and M2 macrophages to suppress immune responses to collagen [12]. Elucidating the mechanism of action of hydrolysed collagen peptides has the potential to allow for the development of targeted interventions for ageing and age-related disease. Further work is required for a complete mechanistic understanding and to elucidate the extent of the beneficial effects of ingested collagen peptides on normal physiology, disease and ageing (Fig. 12.8).

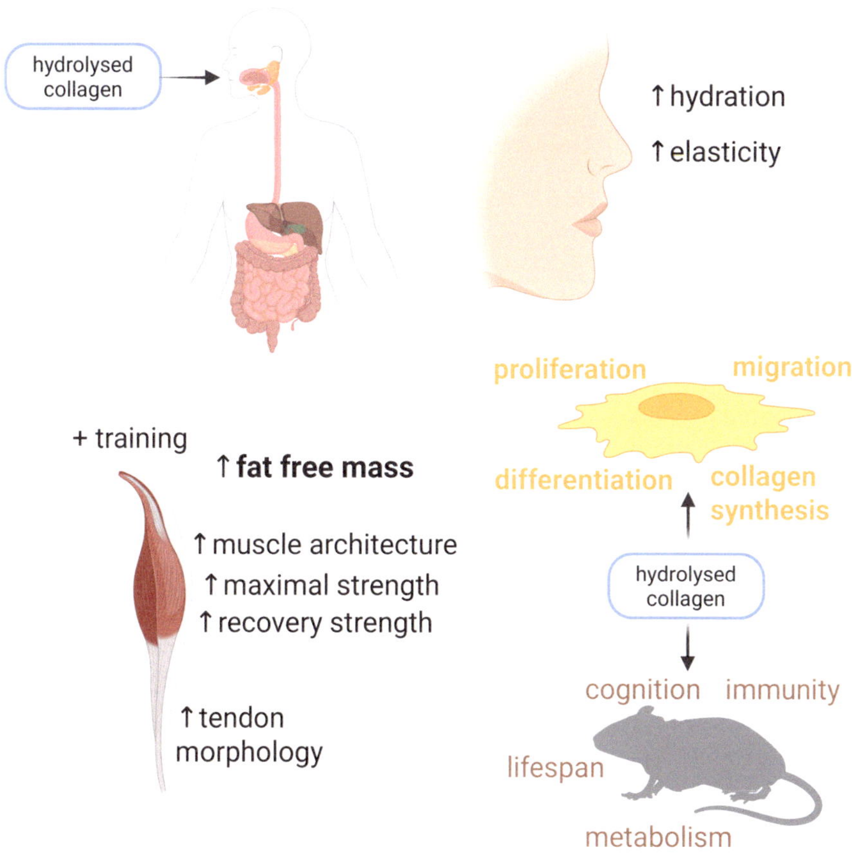

Fig. 12.8 Summary of evidence to date for the beneficial effects of hydrolysed collagens. (Figure created in Biorender)

Conclusion and Future Perspectives

Although the protein collagen was recognised almost 200 years ago and much ground-breaking work was carried out in the 1950s, 1960s and 1970s to elucidate the chemical structure, there is still much to be discovered. Modern techniques such as RNA sequencing and large-scale proteomics have taken the focus away from furthering the understanding of collagen-specific post-translational modifications. More recent findings of collagen peptide influence on cell metabolism and the role that ageing collagen may have on cell senescence, and the finding that maintaining collagen turnover may be key to youthfulness has revitalised interest in collagen. Future research should be directed towards harnessing the power of modern mass spectrometry techniques to further the understanding of post-translational modification of collagen. The impact that these age-related changes have on cell behaviour, ultrastructure and nano- and macromechanics deserve further attention and will undoubtably translate into improving human health.

Acknowledgements EGC-L received funding from BBSRC and MRC to support the extracellular matrix ageing across the life course interdisciplinary research network, ECMage (BB/W018314/1) which supported initial discussions between the authors.

Compliance with Ethical Standards No authors have declared any conflicts of interest.

References

1. Acevedo C, Sylvia M, Schaible E, Graham JL, Stanhope KL, Metz LN, Gludovatz B, Schwartz AV, Ritchie RO, Alliston TN, Havel PJ, Fields AJ (2018) Contributions of material properties and structure to increased bone fragility for a given bone mass in the ucd-t2dm rat model of type 2 diabetes. J Bone Miner Res 33(6):1066
2. Adamska O, Stolarczyk A, Gondek A, Maciąg B, Świderek J, Czuchaj P, Modzelewski K (2022) Ligament alteration in diabetes mellitus. J Clin Med 11:5719
3. Ahmed T, Nash A, Clark KEN, Ghibaudo M, De Leeuw NH, Potter A, Stratton R, Birch HL, Casse RE, Bozec L (2017) Combining nano-physical and computational investigations to understand the nature of "aging" in dermal collagen. Int J Nanomedicine 12:3303
4. Amar S, Smith L, Fields GB (2017) Matrix metalloproteinase collagenolysis in health and disease. Biochim Biophys Acta Mol Cell Res 1864(11 Pt A):1940–1951
5. Annoni G, Luvarà G, Arosio B, Gagliano N, Fiordaliso F, Santambrogio D, Jeremic G, Mircoli L, Latini R, Vergani C, Masson S (1998) Age-dependent expression of fibrosis-related genes and collagen deposition in the rat myocardium. Mech Ageing Dev 101(1):57–72
6. Ariosa-Morejon Y, Santos A, Fischer R, Davis S, Charles P, Thakker R, Wann A, Vincent TL (2021) Age-dependent changes in protein incorporation into collagen-rich tissues of mice by in vivo pulsed silac labelling. bioRxiv.2021.2001.2013.426496
7. Babraj JA, Cuthbertson DJ, Smith K, Langberg H, Miller B, Krogsgaard MR, Kjaer M, Rennie MJ (2005) Collagen synthesis in human musculoskeletal tissues and skin. Am J Physiol Endocrinol Metab 289(5):E864–E869
8. Bailey AJ, Shimokomaki MS (1971) Age related changes in the reducible cross-links of collagen. FEBS Lett 16(2):86–88
9. Bailey AJ, Paul RG, Knott L (1998) Mechanisms of maturation and ageing of collagen. Mech Ageing Dev 106(1–2):1–56
10. Bailey AJ, Sims TJ, Ebbesen EN, Mansell JP, Thomsen JS, Mosekilde L (1999) Age-related changes in the biochemical properties of human cancellous bone collagen: relationship to bone strength. Calcif Tissue Int 65(3):203–210
11. Bailey AJ, Sims TJ, Knott L (2002) Phenotypic expression of osteoblast collagen in osteoarthritic bone: production of type i homotrimer. Int J Biochem Cell Biol 34(2):176–182
12. Barati M, Jabbari M, Navekar R, Farahmand F, Zeinalian R, Salehi-Sahlabadi A, Abbaszadeh N, Mokari-Yamchi A, Davoodi SH (2020) Collagen supplementation for skin health: a mechanistic systematic review. J Cosmet Dermatol 19(11):2820–2829
13. Barnes MJ, Constable BJ, Morton LF, Royce PM (1974) Age-related variations in hydroxylation of lysine and proline in collagen. Biochem J 139(2):461–468

14. Besse S, Robert V, Assayag P, Delcayre C, Swynghedauw B (1994) Nonsynchronous changes in myocardial collagen mrna and protein during aging: effect of doca-salt hypertension. Am J Phys Heart Circ Phys 267(6):H2237–H2244
15. Biemel KM, Reihl O, Conrad J, Lederer MO (2001) Formation pathways for lysine-arginine cross-links derived from hexoses and pentoses by maillard processes: unraveling the structure of a pentosidine precursor. J Biol Chem 276(26):23405–23412
16. Birch HL (2018) Extracellular matrix and ageing. Subcell Biochem 90:169–190
17. Bischof K, Moitzi AM, Stafilidis S, König D (2024) Impact of collagen peptide supplementation in combination with long-term physical training on strength, musculotendinous remodeling, functional recovery, and body composition in healthy adults: a systematic review with meta-analysis, vol 54. Sports Med, p 2865
18. Bizot-Foulon V, Bouchard B, Hornebeck W, Dubertret L, Bertaux B (1995) Uncoordinate expressions of type i and iii collagens, collagenase and tissue inhibitor of matrix metalloproteinase 1 along in vitro proliferative life span of human skin fibroblasts. Regulation by all-trans retinoic acid. Cell Biol Int 19(2):129–135
19. Bornstein P (1967) Comparative sequence studies of rat skin and tendon collagen. I. Evidence for incomplete hydroxylation of individual prolyl residues in the normal proteins*. Biochemistry 6(10):3082–3093
20. Bowers SLK, Meng Q, Kuwabara Y, Huo J, Minerath R, York AJ, Sargent MA, Prasad V, Saviola AJ, Galindo DC, Hansen KC, Vagnozzi RJ, Yutzey KE, Molkentin JD (2023) Col1a2-deleted mice have defective type i collagen and secondary reactive cardiac fibrosis with altered hypertrophic dynamics. Cells 12(17):2174
21. Brandenberger C, Mühlfeld C (2017) Mechanisms of lung aging. Cell Tissue Res 367(3): 469–480
22. Brull DJ, Murray LJ, Boreham CA, Ralston SH, Montgomery HE, Gallagher AM, McGuigan FE, Davey Smith G, Savage M, Humphries SE, Young IS (2001) Effect of a col1a1 sp1 binding site polymorphism on arterial pulse wave velocity: an index of compliance. Hypertension 38(3):444–448
23. Chaib S, Tchkonia T, Kirkland JL (2022) Cellular senescence and senolytics: the path to the clinic. Nat Med 28(8):1556–1568
24. Chang Z, Paoletti P, Barrett SD, Chim YH, Caamaño-Gutiérrez E, Hansen ML, Beck HC, Rasmussen LM, Akhtar R (2018) Nanomechanics and ultrastructure of the internal mammary artery adventitia in patients with low and high pulse wave velocity. Acta Biomater 73:437
25. Chang J, Garva R, Pickard A, Yeung CC, Mallikarjun V, Swift J, Holmes DF, Calverley B, Lu Y, Adamson A, Raymond-Hayling H, Jensen O, Shearer T, Meng QJ, Kadler KE (2020) Circadian control of the secretory pathway maintains collagen homeostasis. Nat Cell Biol 22(1):74–86
26. Chretien A, Couchot M, Mabilleau G, Behets C (2022) Biomechanical, microstructural and material properties of tendon and bone in the young oim mice model of osteogenesis imperfecta. Int J Mol Sci 23(17):9928
27. Chung JH, Seo JY, Choi HR, Lee MK, Youn CS, Rhie G-e, Cho KH, Kim KH, Park KC, Eun HC (2001) Modulation of skin collagen metabolism in aged and photoaged human skin in vivo. J Invest Dermatol 117(5):1218–1224
28. Chung L, Dinakarpandian D, Yoshida N, Lauer-Fields JL, Fields GB, Visse R, Nagase H (2004) Collagenase unwinds triple-helical collagen prior to peptide bond hydrolysis. EMBO J 23(15):3020–3030
29. Clarke BL, Ebeling PR, Jones JD, Wahner HW, O'Fallon WM, Riggs BL, Fitzpatrick LA (1996) Changes in quantitative bone histomorphometry in aging healthy men. J Clin Endocrinol Metab 81(6):2264–2270
30. Connizzo BK, Grodzinsky AJ (2018) Multiscale poroviscoelastic compressive properties of mouse supraspinatus tendons are altered in young and aged mice. J Biomech Eng 140(5): 0510021

31. Corps AN, Robinson AH, Harrall RL, Avery NC, Curry VA, Hazleman BL, Riley GP (2012) Changes in matrix protein biochemistry and the expression of mrna encoding matrix proteins and metalloproteinases in posterior tibialis tendinopathy. Ann Rheum Dis 71(5):746–752
32. Couppe C, Hansen P, Kongsgaard M, Kovanen V, Suetta C, Aagaard P, Kjaer M, Magnusson SP (2009) Mechanical properties and collagen cross-linking of the patellar tendon in old and young men. J Appl Physiol (1985) 107(3):880–886
33. de Magalhães JP (2024) Cellular senescence in normal physiology. Science 384(6702): 1300–1301
34. de Miranda RB, Weimer P, Rossi RC (2021) Effects of hydrolyzed collagen supplementation on skin aging: a systematic review and meta-analysis. Int J Dermatol 60(12):1449–1461
35. de Souza RR (2002) Aging of myocardial collagen. Biogerontology 3(6):325–335
36. Debessa CRG, Maifrino LBM, Rodrigues de Souza R (2001) Age related changes of the collagen network of the human heart. Mech Ageing Dev 122(10):1049–1058
37. Delire B, Lebrun V, Selvais C, Henriet P, Bertrand A, Horsmans Y, Leclercq IA (2016) Aging enhances liver fibrotic response in mice through hampering extracellular matrix remodeling. Aging (Albany NY) 9(1):98–113
38. Dewi DAR, Arimuko A, Norawati L, Yenny SW, Setiasih NL, Perdiyana A, Arkania N, Nadhira F, Wiliantari N (2023) Exploring the impact of hydrolyzed collagen oral supplementation on skin rejuvenation: a systematic review and meta-analysis. Cureus 15(12):e50231
39. Diekmann J, Alili L, Scholz O, Giesen M, Holtkötter O, Brenneisen P (2016) A three-dimensional skin equivalent reflecting some aspects of in vivo aged skin. Exp Dermatol 25(1):56–61
40. Ding J-F, Li Y-Y, Xu J-J, Su X-R, Gao X, Yue F-P (2011) Study on effect of jellyfish collagen hydrolysate on anti-fatigue and anti-oxidation. Food Hydrocoll 25(5):1350–1353
41. Dudek M, Swift J, Meng QJ (2023) The circadian clock and extracellular matrix homeostasis in aging and age-related diseases. Am J Physiol Cell Physiol 325(1):C52–c59
42. Eghbali M, Eghbali M, Robinson TF, Seifter S, Blumenfeld OO (1989) Collagen accumulation in heart ventricles as a function of growth and aging. Cardiovasc Res 23(8):723–729
43. El-Domyati M, Attia S, Saleh F, Brown D, Birk DE, Gasparro F, Ahmad H, Uitto J (2002) Intrinsic aging vs. Photoaging: a comparative histopathological, immunohistochemical, and ultrastructural study of skin. Exp Dermatol 11(5):398
44. Exposito J-Y, Valcourt U, Cluzel C, Lethias C (2010) The fibrillar collagen family. Int J Mol Sci 11(2):407–426
45. Eyre DR, Dickson IR, Van Ness K (1988) Collagen cross-linking in human bone and articular cartilage. Age-related changes in the content of mature hydroxypyridinium residues. Biochem J 252(2):495–500
46. Eyre DR, Weis M, Hudson DM, Wu JJ, Kim L (2011) A novel 3-hydroxyproline (3hyp)-rich motif marks the triple-helical c terminus of tendon type i collagen. J Biol Chem 286(10): 7732–7736
47. Fede C, Fan C, Pirri C, Petrelli L, Biz C, Porzionato A, Macchi V, De Caro R, Stecco C (2022) The effects of aging on the intramuscular connective tissue. Int J Mol Sci 23(19):11061
48. Fernandes RJ, Farnand AW, Traeger GR, Weis MA, Eyre DR (2011) A role for prolyl 3-hydroxylase 2 in post-translational modification of fibril-forming collagens. J Biol Chem 286(35):30662–30669
49. Fietzek PP, Kühn K (1975) Information contained in the amino acid sequence of theα1(i)-chain of collagen and its consequences upon the formation of the triple helix, of fibrils and crosslinks. Mol Cell Biochem 8(3):141–157
50. Fisher GJ, Varani J, Voorhees JJ (2008) Looking older: fibroblast collapse and therapeutic implications, vol 144. Arch Dermatol, p 666
51. Fisher GJ, Quan T, Purohit T, Shao Y, Cho MK, He T, Varani J, Kang S, Voorhees JJ (2009) Collagen fragmentation promotes oxidative stress and elevates matrix metalloproteinase-1 in fibroblasts in aged human skin. Am J Pathol 174(1):101–114

52. Forrester A, De Leonibus C, Grumati P, Fasana E, Piemontese M, Staiano L, Fregno I, Raimondi A, Marazza A, Bruno G, Iavazzo M, Intartaglia D, Seczynska M, van Anken E, Conte I, De Matteis MA, Dikic I, Molinari M, Settembre C (2019) A selective er-phagy exerts procollagen quality control via a calnexin-fam134b complex. EMBO J 38(2):e99847
53. Fratzl P (2008) Collagen: structure and mechanics. Springer US
54. Fratzl P, Weinkamer R (2007) Nature's hierarchical materials. Prog Mater Sci 52:1263–1334
55. Freitas-Rodríguez S, Folgueras AR, López-Otín C (2017) The role of matrix metalloproteinases in aging: tissue remodeling and beyond. Biochim Biophys Acta (BBA) – Mol Cell Res 1864(11, Part A):2015–2025
56. Giannini C, Siliqi D, Ladisa M, Altamura D, Diaz A, Beraudi A, Sibillano T, De Caro L, Stea S, Baruffaldi F, Bunk O (2014) Scanning saxs-waxs microscopy on osteoarthritis-affected bone – an age-related study. J Appl Crystallogr 47:110–117
57. Giannini C, Terzi A, Fusaro L, Sibillano T, Diaz A, Ramella M, Lutz-Bueno V, Boccafoschi F, Bunk O (2019) Scanning x-ray microdiffraction of decellularized pericardium tissue at increasing glucose concentration. J Biophotonics 12(10):e201900106
58. Gonzalez AD, Gallant MA, Burr DB, Wallace JM (2014) Multiscale analysis of morphology and mechanics in tail tendon from the zdsd rat model of type 2 diabetes. J Biomech 47(3):681
59. Gorisse L, Pietrement C, Vuiblet V, Schmelzer CE, Köhler M, Duca L, Debelle L, Fornès P, Jaisson S, Gillery P (2016) Protein carbamylation is a hallmark of aging. Proc Natl Acad Sci USA 113(5):1191–1196
60. Greenwald SE (2007) Ageing of the conduit arteries. J Pathol 211(2):157–172
61. Guarnieri V, Morlino S, Di Stolfo G, Mastroianno S, Mazza T, Castori M (2019) Cardiac valvular ehlers-danlos syndrome is a well-defined condition due to recessive null variants in col1a2. Am J Med Genet A 179(5):846–851
62. Gupta HS, Seto J, Krauss S, Boesecke P, Screen HRC (2010) In situ multi-level analysis of viscoelastic deformation mechanisms in tendon collagen. J Struct Biol 169(2):183–191
63. Gustavson KH (1955) The function of hydroxyproline in collagens. Nature 175(4445):70–74
64. Hammond MA, Gallant MA, Burr DB, Wallace JM (2013) Nanoscale changes in collagen are reflected in physical and mechanical properties of bone at the microscale in diabetic rats. Bone 60:26
65. Han S, Makareeva E, Kuznetsova NV, DeRidder AM, Sutter MB, Losert W, Phillips CL, Visse R, Nagase H, Leikin S (2010) Molecular mechanism of type i collagen homotrimer resistance to mammalian collagenases. J Biol Chem 285(29):22276–22281
66. Harmel R, Fiedler D (2018) Features and regulation of non-enzymatic post-translational modifications. Nat Chem Biol 14(3):244–252
67. Horn MA, Trafford AW (2016) Aging and the cardiac collagen matrix: novel mediators of fibrotic remodelling. J Mol Cell Cardiol 93:175–185
68. Horn MA, Graham HK, Richards MA, Clarke JD, Greensmith DJ, Briston SJ, Hall MCS, Dibb KM, Trafford AW (2012) Age-related divergent remodeling of the cardiac extracellular matrix in heart failure: collagen accumulation in the young and loss in the aged. J Mol Cell Cardiol 53(1):82–90
69. Hoyle A, Chang J, Cutiongco MFA, O'Cualain R, Warwood S, Knight D, Meng Q-J, Kadler KE, Swift J (2024) Circadian and permanent pools of extracellular matrix co-exist in tendon tissue, but have distinct rates of turnover and differential responses to ageing. bioRxiv.2024.2008.2009.607297
70. Huang S, Strange A, Maeva A, Siddiqui S, Bastien P, Aguayo S, Vaez M, Montagu-Pollock H, Ghibaudo M, Potter A, Pageon H, Bozec L (2023) Quantitative nanohistology of aging dermal collagen. Front Aging 4:1178566
71. Ide K, Takahashi S, Sakai K, Taga Y, Ueno T, Dickens D, Jenkins R, Falciani F, Sasaki T, Ooi K, Kawashiri S, Mizuno K, Hattori S, Sakai T (2021) The dipeptide prolyl-hydroxyproline promotes cellular homeostasis and lamellipodia-driven motility via active β1-integrin in adult tendon cells. J Biol Chem 297(1):100819

72. Ihme A, Krieg T, Nerlich A, Feldmann U, Rauterberg J, Glanville RW, Edel G, Müller PK (1984) Ehlers-danlos syndrome type vi: collagen type specificity of defective lysyl hydroxylation in various tissues. J Invest Dermatol 83(3):161–165
73. IMPC. International mouse phenotyping consortium: Col1a2. https://www.mousephenotype.org/data/genes/MGI:88468. Accessed 4 Apr 2025
74. Inamdar SR, Knight DP, Terrill NJ, Karunaratne A, Cacho-Nerin F, Knight MM, Gupta HS (2017) The secret life of collagen: temporal changes in nanoscale fibrillar pre-strain and molecular organization during physiological loading of cartilage. ACS Nano 11(10): 9728–9737
75. Inamdar SR, Barbieri E, Terrill NJ, Knight MM, Gupta HS (2019) Proteoglycan degradation mimics static compression by altering the natural gradients in fibrillar organisation in cartilage. Acta Biomater 97:437–450
76. Inamdar SR, Prévost S, Terrill NJ, Knight MM, Gupta HS (2021) Reversible changes in the 3d collagen fibril architecture during cyclic loading of healthy and degraded cartilage. Acta Biomater 136:314–326
77. Kadler KE, Baldock C, Bella J, Boot-Handford RP (2007) Collagens at a glance. J Cell Sci 120 (Pt 12):1955–1958
78. Khatri M, Naughton RJ, Clifford T, Harper LD, Corr L (2021) The effects of collagen peptide supplementation on body composition, collagen synthesis, and recovery from joint injury and exercise: a systematic review. Amino Acids 53(10):1493–1506
79. Khorramizadeh MR, Tredget EE, Telasky C, Shen Q, Ghahary A (1999) Aging differentially modulates the expression of collagen and collagenase in dermal fibroblasts. Mol Cell Biochem 194(1–2):99–108
80. Kwok J, Grogan S, Meckes B, Arce F, Lal R, D'Lima D (2014) Atomic force microscopy reveals age-dependent changes in nanomechanical properties of the extracellular matrix of native human menisci: implications for joint degeneration and osteoarthritis. Nanomedicine 10(8):1777
81. Lago JC, Puzzi MB (2019) The effect of aging in primary human dermal fibroblasts. PLoS One 14(7):e0219165
82. Lee JM, Veres SP (2019) Advanced glycation end-product cross-linking inhibits biomechanical plasticity and characteristic failure morphology of native tendon. J Appl Physiol 126(4): 832
83. Lee KJ, Rambault L, Bou-Gharios G, Clegg PD, Akhtar R, Czanner G, van't Hof R, Canty-Laird EG (2022) Collagen (i) homotrimer potentiates the osteogenesis imperfecta (oim) mutant allele and reduces survival in male mice. Dis Model Mech 15(9):dmm049428
84. León-López A, Morales-Peñaloza A, Martínez-Juárez VM, Vargas-Torres A, Zeugolis DI, Aguirre-Álvarez G (2019) Hydrolyzed collagen-sources and applications. Molecules 24(22): 4031
85. Li Y, Fessel G, Georgiadis M, Snedeker JG (2013) Advanced glycation end-products diminish tendon collagen fiber sliding. Matrix Biol 32(3–4):169
86. Liang J, Pei XR, Wang N, Zhang ZF, Wang JB, Li Y (2010) Marine collagen peptides prepared from chum salmon (oncorhynchus Keta) skin extend the life span and inhibit spontaneous tumor incidence in sprague-dawley rats. J Med Food 13(4):757–770
87. Lin Y, Li Y, Liang G, Yang X, Yang J, Hu Q, Sun J, Zhang C, Fang H, Liu A (2024) Single-cell transcriptome analysis of aging mouse liver. FASEB J 38(4):e23473
88. Lindsey ML, Goshorn DK, Squires CE, Escobar GP, Hendrick JW, Mingoia JT, Sweterlitsch SE, Spinale FG (2005) Age-dependent changes in myocardial matrix metalloproteinase/tissue inhibitor of metalloproteinase profiles and fibroblast function. Cardiovasc Res 66(2):410–419
89. Liu W, Tian Z, Li C, Li G (2014) Thermal denaturation of fish collagen in solution: a calorimetric and kinetic analysis. Thermochim Acta 581:32–40
90. Liu F, Hu Y, Zhang Y, Ren C, Qiao F, Yang H, Xu H, Yang P (2025) Aging-induced discrepant response of fracture healing is initiated from the organization and mineralization of collagen fibrils in callus. ACS Biomater Sci Eng 11(2):1038–1050

91. Lopez-Clavijo AF, Nash N, Barrow MP, O'Connor PB, Bozec L, De Leeuw N, Birch HL (2015) Glucosepane: investigating formation and structure in tendon collagen. Int J Exp Pathol 96(2):A15–A15
92. López-Otín C, Blasco MA, Partridge L, Serrano M, Kroemer G (2013) The hallmarks of aging. Cell 153(6):1194–1217
93. Ma S, Goh EL, Tay T, Wiles CC, Boughton O, Churchwell JH, Wu Y, Karunaratne A, Bhattacharya R, Terrill N, Cobb JP, Hansen U, Abel RL (2020) Nanoscale mechanisms in age-related hip-fractures. Sci Rep 10(1):14208
94. Malfait F, Symoens S, Coucke P, Nunes L, De Almeida S, De Paepe A (2006) Total absence of the alpha2(i) chain of collagen type i causes a rare form of ehlers-danlos syndrome with hypermobility and propensity to cardiac valvular problems. J Med Genet 43(7):e36
95. Mann V, Hobson EE, Li B, Stewart TL, Grant SFA, Robins SP, Aspden RM, Ralston SH (2001) A col1a1 sp1 binding site polymorphism predisposes to osteoporotic fracture by affecting bone density and quality. J Clin Invest 107(7):899–907
96. Mavrogonatou E, Pratsinis H, Papadopoulou A, Karamanos NK, Kletsas D (2019) Extracellular matrix alterations in senescent cells and their significance in tissue homeostasis. Matrix Biol 75–76:27–42
97. Mayne R, Vail MS, Miller EJ (1975) Analysis of changes in collagen biosynthesis that occur when chick chondrocytes are grown in 5-bromo-2′-deoxyuridine. Proc Natl Acad Sci USA 72(11):4511–4515
98. Mayne R, Vail MS, Mayne PM, Miller EJ (1976) Changes in type of collagen synthesized as clones of chick chondrocytes grow and eventually lose division capacity. Proc Natl Acad Sci USA 73(5):1674–1678
99. Mays PK, McAnulty R, Laurent GJ (1991a) Age-related changes in total protein and collagen metabolism in rat liver. Hepatology 14(6):1224–1229
100. Mays PK, McAnulty RJ, Campa JS, Laurent GJ (1991b) Age-related changes in collagen synthesis and degradation in rat tissues. Importance of degradation of newly synthesized collagen in regulating collagen production. Biochem J 276(Pt 2):307–313
101. McNulty M, Spiers P, McGovern E, Feely J (2005) Aging is associated with increased matrix metalloproteinase-2 activity in the human aorta*. Am J Hypertens 18(4):504–509
102. Millis AJ, Sottile J, Hoyle M, Mann DM, Diemer V (1989) Collagenase production by early and late passage cultures of human fibroblasts. Exp Gerontol 24(5–6):559–575
103. Millis AJ, Hoyle M, McCue HM, Martini H (1992) Differential expression of metalloproteinase and tissue inhibitor of metalloproteinase genes in aged human fibroblasts. Exp Cell Res 201(2):373–379
104. Misof K, Landis WJ, Klaushofer K, Fratzl P (1997) Collagen from the osteogenesis imperfecta mouse model (oim) shows reduced resistance against tensile stress. J Clin Invest 100(1):40–45
105. Mizushige T, Nogimura D, Nagai A, Mitsuhashi H, Taga Y, Kusubata M, Hattori S, Kabuyama Y (2019) Ginger-degraded collagen hydrolysate exhibits antidepressant activity in mice. J Nutr Sci Vitaminol (Tokyo) 65(3):251–257
106. Molnar J, Fong KS, He QP, Hayashi K, Kim Y, Fong SF, Fogelgren B, Szauter KM, Mink M, Csiszar K (2003) Structural and functional diversity of lysyl oxidase and the lox-like proteins. Biochim Biophys Acta 1647(1–2):220–224
107. Monnier VM, Bautista O, Kenny D, Sell DR, Fogarty J, Dahms W, Cleary PA, Lachin J, Genuth S (1999) Skin collagen glycation, glycoxidation, and crosslinking are lower in subjects with long-term intensive versus conventional therapy of type 1 diabetes: relevance of glycated collagen products versus hba1c as markers of diabetic complications. Dcct skin collagen ancillary study group. Diabetes control and complications trial. Diabetes 48(4):870–880
108. Monnier VM, Sell DR, Strauch C, Sun W, Lachin JM, Cleary PA, Genuth S (2013) The association between skin collagen glucosepane and past progression of microvascular and neuropathic complications in type 1 diabetes. J Diabetes Complicat 27(2):141–149
109. Monnier VM, Genuth S, Sell DR (2016) The pecking order of skin advanced glycation endproducts (ages) as long-term markers of glycemic damage and risk factors for micro- and subclinical macrovascular disease progression in type 1 diabetes. Glycoconj J 33(4):569–579

110. Morikiri Y, Matsuta E, Inoue H (2018) The collagen-derived compound collagen tripeptide induces collagen expression and extends lifespan via a conserved p38 mitogen-activated protein kinase cascade. Biochem Biophys Res Commun 505(4):1168–1173
111. Mosekilde L (1989) Sex differences in age-related loss of vertebral trabecular bone mass and structure--biomechanical consequences. Bone 10(6):425–432
112. Nishikimi A, Koyama YI, Ishihara S, Kobayashi S, Tometsuka C, Kusubata M, Kuwaba K, Hayashida O, Hattori S, Katagiri K (2018) Collagen-derived peptides modulate cd4(+) t-cell differentiation and suppress allergic responses in mice. Immun Inflamm Dis 6(2):245–255
113. Nogimura D, Mizushige T, Taga Y, Nagai A, Shoji S, Azuma N, Kusubata M, Adachi SI, Yoshizawa F, Kabuyama Y (2020) Prolyl-hydroxyproline, a collagen-derived dipeptide, enhances hippocampal cell proliferation, which leads to antidepressant-like effects in mice. FASEB J 34(4):5715–5723
114. Norris GE, Parry DAD (2025) Comparative study of the positions of glycosylated residues in the chains of fibril-forming collagens and the role that these residues play in axial and lateral assembly. In: Parry DAD, Luther PK (eds) Subcellular biochemistry: fibrous proteins: structures and mechanisms, 2nd edn. Springer International Publishing, Switzerland
115. Notbohm H, Mosler S, Bodo M, Yang C, Lehmann H, Bätge B, Müller PK (1992) Comparative study on the thermostability of collagen i of skin and bone: influence of posttranslational hydroxylation of prolyl and lysyl residues. J Protein Chem 11(6):635–643
116. Ogle JD, Arlinghaus RB, Lgan MA (1962) 3-hydroxyproline, a new amino acid of collagen. J Biol Chem 237:3667–3673
117. Oguma Y, Alessio N, Aprile D, Dezawa M, Peluso G, Di Bernardo G, Galderisi U (2023) Meta-analysis of senescent cell secretomes to identify common and specific features of the different senescent phenotypes: a tool for developing new senotherapeutics. Cell Commun Signal 21(1):262
118. Padilla Colón CJ, Molina-Vicenty IL, Frontera-Rodríguez M, García-Ferré A, Rivera BP, Cintrón-Vélez G, Frontera-Rodríguez S (2018) Muscle and bone mass loss in the elderly population: advances in diagnosis and treatment. J Biomed (Syd) 3:40–49
119. Panwar P, Lamour G, Mackenzie NC, Yang H, Ko F, Li H, Brömme D (2015) Changes in structural-mechanical properties and degradability of collagen during aging-associated modifications. J Biol Chem 290(38):23291–23306
120. Panwar P, Butler GS, Jamroz A, Azizi P, Overall CM, Brömme D (2018) Aging-associated modifications of collagen affect its degradation by matrix metalloproteinases. Matrix Biol 65: 30–44
121. Pei X, Yang R, Zhang Z, Gao L, Wang J, Xu Y, Zhao M, Han X, Liu Z, Li Y (2010) Marine collagen peptide isolated from chum salmon (oncorhynchus Keta) skin facilitates learning and memory in aged c57bl/6j mice. Food Chem 118(2):333–340
122. Pfeiffer BJ, Franklin CL, Hsieh FH, Bank RA, Phillips CL (2005) Alpha 2(i) collagen deficient oim mice have altered biomechanical integrity, collagen content, and collagen crosslinking of their thoracic aorta. Matrix Biol 24(7):451–458
123. Pham TX, Lee J, Guan J, Caporarello N, Meridew JA, Jones DL, Tan Q, Huang SK, Tschumperlin DJ, Ligresti G (2022) Transcriptional analysis of lung fibroblasts identifies pim1 signaling as a driver of aging-associated persistent fibrosis. JCI Insight 7(6):e153672
124. Phillips SA, Thornalley PJ (1993) The formation of methylglyoxal from triose phosphates. Investigation using a specific assay for methylglyoxal. Eur J Biochem 212(1):101–105
125. Philp AM, Collier RL, Grover LM, Davis ET, Jones SW (2017) Resistin promotes the abnormal type i collagen phenotype of subchondral bone in obese patients with end stage hip osteoarthritis. Sci Rep 7(1):4042
126. Podolsky MJ, Yang CD, Valenzuela CL, Datta R, Huang SK, Nishimura SL, Dallas SL, Wolters PJ, Le Saux CJ, Atabai K (2020) Age-dependent regulation of cell-mediated collagen turnover. JCI Insight 5(10):e137519
127. Popenoe EA, Aronson RB, Van Slyke DD (1965) The course of lysine hydroxylation during collagen biosynthesis. Paper presented at: Conference: international symposium on the biochemistry and physiology of connective tissue, Lyon (France), 31 Aug–3 Sep 1965. Brookhaven National Lab. (BNL), Upton

128. Prockop DJ, Weinstein E, Mulveny T (1966) Hydroxylation of lysine in a polypeptide precursor of collagen. Biochem Biophys Res Commun 22(1):124–128
129. Pu SY, Huang YL, Pu CM, Kang YN, Hoang KD, Chen KH, Chen C (2023) Effects of oral collagen for skin anti-aging: a systematic review and meta-analysis. Nutrients 15(9):2080
130. Quan T, Fisher GJ (2015) Role of age-associated alterations of the dermal extracellular matrix microenvironment in human skin aging: a mini-review. Gerontology 61:427
131. Quan T, Shao Y, He T, Voorhees JJ, Fisher GJ (2010) Reduced expression of connective tissue growth factor (ctgf/ccn2) mediates collagen loss in chronologically aged human skin. J Invest Dermatol 130(2):415–424
132. Quan T, Qin Z, Voorhees JJ, Fisher GJ (2012) Cysteine-rich protein 61 (ccn1) mediates replicative senescence-associated aberrant collagen homeostasis in human skin fibroblasts. J Cell Biochem 113(9):3011–3018
133. Ramachandran GN, Kartha G (1954) Structure of collagen. Nature 174(4423):269–270
134. Ren LL, Miao H, Wang YN, Liu F, Li P, Zhao YY (2023) Tgf-β as a master regulator of aging-associated tissue fibrosis. Aging Dis 14(5):1633–1650
135. Rennard SI, Stier LE, Crystal RG (1982) Intracellular degradation of newly synthesized collagen. J Invest Dermatol 79(Suppl 1):77s–82s
136. Ricard-Blum S (2011) The collagen family. Cold Spring Harb Perspect Biol 3(1):a004978
137. Ricciarelli R, Maroni P, Ozer N, Zingg JM, Azzi A (1999) Age-dependent increase of collagenase expression can be reduced by alpha-tocopherol via protein kinase c inhibition. Free Radic Biol Med 27(7–8):729–737
138. Rosenberg JL, Woolley W, Elnunu I, Kamml J, Kammer DS, Acevedo C (2023) Effect of non-enzymatic glycation on collagen nanoscale mechanisms in diabetic and age-related bone fragility. Biocell 47(7):1651
139. Sasaki N, Odajima S (1996) Elongation mechanism of collagen fibrils and force-strain relations of tendon at each level of structural hierarchy. J Biomech 29(9):1131–1136
140. Schafer MJ, White TA, Iijima K, Haak AJ, Ligresti G, Atkinson EJ, Oberg AL, Birch J, Salmonowicz H, Zhu Y, Mazula DL, Brooks RW, Fuhrmann-Stroissnigg H, Pirtskhalava T, Prakash YS, Tchkonia T, Robbins PD, Aubry MC, Passos JF, Kirkland JL, Tschumperlin DJ, Kita H, LeBrasseur NK (2017) Cellular senescence mediates fibrotic pulmonary disease. Nat Commun 8(1):14532
141. Schalkwijk CG, Stehouwer CD, van Hinsbergh VW (2004) Fructose-mediated non-enzymatic glycation: sweet coupling or bad modification. Diabetes Metab Res Rev 20(5):369–382
142. Schurman CA, Kaya S, Dole N, Luna NMM, Castillo N, Potter R, Rose JP, Bons J, King CD, Burton JB, Schilling B, Melov S, Tang S, Schaible E, Alliston T (2024) Aging impairs the osteocytic regulation of collagen integrity and bone quality. Bone Res 12(1):13
143. Segrest JP, Cunningham LW (1970) Variations in human urinary o-hydroxylysyl glycoside levels and their relationship to collagen metabolism. J Clin Invest 49(8):1497–1509
144. Sell DR, Biemel KM, Reihl O, Lederer MO, Strauch CM, Monnier VM (2005) Glucosepane is a major protein cross-link of the senescent human extracellular matrix. Relationship with diabetes. J Biol Chem 280(13):12310–12315
145. Shuster S (2005) Osteoporosis, a unitary hypothesis of collagen loss in skin and bone. Med Hypotheses 65(3):426–432
146. Shuster SAM, Black MM, McVitie EVA (1975) The influence of age and sex on skin thickness, skin collagen and density. Br J Dermatol 93(6):639–643
147. Skorupski P, Krol J, Starega J, Adamiak A, Jankiewicz K, Rechberger T (2006) An alpha-1 chain of type i collagen sp1-binding site polymorphism in women suffering from stress urinary incontinence. Am J Obstet Gynecol 194(2):346–350
148. Spielmann N, Miller G, Oprea TI, Hsu C-W, Fobo G, Frishman G, Montrone C, Haseli Mashhadi H, Mason J, Munoz Fuentes V, Leuchtenberger S, Ruepp A, Wagner M, Westphal DS, Wolf C, Görlach A, Sanz-Moreno A, Cho Y-L, Teperino R, Brandmaier S, Sharma S, Galter IR, Östereicher MA, Zapf L, Mayer-Kuckuk P, Rozman J, Teboul L, Bunton-Stasyshyn RKA, Cater H, Stewart M, Christou S, Westerberg H, Willett AM, Wotton JM, Roper WB, Christiansen AE, Ward CS, Heaney JD, Reynolds CL, Prochazka J, Bower L, Clary D, Selloum M, Bou About G, Wendling O, Jacobs H, Leblanc S, Meziane H, Sorg T,

Audain E, Gilly A, Rayner NW, Aguilar-Pimentel JA, Becker L, Garrett L, Hölter SM, Amarie OV, Calzada-Wack J, Klein-Rodewald T, da Silva-Buttkus P, Lengger C, Stoeger C, Gerlini R, Rathkolb B, Mayr D, Seavitt J, Gaspero A, Green JR, Garza A, Bohat R, Wong L, McElwee ML, Kalaga S, Rasmussen TL, Lorenzo I, Lanza DG, Samaco RC, Veeraragaven S, Gallegos JJ, Kašpárek P, Petrezsélyová S, King R, Johnson S, Cleak J, Szkoe-Kovacs Z, Codner G, Mackenzie M, Caulder A, Kenyon J, Gardiner W, Phelps H, Hancock R, Norris C, Moore MA, Seluke AM, Urban R, Kane C, Goodwin LO, Peterson KA, McKay M, Cook JJ, Lowy JP, McFarland M, Wood JA, Willis BJ, Tolentino H, Tolentino T, Schuchbauer M, Salazar J, Johnson J, Munson R, Ayadi A, Pavlovic G, Birling M-C, Jacquot S, Ali-Hadji D, Charles P, Andre P, Champy M-F, Riet F, Vukobradovic I, Berberovic Z, Qu D, Guo R, D'Souza A, Huang Z, Camilleri S, Ganguly M, Adissu H, Eskandarian M, Shang X, Duffin K, Xu C, Roberton K, Laurin V, Lan Q, Sleep G, Creighton A, Lintott L, Gertsenstein M, Pereira M, Tondat S, Patel A, Cruz M, Bezginov A, Miller D, Hy W, Yoshiki A, Tanaka N, Tamura M, Liu Z, Ermakova O, Ferrara A, Fruscoloni P, Seisenberger C, Bürger A, Giesert F, Ambrose JC, Arumu gam P, Bevers R, Bleda M, Boardman-Pretty F, Boustred CR, Brittain H, Caulfield MJ, Chan GC, Fowler T, Giess A, Hamblin A, Henderson S, Hubbard TJP, Jackson R, Jones LJ, Kasperaviciute D, Kayikci M, Kousathanas A, Lahnstein L, Leigh SEA, Leong IUS, Lopez FJ, Maleady-Crowe-F, McEntagart M, Minneci F, Moutsianas L, Mueller M, Murugaesu N, Need AC, O'Donovan P, Odhams CA, Patch C, Perez-Gil D, Pereira MB, Pullinger J, Rahim T, Rendon A, Rogers T, Savage K, Sawant K, Scott RH, Siddiq A, Sieghart A, Smith SC, Sosinsky A, Stuckey A, Tanguy M, Taylor-Tavares AL, Thomas ERA, Thompson SR, Tucci A, Welland MJ, Williams E, Witkowska K, Wood SM, Hitz M-P, Zeggini E, Wolf E, Sedlacek R, Murray SA, Svenson KL, Braun RE, White JK, Kelsey L, Gao X, Shiroishi T, Xu Y, Seong JK, Mammano F, Tocchini-Valentini GP, Beaudet AL, Meehan TF, Parkinson H, Smedley D, Mallon A-M, Wells SE, Grallert H, Wurst W, Marschall S, Fuchs H, Brown SDM, Flenniken AM, Nutter LMJ, McKerlie C, Herault Y, Lloyd KCK, Dickinson ME, Gailus-Durner V, Hrabe de Angelis M, consortium I, Genomics England Research C (2022) Extensive identification of genes involved in congenital and structural heart disorders and cardiomyopathy. Nat Cardiovasc Res 1(2):157–173

149. Sprangers S, Everts V (2019) Molecular pathways of cell-mediated degradation of fibrillar collagen. Matrix Biol 75-76:190–200
150. Sprenger CC, Plymate SR, Reed MJ (2010) Aging-related alterations in the extracellular matrix modulate the microenvironment and influence tumor progression. Int J Cancer 127(12):2739–2748
151. Sricholpech M, Perdivara I, Nagaoka H, Yokoyama M, Tomer KB, Yamauchi M (2011) Lysyl hydroxylase 3 glucosylates galactosylhydroxylysine residues in type i collagen in osteoblast culture. J Biol Chem 286(11):8846–8856
152. Taga Y, Kusubata M, Ogawa-Goto K, Hattori S (2016) Developmental stage-dependent regulation of prolyl 3-hydroxylation in tendon type i collagen. J Biol Chem 291(2):837–847
153. Taga Y, Tanaka K, Hamada C, Kusubata M, Ogawa-Goto K, Hattori S (2017) Hydroxyhomocitrulline is a collagen-specific carbamylation mark that affects cross-link formation. Cell Chem Biol 24(10):1276–1284.e1273
154. Taga Y, Hayashida O, Ashour A, Amen Y, Kusubata M, Ogawa-Goto K, Shimizu K, Hattori S (2018a) Characterization of angiotensin-converting enzyme inhibitory activity of x-hyp-gly-type tripeptides: importance of collagen-specific prolyl hydroxylation. J Agric Food Chem 66(33):8737–8743
155. Taga Y, Kusubata M, Ogawa-Goto K, Hattori S, Funato N (2018b) Collagen-derived x-hyp-gly-type tripeptides promote differentiation of mc3t3-e1 pre-osteoblasts. J Funct Foods 46:456–462
156. Talwar HS, Griffiths CEM, Fisher GJ, Hamilton TA, Voorhees JJ (1995) Reduced type i and type iii procollagens in photodamaged adult human skin. J Invest Dermatol 105(2):285
157. Thomas DP, Zimmerman SD, Hansen TR, Martin DT, McCormick RJ (2000) Collagen gene expression in rat left ventricle: interactive effect of age and exercise training. J Appl Physiol 89(4):1462–1468

158. Thorpe CT, Streeter I, Pinchbeck GL, Goodship AE, Clegg PD, Birch HL (2010) Aspartic acid racemization and collagen degradation markers reveal an accumulation of damage in tendon collagen that is enhanced with aging. J Biol Chem 285(21):15674–15681
159. Tometsuka C, Funato N, Mizuno K, Taga Y (2021) Long-term intake of ginger protease-degraded collagen hydrolysate reduces blood lipid levels and adipocyte size in mice. Curr Res Food Sci 4:175–181
160. Tsuchida T, Friedman SL (2017) Mechanisms of hepatic stellate cell activation. Nat Rev Gastroenterol Hepatol 14(7):397–411
161. Tsuruoka N, Yamato R, Sakai Y, Yoshitake Y, Yonekura H (2007) Promotion by collagen tripeptide of type i collagen gene expression in human osteoblastic cells and fracture healing of rat femur. Biosci Biotechnol Biochem 71(11):2680–2687
162. Tvaroška I (2024) Glycosylation modulates the structure and functions of collagen: a review. Molecules 29(7):1417
163. Uzawa K, Grzesik WJ, Nishiura T, Kuznetsov SA, Robey PG, Brenner DA, Yamauchi M (1999) Differential expression of human lysyl hydroxylase genes, lysine hydroxylation, and cross-linking of type i collagen during osteoblastic differentiation in vitro. J Bone Miner Res 14(8):1272–1280
164. Vafaie F, Yin H, O'Neil C, Nong Z, Watson A, Arpino JM, Chu MW, Wayne Holdsworth D, Gros R, Pickering JG (2014) Collagenase-resistant collagen promotes mouse aging and vascular cell senescence. Aging Cell 13(1):121–130
165. Valtavaara M, Papponen H, Pirttilä AM, Hiltunen K, Helander H, Myllylä R (1997) Cloning and characterization of a novel human lysyl hydroxylase isoform highly expressed in pancreas and muscle. J Biol Chem 272(11):6831–6834
166. Valtavaara M, Szpirer C, Szpirer J, Myllylä R (1998) Primary structure, tissue distribution, and chromosomal localization of a novel isoform of lysyl hydroxylase (lysyl hydroxylase 3). J Biol Chem 273(21):12881–12886
167. Varani J, Warner RL, Gharaee-Kermani M, Phan SH, Kang S, Chung JH, Wang ZQ, Datta SC, Fisher GJ, Voorhees JJ (2000) Vitamin a antagonizes decreased cell growth and elevated collagen-degrading matrix metalloproteinases and stimulates collagen accumulation in naturally aged human skin. J Invest Dermatol 114(3):480–486
168. Varani J, Dame MK, Rittie L, Fligiel SE, Kang S, Fisher GJ, Voorhees JJ (2006) Decreased collagen production in chronologically aged skin: roles of age-dependent alteration in fibroblast function and defective mechanical stimulation. Am J Pathol 168(6):1861–1868
169. Vidal R, Wagner JUG, Braeuning C, Fischer C, Patrick R, Tombor L, Muhly-Reinholz M, John D, Kliem M, Conrad T, Guimarães-Camboa N, Harvey R, Dimmeler S, Sauer S (2019) Transcriptional heterogeneity of fibroblasts is a hallmark of the aging heart. JCI Insight 4(22): e131092
170. Vouyouka AG, Pfeiffer BJ, Liem TK, Taylor TA, Mudaliar J, Phillips CL (2001) The role of type i collagen in aortic wall strength with a homotrimeric. J Vasc Surg 33(6):1263–1270
171. Wang W-J, Chen X-M, Cai G-Y (2021) Cellular senescence and the senescence-associated secretory phenotype: potential therapeutic targets for renal fibrosis. Exp Gerontol 151:111403
172. Watanabe T, Nishimura K, Takeuchi R, Koyama Y, Kusubata M, Takehana K, Hiramatsu K (2015) Oral ingestion of collagen peptide causes change in width of the perimysium of the chicken iliotibialis lateralis muscle. J Vet Med Sci 77(11):1413–1417
173. Watson RE, Gibbs NK, Griffiths CE, Sherratt MJ (2014) Damage to skin extracellular matrix induced by uv exposure. Antioxid Redox Signal 21(7):1063–1077
174. Weis SM, Emery JL, Becker KD, McBride DJ Jr, Omens JH, McCulloch AD (2000) Myocardial mechanics and collagen structure in the osteogenesis imperfecta murine (oim). Circ Res 87(8):663–669
175. Weis MA, Hudson DM, Kim L, Scott M, Wu JJ, Eyre DR (2010) Location of 3-hydroxyproline residues in collagen types i, ii, iii, and v/xi implies a role in fibril supramolecular assembly. J Biol Chem 285(4):2580–2590
176. Wen CY, Wu CB, Tang B, Wang T, Yan CH, Lu WW, Pan H, Hu Y, Chiu KY (2012) Collagen fibril stiffening in osteoarthritic cartilage of human beings revealed by atomic force microscopy. Osteoarthr Cartil 20(8):916

177. Wen XX, Wang FQ, Xu C, Wu ZX, Zhang Y, Feng YF, Yan YB, Lei W (2015) Time related changes of mineral and collagen and their roles in cortical bone mechanics of ovariectomized rabbits. PLoS One 10(6):e0127973
178. WHO. World health organization: ageing and health factsheet. https://www.who.int/newsroom/fact-sheets/detail/ageing-and-health. Accessed 1 May 2025
179. Williamson KA, Katie JL, Beamish E, Carter A, Gumbs JA, Cooper G, Cheung G, Brown D, Pettitt R, Comerford EV, Clegg PD, Canty-Laird EG (2023) Active synthesis of type i collagen homotrimer in dupuytren's fibrosis is unaffected by anti-tnf-α treatment. bioRxiv.2020.2007.2013.195107
180. Woo M, Song YO, Kang KH, Noh JS (2018) Anti-obesity effects of collagen peptide derived from skate (raja kenojei) skin through regulation of lipid metabolism. Mar Drugs 16(9):306
181. Worthen CA, Cui Y, Orringer JS, Johnson TM, Voorhees JJ, Fisher GJ (2020) Cd26 identifies a subpopulation of fibroblasts that produce the majority of collagen during wound healing in human skin. J Invest Dermatol 140(12):2515–2524.e2513
182. Wu J, Yu M, Zhou Y (2017) Association of collagen type i alpha 1 +1245g/t polymorphism and osteoporosis risk in post-menopausal women: a meta-analysis. Int J Rheum Dis 20(7): 903–910
183. Xia W, Hammerberg C, Li Y, He T, Quan T, Voorhees JJ, Fisher GJ (2013) Expression of catalytically active matrix metalloproteinase-1 in dermal fibroblasts induces collagen fragmentation and functional alterations that resemble aged human skin. Aging Cell 12(4): 661–671
184. Yamauchi M, Sricholpech M (2012) Lysine post-translational modifications of collagen. Essays Biochem 52:113–133
185. Yamauchi M, Terajima M, Shiiba M (2019) Lysine hydroxylation and cross-linking of collagen. Methods Mol Biol 1934:309–324
186. Yeowell HN, Walker LC (1999) Tissue specificity of a new splice form of the human lysyl hydroxylase 2 gene. Matrix Biol 18(2):179–187
187. Yu T-Y, Pang J-HS, Wu KP-H, Chen MJL, Chen C-H, Tsai W-C (2013) Aging is associated with increased activities of matrix metalloproteinase-2 and -9 in tenocytes. BMC Musculoskelet Disord 14(1):2
188. Zhong B, Huang D, Ma K, Deng X, Shi D, Wu F, Shao Z (2017) Association of col1a1 rs1800012 polymorphism with musculoskeletal degenerative diseases: a meta-analysis. Oncotarget 8(43):75488–75499
189. Zimmermann EA, Schaible E, Bale H, Barth HD, Tang SY, Reichert P, Busse B, Alliston T, Ager JW, Ritchie RO (2011) Age-related changes in the plasticity and toughness of human cortical bone at multiple length scales. Proc Natl Acad Sci USA 108(35):14416–14421
190. Makareeva E, Han S, Vera JC, Sackett DL, Holmbeck K, Phillips CL, Visse R, Nagase H, Leikin S (2010) Carcinomas contain a matrix metalloproteinase–resistant isoform of type I collagen exerting selective support to invasion. Cancer Res 70(11):4366–4374. https://doi.org/10.1158/0008-5472.can-09-4057
191. Chen Y, Yang S, Tavormina J, Tampe D, Zeisberg M, Wang H, Mahadevan KK, Wu CJ, Sugimoto H, Chang CC, Jenq RR, McAndrews KM, Kalluri R (2022) Oncogenic collagen I homotrimers from cancer cells bind to α3β1 integrin and impact tumor microbiome and immunity to promote pancreatic cancer. Cancer Cell 40(8):818–834.e819. https://doi.org/10.1016/j.ccell.2022.06.011

Chapter 13
Fibrin as a Versatile Fibrous Biopolymer

John W. Weisel and Rustem I. Litvinov

Abstract Fibrin is a fibrous biopolymer that plays a crucial role in hemostasis, thrombosis, wound healing, and various other biological functions and pathological conditions. The X-ray crystallographic structure of fibrinogen, along with computational reconstructions of missing regions and extensive biochemical and biophysical studies, has provided significant insights into the molecular mechanisms of fibrin formation, its structural organization, and its biological and mechanical properties. Upon cleavage of fibrinopeptides by thrombin, the blood protein fibrinogen is converted into fibrin monomers, which then interact through "knobs" exposed by the removal of fibrinopeptides in the central region and "holes" that are constitutively available at the ends of the molecules. The result is half-staggered, double-stranded oligomers that elongate into protofibrils, which then aggregate laterally to form fibers, and branch to create a three-dimensional network. Much has been learned about how the structure of fibrin contributes to the mechanical properties of the clot, including changes in fiber orientation, stretching, buckling, and the forced unfolding of molecular domains. Recent research into the mechanical stability of fibrin has enhanced our understanding of its rupture resistance, which is relevant to thrombotic embolization and mechanical thrombectomy. The fibrinolytic system, in which plasminogen, along with tissue-type plasminogen activator, binds to fibrin and is activated to plasmin, leads to the digestion of fibrin at specific lysine residues. Fibrin has been utilized in hemostatic fibrin sealants and as a biomaterial in tissue engineering and regenerative medicine. Despite significant advances in our understanding of these interconnected processes, much remains unknown about the molecular mechanisms underlying fibrin's biological functions, particularly concerning the molecular origins of its mechanical properties and the more complex structure and properties of hemostatic clots and pathological thrombi and their clinical implications.

J. W. Weisel (✉) · R. I. Litvinov
Department of Cell and Developmental Biology, University of Pennsylvania School of Medicine, Philadelphia, PA, USA
e-mail: weisel@pennmedicine.upenn.edu; litvinov@pennmedicine.upenn.edu

D. A. D. Parry, P. K. Luther (eds.), *Fibrous Proteins: Structures and Mechanisms*, Subcellular Biochemistry 113, https://doi.org/10.1007/978-3-032-05273-5_13

Keywords Fibrin formation · Fibrin structure · Fibrin properties · Fibrinogen composition · α-Helical coiled-coil · Blood clot · Thrombus · Fibrin polymerization · Fibrin mechanical properties · Fibrin rupture resistance · Fibrinolysis · Fibrin biomaterials · Fibrin sealants

Abbreviations

α2-AP	α2-antiplasmin
CGS	centimeter-gram-second system of units
FDP	fibrin(ogen) degradation products
FGA	fibrinogen Aα chain gene
FGB	fibrinogen Bβ chain gene
FGG	fibrinogen γ chain gene
FpA and FpB	fibrinopeptides A and B
GHRP	Gly-His-Arg-Pro
GPRP	Gly-Pro-Arg-Pro
Lp(a)	lipoprotein(a)
PAI-1	Plg activator inhibitor-1
PDGF	platelet-derived growth factor
Plg	plasminogen
Pn	plasmin
PRF	platelet-rich fibrin
RBC	red blood cell
SARS-CoV-2	Severe Acute Respiratory Syndrome Coronavirus 2
SI	International System of Units
TAFI	thrombin-activatable fibrinolysis inhibitor
TGF-β	transforming growth factor beta
TM	thrombomodulin
t-PA	tissue-type Plg activator
u-PA	urokinase-type Plg activator
VEGF	vascular endothelial growth factor
VLDLR	very-low-density lipoprotein receptor

Introduction

Fibrin is a fibrous, insoluble protein biopolymer derived from fibrinogen through enzymatic cleavage by thrombin, serving as the final product of blood clotting. It polymerizes into a mesh-like structure that includes platelets and captures red blood cells, forming the framework of blood clots. In the body, this process either stops bleeding at the injury site (a protective mechanism known as hemostasis) or can obstruct blood vessels, leading to impaired blood flow (a pathological condition called thrombosis). Fibrin gives blood clots and thrombi their viscoelastic properties, providing unique mechanical properties crucial for their biological functions. Beyond hemostasis and thrombosis, fibrin is a key component of the extracellular matrix, contributing to cellular processes such as adhesion, migration, proliferation, differentiation, wound healing, tissue repair, angiogenesis, immune defense, inflammation, and more. As such, any structural or functional changes to fibrin can result in diverse pathological conditions of various etiologies.

The assembly of fibrin creates a multiscale structure that offers adaptable mechanical and biochemical properties, making it well-suited for various biological functions. Fibrin displays both elasticity and plasticity, enabling clots to stretch and deform under mechanical stress while preserving their structural integrity. As a viscoelastic biopolymer, fibrin governs the deformability and rupture resistance of blood clots and fibrin-based biomaterials in response to biomechanical forces. The mechanical behavior of fibrin in reaction to stresses from blood flow, deformations of the pulsating heart and vessel wall, platelet-driven clot contraction, skeletal muscles surrounding blood vessels, and other forces influences the development and outcomes of thrombotic and hemostatic conditions, such as venous thromboembolism, heart attacks, and ischemic strokes.

The diverse properties of fibrin make it not only a crucial player in physiology but also a valuable biomaterial for medical and bioengineering applications. In biomedicine, fibrin is used as a versatile polymer with unique mechanical properties, serving a range of clinical and laboratory purposes, including hemostatic fibrin sealants and scaffolds for tissue engineering. Fibrin is naturally cleaved by plasmin during fibrinolysis, ensuring that blood clots and the fibrin matrix are removed when no longer necessary. Despite the significant clinical importance of fibrin, both as the framework of blood clots and thrombi and as a widely used biomaterial, many aspects of its structural mechanisms and properties, which underpin its biomedical value, remain insufficiently explored.

Fibrinogen and fibrin have been previously reviewed in this series of books focused on fibrous proteins [281, 285]. Numerous recent reviews have addressed various aspects of fibrin(ogen) structure and biology, each with varying levels of depth ([15, 18, 38, 61, 63, 65, 74, 93, 95, 104, 119, 139, 146, 151, 152, 169, 171, 182, 188, 202, 229, 259, 266, 273, 292, 296, 300, 303, 304], and more). Over the past decade, significant progress has been made in our understanding of fibrin formation and structure, modulation of clot properties, and fibrinolysis. While this review primarily focuses on human fibrinogen and fibrin, the core principles discussed are applicable beyond species-specific differences.

Fibrinogen, the Fibrin Precursor

Fibrinogen is a soluble protein, present in the blood plasma of healthy humans at the level of 2–4 g/L, which is necessary for blood clotting and many other biological functions. Since fibrin is derived from fibrinogen as described below (section

"Conversion of fibrinogen to fibrin") and is structurally similar, it is important to describe the molecular structure of fibrinogen. The primary, secondary, tertiary, and quaternary structures of fibrinogen are complex and directly underlie its interactions and properties. Divalent calcium ions are important in blood clotting, and fibrinogen has binding sites for calcium. Fibrinogen is a glycoprotein, and its carbohydrate moieties influence its functions.

Polypeptide Composition and Conformation of Fibrinogen Molecules

Human fibrinogen consists of three pairs of polypeptide chains: Aα, Bβ, and γ, with molecular weights of 66,500, 52,000, and 46,500 Da, respectively (Fig. 13.1). The addition of N-linked carbohydrates to the Bβ and γ chains increases the overall

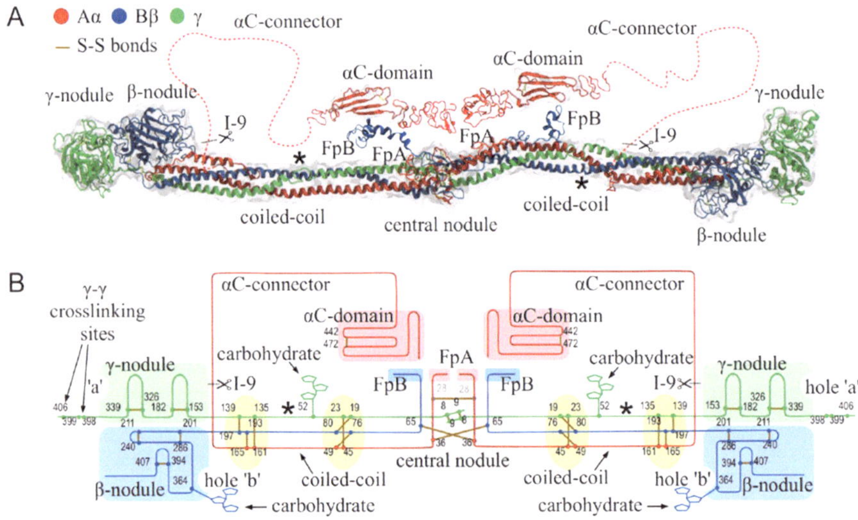

Fig. 13.1 Human fibrinogen structure (PDB entry: 3GHG; color denotation is presented in the figure). (**a**) Structural details showing the central nodule, γ-nodules, β-nodules, disulfide rings, αC-connectors, αC-domains, and fibrinopeptides A (FpA) and B (FpB). The fibrinogen molecule is shown in the compact crystal structure, with the truncation sites for I-9 fibrinogen variant indicated by the scissors. Addition of flexible C-terminal portions of the Aα chains (αC regions) is reconstructed computationally with respect to the NMR structure of isolated αC domains [30, 31] and the high-resolution atomic force microscope images of fibrinogen [211]. (**b**) A schematic representation of the fibrinogen molecule in the naturally folded state with the addition of γ−γ cross-linking sites and carbohydrate moieties. The schematic also shows the location of holes "a" and holes "b" in the γ-nodules and β-nodules, respectively. Both panels also show the location of the cleavage site for the I-9 variant of fibrinogen (scissors) and the location of the plasmin cleavage site, which coincide with the flexible hinge in the central part of the coiled coils (asterisk). With permission from Zuev et al. [313]

molecular weight to approximately 340 kDa. The notation for fibrinogen's polypeptide structure, $(A\alpha\ B\beta\ \gamma)_2$, refers to the small fibrinopeptides A and B (FpA and FpB), which are present at the N-terminal ends of the Aα and Bβ chains, respectively. These fibrinopeptides are cleaved by thrombin, resulting in the α and β chains without the fibrinopeptides. No thrombin cleavage occurs in the γ chains. Thus, the structure of the fibrin monomer is $(\alpha\ \beta\ \gamma)_2$, and the transformation of fibrinogen into fibrin monomer can be represented as: $(A\alpha\ B\beta\ \gamma)_2 \rightarrow (\alpha\ \beta\ \gamma)_2 + 2FpA + 2FpB$.

The six polypeptide chains are connected by 29 disulfide bonds, including 17 interchain and 12 intrachain bonds, forming two symmetrical half-molecules (Fig. 13.1b). The Aα, Bβ, and γ chains contain 8, 11, and 10 cysteine residues, respectively. The amino-terminal portions of all six chains are linked via disulfide bonds in the central globule. The chains each feature Cys-Pro-X-X-Cys sequences, forming "disulfide ring" structures where all three chains are connected at the ends of their α-helical coiled coils. The two halves of the molecule are linked by three interchain disulfide bonds, one between the Aα chains and two between the γ chains. Additionally, an interchain disulfide bond connects the Aα and Bβ chains within each half-molecule. The remaining Aα chain contains one intrachain disulfide, the Bβ chain has three intrachain disulfides, and the γ chain has two. Remarkably, some of these disulfide bonds are not entirely stable and are present in reduced form under some conditions [32].

Fibrinogen molecule has an elongated structure, approximately 45 nm in length and 2–5 nm in diameter, as determined by transmission electron microscopy, atomic force microscopy, and X-ray crystallography (Fig. 13.1a). Complete plasmin cleavage of fibrinogen produces core fragments, including fragment E, which originates from the central region, and fragment D, which contains one of the two identical terminal regions of fibrinogen. According to the nomenclature for the proteolytic fragments, fibrinogen consists of two distal globular D regions and one central globular E region, each containing part of α-helical coiled-coils [170]. X-ray crystallography has provided an atomic-resolution structure of more than two-thirds of fibrinogen, though the full structure has yet to be resolved due to unstructured portions that remain unclear in crystallographic analysis. These unresolved regions include residues 1–26, 1–57, and 1–13 at the N-termini of the Aα, Bβ, and γ chains, respectively, and residues 201–610, 459–461, and 395–411 at the C-termini of the Aα, Bβ, and γ chains, respectively. These missing segments have been computationally reconstructed [169, 309] to create a more complete molecular structure of fibrinogen. Unlike other unresolved regions, the 390-residue C-termini of the Aα chains (residues 221–610), referred to as the αC regions, have been visualized using transmission electron microscopy [285] and atomic force microscopy [212]. These αC regions fold back from the distal ends of the triple coiled-coils to form a fourth strand, then extend outward via a flexible connector to compact C-terminal domains that interact with the central globule of fibrinogen [143]. Truncation of the αC regions significantly alters the hydrodynamic properties of fibrinogen [220]. Studies of fibrinogen in solution have demonstrated its considerable flexibility [313], and systematic mapping of its conformational landscape has begun [169, 207, 311].

Domains of Fibrinogen

Fibrinogen is organized into domains, which are distinct, independently folded structural units. According to the crystal structure, the E region consists of four structural domains, while each D region contains seven structural domains [170]. In each of the two halves of the E region, the N-terminal portions of all three chains form a triple-helical coiled-coil-E domain. The N-terminal portions of both γ chains come together at the center of the E region to form a single asymmetric domain, known as the γN-domain. On the opposite side of the E region, parts of the two Aα and two Bβ chains combine to form another structural domain called the funnel-shaped domain. The globular part of the E region, excluding the coiled-coil-E domains, is referred to as the central nodule. In each D region, the C-terminal portions of all three chains form a triple helical coiled-coil-D domain. The remaining C-terminal portions of the Bβ and γ chains form the β-nodule (or β-module) and γ-nodule (or γ-module), respectively, each containing three structural domains. These domains were identified in the crystal structure of the recombinant γ-nodule and named the N-terminal A-domain, central B-domain, and C-terminal P-domain.

While the structure of the regions of fibrinogen not included in the crystal structure is less clearly defined, it is well understood that each αC region (residues Aα221–610) consists of two distinct portions, visualized by atomic force microscopy [212]. Specifically, the C-terminal portion (residues Aα392–610) has a relatively compact structure, connected to the rest of the molecule by the N-terminal portion (residues Aα221–391), which forms a flexible, unstructured tether. Consequently, the compact part is referred to as the αC-domain, while the flexible part is known as the αC-connector. The two unstructured BβN regions, consisting of the flexible N-terminal portions of the Bβ chains, contain several functionally important binding sites [88]. Structures of other regions of fibrinogen not seen in the crystal structures have been determined by molecular dynamics simulations [169, 312] and cross-linking mass spectrometry and structural modeling [123].

The central nodule and the globular end regions of fibrin(ogen) are connected by 17-nm-long triple (and partially quadruple) α-helical coiled-coil connectors, which consist of 111 or 112 amino acid residues from each of the Aα, Bβ, and γ chains, bordered by “disulfide rings.” In these α-helical coiled-coils, three right-handed α-helices twist around one another to form a left-handed supercoil. Unlike most triple helices, fibrinogen’s coiled-coils feature a fourth helix in the bundle, which contains 30 residues (AαSer166 to AαPro195). This helix starts at the lateral disulfide ring, where the Aα chain makes a U-turn and stretches back in the opposite direction for about a quarter of the coiled-coil connector’s length. The coiled-coils can also bend at a central hinge point located in a non-helical segment of the γ chain near the carbohydrate attachment site γAsn52 [164]. This bent section of the coiled-coil exposes cleavage sites that are susceptible to hydrolysis by plasmin and other proteases, leading to the degradation of fibrinogen into proteolytic fragments. The bending also accounts for the conformational flexibility of fibrinogen observed both in solution and at interfaces [124]. Recent studies suggest that the coiled-coil region plays a functional role in the tensile deformation of fibrin fibers, specifically enabling partial unraveling and spring-like reversible extension-contraction, which helps accommodate and propagate tensile stress along the fiber axis [309]. At high levels of tensile or compressive deformation, the α-helices can undergo a mechanical conversion into β-sheets [144, 310].

Ca^{2+}-Binding Sites of Fibrinogen

Fibrinogen contains both strong and weak binding sites for calcium ions, which are essential for its functions, such as fibrin polymerization and lytic stability. High-affinity calcium-binding sites, known as γ1, are located in the γ chains and are coordinated by four amino acid residues—γAsp318, γAsp320, γGly324, and γPhe322. Another set of high-affinity Ca^{2+}-binding sites, termed β1, is found in the β-nodules, specifically in loop β381–385, where each site is coordinated by a single water molecule. The dissociation constant for calcium ions at these binding sites is low enough that both types of sites are fully occupied under physiological Ca^{2+} concentrations in fibrinogen.

Two additional calcium-binding sites, γ2 and β2, exhibit much lower affinities. The β2 sites anchor the β-nodules to the coiled-coil connector and are involved in the lateral aggregation of protofibrils. It is suggested that β2 sites play a role in regulating the accessibility of the tissue-type plasminogen activator (t-PA)-binding site in fibrin(ogen) [64]. There may also be low-affinity Ca^{2+}-binding sites associated with sialic acid residues on the carbohydrate chains.

Binding of calcium ions to the γ chain's high-affinity sites helps protect the γ chains from enzymatic degradation, similar to how the peptide Gly-Pro-Arg-Pro protects the molecule from plasmin digestion. Ca^{2+} binding does not impact FpA cleavage by thrombin or batroxobin, but is crucial for modulating fibrin polymerization by enhancing lateral aggregation, which leads to thicker fibers. Mutations that affect the high-affinity Ca^{2+}-binding sites have significant functional consequences [23]. A Ca^{2+}-dependent conformational change in fibrin occurs during FpB cleavage. Additionally, the affinity of Ca^{2+}-binding sites changes with FpB release, along with the associated conformational change. The peptide Gly-His-Arg-Pro, a mimetic of the knob 'B' structure, binds ten times more strongly to fibrinogen in the presence of Ca^{2+} compared to its absence. This stronger binding may be linked to the β2 site involved in the conformational change associated with the binding of Gly-His-Arg-Pro to the β-nodule.

Fibrinogen's Carbohydrate Moieties

Each fibrinogen molecule is linked to four biantennary oligosaccharide chains via N-glycosidic bonds: two are attached to the BβAsn364 residues in the β-nodule (as shown in the crystal structure 3GHG), and the other two are connected to γAsn52 in the coiled-coils (which remain unresolved crystallographically but could be reproduced computationally as in [312]) (Fig. 13.1). These carbohydrate attachment sites feature the typical NXS or NXT sequences, characteristic of N-glycosylation. In contrast, the Aα chains do not contain any carbohydrate, despite having two NXS sequences. The heterogeneity of circulating fibrinogen is partly attributed to variable desialylation, or the removal of the terminal N-acetylneuraminic acid (sialic acid) residue. Fibrinogen isolated from human plasma contains equal amounts of mono- and di-sialylated carbohydrate chains but no asialo-chains.

The carbohydrate portion of fibrinogen plays a significant role in fibrin polymerization and clot structure. For example, patients with cirrhosis or other liver diseases have fibrinogen with highly sialylated carbohydrates, which leads to the formation of clots with thin fibers and many branching points [285]. This finding is consistent with experiments where neuraminidase was used to remove sialic acid from normal fibrinogen, resulting in clots made of thicker fibers. Fibrinogen, produced during inflammatory conditions (acute-phase protein), has a markedly different oligosaccharide structure [24]. Complete removal of carbohydrate results in even more dramatic changes to clot structure, producing clots composed of very thick fibers. These observations suggest that both the charge and mass of the carbohydrate influence the extent of lateral aggregation and that the carbohydrate moieties significantly enhance fibrinogen's solubility.

Conversion of Fibrinogen to Fibrin

The conversion of fibrinogen to fibrin is a key outcome of the enzymatic cascade in blood coagulation, playing a crucial role in stopping bleeding (hemostasis) and in conditions like vascular obstruction or thrombosis. This process occurs in two main phases: enzymatic and nonenzymatic (Fig. 13.2). During the enzymatic phase, thrombin catalyzes the cleavage of fibrinopeptides from fibrinogen, resulting in the formation of fibrin monomers. Thrombin, a highly specific serine protease, is activated from its zymogen, prothrombin, which is naturally found in the blood. Moreover, fibrin clots also sequester thrombin, which has significant effects on prothrombotic conditions in blood [55]. In the nonenzymatic phase, the fibrin monomers spontaneously self-assemble into fibrin oligomers, which elongate to form two-stranded protofibrils. These protofibrils aggregate both laterally and longitudinally to create fibers, which branch and form a three-dimensional network known as a clot. Lastly, the fibrin polymer is stabilized through covalent crosslinking by plasma transglutaminase, Factor XIIIa, resulting in a mature fibrin clot that is more stable both mechanically and chemically.

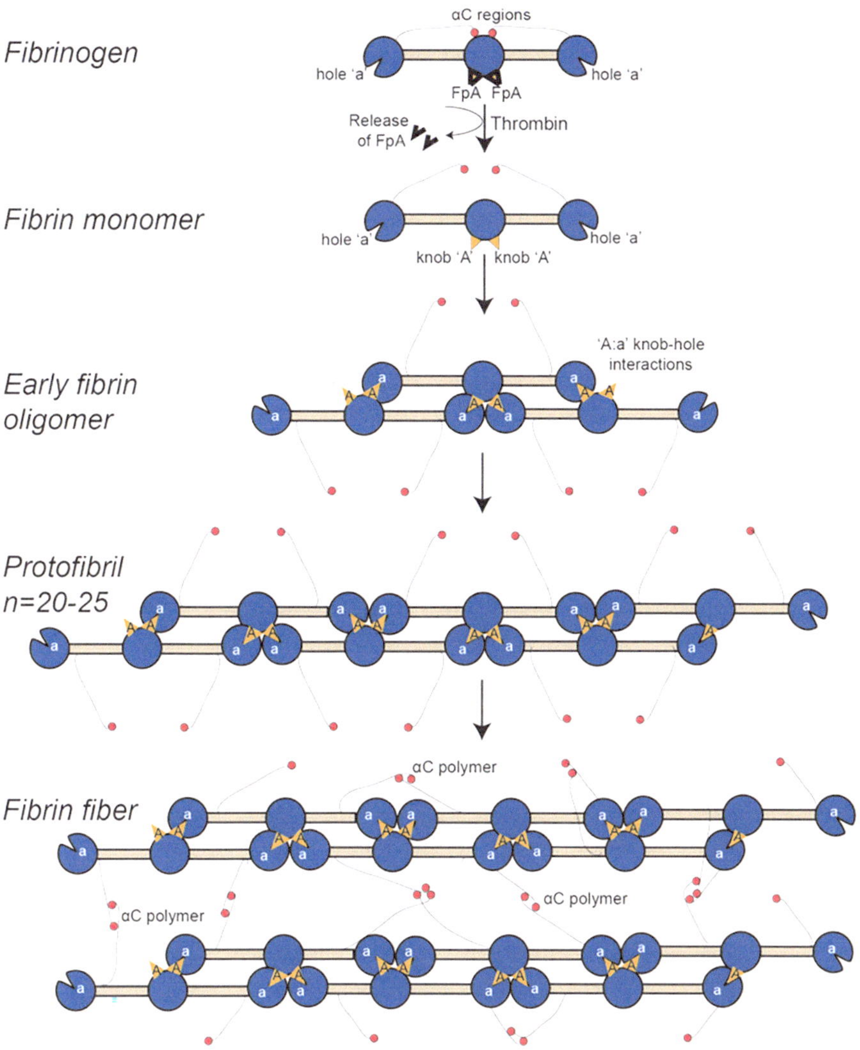

Fig. 13.2 Schematic diagram of fibrin polymerization. Fibrinopeptides in the central nodule cover knobs that are complementary to holes that are always exposed at the ends of the protein. When the fibrinopeptides are removed by thrombin, 'A:a' knob-hole interactions occur, giving rise to oligomers (a trimer is shown), which elongate to produce the two-stranded protofibrils made up of 20–25 half-staggered monomeric molecules. The protofibrils aggregate laterally to make fibers, a process enhanced by interactions of the αC regions and formation of αC-polymers. (Published with permission and thanks to Dr. Alina Peshkova)

Thrombin-Catalyzed Cleavage of Fibrinopeptides

Fibrin formation begins with thrombin cleaving FpA and FpB from the N-termini of the Aα and Bβ chains of fibrinogen, respectively, resulting in the formation of fibrin monomers (Fig. 13.2). The thrombin-fibrin binding during fibrin polymerization has been modeled [120]. FpA (residues 1–16) is cleaved at the AαArg16-Gly17 bond, while FpB (residues 1–14) is cleaved at the BβArg14-Gly15 bond, though at a slower rate than FpA. As fibrin polymerization progresses, the rate of FpB release increases, reaching its peak when polymerization is nearly complete, suggesting that FpB is preferentially released from the fibrin polymer. In surface-bound fibrinogen, unlike in solution, FpBs are cleaved more rapidly than FpAs, depending on the surface density and orientation of fibrinogen [222], indicating that the conformation of fibrinogen influences the accessibility of fibrinopeptides to thrombin. The sequential cleavage of FpA and FpB is driven by steric factors in the interaction between thrombin and fibrinogen, where the FpA-containing regions are positioned more favorably to bind to thrombin's active site cleft [200]. In cases of naturally occurring dysfibrinogenemias or recombinant fibrinogen variants with substitutions in AαArg16 or BβArg14, the cleavage of FpA or FpB is impaired, leading to defective fibrin polymerization [174]. Studies with a recombinant fibrinogen with fibrinopeptide A not cleavable by thrombin have allowed the dissection of pathophysiological functions of fibrinogen versus those requiring fibrin [108, 209].

'A-a' Knob-Hole Interactions: Main Driving Force for Fibrin Polymerization

The release of FpA exposes a new N-terminal α chain motif, Gly-Pro-Arg- (GPR), referred to as knob 'A'. Knob 'A' interacts with complementary pockets or holes 'a' found in the γ-nodules of another fibrin molecule, and this interaction is known as the 'A-a' bond (Figs. 13.2 and 13.3). Studies on these knob-hole interactions at the single-molecule level have shown that the 'A-a' bonds are strong and stable [141]. Moreover, these bonds have an unusual property in that they become stronger with increased force; i.e., they are so-called catch bonds [145]. In addition, these catch bonds affect protofibril packing and clot stability [8].

Crystallographic structures of fragment D (from the fibrinogen's lateral D regions containing the γ-nodule) co-crystallized with the peptide GPRP (a synthetic mimetic of knob 'A') have demonstrated binding of the peptide to pockets in the γ-nodule, indicating the presence of holes 'a' (Fig. 13.3). The binding sites for the GPRP peptide in holes 'a' were identified as residues γTrp315-Trp330, γTrp335-Asn365, and γPhe295-Thr305. Furthermore, molecular dynamics simulations revealed the basis of the catch bonds as a moveable flap that changes conformation in response to force to increase the bond strength [145]. Molecular dynamics simulations further revealed that during fibrin oligomerization, the D:E interface includes binding sites

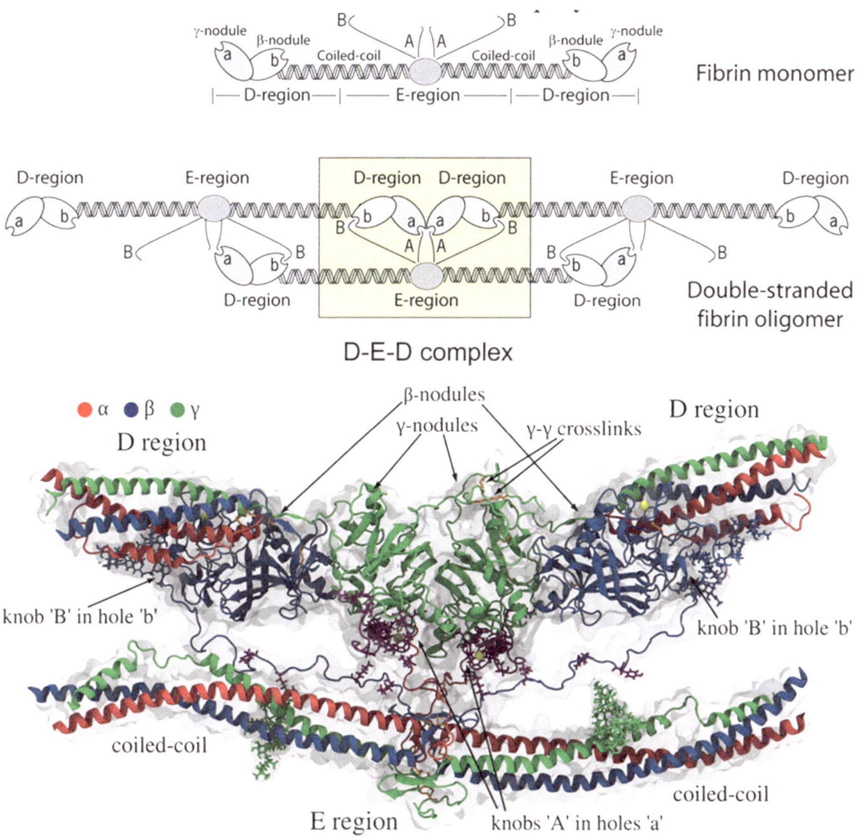

Fig. 13.3 Complementary binding sites or knob-hole interactions in fibrin polymerization. *Top*. Schematic diagram of knob-hole interactions. Knobs 'A' and 'B' in the central domain of a fibrin monomer are complementary to holes 'a' and 'b' that are always exposed at the ends of the protein. When the fibrinopeptides are removed by thrombin, exposing the knobs, knob-hole interactions occur, giving rise to the trimer shown and eventually to the two-stranded protofibril made up of half-staggered molecules. *Bottom*. Atomic resolution structure of the knob-hole interactions. The γ- and β-nodules near the ends of the molecule contain the holes 'a' and 'b', respectively, that are complementary to the knobs 'A' and 'B' in the central nodule. Most of these structures were derived from X-ray crystallographic data, although the disordered and/or flexible N-terminal regions of the α and β chains were derived from computational modeling

beyond the 'A-a' knob-hole interactions, confirming the presence of additional intermolecular 'A' and 'a' binding sites [127]. However, the cleavage of FpA and the exposure of knobs 'A' are both necessary and sufficient for fibrin clot formation, referred to as desA-fibrin. Additionally, the constitutive exposure of fibrin(ogen) holes 'a' is essential for clot formation. If these holes are blocked by the GPRP peptide or altered through a mutation of the key residue γAsp364 [195], fibrin polymerization is inhibited. These findings, along with others, suggest that the 'A-a' knob-hole interactions drive fibrin polymerization and clot formation.

Fibrin Oligomers and Protofibrils: Intermediate Fibrin Precursors

Fibrin polymerization begins when two fibrin monomers, produced by the release of FpA, interact with each other in a half-staggered manner. In this configuration, knob 'A' fits into hole 'a', creating two sets of 'A-a' knob-hole bonds that hold the monomers together. The addition of a third molecule to the half-staggered dimer forms an end-to-end junction, where the ends of two molecules come together, and the lateral D regions of the two molecules create the D:D interface, linking monomers in each of the two strands of fibrin oligomers (Fig. 13.3). The D:D interface, which spans residues γ275–309, is weak and is the first to break when fibrin(ogen) oligomers are subjected to mechanical stretching [309]. In addition, the D:D intermonomer junction is a hinge region that provides a structural basis for the intermolecular flexibility of fibrin(ogen) oligomers [311]. Studies on natural and recombinant fibrinogen variants have shown that residues γ275, γ308, and γ309 are critical for D:D interactions and fibrin strand elongation [20, 161]. Fibrin monomers can continue to add longitudinally, forming longer two-stranded fibrin oligomers of varying lengths. Lateral interactions between two strands of fibrin oligomers occur through the central E region of one fibrin molecule and the lateral D regions of two other molecules. The D-E-D complex is primarily held together by 'A-a' knob-hole bonds and additional interactions at the D:E and D:D interfaces, with additional details revealed by atomic structural models [127, 312].

As fibrin monomers keep adding longitudinally to the oligomers, they elongate further to form two-stranded protofibrils, which are an essential intermediate in fibrin polymerization (Fig. 13.2). Protofibrils are defined as fibrin oligomers long enough to aggregate laterally, typically around 0.5–0.6 μm in length, corresponding to approximately 20–25 monomers [42]. Protofibril growth is partially hindered by the presence of the γ chain splice variant, γ′, likely due to electrostatic repulsion [6, 84]. Since fibrinogen has constitutively open holes, it can bind to fibrin monomer and oligomers through the exposed knobs; such binding can have profound effects on polymerization, especially at early stages [187]. Excess of fibrinogen over slowly polymerizing fibrin can even block polymerization, leading to formation of stable soluble fibrin precursors referred to as "soluble fibrin complexes" or "soluble fibrin" similar to the oligomeric structures produced by the partial enzymatic lysis of fibrin (see section "Fibrinolysis: enzymatic lysis of fibrin").

Lateral Aggregation of Protofibrils Forming Fibrin Fibers

Protofibrils laterally aggregate to form fibers of varying thickness, with the half-staggered molecular arrangement leading to a consistent 22.5-nm repeat, which corresponds to half the length of a fibrin monomer (Fig. 13.2). This ordered longitudinal packing of fibrin fibers has been visualized using transmission electron microscopy with negative contrast and atomic force microscopy. The exact mechanisms, structural motifs, and forces that support the lateral aggregation of protofibrils remain largely unclear. Protofibrils only laterally associate once they reach a certain length, indicating that the interactions responsible for lateral aggregation are relatively weak and accumulate along the protofibril's length. Currently, several structures have been identified or suggested to contribute to inter-protofibril lateral interactions, including knobs 'B' and holes 'b', αC regions, the C-terminal portions of the γ chains, adjacent β-nodules [297], coiled-coils [194], and N-glycans at βAsn364 and γAsn52 [164].

Because structural proteins that self-assemble typically show similar intermolecular interactions in vitro and in vivo, a plausible model for lateral aggregation of protofibrils has been derived from the study of D-region interactions in fibrin crystals. One proposed mechanism for lateral aggregation involves interactions between the β-nodules of adjacent protofibrils, particularly residues β330–375 [297]. Additionally, certain residues, such as βAla68 in the N-terminal part of the β chain [180] and βGly15, the terminal residue of knob **'B'**, play a significant role in lateral aggregation, regardless of whether FpB has been cleaved [101].

Fibrin packing appears to be paracrystalline, with fibrin molecules arranged more regularly along the longitudinal axis than laterally, although regularity in lateral packing has also been observed [35, 139]. The molecular packing of protofibrils in fibrin fibers has been determined by a combination of small-angle X-ray scattering and computational molecular modeling [113]. Packing of protofibrils in fibers differs significantly in fibers of different diameters [134]. Furthermore, twisted protofibrils appear to associate in a specific pattern that results in twisted fibers. The 22.5-nm repeat is maintained as new protofibrils are added to the outer edges of fibers, and these protofibrils must stretch as their path length increases [139]. This stretching could provide a mechanism to limit fiber thickness, as lateral growth halts when the energy needed to stretch an added protofibril exceeds the energy available from bonding.

An alternative model of fibrin polymerization and structure has been proposed based on the formation of ultrathin fibrin sheets across channels on a plastic substrate, though the physiological relevance of these findings is still unclear [193]. Another proposed model suggests that in the early stages of fibrin assembly, short, thin fibrin branches form, and single-bonded "Y-ladder" polymers quickly elongate before transitioning to double-stranded fibrils [232].

'B-b' Knob-Hole Interactions: Auxiliary Mechanism

The cleavage of FpB reveals the N-terminal β-chain sequence Gly-His-Arg-Pro (GHRP), known as knob 'B', which complements hole 'b' found in the globular β-nodule. The affinity of the knob 'B'-mimicking GHRP peptide for fibrinogen ($K_d = 140$ μM) is lower compared to the knob 'A'-mimicking GPRP ($K_d = 25$ μM). Moreover, a larger fragment corresponding to the central nodule, with exposed knobs ‘A’ binding to surface-attached fibrinogen (containing holes ‘a’), has a K_d of 5.8 ± 1.1 μM [83], suggesting that the binding site for knob ‘A’ in fibrin is not limited to the GPR sequence and exhibits a substantially higher affinity. This has been further supported by structural modeling [127, 311].

While direct evidence exists for ‘B-b’ interactions [142], their physiological role remains uncertain and is primarily supported by indirect evidence. Cleaving FpA without cleaving FpB results in clots with thinner fibers than those formed by thrombin cleavage of both fibrinopeptides, implying that ‘B-b’ interactions play a role in the lateral aggregation of protofibrils. However, lateral aggregation and fiber formation can occur even without cleavage of FpB. Further evidence for ‘B-b’ interactions comes from congenital fibrinogen variants or dysfibrinogenemias (such as the Metz and Frankfurt XIII fibrinogens), where all fibrinogen molecules are homodimers, and only FpB is cleaved. These clots exhibit thin fibers at low temperatures. Additionally, recombinant mutants of the γAsp364 residue, which lack functional holes ‘a’ and cannot form ‘A-a’ bonds, formed clots slowly with thrombin but not with reptilase, an enzyme that cleaves FpA but not FpB [195]. On the other hand, the BβAsp432Ala fibrinogen variant, with impaired hole ‘b’, displayed normal polymerization [20]. These findings suggest that ‘B-b’ interactions may occur when ‘A-a’ interactions are disrupted, but their normal functional role remains unclear. Molecular dynamics simulations have provided insights into the structural and thermodynamic properties of ‘B-b’ interactions [127]. It has been proposed that these interactions may influence the susceptibility of clots to proteolytic digestion [64].

In addition to the established ‘A-a’ and ‘B-b’ knob-hole interactions, there is some evidence for the existence of ‘A:b’ interactions and suggestions that ‘B:a’ interactions may not occur [141, 142]. Fibrin structure and properties are influenced by the addition of a knob ‘B’ mimetic combined with polyethylene glycol, indirectly confirming the functional importance of ‘B-b’ knob-hole interactions [28].

The Role of the αC Regions in Fibrin Formation

During fibrin polymerization, the protruding and flexible αC regions (discussed in section “Domains of fibrinogen”) interact intermolecularly, both within and between protofibrils (Fig. 13.2), and are further crosslinked by Factor XIIIa, leading to the formation of αC polymers. The αC-domains adopt an active physiological conformation when they self-associate through their N-terminal subdomains, likely involving β-hairpin swapping and interaction of the C-terminal subdomains with the αC-connector [171]. While the αC regions are not essential for lateral aggregation [68], they play a role in enhancing it [143, 171] and also affect clot mechanical properties [51, 75]. Clots formed from fibrinogen lacking the αC regions consist of thinner fibers with more branching points compared to control clots [51]. Additionally, a recombinant hybrid fibrinogen variant, where the human αC regions were replaced with the shorter chicken αC regions, exhibited selective impairment in the lateral aggregation of protofibrils [208].

Fibrin Branching and Three-Dimensional Network Architecture

As fibrin fibers thicken through lateral aggregation and increase in length, they also branch, a necessary process for forming a three-dimensional network. Electron microscopy has provided evidence for at least two distinct molecular mechanisms by which branch points can form. One mechanism, known as a "bilateral junction," occurs when two protofibrils laterally aggregate to form a four-stranded fibril, and then separate again into two individual protofibrils. The second type of branch point, called a "trimolecular junction" or "equilateral junction," forms when a fibrin molecule binds at the end of a protofibril through a single γ-nodule, allowing both the attached fibrin molecule and the protofibril to elongate as two-stranded protofibrils [78]. In both cases, most branch points in clots consist of three fibers of similar diameters joined together, implying that additional protofibrils are added fairly equally to the original structure, regardless of the type of branch point. Branching has been directly observed in imaging studies of the early stages of fibrin polymerization [42]. Overall, as the number of branch points in a clot increases, the fiber diameters tend to decrease. These observations suggest that branching and lateral aggregation compete: conditions favoring lateral aggregation tend to produce clots with thicker fibers and fewer branch points, while conditions that inhibit lateral aggregation tend to result in clots with thinner fibers and more branch points. A kinetic model was developed to explore the spatial-temporal dynamics of branching [79].

A fibrin clot is a gel that forms once branching fibers create a three-dimensional, space-filling network (Fig. 13.4a). The gelation point, or blood/plasma clotting time, is commonly used in clinical assays to detect coagulation disorders. This point occurs relatively early in polymerization, when only about 15–20% of the fibrinogen has been incorporated into the gel [41]. Some correlations between gelation time and final clot structure have been established, but the complete branching network is not fully formed at the gelation point; new fibers and branch points continue to form after this stage [41]. The structure of clots can be characterized by parameters such as fiber diameter, density, number of branch points, distances between them (fiber length), and pore size, which can be determined using scanning electron microscopy [57] or turbidity [17]. All of these parameters are significantly influenced by variations in the kinetics of fibrin polymerization. The ultrastructure of fibrin clots at and after the gelation point has been studied using various biophysical techniques, revealing the dynamics and complex hierarchy at different spatial scales. While the fiber network does not greatly affect the diffusion of proteins due to the large spaces between fibers, the perfusion of fluid or nanoparticles through the gel is a sensitive

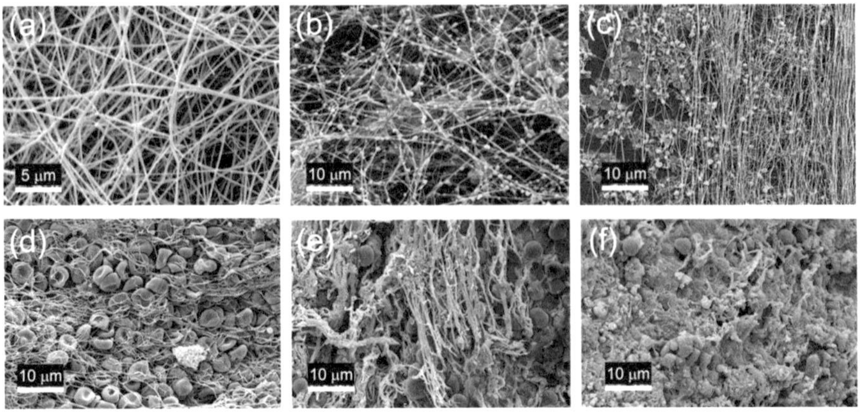

Fig. 13.4 Scanning electron microscope images of clots and thrombi. (**a**–**c**) In vitro clots made from human (**a**) platelet-poor plasma with an isotropic (random) fibrin network structure and (**b**) platelet-rich plasma showing oriented fibrin fibers stretched between platelet aggregates; many platelet-derived microvesicles are attached to fibrin fibers. (**c**) The fibrin formed on a surface coated with collagen adherent platelets is made up of many fibers aligned along the direction of flow. (**d**–**f**) The images show major types of fibrin structures revealed in ex vivo human cerebral artery thrombi: (**d**) fibers, (**e**) bundles, and (**f**) sponge

measure of pore size and overall clot structure. The diffusivity of the fibrin network decreases significantly when it is embedded with blood cells and compressed, as occurs in a contracted whole blood clot [44].

Factor XIIIa-Catalyzed Covalent Crosslinking of Fibrin

To stabilize the clot against proteolytic and mechanical damage, fibrin is covalently crosslinked by the plasma transglutaminase, Factor XIIIa, the active form of the Factor XIII zymogen, which is activated by thrombin in the presence of Ca^{2+}. Molecular mechanisms of this process are being identified [152, 175, 213]. Transglutaminase catalyzes the formation of an isopeptide bond between the ε-amino group of a lysine residue and the γ-carboxamide group of a glutamine residue. The C-terminal portion of each γ chain in fibrin(ogen) contains a crosslinking site where two adjacent molecules form an intermolecular ε-(γ-glutamyl)-lysyl covalent isopeptide bond between γLys406 of one γ chain and γGln398/399 of another γ chain. There has been some debate over whether the γ-γ crosslinking occurs longitudinally within a strand of a protofibril or transversely between fibrin strands [176, 280], but recent evidence supports the longitudinal orientation of these bonds [148, 234, 311]. The γ chain crosslinks stabilize clots against embolization [69]. Similarly, the same intermolecular ε-(γ-glutamyl)-lysyl bonds form more slowly between the C-terminal portions of fibrin α chains (αC regions), creating αC polymers with multiple donor and acceptor sites for crosslinking within the αC regions. Some of these crosslinking pairs have been identified, as well as their susceptibility to crosslinking [179, 238, 241]. Crosslinking also occurs between α and γ chains, resulting in the formation of α-γ-heterodimers [252]. Prior to crosslinking, fibrin polymerization is reversible, with fibrin existing as an equilibrium polymer [43]. However, after crosslinking within and between protofibrils, polymerization becomes irreversible, making the clot more stable, mechanically stronger, and resistant to fibrinolysis. Crosslinked fibrin can be dissolved through the reduction of disulfide bonds holding polypeptide chains together or by chemical/enzymatic hydrolysis of peptide bonds. A genetic polymorphism of Factor XIII, where Val34 is replaced by Leu, alters fibrin clot structure, leading to either coarse, highly permeable clots or dense clots with reduced permeability, depending on fibrinogen concentration [7]. A synthetic hemostatic polymer has recently been developed to stabilize fibrin clots chemically, mimicking the crosslinking effect of Factor XIIIa [39].

Variations of Fibrin(ogen) Structure and Properties

Fibrinogen is a highly heterogeneous protein that largely determines the structural and functional diversity of fibrin. There are fibrinogen variants present from several common genetic polymorphisms or naturally occurring alternative primary structures. There are several dozen different types of post-translational modifications of fibrinogen that can occur in a huge variety of combinations. Furthermore, depending on the environmental conditions during clotting, fibrin network structures and mechanical properties are tremendously variable. Fibrin(ogen) binds a great many other proteins and cells as part of its various functions. Finally, when blood cells are included and especially when clots are formed in vivo under flow as a function of time and space, fibrin structures and properties are substantially modified.

Genetic Polymorphisms

Fibrinogen variants are present in the blood due to several common polymorphisms or naturally occurring alternative primary structures. The most prevalent variants include two types of γ chains, γA and γ′, which result from an alternative polyadenylation signal in intron 9 of the FGG gene. Around 8–15% of total fibrinogen contains the γ′ chain, with the majority existing in the heterodimeric γA/γ′ form, while the homodimer γ′/γ′ makes up only about 1%. The presence of γ′ chains has been shown to slow down the lateral aggregation of protofibrils and alter the formation and structure of fibrin [62]. Increased plasma γ′ fibrinogen concentration is linked to a higher risk of myocardial infarction and other thrombotic conditions [154]. Functional fibrinogen polymorphisms can also occur in polypeptide chains other than the γ chain. For example, αE fibrinogen (see section "Polypeptide composition and conformation of fibrinogen molecules") has a reduced rate of fibrin polymerization and forms thinner, more branched fibers than fibrin networks containing the α chains [178]. The FGA gene polymorphism 2224G/A has been linked to reduced clot permeability [160], while the Ala312 allele of FGA 6534/Thr312Ala is associated with increased clot stiffness [251]. The Lys448 allele of the fibrinogen Bβ chain gene (FGB) polymorphism BβArg448Lys results in a compact fibrin network that is resistant to lysis [2]. There are numerous other examples of correlations between fibrinogen polymorphisms, clot structure and properties, and disease. Other common polymorphisms in the fibrinogen genes occur in noncoding regions and can influence plasma fibrinogen levels.

Post-translational Modifications

Fibrinogen exists in many molecular forms in the blood, initially identified by variations in biochemical properties and gel electrophoretic behavior. It is estimated that there are over a million nonidentical forms of fibrinogen in a healthy individual, resulting from numerous combinations of modified or inherently polymorphic sites [99]. While several genetically determined variations have been discussed, post-translational heterogeneities also contribute to fibrinogen diversity [59, 188, 247, 259]. These heterogeneities arise from various biochemical reactions associated with physiological and pathological conditions, such as inflammation or ischemia. These modifications can include phosphorylation at specific seryl and threonyl sites, prolyl hydroxylation, tyrosyl sulfation, asparaginyl or glutaminyl deamidation, N-terminal pyroglutamate formation from glutaminyl precursors, oxidation of methionine, histidine, and tryptophan residues, tyrosyl nitration, cysteine residue modifications, dityrosine formation, the introduction of carbonyl groups, and more.

The C-terminal portion of the Aα chain is particularly susceptible to limited cleavage by intracellular and extracellular proteolytic enzymes resulting in the formation of I-9 fibrinogen (Fig. 13.1), but specific sites in the Bβ and γ chains can also be digested, leading to the presence of lower molecular weight forms in plasma fibrinogen. Alternative N-glycosylation can contribute to fibrinogen heterogeneity by producing oligosaccharides with variable structures [24]. Acquired abnormal fibrinogen variants can arise in patients with various conditions through nonenzymatic reactions, such as lysine glycation in uncontrolled diabetes mellitus [66] or homocysteinylation in hyperhomocysteinemia [236, 267]. Fibrinogen derivatives can also form under pathological conditions linked to thrombin generation, fibrinolytic enzyme activation, or immune reactions. For example, fibrin(ogen) degradation products are produced when excessive fibrinolytic activity occurs in the blood, and antibody-fibrinogen complexes are formed in certain autoimmune diseases. Individuals treated with acetylsalicylic acid (aspirin) exhibit acetylated lysine residues [3].

Oxidative stress, which is commonly involved in physiological processes such as aging and various disease pathologies—including arterial and venous thrombosis—

can lead to post-translational modifications of plasma proteins, including fibrinogen [118, 126, 165, 302]. Neutrophils and monocytes, prevalent leukocytes in sites of inflammation and venous thrombi, generate nitrating intermediates capable of nitrating fibrinogen [97]. Tobacco smokers often experience nitration of tyrosine and oxidation of methionine, histidine, or tryptophan residues [198]. Many of these chemically modified forms of fibrinogen are linked to alterations in its functional and structural properties, contributing to a thrombogenic phenotype that increases the risk of both arterial and venous thrombosis [198, 279]. Fibrinogen is susceptible to citrullination by peptidyl arginine deiminases, secreted from neutrophils during the formation of neutrophil extracellular traps, affecting clot structure and properties [56, 60, 271].

Hereditary Fibrinogen Abnormalities

Dysfibrinogenemias are characterized by inherent structural changes in the fibrinogen molecule, which often lead to altered clotting or other functional abnormalities of the protein, while immunochemically, fibrinogen as an antigen is determined in blood at a normal level [38]. Traditionally, defective fibrinogen variants are named after the city of their discovery or the patient's place of residence. Although most congenital defects are rare, they have provided significant insights into the structure-function relationships of fibrin(ogen).

Mutations that lead to dysfibrinogenemias are commonly single base mutations that cause the substitution of a single amino acid, although other mutations may create stop codons, resulting in truncation of one of the fibrinogen chains. In some cases, base additions or deletions may occur, impacting fibrinogen structure and function. These mutations can lead to a predisposition for either thrombosis or bleeding or may be asymptomatic, depending on their functional effects [162]. Mutations can influence processes such as fibrinopeptide cleavage, fibrin polymerization, Factor XIIIa-catalyzed crosslinking, susceptibility to fibrinolysis, or integrin αIIbβ3-mediated platelet aggregation. In homozygous forms of dysfibrinogenemias, all fibrinogen molecules are mutated, resulting in only abnormal homodimers in the blood. However, such mutations are rare, and most dysfibrinogenemias are heterozygous. As a result, individuals with dysfibrinogenemia typically have a fibrinogen pool consisting of varying proportions of normal homodimers, mutant homodimers, and heterodimers.

The most common alterations in dysfibrinogenemias occur in the N-terminal part of the Aα chain, in regions responsible for thrombin binding or cleavage, or in or near knob 'A,' such as the AαGly17-Pro18-Arg19 motif. A specific mutation involves the substitution of AαArg16 (the last residue of FpA) with His or Cys, with the former causing delayed release of FpA and the latter preventing its release entirely. Without FpA cleavage, FpB is released slowly, leading to the formation of desB-fibrin clots, but only at lower temperatures. Mutations in AαGly17, AαPro18, AαArg19, or AαVal20 result in defective polymerization due to changes in knobs

'A.' A mutation that results in fibrinogen molecules in plasma lacking both FpA and knobs 'A' makes clots with thin fibers and frequent ends [274].

Mutations in the Bβ chain are less frequent. For example, substituting Cys for BβGly at position 15 results in slow polymerization due to delayed release of FpB [254]. Another mutation, changing BβAla68 to Thr, impairs thrombin binding to fibrin, leading to thrombosis, as fibrin is a natural thrombin absorber and was originally termed antithrombin I [177].

Several mutations have been identified in the C-terminal portion of the γ chain, especially at position γArg275. This residue interacts at the D:D interface, and mutations here disrupt fibrin polymerization. Mutations near hole 'a' are common, but patients with most of these substitutions are asymptomatic, likely due to being heterozygous.

Point mutations can also cause congenital hypo- and afibrinogenemia due to defects in fibrinogen's molecular processing, assembly, secretion, and domain stability. Afibrinogenemia, an autosomal recessive disorder, is characterized by the complete absence of detectable fibrinogen in the blood, while hypofibrinogenemia involves a reduced fibrinogen level. Analysis of the three fibrinogen genes in affected individuals has led to the identification of several causative mutations [22, 184], most of which are linked to mutations in the FGA gene. The severity of hypo- and afibrinogenemia associated with homozygous Aα chain truncations suggests that the minimal requirement for fibrinogen's molecular assembly and secretion is the formation of the distal disulfide ring of the coiled-coil.

Modulation of Fibrin Structure

Fibrin formation and clot structure are influenced by numerous external factors, both in vitro and in vivo, including ionic conditions such as pH, ionic strength, and electrolyte composition, as well as the presence of various endogenous and exogenous compounds [283]. These include polyphosphates, mono-, oligo-, and polysaccharides, peptides, lipids, proteins, nucleic acids, medications, and other normal and pathological substances found in blood and injured tissues. The concentration of active thrombin, determined by the balance of its generation and inhibition/elimination, plays a significant role in clot structure. High thrombin levels lead to clots with many thin fibers, multiple branch points, and small pores, while low thrombin levels result in fewer, thicker fibers with fewer branch points and larger pores [283, 299]. These differences are largely explained by their effects on the kinetics of fibrin polymerization [287], but fibrinogen aggregates also affect clot structure [81]. Surprisingly, the effects of fibrinogen concentration on clot structure have not been fully explored until recently [226]. Additionally, physical factors such as external forces or a strong magnetic field can affect fibrin structure, leading to the formation of oriented, anisotropic fibrin fibers [34, 72, 85]. The presence of blood cells, particularly activated platelets and erythrocytes, significantly impacts fibrin network formation and properties [4], creating a dynamic environment for clot formation (Fig. 13.4b). Circulating cell-derived microvesicles can alter fibrin clot structure and properties [314]. Fibrin interacts with various components of the extracellular matrix, both filamentous and nonfilamentous, which not only affects fibrin structure but also adds mechanical and chemical stability. The structure of plasma clots in vitro is closely linked to clinical conditions associated with thrombosis [52, 172, 246, 304].

Blood flow in vivo significantly impacts fibrin network formation, its structure, and properties [255]. In static conditions, clotting halts once all fibrinogen has been converted to fibrin and polymerized, so clots in normal blood plasma (with fibrinogen concentrations of about 3 g/L) consist of only 0.3% protein and 99.7% liquid. However, in flowing blood, fibrinogen continues to be delivered to the forming clot, resulting in a denser clot with thicker, bundled fibers [185, 245]. Fluid flow can also align fibrin fibers in the direction of flow [34, 85, 290] (Fig. 13.4c), which significantly affects the mechanical properties of the clot and its susceptibility to fibrinolysis [34, 72, 269]. Additionally, microfluidic experiments suggest that shear forces from blood flow influence the likelihood of embolization, the process by which a clot fragment is carried through the bloodstream and potentially blocks another vessel [21, 45] (see section "Mechanical and chemical stability of fibrin").

Interactions of Fibrin(ogen) with Plasma and Cellular Proteins

Many proteins and compounds bind specifically to fibrinogen or fibrin [257]: actin, albumin, α1-antitrypsin, α2-antiplasmin, apolipoproteins A-IV, A-I, J, E, carboxypeptidase N, CD44, coagulation Factor XIII, SARS CoV-2, ferritin, fibroblast growth factor-2, fibulin, haptoglobin, heparin, hepatocyte-derived fibrinogen-related protein-1, immunoglobulins, interleukin-1β, lipoprotein(a), α2-macroglobulin, myosin, plasma fibronectin, plasminogen, plasminogen activator inhibitor-1, serum amyloid P, thrombin, thrombospondin, tissue-type plasminogen activator (t-PA), vascular endothelial growth factor, VE-cadherin, very low-density lipoprotein receptor, vitronectin, von Willebrand factor, and more [285].

Fibrin(ogen) is a ligand for a host of cell surface receptors. Most importantly, both fibrinogen and fibrin bind to the integrin αIIbβ3 on platelets, which is necessary for both platelet aggregation and clot contraction [140]. In addition, both fibrin and immobilized fibrinogen bind to GPVI on platelets, which is important for platelet adhesion and activation [159]. Fibrin(ogen) has a multifaceted role in inflammation, tissue injury, and many neurological diseases via binding to other cells. Fibrin can modify multiple aspects of inflammatory cell function and antimicrobial host defense by engaging leukocytes through a variety of cellular receptors. Converting fibrinogen into fibrin exposes cryptic epitopes that promote inflammation. Fibrin or specific fibrin degradation products support a bridging mechanism through leukocyte αXβ2 and endothelial cell VE-cadherin to promote leukocyte transmigration across endothelial monolayers [151, 296]. Fibrin or fibrin degradation products (but not fibrinogen) can also be anchored to the endothelial surface through very-low-density lipoprotein receptor (VLDLR), which also promotes leukocyte transmigration. The inflammatory functions of fibrinogen are mediated primarily via binding to the integrin receptor αMβ2 (also known as Mac-1 or CD11b/CD18 or complement receptor 3) on leukocytes, microglia, and macrophages. Fibrinogen is a molecular mediator entering the central nervous system after blood–brain barrier disruption that is causally linked with neuroinflammation, neuronal damage, and immune cell recruitment in the nervous system [202]. At the same time, fibrinogen binding to proteins such as latent transforming growth factor-β (TGFβ) or Aβ contributes to brain trauma and Alzheimer's disease, respectively.

Fibrin in the Context of In Vivo Blood Clots and Thrombi

Fibrin is the structural and mechanical scaffold of in vivo hemostatic clots and thrombi. Since clot and thrombus structure and composition determine the clinical consequences of both hemostasis and thrombosis, it is important to understand fibrin in its pathophysiological context. Mechanical stability of blood clots is necessary for hemostasis, and the mechanical properties of thrombi determine their obstructiveness, susceptibility to embolization, and treatment outcomes (see section "Mechanical and chemical stability of fibrin"). Hemostatic clots and thrombi are complex in structure, since they are formed over time in a blood vessel, usually with the forces of flowing blood and/or surrounding musculature.

Hemostasis has been studied in vivo primarily through the use of murine models, but there have been some larger animal models. Although most early models merely characterized clotting by measuring bleeding times, now various types of microscopy have been used to study hemostatic clot structure as a function of time. It should be noted that there are more murine thrombosis models than hemostasis models. However, puncture of blood vessels either with a laser or a physical object simulates adequately bleeding injuries [250]. Studies with fluorescently labeled platelets and fibrin have shown the accumulation of these components as a function of time, from the initial adhesion of platelets to the injury site, platelet aggregation to form a plug with a dense core and loose shell, and fibrin polymerization near the vessel wall and in the extravascular space [250]. Newer approaches, such as super-resolution microscopy, cryo-electron tomography, and scanning electron microscopy of serial block face sections, are now providing more detailed structural information [261].

Thrombus morphology has been studied as a result of mechanical thrombectomy for the treatment of myocardial infarction and ischemic stroke, which provides fresh thrombi from patients for histology or scanning electron microscopy for more detail [286]. The major structural elements of a thrombus are platelets, erythrocytes, and fibrin, each playing a critical role in determination of biological and physical properties of thrombi, such as permeability, stiffness, lytic, and mechanical stability. The minor components include neutrophils, monocytes, von Willebrand factor, cellular microvesicles, plasma proteins, cholesterol crystals, and other structures.

There are interactions between fibrin(ogen) and many of these components. For example, platelet microvesicles bind strongly to fibrin fibers via the integrin $\alpha IIb\beta 3$. In addition, fibrin regulates neutrophil function at the oral mucosal barrier [244]. The relative volume fractions of thrombus components, their functional and structural forms vary substantially, providing a basis for the diverse pathogenic mechanisms and clinical manifestations of thrombosis.

One of the newly observed features of the structure of thrombi is that there is very little open space in thrombi, making them almost impermeable for most drug treatments, including thrombolysis [286]. The main reason for the lack of space between structures making up thrombi is intravital contraction (or retraction), which is the volume shrinkage of clots or thrombi caused by activated platelets pulling on fibrin fibers [140]. Clot contraction results in compaction of the fibrin network and tight packing of erythrocytes. This is important for creating a better seal to stem bleeding in hemostasis, for reducing the obstructiveness of thrombi, and for wound healing. During clot contraction, there is a redistribution of components from a homogeneous distribution to most platelets and fibrin on the exterior and RBCs on the interior [140]. The effects of redistribution have been observed in ex vivo thrombi. Furthermore, fibrin forms a thin film at the liquid–air interface during the process of clot formation, which helps to protect wounds from bacterial invasion [155]. A similar fibrin film has also been observed on the surface of thrombi [5], but its pathophysiological significance is mostly unknown.

Because of platelets pulling on fibrin during clot contraction, erythrocytes, which are prevalent in all types of thrombi, undergo compressive deformation to polyhedral (polyhedrocytes) and polyhedral-like cells [140]. As a consequence, there are very few biconcave RBCs. The retention of RBCs is controlled by Factor XIIIa-catalyzed crosslinking of fibrin α-chains [33], which affects thrombus stability, composition, and size [152].

The high prevalence of fibrin in thrombi is quite striking [286]. Arterial thrombi contain a larger amount of fibrin than expected. There are different forms of fibrin, fibrin fibers (Fig. 13.4d) with fibrin bundles and fibrin network densification arising from clot contraction (Fig. 13.4e) as well as fibrin sponge (Fig. 13.4f), likely coming from the effects of partial fibrinolysis (see section "Fibrinolysis: enzymatic lysis of fibrin"). Fibrin in thrombi changes in time and space. The composition of thrombi from acute ST-elevation myocardial infarction patients as a function of time (meaning the time from the start of chest pain until extraction of the thrombus) showed large changes, with a relative two-fold increase in fibrin content for each hour of ischemic time and a 50% reduction in the relative amount of platelets [245]. Intact thrombi extracted from these patients showed that thrombi removed at times less than an hour were almost entirely composed of platelets, while thrombi removed at longer times consisted of mostly platelets at the site of initiation, with a gradual transition to almost entirely polyhedrocytes at the other end, surrounded by fibrin [286]. Ex vivo thrombi from patients with acute ischemic stroke displayed a dense, compact fibrin- and platelet-rich shell, and a core rich in polyhedrocytes, likely a result of clot contraction, in which redistribution occurs such that most fibrin and platelets are on the surface and RBCs are in the interior [121].

Fibrin Mechanical Properties and Their Structural Origins

Fibrin mechanics is a crucial and rapidly advancing area of hemostasis and thrombosis, as fibrin forms the scaffold of blood clots and thrombi that undergo dramatic deformations under (patho)physiological conditions, such as hydrodynamic blood shear, platelet-induced contraction, compression by extravascular muscle, and vessel wall pulsation. As a result, the mechanical properties of fibrin networks significantly influence the outcomes of various bleeding and thrombotic disorders [74, 139]. Additionally, mechanical and rheological properties of fibrin clots are extremely important for their clinical use as hemostatic tissue sealants (see section "Fibrin sealants for hemostasis"). A new impetus to study fibrin mechanics came from biomedical engineering, since fibrin has been widely utilized as a biomaterial for forming a temporary matrix that supports blood vessel growth and tissue regeneration, which rely markedly on the mechanical characteristics of gels like fibrin (see section "Fibrin-based scaffolds for tissue engineering and regenerative medicine").

Fibrin Is a Viscoelastic Biopolymer

Fibrin is a viscoelastic polymer, meaning it exhibits both elastic and viscous characteristics. Elasticity (or stiffness) refers to the ability to undergo reversible mechanical deformation, while viscosity and plasticity involve slower, irreversible deformation (creep) under applied force. Viscoelastic biomaterials vary in their balance of bulk elastic and viscous properties measured using various types of rheometers. The small-strain elastic behavior of a fibrin clot is usually quantified by the shear storage modulus (G'), which represents the portion of shear stress that aligns with the strain. The viscous behavior is captured by the shear loss modulus (G''), where strain lags behind stress. Together, the storage and loss moduli determine the clot's response to external forces. The ratio of G''/G', called tangent delta (tan δ), is often used to describe the relative viscosity and elasticity of a fibrin clot. In stretching and compression deformations, the slope of the stress-strain curve reflects the stiffness of the material. The units for elasticity, also known as the modulus of elasticity or Young's modulus, are force per unit area (N/m^2) equivalent to the Pascal (Pa) in SI or dyne per square centimeter ($dyne/cm^2$) in CGS. Compared to other fibrous proteins, fibrin clots are relatively soft and highly deformable biopolymers with a low Young's modulus generally on the order of kPa [139], but they have highly variable mechanical properties depending on the conditions of formation and structure, so the moduli can be cPa or even MPa (for fibrin sealants). Fibrin fibers have Young's moduli of about 1–10 MPa and are highly extensible [50, 92, 147].

A variety of environmental factors influence the stiffness of fibrin clots both in vitro and in vivo. One of the primary physiological modulators of fibrin mechanics is Factor XIIIa, which catalyzes fibrin covalent crosslinking and enhances the elastic modulus of fibrin several times [131, 139]. The viscoelastic properties of fibrin depend on fibrinogen properties, the conditions of polymerization, and structure (see section "Variations of fibrin(ogen) structure and properties"). The responses of platelet-rich plasma clots, whole blood clots, and human thrombi to forces are all similar, demonstrating that fibrin is the main determinant of the mechanical properties of clots and thrombi [136]. However, these studies also show how platelets and RBCs affect clot mechanical properties. Moreover, clots formed under flow have higher stiffness values associated with thinner fibrin fibers, smaller pores, and a denser fibrin network compared to clots formed under static conditions [72]. One of the main causes of fibrin clot stiffness and a major determinant of mechanical response to applied loads is natural pre-stress in fibrin fibers [248], which can be increased by platelet contraction and platelet-fibrin interactions [199]. Based on in vitro findings, it has been suggested that zinc, which decreases fibrin clot stiffness, might be released from activated platelets to help modulate clot strength and stability [295]. Fibrin elasticity can also be influenced by other physical factors, biochemical modifications, and by the blood components and cells integrated into the fibrin network [98, 112, 139, 181, 279].

The viscoelastic properties of fibrin are closely related to pathological conditions and can provide useful information for disease diagnosis and therapies. Clots made from the plasma of patients who suffered heart attacks at a young age have stiffness levels that are 50% higher compared to controls [52] and generally clots made from the plasma of (pro)thrombotic patients are made up of thin, highly branched fibers, and are less permeable and stiff [303, 304]. Moreover, blood clots from individuals with pulmonary embolism have been found to establish ultimate viscoelastic properties faster than those from healthy individuals [166]. The fractal dimension of clots, based on their viscoelastic properties, has been identified as a potential biomarker for the microstructure of prothrombotic clots [58]. Platelets are sensitive to the stiffness of the fibrin/fibrinogen substrate, with increased stiffness leading to enhanced platelet activation, adhesion, and spreading [215, 293].

Nonlinear Elasticity of Fibrin

The elasticity of fibrin clots is often described using a stress-strain curve, where stress (force per area) is plotted against the amount of deformation (strain) induced. At low levels of strain, the relationship between stress and strain is linear, and the slope of the curve characterizing the gel's elasticity remains constant. However, as strain increases, this linear relationship breaks down, and the slope steepens significantly, causing the fibrin clot's stiffness or elastic modulus to increase by up to an order of magnitude [117, 139]. This nonlinear behavior is referred to as strain hardening or strain stiffening, which is a key characteristic of many biological fibrous networks [268].

The propagation of strain throughout the fibrin gel is influenced by the fine structure and the nonlinear elastic behavior of individual fibrin fibers [107]. As the fibers stretch, they become stiffer than the surrounding fibers at lower strains, allowing these stiffer fibers to redistribute the strain to the less strained fibers and reduce strain concentrations [105]. In addition to shear and tension, the nonlinear elasticity of fibrin has also been observed in response to compressive deformations [122, 136, 275]. Stiffening during repeated compressive cycles arises due to the formation of new bonds between fibers in the compressed state, which develop tensile stress when the network is decompressed [122, 275]. A similar mechanism for strain-stiffening of a fibrin network under tension is based on the nascent cohesive interactions between individual fibers (crisscrossing) developing under tensile load [25]. An important mechanism that underlies the strain-stiffening response of fibrin gels is Factor XIIIa, which increases fibrin stiffness via γ-γ-chain cross-linking at the level of protofibrils, whereas α-chain cross-linking stiffens fibrin fibers because of tighter coupling between the constituent protofibrils [139, 205]. The importance of molecular heterogeneity in determining clot mechanics has been demonstrated by comparing the strain stiffening of fibrin gels formed from two recombinant fibrinogens differing in the length of the Aα-chain [167].

The adaptable nonlinear elasticity of fibrin may serve a crucial biological role by allowing fibrin clots to remain flexible at smaller strains while becoming stiffer at higher deformations, which helps prevent clot disruption and reduces the risk of embolization (see section "Rupture resistance of fibrin, thrombotic embolism, and mechanical thrombectomy". Additionally, the complex mechanical behavior of fibrin is vital for cell interactions with the extracellular matrix [289]. As mechanical stress increases, fibrin becomes more resistant to fibrinolysis [269], suggesting that fibrin elasticity could play a significant role in determining the susceptibility of clots and thrombi to enzymatic breakdown [150] (see section "Fibrinolysis: enzymatic lysis of fibrin").

Multiscale Structural Mechanics of Fibrin Clots

Fibrin mechanics are influenced by a structural hierarchy, suggesting that fibrin deformation involves multiple structural changes at various scales, from the submolecular and molecular level to individual fibers, the fiber network, and the entire macroscopic clot [27, 132, 204, 214, 298, 300].

At the macroscopic scale ($\sim 10^{-2}$ m), fibrin clots, in addition to deformation in the direction of strain, undergo a substantial volume reduction when stretched [27]. This shrinkage is attributed to water expulsion and the densification of the network, as shown by a tenfold increase in protein content in clots stretched threefold. This phenomenon may be linked to the negative normal stress observed in networks of semiflexible polymers, including fibrin [117]. Alternatively, it could be due to a molecular structural transition in the fibrin fibers as a result of molecular unfolding during stretching, followed by exposure of hydrophobic chemical groups normally on the interior of proteins [27, 214] (see section "High extensibility and forced molecular unfolding of fibrin"). Characterization of fibrin stress-strain response and the deformation at the macroscale provides information about the strain fields that develop within heterogeneous fibrin gels with spatial variations in material properties [115].

At the network scale or mesoscale (~10^{-5} m), when strain is applied, the fibrin fibers begin to align with the strain direction [27, 214]. The fibers become thinner, move closer together, and bundle [27]. A significant reorganization of the fibrin network is also observed under compressive deformation [122], where network density increases, fibers perpendicular to applied compressive load buckle, and shorter fiber segments are formed due to fiber crisscrossing. These initial network structural changes of buckling are accompanied by clot softening, and the later fiber crisscrossing and sticking together result in a marked increase in fibrin elasticity. The elastic behavior of fibrin networks studied using experiments and modeling provides a framework for predicting and comprehending the nonlinear elastic behavior of fibrin clots and other active biopolymer networks [305].

At the level of individual fibers (~10^{-6} m), studies of the unusual viscoelastic properties of fibrin have provided important mechanistic insights. For example, when a fibrin fiber is stretched laterally using an atomic force microscope, both crosslinked and uncrosslinked fibrin fibers can extend to approximately 2.5 and 3.3 times their original length, respectively, before rupturing [147, 148]. The variability in individual fibrin fiber mechanics has been shown to depend not only on the Factor XIIIa-catalyzed crosslinking [156], but also on fibrin(ogen) mass and volume fraction [95]. At the fiber scale, when subjected to compression or shear, individual fibrin fibers begin to buckle in the direction of the deformation [122]. Since buckling and bending increase the fibers' compliance, the stiffness of the network in the transverse shear direction decreases gradually under compression before crisscrossing occurs. Comparing the fibrin response to shear and tensile loads, studied at the fiber and molecular scales, highlights the importance of fibrin's molecular response to different types of deformation. This may be related to the functionally essential fibrin's resistance to shear and tensile forces [278]. The contribution of molecular structure of fibrin to its unique mechanical properties has been studied using natural [214] or engineered proteins with alterations in αC-region, which has led to a new structural model of protofibril organization within fibrin fibers that fully explains their strain stiffening, relatively low elastic modulus and their high extensibility [75]. Converting a phenomenological model based on experimental fibrin fiber force-strain curves to a mechanistic model provides important structural and thermodynamic insights into deformability of fibrin fibers [77].

High Extensibility and Forced Molecular Unfolding of Fibrin

Fibrin is an exceptionally extensible polymer, meaning that blood clots tend to elongate under stress rather than fracture. Fibrin clots that are stabilized by Factor XIIIa can stretch to more than three times their original length before breaking [27]. Fibrin is capable of significant elongation due to its flexible and partially disordered structure, particularly in the αC-regions. This remarkable extensibility

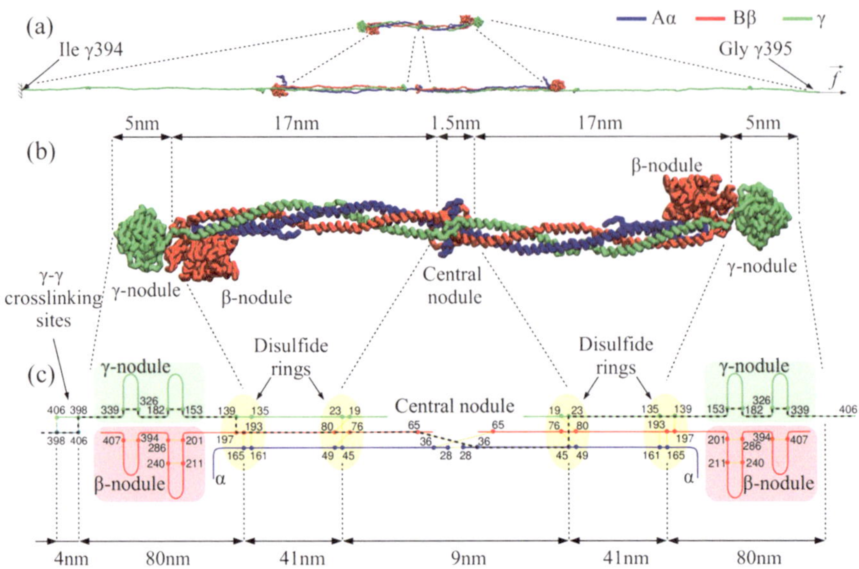

Fig. 13.5 Unfolding of fibrin(ogen). A schematic representation of the fibrin(ogen) molecule in the naturally folded (**b**) and fully unfolded states (**a**) and (**c**). The molecule is constrained at the C-terminal part of one γ chain, and mechanical force is applied to the C-terminal part of the other γ chain. (**a**) Shows full unfolding without much detail, while the structural details are given in (**c**), showing schematically the lengths of the central nodule, γ-nodules, β-nodules, coiled-coils, taking into account the disulfide bonds. Dimensions are shown in the compact crystal structure (**b**), and the contour lengths of various structural elements are shown in the fully unfolded state (**c**), assuming a contour length per residue of 0.38 nm. (With permission from Zhmurov et al. [309])

allows fibrin to stretch, deform, and adapt, which is critical for the proper functioning of hemostatic blood clots formed in dynamic conditions. The cutting edge of fibrin mechanics is identifying parts of fibrin fibers, including protofibrils and monomers that undergo structural transitions in response to deformations of clots.

At the molecular level, fibrin can potentially elongate more than five times its original length, from 45 nm to around 250 nm [309] (Fig. 13.5). Key structural changes during fibrin deformation include the unfolding of coiled-coils [26, 137, 144], γ-nodules [9, 10], and the extension of αC regions [11]. Other studies have highlighted the role of network branch points [36] and the γA/γ′ splice variants [6, 84]. Based on the known crystal structure of a folded fibrin(ogen) molecule, it is predicted that full unfolding of compact structures would result in approximately a 4.7-fold elongation [106, 309], which is unlikely but emphasizes the significant contribution of molecular unfolding to large fibrin deformations. Molecular deformation of fibrin monomers in the network in response to tension was observed using Förster resonance energy transfer [94].

The first direct experimental observation of an α-helix to β-strand conversion in the coiled-coils during fibrin extension was made using wide-angle X-ray scattering of squeezed fibrin films [12]. Further studies by Fourier Transform infrared spectroscopy have shown that both extension and compression of hydrated fibrin clots cause a transition from an α-helical coiled-coil to β-sheet [144]. This structural change has been further confirmed and analyzed through full-atom molecular dynamics simulations of the extensional mechanical unfolding of a fibrin(ogen) molecule [310] and generalized for other polypeptide coiled-coil structures [173].

Fibrin fibers exhibit paracrystalline molecular packing with a longitudinal periodicity of 22.5 nm, and it has been hypothesized that forced elongation and unfolding of fibrin molecules during stretching would increase this periodicity. However, small-angle X-ray scattering revealed that the peak corresponding to this 22.5-nm spacing remains relatively unchanged as the clot is stretched [27]. Instead, the sample showed an increase in disorder, indicating that molecules unfold nonuniformly in response to large strain, as expected in a two-state system where some molecules fully extend while others remain folded. A negligibly small increase in axial periodicity, only 0.7% as observed in another study [276], is also indicative of molecular unfolding. However, this increase was much smaller than the 80% length increase of the fibers, suggesting that fiber elongation primarily may result from the uncoiling of the natively disordered αC-polypeptide linkers that laterally bond the molecules.

Single-stranded fibrinogen oligomers have been studied using single-molecule atomic force microscopy to measure the forced unfolding of fibrinogen domains [26, 309]. Additionally, computer modeling of these experimental data has allowed visualization of the various structural transitions and provided insight into the forces observed in experimental or simulated force-extension curves. Molecular elongation of fibrin(ogen) is primarily driven by sequential unfolding transitions of the domains in the C-terminal γ-chain nodules (Fig. 13.6), along with the reversible extension and contraction of the α-helical coiled-coil connectors. These coiled-coils act as molecular springs, absorbing slack as other domains unfold.

The role of αC polymers in fibrin mechanics has received much attention [11, 73, 96, 102]. It has been shown that transglutaminase-catalyzed covalent crosslinking of the α-chains significantly contributes to the stiffness and elasticity of the fibrin clot [46, 67, 96, 131, 204], including the rapid elastic recoil of stretched fibrin fibers [106]. The αC-connector region of fibrin plays a crucial role in its extensibility. This region, found in the C-terminal part of the α-chain, is highly disordered in its native state, which enables it to stretch when mechanical tension is applied. The unfolding of these disordered αC-polypeptides contributes significantly to the overall elongation of the fibrin fibers. Given the critical role of the αC regions in fibrin mechanics, a structural model of fibrinogen has been proposed to capture aspects of the stress-strain behavior of individual fibrin fibers [11], but much remains to be discovered.

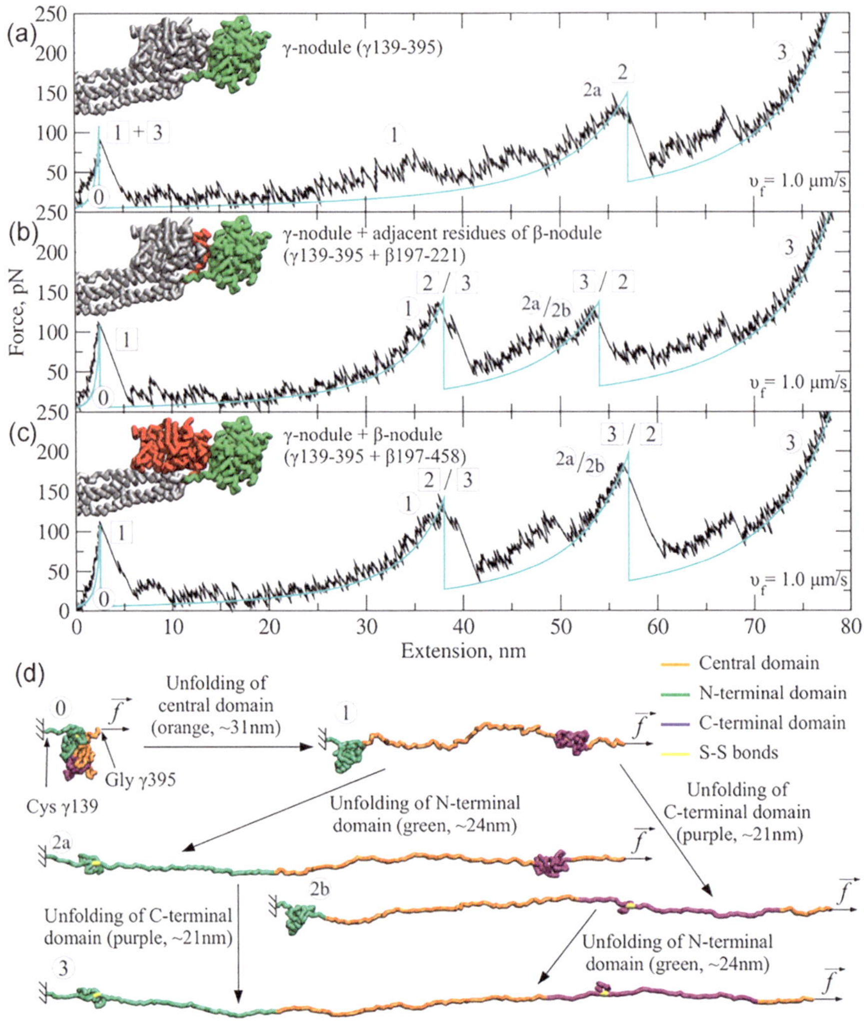

Fig. 13.6 Computational modeling of stepwise forced unfolding of the central, N-terminal, and C-terminal domains of the fibrin's γ-nodule. (**a**) Coarse-grained structure of a fibrin monomer with a highlighted lateral D-region containing the γ-nodule, β-nodule, and a portion of the coiled-coil. (**a**–**c**) Panels show force-extension profiles for the γ-nodule alone (**a**), for the γ-nodule and adjacent portion of the β-chain (**b**), and for the complex of the γ- and β-nodules (**c**). In each structure, the simulated portion is colorized, and the "truncated" portion, not used in the simulations, is shown in gray. The force peaks correspond to the structural transitions of types 1–3. (**d**) The unfolding steps and intermediate conformations observed for the γ-nodule. Structures 0, 1, 2a, 2b, and 3 correspond to the accordingly numbered force peaks in the force-distance curves (**a**–**c**). In parenthesis, the maximal extensions are shown, corresponding to the peak-to-peak distances in the force-distance curves. (Modified from Zhmurov et al. [309] with permission)

Mechanical and Lytic Stability of Fibrin

The mechanical and lytic stability of fibrin is crucial for its functions in hemostasis and thrombosis. The mechanical stability or rupture resistance of fibrin ensures the clot's strength and structural integrity, which determines the ability to withstand forces that could cause it to break apart. Understanding the physical mechanisms of fibrin toughness (ability to deform without fracture) would help to explain, predict, and prevent the embolization of thrombi and provide the potential for developing new devices and protocols for therapeutic removal of thrombi, i.e., mechanical thrombectomy. Fibrin toughness also has relevance to rupture in the broad class of materials called fibrous hydrogels. The ability of fibrin gels to break determines the mechanical stability of fibrin-based engineered tissues, grafts, and hemostatic sealants widely used in clinical medicine (see section "Fibrin as a biomaterial"). Lytic stability refers to the resistance of fibrin against degradation by various proteolytic enzymes, mostly plasmin in vivo, which allows the clot to last long enough to prevent bleeding and promote wound healing. Either mechanical or proteolytic instability of fibrin would cause thrombotic embolism or premature clot dissolution, respectively, with adverse pathophysiological consequences.

Rupture Resistance of Fibrin, Thrombotic Embolism, and Mechanical Thrombectomy

Rupture resistance is unrelated to viscoelastic properties of a fibrin gel. While most of the studies in fibrin biomechanics have been focused on viscoelasticity and deformability of intact fibrin gels or fibers (see section "Fibrin mechanical properties and their structural origins "), the mechanisms of fibrin mechanical disintegration or rupture remain largely unknown, despite their tremendous theoretical and practical importance. As fibrin determines the mechanical properties of blood clots and thrombi in the highly dynamic intra- and extravascular environment, it is critical to examine the mechanical stability of fibrin gels, i.e., their rupture resistance, also called toughness. From the theory of fracture mechanics, polymer toughness is distinct from the mechanical characteristics of deformability and viscoelasticity. Toughness of a material is a property that is largely independent of bulk properties such as storage or loss moduli and is characterized by the energy released per unit increase in crack face area, termed the critical energy release rate, G_c [264].

The mechanical stability or rupture resistance of polymeric fibrin gels is governed by several factors, of which the structure of a fibrin network is the main determinant. It has been known that the fibrin network architecture, including the density, length, and thickness of fibrin fibers, determines the viscoelastic mechanical properties of clots, but its importance for fracture mechanics has been revealed only recently. A variable shown to modulate rupture resistance of fibrin gels as a result of structural alterations is the concentration of tissue factor, which determines the kinetics of thrombin generation and fibrin formation [218]. Another important modulator of fibrin toughness is fibrinogen concentration in the initial blood plasma, equivalent to the mass and volume fraction of fibrin in the clot formed [82, 216]. Fracture tests performed on blood clots with distinct but physiologically relevant compositions have shown that clot fracture toughness increases significantly with increasing fibrin content, while it is inversely related to hematocrit or the volume fraction of RBCs [76]. There is interplay between mechanical deformation in the entire fibrin fiber networks and individual fibers, offering new insights into rupture resistance and failure mechanisms in fibrin clots [90].

Mechanisms of fibrin rupture. The fracture of fibrin gels can be considered in terms of thermodynamic aspects as well as dynamic structural alterations in the fibrin network associated with water expulsion, in addition to breakage of fibrin fibers and molecular unfolding [253, 265]. To study the mechanisms of fibrin rupture, macroscopic clots made of human blood plasma are stretched or sheared to high strains without or with linear incisions (cracks) of varying length at an edge (Fig. 13.7).

In experiments on strain-driven rupture of fibrin gels made from plasma, the resistance to rupture quantified by the critical energy release rate was determined to be 7.6 ± 0.45 J/m^2 [264]. Fibrin gels have a relatively low toughness for a ductile material, as most hydrogels that have been tested for rupture are synthetic rather than biological, with toughness values of >10 up to 1000 J/m^2 [306]. The free energy changes for fibrin rupture have been obtained by performing the numerical integration of experimental stress-strain curves that are dominated by the energy contribution associated with the expulsion of water from the fibrin gel, with the energy changes for stretching of fibrin fibers playing a minor role [265]. The rupture resistance of fibrin clots is enhanced by mechanical cooperativity among individual fibrin fibers, which has important physiological implications related to the varying density of clots. Using experimental and theoretical approaches, it has been shown that fracture toughness is relatively independent of the type of loading (stretching or shear) and is therefore a fundamental property of the gel [217, 265]. By coupling bulk mechanical testing, microstructural measurements, and theoretical modeling to capture the force-stretch response of fibrin clots, it has been shown that increased fibrin content in clots reduces porosity and limits fluid motion, and that fluid motion drastically alters the clot toughness [82]. Experiments and simulations suggest that stretching and alignment of the fibrin network in tensile regions is a key part of the mechanical behavior of the clot. In particular, alignment of the fibrin network in front of the crack tip significantly enhances the fracture toughness of fibrin-rich clots, which provides a mechanistic explanation for the observed dependence of clot fracture toughness on fibrin content [76, 82]. New insights into the fatigue and fracture of fibrin clots and fibrous materials have been gleaned from the fracture behavior of fibrin clots under cyclic fatigue and monotonic variable rate loads. The cyclic fatigue tests show the fatigue threshold of fibrin clots at 1.66 J/m^2, compared to the overall fracture toughness of 15.8 J/m^2 [149].

The presumed additional toughening mechanisms in fibrin clots are the following: (i) unfolding of individual fibrin monomers [309], (ii) stretching of long flexible αC-polymers, connecting protofibrils within a fibrin fiber [148]; (iii) the formation

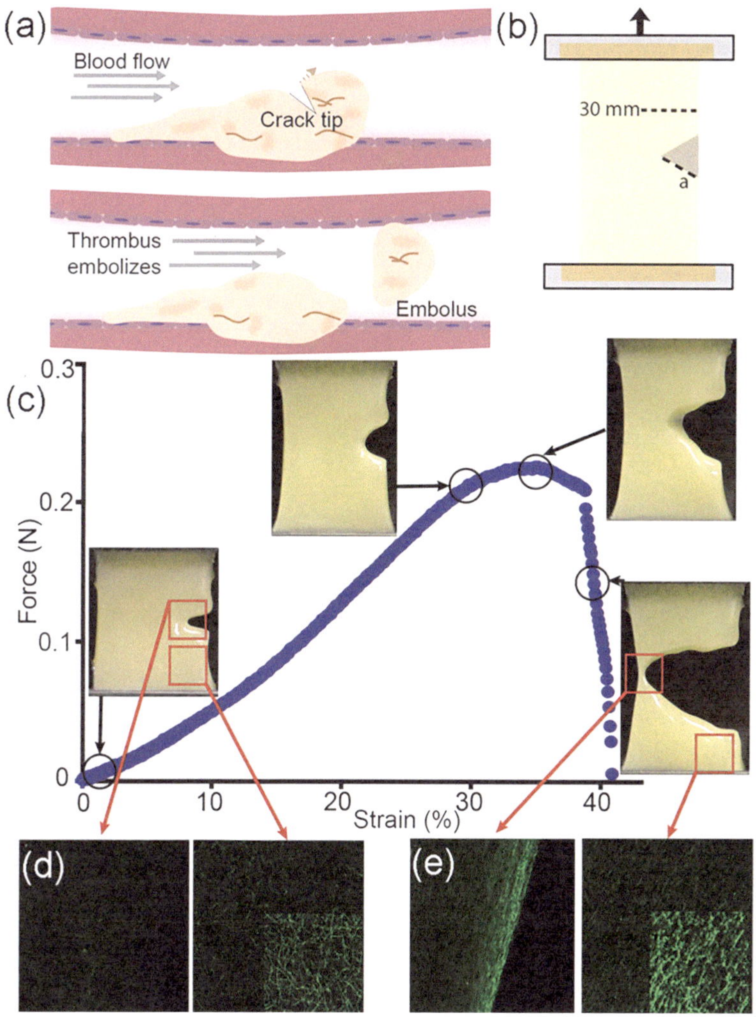

Fig. 13.7 To study the mechanisms of fibrin rupture, macroscopic clots made of human blood plasma are stretched or sheared to high strains without or with linear incisions (cracks) of varying length at an edge. (**a**) Schematic of incipient rupture and embolization of an intravascular blood clot due to shear tractions of blood flow acting on its surface (*dashed arrow*), causing opening of a tensile crack. (**b**) Schematic of crack geometry showing a 30-mm-wide fibrin clot sample with an edge crack (**a**). (**c**) Crack propagation in a stretched plasma gel made with human blood plasma. Samples were stretched at 3 mm/min while recording force-displacement curves. (**d** and **e**) Representative fluorescence confocal microscopy images at the crack tip or far from the crack in the fibrin network that was unstrained (**d**) and stretched at 40% strain. (**e**) Images are all at the same gain and other microscope settings for comparison, showing the densification with stretching, except for the insets, where the brightness was increased to show fiber orientation. Fibrin gels were strained to 10, 20, and 40%. While still under tension, samples were immersed in fixative and then excised from the stretching device. (Modified from Tutwiler et al. [264] with permission)

and breakage of adhesive interactions in fibrin that leads to reversible bundling of fibers upon clot deformation [25]; and (iv) dissipation due to motion of fluid (volume decrease) in fibrin gels [191], although its contribution depends on deformation rates, and is relatively small because of the large pore sizes in many fibrin clots. The low toughness of fibrin despite these mechanisms must be due to the relatively low physiological volume fraction of fibrin in blood plasma-derived clots. Model-based calculation of fracture toughness (G_c) as a function of protein volume fraction in fibrin gels shows that at extremely high-fibrin content (as in ex vivo thrombi and fibrin sealants), exceeding physiological plasma values 1000-fold, the toughness increases greatly [264]. Taken together, this information provides a fundamental understanding of fibrin rupture and the structural biomechanics underlying the mechanical (in)stability of blood clots and thrombi [262].

Rupture resistance of fibrin and the risk of thrombotic embolism. The rupture resistance of fibrin likely determines the mechanical stability of blood clots because of the similarities in the mechanical response of human platelet-poor plasma clots, platelet-rich plasma clots, and whole-blood clots [136]. Plasma clot rupture occurs at the same strain as paired whole-blood clots [264]. Under shear loading (e.g., as in the flow of blood past a wall-attached clot in a blood vessel), cracks in bulk materials tend to propagate at that same value of G_c but at an angle ~45° to the shear, so that the crack-tip stresses are predominantly tensile [71]. Therefore, the experimental parameters obtained under tension and mechanisms of rupture resistance of fibrin have more or less direct and indirect clinical implications related to the mechanical stability of blood clots and thrombi.

Intravascular blood clots are subjected to various mechanical forces that can cause them to rupture, releasing fragments called thrombotic emboli that block downstream vessels. The rupture resistance of fibrin gels plays a crucial role in thrombus stability, with weaker fibrin networks increasing the risk of clot fragmentation and embolization. When a clot is unstable, parts of it can detach, travel through the bloodstream, and obstruct smaller blood vessels, leading to conditions like pulmonary embolism, complicating deep vein thrombosis. This underscores the importance of the balance between fibrin clot formation, stability, and breakdown in preventing embolic events, as fibrin gels with higher rupture resistance form more stable thrombi, while weaker networks are more susceptible to disruption, releasing embolic material. A good example is elimination of fibrin γ-chain cross-linking, which results in the formation of thrombi in vivo with reduced strength that are prone to fragmentation and embolization [69].

It has been demonstrated that a fibrin clot without cracks can sustain much larger deformations than one with a crack, and that as crack length increases, the loads at which crack advance occurs decrease [264]. Physiologically, this suggests that an intact blood clot remains relatively stable until a crack forms, at which point it becomes more susceptible to tearing. In other words, thrombotic embolization occurs when a physical defect forms and is subsequently enlarged through mechanical forces, ultimately leading to clot rupture, as illustrated in Fig. 13.7.

There are at least two possible explanations for the initial formation of this defect in the blood clot. The first possibility is that the initial defect may have a purely mechanical origin, resulting from initial structural heterogeneity and/or localized stress applied to the blood clot. Various intra- and extravascular mechanical forces can act on blood clots and thrombi, including blood flow shear, muscle contractions outside blood vessels, smooth muscle contractions in vessel walls, rhythmic arterial wall expansion (pulsation), platelet-driven clot contraction, respiratory movements (diaphragmatic), and heart contractions (in the case of cardiac thrombi). Due to the pronounced structural nonuniformity of thrombus surfaces, certain areas may experience higher localized stresses that can reach the threshold required to break fibrin fibers, leading to a mechanical tear or crack in the fibrin network, compromising its integrity.

Another possible, and not mutually exclusive, mechanism for the formation of the initial defect in a blood clot is partial local proteolytic digestion or fibrinolysis (see section "Fibrinolysis: enzymatic lysis of fibrin"). Since plasmin cleaves covalent bonds within fibrin, this process can create an initial defect that weakens blood clots and thrombi, making them more prone to rupture and embolization. Fibrinolysis may work synergistically with mechanical forces, enhancing the risk of thrombotic embolization [129, 294]. Partial cleavage, acting like a "proteolytic razor blade," can initiate a crack on or within the clot or thrombus, which may trigger or facilitate further mechanical rupture due to hydrodynamic shear forces from blood flow.

Irrespective of the underlying mechanisms, thrombotic embolism is crucial because it can lead to life-threatening conditions like stroke and pulmonary embolism. The study of rupture resistance of blood clots and thrombi is a prospective avenue and a new (sub)field in the biomechanics of hemostasis and thrombosis. This research may eventually enable clinicians to predict and reduce the probability of thrombotic embolization based on structural and mechanical studies of fibrin.

Rupture resistance of fibrin and effectiveness of mechanical thrombectomy. Apart from providing the mechanistic basis for thrombotic embolization, the rupture resistance of blood clots has a number of medical implications, such as for mechanical thrombectomy, an interventional procedure used to remove obstructive thrombi from blood vessels, typically in cases of acute ischemic stroke or severe peripheral artery disease. Thrombus composition in stroke is highly heterogeneous, but thrombi could roughly be divided into fibrin- and RBC-rich clots. Fibrin-rich clots are associated with increased recanalization maneuvers, longer procedure time, and less favorable clinical outcomes compared with red blood cell-rich clots [116]. Accordingly, in in vitro experiments, fracture toughness of blood clots with physiologically relevant compositions is increased significantly with increasing fibrin content, i.e., RBC-rich clots are more prone to tear during loading compared to the fibrin-rich clots [76].

The content and structure of the fibrin scaffold in occlusive thrombi is a major determinant for the effectiveness of their mechanical damage and removal to restore the impaired blood flow [288]. The mechanical resistance to clot removal is related to the structure and physical properties of the fibrin network, such that clots that have a higher density and more cross-linked fibrin fibers tend to be more difficult to remove. This means that for thrombectomy devices, which rely on capturing and physically removing the clot, the mechanical properties of the fibrin network are a critical factor in determining how easily the clot can be aspirated, fragmented, or pulled out. These thrombi may require more extensive manipulation or multiple attempts to fully remove them. On the other hand, thrombi with weaker fibrin networks are more likely to fragment under the stress of the mechanical thrombectomy procedure. These clots may be easier to remove, but their fragments could pose a risk of causing secondary emboli. This is particularly concerning in procedures like thrombectomy for ischemic stroke, where the goal is to clear the clot without causing additional brain damage through embolization.

An important mechanical property that contributes to the efficacy of thrombus removal is static friction between the clot and the vessel or catheter wall. Fibrin-rich clots (<20% RBC content) have a significantly higher coefficient of friction than clots with an RBC content >20%. Therefore, intravital imaging techniques that can show fibrin and RBC content might help interventional neuroradiologists to predict the stickiness of a clot [91].

Despite significant advancements in the procedural efficacy of mechanical thrombectomy in recent years, there still remains a large fraction of clinical cases that do not achieve good vessel recanalization. The mechanical properties of fibrin clots must be taken into account during the design of devices performing various types of mechanical thrombectomy (aspiration or extraction). For instance, stent retrievers, which expand inside the clot and trap it for removal, must be able to apply enough force to grasp the clot without significant disruption. In cases of very dense, fibrin-rich thrombi, the device may need to apply higher forces or use additional techniques (e.g., cyclic aspiration or multiple retrieval attempts) to successfully remove the clot. New devices are coming along that have increased capability in removing tough clots rich in fibrin [114, 190].

Fibrinolysis: Enzymatic Lysis of Fibrin

Physiological importance. The primary physiological role of fibrinolysis is to resolve hemostatic blood clots after they have served their purpose in stopping bleeding. By cleaving fibrin, fibrinolysis also helps prevent inappropriate clot formation and restores normal blood flow in obstructive thromboses. Fibrinolysis is essential for wound healing as it breaks down the fibrin clot that forms at the injury site. This process enables the movement of cells required for tissue repair and the subsequent remodeling of the wound matrix, creating space for new tissue growth and promoting the healing process [133]. In addition, fibrinolytic (pro)enzymes are involved in several other (patho)physiological processes, such as inflammation, cell migration, angiogenesis, embryogenesis, ovulation, tumor progression, metastasis, and atherosclerosis [284]. Understanding of the fibrinolytic system and its interaction with infection and immunity, and beyond, is necessary for new developments in novel therapeutics and applications [183].

Molecular mechanisms of fibrinolysis. Fibrinolysis is a complex series of biochemical events, along with its regulatory mechanisms (Fig. 13.8). It begins with the

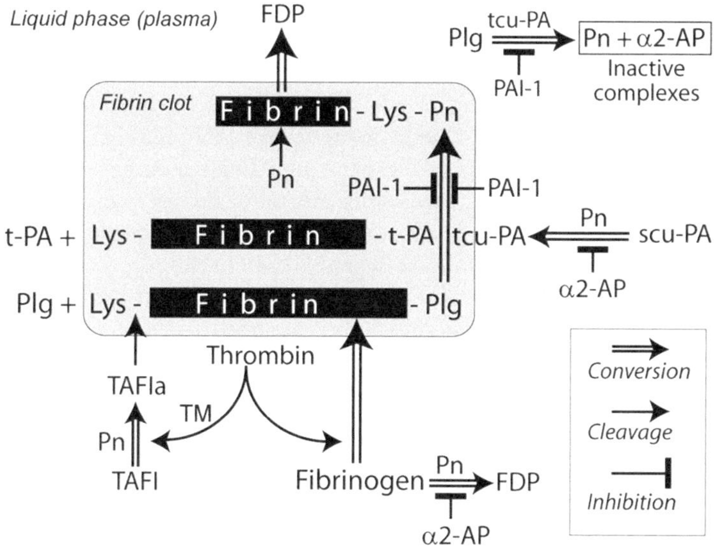

Fig. 13.8 Schematic representation of the major reactions of fibrinolysis and their regulation on a fibrin clot surface and in the surrounding liquid plasma milieu. The grey highlighted area represents a fibrin clot surface (*solid phase*) surrounded by the blood plasma (*liquid phase*). Black arrows show the biochemical conversions involving proteolytic cleavage. T-like symbols indicate inhibitory effects. Lys-Fibrin—C-terminal lysine residues on fibrin to which Plg and t-PA bind selectively; Plg—plasminogen, bound to the C-terminal lysine residues on fibrin and free in plasma; Pn—plasmin, formed on fibrin (by the action of t-PA) and in plasma (by the action of tcu-PA) from Plg. Pn cleaves fibrin and fibrinogen, activates scu-PA into tcu-PA and TAFI to TAFIa; t-PA—tissue-type Plg activator, fibrin-selective Plg activator, bound to fibrin via the C-terminal lysine residues on fibrin; scu-PA—single-chain urokinase-type Plg activator (inactive); tcu-PA—two-chain uPA (active), non-fibrin-selective Plg activator; PAI-1—plasminogen activator inhibitor-1, blocks both t-PA and tcu-PA; TAFI(a)—thrombin-activatable fibrinolysis inhibitor (enzymatically active form) that splits off the C-terminal lysine residues from fibrin, thus preventing binding of Plg and t-PA to fibrin; α2-AP—α2-antiplasmin, direct Pn inhibitor, forms circulating (Pn-α2-AP) complexes; TM—thrombomodulin; FDP—fibrin(ogen) degradation products, resulting from cleavage of fibrin or fibrinogen by Pn

conversion of a proenzyme plasminogen (Plg) to an active serine protease plasmin (Pn) by a Plg activator, primarily on the fibrin surface, followed by the breakdown of fibrin by Pn. Any Pn that dissociates from the fibrin surface is rapidly neutralized by inhibitors. Plg activators include serine proteases like tissue-type plasminogen activator (t-PA) or urokinase-type plasminogen activator (u-PA), as well as bacterial proteins like streptokinase and staphylokinase, which acquire proteolytic activity upon interacting with human Plg or Pn. t-PA and staphylokinase are selective for fibrin, remaining attached to it and protected from rapid inhibition, while streptokinase and two-chain u-PA are not fibrin-specific, activating both Plg in the blood and Plg bound to fibrin. The attachment of Plg and t-PA to fibrin is facilitated by the C-terminal lysine residues of fibrin and the specific lysine-binding kringle domains on the Plg and t-PA molecules. Since Pn cleaves fibrin at lysine residues, the newly exposed C-terminal lysines provide additional binding sites for Plg and t-PA, acting as a positive feedback mechanism [224, 284]. Alternative fibrin-cleaving proteolytic enzymes are released from inflammatory cells and metalloproteases of the extracellular matrix.

Plasmin molecules digest fibers laterally, but the rates of digestion are nonuniform, which may result in cleavage at a single location along the fiber [153, 284]. The lysis rates estimated by theoretical modelling compared to experiments on fiber degradation suggest that plasmin likely crawls between protofibrils, and a certain fraction of a fiber must be degraded prior to cleavage [197, 284]. In line with this hypothesis, protofibril packing density within a fiber has been shown to play a critical role in fibrinolysis [230].

Enzymatic breakdown of fibrin leads to the formation of soluble degradation products. The fibrin degradation products generated by plasmin are similar to the proteolytic fragments of fibrinogen, such as intermediate fragments X and Y, and terminal fragments D and E, with the distinction being the presence or absence of fibrinopeptides [163]. What sets apart the fibrin degradation products formed in vivo from those of fibrinogen is the noncovalent interactions and the covalent crosslinking by Factor XIIIa, which leads to the formation of various soluble oligomeric structures [272]. The smallest crosslinked fibrin degradation product is D-dimer, which consists of two D regions from adjacent fibrin molecules joined by γ-γ bonds, always associated with fragment E because of the strong knob-hole bonds. Since D-dimer only comes from crosslinked fibrin, elevated levels of D-dimer-containing products in the blood, often much larger than DED, are commonly used as a clinical indicator of intravascular fibrin deposition and fibrinolysis, seen in conditions like local thrombosis or disseminated intravascular coagulation [138]. Fibrin degradation products consist of fragments with incomplete binding sites, which competitively inhibit fibrin polymerization, potentially leading to the formation of large oligomeric structures, referred to as "soluble fibrin complexes" or "soluble fibrin."

Physiologic regulation of fibrinolysis. A variety of biochemical processes regulate the activity of pro-fibrinolytic components, including Plg activator inhibitor-1 (PAI-1), a strong inhibitor of t-PA and u-PA, and α2-antiplasmin, which directly inhibits Pn (Fig. 13.8). Thrombin-activatable fibrinolysis inhibitor (TAFI) works by cleaving C-terminal lysine residues from partially degraded fibrin, preventing fibrinolysis by hindering the lysine-dependent binding of Plg to fibrin. Additionally, lipoprotein(a) [Lp(a)] shares structural similarities with Plg and can compete for binding to lysine residues, thus impairing fibrinolysis [284].

The efficiency of fibrinolysis is influenced by a combination of regulated enzymatic activity and the physical properties of the fibrin scaffold, such as fiber density, branch point distribution, pore size, and fiber thickness. Generally, clots with thicker fibers tend to undergo lysis more quickly than those composed of thinner fibers, although other biophysical characteristics of the clot also play a role [282]. A three-dimensional multiscale model of fibrinolysis has been developed to study the effect of plasmin-mediated degradation of fibrin on tissue plasminogen activator (tPA) diffusion parameters. It has been shown that lysis rates depend directly on the ability of tPA to diffuse into a clot and on the dissociation constants between tPA and fibrin [13, 14]. Fibrinolysis is a physical process as it occurs amid fluid flow and other dynamic biophysical mechanisms that regulate fibrinolytic rates [104]. At the molecular level, the susceptibility of fibrin to lysis depends substantially on the accessibility of plasmin cleavage sites as well as on the presence of the αC-regions and conformational transitions of its domains [168].

Internal and external (therapeutic) fibrinolysis. The process that most closely resembles the natural physiological process of fibrinolysis is referred to as internal or intrinsic lysis. Internal fibrinolysis is the degradation of fibrin by the endogenous or innate presence or release of lytic enzymes and activators in the bloodstream; under healthy conditions, this process regulates hemostasis and prevents bleeding or excessive clotting. To model this in vitro, t-PA is mixed with blood plasma, or both t-PA and Plg are mixed with fibrinogen, resulting in thrombin-induced clot formation followed by dissolution as t-PA activates Plg to Pn on the fibrin surface. Lysis is monitored by tracking the decrease in turbidity, or using confocal or other light microscopy, or by detecting the appearance of lysis products biochemically [49, 206]. In clinical laboratory settings, the array of methods to assess fibrinolysis is different [307]. Analysis of simulations and experiments indicates that fibrinolysis is driven by pore expansion of the fibrin network, and this effect is strongly influenced by the ratio of fibrin:tPA when compared to absolute tPA concentration [228].

In contrast, the process that mimics clinical or therapeutic thrombolysis is termed external or extrinsic fibrinolysis. In therapeutic thrombolysis, t-PA is administered into the bloodstream, where it circulates and binds to fibrin in thrombi, activating Plg to Pn on the fibrin surface. As an in vitro model, t-PA can be applied to the edge of a pre-formed clot within a chamber, typically made from a microscope slide and cover slip, allowing lysis to be observed either visually or via light microscopy [47, 224]. In this external lysis model, t-PA enters the clot solely through diffusion. However, in thrombolysis, t-PA is delivered through flowing blood, often resulting in permeation or perfusion into the thrombus, depending on the thrombus structure and density. Importantly, the kinetics of flow-driven fibrinolysis are different from the static conditions [203]. The efficacy of therapeutic thrombolysis can depend strongly on the altered properties and reduced lytic susceptibility of pathologically modified fibrin; e.g., clots made of citrullinated fibrinogen [270] or clots containing histones [125] are lysis-resistant. In type 2 diabetes, the reduced lytic susceptibility of blood clots is associated with the fibrin(ogen) polymorphism BβArg448Lys [89], and other abnormal structural features [130]. On the other hand, anticoagulants, such as apixaban, a factor Xa inhibitor, can facilitate fibrinolysis [227]. In some cases, fibrinolytic agents (t-PA or other Plg activators) may be administered before or during the mechanical thrombectomy (see section "Rupture resistance of fibrin, thrombotic embolism, and mechanical thrombectomy") to weaken the fibrin network and reduce the rupture resistance of the clot, given that lysis reduces the clot's stiffness and toughness [16]. This can make the thrombus easier to remove with less mechanical force, potentially reducing the risk of embolization and making the procedure more efficient.

Altered susceptibility to fibrinolysis in contracted blood clots. Clot contraction or retraction significantly affects the clot's susceptibility to fibrinolysis by altering the structure and physical properties of the clot. Importantly, the contraction of a blood clot appears to have differing effects on internal and external fibrinolysis. In vitro studies indicate that contracted blood clots exhibit greater resistance to external fibrinolysis compared to uncontracted clots lacking platelets [48, 263, 291]. This resistance is likely due to a number of reasons. First, as the clot contracts, the surface area available for Plg and t-PA binding is reduced. Since fibrinolysis depends on the activation of plasminogen to plasmin (Pn) on the fibrin surface, a smaller surface area means fewer binding sites for Plg and t-PA, which can slow down the rate of fibrinolysis. Second, contraction leads to a denser, more compact clot with tighter fibrin networks. This can make it more difficult for t-PA to penetrate the clot and reach the fibrin fibers, further hindering fibrinolysis.

Conversely, when both Pn and Plg activators are present in the blood before clot formation, contracted clots dissolve more rapidly than their uncontracted counterparts in vitro [37, 258, 263]. This accelerated internal fibrinolysis can be attributed to higher local concentrations of fibrinolytic enzymes bound to fibrin, such as Pn and Plg activator, along with their fibrin substrate. Additionally, Pn, through its kringle domains that bind to fibrin's C-terminal lysines, can transfer between fibers without being neutralized by α2-antiplasmin and other plasma inhibitors. However, there is limited literature on measuring internal fibrinolysis of contracted clots in vivo.

Beyond the overall porosity and stiffness of the fibrin network, fibrinolysis is also influenced by the structural and mechanical properties of individual fibrin fibers. The specific mechanical tension of these fibers can either inhibit [135, 269] or promote [29, 53] fibrinolysis depending on various conditions. The process is further modulated by local concentrations and ratios of pro- and anti-fibrinolytic factors, their covalent binding to fibrin, and additional systemic and localized elements that regulate clot contraction and cleavage [223, 225]. Interestingly, studies suggest that endogenous fibrinolysis supports clot contraction both in vitro and in vivo, likely due to partial fibrin degradation leading to reduced clot stiffness [235].

Regardless of the exact mechanisms, clot contraction alters the clot's physical structure and can make the clot either more or less resistant to fibrinolysis, slowing down or accelerating the overall process. Importantly, the interplay between clot contraction and fibrinolysis holds potential clinical relevance for thrombotic patients, as these interactions can influence the effectiveness of natural or therapeutic clot dissolution, depending on factors such as the timing of clot formation, platelet count and contractility, blood composition, and other elements that impact clot contraction.

Fibrin as a Biomaterial

Recent interest in studying fibrin has been driven by its applications in biomedicine, particularly as a hydrogel [111]. Fibrin has become a popular biopolymer scaffold in tissue engineering, due to its distinctive biological and physical properties, such as porosity, flexibility, elasticity, and biodegradability [1]. It has been used alone or combined with other materials as a scaffold for stem or primary cells to regenerate various tissues, including adipose, bone, cardiac, cartilage, liver, nerve, ocular, skin, tendons, and ligaments. While fibrin can function as a standalone biomaterial, research has shown that incorporating collagen and other biopolymers into fibrin enhances its stiffness and durability [277]. Fibrin-based gels, whether used alone or in combination with gelatin, collagen, or elastin, have been designed to optimize cellular processes like differentiation, proliferation, and morphological changes. The concept of mechanosensing—where cells adjust their phenotype and behavior in response to variations in the stiffness of fibrin-based substrates—has been utilized in applications such as promoting neurite outgrowth in fibrin gels [157]. Additionally, fibrin clots with longitudinally aligned fibers could serve as scaffolds for nerve regeneration in cases of traumatic peripheral nerve injuries [86]. Fibrin sealants are another important aspect of using fibrin as a biomaterial created to control bleeding [54].

Fibrin Sealants for Hemostasis

Fibrin sealants are now commonly used in clinical settings for hemostasis due to their ability to promote clot formation and close a wound, in Europe since 1972 and since 1998 in the USA [158, 231, 249]. Fibrin sealants are typically composed of two key reacting components, fibrinogen and thrombin. When these compounds are mixed and applied to the site of bleeding, thrombin converts fibrinogen into fibrin, which forms a gel and adheres to the tissue, effectively sealing the wound. This mechanism mimics the natural blood clotting process, providing a temporary seal to prevent excessive blood loss. Autologous fibrin sealants are derived from the patient's own blood, reducing the risk of immune rejection or disease transmission. Allogeneic or recombinant fibrin sealants are derived from donor blood or manufactured through recombinant DNA technology, offering alternatives in cases where autologous blood is unavailable or impractical. Some formulations may also include additional components, such as factor XIII, which covalently crosslinks the fibrin fibers, or antifibrinolytic agents to prevent premature clot lysis and enhance hemostatic efficacy. It is important to note that commercially available fibrin sealants exhibit varying mechanical properties [100], and the structure and stiffness of different fibrin-based glues can lead to differing levels of hemostatic efficacy [80]. The mechanical properties of fibrin sealants are mainly governed by the fibrinogen concentration, with minimal impact of other components [70].

Compared to traditional hemostatic procedures, fibrin sealants have a number of advantages [18]: (i) Fibrin sealants are formed quickly and act immediately upon application, providing instant hemostasis, which is crucial in emergency situations. (ii) Biocompatibility is another benefit as fibrin sealants are generally well tolerated, especially autologous versions. (iii) There is a reduced need for suturing or stapling the wound, particularly in areas that are difficult to reach. (iv) Besides stopping bleeding, fibrin sealants can also support tissue repair and regeneration by providing a scaffold for cells to migrate and proliferate. The shortcomings of using fibrin sealants include a risk of infection, particularly with allogeneic or recombinant sealants, and inconsistent efficacy, which can vary depending on factors like clot quality, application technique, and the specific formulation used.

Fibrin sealants are widely used in a variety of clinical situations to control bleeding. The most common areas include soft tissue surgery [189, 308], treatment of burns [221], cardiac surgery [192], neurosurgery [87], plastic surgery [256], as well as traumatic injuries and minor invasive procedures, such as endoscopic or laparoscopic surgeries, when fibrin sealants can be applied to provide a barrier against infection and accelerate wound healing.

In summary, fibrin sealants are effective tools for achieving hemostasis in a variety of clinical settings, with their ability to rapidly form a stable clot, support tissue healing, and reduce bleeding. However, their use requires careful consideration of the clinical context, and they may be best suited for specific situations where traditional methods are less effective or feasible. Ongoing research is focused on improving fibrin sealant formulations, increasing their durability, and enhancing their ability to promote tissue healing. New formulations aim to address issues like faster clot formation, longer-lasting hemostasis, and reducing the potential for unwanted fibrinolysis.

Fibrin-Based Scaffolds for Tissue Engineering and Regenerative Medicine

Fibrin scaffolds play an important role in tissue engineering and regenerative medicine, offering a versatile material that can support cell growth, tissue repair, and regeneration. Fibrin offers several important properties that make it well-suited for tissue engineering applications. Fibrin is inherently *biocompatible*, meaning it is well-tolerated by the body with minimal immune response. Fibrin scaffolds are biodegradable because they naturally dissolve over time through the action of enzymes like plasmin, making them ideal for tissue regeneration, as the scaffold can gradually break down as new tissue forms. Fibrin has a high degree of elasticity and deformability (see section "Fibrin is a viscoelastic biopolymer"), making it suitable for use in various types of tissues that experience mechanical stress, such as skin, cartilage, or muscle. The fibrin network forms a porous structure that supports cell migration and growth, facilitating the formation of new tissue. Finally, fibrin is a highly adhesive biomaterial that provides a surface for cell attachment, promoting cell infiltration, proliferation, and differentiation.

Fibrin scaffolds have been used in numerous tissue engineering and regenerative medicine applications. One of the oldest areas is the use of fibrin for nerve regeneration, particularly in peripheral nerve injuries [128, 301]. The fibrin scaffold can guide nerve growth by providing a physical structure for neurons to grow along and enhancing nerve repair. Fibrin scaffolds are used in cardiac tissue engineering to repair necrotic heart tissue following myocardial infarction or heart attack [40, 219]. The scaffold helps to promote cardiac muscle cell growth and angiogenesis, supporting the repair of damaged heart tissue [260].

Fibrin scaffolds are used in wound healing due to their ability to support the formation of new tissue and accelerate the healing process [15, 119]. They are used in chronic wounds (such as diabetic ulcers) or in skin grafting. Surprisingly, fibrin scaffolds sourced from neonatal plasma may improve healing outcomes compared to scaffolds from adult plasma [186]. In cartilage tissue engineering, fibrin scaffolds are used to support chondrocyte (cartilage cell) growth and help repair damaged cartilage, especially in cases of osteoarthritis or trauma [239]. When used in combination with stem cells (e.g., mesenchymal stem cells, endothelial progenitor cells), fibrin scaffolds provide a supportive structure for the cells to grow and differentiate into the required tissue type by modulating gene expression [103]. A promising approach to wound healing therapy is the use of fibrin gels made with sonicated fibrin nanoparticles that support cell migration in an in vitro angiogenic sprouting assay, which is relevant for wound healing [233]. Loading fibrin nanoparticles with antimicrobials such as vancomycin enabled a targeted and sustained release of antibiotic that increased drug contact time and reduced the therapeutic dose required for eradicating bacteria from the wound [240].

Fibrin may not be mechanically strong and stable enough to act as a singular wound repair material. To alleviate this problem, fibrin can be combined with other biomaterials to improve its performance. Collagen and elastin are added to fibrin scaffolds to provide additional structural support, as they form major components of the extracellular matrix in many tissues [237]. Fibrin is incorporated into other hydrogels, which provide a hydrated environment that mimics the natural extracellular matrix, further supporting cell viability and differentiation [15]. Fibrin scaffolds are combined with synthetic materials (such as hydroxyapatite) to create composite scaffolds for bone regeneration [196] in bone defects or fractures. Fibrin scaffolds have also been investigated in organ regeneration, such as liver tissue engineering [109].

A relatively new variation of a fibrin-based biomaterial is platelet-rich fibrin (PRF), which is derived from autologous blood plasma and contains a high concentration of platelets and fibrin. PRF is not the same as platelet-rich plasma (PRP). While PRP is prepared with an anticoagulant, PRF is prepared without any anticoagulant, so that platelets are activated and a fibrin clot is formed containing a high platelet count, as well as white blood cells. Thus, in PRF, platelets have released their contents and are present in a fibrin matrix, and PRF is commonly used topically as a novel but largely unproven therapy. PRF allegedly offers several advantages over traditional fibrin scaffolds, particularly due to release of growth factors from platelets like PDGF, TGF-β, and VEGF [19, 110, 201, 210, 242, 243]. However, because of the complexity of platelet functions and contents, the exact mechanisms of any effects of PRF remain largely indefinable. The most common uses of PRF include dental applications (implantology, periodontal treatment, and tooth extraction), orthopedic surgery (joint surgery and soft tissue injuries), plastic and reconstructive surgery (to reduce scarring after procedures like facelifts, breast reconstruction, or skin grafts), otolaryngology, chronic wounds (like diabetic ulcers), and cosmetic procedures (e.g., skin rejuvenation). Some PRF samples contain concentrated leukocytes, which release cytokines and growth factors, promoting phagocytosis of debris, and also modulating the immune response to prevent excessive inflammation and aid the tissue regeneration process. The presence of leukocytes also contributes to antibacterial properties, helping to reduce the risk of infection at the wound site. Due to simplicity of the preparation process for PRF, requiring just blood draw and centrifugation followed by layer separation, it is often prepared in-office or in a clinical setting just before use. Since PRF is derived from the patient's own blood, it is biocompatible and non-immunogenic, so the risk of infection, allergic reactions, or other complications is notably reduced. Some physicians believe that PRFs have a significant advantage over traditional fibrin scaffolds, especially in terms of promoting faster, more effective healing with fewer complications, but controlled studies are lacking.

In summary, fibrin scaffolds are a promising material in tissue engineering and regenerative medicine, offering unique advantages such as biocompatibility, biodegradability, and support for cell growth. They have been successfully applied in the regeneration of various tissues, making fibrin a promising therapeutic tool. Research into fibrin scaffolds for tissue engineering is ongoing, with efforts focused on improving scaffold properties, such as the mechanical strength, degradation rate, and bioactivity of fibrin scaffolds through novel crosslinking methods, composites, and functionalization with bioactive molecules.

Conclusions and Future Perspectives

The most basic aspects of the structure and function of fibrinogen and its conversion to the polymer fibrin are still necessary for understanding this versatile fibrous biopolymer, and so these topics are fully included in this review. Yet, there is so much new information about fibrinogen and fibrin that it is difficult to encompass it all here. New technologies have given us submolecular insight as well as changes in structure with application of force, especially aspects of molecular unfolding. There has been an intensive focus on the mechanics of fibrin as a biopolymer, not only on viscoelastic properties but also on rupture mechanics, which is important for both biomaterial applications and for understanding embolization in cardiovascular disease. The structure and properties of thrombi have been studied, including the structural and mechanical roles of fibrin. Finally, fibrin has become increasingly used as a versatile and unique biomaterial.

Despite intensive research on fibrin(ogen), much remains unknown. The molecular mechanisms of lateral aggregation and branching in fibrin polymerization are largely still mysteries. Although the zebrafish has been established as an important vertebrate model for hemostasis and thrombosis, the predominant focus of research has been on mammalian fibrinogen and fibrin, although comparative studies can be quite informative. The great multitude of fibrinogen variants, and even more so diverse fibrin clot structures, suggests that modulation of clot structure could become a treatment approach for bleeding and thrombotic disorders. Fibrin(ogen) binds to many other proteins, but most of the functions of these interactions are unknown. The study of fibrin mechanics is only in the beginning stages, and the applications of these results to both thrombectomy and other treatment approaches in thrombosis and biomaterials will multiply. Studies of the process of thrombosis and the structure of thrombi, together with an increasing understanding of fibrinolysis, may lead to renewed efforts in therapeutic thrombolysis. The biological roles of fibrinogen in complex biological processes, such as inflammation, infection, and wound healing, will become more apparent. New and powerful technologies in genetics, proteomics, biomechanics, and imaging will continue to proliferate and drive this research. As much as we have learned, even more about fibrinogen and fibrin remains unknown.

Acknowledgments Some of the authors' work mentioned here was supported by National Institutes of Health grants PO1-HL146373 and RO1-148227. We thank Dr. Prashant Purohit (University of Pennsylvania) for careful reading of portions of the manuscript and helpful suggestions.

Compliance with Ethical Standards Some of the research described here was supported by grants from the National Institutes of Health, USA. The authors have no other conflicts. We confirm that experiments described in the chapter involving human blood products were conducted by following all appropriate ethical guidelines and that informed consent was obtained from donors.

References

1. Ahmed TE, Dare EV, Hincke M (2008) Fibrin: a versatile scaffold for tissue engineering applications. Tissue Eng Part B-Rev 14:199–215
2. Ajjan R, Lim BC, Standeven KF, Harrand R, Dolling S, Phoenix F, Greaves R, Abou-Saleh RH, Connell S, Smith DA, Weisel JW, Grant PJ, Ariens RA (2008) Common variation in the C-terminal region of the fibrinogen beta-chain: effects on fibrin structure, fibrinolysis and clot rigidity. Blood 111:643–650
3. Ajjan RA, Standeven KF, Khanbhai M, Phoenix F, Gersh KC, Weisel JW, Kearney MT, Ariens RA, Grant PJ (2009) Effects of aspirin on clot structure and fibrinolysis using a novel in vitro cellular system. Arterioscler Thromb Vasc Biol 29:712–717
4. Aleman MM, Walton BL, Byrnes JR, Wolberg AS (2014) Fibrinogen and red blood cells in venous thrombosis. Thromb Res 133(Suppl 1):S38–S40
5. Alkarithi G, Duval C, McPherson HR, Stewart L, De Simone I, Macrae FL, Ariëns RAS (2025) Fibrin film on clots is increased by hematocrit but reduced by inflammation: implications for platelets and fibrinolysis. J Thromb Haemost. https://doi.org/10.1016/j.jtha.2024.12.023
6. Allan P, Uitte De Willige S, Abou-Saleh RH, Connell SD, Ariens RA (2012) Evidence that fibrinogen gamma' directly interferes with protofibril growth: implications for fibrin structure and clot stiffness. J Thromb Haemost 10:1072–1080
7. Ariëns RAS, Lai T-S, Weisel JW, Greenberg CS, Grant PJ (2002) Role of factor XIII in fibrin clot formation and effects of genetic polymorphisms. Blood 100:743–754
8. Asquith NL, Duval C, Zhmurov A, Baker SR, McPherson HR, Domingues MM, Connell SDA, Barsegov V, Ariëns RAS (2022) Fibrin protofibril packing and clot stability are enhanced by extended knob-hole interactions and catch-slip bonds. Blood Adv 6:4015–4027

9. Averett LE, Geer CB, Fuierer RR, Akhremitchev BB, Gorkun OV, Schoenfisch MH (2008) Complexity of "A-a" knob-hole fibrin interaction revealed by atomic force spectroscopy. Langmuir 24:4979–4988
10. Averett LE, Schoenfisch MH, Akhremitchev BB, Gorkun OV (2009) Kinetics of the multistep rupture of fibrin 'A-a' polymerization interactions measured using atomic force microscopy. Biophys J 97:2820–2828
11. Averett RD, Menn B, Lee EH, Helms CC, Barker T, Guthold M (2012) A modular fibrinogen model that captures the stress-strain behavior of fibrin fibers. Biophys J 103:1537–1544
12. Bailey K, Astbury WT, Rudall KM (1943) Fibrinogen and fibrin as members of the keratin-myosin group. Nature 151:716–717
13. Bannish BE, Chernysh IN, Keener JP, Fogelson AL, Weisel JW (2017) Molecular and physical mechanisms of fibrinolysis and thrombolysis from mathematical modeling and experiments. Sci Rep 7:6914
14. Bannish BE, Paynter B, Risman RA, Shroff M, Tutwiler V (2024) The effect of plasmin-mediated degradation on fibrinolysis and tissue plasminogen activator diffusion. Biophys J 123:610–621
15. Bayer IS (2022) Advances in fibrin-based materials in wound repair: a review. Molecules 27: 4504
16. Bechtel GN, Sugerman GP, Eades T, Malinowska Z, Bush AM, Saber H, Parekh SH, Rausch MK (2025) Mechano-lysis in whole blood clots: on how mechanics affect clot lysis, and how lysis affects clot mechanics. Adv Healthc Mater 14:e2403389
17. Belcher HA, Litwa K, Guthold M, Hudson NE (2022) The applicability of current turbidimetric approaches for analyzing fibrin fibers and other filamentous networks. Biomolecules 12:807
18. Beudert M, Gutmann M, Lühmann T, Meinel L (2022) Fibrin sealants: challenges and solutions. ACS Biomater Sci Eng 8:2220–2231
19. Blanco J, García A, Hermida-Nogueira L, Castro AB (2025) How to explain the beneficial effects of leukocyte- and platelet-rich fibrin. Periodontol 97:74–94
20. Bowley SR, Okumura N, Lord ST (2009) Impaired protofibril formation in fibrinogen gamma N308K is due to altered D:D and "A:a" interactions. Biochemistry 48:8656–8663
21. Brass LF, Diamond SL (2016) Transport physics and biorheology in the setting of hemostasis and thrombosis. J Thromb Haemost 14:906–917
22. Brennan SO, Fellowes AP, George PM (2001) Molecular mechanisms of hypo- and afibrinogenemia. Ann N Y Acad Sci 936:91–100
23. Brennan SO, Davis RL, Mosesson MW, Hernandez I, Lowen R, Alexander SJ (2007) Congenital hypodysfibrinogenaemia (Fibrinogen Des Moines) due to a gamma320Asp deletion at the Ca2+ binding site. Thromb Haemost 98:467–469
24. Brennan SO (2015) Variation of fibrinogen oligosaccharide structure in the acute phase response: possible haemorrhagic implications. BBA Clin 3:221–226
25. Britton S, Kim O, Pancaldi F, Xu Z, Litvinov RI, Weisel JW, Alber M (2019) Contribution of nascent cohesive fiber-fiber interactions to the non-linear elasticity of fibrin networks under tensile load. Acta Biomater 94:514–523
26. Brown AE, Litvinov RI, Discher DE, Weisel JW (2007) Forced unfolding of coiled-coils in fibrinogen by single-molecule AFM. Biophys J 92:L39–L41
27. Brown AE, Litvinov RI, Discher DE, Purohit PK, Weisel JW (2009) Multiscale mechanics of fibrin polymer: gel stretching with protein unfolding and loss of water. Science 325:741–744
28. Brown AC, Baker SR, Douglas AM, Keating M, Alvarez-Elizondo MB, Botvinick EL, Guthold M, Barker TH (2015) Molecular interference of fibrin's divalent polymerization mechanism enables modulation of multiscale material properties. Biomaterials 49:27–36
29. Bucay I, O'Brien ET 3rd, Wulfe SD, Superfine R, Wolberg AS, Falvo MR, Hudson NE (2015) Physical determinants of fibrinolysis in single fibrin fibers. PLoS One 10:e0116350

30. Burton RA, Tsurupa G, Medved L, Tjandra N (2006) Identification of an ordered compact structure within the recombinant bovine fibrinogen αC-domain fragment by NMR. Biochemistry 45:2257–2266
31. Burton RA, Tsurupa G, Hantgan RR, Tjandra N, Medved L (2007) NMR solution structure, stability, and interaction of the recombinant bovine fibrinogen αC-domain fragment. Biochemistry 46:8550–8560
32. Butera D, Hogg PJ (2020) Fibrinogen function achieved through multiple covalent states. Nat Commun 11:5468
33. Byrnes JR, Duval C, Wang Y, Hansen CE, Ahn B, Mooberry MJ, Clark MA, Johnsen JM, Lord ST, Lam WA, Meijers JC, Ni H, Ariëns RA, Wolberg AS (2015) Factor XIIIa-dependent retention of red blood cells in clots is mediated by fibrin α-chain crosslinking. Blood 126: 1940–1948
34. Campbell RA, Aleman M, Gray LD, Falvo MR, Wolberg AS (2010) Flow profoundly influences fibrin network structure: implications for fibrin formation and clot stability in haemostasis. Thromb Haemost 104:1281–1284
35. Caracciolo G, De Spirito M, Castellano AC, Pozzi D, Amiconi G, De Pascalis A, Caminiti R, Arcovito G (2003) Protofibrils within fibrin fibres are packed together in a regular array. Thromb Haemost 89:632–636
36. Carlisle CR, Sparks EA, Der Loughian C, Guthold M (2010) Strength and failure of fibrin fiber branchpoints. J Thromb Haemost 8:1135–1138
37. Carroll RC, Gerrard JM, Gilliam JM (1981) Clot retraction facilitates clot lysis. Blood 57:44–48
38. Casini A, Moerloose P, Neerman-Arbez M (2025) Clinical, laboratory, and molecular aspects of congenital fibrinogen disorders. Semin Thromb Hemost 51:103–110
39. Chan LW, Wang X, Wei H, Pozzo LD, White NJ, Pun SH (2015) A synthetic fibrin crosslinking polymer for modulating clot properties and inducing hemostasis. Sci Transl Med 7: 277ra229
40. Chang T, Liu C, Yang H, Lu K, Han Y, Zheng Y, Huang H, Wu Y, Song Y, Yu Q, Shen Z, Jiang T, Zhang Y (2022) Fibrin-based cardiac patch containing neuregulin-1 for heart repair after myocardial infarction. Colloids Surf B Biointerfaces 220:112936
41. Chernysh IN, Weisel JW (2008) Dynamic imaging of fibrin network formation correlated with other measures of polymerization. Blood 111:4854–4861
42. Chernysh IN, Nagaswami C, Weisel JW (2011) Visualization and identification of the structures formed during early stages of fibrin polymerization. Blood 117:4609–4614
43. Chernysh IN, Nagaswami C, Purohit PK, Weisel JW (2012) Fibrin clots are equilibrium polymers that can be remodeled without proteolytic digestion. Sci Rep 2:879
44. Cines DB, Lebedeva T, Nagaswami C, Hayes V, Massefski W, Litvinov RI, Rauova L, Lowery TJ, Weisel JW (2014) Clot contraction: compression of erythrocytes into tightly packed polyhedra and redistribution of platelets and fibrin. Blood 123:1596–1603
45. Colace TV, Muthard RW, Diamond SL (2012) Thrombus growth and embolism on tissue factor-bearing collagen surfaces under flow: role of thrombin with and without fibrin. Arterioscler Thromb Vasc Biol 32:1466–1476
46. Collet J-P, Veklich Y, Mullin JL, Gorkun OV, Lord ST, Weisel JW (1999) The αC domains of fibrinogen affect the structure of the clot and its physical and biochemical properties. Thromb Haemost 82(Suppl):692
47. Collet J-P, Park D, Lesty C, Soria J, Soria C, Montalescot G, Weisel JW (2000) Influence of fibrin network conformation and fibrin fiber diameter on fibrinolysis speed: dynamic and structural approaches by confocal microscopy. Arterioscler Thromb Vasc Biol 20:1354–1361
48. Collet JP, Montalescot G, Lesty C, Weisel JW (2002) A structural and dynamic investigation of the facilitating effect of glycoprotein IIb/IIIa inhibitors in dissolving platelet-rich clots. Circ Res 90:428–434

49. Collet JP, Lesty C, Montalescot G, Weisel JW (2003) Dynamic changes of fibrin architecture during fibrin formation and intrinsic fibrinolysis of fibrin-rich clots. J Biol Chem 278:21331–21335
50. Collet JP, Shuman H, Ledger RE, Lee ST, Weisel JW (2005a) The elasticity of an individual fibrin fiber in a clot. Proc Natl Acad Sci USA 102:9133
51. Collet J-P, Moen JL, Veklich YI, Gorkun OV, Lord ST, Montalescot G, Weisel JW (2005b) The alphaC domains of fibrinogen affect the structure of the fibrin clot, its physical properties, and its susceptibility to fibrinolysis. Blood 106:3824–3830
52. Collet J-P, Allali Y, Lesty C, Tanguy ML, Silvain J, Ankri A, Blanchet B, Dumaine R, Giannetti J, Payot L, Weisel JW, Montalescot G (2006) Altered fibrin architecture is associated with hypofibrinolysis and premature coronary artery atherothrombosis. Arterioscler Thromb Vasc Biol. 26:2567–2573
53. Cone SJ, Fuquay AT, Litofsky JM, Dement TC, Carolan CA, Hudson NE (2020) Inherent fibrin fiber tension propels mechanisms of network clearance during fibrinolysis. Acta Biomater 107:164–177
54. Corral M, Ferko N, Hollmann S, Hogan A, Jamous N, Batiller J, Shen J (2016) Clinician reported ease of use for a novel fibrin sealant patch for hemostasis: results from four randomized controlled trials. Curr Med Res Opin 32:367–375
55. Crossen J, Shankar KN, Diamond SL (2023) Investigating thrombin-loaded fibrin in whole blood clot microfluidic assay via fluorogenic peptide. Biophys J 122:697–712
56. Damiana T, Damgaard D, Sidelmann JJ, Nielsen CH, de Maat MPM, Münster AB, Palarasah Y (2020) Citrullination of fibrinogen by peptidylarginine deiminase 2 impairs fibrin clot structure. Clin Chim Acta 501:6–11
57. Daraei A, Pieters M, Baker SR, de Lange-Loots Z, Siniarski A, Litvinov RI, Veen CSB, de Maat MPM, Weisel JW, Ariëns RAS, Guthold M (2021) Automated fiber diameter and porosity measurements of plasma clots in scanning electron microscopy images. Biomolecules 11:1536
58. Davies NA, Harrison NK, Morris RH, Noble S, Lawrence MJ, D'silva LA, Broome L, Brown MR, Hawkins KM, Williams PR, Davidson S, Evans PA (2015) Fractal dimension (df) as a new structural biomarker of clot microstructure in different stages of lung cancer. Thromb Haemost 114:1251–1259
59. de Vries JJ, Snoek CJM, Rijken DC, de Maat MPM (2020a) Effects of post-translational modifications of fibrinogen on clot formation, clot structure, and fibrinolysis: a systematic review. Arterioscler Thromb Vasc Biol 40:554–569
60. de Vries JJ, Hoppenbrouwers T, Martinez-Torres C, Majied R, Özcan B, van Hoek M, Leebeek FWG, Rijken DC, Koenderink GH, de Maat MPM (2020b) Effects of diabetes mellitus on fibrin clot structure and mechanics in a model of acute neutrophil extracellular traps (NETs) formation. Int J Mol Sci 21:7107
61. de Vries JJ, Laan DM, Frey F, Koenderink GH, de Maat MPM (2023) A systematic review and comparison of automated tools for quantification of fibrous networks. Acta Biomater 157:263–274
62. Domingues MM, Macrae FL, Duval C, McPherson HR, Bridge KI, Ajjan RA, Ridger VC, Connell SD, Philippou H, Ariëns RA (2016) Thrombin and fibrinogen γ' impact clot structure by marked effects on intrafibrillar structure and protofibril packing. Blood 127:487–495
63. Doolittle RF (2017) The conversion of fibrinogen to fibrin: a brief history of some key events. Matrix Biol 60-61:5–7
64. Doolittle LR, Pandi L (2006) Binding of synthetic B knobs to fibrinogen changes the character of fibrin and inhibits its ability to activate tissue plasminogen activator and its destruction by plasmin. Biochemist 45:2657–2667
65. Driever EG, Lisman T (2023) Fibrin clot properties and thrombus composition in cirrhosis. Res Pract Thromb Haemost 7:100055
66. Dunn EJ, Ariens RA (2004) Fibrinogen and fibrin clot structure in diabetes. Herz 29:470–479

67. Duval C, Allan P, Connell SD, Ridger VC, Philippou H, Ariens RA (2014) Roles of fibrin alpha- and gamma-chain specific cross-linking by FXIIIa in fibrin structure and function. Thromb Haemost 111:842–850
68. Duval C, Profumo A, Aprile A, Salis A, Millo E, Damonte G, Gauer JS, Ariëns RAS, Rocco M (2020) Fibrinogen αC-regions are not directly involved in fibrin polymerization as evidenced by a "Double-Detroit" recombinant fibrinogen mutant and knobs-mimic peptides. J Thromb Haemost 18:802–814
69. Duval C, Baranauskas A, Feller T, Ali M, Cheah LT, Yuldasheva NY, Baker SR, McPherson HR, Raslan Z, Bailey MA, Cubbon RM, Connell SD, Ajjan RA, Philippou H, Naseem KM, Ridger VC, Ariëns RAS (2021) Elimination of fibrin γ-chain cross-linking by FXIIIa increases pulmonary embolism arising from murine inferior vena cava thrombi. Proc Natl Acad Sci USA 118:e2103226118
70. Elbaz U, Berliner O, Tabo S, Yeshayahu S, Kesner R, Cohen-Gerassi D, Adler-Abramovich L, Halperin-Sternfeld M, Aviv M (2025) In vitro evaluation of a semi-autologous fibrin sealant for surgical applications. Macromol Biosci 20:e2400165
71. Erdogan F, Sih GC (1963) On the crack extension in plates under plane loading and transverse shear. J Basic Eng 85:519–525
72. Eyisoylu H, Hazekamp ED, Cruts J, Koenderink GH, de Maat MPM (2024) Flow affects the structural and mechanical properties of the fibrin network in plasma clots. J Mater Sci Mater Med 35:8
73. Falvo MR, Gorkun OV, Lord ST (2010) The molecular origins of the mechanical properties of fibrin. Biophys Chem 152:15–20
74. Feller T, Connell SDA, Ariëns RAS (2022) Why fibrin biomechanical properties matter for hemostasis and thrombosis. J Thromb Haemost 20:6–16
75. Feller T, McPherson HR, Connell SD, Ariëns RAS (2024) Fibrinogen αC-region acts as a functional safety latch: implications for a fibrin biomechanical behaviour model. Acta Biomater 189:179–191
76. Fereidoonnezhad B, Dwivedi A, Johnson S, McCarthy R, McGarry P (2021) Blood clot fracture properties are dependent on red blood cell and fibrin content. Acta Biomater 127: 213–228
77. Filla N, Zhao Y, Wang X (2023) Fibrin fiber deformation mechanisms: insights from phenomenological modeling to molecular details. Biomech Model Mechanobiol 22:851–869
78. Fogelson AL, Keener JP (2010) Toward an understanding of fibrin branching structure. Phys Rev E Stat Nonlinear Soft Matter Phys 81:051922
79. Fogelson AL, Nelson AC, Zapata-Allegro C, Keener JP (2022) Development of fibrin branch structure before and after gelation. SIAM J Appl Math 82:267–293
80. Fortelny RH, Petter-Puchner AH, Ferguson J, Gruber-Blum S, Brand J, Mika K, Redl H (2011) A comparative biomechanical evaluation of hernia mesh fixation by fibrin sealant. J Surg Res 171:576–581
81. García X, Seyve L, Tellier Z, Chevreux G, Bihoreau N, Polack B, Caton F (2020) Aggregates dramatically alter fibrin ultrastructure. Biophys J 118:172–181
82. Garyfallogiannis K, Ramanujam RK, Litvinov RI, Yu T, Nagaswami C, Bassani JL, Weisel JW, Purohit PK, Tutwiler V (2023) Fracture toughness of fibrin gels as a function of protein volume fraction: mechanical origins. Acta Biomater 159:49–62
83. Geer CB, Tripathy A, Schoenfisch MH, Lord ST, Gorkun OV (2007) Role of 'B-b' knob-hole interactions in fibrin binding to adsorbed fibrinogen. J Thromb Haemost 5:2344–2351
84. Gersh KC, Nagaswami C, Weisel JW, Lord ST (2009) The presence of gamma' chain impairs fibrin polymerization. Thromb Res 124:356–363
85. Gersh KC, Edmondson KE, Weisel JW (2010) Flow rate and fibrin fiber alignment. J Thromb Haemost 8:2826–2828
86. Gessmann J, Seybold D, Peter E, Schildhauer TA, Koller M (2016) Alignment of the fibrin network within an autologous plasma clot. Tissue Eng Part C Methods 22:30–37

87. Giammalva RG, Brunasso L, Costanzo R, Paolini S, Umana G, Yağmurlu K, Chaurasia B, Cicero S, Scalia G, Basile L, Maria Gerardi R, Gulì C, Angela Pino M, Graziano F, Federico Nicoletti G, Tumbiolo S, Gerardo Iacopino D, Maugeri R (2021) The role of hemostatic devices in neurosurgery. A systematic review. J Clin Neurosci 89:151–157
88. Gorkun OV, Litvinov RI, Veklich YI, Weisel JW (2006) Interactions mediated by the N-terminus of fibrinogen's Bbeta chain. Biochemistry 45:14843–14852
89. Greenhalgh KA, Strachan MW, Alzahrani S, Baxter PD, Standeven KF, Storey RF, Ariens RA, Grant PJ, Price JF, Ajjan RA (2017) BβArg448Lys polymorphism is associated with altered fibrin clot structure and fibrinolysis in type 2 diabetes. Thromb Haemost 117:295–302
90. Gu B, Hou J, Filla N, Li H, Wang X (2025) Rupture mechanics of blood clot fibrin fibers: a coarse-grained model study. J Mech Phys Solids 196:105998
91. Gunning GM, McArdle K, Mirza M, Duffy S, Gilvarry M, Brouwer PA (2018) Clot friction variation with fibrin content; implications for resistance to thrombectomy. J Neurointerv Surg 10:34–38
92. Guthold M, Liu W, Sparks EA, Jawerth LM, Peng L, Falvo M, Superfine R, Hantgan RR, Lord ST (2007) A comparison of the mechanical and structural properties of fibrin fibers with other protein fibers. Cell Biochem Biophys 49:165–181
93. Güven B, Can M (2024) Fibrinogen: structure, abnormalities and laboratory assays. Adv Clin Chem 120:117–143
94. Hedayati M, Chen YI, Houser JR, Wang Y, Norouzi S, Yeh HC, Parekh SH (2023) Visualizing molecular deformation in fibrin networks under tensile loading via FLIM-FRET. Chem Commun (Camb) 59:14575–14578
95. Helms CC (2024) Variability in individual native fibrin fiber mechanics. Phys Biol 21:6
96. Helms CC, Ariens RA, Uitte De Willige S, Standeven KF, Guthold M (2012) α-α cross-links increase fibrin fiber elasticity and stiffness. Biophys J 102:168–175
97. Helms CC, Kapadia S, Gilmore AC, Lu Z, Basu S, Kim-Shapiro DB (2017) Exposure of fibrinogen and thrombin to nitric oxide donor ProliNONOate affects fibrin clot properties. Blood Coagul Fibrinolysis 28:356–364
98. Henderson SJ, Xia J, Wu H, Stafford AR, Leslie BA, Fredenburgh JC, Weitz DA, Weitz JI (2016) Zinc promotes clot stability by accelerating clot formation and modifying fibrin structure. Thromb Haemost 115:533–542
99. Henschen-Edman AH (2001) Fibrinogen non-inherited heterogeneity and its relationship to function in health and disease. Ann N Y Acad Sci 936:580–593
100. Hickerson WL, Nur I, Meidler R (2011) A comparison of the mechanical, kinetic, and biochemical properties of fibrin clots formed with two different fibrin sealants. Blood Coagul Fibrinolysis 22:19–23
101. Hirota-Kawadobora M, Terasawa F, Yonekawa O, Sahara N, Shimizu E, Okumura N, Katsuyama T, Shigematsu H (2003) Fibrinogens Kosai and Ogasa: Bbeta15Gly→Cys (GGT→TGT) substitution associated with impairment of fibrinopeptide B release and lateral aggregation. J Thromb Haemost 1:275–283
102. Houser JR, Hudson NE, Ping L, O'brien ET 3rd, Superfine R, Lord ST, Falvo MR (2010) Evidence that alphaC region is origin of low modulus, high extensibility, and strain stiffening in fibrin fibers. Biophys J 99:3038–3047
103. Huang NF, Chu J, Lee RJ, Li S (2010) Biophysical and chemical effects of fibrin on mesenchymal stromal cell gene expression. Acta Biomater 6:3947–3956
104. Hudson NE (2017) Biophysical mechanisms mediating fibrin fiber lysis. Biomed Res Int 2017: 2748340
105. Hudson NE, Houser JR, O'brien ET 3rd, Taylor RM 2nd, Superfine R, Lord ST, Falvo MR (2010) Stiffening of individual fibrin fibers equitably distributes strain and strengthens networks. Biophys J 98:1632–1640
106. Hudson NE, Ding F, Bucay I, O'brien ET 3rd, Gorkun OV, Superfine R, Lord ST, Dokholyan NV, Falvo MR (2013) Submillisecond elastic recoil reveals molecular origins of fibrin fiber mechanics. Biophys J 104:2671–2680

107. Hudson NE, Houser JR, O'Brien ET 3rd, Taylor RM 2nd, Superfine R, Lord ST, Falvo MR (2015) Stiffening of individual fibrin fibers equitably distributes strain and strengthens networks. Biophys J 98:1632–1640
108. Hur WS, Kawano T, Mwiza JMN, Paul DS, Lee RH, Clark EG, Bouck EG, Dutta A, Cai C, Baker SR, Guthold M, Mackman N, Mangin P, Wolberg AS, Bergmeier W, Flick MJ (2024) Mice expressing nonpolymerizable fibrinogen have reduced arterial and venous thrombosis with preserved hemostasis. Blood 143:105–117
109. Imamura H, Tomimaru Y, Kobayashi S, Harada A, Kita S, Sasaki K, Iwagami Y, Yamada D, Noda T, Takahashi H, Hokkoku D, Kado T, Toya K, Kodama T, Saito S, Shimomura I, Miyagawa S, Doki Y, Eguchi H (2025) Adipose-derived stem cells using fibrin gel as a scaffold enhances post-hepatectomy liver regeneration. Sci Rep 15:6334
110. Jamjoom AG (2024) From healing to regeneration: a comprehensive review of the efficacy of platelet-rich fibrin in periodontal plastic surgery procedures. Cureus 16:e69287
111. Janmey PA, Winer JP, Weisel JW (2009) Fibrin gels and their clinical and bioengineering applications. J R Soc Interface 6:1–10
112. Jansen KA, Bacabac RG, Piechocka IK, Koenderink GH (2013) Cells actively stiffen fibrin networks by generating contractile stress. Biophys J 105:2240–2251
113. Jansen KA, Zhmurov A, Vos BE, Portale G, Hermida-Merino D, Litvinov RI, Tutwiler V, Kurniawan NA, Bras W, Weisel JW, Barsegov V, Koenderink GH (2020) Molecular packing structure of fibrin fibers resolved by X-ray scattering and molecular modeling. Soft Matter 16: 8272–8283
114. Jazayeri SB, Al-Janabi OM, Alateya A, Muhammad S, Ghozy S, Rabinstein AA, Kadirvel R, Kallmes DF (2025) Efficacy and safety of mechanical thrombectomy in patients with acute ischemic stroke and left ventricular assist device: review of the literature and meta-analysis. J Neurointerv Surg. https://doi.org/10.1136/jnis-2024-022792
115. Jimenez JM, Tuttle T, Guo Y, Miles D, Buganza-Tepole A, Calve S (2023) Multiscale mechanical characterization and computational modeling of fibrin gels. Acta Biomater 162: 292–303
116. Jolugbo P, Ariëns RAS (2021) Thrombus composition and efficacy of thrombolysis and thrombectomy in acute ischemic stroke. Stroke 52:1131–1142
117. Kang H, Wen Q, Janmey PA, Tang JX, Conti E, MacKintosh FC (2009) Nonlinear elasticity of stiff filament networks: strain stiffening, negative normal stress, and filament alignment in fibrin gels. J Phys Chem B 113:3799–3805
118. Kaufmanova J, Stikarova J, Hlavackova A, Chrastinova L, Maly M, Suttnar J, Dyr JE (2021) Fibrin clot formation under oxidative stress conditions. Antioxidants (Basel) 10:923
119. Kearney KJ, Ariëns RAS, Macrae FL (2022) The role of fibrin(ogen) in wound healing and infection control. Semin Thromb Hemost 48:174–187
120. Kelley MA, Leiderman K (2022) Mathematical modeling to understand the role of bivalent thrombin-fibrin binding during polymerization. PLoS Comput Biol 18:e1010414
121. Khismatullin RR, Nagaswami C, Shakirova AZ, Vrtkova A, Prochazka V, Gumulec J, Macak J, Litvinov RI, Weisel JW (2020) Quantitative morphology of cerebral thrombi related to intravital contraction and clinical features of ischemic stroke. Stroke 51:3640–3650
122. Kim OV, Litvinov RI, Weisel JW, Alber MS (2014) Structural basis for the nonlinear mechanics of fibrin networks under compression. Biomaterials 35:6739–6749
123. Klykov O, van der Zwaan C, Heck AJR, Meijer AB, Scheltema RA (2020) Missing regions within the molecular architecture of human fibrin clots structurally resolved by XL-MS and integrative structural modeling. Proc Natl Acad Sci USA 117:1976–1987
124. Kohler S, Schmid F, Settanni G (2015) The internal dynamics of fibrinogen and its implications for coagulation and adsorption. PLoS Comput Biol 11:e1004346
125. Komorowicz E, Balázs N, Tanka-Salamon A, Varga Z, Szabó L, Bóta A, Longstaff C, Kolev K (2021) Size- and charge-dependent modulation of the lytic susceptibility and mechanical stability of fibrin-histone clots by heparin and polyphosphate variants. J Thromb Haemost 19: 1307–1318

126. Konieczyńska M, Natorska J, Undas A (2024) Thrombosis and aging: fibrin clot properties and oxidative stress. Antioxid Redox Signal 41:233–254
127. Kononova O, Litvinov RI, Zhmurov A, Alekseenko A, Cheng CH, Agarwal S, Marx KA, Weisel JW, Barsegov V (2013) Molecular mechanisms, thermodynamics, and dissociation kinetics of knob-hole interactions in fibrin. J Biol Chem 288:22681–22692
128. Koopman JE, Duraku LS, de Jong T, de Vries RBM, Zuidam MJ, Hundepool CA (2022) A systematic review and meta-analysis on the use of fibrin glue in peripheral nerve repair: Can we just glue it? J Plast Reconstr Aesthet Surg 75:1018–1033
129. Kucinsky R, Goldhaber S, Tavel ME (2005) Acute myocardial infarction complicated by pulmonary embolism after thrombolytic therapy: problems in clinical management. Chest 128: 3572–3575
130. Kurdee Z, King R, Ajjan RA (2014) The fibrin network in diabetes: its role in thrombosis risk. Pol Arch Med Wewn 124:617–627
131. Kurniawan NA, Grimbergen J, Koopman J, Koenderink GH (2014) Factor XIII stiffens fibrin clots by causing fiber compaction. J Thromb Haemost 12:1687–1696
132. Kurniawan NA, Vos BE, Biebricher A, Wuite GJ, Peterman EJ, Koenderink GH (2016) Fibrin networks support recurring mechanical loads by adapting their structure across multiple scales. Biophys J 111:1026–1034
133. Laurens N, Koolwijk P, de Maat MP (2006) Fibrin structure and wound healing. J Thromb Haemost 4:932–939
134. Li W, Sigley J, Baker SR, Helms CC, Kinney MT, Pieters M, Brubaker PH, Cubcciotti R, Guthold M (2017a) Nonuniform internal structure of fibrin fibers: protein density and bond density strongly decrease with increasing diameter. Biomed Res Int 2017:6385628
135. Li W, Lucioni T, Li R, Bonin K, Cho SS, Guthold M (2017b) Stretching single fibrin fibers hampers their lysis. Acta Biomater 60:264–274
136. Liang X, Chernysh I, Purohit PK, Weisel JW (2017) Phase transitions during compression and decompression of clots from platelet-poor plasma, platelet-rich plasma and whole blood. Acta Biomater 60:275–290
137. Lim BCB, Lee EH, Sotomayor M, Schulten K (2008) Molecular basis of fibrin clot elasticity. Structure 16:449–459
138. Lippi G, Mullier F, Favaloro EJ (2022) D-dimer: old dogmas, new (COVID-19) tricks. Clin Chem Lab Med 61:841–850
139. Litvinov RI, Weisel JW (2017) Fibrin mechanical properties and their structural origins. Matrix Biol 60–61:110–123
140. Litvinov RI, Weisel JW (2023) Blood clot contraction: mechanisms, pathophysiology, and disease. Res Pract Thromb Haemost 7:100023
141. Litvinov RI, Gorkun OV, Owen SF, Shuman H, Weisel JW (2005) Polymerization of fibrin: specificity, strength, and stability of knob-hole interactions studied at the single-molecule level. Blood 106:2944–2951
142. Litvinov RI, Gorkun OV, Galanakis DK, Yakovlev S, Medved L, Shuman H, Weisel JW (2007) Polymerization of fibrin: direct observation and quantification of individual B:b knob-hole interactions. Blood 109:130–138
143. Litvinov RI, Yakovlev S, Tsurupa G, Gorkun OV, Medved L, Weisel JW (2007) Direct evidence for specific interactions of the fibrinogen alphaC-domains with the central E region and with each other. Biochemistry 46:9133–9914
144. Litvinov RI, Faizullin DA, Zuev YF, Weisel JW (2012) The α-helix to β-sheet transition in stretched and compressed hydrated fibrin clots. Biophys J 103:1020–1027
145. Litvinov RI, Kononova O, Zhmurov A, Marx KA, Barsegov V, Thirumalai D, Weisel JW (2018) Regulatory element in fibrin triggers tension-activated transition from catch to slip bonds. Proc Natl Acad Sci USA 115:8575–8580
146. Litvinov RI, Pieters M, de Lange-Loots Z, Weisel JW (2021) Fibrinogen and fibrin. Subcell Biochem 96:471–501

147. Liu W, Jawerth LM, Sparks EA, Falvo MR, Hantgan RR, Superfine R, Lord ST, Guthold M (2006) Fibrin fibers have extraordinary extensibility and elasticity. Science 313:634
148. Liu W, Carlisle CR, Sparks EA, Guthold M (2010) The mechanical properties of single fibrin fibers. J Thromb Haemost 8:1030–1036
149. Liu S, Bahmani A, Ghezelbash F, Li J (2024) Fibrin clot fracture under cyclic fatigue and variable rate loading. Acta Biomater 177:265–277
150. Longstaff C, Varju I, Sotonyi P, Szabo L, Krumrey M, Hoell A, Bota A, Varga Z, Komorowicz E, Kolev K (2013) Mechanical stability and fibrinolytic resistance of clots containing fibrin, DNA, and histones. J Biol Chem 288:6946–6956
151. Luyendyk JP, Schoenecker JG, Flick MJ (2019) The multifaceted role of fibrinogen in tissue injury and inflammation. Blood 133:511–520
152. Luyendyk JP, Flick MJ, Wolberg AS (2025) Factor XIII: driving (cross-)links in hemostasis, thrombosis, and disease. Blood. https://doi.org/10.1182/blood.2024025321
153. Lynch SR, Laverty SM, Bannish BE, Hudson NE (2022) Microscale structural changes of individual fibrin fibers during fibrinolysis. Acta Biomater 141:114–122
154. Macrae FL, Domingues MM, Casini A, Ariëns RA (2016) The (patho)physiology of fibrinogen γ'. Semin Thromb Hemost 42:344–355
155. Macrae FL, Duval C, Papareddy P, Baker SR, Yuldasheva N, Kearney KJ, McPherson HR, Asquith N, Konings J, Casini A, Degen JL, Connell SD, Philippou H, Wolberg AS, Herwald H, Ariëns RA (2018) A fibrin biofilm covers blood clots and protects from microbial invasion. J Clin Invest 128:3356–3368
156. Maksudov F, Daraei A, Sesha A, Marx KA, Guthold M, Barsegov V (2021) Strength, deformability and toughness of uncrosslinked fibrin fibers from theoretical reconstruction of stress-strain curves. Acta Biomater 136:327–342
157. Man AJ, Davis HE, Itoh A, Leach JK, Bannerman P (2011) Neurite outgrowth in fibrin gels is regulated by substrate stiffness. Tissue Eng A 17:2931–2942
158. Mandell SP, Gibran NS (2014) Fibrin sealants: surgical hemostat, sealant and adhesive. Expert Opin Biol Ther 14:821–830
159. Mangin PH, Gardiner EE, Ariëns RAS, Jandrot-Perrus M (2023) Glycoprotein VI interplay with fibrin(ogen) in thrombosis. J Thromb Haemost 21:1703–1713
160. Mannila MN, Eriksson P, Ericsson CG, Hamsten A, Silveira A (2006) Epistatic and pleiotropic effects of polymorphisms in the fibrinogen and coagulation factor XIII genes on plasma fibrinogen concentration, fibrin gel structure and risk of myocardial infarction. Thromb Haemost 95:420–427
161. Marchi RC, Carvajal Z, Boyer-Neumann C, Angles-Cano E, Weisel JW (2006) Functional characterization of fibrinogen Bicetre II: a gamma 308 Asn→Lys mutation located near the fibrin D:D interaction sites. Blood Coagul Fibrinolysis 17:193–201
162. Marchi R, Vilar R, Durual S, Goodyer M, Gay V, Neerman-Arbez M, Casini A (2021) Fibrin clot properties to assess the bleeding phenotype in unrelated patients with hypodysfibrinogenemia due to novel fibrinogen mutations. Thromb Res 197:56–64
163. Marder VJ, Budzynski AZ (1975) Data for defining fibrinogen in its plasmic degradation products. Thromb Diath Hemorrh 33:199–207
164. Marsh JJ, Guan HS, Li S, Chiles PG, Tran D, Morris TA (2013) Structural insights into fibrinogen dynamics using amide hydrogen/deuterium exchange mass spectrometry. Biochemistry 52:5491–5502
165. Martinez M, Weisel JW, Ischiropoulos H (2013) Functional impact of oxidative posttranslational modifications on fibrinogen and fibrin clots. Free Radic Biol Med 65:411–418
166. Martinez MR, Cuker A, Mills AM, Crichlow A, Lightfoot RT, Chernysh IN, Nagaswami C, Weisel JW, Ischiropoulos H (2014) Enhanced lysis and accelerated establishment of viscoelastic properties of fibrin clots are associated with pulmonary embolism. Am J Physiol Lung Cell Mol Physiol 306:L397–L404

167. Martinez-Torres C, Grimbergen J, Koopman J, Koenderink GH (2014) Interplay of fibrinogen αEC globular domains and factor XIIIa cross-linking dictates the extensibility and strain stiffening of fibrin networks. J Thromb Haemost 22:715–726
168. McPherson HR, Duval C, Baker SR, Hindle MS, Cheah LT, Asquith NL, Domingues MM, Ridger VC, Connell SD, Naseem KM, Philippou H, Ajjan RA, Ariëns RA (2021) Fibrinogen αC-subregions critically contribute blood clot fibre growth, mechanical stability, and resistance to fibrinolysis. eLife 10:e68761
169. Medeiros R, Cantero J, Borthagaray G, Paulino M (2025) Unveiling the architecture of human fibrinogen: a full-length structural model. Chembiochem. https://doi.org/10.1002/cbic.202400425
170. Medved L, Weisel JW (2009) Fibrinogen and Factor XIII Subcommittee of Scientific Standardization Committee of International Society on Thrombosis and Haemostasis. Recommendations for nomenclature on fibrinogen and fibrin. J Thromb Haemost 7:355–359
171. Medved L, Weisel JW (2022) The story of the fibrin(ogen) αC-domains: evolution of our view on their structure and interactions. Thromb Haemost 122:1265–1278
172. Mihalko E, Brown AC (2020) Clot structure and implications for bleeding and thrombosis. Semin Thromb Hemost 46:96–104
173. Minin KA, Zhmurov A, Marx KA, Purohit PK, Barsegov V (2017) Dynamic transition from α-helices to β-sheets in polypeptide coiled-coil motifs. J Am Chem Soc 139:16168–16177
174. Moen JL, Gorkun OV, Weisel JW, Lord ST (2003) Recombinant BbetaArg14His fibrinogen implies participation of N-terminus of Bbeta chain in desA fibrin polymerization. Blood 102: 2466–2471
175. Mohammed RDS, Piell KM, Maurer MC (2024) Identification of factor XIII β-sandwich residues mediating glutamine substrate binding and activation peptide cleavage. Thromb Haemost 124:408–422
176. Mosesson MW (2004) Cross-linked gamma chains in a fibrin fibril are situated transversely between its strands: yes. J Thromb Haemost 2:388–393
177. Mosesson MW (2007) Update on antithrombin I (fibrin). Thromb Haemost 98:105–108
178. Mosesson MW, Diorio JP, Hernandez I, Hainfeld JF, Wall JS, Grieninger G (2004) The ultrastructure of fibrinogen-420 and the fibrin-420 clot. Biophys Chem 112:209–214
179. Mouapi KN, Bell JD, Smith KA, Ariëns RA, Philippou H, Maurer MC (2016) Ranking reactive glutamines in the fibrinogen αC region that are targeted by blood coagulant factor XIII. Blood 127:2241–2248
180. Mullin JL, Gorkun OV, Lord ST (2000) Decreased lateral aggregation of a variant recombinant fibrinogen provides insight into the polymerization mechanism. Biochemistry 39:9843–9849
181. Münster S, Jawerth LM, Fabry B, Weitz DA (2013) Structure and mechanics of fibrin clots formed under mechanical perturbation. J Thromb Haemost 11:557–560
182. Murano G (2024) The molecular structure of fibrinogen. Semin Thromb Hemost 50:131–147
183. Mutch NJ, Medcalf RL (2023) The fibrinolysis renaissance. J Thromb Haemost 21:3304–3316
184. Neerman-Arbez M (2001) Fibrinogen gene mutations accounting for congenital afibrinogenemia. Ann N Y Acad Sci 936:496–508
185. Neeves KB, Illing DA, Diamond SL (2010) Thrombin flux and wall shear rate regulate fibrin fiber deposition state during polymerization under flow. Biophys J 98:1344–1352
186. Nellenbach K, Nandi S, Peeler C, Kyu A, Brown AC (2020) Neonatal fibrin scaffolds promote enhanced cell adhesion, migration, and wound healing in vivo compared to adult fibrin scaffolds. Cell Mol Bioeng 13:393–404
187. Nelson AC, Fogelson AL (2023) Towards understanding the effect of fibrinogen interactions on fibrin gel structure. Phys Rev E 107:024413
188. Nencini F, Bettiol A, Argento FR, Borghi S, Giurranna E, Emmi G, Prisco D, Taddei N, Fiorillo C, Becatti M (2024) Post-translational modifications of fibrinogen: implications for clotting, fibrin structure and degradation. Mol Biomed 5:45

189. Nguyen M, Foreman A, Lockwood C (2024) Effectiveness of fibrin sealants in head and neck surgery: a systematic review protocol. JBI Evid Synth 22:1151–1160
190. Nogueira RG, Andersson T, Haussen DC, Yoo AJ, Hanel RA, Zaidat OO, Hacke W, Jovin TG, Fiehler J, De Meyer SF, Brinjikji W, Doyle KM, Kallmes DF, Liebeskind DS, Virmani R, Kokoszka MA, Inoa V, Humphries W, Woodward KB, Jabbour PM, François O, Levy EI, Bozorgchami H, Boor S, Cohen JE, Dashti SR, Taqi MA, Budzik RF, Schirmer CM, Hussain MS, Estrade L, De Leacy RA, Puri AS, Chitale RV, Brekenfeld C, Siddiqui AH (2024) EXCELLENT registry: a prospective, multicenter, global registry of endovascular stroke treatment with the EMBOTRAP device. Stroke 55:2804–2814
191. Noselli G, Lucantonio A, McMeeking RM, DeSimone A (2016) Poroelastic toughening in polymer gels: a theoretical and numerical study. J Mech Phys Solids 94:33–46
192. Novotny R, Hlubocký J, Mitáš P, Lindner J (2018) Fibrin sealants in cardiac surgery: the last five years of their development and application. Adv Clin Exp Med 27:857–862
193. O'Brien ET, Falvo MR, Millard D, Eastwood B, Taylor RM, Superfine R (2008) Ultrathin self-assembled fibrin sheets. Proc Natl Acad Sci USA 105:19438–19443
194. Okumura N, Terasawa F, Hirota-Kawadobora M, Yamauchi K, Nakanishi K, Shiga S, Ichiyama S, Saito M, Kawai M, Nakahata T (2006) A novel variant fibrinogen, deletion of Bbeta111Ser in coiled-coil region, affecting fibrin lateral aggregation. Clin Chim Acta 365: 160–167
195. Okumura N, Terasawa F, Haneishi A, Fujihara N, Hirota-Kawadobora M, Yamauchi K, Ota H, Lord ST (2007) B:b interactions are essential for polymerization of variant fibrinogens with impaired holes 'a'. J Thromb Haemost 5:2352–2359
196. Osathanon T, Linnes ML, Rajachar RM, Ratner BD, Somerman MJ, Giachelli CM (2008) Microporous nanofibrous fibrin-based scaffolds for bone tissue engineering. Biomaterials 29: 4091–4099
197. Ouedraogo RR, Sowers HK, Lynch SR, Hudson NE, Bannish BE (2024) A mathematical model of plasmin-mediated fibrinolysis of single fibrin fibers. PLoS Comput Biol 20: e1012684
198. Parastatidis I, Thomson L, Burke A, Chernysh I, Nagaswami C, Visser J, Stamer S, Liebler DC, Koliakos G, Heijnen HF, Fitzgerald GA, Weisel JW, Ischiropoulos H (2008) Fibrinogen beta-chain tyrosine nitration is a prothrombotic risk factor. J Biol Chem 283:33846–33853
199. Pathare SJ, Eng W, Lee SJ, Ramasubramanian AK (2021) Fibrin prestress due to platelet aggregation and contraction increases clot stiffness. Biophys Rep (NY) 1:100022
200. Pechik I, Yakovlev S, Mosesson MW, Gilliland GL, Medved L (2006) Structural basis for sequential cleavage of fibrinopeptides upon fibrin assembly. Biochemistry 45:3588–3597
201. Pensato R, Al-Amer R, La Padula S (2024) Clinical application of platelet-rich fibrin in plastic and reconstructive surgery: a systematic review. Aesth Plast Surg 48:3047–3048
202. Petersen MA, Ryu JK, Peterse K (2018) Fibrinogen in neurological diseases: mechanisms, imaging and therapeutics. Nat Rev Neurosci 19:283–301
203. Petkantchin R, Rousseau A, Eker O, Zouaoui Boudjeltia K, Raynaud F, Chopard B, INSIST investigators (2023) A simplified mesoscale 3D model for characterizing fibrinolysis under flow conditions. Sci Rep 13:13681
204. Piechocka IK, Bacabac RG, Potters M, Mackintosh FC, Koenderink GH (2010) Structural hierarchy governs fibrin gel mechanics. Biophys J 98:2281–2289
205. Piechocka IK, Kurniawan NA, Grimbergen J, Koopman J, Koenderink GH (2017) Recombinant fibrinogen reveals the differential roles of α- and γ-chain cross-linking and molecular heterogeneity in fibrin clot strain-stiffening. J Thromb Haemost 15:938–949
206. Pieters M, Philippou H, Undas A, de Lange Z, Rijken DC, Mutch NJ (2018) Subcommittee on Factor XIII and Fibrinogen, and the Subcommittee on Fibrinolysis. An international study on the feasibility of a standardized combined plasma clot turbidity and lysis assay: communication from the SSC of the ISTH. J Thromb Haemost 16:1007–1012

207. Pinelo JEE, Manandhar P, Popovic G, Ray K, Tasdelen MF, Nguyen Q, Iavarone AT, Offenbacher AR, Hudson NE, Sen M (2023) Systematic mapping of the conformational landscape and dynamism of soluble fibrinogen. J Thromb Haemost 21:1529–1543
208. Ping L, Huang L, Cardinali B, Profumo A, Gorkun OV, Lord ST (2011) Substitution of the human alphaC region with the analogous chicken domain generates a fibrinogen with severely impaired lateral aggregation: fibrin monomers assemble into protofibrils but protofibrils do not assemble into fibers. Biochemistry 50:9066–9075
209. Prasad JM, Gorkun OV, Raghu H, Thornton S, Mullins ES, Palumbo JS, Ko YP, Höök M, David T, Coughlin SR, Degen JL, Flick MJ (2015) Mice expressing a mutant form of fibrinogen that cannot support fibrin formation exhibit compromised antimicrobial host defense. Blood 126:2047–2058
210. Proffer SL, Wyles SP, Hausauer AK (2024) Update on platelet-rich plasma and platelet-rich fibrin for dermatologic surgery: addressing knowns and unknowns. Dermatologic Surg 50: S149–S159
211. Protopopova AD, Barinov NA, Zavyalova EG, Kopylov AM, Sergienko VI, Klinov DV (2015) Visualization of fibrinogen αC regions and their arrangement during fibrin network formation by high-resolution AFM. J Thromb Haemost 13:570–579
212. Protopopova AD, Litvinov RI, Galanakis DK, Nagaswami C, Barinov NA, Mukhitov AR, Klinov DV, Weisel JW (2017) Morphometric characterization of fibrinogen's αC regions and their role in fibrin self-assembly and molecular organization. Nanoscale 9:13707–13716
213. Protopopova AD, Ramirez A, Klinov DV, Litvinov RI, Weisel JW (2019) Factor XIII topology: organization of B subunits and changes with activation studied with single-molecule atomic force microscopy. J Thromb Haemost 17:737–748
214. Purohit PK, Litvinov RI, Brown AE, Discher DE, Weisel JW (2011) Protein unfolding accounts for the unusual mechanical behavior of fibrin networks. Acta Biomater 7:2374–2323
215. Qiu Y, Brown AC, Myers DR, Sakurai Y, Mannino RG, Tran R, Ahn B, Hardy ET, Kee MF, Kumar S, Bao G, Barker TH, Lam WA (2014) Platelet mechanosensing of substrate stiffness during clot formation mediates adhesion, spreading, and activation. Proc Natl Acad Sci USA 111:14430–14435
216. Ramanujam RK, Maksudov F, Litvinov RI, Nagaswami C, Weisel JW, Tutwiler V, Barsegov V (2023) Biomechanics, energetics, and structural basis of rupture of fibrin networks. Adv Healthc Mater 12:e2300096
217. Ramanujam RK, Garyfallogiannis K, Litvinov RI, Bassani JL, Weisel JW, Purohit PK, Tutwiler V (2024a) Mechanics and microstructure of blood plasma clots in shear driven rupture. Soft Matter 20:4184–4196
218. Ramanujam RK, Maksudov F, Risman RA, Litvinov RI, Weisel JW, Bassani JL, Barsegov V, Purohit PK, Tutwiler V (2024b) Rupture mechanics of blood clots: influence of fibrin network structure on the rupture resistance. Acta Biomater 190:329–343
219. Rane AA, Christman KL (2011) Biomaterials for the treatment of myocardial infarction: a 5-year update. J Am Coll Cardiol 58:2615–2629
220. Raynal B, Cardinali B, Grimbergen J, Profumo A, Lord ST, England P, Rocco M (2013) Hydrodynamic characterization of recombinant human fibrinogen species. Thromb Res 132: e48–e53
221. Reuland C, Chidiac C, Cappiello C, Hodgman E (2024) Artiss fibrin sealant for the fixation of autografts in pediatric burn care. Burns 50:1848–1852
222. Riedel T, Suttnar J, Brynda E, Houska M, Medved L, Dyr JE (2011) Fibrinopeptides A and B release in the process of surface fibrin formation. Blood 117:1700–1706
223. Rijken DC, Uitte de Willige S (2017) Inhibition of fibrinolysis by coagulation factor XIII. Biomed Res Int 2017:1209676
224. Risman RA, Tutwiler V (2024) Visualizing the degradation of fibrin fibers. Biophys Rev (Melville) 5:032101

225. Risman RA, Abdelhamid A, Weisel JW, Bannish BE, Tutwiler V (2022) Effects of clot contraction on clot degradation: a mathematical and experimental approach. Biophys J 121: 3271–3285
226. Risman RA, Belcher HA, Ramanujam RK, Weisel JW, Hudson NE, Tutwiler V (2024) Comprehensive analysis of the role of fibrinogen and thrombin in clot formation and structure for plasma and purified fibrinogen. Biomolecules 14:230
227. Risman RA, Shroff M, Goswami J, Tutwiler V (2024) Dependence of clot structure and fibrinolysis on apixaban and clotting activator. Res Pract Thromb Haemost 8:102614
228. Risman RA, Paynter B, Percoco V, Shroff M, Bannish BE, Tutwiler V (2024) Internal fibrinolysis of fibrin clots is driven by pore expansion. Sci Rep 14:2623
229. Risman RA, Sen M, Tutwiler V, Hudson NE (2025) Deconstructing fibrin(ogen) structure. J Thromb Haemost 23:368–380
230. Risman RA, Percoco V, Paynter B, Bannish BE, Tutwiler V (2025) Protofibril packing density of individual fibers alters fibrinolysis. Res Pract Thromb Haemost 9:e102708
231. Roberts IV, Bukhary D, Valdivieso CYL, Tirelli N (2020) Fibrin matrices as (injectable) biomaterials: formation, clinical use, and molecular engineering. Macromol Biosci 20: e1900283
232. Rocco M, Molteni M, Ponassi M, Giachi G, Frediani M, Koutsioubas A, Profumo A, Trevarin D, Cardinali B, Vachette P, Ferri F, Perez J (2014) A comprehensive mechanism of fibrin network formation involving early branching and delayed single- to double-strand transition from coupled time-resolved X-ray/light-scattering detection. J Am Chem Soc 136: 5376–5384
233. Roosa CA, Muhamed I, Young AT, Nellenbach K, Daniele MA, Ligler FS, Brown AC (2021) Synthesis of sonicated fibrin nanoparticles that modulate fibrin clot polymerization and enhance angiogenic responses. Colloids Surf B Biointerfaces 204:111805
234. Rosenfeld MA, Leonova VB, Shchegolikhin AN, Bychkova AV, Kostanova EA, Biryukova MI (2015) Covalent structure of single-stranded fibrin oligomers cross-linked by FXIIIa. Biochem Biophys Res Commun 461:408–412
235. Samson AL, Alwis I, Maclean JAA, Priyananda P, Hawkett B, Schoenwaelder SM, Jackson SP (2017) Endogenous fibrinolysis facilitates clot retraction in vivo. Blood 130:2453–2462
236. Sauls DL, Lockhart E, Warren ME, Lenkowski A, Wilhelm SE, Hoffman M (2006) Modification of fibrinogen by homocysteine thiolactone increases resistance to fibrinolysis: a potential mechanism of the thrombotic tendency in hyperhomocysteinemia. Biochemistry 45:2480–2487
237. Sawadkar P, Mandakhbayar N, Patel KD, Owji N, Rajasekar P, Sarama R, Lee JH, Kim HW, Knowles J, García-Gareta E (2024) 3D porous binary composites of collagen, elastin, and fibrin proteins orchestrate adipose tissue regeneration. Macromol Biosci 24:e2400073
238. Schmitt LR, Henderson R, Barrett A, Darula Z, Issaian A, D'Alessandro A, Clendenen N, Hansen KC (2019) Mass spectrometry-based molecular mapping of native FXIIIa cross-links in insoluble fibrin clots. J Biol Chem 294:8773–8778
239. Scull G, Fisher MB, Brown AC (2021) Fibrin-based biomaterial systems to enhance anterior cruciate ligament healing. Med Devices Sens 4:e10147
240. Scull G, Aligwekwe A, Rey Y, Koch D, Nellenbach K, Sheridan A, Pandit S, Sollinger J, Pierce JG, Flick MJ, Gilbertie J, Schnabel L, Brown AC (2024) Fighting fibrin with fibrin: vancomycin delivery into coagulase-mediated Staphylococcus aureus biofilms via fibrin-based nanoparticle binding. J Biomed Mater Res A 112:2071–2085
241. Semkova ME, Hsuan JJ (2021) Mass spectrometric identification of a novel Factor XIIIa cross-linking site in fibrinogen. Proteomes 9:43
242. Sharifi A, Kouhi A, Patel ZM (2024) Utilization and efficacy of platelet-rich plasma and platelet-rich fibrin in otolaryngology: a systematic evidence-based review. Eur Arch Otorrinolaringol 281:5091–5110

243. Shirbhate U, Bajaj P, Wanjari M, Patil M (2025) Role of injectable platelet-rich fibrin in the management of soft and hard tissue periodontal regeneration in dentistry: protocol for a systematic review. JMIR Res Protoc 14:e65137
244. Silva LM, Doyle AD, Greenwell-Wild T, Dutzan N, Tran CL, Abusleme L, Juang LJ, Leung J, Chun EM, Lum AG, Agler CS, Zuazo CE, Sibree M, Jani P, Kram V, Martin D, Moss K, Lionakis MS, Castellino FJ, Kastrup CJ, Flick MJ, Divaris K, Bugge TH, Moutsopoulos NM (2021) Fibrin is a critical regulator of neutrophil effector function at the oral mucosal barrier. Science 374:eabl5450
245. Silvain J, Collet JP, Nagaswami C, Beygui F, Edmondson KE, Bellemain-Appaix A, Cayla G, Pena A, Brugier D, Barthelemy O, Montalescot G, Weisel JW (2011) Composition of coronary thrombus in acute myocardial infarction. J Am Coll Cardiol 57:1359–1367
246. Siniarski A, Baker SR, Duval C, Malinowski KP, Gajos G, Nessler J, Ariëns RAS (2021) Quantitative analysis of clot density, fibrin fiber radius, and protofibril packing in acute phase myocardial infarction. Thromb Res 205:110–119
247. Sovová Ž, Štikarová J, Kaufmanová J, Májek P, Suttnar J, Šácha P, Malý M, Dyr JE (2020) Impact of posttranslational modifications on atomistic structure of fibrinogen. PLoS One 15: e0227543
248. Spiewak R, Gosselin A, Merinov D, Litvinov RI, Weisel JW, Tutwiler V, Purohit PK (2022) Biomechanical origins of inherent tension in fibrin networks. J Mech Behav Biomed Mater 133:105328
249. Spotnitz WD (2010) Fibrin sealant: past, present, and future: a brief review. World J Surg 34: 632–634
250. Stalker TJ (2020) Mouse laser injury models: variations on a theme. Platelets 31:423–431
251. Standeven KF, Grant PJ, Carter AM, Scheiner T, Weisel JW, Ariens RA (2003) Functional analysis of the fibrinogen Aalpha Thr312Ala polymorphism: effects on fibrin structure and function. Circulation 107:2326–2330
252. Standeven KF, Carter AM, Grant PJ, Weisel JW, Chernysh I, Masova L, Lord ST, Ariens RA (2007) Functional analysis of fibrin {gamma}-chain cross-linking by activated factor XIII: determination of a cross-linking pattern that maximizes clot stiffness. Blood 110:902–907
253. Sugerman GP, Parekh SH, Rausch MK (2020) Nonlinear, dissipative phenomena in whole blood clot mechanics. Soft Matter 16:9908–9916
254. Sugo T, Nakamikawa C, Yoshida N, Niwa K, Sameshima M, Mimuro J, Weisel JW, Nagita A, Matsuda M (2000) End-linked homodimers in fibrinogen Osaka VI with a B beta-chain extension lead to fragile clot structure. Blood 96:3779–3785
255. Swieringa F, Baaten CC, Verdoold R, Mastenbroek TG, Rijnveld N, van der Laan KO, Breel EJ, Collins PW, Lancé MD, Henskens YM, Cosemans JM, Heemskerk JW, van der Meijden PE (2016) Platelet control of fibrin distribution and microelasticity in thrombus formation under flow. Arterioscler Thromb Vasc Biol 36:692–699
256. Szczyt M, Batko J, Pasternak A (2024) A systematic review: fibrin glue in drainless rhytidoplasty. Aesth Plast Surg 48:2224–2230
257. Talens S, Leebeek FW, Demmers JA, Rijken DC (2012) Identification of fibrin clot-bound plasma proteins. PLoS One 7:e41966
258. Taylor FB Jr, Muller-Eberhard HJ (1970) Qualitative description of factors involved in the retraction and lysis of dilute whole blood clots and in the aggregation and retraction of platelets. J Clin Invest 49:2068–2085
259. Tenopoulou M (2025) Fibrinogen post-translational modifications are biochemical determinants of fibrin clot properties and interactions. FEBS J 292:11–27
260. Thomson KS, Korte FS, Giachelli CM, Ratner BD, Regnier M, Scatena M (2013) Prevascularized microtemplated fibrin scaffolds for cardiac tissue engineering applications. Tissue Eng Part A 19:967–977
261. Tomaiuolo M, Litvinov RI, Weisel JW, Stalker TJ (2020) Use of electron microscopy to study platelets and thrombi. Platelets 31:580–588

262. Tran R, Myers DR, Ciciliano J, Trybus Hardy EL, Sakurai Y, Ahn B, Qiu Y, Mannino RG, Fay ME, Lam WA (2013) Biomechanics of haemostasis and thrombosis in health and disease: from the macro- to molecular scale. J Cell Mol Med 17:579–596
263. Tutwiler V, Peshkova AD, Le Minh G et al (2019) Blood clot contraction differentially modulates internal and external fibrinolysis. J Thromb Haemost 17(2):361–370
264. Tutwiler V, Singh J, Litvinov RI, Bassani JL, Purohit PK, Weisel JW (2020) Rupture of blood clots: mechanics and pathophysiology. Sci Adv 6:eabc0496
265. Tutwiler V, Maksudov F, Litvinov RI, Weisel JW, Barsegov V (2021) Strength and deformability of fibrin clots: biomechanics, thermodynamics, and mechanisms of rupture. Acta Biomater 131:355–369
266. Undas A (2024) Reviewing the rich history of fibrin clot research with a focus on clinical relevance. Semin Thromb Hemost 50:751–759
267. Undas A, Brozek J, Jankowski M, Siudak Z, Szczeklik A, Jakubowski H (2006) Plasma homocysteine affects fibrin clot permeability and resistance to lysis in human subjects. Arterioscler Thromb Vasc Biol 26:1397–1404
268. Vahabi M, Sharma A, Licup AJ, van Oosten AS, Galie PA, Janmey PA, MacKintosh FC (2016) Elasticity of fibrous networks under uniaxial prestress. Soft Matter 12:5050–5060
269. Varjú I, Sótonyi P, Machovich R, Szabó L, Tenekedjiev K, Silva MM, Longstaff C, Kolev K (2011) Hindered dissolution of fibrin formed under mechanical stress. J Thromb Haemost 9: 979–986
270. Varjú I, Sorvillo N, Cherpokova D, Farkas ÁZ, Farkas VJ, Komorowicz E, Feller T, Kiss B, Kellermayer MZ, Szabó L, Wacha A, Bóta A, Longstaff C, Wagner DD, Kolev K (2021) Citrullinated fibrinogen renders clots mechanically less stable, but lysis-resistant. Circ Res 129:342–344
271. Varjú I, Tóth E, Farkas ÁZ, Farkas VJ, Komorowicz E, Feller T, Kiss B, Kellermayer MZ, Szabó L, Wacha A, Bóta A, Longstaff C, Kolev K (2022) Citrullinated fibrinogen forms densely packed clots with decreased permeability. J Thromb Haemost 20:2862–2872
272. Veklich Y, Francis CW, White J, Weisel JW (1998) Structural studies of fibrinolysis by electron microscopy. Blood 92:4721–4729
273. Vilar R, Fish RJ, Casini A, Neerman-Arbez M (2020) Fibrin(ogen) in human disease: both friend and foe. Haematologica 105:284–296
274. Vorjohann S, Fish RJ, Biron-Andréani C, Nagaswami C, Weisel JW, Boulot P, Reyftmann L, de Moerloose P, Neerman-Arbez M (2010) Hypodysfibrinogenaemia due to production of mutant fibrinogen alpha-chains lacking fibrinopeptide A and polymerisation knob 'A'. Thromb Haemost 104:990–997
275. Vos BE, Liebrand LC, Vahabi M, Biebricher A, Wuite GJL, Peterman EJG, Kurniawan NA, MacKintosh FC, Koenderink GH (2017) Programming the mechanics of cohesive fiber networks by compression. Soft Matter 13:8886–8893
276. Vos BE, Martinez-Torres C, Burla F, Weisel JW, Koenderink GH (2020) Revealing the molecular origins of fibrin's elastomeric properties by in situ X-ray scattering. Acta Biomater 104:39–52
277. Wachendörfer M, Buhl EM, Messaoud GB, Richtering W, Fischer H (2023) pH and thrombin concentration are decisive in synthesizing stiff, stable, and open-porous fibrin-collagen hydrogel blends without chemical cross-linker. Adv Healthc Mater 12:e2203302
278. Wang Y, Kumar S, Nisar A, Bonn M, Rausch MK, Parekh SH (2021) Probing fibrin's molecular response to shear and tensile deformation with coherent Raman microscopy. Acta Biomater 121:383–392
279. Weigandt KM, White N, Chung D, Ellingson E, Wang Y, Fu X, Pozzo DC (2012) Fibrin clot structure and mechanics associated with specific oxidation of methionine residues in fibrinogen. Biophys J 103:2399–2407
280. Weisel JW (2004) Cross-linked gamma chains in a fibrin fibril are situated transversely between its strands: no. J Thromb Haemost 2:394–399
281. Weisel JW (2005) Fibrinogen and fibrin. Adv Protein Chem 70:247–299

282. Weisel JW (2007) Structure of fibrin: impact on clot stability. J Thromb Haemost 5(Suppl. 1): 116–124
283. Weisel JW, Litvinov RI (2013) Mechanisms of fibrin polymerization and clinical implications. Blood 121:1712–1719
284. Weisel JW, Litvinov RI (2014) Mechanisms of fibrinolysis and basic principles of management. In: Saba HI, Roberts HR (eds) Handbook of hemostasis and thrombosis: practical guidelines in clinical management. Wiley-Blackwell, Oxford, pp 169–179
285. Weisel JW, Litvinov RI (2017) Fibrin formation, structure and properties. Subcell Biochem 82:405–456
286. Weisel JW, Litvinov RI (2025) Exploring the thrombus niche: lessons learned and potential therapeutic opportunities. Blood. https://doi.org/10.1182/blood.2024025319
287. Weisel JW, Nagaswami C (1992) Computer modeling of fibrin polymerization kinetics correlated with electron microscope and turbidity observations: clot structure and assembly are kinetically controlled. Biophys J 63:111–128
288. Weiss HL, Selvarah P, Okita K, Matsumoto Y, Voie A, Hoelscher T, Szeri AJ (2013) Mechanical clot damage from cavitation during sonothrombolysis. J Acoust Soc Am 133: 3159–3174
289. Wen Q, Janmey PA (2013) Effects of non-linearity on cell-ECM interactions. Exp Cell Res 319:2481–2489
290. Whittaker P, Przyklenk K (2009) Fibrin architecture in clots: a quantitative polarized light microscopy analysis. Blood Cells Mol Dis 42:51–56
291. Whyte CS, Mitchell JL, Mutch NJ (2017) Platelet-mediated modulation of fibrinolysis. Semin Thromb Hemost 43:115–128
292. Wolberg AS (2023) Fibrinogen and fibrin: synthesis, structure, and function in health and disease. J Thromb Haemost 21:3005–3015
293. Wufsus AR, Rana K, Brown A, Dorgan JR, Liberatore MW, Neeves KB (2015) Elastic behavior and platelet retraction in low- and high-density fibrin gels. Biophys J 108:173–183
294. Xia Z, Heng Y, Ma B, Zhu SH, Tang HT, Ben D (2010) Acute pulmonary embolism complicated by thrombolytic therapy. J Trauma 69:E109
295. Xia J, Cai LH, Wu H, MacKintosh FC, Weitz DA (2021) Anomalous mechanics of Zn^{2+}-modified fibrin networks. Proc Natl Acad Sci USA 118:e2020541118
296. Yakovlev S, Strickland DK, Medved L (2022) Current view on the molecular mechanisms underlying fibrin(ogen)-dependent inflammation. Thromb Haemost 122:1858–1868
297. Yang Z, Mochalkin I, Doolittle LR (2000) A model for fibrin formation based on crystal structures of fibrinogen and fibrin fragments complexed with specific peptides. Proc Natl Acad Sci USA 97:14156–14161
298. Yeromonahos C, Polack B, Caton F (2010) Nanostructure of the fibrin clot. Biophys J 99: 2018–2027
299. Yesudasan S, Wang X, Averett RD (2018) Coarse-grained molecular dynamics simulations of fibrin polymerization: effects of thrombin concentration on fibrin clot structure. J Mol Model 24:109
300. Yesudasan S, Averett RD (2020) Multiscale network modeling of fibrin fibers and fibrin clots with protofibril binding mechanics. Polymers (Basel) 12:1223
301. Young JZ, Oxfd MA, Medawar PB, Oxfd MA (1940) Fibrin suture of the peripheral nerves. Lancet 236:126–128
302. Yurina LV, Vasilyeva AD, Gavrilina ES, Ivanov VS, Obydennyi SI, Chabin IA, Indeykina MI, Kononikhin AS, Nikolaev EN, Rosenfeld MA (2024) A role of methionines in the functioning of oxidatively modified fibrinogen. Biochim Biophys Acta Proteins Proteom 1872:141013
303. Ząbczyk M, Undas A (2017) Plasma fibrin clot structure and thromboembolism: clinical implications. Pol Arch Intern Med 127:873–881
304. Ząbczyk M, Ariëns RAS, Undas A (2023) Fibrin clot properties in cardiovascular disease: from basic mechanisms to clinical practice. Cardiovasc Res 119:94–111

305. Zakharov A, Awan M, Cheng T, Gopinath A, Lee SJ, Ramasubramanian AK, Dasbiswas K (2024) Clots reveal anomalous elastic behavior of fiber networks. Sci Adv 10:eadh1265
306. Zhang W, Hu J, Tang J, Wang Z, Wang J, Lu T, Suo Z (2019) Fracture toughness and fatigue threshold of tough hydrogels. ACS Macro Lett 8:17–23
307. Zheng Z, Mukhametova L, Boffa MB, Moore EE, Wolberg AS, Urano T, Kim PY (2023) Assays to quantify fibrinolysis: strengths and limitations. Communication from the International Society on Thrombosis and Haemostasis Scientific and Standardization Committee on fibrinolysis. J Thromb Haemost 21:1043–1054
308. Zheng X, Wang F, Su YC, Xu CY, Wang MZ (2024) Efficacy and safety of fibrin sealant application in patients undergoing thyroidectomy: a systematic review and meta-analysis. BMC Surg 24:122
309. Zhmurov A, Brown AE, Litvinov RI, Dima RI, Weisel JW, Barsegov V (2011) Mechanism of fibrin(ogen) forced unfolding. Structure 19:1615–1624
310. Zhmurov A, Kononova O, Litvinov RI, Dima RI, Barsegov V, Weisel JW (2012) Mechanical transition from alpha-helical coiled coils to beta-sheets in fibrin(ogen). J Am Chem Soc 134: 20396–20402
311. Zhmurov A, Protopopova AD, Litvinov RI, Zhukov P, Mukhitov AR, Weisel JW, Barsegov V (2016) Structural basis for interfacial flexibility in fibrin oligomers. Structure 24:1907–1917
312. Zhmurov A, Protopopova AD, Litvinov RI, Zhukov P, Weisel JW, Barsegov V (2018) Atomic structural models of fibrin oligomers. Structure 26:857–868.e4
313. Zuev YF, Litvinov RI, Sitnitsky AE, Idiyatullin BZ, Bakirova DR, Galanakis DK, Zhmurov A, Barsegov V, Weisel JW (2017) Conformational flexibility and self-association of fibrinogen in concentrated solutions. J Phys Chem B 121:7833–7843
314. Zubairova LD, Nabiullina RM, Nagaswami C, Zuev YF, Mustafin IG, Litvinov RI, Weisel JW (2015) Circulating microparticles alter formation, structure, and properties of fibrin clots. Sci Rep 5:17611

Chapter 14
Biology and Molecular Structure of the α-Fibrous Silks in Bees, Ants, Wasps and Hornets (*Hymenoptera*)

Caitlin Johnston, Lyndall Briggs, Michelle Michie, and Tara D. Sutherland

Abstract Larval aculeate insects (bees, ants and hornets) have evolved a silk for cocoon and nest construction based on entirely different design principles from the convergently evolved silk of spiders and silkworms. The silk is based on the coiled coil protein structure: a structure proposed in the 1950s by Francis Crick, where multiple α-helical proteins wrap around each other to form a supercoil. Whilst now recognised as a ubiquitous protein assembly motif, the purity of silk pulled from the honeybee silk gland made it one of the earliest experimentally validated examples of this structure. Here, we describe the current state of understanding of the relationship between the coiled coil structure and aculeate silk biology gained from early studies dating from the mid-twentieth century up to the present time. Further, efforts to replicate the natural material and functional materials generated from the silk are outlined. Finally, we consider the future of research in this area, focusing on mimicking natural silk and the use of recombinant silk for rational design of functional materials.

Keywords Coiled coil · Silk · Hymenoptera · Cross-link · Bee · Ant · Hornet · Rational design · Protein materials

Introduction

Bees, ants, wasps, and hornets—the social Aculeata—are a spectacularly successful subclade of Hymenoptera that dominate the terrestrial ecosystem both in terms of biomass and energy consumption. There are three superfamilies within this subclade: Apoidea (bees and sphecid wasps), Vespoidea (wasps and ants), and Chrysidoidea (parasitic wasps). Species within Apoidea and Vespoidea produce a silk that is made up of proteins in a coiled coil structure, whereby α-helical proteins twist around each other to form a supercoil. This molecular structure is very distinct

C. Johnston · L. Briggs · M. Michie · T. D. Sutherland (✉)
CSIRO, Black Mountain, Canberra, ACT, Australia
e-mail: Tara.Sutherland@CSIRO.AU

D. A. D. Parry, P. K. Luther (eds.), *Fibrous Proteins: Structures and Mechanisms*, Subcellular Biochemistry 113, https://doi.org/10.1007/978-3-032-05273-5_14

Fig. 14.1 Silk with the coiled coil molecular structure (white fibrous material) covering the brood cells of a *Vespula pensylvanica* wasp nest

from the well-studied dragline silk of spiders and the cocoon silk of silkworms. Spiders and silkworms independently evolved very large silk proteins (>10 times that of the bee, ant, and hornet silks). These proteins have highly repetitive amino acid sequences that, in the final material, are found in β-sheets: closely aligned extended protein structures. Whilst β-sheet structures are also present in the bee, ant, and hornet silks, they are less prominent compared to the coiled coil structure.

Coiled coil silk is produced by final instar larvae and used by itself or as a composite with other natural materials (Fig. 14.1). Wasps and hornets use the silk in large, elaborate nests and to seal brood cells. Successive generations of honeybee larvae coat the brood cell walls immediately before pupation to produce a silk/wax composite hive. Bumblebees reuse their silken cocoons to store pollen and honey. Bulldog ants and many other species produce silken cocoons for protection during pupation. Green ants (weaver ants) use the silk to weave together leaves, forming large communal nests in which the larvae pupate without a cocoon.

The mechanical properties, biocompatibility, and biodegradability of all types of silk have, for decades, attracted researchers interested in replicating these features for various applications. Native silkworm silk can be harvested, dissolved, and reprocessed into new materials but, to date, these materials do not replicate the physical properties of the natural materials. The large size and repetitive nature of the silk proteins of silkworms and spiders have made it impossible to replicate them in existing recombinant systems. Instead, studies using recombinant spider or silkworm silks have thus far been able to produce only fractions of the natural silk protein, and at levels that are below what is considered commercially viable for most applications. A further challenge with spider and silkworm silks is the natural process of silk production that involves several molecular structural transitions, each of which is complex, poorly understood at present, and unlikely to be scalable in existing commercial facilities. The coiled coil silk proteins, however, are attracting research interest as they are small and therefore compatible with production in recombinant systems, and the fabrication process does not involve molecular structural transitions, yet they produce similar mechanical properties, biocompatibility, and biodegradability to those of the spider and silkworm silks.

In addition to the mechanical, biocompatibility, and biodegradability properties of this silk, there is interest in what other properties these silks might offer to the aculeate insects. The term 'aculeate' refers to the sting that is a characteristic feature of this group. Within the aculeate insects, eusociality (the highest organisation of sociality) has arisen independently at least 12 times. It is argued that the sting was an enabling mechanism for evolution of eusociality in this subclade [44]. However, social behaviour is found only within the two aculeate superfamilies that produce the coiled coil silk: Apoidea and Vespoidea. Insects from Chrysidoidea have evolved the sting but have never evolved sociality. Also critical to social behaviour is a secure habitat to house young and food supplies. The coincidence of the use of the coiled coil silk and evolution of sociality is an intriguing observation that drives an appetite to better understand what other properties this silk has, which led to its evolution.

In this chapter, we review published research on this silk, starting from the early history up to the present efforts. Finally, we discuss the potential applications that this material may play in the future as a template for new and exciting materials.

Early Research into Aculeate Silk (1960–2000)

The Coiled Coil Structure of Bee Silk

Underpinning the early research interest in the honeybee silk was Kenneth Rudall's X-ray diffraction data from 1962 showing that silk fibres of bees, wasps, and ants were in an α-helical structure, in marked contrast to the antiparallel pleated sheets of silkworms and spiders [38]. In 1967, Atkins [1] interpreted this data as indicative of a tetrameric coiled coil, with the coiled coils orientated parallel to the fibre axis. Around the same time, amino acid analysis of honeybee silk found high levels of alanine and the small polar residue serine [26], a composition not associated with canonical coiled coils and implying the silk proteins formed an unusual coiled coil structure. To date, details of the atomic structure of this silk have remained elusive.

Studies of the Functional Properties of Bee, Ant, and Hornet Silk

In general, it is widely assumed that the key role for cocoons is to protect the animal during metamorphosis from environmental damage (dryness, flooding, cold) and for nests is to protect the animal from predators and pathogens. Whilst there have been limited studies verifying this assumption, insects invest heavily in this process, with studies showing that 10% of prepupal weight of solitary bees goes into cocoon construction [2]. The studies below outline other roles for the bee and hornet silk.

Hepburn and Kurstjens [11] investigated the function of bee silk by comparing the mechanical properties of African honeybee (*Apis mellifera scutellata*) nests at different temperatures and stages of maturation. In early spring, young nests were templated with wax, an isotropic plastic with mechanical properties that are highly temperature dependent. As spring turned to summer, layers of silk added by successive generations of larvae gradually converted the nest into a silk/wax composite containing up to 34% silk. The silk greatly enhanced the stiffness and strength of the hive and, in contrast to the wax component, the mechanical properties (stiffness, strength) did not change over the natural temperature range of the hive (25–45 °C). The study concluded that the silk was critical for tensile strength and mechanical integrity of the hive in warmer temperatures. A similar conclusion was reached by Zhang et al. [58].

The mechanical properties of bee silk threads have been described in two contradictory studies. Hepburn et al. [12] hand-drew silk fibres from African honeybee (*Apis mellifera adansonii*) larvae and measured their tensile properties at 65% humidity. They observed stress-strain curves with an initial linear region, followed by yield point, and then maximum true stress at break of around 400 MPa (true stress is defined as the force divided by the cross-sectional area existing at a given point in the test) and strain at break of 110%. As other publications have presented data as engineering stress (the force divided by the cross-sectional area at the start of the test), Weisman et al. [54] back-calculated this to an engineering strain of 132 MPa. A later study from Zhang et al. [59], also on the silk of *A. mellifera* but at 30% humidity and from larvae that had been allowed to spin silk on glass slides, found the silk had a tensile strength of 164 $\pm$ 3 MPa (engineering stress) but had a much lower breaking strain than reported by Hepburn et al. (4 $\pm$ 1%). The stress-strain curves from Zhang et al. [59] had a very different shape that lacked the initial linear region or yield point observed by Hepburn et al. [12]. Whether natural silk fibres are brittle as measured by Zhang et al. [59] or tough as measured by Hepburn et al. [12], or very condition-dependent, is still to be determined.

Hornets produce multicellular nests from a range of natural building materials mixed with saliva from adults. Similar to bees, the larvae spin a cocoon within the individual cells. Shabtai and Ishay [40] demonstrated that the end caps of these cocoons allow gas exchange but filter out physical and biological contaminants, with the interior of the comb cells being essentially sterile. These end caps have also been found to act as thermoregulators [15] and have been proposed to act as electric (Joule) heaters where the silk stores heat as electrical charge during the day and then releases it during the night when the temperature drops.

How the Insects Produce the Silk

The coiled coil silk is made from proteins produced in the labial glands of larval insects (Fig. 14.2a) and fabricated using a spinneret that is a long, thin, hardened pore (Fig. 14.2b). Jay [16] observed that fibres formed after the insect placed the spinneret on a substrate and then moved away; conversely, films were formed when the insect dragged the spinneret over the substrate. Assuming a sufficiently viscous silk solution, the slit-like structure of the spinneret is ideally suited to the formation of these two structures. The surface tension of aqueous silk extruded as a ribbon will lead to capillary breakup of the solution and fibre formation, which can be prevented by casting the ribbon on a substrate to form a film (Fig. 14.2c; [47]).

So, how does the insect generate silk protein solutions with sufficient viscosity to allow silk formation but also allow the silk solution to pass through the spinneret? In silkworm silk glands, the silk proteins reach concentrations of around 30% with solution viscosity of over three million times that of water [24]. Early work by Flower and Kenchington [8] provides the most likely explanation. With an elegant series of light and electron microscope images combined with the use of various stains, they demonstrated that the silk is organised as strongly birefringent cigar-shaped bodies a few microns long (Fig. 14.2d). This structural organisation was later confirmed by Silva-Zacarin et al. [42]. These tactoids break down to fibrils of 250–500 × 10 nm at low potassium chloride concentrations, then smaller fibrils of

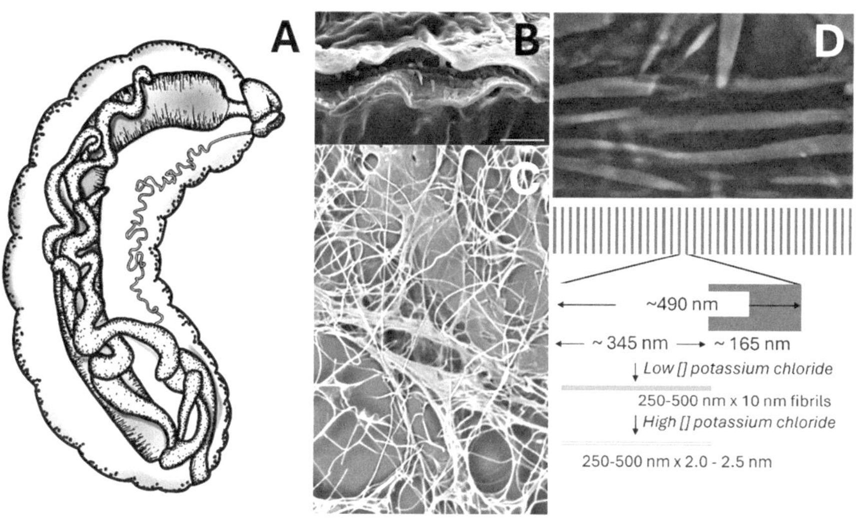

Fig. 14.2 Silk production in honeybee larvae. (**a**) Schematic of a honeybee larvae showing morphology of the silk gland (redrawn from [29]); (**b**) spinneret of a bumble bee [48]; (**c**) scanning electron microscopy image of the silken end cap of a honeybee broad cell showing sheets and fibres; (**d**) Image of 'tactoids' observed in the silk gland of larval honeybees [8] and schematics of the various structures identified after potassium chloride treatments, with the finest structures identified being fibrils of 2–2.5 nm diameter

250–500 × 2 nm at higher potassium chloride concentrations (Fig. 14.2d). It is likely that it is this supramolecular organisation within the silk gland that enables the silk solution to flow out of the spinneret at a concentration where it can almost immediately dehydrate to form solid films or fibres.

The concept that the critical role of the coiled coils is in allowing biologically compatible rheology in the silk solution is supported by the finding that, at least in honeybees, the coiled coil is not retained in the natural material [20]. Whilst the coiled coil structure is found in fibres directly isolated from larvae [38], high-resolution ^{13}C solid-state NMR studies of silk in the presence of wax, akin to what is found in the native hive, identified alanine chemical shifts indicative of isolated α-helices rather than coiled coils [20].

The Present: Bee Silk in the Age of Modern Molecular Biology

Modern molecular biology has opened the door to understanding and replicating the biological structure and function of proteins. In this section, we discuss the efforts to date to: replicate the coiled coil silks; understand how and why these silks evolved; and generate functional materials based on these silk proteins.

Replicating the Natural Material

Identification of the Silk Gene Sequences

The DNA sequences encoding the silk proteins of bees, ants, and hornets were first identified by two separate groups in 2006 [39, 45]. By matching sequences encoded in cDNA clones from a library constructed from the silk gland of late final instar honeybee larvae with mass spectrometry peptide fragmentation patterns obtained after tryptic digestion of the natural silk, Sutherland et al. [45] identified four homologous silk proteins that were predicted to fold into a coiled coil structure. Independently, and using a similar method, Sezutsu et al. [39] identified the four homologous proteins from the Asian hornet. A year later, the homologous silk proteins from bumble bees, green (weaver) ants, and the Australian bulldog ant [46] were identified. Through genome projects and other studies, the number of available homologous aculeate silk protein sequences has dramatically expanded, and these can be identified by sequence homology in gene or protein databases, although at this stage many are misannotated.

Recombinant Production of Silk Proteins

The availability of gene sequences has allowed recombinant production of the silk proteins, with bacterial expression in *Escherichia coli* being the preferred method of production. Expression levels are generally high as far as recombinant expression levels go: *Apis cerana* silk: 10–60 mg/L [41]; *Apis mellifera* silk: 200–2500 mg/L [54]; *Vespa simillima* (hornet) silk: 160–330 mg/L [18]; *Apis dorsata* silk: 1600–2600 mg/L [27]. Sutherland et al. [49] report yields from a commercial partner who optimised the protein production process and were able to achieve yields of the AmelF3 silk protein from *A. mellifera* of greater than 10 g/L.

Silk Mimics Made from Recombinant Proteins

Bee, ant, and hornet silk is produced by the insect as fibres (see [5] for images), with bee silk also produced as films [16]. Fibres have been air-drawn from solutions containing all four recombinantly produced honeybee silk proteins, and then further drawn to twice their length in aqueous methanol [48, 54]. These fibres had breaking stresses similar to those of native fibres [12, 59], and were less brittle than observed in native fibres by Zhang et al. [59] but were not as extensible as those described by Hepburn et al. [12]. Later studies produced fibres from single, recombinantly produced honeybee silk proteins either by continuous extrusion into aqueous methanol followed by post-fabrication drawing [30] or by electrospinning [56]. The fibres produced by the optimised extrusion method had similar stress and strain properties to the hand-drawn fibres from the 2010 study, again not mimicking the toughness or brittleness observed in the native fibres by Hepburn et al. [12] or Zhang et al. [59].

Films have been made by casting concentrated solutions of recombinant AmelF3 proteins onto Teflon surfaces [31], but to date, the properties of these have not been compared to natural materials. Both films and fibres can be dissolved in aqueous solutions when first fabricated and can be stabilised by either dry heat treatment, which generates covalent cross-links, or by aqueous methanol treatment, which increases β-sheet structure and hydrogen bonding between proteins [14]. Films have also been generated from recombinant hornet silk proteins after protein solutions were freeze-dried and then dissolved in hexafluoroisopropanol (HFIP) [18]. These films had mechanical properties that equalled or surpassed those of films made after dissolving and reconstituting either native hornet silk or silkworm silk. To date, there are no studies that have measured the mechanical properties of natural hornet silk material.

Whilst these materials mimic the natural materials in terms of chemical composition, so far there have been no reports where the macromolecular liquid crystal structure of the silk proteins found in natural silk glands (see section "How the insects produce the silk") has been replicated prior to material fabrication. It is likely that replication of the native properties will require this macromolecular assembly process to be mimicked.

Functional Materials Produced Using Recombinant Bee Silk

Functional materials have been produced from recombinant honeybee silk proteins by entrapping heme within silk protein materials [32–35], with the materials' functionality coming from the reactivity of the heme within the silk protein environment. The α-helical structure of the silk proteins is similar to that found in many naturally occurring heme proteins, and residues within the coiled coil coordinate to the iron centre of heme, allowing control of the reactivity of the metal centre [32].

The first functional materials generated using this approach were optical biosensors for nitric oxide detection [32]. Nitric oxide levels in breath are currently used clinically to support asthma diagnosis and management, and these sensors were designed to target home management of this disease [35]. The heme-binding capacity of the silk protein was increased by the addition of histidine residues to key positions in the silk [33]. Finally, a third-generation nitric oxide biosensor (a sensor based on direct electron transfer between a metal centre in a protein and an electrode) was designed by Musameh et al. [28]. The sensor was capable of nanomolar nitric oxide detection in the presence or absence of oxygen, nitrite, nitrate, dopamine, or ascorbic acid and retained nearly 90% activity over 2 months.

A similar approach was used to generate oxygen reduction reaction catalysts as an alternative catalyst to platinum in fuel cells [34, 36]. Materials generated after entrapping hemin within silk scaffolds were capable of oxygen reduction reactions with turnover numbers $>10^7$ and could be operated for at least 96 h. These materials had very high activity—3 orders of magnitude better than previously engineered iron porphyrin oxygen reduction catalysts—which was attributed to the silk isolating each heme catalytic centre, preventing them from stacking against each other.

In addition to hemin, the approach has been extended to entrap other porphyrin-based macrocycles. Horgan et al. [13] produced materials for photodynamic therapy by entrapping different macrocycles within recombinant silk that were able to undergo transition from singlet state to triplet state following excitation with visible light, leading to phosphorescence and the production of cytotoxic oxygen species. These materials could also be used as phosphorescent sensors for oxygen. Aimed at achieving clean sources of hydrogen, Rapson et al. [37] designed recombinant honeybee silk materials containing cobalt protoporphyrin to generate a catalytic material capable of hydrogen evolution. Activity was improved by modifying amino acids in the silk protein to improve coordination of the metal centre of the protoporphyrin. The viability of the catalyst was demonstrated after it was incorporated into a proton exchange water electrolysis cell and demonstrated hydrogen production with 98% Faradic efficiency [37].

Recombinant silk materials with entrapped silver nanoparticles have antimicrobial activity [52]. The activity was attributed to silver ions dissolving from the nanoparticles, with the silk preventing aggregation of nanoparticles and thereby enhancing the rate of dissolution of the particles to provide an antimicrobial effect.

Generation of Materials from Natural Silks

Use of Natural Silk to Make New Materials

Whilst it is difficult to imagine commercially viable methods to harvest natural bee or ant silk, there have been a few studies looking at potential materials from natural silks. It has been suggested that weaver ant silk could be used as a biomaterial based on evidence that it supported cell adhesion and proliferation of NIH 3T3 cells [43]. Native *A. cerana* silk has been proposed as a material for tissue engineering and regenerative medicine because the silk is cytocompatible and supports the proliferation of various human cell lines [25]. Natural *A. mellifera* silk, blended with dissolved chitosan and freeze-dried to generate scaffolds, has been shown to be antimicrobial [51], although the activity was due to the chitosan component.

Materials Made from Reconstituted Natural Silk

Whereas bee and ant silks are stabilised by covalent cross-links and cannot be dissolved [5], hornet silks are stabilised by hydrogen bonding between β-sheet regions and can be readily dissolved and processed back into different material forms. The volume and cleanliness of hornet silk make it relatively easy to collect, and a number of studies have processed this silk into various material forms. Silk from the hornet *V. simillima* was dissolved in lithium bromide, dialysed against water and fabricated into films [23] and tubes [21]; in a separate study dissolved silk was dialysed against basic ammonia and cast into films [19]; and in another study the silk was resolubilised into HFIP and cast into transparent films [22]. This silk has been developed into a nail polish and a nail lotion, both of which are sold in Japan.

Evolution of the Coiled Coil Silks

Sutherland et al. [46] established that within Hymenoptera the coiled coil silks were produced by members of the Apoidea and the Vespoidea, but not by insects of the Chrysidoidea, the oldest lineage of the Aculeata, which produce silk with a β-sheet structure. The paraphyly of Chrysidoidea with respect to the aculeate Hymenoptera suggests that the β-sheet silks are plesiomorphic and that coiled coil silks represent the derived condition within this insect order [46]. Sociality has evolved multiple times within the aculeate insects, but not in any other Hymenopteran lineage. Disease and parasitism are major threats to social insects. Sutherland et al. [46] proposed that this radically different building material that replaced the plesiomorphic β-sheet silk may have played a role in the evolution of social behaviour, allowing the insects to build nests and protect juveniles. Sutherland et al. [50] also demonstrated that silk peptides have antimicrobial activity, and they suggested that the silks evolved as an antifouling material and that it was this feature that facilitated the evolution of sociality.

The concept that the silk may provide a beneficial effect in reducing fungal disease has been tested in weaver ants [9]. These social ants construct elaborate silken nests, but they have lost the antibiotic-producing metapleural gland, which is a primary mechanism of disease defence. However, the silk was found not to improve the already high resistance of ants to the fungi *Metarhizium*, and *Aspergillus* was found widely in ant nests, leading to the conclusion that the silk did not improve resistance to these fungi. Further testing is thus required to determine if the silks do indeed provide a role in disease resistance.

Silk Molecular Structure

Genetic Structure

The genes encoding the four distinct proteins that form the main components of silk materials are conserved across bees, ants and hornets [5]. It is hypothesised that these genes originated from an ancient duplication event that occurred before the divergence of the Apoidea (bees) and Vespoidea (ants and hornets) superfamilies around 150 million years ago [46]. In honeybees (Apoidea), the genes are arranged sequentially in a high GC region, a feature thought to be linked to increased gene expression [45]. In contrast, the silk genes in the Indian jumping ant are more dispersed through the genome and interrupted by non-silk genes [48]. Despite these differences in genomic arrangement, all species retain a single copy of each of the four homologous genes, implying an essential role for each protein in silk production. Quantitative PCR has shown that the four silk genes are expressed in roughly equimolar amounts within the silk glands of bumblebees, weaver ants, bulldog ants, and honeybees [46], suggesting that maintaining stoichiometric balance among these proteins may be critical for the proper formation of the coiled coil structure. The silk proteins exhibit low primary sequence similarity, averaging only 20–30% in the central coiled coil domain, with the N- and C-terminal regions being more variable [39, 46, 48].

Protein Structure

Ample evidence indicates that the larval silk proteins from bees, ants, and hornet species adopt a coiled coil structure as proposed by Atkins [1] (section "The coiled coil structure of bee silk"). Experimentally, the structure has been confirmed multiple times in further X-ray diffraction, FTIR, and solid-state NMR studies from a number of species [6, 19, 20, 22, 46]. The importance of all four proteins in forming the structure was highlighted in a study on silk proteins from the giant Asian honeybee (*A. dorsata*): circular dichroism (CD) analysis of solutions of the four recombinant silk proteins readily formed stable coiled coil assemblies, whereas solutions of the individual proteins required much higher concentrations to associate into coiled coils [27].

The four predicted silk proteins share key structural similarities. Each protein is small, ranging from 29 to 50 kDa [5, 48], with a central alanine-rich region of about 210 residues consisting of continuous repeating heptads of amino acid character HPPHPPP (H: hydrophobic; P: polar). Analysis using bioinformatic prediction programs [39, 46] and, more recently, the protein structural prediction program, AlphaFold [17], predicts that this central region forms the coiled coil structure. The model of the four honeybee silk proteins (F1-F4) predicted by AlphaFold was experimentally validated using crosslinking mass spectrometry and medium-resolution cryo-electron microscopy.

The proteins are orientated with F1 and F2 running antiparallel to F3 and F4, arranged with the canonical 'knobs-into-holes' packing that is common to the hydrophobic core of many coiled coils [17]. Whilst both the radius and the pitch of the coiled coil in the validated AlphaFold model are larger than proposed by Atkins [1], Atkins did concede that his proposed values could be increased whilst still maintaining good agreement with the X-ray diffraction data. The radius is smaller than typical coiled coils, which has been attributed to the high level of alanine in the core positions. The high levels of alanine in the predicted core positions of the coiled coil are unusual: coiled coils most commonly contain large hydrophobic residues at these positions to maximise the hydrophobic forces stabilising the core [57]. It has been proposed that the atypical composition of this core has arisen as a consequence of metabolic constraints: the silks are produced in large quantities over a short time period and are not recycled, so insects are constrained to using nonessential amino acids [46]. Of residues able to form α-helices, the most hydrophobic of the nonessential amino acids for bees is alanine [7]. The length and continuity of the coiled coil region, combined with a structure comprised of four proteins, allows use of alanine over larger, more hydrophobic but essential amino acids.

Analysis of protein-protein interactions between the silk proteins using blue-native PAGE and circular dichroism spectroscopy has indicated that coiled coil interactions between specific proteins drive the assembly process [17]. Initially, F2 and F4 interact to form a dimeric coiled coil, F3 then binds to this complex, and finally F1 binds to form a highly stable coiled coil structure. It is speculated that each intermediate complex is stabilised by selective heptad complementarity and electrostatic interactions, which together contribute to the accurate and efficient assembly of the final heterotetrameric complex.

Outside the central coiled coil domain, the length of the terminal regions differs across species; for instance, the terminal regions in hornets are notably longer than those in Apoidea species [19, 22, 39]. These regions are generally considered to form β-sheets in the final material, with all the aforementioned structural studies observing significant levels of this structure in native silk materials. There is little or no sequence homology between homologs in these regions, and AlphaFold was unable to resolve a structure for them. It has been proposed that, as the β-sheet cross-linking in this material does not require sequence-specific protein interactions, weaker coevolutionary signal between the proteins would result in low confidence structural predictions using AlphaFold for these regions [17].

Supramolecular Structure

General consensus is that the tetrameric coiled coil structures serve as core building blocks, underpinning the supramolecular organisation and assembly of silk fibres. The silk gland contains birefringent cigar-shaped bodies a few microns long [8, 42]. These structures break down to 250–500 $\times$ 10 nm fibrils at low potassium chloride, then smaller fibrils of 250–500 $\times$ 2 nm at higher potassium chloride concentrations. Models based on X-ray diffraction data, proposed by Fraser and Parry [10], suggest that tetrameric coiled coil subunits may organise into a 'four-rope' model, where four coiled coil tetramers twist together to form a tightly packed filament of around 4–5 nm. This model aligns with electron microscopy images of the size of the intermediate filament structures from honeybee silk gland, and these could come apart to form the smaller fibrils observed by electron microscopy [8].

The coiled coils remain aligned through the spinning process with studies on native aculeate silks, including honeybee, weaver ant, and hornet silks, showing consistent alignment of coiled coil subunits parallel to the fibre axis in both natural silk materials and in artificially produced honeybee silk fibres, through techniques including X-ray diffraction analysis [1, 6, 19, 38] and Raman spectroscopy [30]. In addition to coiled coil alignment, which may involve salt bridges, ionic interactions, and hydrogen bonding, there are a number of other cross-links that stabilise the final silk material, with some variation between species. β-sheet crosslinks are notably present within the N- and C-terminal regions of certain silk proteins, such as those in hornet silk [19]. In contrast, bee silk remains predominantly α-helical, with significantly less β-sheet in its terminal regions [1, 19]. Iso-peptide bonds between glutamine and lysine residues are present in bee and ant silks but not hornet silks [5]. Lastly, tanning reactions in bee silk, such as those induced by laccase enzymes, further modify the silk material by forming additional covalent cross-links [5, 6]. This process is visible as the silk transitions from white to brown. Together, these cross-linking mechanisms provide a robust structural network that stabilises the protein assemblies, enabling aculeate silks to achieve their unique mechanical properties.

The Future of Aculeate Coiled Coil Silk Research

Mimicking Natural Silk

Mimicking natural silks has attracted attention both for an interest in replicating the natural production process and producing materials with the extraordinary properties found in these natural materials. The two interests—production process and biophysical properties—are interlinked with the natural process having evolved to optimise the structure-property-function relationship of the silk proteins. To date, the properties of artificial silks have not replicated those of native silks, and this is generally attributed to an inability to reproduce the complexity of the natural production process.

Mimicking the Materials Fabrication Process

All natural silk production processes occur from aqueous solutions at ambient temperatures and pressures and chemically benign conditions. As materials fabrication processes go, these are the gold standard in terms of sustainability, but have yet to be achieved synthetically. Successful replication of these processes in artificial systems would allow a sustainable supply of consistent quality materials, a feature that is not available with other biological materials or many synthetics. However, to date, efforts that replicate the pH, ion environment, protein concentration, and/or spinneret geometries of silkworm and spider silk glands have not been able to generate material with native silk properties.

A key feature common to all types of native silk production is the organisation of the silk proteins into liquid crystals (supramolecular assemblies that flow like a liquid) in the silk gland [53]. This hierarchical structure and liquid crystalline behaviour are critical in allowing flow of the silk protein solution along and out of the silk gland.

What is particularly interesting about the bee, ant, and hornet silks is that the protein structures believed to underpin the supramolecular assemblies in the silk gland are retained through the spinning process and in the final material (see section "Supramolecular structure"). Silkworm and spider silk production is entirely different: the silk proteins undergo multiple structural transformations as they move through the silk gland, which involves a complete transformation of their secondary structure (for review of the silkworm process, see [3]). Mimicking the spider or silkworm silk process will require understanding and replicating each of these transitions, and it is likely that, once understood, they will be challenging to scale from an engineering perspective. In contrast, if conditions, whether native-like or not, that allow self-assembly of supramolecular structures with the coiled coil silk proteins can be determined, then it is likely that the fabrication of materials with native-like properties will quickly follow. Being able to disconnect the mechanisms for the formation of supramolecular structures from the native process that leads to the same structures allows design of processes that are amenable to existing engineering solutions and is arguably a simpler approach to what is required to mimic the production method of spider or silkworm silk.

Mimicking the Material Properties of Silk

Silks are generally strong and tough materials that are biocompatible and therefore of particular interest for biomedical applications. Many of the desired material properties of silk result from the hierarchical structure or close and ordered packing of silk proteins in the final material. In contrast to most engineered materials that do not have hierarchy, where the presence of cracks leads to severe reduction in strength, the hierarchy, coupled with high levels of hydrogen bonds in silk materials, allows the materials to tolerate structural flaws of virtually any size [4]. Silk materials are also transparent, insulating, and have useful optical properties. However, these are not reliant on the hierarchical molecular structure and have been readily mimicked in artificial systems.

Given the role that the production process plays in forming the atomic structure of the final materials, it is likely that once fabrication processes akin to native processes are developed, then the properties of the resultant material will more closely mimic those of natural materials. A massive advantage of the coiled coil silks is that the silk proteins are small and can be readily produced in recombinant systems (see section "Recombinant production of silk proteins"). A further advantage of the coiled coil silks is that the silk proteins readily self-assemble into the coiled coil structure [54] that forms the basis of the material properties of that silk. Simple recombinant production and self-assembly of the native atomic structure make these proteins considerably easier to work with than other silk systems. Consequently, it is likely that artificial versions of these silks will be achieved in the near future.

Use of Silks for Rational Design of New Materials

Our interest in the coiled coil silks stems from an interest in the rational design of stimuli-responsive or multifunctional materials. It is generally considered that polymeric materials, when controlled production of sequence-defined polymers is possible, will allow rational design of such materials. Structural proteins are encoded by a DNA sequence and therefore are sequence-defined polymers. There are many precedents for stimuli-responsive or multifunctional materials in nature: for example, natural collagen molecules are known to be capable of over 50 different molecular interactions [55]. However, structural proteins are underrepresented in material science because, whilst the sequence can be changed, the structure-function relationship of most changes is not yet predictable to an extent that allows rational design of new function into those materials.

The constraints limiting rational design of recombinant structural proteins are described in Sutherland et al. [49] and include the following:

- Proteins have limited tolerance of change without disrupting their structure or function.

- Some structural proteins, such as spider and silkworm silks, do not have a precise sequence-structure relationship.
- The sequence-structure relationship of many structural proteins is not yet sufficiently understood.

The properties of the coiled coil silks go a long way to addressing these constraints. Regarding tolerance to change, there is only very low sequence homology between the proteins from different species, indicating that the identity of the amino acids in the primary sequence is less important than the character of the amino acids. The alignments of the known silk proteins show conservation of the protein architecture with a central coiled coil region, predicted to be around 30 repeats of an HPPHPPP sequence, where H is generally a hydrophobic residue, and P is generally a polar residue [5, 46]. This central region is flanked by shorter unstructured regions and preceded by a highly conserved signal peptide. This flexibility in the required amino acid sequence indicates that the proteins are tolerant of change without disrupting their ability to form a coiled coil structure.

The extensive understanding of the aculeate silks, combined with their amenability for recombinant production, capacity to be modified, natural biocompatibility, mechanical properties, and that they fold into their final structure in solution positions them as an ideal candidate for further development into functional materials.

References

1. Atkins EDT (1967) A four-strand coiled-coil model for some insect fibrous proteins. J Mol Biol 24:139–141
2. Bosch J, Vicens N (2002) Body size as an estimator of production costs in a solitary bee. Ecol Entomol 27:129–137
3. Brookstein O, Shimoni E, Eliaz D, Dezorella N, Biran I, Rechav K, Sivan E, Kozell A, Shimanovich U (2024) The natural material evolution and stage-wise assembly of silk along the silk gland. bioRxiv:2024.04.16.589504
4. Buehler MJ, Yung YC (2010) How protein materials balance strength, robustness, and adaptability. HFSP J 4:26–40
5. Campbell PM, Trueman HE, Zhang Q, Koijima K, Kameda T, Sutherland TD (2014) Cross-linking in the silks of bees, ants and hornets. Insect Biochem Mol Biol 48:40–50
6. Crewe RM, Thompson PR (1979) Oecophylla silk: functional adaptation in a biopolymer. Sci Nat 66:57–58
7. DeGroot AP (1953) Protein and amino acid requirements of the honeybee (Apis mellifera). Physiologia Comparata et D'Ecologia 3:197–285
8. Flower NE, Kenchington W (1967) Studies on insect fibrous proteins: the larval silk of Apis, Bombus and Vespa (Hymenoptera: Aculeata). J R Microsc Soc 86:297–310
9. Fountain T, Hughes WHO (2011) Weaving resistance: silk and disease resistance in the weaver ant *Polyrhachis dives*. Insect Soc 58:453–458
10. Fraser RDB, Parry DAD (2015) The molecular structure of the silk fibres from Hymenoptera aculeata (bees, wasps, ants). J Struct Biol 192:528–538
11. Hepburn HR, Kurstjens S (1988) The combs of honeybees as composite materials. Apidologie 19:25–36

12. Hepburn HR, Chandler HD, Davidoff MR (1979) Extensometric properties of insect fibroins: the green lacewing cross-b, honeybee a-helical and greater waxmoth parallel-b conformations. Insect Biochem 9:69–77
13. Horgan CC, Han Y-S, Trueman H, Jackson CJ, Sutherland TD, Rapson TD (2016) Phosphorescent oxygen-sensing and singlet oxygen production by a biosynthetic silk. RSC Adv 6: 39530–39533
14. Huson MG, Church JS, Poole JM, Weisman S, Sriskantha S, Warden AC, Ramshaw JAM, Sutherland TD (2012) Structural and physical changes of honeybee silk materials induced by heating or by immersion in aqueous methanol solutions. PLoS One 7:e52308
15. Ishay JS, Barenholz-Paniry V (1995) Thermoelectric effect in hornet (*Vespa orientalis*) silk and thermoregulation in a hornet's nest. J Insect Physiol 41:753–759
16. Jay SC (1964) The cocoon of the honey bee, *Apis mellifera* L. Can Entomol 96:784–792
17. Johnston CL, Jobichen C, Briggs LJ, Michie M, Liu J-W, Morton CJ, Warden AC, Sutherland TD (2025) Resolving the structure and assembly of the honeybee silk heterotetrameric coiled coil. Protein Sci 34:e70230
18. Kambe Y, Sutherland TD, Kameda T (2014) Recombinant production and film properties of full-length hornet silk proteins. Acta Biomater 10:3590–3598
19. Kameda T (2014) Influence of pH, temperature and concentration on stabilization of aqueous hornet silk solution and fabrication of salt-free materials. Biopolymers 103:41–52
20. Kameda T, Tamada Y (2009) Variable-temperature ^{13}C solid-state NMR study of the molecular structure of honeybee wax and silk. Int J Biol Macromol 44:64–69
21. Kameda T, Aratani E (2012) Production and characterizations of tubes from hornet (*vespa*) silk. J Insect Biotechnol Sericol 80:109–116
22. Kameda T, Kojima K, Miyazawa M, Fujiwara S (2005) Film formation and structural characterization of silk of the hornet *Vespa simillima* xanthoptera Cameron. Z Naturforsch C J Biosci 60:906–914
23. Kameda T, Kojima K, Togawa E, Sezutsu H, Zhang Q, Teramoto H, Tamada Y (2010) Drawing-induced changes in morphology and mechanical properties of hornet silk gel films. Biomacromolecules 11:1009–1018
24. Kojić N, Bico J, Clasen C, McKinley GH (2006) Ex vivo rheology of spider silk. J Exp Biol 209:4355–4362
25. Kumar M, Jain D, Bhardwaj N, Gupta P, Nandi SK, Mandal BB (2016) Native honeybee silk membrane: a potential matrix for tissue engineering and regenerative medicine. RSC Adv 6: 54394–54403
26. Lucas F, Rudall KM (1968) Extracellular fibrous proteins: the silks. In: Florkin M, Stotz EH (eds) Comprehensive biochemistry, vol 26. Elsevier, Amsterdam, pp 475–558
27. Maitip J, Trueman HE, Kaehler BD, Huttley GA, Chantawannakui P, Sutherland TD (2015) Folding behaviour of four silks of giant honeybee reflects the evolutionary conservation of aculeate silk proteins. Insect Biochem Mol Biol 48:40–50
28. Musameh MM, Dunn CJ, Uddin MH, Sutherland TD, Rapson TD (2018) Silk provides a new avenue for third generation biosensors: sensitive, stable and selective electrochemical detection of nitric oxide. Biosens Bioelectron 103:26–31
29. Nelson JA (1924) Morphology of the honeybee larvae. J Agric Res 18:1167–1213
30. Poole JM, Church JS, Woodhead AL, Huson MG, Sriskantha A, Kyratzis IL, Sutherland TD (2013) Continuous production of flexible fibers from transgenically-produced honeybee silk proteins. Macromol Biosci 13:1321–1326
31. Rapson TD, Church JS, Trueman HE, Hacres H, Sutherland TD, Trowell SC (2014) Micromolar biosensing of nitric oxide using myoglobin immobilised in a synthetic silk film. Biosens Bioelectron 62:214–220
32. Rapson TD, Sutherland TD, Church JS, Trueman HE, Dacres H, Trowell SC (2015) *De novo* engineering of solid-state metalloproteins using recombinant coiled-coil silk. ACS Biomater Sci Eng 1:1114–1120

33. Rapson TD, Liu JW, Sriskantha A, Musameh M, Dunn CJ, Church JS, Woodhead A, Warden A, Riley MJ, Harmer JR, Noble CJ, Sutherland TD (2017a) Design of silk proteins with increased heme binding capacity and fabrication of silk-heme materials. J Inorg Biochem 117:219–227
34. Rapson TD, Kusuoka R, Butcher J, Musameh M, Dunn CJ, Church JS, Warden A, Blanford CF, Nakamura N, Sutherland TD (2017b) Bioinspired electrocatalysts for oxygen reduction using recombinant silk films. J Mat Chem A 5:10236–10243
35. Rapson TD, Hall GL, Sutherland TD (2019) Could home based FeNO measurements breathe new life into asthma management? J Asthma 56:910–913
36. Rapson TD, Christley-Balcomb AM, Jackson CJ, Sutherland TD (2020a) Enhancement of metallomacrocycle-based oxygen reduction catalysis through immobilization in a tunable silk-protein scaffold. J Inorg Biochem 204:110960
37. Rapson TD, Ju H, Paul Marshall P, Devilla R, Jackson CJ, Giddey S, Sutherland TD (2020b) Engineering a solid-state metalloprotein hydrogen evolution catalyst. Sci Rep 10:3774
38. Rudall KM (1962) Silk and other cocoon proteins. In: Florkin M, Mason HS (eds) Comparative biochemistry, vol IV. Academic, New York, pp 397–433
39. Sezutsu H, Kajiwara H, Kojima K, Mita K, Tamura T, Tamada Y, Kameda T (2007) Identification of four major hornet silk genes with a complex of alanine-rich and serine-rich sequences in *Vespa simillima* xanthoptera Cameron. Biosci Biotechnol Biochem 71:2725–2734
40. Shabtai Y, Ishay J (1998) Hornet silk caps maintain a clean room environment: a device for filtering out bacteria and dust particles. Comp Biochem Physiol A 120:565–570
41. Shi J, Lua S, Du N, Liu X, Song J (2008) Identification, recombinant production and structural characterization of four silk proteins from the Asiatic honeybee *Apis cerana*. Biomaterials 29: 2820–2828
42. Silva-Zacarin ECM, Silva de Moraes RLM, Taboga SR (2004) Silk formation mechanisms in the larval salivary glands of *Apis mellifera* (Hymenoptera: Apidae). J Biosci 28:753–764
43. Siri S, Maensiri S (2010) Alternative biomaterials: natural, non-woven, fibroin-based silk nanofibers of weaver ants (*Oecophylla smaragdina*). Int J Biol Macromol 46:529–534
44. Starr CK (1985) Enabling mechanisms in the origin of sociality in the hymenoptera—the stings the thing. Ann Entomol Soc Am 78:836–840
45. Sutherland TD, Campbell PM, Weisman S, Trueman HE, Sriskantha A, Wanjura WJ, Haritos VS (2006) A highly divergent gene cluster in honey bees encodes a novel silk family. Gen Res 16:1414–1421
46. Sutherland TD, Weisman S, Trueman HE, Sriskantha A, Trueman JW, Haritos VS (2007) Conservation of essential design features in coiled coil silks. Mol Biol Evol 24:2424–2432
47. Sutherland TD, Young J, Weisman S, Hayashi CY, Merrit D (2010) Insect silk: one name, many materials. Annu Rev Entomol 55:171–188
48. Sutherland TD, Weisman S, Walker AA, Mudie ST (2011) The coiled coil silk of bees, ants and hornets. Biopolymers 97:446–454
49. Sutherland TD, Rapson TD, Huson MG, Church JS (2017) Recombinant structural proteins and their use in future materials. In: Parry DAD, Squire JM (eds) Fibrous proteins: structures and mechanisms, vol 82. Springer International Publishing, pp 491–526
50. Sutherland TD, Sriskantha A, Rapson TD, Kaehler BD, Huttley GA (2018) Did aculeate silk evolve as an antifouling material? PLoS One 13:e0203948
51. Tatlisulu S, Ozgor E, Kavaz D, Djamgoz MBA (2024) Honeybee silk and chitosan: a promising biocomposite for wound healing applications. J Med Biol Eng. https://doi.org/10.1007/s40846-024-00853-z
52. Trueman HE, Sriskantha A, Qu Y, Rapson TD, Sutherland TD (2017) Modification of honeybee silk by addition of antimicrobial agents. Omega 2:4456–4463
53. Walker AA, Holland C, Sutherland TD (2015) More than one way to spin a crystallite: multiple trajectories through liquid crystallinity to solid silk. Proc R Soc B 282:20150259

54. Weisman S, Haritos VS, Church JS, Huson MG, Mudie ST, Rodgers AJW, Dumsday GJ, Sutherland TD (2010) Honeybee silk: recombinant protein production, assembly and fiber spinning. Biomaterials 31:2695–2700
55. Werkmeister JA, Ramshaw JAM (2012) Recombinant protein scaffolds for tissue engineering. Biomed Mater 7:012002
56. Wittmer CR, Hu X, Gauthier P-C, Weisman S, Kaplan DL, Sutherland TD (2011) Production, structure and *in vitro* degradation of electrospun honeybee silk nanofibres. Acta Biomater 7: 3789–3795
57. Woolfson DN (2005) The design of coiled coil structures and assemblies. In: Parry DAD, Squire JM (eds) Fibrous proteins: coiled-coils, collagen and elastomers. Elsevier, San Diego, pp 79–112
58. Zhang K, Duan H, Karihaloo BL, Wang J (2010a) Hierarchical, multilayered cell walls reinforced by recycled silk cocoons enhance the structural integrity of honeybee combs. PNAS 107:9502–9506
59. Zhang K, Si FW, Duan HL, Wang J (2010b) Microstructures and mechanical properties of silks of silkworm and honeybee. Acta Biomater 6:2165–2171

Chapter 15
Designability of α-Helical Protein Filaments

Abhinaba Das and Vincent Conticello

Abstract Helix-helix interactions are mediated through highly designable interfaces within tertiary and quaternary structures of proteins and protein assemblies. The structural regularity of these interfaces suggests that ordered self-assembled structures could be constructed from implementation of these interactions between appropriately designed helical protomers. This review summarizes the current understanding of helix-helix interactions within different classes of naturally occurring α-helical protein filaments. The implications of this structural information for the de novo design of synthetic filamentous nanomaterials will be discussed with reference to examples in which these principles have been successfully implemented. A specific case study will focus on the designability of cross-α helical filaments, a recently discovered structural class in which the helical protomers are arranged in a perpendicular orientation with respect to the protofilament axis. This discussion will include an evaluation of the frequency of occurrence of cross-α interfaces in the PDB, the effectiveness of structural prediction from sequence information, and the potential for de novo design of interfaces that promote cross-α interactions.

Keywords Self-assembly · Helical interfaces · Designability · Protein · Peptide · Coiled-coil · Cross-α filament

Introduction: Helical Interfaces as Designable Structural Elements

The α-helix is the most common secondary structure element, occurring at a frequency of about 33% within the sequences of structurally characterized proteins in the Protein Data Bank (PDB) [6, 42]. The periodic arrangement of amino acid side chains along the contour length of the helix promotes formation of regular packing

A. Das · V. Conticello (✉)
Department of Chemistry, Emory University, Atlanta, Georgia
e-mail: vcontic@emory.edu

D. A. D. Parry, P. K. Luther (eds.), *Fibrous Proteins: Structures and Mechanisms*, Subcellular Biochemistry 113, https://doi.org/10.1007/978-3-032-05273-5_15

interfaces that are commensurate with its underlying helical symmetry (twist ~100° and rise ~1.5 Å) [60, 61]. Statistical [89, 103] and geometrical [90] analyses of helix-helix packing interfaces have identified inter-helix crossing angles that promote regular knobs-into-holes (KIH) [19, 55] packing interactions between helices. The conservation of these packing interfaces between soluble and transmembrane helical bundles [103] suggests that these interfaces are richly designable and mediate inter-helical interactions within a diverse range of protein sequences and structures. Designability in the context of protein structural prediction and de novo design refers to robustness of a protein fold in sequence space [31, 46]. One measure of designability is the frequency of occurrence of a structural motif within the PDB. Designable protein motifs have higher levels of representation within the PDB, while non-designable ones are sparsely represented or absent [49, 106]. The high level of representation of regular helix-helix packing interfaces within high-resolution protein structures in the PDB is consistent with rich native designability [43].

The most observed helical interaction motifs are based on left-handed coiled-coil [18, 19] heptad repeats (ZxxZxxx, Z = large hydrophobic residue) [77, 80] and right-handed glycine zipper-like tetrad repeats (Zxxx, Z = small side-chain residue) [39, 66, 71]. The respective repeat-motifs display different inter-helix crossing angles (~ +20° versus ~ –40°, respectively) and distinct residue preferences (large (e.g., Val, Leu, Ile, Phe) versus small (e.g., Gly, Ala, Ser), respectively) at the helix packing interfaces. Given the regularity of these packing interfaces, helical proteins lend themselves well to parametric design [32, 34, 54, 84], especially if the corresponding tertiary or quaternary structures display a high degree of symmetry. In particular, the coiled-coil motif has played a critical role in understanding the relationship between sequence and structure in de novo protein design [97]. The high designability and thermodynamic stability of coiled-coils has made them an attractive scaffold for the rational design of novel catalysts [13], transporters [37, 70], protein-protein interaction inhibitors [25], and biomaterials [3], as well as being a convenient proxy to structurally explore the sparsely sampled "dark matter" of the protein universe [33, 96]. While glycine zipper motifs have been employed to a lesser extent in protein design, peptides based on this motif have been designed to act as binders to membrane proteins [53, 101] and mediate interactions in self-assembled protein-based materials [45, 85, 104].

The high thermodynamic stability and native designability of helix-helix packing interfaces suggests that propagation of these local interactions in an extended assembly would represent a potentially effective method for the reliable design of filamentous nanomaterials. Regular helical packing interfaces are commonly observed between protomers within naturally occurring protein filaments [3, 26, 51, 110]. A number of filamentous α-helical nanostructures have been characterized at near-atomic resolution in the native state (vide infra). A range of helix-helix interfaces have been observed within these filaments, which suggests that long-range order can be designed into filamentous protein assemblies using helical protomers of diverse tertiary structure. The most easily distinguishable structural variation between different classes of biologically derived assemblies corresponds to the relative orientation of the helical protomers with respect to the long axis of the filament. Helical protomers in filaments commonly display shape anisotropy as a consequence of their extended periodic structure [14]. The helix-helix interface is usually oriented at a right angle to the long (helical) axis of a protomer. However, depending on the packing geometry, individual helical protomers, and, consequently, the helix-helix interfaces that arise from interactions between them, can display either a parallel, skew (tilted), or perpendicular orientation with respect to the propagation axis of a filament or protofilament. While it may be more appropriate to consider the arrangement of the protomers as representing a continuum of orientations, different classes of filamentous nanomaterials often display preferences for specific packing arrangements depending upon the biological function of the assembly (e.g., mechanical support, binding, transport, storage, etc.).

Longitudinal α-Helical Protein Filaments

The most common helix-helix interaction observed in filamentous α-helical proteins derives from parallel or antiparallel packing of coiled-coil structural motifs into a longitudinal oligomers or polymers. Historically, coiled-coil packing interfaces were first identified from X-ray fiber diffraction analyses of oriented fibers of k-m-e-f (keratin-myosin-epidermin-fibrinogen) family proteins and α-helical silks [78]. Subsequent crystallographic analyses confirmed the presence of coiled-coil packing interfaces at atomic-level resolution in discrete protein-protein interaction modules, e.g., the leucine zipper motif of the GCN4 transcription factor [55], as well as in extended protein filaments (Fig. 15.1) [3]. In most of these structures, the long axis of the helical chains is aligned in a nearly parallel orientation to the central axis of the filament, which places the helix-helix interfaces in an orthogonal orientation to the fiber. Therefore, the helix-helix interface can wind along the entire contour length of the filament. Individual protein chains can be quite long with the coiled-coil domain extending over nearly the entire sequence, e.g., as in dimeric coiled-coil structure of porcine tropomyosin (PDB: 1C1G), which spans 284 amino acids and 420 Å in length [95]. The periodicity of the filamentous structures can be disrupted through insertions that result in local structural distortion without perturbing the modularity of the coiled-coil segments (Fig. 15.2) [48]. The formation of higher-order structure is mediated through specific packing interactions that promote lateral and longitudinal assembly between coiled-coil protomers, which results in the formation of meso- to microscale fibrous protein materials.

The sequence-structure correlations for the coiled-coil motif are well understood and can be applied to the design of synthetic peptides that can self-assemble under controllable conditions into filamentous nanomaterials that mimic the corresponding native protein fibers. However, a critical challenge to this process has been the design of protomers that can align in a specific registry that reinforces the formation of filaments having a defined supramolecular architecture. Native α-helical assemblies, e.g., keratin intermediate filaments [17], adopt the correct registry through the

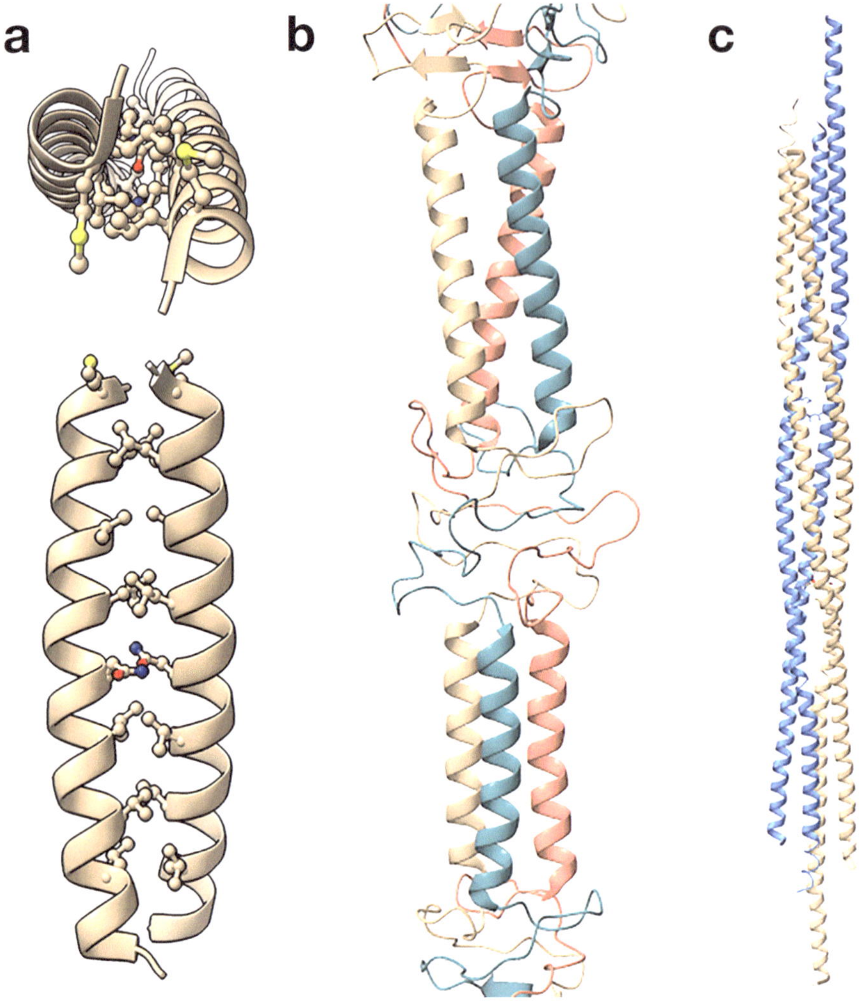

Fig. 15.1 Representative oligomeric coiled-coil structures. (**a**) Side and top view of the structure of the coiled-coil domain of GCN4 (PDB: 2ZTA) with residues at the hydrophobic (a/d) interface rendered in the ball and stick depiction [55]. (**b**) Structural segment of the trimeric autotransporter adhesin AtaA (PDB: 3WPA) from *Acinetobacter* sp. strain Tol 5 indicating the modularity of different coiled-coil domains within the filament [40]. The AtaA monomer comprises 3630 amino acids and contains many coiled-coil domains that are disrupted by intervening sequences that display distinct structures in the trimer. (**c**) Antiparallel coiled-coil tetramer corresponding to a sub-structure of the bipolar assembly (BASS) domain of kinesin-5 (PDB: 4PXU) [69]

influence of structurally unique N- and C-terminal domains that promote alignment between the coiled-coil domains in the proteins and mediate higher-order assembly of the structural subunits into mesoscale fibers. However, the structural role of these terminal domains is complex and incompletely understood from the perspective of

Fig. 15.2 Representative interactions between coiled-coil domains promote the formation of higher-order structures. (**a**) Dimeric coiled-coils in the crystal structure of porcine tropomyosin (PDB: 1C1G) [95]. The two coiled-coils in the asymmetric unit display an antiparallel orientation, which may be representative of the packing in the tropomyosin filament. (**b**) Dimer of tetramers packed within the crystal structure of intermediate filament (IF) protein vimentin 1B (PDB: 5WHF) [58]. The lateral interaction between tetramers (dimers of dimers) is observed in the crystal structure as well as head-to-tail stacking of tetramers. (**c**) Cryo-EM reconstruction of vimentin intermediation filaments (PDB: 8RVE) [28] indicating the presence of five protofibrils (single protofibril highlighted in red) with forty subunits in cross-sections. The low complexity head domains localize to the lumen of the assembling mediating interactions between protofibrils

de novo design. Therefore, the engineering of similar interactions would be challenging to implement in designed peptide materials. Despite this difficulty, synthetic oligopeptides, consisting of 3–6 heptad repeats, can be designed that self-associate in the appropriate registry to form quasi-infinite coiled-coil polymers having

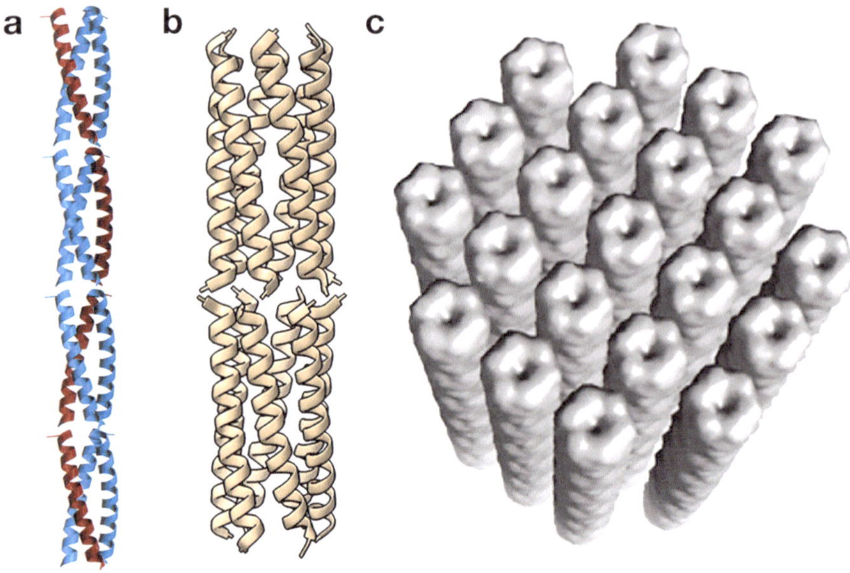

Fig. 15.3 Designed coiled-coil filaments based on stacking of helical protomers. (**a**) Crystal structure of a designed trimeric coiled-coil (PDB: 1COI) in which the protomers stack head-to-tail in the crystal lattice. The stacked trimers maintain a continuous superhelix in the structure. An individual protofilament is highlighted in red. The pitch of the 3-start helix (121.6 Å) corresponds to 12 heptad repeats or three coiled-coil stacks of 4 heptads. (Reprinted with permission from Ogihara et al. [56] Copyright (1997) Wiley). (**b**) Stacked coiled-coil hexamers in the crystal structure of CC-Hex (PDB: 3R3K) [102] (**c**) Structural model of para-crystalline filaments derived from self-assembly of CC-Hex-T, a derivative of CC-Hex. The model is based on in-register stacking of hexameric coiled-coil. The predicted pitch of the 6-start helices (25.8 nm), corresponding to the stacking of six hexamers, was closely related to the corresponding value predicted from crystallographic analysis of CC-Hex (25.9 nm). (Reprinted with permission from Burgess et al. [12] Copyright (2015) American Chemical Society)

mega-Dalton mass [3]. The latter process can involve blunt-ended [12] or lock-washer [99] stacking of helical bundles into extended filaments. Such axial (end-to-end) stacking of helices into "filament"-like extended arrays has been observed in the crystallographic analysis of discrete coiled-coil oligomers (Fig. 15.3) [56, 102]. This structural principle has been employed to guide the design of novel filamentous nanomaterials and assist in the structural analysis. An alternative approach has focused on the design of sequences that associate into filaments through sticky-ended self-assembly [24, 50, 57, 59, 63, 72, 75, 111]. The protomer sequences are designed to align in a staggered orientation in which adjacent peptides are offset along their length by an integral number of heptad repeats. An atomic model built into a cryo-EM density map at medium resolution confirmed the proposed packing arrangement for a filament derived from self-assembly of a heterodimeric pair of coiled-coils (Fig. 15.4) [72]. Molecular design principles have been implemented based on this model, which have resulted in synthetic

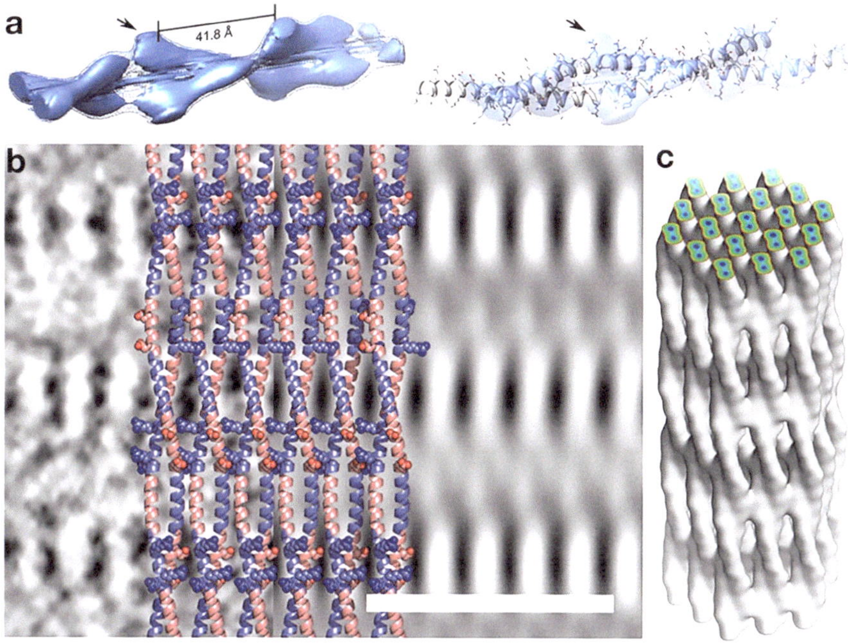

Fig. 15.4 Designed coiled-coil filaments based on staggered (sticky-ended) alignment of helical protomers. (**a**) Cryo-EM density map corresponding to a hypothetical sticky-ended coiled-coil from reconstruction of the SAF-p1/SAF-p2a heterodimeric assembly (left). Fit of the atomic model of the coiled-coil heterodimer into the density indicating the position of a structurally critical salt bridge between Asp and Arg residues. (**b**) Atomic model of the SAF-p1/SAF-p2a heterodimeric filament fit into the densities of the experimental (left) and processed cryo-EM images. The putative interactions between Asp and Arg residues are depicted. For each peptide, the N-terminal two heptads and C-terminal two heptads are highlighted in the respective ribbon diagrams in blue and pink. The two peptides interact in parallel through a half-length stagger along the direction of the central axis of the coiled-coil. (**c**) All-atom model of the SAF-p1/SAF-p2a filament filtered to 10 Å to match the resolution of the cryo-EM density map (EMD: 1995). (Reprinted with permission from Sharp et al. [72] Copyright (2012) National Academy of Sciences)

peptide assemblies displaying a range of materials behavior from elastic hydrogels [2] to mechanically rigid para-crystalline filaments [59].

Tilted α-Helical Protein Filaments

Skew helix-helix packing interfaces mediate noncovalent interactions between protomers in the structures of a range of biological-derived assemblies, including filamentous bacteriophage [81, 92, 100] and extracellular filamentous organelles in bacteria and archaea (Fig. 15.3) [4, 16, 22, 47, 107]. These assemblies encompass a wide range of functions, including controlled delivery and release, macromolecular transport, adhesion, and locomotion. While the structures of a number of these filamentous assemblies have been determined at near-atomic resolution, primarily by cryo-EM helical reconstruction [93], the relationship between sequence and supramolecular structure for these assemblies is not understood to the same extent as for coiled-coil filaments. Since the optimal geometries of the helix-helix packing interfaces are structurally conserved, one would expect that synthetic analogues of these assemblies would be amenable to de novo design, especially given the relatively small size of the protomers. However, several factors complicate the application of computational methods to the de novo design of synthetic filaments that mimic these native assemblies. Unlike coiled-coils, which self-assemble readily in aqueous buffers, the assembly of helical pili, fimbriae, and phage usually involves the action of dedicated secretion systems that process the membrane-bound precursors into the corresponding filaments [15]. The precursor proteins often comprise hydrophobic sequences that bury significant solvent accessible surface in the assembled state and, consequently, display a pronounced tendency toward aggregation upon in vitro reconstitution in aqueous buffers. The protein precursors often co-assemble with other components, such as lipid and nucleic acids, that are retained in the filament post-assembly and may be responsible, in part, for its structural integrity as well as its biological function. In comparison to coiled-coil filaments, the skew (tilted) arrangement of the helices with respect to the central axis of the filament places the protomers in the appropriate orientation for performance of their native function, e.g., DNA binding into or transport through the lumen of the filament for bacteriophage and conjugative pili, respectively. In addition, the individual helices within protomers are often bent or curved rather than straight and can contain globular domains of different lengths and structural complexity. These considerations further complicate efforts to redesign the native filaments (Fig. 15.5).

While at an early stage in the process, rational design has been employed to generate synthetic filaments that mimic the skew arrangement of protomers within extracellular filamentous organelles and bacteriophage. This approach has been based on the use of stable, folded structural motifs as protein building blocks ("tectons") for the creation of filamentous nanostructures. The underlying protein structure of the protomer provides a stable scaffold that enables redesign of surface residues to optimize inter-protomer interactions. Self-association of appropriately designed interfaces drives self-assembly into a helical filament through structurally complementary interactions between protomers. This approach has been employed to design helical filaments from a range of structural motifs, including helical hairpins [35], helical bundles [74], and tandem repeat proteins [7, 73] (Fig. 15.6). However, the practical necessity to achieve a balance between filament stability, usually interpreted in terms of buried solvent accessible surface area at the inter-protomer interfaces [5, 62], and the solubility of the protomers has complicated these design efforts. Often, this process necessitates the introduction of hydrophobic residues at surface positions on the protomers, which can result in folding frustration and structural polymorphism, as is commonly observed in structures of amyloid filaments [68]. Several researchers have successfully implemented alternative mechanisms to promote more selective formation of interfacial interactions between

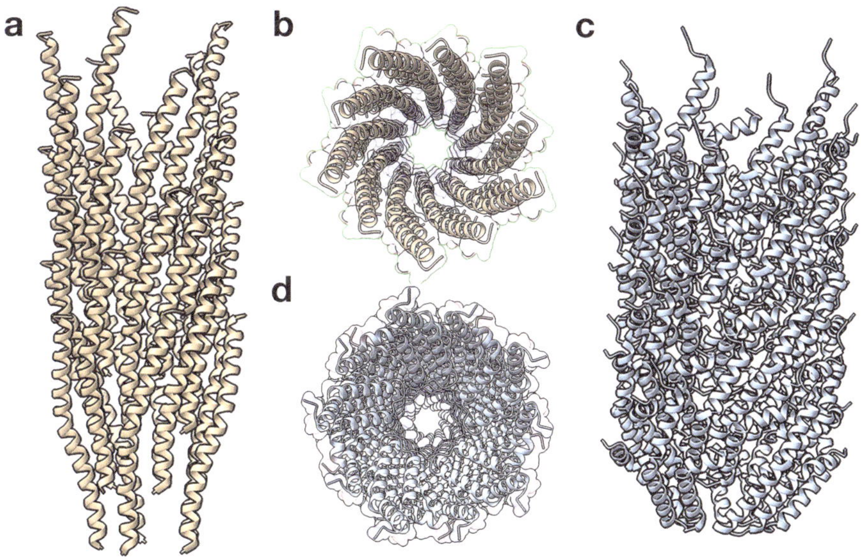

Fig. 15.5 Representative naturally occurring filamentous assemblies derived from packing of helical protomers in a tilted (skew) orientation. (**a, b**) Side and top views of the atomic model derived from cryo-EM helical reconstruction of the filamentous bacteriophage Ike capsid (PDB: 6A7F) [100]. (**c, d**) Side and top views of the conjugative Ti pilus from *Agrobacterium tumefaciens* (PDB: 9EKC) [4]

helical protomers. Residues have been designed into specific surface sites that respond to environmental triggers, e.g., a pH change [74] or the presence of specific metal ions [8, 9], to afford greater control over the self-assembly process and the structure of the resultant filament.

Cross-α Helical Protein Filaments

Recent crystallographic [67, 82, 83] and cryo-EM [41] structural analyses of filamentous peptide assemblies have revealed that α-helical protomers can stack in a perpendicular orientation with respect to the long axis of a protein filament or protofilament. This structural arrangement has been designated as a cross-α assembly in analogy to the cross-β architecture observed in the structures of amyloid fibrils [68]. The cross-α architecture was first described in a crystallographic analysis of the *S. aureus* peptide virulence factor PSMα3 (Fig. 15.7) [82]. Unlike cross-β amyloids, in which inter-strand hydrogen-bonding mediates longitudinal association of the protomers, the stability of the cross-α axial stacking interfaces depends on hydrophobic interactions between solvent-exposed residues. The cross-α stacking interaction necessitates that the peptide protomers adopt an amphipathic conformation in which the hydrophobic residues are arranged on the same helical face. The

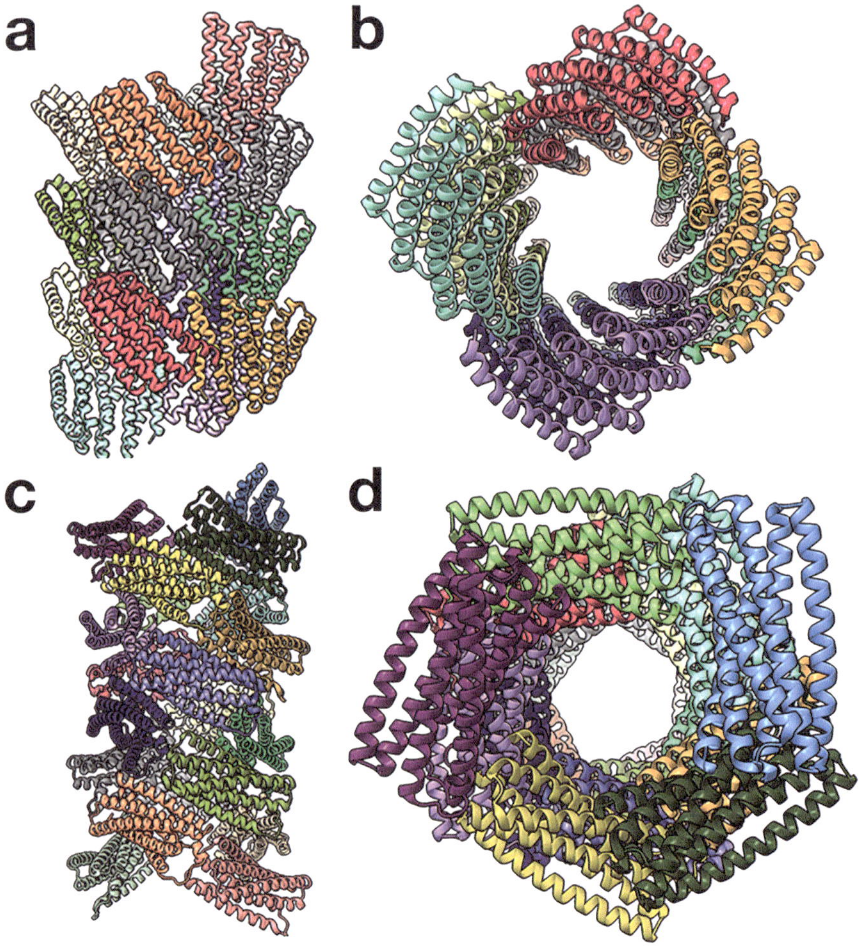

Fig. 15.6 Representative designed helical filaments derived from folded helical protomers. (**a, b**) Side and top views of DHF38 (PDB: 6E9Y) [73], a computationally designed filament derived from self-assembly of a helical repeat protein. (**c, d**) Side and top views of DpH19 (PDB: 8UBG) [74], a computationally designed filament derived from self-assembly of a four-helix bundle protein

hydrophobic face of PSMα3, corresponding to the core structural residues (2–19) of the 22 amino acid peptide, subtends an angle of ~200° with respect to the central axis of the helix, which should be sufficient to mediate bilateral stacking at the axial interface. Cryo-EM helical reconstruction [41] of two different nanotubes derived from solution self-assembly of PSMα3 indicated that the stacking interactions between helical interfaces that were observed in the corresponding crystal structure were maintained within protofilaments in the atomic model of the respective nanotubes. In addition, cryo-EM structural analysis of a nanotube derived from self-assembly of PSMβ2, a 44 amino acid peptide that adopted a helix-turn-helix conformation, demonstrated that the protomers were arranged in a structurally

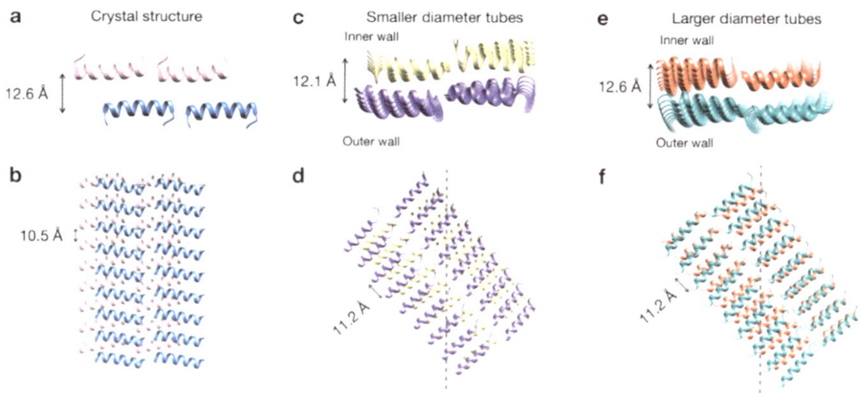

Fig. 15.7 Structures of cross-α helical filaments derived from the PSMα3 peptide. (**a, b**) Top and side views of the peptide packing in the crystal structure of (PDB: 5I55) [82]. (**c, d**) Top and side views of peptide packing in a cross-α protofilament in the smaller diameter PSMα3 nanotube (PDB: 7SZZ). (**e, f**) Top and side views of peptide packing in a cross-α protofilament in the larger diameter PSMα3 nanotube (PDB: 7T0X). In the atomic models of the PSMα3 nanotubes, the peptides are arranged in a perpendicular orientation to the axis of the cross-a protofilament but are oriented at an angle with respect to the central axis of the nanotube. (Reproduced with permission from Kreutzberger et al. [41] Copyright (2022) National Academies of Science)

similar cross-α orientation. Hydrophobic interactions between amphipathic protomers mediated axial association within the 12 protofilaments that defined the nanotube structure. A crystallographic determination of the structure of uperin-3.5 [67], an antimicrobial peptide derived from the amphibian *Uperoleia mjobergii*, revealed that the assemblies adopted a cross-α structure. Like PSMα3, the hydrophobic face of the amphipathic peptide subtended a relatively large angle of 180° within a helical conformation, which promoted axial stacking of helices along the fibril axis.

Given the structural conservation of interfacial helical packing geometries, one would expect that the axial stacking interactions that drove assembly of these peptides into a cross-helical structure would be representative of those observed for helix-helix interactions in proteins. The high designability of helix-helix packing interfaces suggests that peptides that adopt cross-α structures should be amenable to de novo design. Structural analyses of several designed peptide assemblies have provided evidence for the formation of cross-α filaments. Zhang et al. [105] reported a series of peptides, αAm_X (X = mem, G, A, S, L, F), based on coiled-coil repeat sequences that self-assembled into helical filaments (Fig. 15.8). Single crystal X-ray diffraction analyses of these peptides revealed the presence of cross-α helical filaments within the crystal lattice that were derived from axial stacking of parallel peptide dimers in an alternating antiparallel arrangement. The filaments exhibited a left-handed helical twist between successive parallel dimeric repeat units within the structure, although helices within a protofilament, corresponding to the stacking direction of antiparallel helical pairs, displayed a right-handed screw sense. The crossing angles between pairs of helices approach the value of ~ +20° observed for

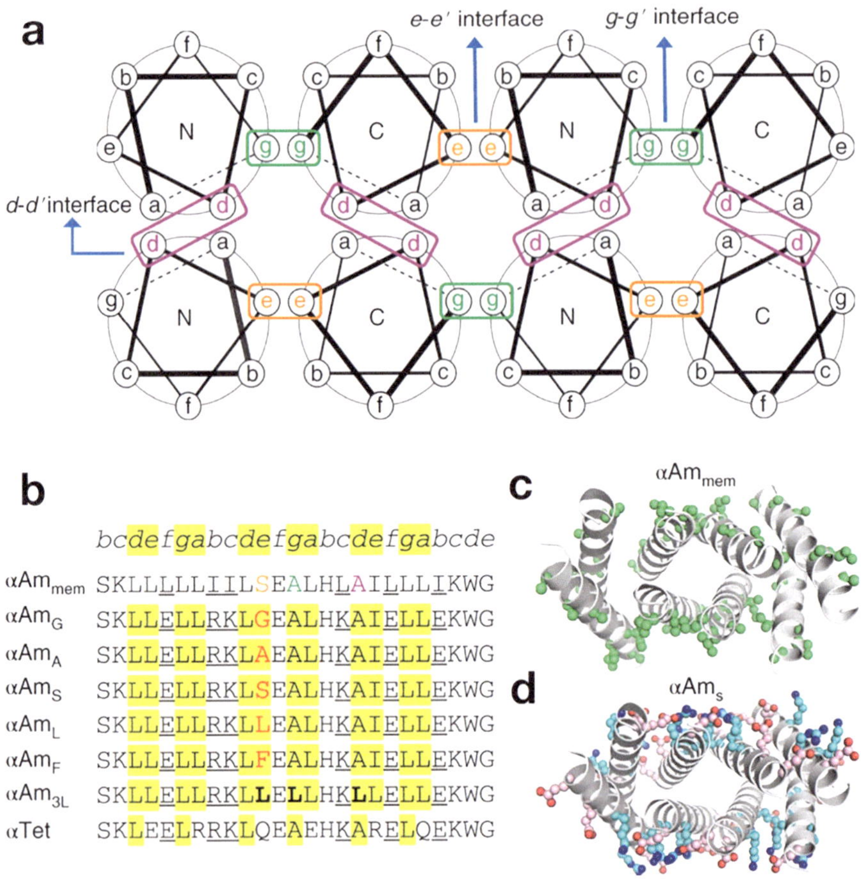

Fig. 15.8 Cross-α filaments derived from αAm_X peptides. (**a**) Helical wheel diagram indicating the relative orientation of packing interfaces between αAm_X peptides in the cross-α filaments. The two hydrophobic faces (*a/e* and *d/g*) are based on a type II polar pattern in which the faces are offset by a single residue. (**b**) Sequences of the αAm_X peptides in which the positions of the residues at the packing interfaces are highlighted in yellow. (**c**) Helical packing in the αAm_{mem} cross-α filament derived from single-crystal X-ray diffraction. (**d**) Helical packing in the αAm_S cross-α filament derived from single-crystal X-ray diffraction. (Reprinted with permission from Zhang et al. [105] Copyright (2018) Springer Nature)

coiled-coil interactions. While knobs-into-holes (KIH) packing occurred over a spatially limited structural interface within these filaments, the amino acid sequences of the heptad repeats approximate that of a type II coiled-coil polar pattern. Walshaw and Woolfson [88] first proposed the latter pattern to rationalize higher-order interactions between extended hydrophobic interfaces in coiled-coils. When projected onto a helical wheel in superhelix space, the hydrophobic face subtends an angle of ~206°, which, in comparison to the structures of the PSMα3 and

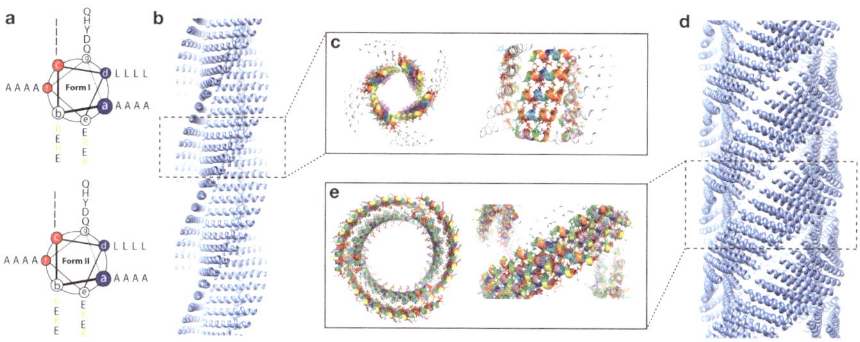

Fig. 15.9 Structure of cross-α helical filaments derived from Form I and II peptide sequences. (**a**) Helical wheel diagrams of the Form I and Form II peptides in which the sequence changes are highlighted in yellow. The two hydrophobic faces (*a/d* and *c/f*) of the type III polar pattern, represented in blue and red, respectively, are offset by two residues. (**b**) An atomic model of the Form I filament from cryo-EM helical reconstruction (PDB: 3 J89) [27]. (**c**) Top (left) and side (right) view of a segment of the Form I filament highlighting consecutive knobs-into-holes interactions (colored segments). (**d**) An atomic model of the Form II filament from cryo-EM helical reconstruction (PDB: 6WL8). (**e**) Top (left) and side (right) view of a segment of the Form II filament highlighting consecutive knobs-into-holes interactions (colored segments). (Reprinted with permission from Egelman et al. Copyright (2015) Elsevier)

uperin-3.5 assemblies, should be sufficient to promote axial stacking over an extended hydrophobic interface.

Egelman, Conticello, and co-workers described the cryo-EM structural analysis of two designed helical peptide nanotubes in which the amino acid sequences of the two peptides, Form I and Form II, were based on a type III coiled-coil repeat pattern [23, 27]. In contrast to the type II repeat of the αAm_X peptides, the two hydrophobic faces were offset from each other in the helical wheel projection (Fig. 15.9). The resulting polar pattern placed the two hydrophobic faces at an angle of ~154° with respect to each other. Each hydrophobic face subtended a cross-sectional angle of ~103° and consequently buried a similar amount of hydrophobic surface area as the type II sequences. The geometrically opposing arrangement of the two hydrophobic faces was predicted to favor axial stacking of helices in the corresponding assemblies. Walshaw and Woolfson originally proposed that the type III heptad pattern should promote lateral association of helices into stacked structures that they designated α-sheets [86]. The computational algorithm SOCKET [87] was employed to query the then current release of Protein Data Bank (PDB). Knobs-into-holes interactions were identified in a limited number of protein structures that were consistent with the proposed α-sheet interfaces. As these interactions occurred within the structural context of globular proteins, the helix-helix interactions did not promote the formation of extended helical filaments.

The Form I and Form II peptides differed solely in the substitution of arginine residues in the former peptide for lysine residues in the latter peptide at identical positions in the respective sequences. Surprisingly, this semi-conservative substitution resulted in the formation of different self-assembled structures (Fig. 15.9). Each nanotube structure was derived from lateral interaction between cross-α protofilaments. However, in the Form I structure, the asymmetric unit (ASU) was derived from a single α-helix, while, in the Form II structure, the ASU comprised a pair of α-helices in a bilayer arrangement. Axial stacking of the respective ASUs resulted in the formation of the cross-α protofilaments. SOCKET analysis of the interfacial interactions between axially stacked helices in the respective filaments revealed that classical coiled-coil knobs-into-holes packing mediated self-association of the helical protomers. The helix-helix crossing angles were significantly more acute (~ +12°) than those observed for classical coiled-coil interfaces (~ +20°), which may reflect the need to maintain extended KIH interactions simultaneously along two extended interfaces [86]. Also, the two nanotubes differed with respect to the number of cross-α protofilaments in the self-assembled structure. Four protofilaments were observed in the Form I structure versus three for the Form II structure. Despite the observed similarity in the axial stacking interfaces, the lateral interactions differed significantly between the two peptide assemblies. In the Form I structure, a pair of arginine residues, R13 and R17, mediated lateral association between cross-α protofilaments through a C-terminal capping interaction. In the Form II structure, a weaker interaction between terminal glutamine residues, Q1 and Q29, mediated lateral association. Mutagenesis of either R13 or R17 to lysine converted the Form I structure to a Form II-like structure. Introduction of arginine at positions 13 and 17 in the Form II sequence resulted in conversion to a Form I-like structure. In either case, the axial stacking interfaces of the cross-α protofilaments were conserved even though the mode of lateral association was altered. These results suggested that cross-α axial stacking interface was thermodynamically stable and richly designable, whereas the lateral interactions were locally weak and sparsely designable (vide infra).

To assess the robustness of the cross-α protofilament structure in sequence space, which can be considered as a measure of the designability of the helix-helix interaction, a series of Form I-like peptides were synthesized [91], and the corresponding structures were determined using cryo-EM helical reconstruction. The $R_{13}xxxR_{17}$ (arginine staple) motif of the Form I peptide was conserved in the resultant sequences; however, the lengths of the peptide were increased or decreased incrementally through addition or deletion of heptad repeats (Fig. 15.9). The design of the peptide sequences retained the structural periodicity of the type III sequence pattern, which was considered to be an important structural determinant for the formation of the cross-α protofilaments. The experimentally determined atomic model of three nanotubes derived from these peptide sequences demonstrated that the cross-α stacking interfaces were conserved in the resultant nanotubes despite a range in peptide length from 15 to 36 amino acids. Like the Form I peptide assemblies, the cross-α protofilaments associated laterally through the interaction between the RxxxR motifs on a given protofilament and the C-termini of peptides in a structurally adjacent helical stack. The lateral interaction occurred between protofilaments that were related in a left-handed screw sense, which differed from the right-handed screw sense of the axial stacking interaction within a protofilament (Fig. 15.10). Despite the similarity in the local structural interfaces within the

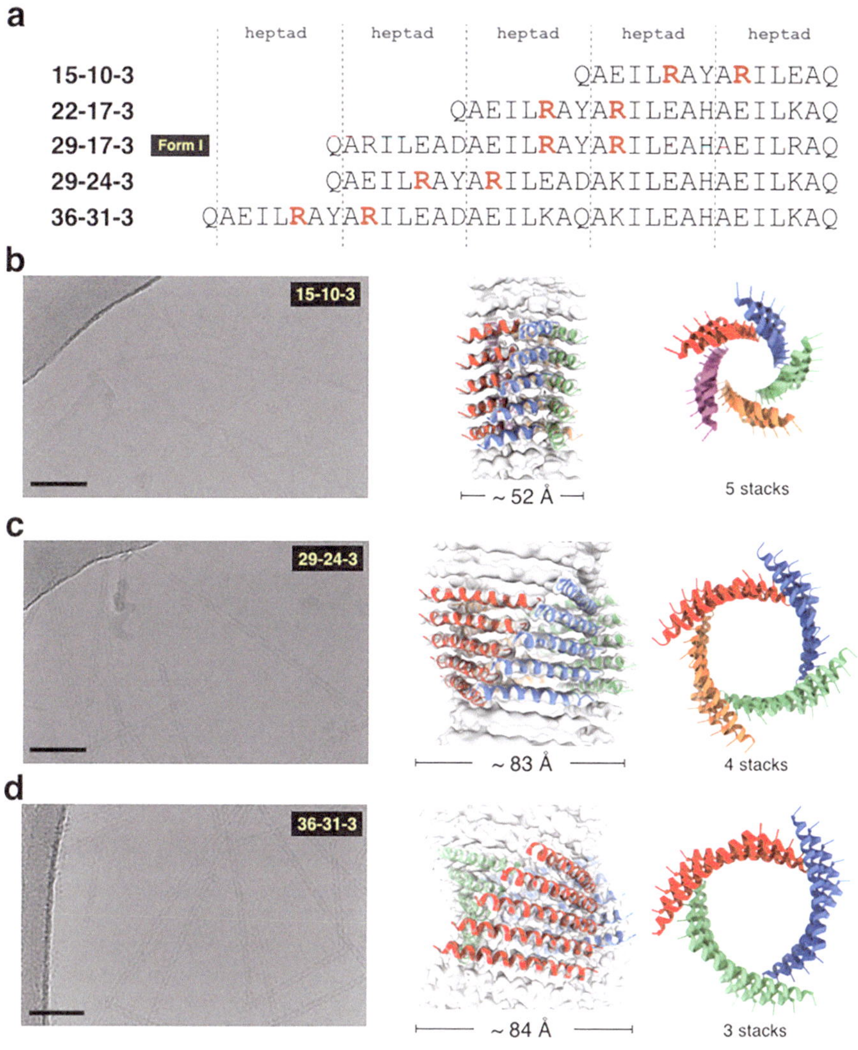

Fig. 15.10 Cryo-EM structural analysis of designed cross-α filaments derived from Form I-like peptide sequences. (**a**) Sequences of the Form I peptide and four designed peptides derived from Form I. (**b–d**) Cryo-EM of peptides 15-10-3, 29-24-3, and 36-31-3 [91]. The cryo-EM reconstructions are shown in the middle, with models fitted into the maps (PDB: 6WKX, 6WKY, and 6WL1). The top views of the atomic models are shown on the right. (Reproduced under the Creative Commons License from [91])

respective assemblies, the global structures of the assemblies differed regarding the number of cross-α protofilaments that self-associated to form the respective nanotubes. A trend was observed in which the number of protofilaments within the nanotube decreased as the length of the Form I-like peptides increased. This observation was rationalized based on the need to simultaneously accommodate the

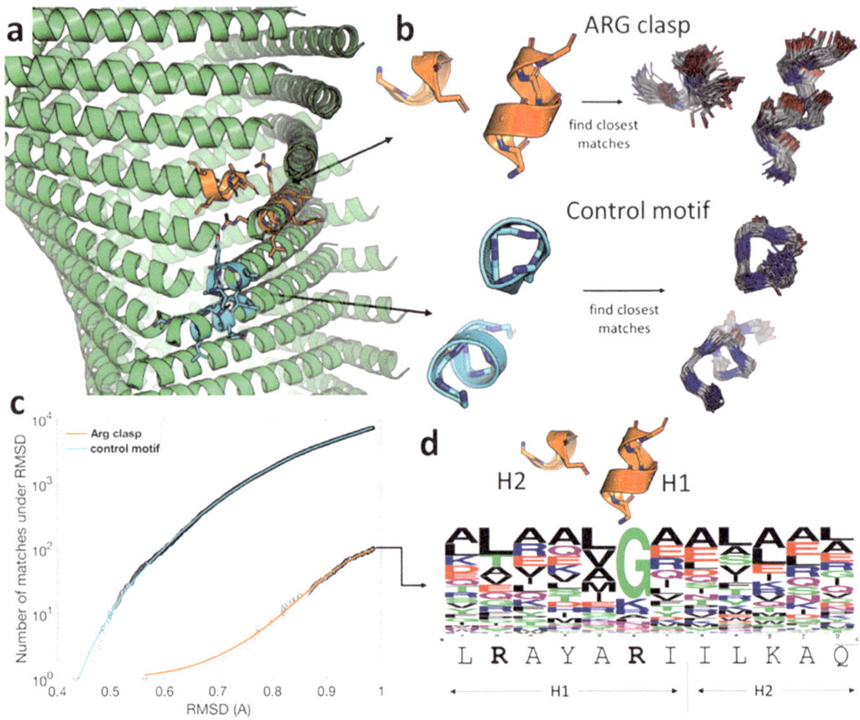

Fig. 15.11 Designability of interfacial interactions within a cross-α nanotube derived from self-assembly of peptide 36-31-3. (**a**) The 36-31-3 assembly showing the definition of the motifs used to interrogate for the presence of the arginine clasp in native structures (orange), as well as a control (axial stacking) motif of the same size and complexity (cyan). (**b**) Structural alignments of the resulting 100 closest matches for both motifs. The control motif, corresponding to the helix stacking interaction within a protofilament, clearly corresponds to a well-populated structural attractor, while the arginine clasp motif brings in much more diverse hits. (**c**) RMSD divergence plots, showing the number of matches below a given RMSD cutoff for both motifs. Matches to the Arg clasp motif are rare as the RMSD required to get up to a certain number of matches grows much more rapidly than for the control. (**d**) Sequence logo of the closest nonredundant sequence matches to the Arg clasp fragment, showing the query sequence underneath. The native statistics do not bear out the Arg clasp sequence motif, and the one conserved position corresponds to a Gly rather than an Arg. (Reproduced under the Creative Commons License from [91])

cross-α helical axial stacking interactions within a protofilament and the lateral arginine clasp interaction between protofilaments to promote the formation of the stable nanotubes.

The difference in designability between the axial stacking interfaces within the cross-α protofilaments and the lateral arginine clasp interface between helices in structurally adjacent protofilaments was explicitly investigated for the longest peptide sequence, 36-31-3 (Fig. 15.11). The program MASTER [108] was employed to search the PDB to identify structurally similar interfaces using a representative axial stacking (control) interface and the arginine clasp interface as queries. The RMSD divergence plot that resulted from this search indicated that the number of hits under a given RMSD rose more sharply for the axial stacking interface than for the arginine clasp interface. The relative frequency of occurrence of structurally similar interfaces within the PDB has been proposed as a proxy for native designability. Therefore, one would consider the axial stacking interaction as richly designable, as would be expected for the ubiquitously observed coiled-coil interface in native protein sequences. In contrast, the increase in the RMSD divergence plot for the arginine clasp interaction was significantly attenuated relative to the helical stacking interaction, which suggested that this interaction was weakly designable with sparse representation in the PDB. The sequence logo for the structural hits to the arginine clasp query indicated a significant divergence from the sequence of the interaction motif in 36-31-3 and related Form I-like peptides, which provided further support for the sparse designability of the latter interface in sequence space.

Designability of Cross-α Helical Protein Filaments

Given the novelty of cross-α filaments and their potential utility as biomaterials [30, 98], questions may be raised regarding whether the cross-α architecture can be reliably and predictably designed from first principles, and, if so, what the nature of the design criteria would be. The high designability of helix-helix packing interfaces suggests that cross-α filaments should be amenable to de novo design; however, the formation of extended filamentous peptide assemblies poses some specific challenges. The preliminary structural analyses described in this chapter indicated that an amphipathic sequence pattern was a necessary but insufficient criterion for the formation of cross-α filaments. The structures of both biologically derived and de novo designed cross-α filaments suggested that the hydrophobic face(s) of the helices should subtend an angle at the helical axis that would expose sufficient hydrophobic surface area in an appropriate spatial orientation to promote axial stacking that would result in the formation of extended filamentous assemblies. A recent theoretical analysis of cross-α filament [20] suggested that the polar face of the helices should subtend an optimal angle of ~95° (and consequently that the hydrophobic face should subtend an angle of ~265°) to promote axial self-assembly of the protomers. However, these estimates were based on binary choices in assignment of residues as hydrophilic versus hydrophobic and don't consider the stereochemical consequences of residue surface area and shape at the helical packing interfaces. The latter consideration has critical consequences for the packing interfaces in coiled-coil proteins in which hydrophobic residue selection has an important influence on oligomerization state and interfacial packing geometry [21, 29, 65]. In addition, the previous theoretical analysis was predicated on the assumption that the packing interface would involve the formation of a bilayer hydrophobic interface as observed for the PSMα3 and αAM_X assemblies. Bilayer packing would require a wide and uninterrupted hydrophobic face within an amphipathic peptide sequence that would expose significant nonpolar surface area. The Form I-like peptide nanotubes provide evidence that alternative polar sequence patterns in designed peptides can support monolayer-like helical stacking within cross-α protofilaments if the inter-helical interaction buries sufficient hydrophobic surface area in an appropriate geometrical orientation to promote axial assembly. However, a recent cryo-EM study of a nanotube assembly derived from an amphipathic antimicrobial peptide indicated that, while the hydrophobic face may subtend a large angle (~170°) [44], it doesn't necessarily guarantee formation of a cross-α structure as lateral interactions between protofilaments may have an influence with respect to the relative orientation of protomers within the assembly.

An important question regarding the designability of cross-α filaments refers to the frequency of their occurrence within structurally characterized proteins. The latter consideration has been employed as a proxy for the native designability of interaction interfaces in tertiary and quaternary structure [106]. The RMSD divergence plot for the helical stacking interface of the 36-31-3 nanotube indicated that it was highly designable. However, the query structure was limited to a dimeric interface between short helical segments, which might not be considered to be representative of the extended, bilateral stacking interface of a cross-α protofilament. Walshaw and Woolfson [86] used the program SOCKET to identify α-sheet interfaces based on type III coiled-coiled heptad repeats within the PDB. Several cross-α interfaces were observed in globular proteins. However, this analysis is nearly 25 years old, and the PDB [6] has grown significantly in terms of the number of deposited structures during the intervening years since this initial report. In addition, the development of more efficient and refined computational methods for conformational searches enables a more effective structural interrogation of the PDB to identify analogues of a query structure. Das and Bhargava [20] developed a computational algorithm to search the PDB for helical segments that had the potential to form cross-α filaments, which were computationally screened for aggregation potential. While this approach may be promising to identify peptides that have potential for adoption of the cross-α architecture, the peptide segments that were identified do not necessarily form cross-α interactions within the context of the native protein structure.

To address this question, we employed the program MASTER [108] to interrogate the PDB (as of 23 May 2024) to identify structures that were homologous to a query derived from three stacked helices (H1, H2, and H3) corresponding to residues 13–29 in structure of the Form I cross-α protofilament (Fig. 15.12). The query encompasses the minimal structural unit of a cross-α filament including a central helix engaging in bilateral stacking interactions with helices above and below it. The MASTER search identified 318 trimeric helical interfaces at RMSD ≤1.5 Å in the 2024 release of the PDB. The structural alignment of the query sequence to the closest matches is consistent with the hypothesis that it is a well-populated structural attractor, as previously postulated for the much smaller query sequence employed in the previous MASTER analysis of the cross-α interface within the Form peptides (Fig. 15.9). While the number of hits rose more slowly than that of the previous RMSD divergence plot, the frequency of recovery of similar structures was consistent with the larger three-helix query that defined a more

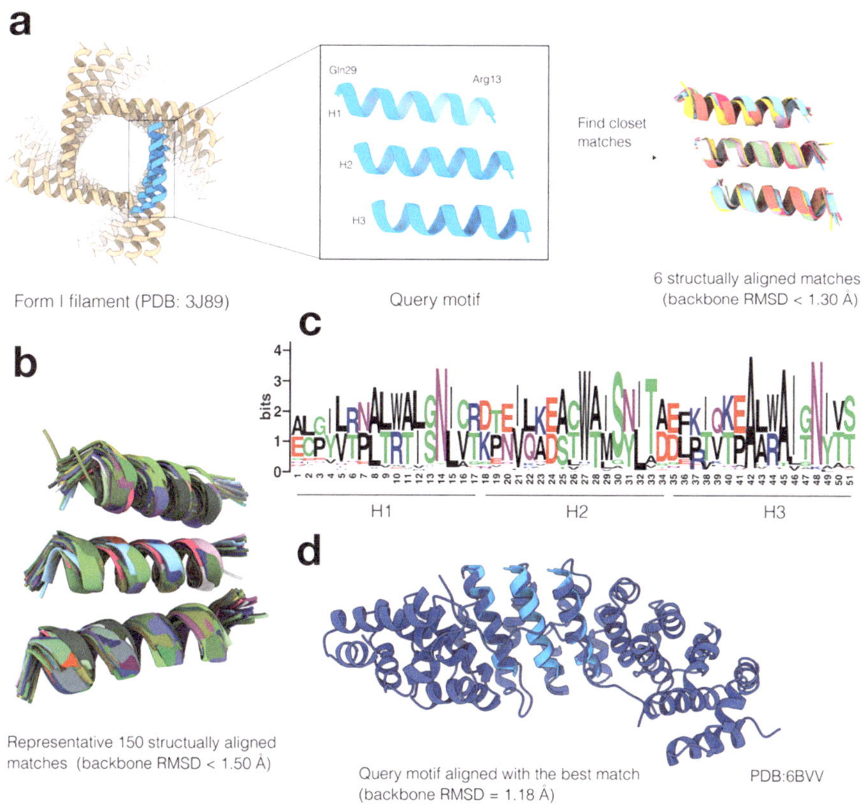

Fig. 15.12 MASTER analysis of the PDB to identify cross-α structural interfaces. (**a**) Summary of the MASTER search workflow starting from the structure of the Form I peptide assembly (PDB: 3 J89). (**b**) Structural alignment of the backbones of a subset of the most closely related hits from the MASTER search to the 3J89-derived query sequence. (**c**) Sequence logo of the closest nonredundant structural matches to the three helices of the 3J89 query motif. (**d**) Structural alignment of a representative segment of the best structural match, importin α3, to the 3J89 structural query

extended structural interface. Notably, the sequence logo of the closest nonredundant sequence matches displayed pronounced residue preferences for specific positions within the three helices of the query. The most notable feature of the analysis was the pronounced preference for tryptophan, asparagine, and serine/threonine at specific positions in all three helices. The closest match to the query structure (RMSD = 1.18 Å) occurred within the armadillo repeat domain of the protein importin α3 (PDB: 6VVB) [76], which mediates transport of protein cargo from the cytoplasm into the nucleus of eukaryotic cells. The armadillo domain consists of short repeats of approximately forty amino acids that are arranged in a three-helix structural motif. However, the query structure in the MASTER search identified stacked helices in three adjacent repeats rather than aligning with helices within the same repeat motif. Tandem repeat proteins based helical hairpin motifs

[38], e.g., TPR, ankyrin, armadillo, and HEAT repeats, are common in eukaryotic systems in which the extended structures that result from concatenation of the motifs are often involved in protein recognition, binding, and transport. While it had been previously speculated that the cross-α helical interaction might resemble the internal helical interfaces in super-helical tandem proteins [82], this MASTER analysis provides direct evidence for the presence of such interactions in native and designed tandem repeat proteins.

Since cross-α helical interfaces appear to be based on a highly designable interaction, a relevant question regards the potential for reliable structural prediction from sequence information and the ability to design novel cross-α assemblies de novo. The Colabfold implementation of AlphaFold-Multimer [52], a state-of-the-art deep-learning method for structural prediction from sequence information, has been investigated recently for its ability to accurately predict the structure of cross-α oligomers of PSMα3 (PDB: 5I55) and uperin 3.5 (PDB: 6GS3) from sequence information. These two systems offer an advantage in that reliable atomic models are available for comparison to structural predictions [64]. While AlphaFold-Multimer could predict the global morphology in both cases, i.e., the bilayer of cross-α helical stacks, the packing of peptides was not realistic in that the inter-helical distances were significantly longer in the predicted structures with less buried surface area than observed in the corresponding structures derived from X-ray diffraction or cryo-EM analyses. The recently released AlphaFold3 [1] provided no significant improvement in the reliability of the predictions based on evaluation of the pLDDT and ipTM metrics for the predictions. For designed cross-α peptides, such as the Form I-like sequences depicted in Fig. 15.8, AlphaFold3 can predict the helical stacking interaction observed for individual oligomers; however, the reliability of the predictions decreases with increasing oligomerization state. For example, the dimeric and tetrameric interfaces of peptide 36-31-3 were predicted with high reliability, but the structures of octameric and dodecameric assemblies were not well predicted and deviate from the structures observed in the cryo-EM atomic models. Since the different versions of AlphaFold were trained using structural information deposited in the PDB, the limited number of filamentous structures based on cross-α interfaces may be responsible for the low reliability of these structural predictions for open-ended extended assemblies.

As regards the second question of de novo design of cross-α filaments, progress has been more limited than for reliable structural prediction. Generative protein design [36, 94] is currently undergoing a revolution based on the implementation of diffusion-based methods in which novel folds and structural interfaces have been designed, fitted with sequences that promote structure formation, and screened with AlphaFold to confirm the structure. The current limitations of AlphaFold for the prediction of filamentous structures represent a significant hurdle for the de novo design of synthetic cross-α filaments (vide supra). However, rather than designing the long-range structure of an extended filament, the local stacking interactions could serve as the focus of design efforts, as these structural interfaces are more robustly predicted than extended one-dimensional lattices. If these interfaces are appropriately designed and accurately predicted, self-association of the designed peptides

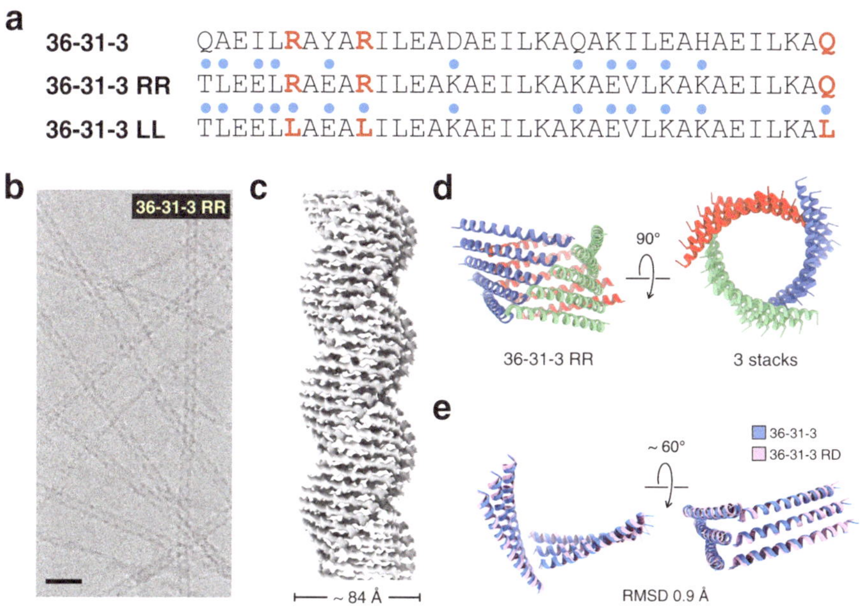

Fig. 15.13 Cryo-EM of the computationally re-designed (36-31-3) peptide nanotube. (**a**) Sequence computationally redesigned peptides 36-31-3_RR and 36-31-3_LL in comparison to the parent sequence. Residues changed from the original 36-31-3 sequences are highlighted with blue dots on top. (**b**) Representative cryo-EM image of 36–31-3_RR peptide filaments out of a total number of 166 images recorded. The scale bar is 20 nm. (**c**) Cryo-EM reconstruction of the 36-31-3_RR filaments (PDB: 6WL0). (**d**) Side view and top view of the atomic model. (**e**) Alignment of two adjacent cross-α stacks between 36-31-3 (blue) and 36-31-3_RR (magenta). (Reproduced under the Creative Commons License from [91])

should result in the formation of extended filamentous structures that maintain the local packing interfaces. Ab initio design of peptide- and protein-based filaments with complete structural control still faces several significant challenges, including the lability of helical symmetry in structural space [93] and potential emergence of weakly designable interfaces [14], e.g., the arginine clasp interaction, that are difficult to predict due to their limited sequence content and sparse representation in structural databases. Therefore, confirmation of the atomic model for a designed peptide filament requires verification of its structure at near-atomic resolution, usually by cryo-EM helical reconstruction.

Despite these potential limitations, computational design has been employed to select sequences that optimize helix-helix packing interfaces in designed cross-α filaments (Fig. 15.13). The computational design engine dTERMen [109] was applied to optimize the sequence of the 36-31-3 peptide, which self-assembled into a nanotube derived from self-association of three cross-α protofilaments (Fig. 15.10). dTERMen scores sequences fitted into the atomic model based on native designability, i.e., the frequency of occurrence of residues within structurally similar interfaces in tertiary motifs (TERMs) in the PDB [49]. Two peptides were synthesized after sequence optimization to maintain the axial and lateral interactions between peptides in the cross-α filament. In the first peptide, 36-31-3_RR, dTERMen was employed to design a sequence in which all positions were redesigned except for those involved in the arginine clasp interaction. In the second peptide, 36-31-3_LL, the entire sequence was redesigned over both interfaces, which resulted in the arginine clasp motif interaction being substituted with a cluster of hydrophobic interactions between leucine residues at structurally homologous lateral packing interfaces. Both sequences self-assembled into nanotubes having similar diameter and morphology to filaments derived from the parent peptide. However, the 36-31-3_LL assemblies unraveled over time into individual protofilaments, which suggested that the re-designed lateral interactions could not maintain the structural integrity of the nanotube. In contrast, the 36-31-L_RR nanotubes were stable, which suggested that the arginine clasp motif could maintain the lateral interaction between protofilaments in a different sequence context. Cryo-EM helical reconstruction of the 36-31-3 nanotubes afforded a 3D density map into which a reliable model could be built. A comparison between the 36-31-3 and 36-31-3_RR nanotubes indicated that the helical symmetry parameters were nearly identical. Structural alignment of three helices in two adjacent stacks, encompassing *both* the axial and lateral interfaces, indicated a high degree of structural similarity (RMSD ~0.9 Å). While generative design of cross-α filaments has not yet been achieved, these latter results provided evidence that optimization of interfacial interactions can be successfully accomplished and represents a viable strategy that can guide future design efforts.

Conclusion

The rich designability of helix-helix packing interfaces suggested that these interactions could be employed for the design of synthetic filaments and assemblies. Broad success has been achieved in the design of artificial filaments based on coiled-coil structural motifs. In the latter case, the coiled-coil protomers align along the central axis of the filament, which mimics the parallel arrangement seen in many α-helical fibrous proteins. Filamentous assemblies are also observed in biological systems in which the helical protomers are oriented in a skew (tilted) arrangement with respect to the central axis of the filament. The tilt angle associated with skew packing of protomers can vary over a wide range, as has been shown in the family of designed helical repeat proteins [11]. While fewer examples of these types of synthetic helical filaments have been reported, they represent an attractive design target for nanomaterial applications. Finally, recent structural analyses have identified a novel packing arrangement for helical protomers, the cross-α filament, in which the peptide backbone is oriented perpendicular to the major axis of a filament or protofilament. This cross-α arrangement results in a high density of peptides along the contour length of the filaments. Since the termini of the peptides are exposed at the edges of the filament, the surface of the assembly can be functionalized with chemical entities that promote the development of novel function. Several examples of designed cross-α filaments have been reported; however, the field is ripe for further exploration, especially through computational design. However, a current impediment to progress in the reliable design of synthetic α-helical filaments is the limited mechanistic insight available into the process of self-assembly and its dependence on peptide sequence. In contrast to designed β-sheet filaments, only a few thermodynamic and kinetic investigations have been reported for the self-assembly of α-helical filaments [10, 79]. Detailed studies into the self-assembly process would promote the development of a more fundamental understanding of the relationship between sequence and supramolecular structure along the energy landscape for α-helical filament formation, which would afford better control over materials fabrication and facilitate incorporation of function.

Acknowledgments The authors acknowledge financial support from the National Science Foundation grants CHE-2108621 and CHE-2404274.

Compliance with Ethical Standards The authors declare no competing interests.

References

1. Abramson J, Adler J, Dunger J, Evans R, Green T, Pritzel A, Ronneberger O, Willmore L, Ballard AJ, Bambrick J et al (2024) Accurate structure prediction of biomolecular interactions with AlphaFold 3. Nature 630(8016):493–500
2. Banwell EF, Abelardo ES, Adams DJ, Birchall MA, Corrigan A, Donald AM, Kirkland M, Serpell LC, Butler MF, Woolfson DN (2009) Rational design and application of responsive alpha-helical peptide hydrogels. Nat Mater 8(7):596–600
3. Beesley JL, Woolfson DN (2019) The de novo design of alpha-helical peptides for supramolecular self-assembly. Curr Opin Biotechnol 58:175–182
4. Beltran LC, Cvirkaite-Krupovic V, Miller J, Wang F, Kreutzberger MAB, Patkowski JB, Costa TRD, Schouten S, Levental I, Conticello VP et al (2023) Archaeal DNA-import apparatus is homologous to bacterial conjugation machinery. Nat Commun 14(1):666
5. Bennett MJ, Sawaya MR, Eisenberg D (2006) Deposition diseases and 3d domain swapping. Structure 14(5):811–824
6. Berman HM, Westbrook J, Feng Z, Gilliland G, Bhat TN, Weissig H, Shindyalov IN, Bourne PE (2000) The protein data bank. Nucleic Acids Res 28(1):235–242
7. Bethel NP, Borst AJ, Parmeggiani F, Bick MJ, Brunette TJ, Nguyen H, Kang A, Bera AK, Carter L, Miranda MC et al (2023) Precisely patterned nanofibres made from extendable protein multiplexes. Nat Chem 15(12):1664–1671
8. Brodin JD, Ambroggio XI, Tang C, Parent KN, Baker TS, Tezcan FA (2012) Metal-directed, chemically tunable assembly of one-, two- and three-dimensional crystalline protein arrays. Nat Chem 4(5):375–382
9. Brodin JD, Smith SJ, Carr JR, Tezcan FA (2015) Designed, helical protein nanotubes with variable diameters from a single building block. J Am Chem Soc 137(33):10468–10471
10. Bromley EH, Channon KJ, King PJ, Mahmoud ZN, Banwell EF, Butler MF, Crump MP, Dafforn TR, Hicks MR, Hirst JD et al (2010) Assembly pathway of a designed alpha-helical protein fiber. Biophys J 98(8):1668–1676

11. Brunette TJ, Parmeggiani F, Huang PS, Bhabha G, Ekiert DC, Tsutakawa SE, Hura GL, Tainer JA, Baker D (2015) Exploring the repeat protein universe through computational protein design. Nature 528(7583):580–584
12. Burgess NC, Sharp TH, Thomas F, Wood CW, Thomson AR, Zaccai NR, Brady RL, Serpell LC, Woolfson DN (2015) Modular design of self-assembling peptide-based nanotubes. J Am Chem Soc 137(33):10554–10562
13. Burton AJ, Thomson AR, Dawson WM, Brady RL, Woolfson DN (2016) Installing hydrolytic activity into a completely de novo protein framework. Nat Chem 8(9):837–844
14. Conticello VP (2023) Peptide-based nanomaterials: building back better & beyond. Curr Opinion Solid State Mater Sci 27(2):101066
15. Costa TR, Felisberto-Rodrigues C, Meir A, Prevost MS, Redzej A, Trokter M, Waksman G (2015) Secretion systems in gram-negative bacteria: structural and mechanistic insights. Nat Rev Microbiol 13(6):343–359
16. Costa TRD, Ilangovan A, Ukleja M, Redzej A, Santini JM, Smith TK, Egelman EH, Waksman G (2016) Structure of the bacterial sex f pilus reveals an assembly of a stoichiometric protein-phospholipid complex. Cell 166(6):1436–1444.e1410
17. Coulombe PA, Fuchs E (1990) Elucidating the early stages of keratin filament assembly. J Cell Biol 111(1):153–169
18. Crick FH (1952) Is alpha-keratin a coiled coil? Nature 170(4334):882–883
19. Crick F (1953) The packing of [alpha]-helices: simple coiled-coils. Acta Crystallogr 6(8–9): 689–697
20. Das M, Bhargava BL (2020) Exploring the candidates for a new protein folding – cross-alpha amyloid – in available protein databases. Phys Chem Chem Phys 22(41):23725–23734
21. Dawson WM, Martin FJO, Rhys GG, Shelley KL, Brady RL, Woolfson DN (2021) Coiled coils 9-to-5: rational de novo design of alpha-helical barrels with tunable oligomeric states. Chem Sci 12(20):6923–6928
22. Demers JP, Habenstein B, Loquet A, Kumar Vasa S, Giller K, Becker S, Baker D, Lange A, Sgourakis NG (2014) High-resolution structure of the Shigella type-III secretion needle by solid-state NMR and cryo-electron microscopy. Nat Commun 5:4976
23. DiMaio F, Song Y, Li X, Brunner MJ, Xu C, Conticello V, Egelman E, Marlovits T, Cheng Y, Baker D (2015) Atomic-accuracy models from 4.5-a cryo-electron microscopy data with density-guided iterative local refinement. Nat Methods 12(4):361–365
24. Dublin SN, Conticello VP (2008) Design of a selective metal ion switch for self-assembly of peptide-based fibrils. J Am Chem Soc 130(1):49–51
25. Eckert DM, Kim PS (2001) Design of potent inhibitors of HIV-1 entry from the gp41 N-peptide region. Proc Natl Acad Sci USA 98(20):11187–11192
26. Egelman EH (2017) Cryo-em of bacterial pili and archaeal flagellar filaments. Curr Opin Struct Biol 46:31–37
27. Egelman EH, Xu C, DiMaio F, Magnotti E, Modlin C, Yu X, Wright E, Baker D, Conticello VP (2015) Structural plasticity of helical nanotubes based on coiled-coil assemblies. Structure 23(2):280–289
28. Eibauer M, Weber MS, Kronenberg-Tenga R, Beales CT, Boujemaa-Paterski R, Turgay Y, Sivagurunathan S, Kraxner J, Koster S, Goldman RD et al (2024) Vimentin filaments integrate low-complexity domains in a complex helical structure. Nat Struct Mol Biol 31(6):939–949
29. Fletcher JM, Boyle AL, Bruning M, Bartlett GJ, Vincent TL, Zaccai NR, Armstrong CT, Bromley EH, Booth PJ, Brady RL et al (2012) A basis set of de novo coiled-coil peptide oligomers for rational protein design and synthetic biology. ACS Synth Biol 1(6):240–250
30. Fries CN, Wu Y, Kelly SH, Wolf M, Votaw NL, Zauscher S, Collier JH (2020) Controlled lengthwise assembly of helical peptide nanofibers to modulate CD8+ T-cell responses. Adv Mater 32(39):2003310
31. Govindarajan S, Goldstein RA (1996) Why are some proteins structures so common? Proc Natl Acad Sci USA 93(8):3341–3345

32. Grigoryan G, Degrado WF (2011) Probing designability via a generalized model of helical bundle geometry. J Mol Biol 405(4):1079–1100
33. Heal JW, Bartlett GJ, Wood CW, Thomson AR, Woolfson DN (2018) Applying graph theory to protein structures: an atlas of coiled coils. Bioinformatics 34(19):3316–3323
34. Huang PS, Oberdorfer G, Xu C, Pei XY, Nannenga BL, Rogers JM, DiMaio F, Gonen T, Luisi B, Baker D (2014) High thermodynamic stability of parametrically designed helical bundles. Science 346(6208):481–485
35. Hughes SA, Wang F, Wang S, Kreutzberger MAB, Osinski T, Orlova A, Wall JS, Zuo X, Egelman EH, Conticello VP (2019) Ambidextrous helical nanotubes from self-assembly of designed helical hairpin motifs. Proc Natl Acad Sci USA 116(29):14456–14464
36. Ingraham JB, Baranov M, Costello Z, Barber KW, Wang W, Ismail A, Frappier V, Lord DM, Ng-Thow-Hing C, Van Vlack ER et al (2023) Illuminating protein space with a programmable generative model. Nature 623(7989):1070–1078
37. Joh NH, Wang T, Bhate MP, Acharya R, Wu Y, Grabe M, Hong M, Grigoryan G, DeGrado WF (2014) De novo design of a transmembrane Zn(2)(+)-transporting four-helix bundle. Science 346(6216):1520–1524
38. Kajava AV (2012) Tandem repeats in proteins: from sequence to structure. J Struct Biol 179(3):279–288
39. Kim S, Jeon TJ, Oberai A, Yang D, Schmidt JJ, Bowie JU (2005) Transmembrane glycine zippers: physiological and pathological roles in membrane proteins. Proc Natl Acad Sci USA 102(40):14278–14283
40. Koiwai K, Hartmann MD, Linke D, Lupas AN, Hori K (2016) Structural basis for toughness and flexibility in the C-terminal passenger domain of an Acinetobacter trimeric autotransporter adhesin. J Biol Chem 291(8):3705–3724
41. Kreutzberger MAB, Wang S, Beltran LC, Tuachi A, Zuo X, Egelman EH, Conticello VP (2022) Phenol-soluble modulins psmalpha3 and psmbeta2 form nanotubes that are cross-alpha amyloids. Proc Natl Acad Sci USA 119(20):e2121586119
42. Kumar P, Paterson NG, Clayden J, Woolfson DN (2022) De novo design of discrete, stable 310-helix peptide assemblies. Nature 607(7918):387–392
43. Kumar P, Petrenas R, Dawson WM, Schweke H, Levy ED, Woolfson DN (2023) Cc(+): a searchable database of validated coiled coils in PDB structures and AlphaFold2 models. Protein Sci 32(11):e4789
44. Laguera B, Golden MM, Wang F, Gnewou O, Tuachi A, Egelman EH, Wuest WM, Conticello VP (2025) Amphipathic antimicrobial peptides illuminate a reciprocal relationship between self-assembly and cytolytic activity. Angew Chem Int Ed Engl 64:e202500040
45. Lanci CJ, MacDermaid CM, Kang SG, Acharya R, North B, Yang X, Qiu XJ, DeGrado WF, Saven JG (2012) Computational design of a protein crystal. Proc Natl Acad Sci USA 109(19): 7304–7309
46. Li H, Helling R, Tang C, Wingreen N (1996) Emergence of preferred structures in a simple model of protein folding. Science 273(5275):666–669
47. Loquet A, Sgourakis NG, Gupta R, Giller K, Riedel D, Goosmann C, Griesinger C, Kolbe M, Baker D, Becker S et al (2012) Atomic model of the type III secretion system needle. Nature 486(7402):276–279
48. Lupas AN, Bassler J (2017) Coiled coils – a model system for the 21st century. Trends Biochem Sci 42(2):130–140
49. Mackenzie CO, Zhou J, Grigoryan G (2016) Tertiary alphabet for the observable protein structural universe. Proc Natl Acad Sci USA 113(47):E7438–E7447
50. Melnik TN, Villard V, Vasiliev V, Corradin G, Kajava AV, Potekhin SA (2003) Shift of fibril-forming ability of the designed alpha-helical coiled-coil peptides into the physiological pH region. Protein Eng 16(12):1125–1130
51. Miller JG, Hughes SA, Modlin C, Conticello VP (2022) Structures of synthetic helical filaments and tubes based on peptide and peptido-mimetic polymers. Q Rev Biophys 55:e2

52. Mirdita M, Schutze K, Moriwaki Y, Heo L, Ovchinnikov S, Steinegger M (2022) Colabfold: making protein folding accessible to all. Nat Methods 19(6):679–682
53. Mravic M, Hu H, Lu Z, Bennett JS, Sanders CR, Orr AW, DeGrado WF (2018) De novo designed transmembrane peptides activating the alpha5beta1 integrin. Protein Eng Des Sel 31(5):181–190
54. Mravic M, Thomaston JL, Tucker M, Solomon PE, Liu L, DeGrado WF (2019) Packing of apolar side chains enables accurate design of highly stable membrane proteins. Science 363(6434):1418–1423
55. O'Shea EK, Klemm JD, Kim PS, Alber T (1991) X-ray structure of the gcn4 leucine zipper, a two-stranded, parallel coiled coil. Science 254(5031):539–544
56. Ogihara NL, Weiss MS, Degrado WF, Eisenberg D (1997) The crystal structure of the designed trimeric coiled coil coil-vald: implications for engineering crystals and supramolecular assemblies. Protein Sci 6(1):80–88
57. Pandya MJ, Spooner GM, Sunde M, Thorpe JR, Rodger A, Woolfson DN (2000) Sticky-end assembly of a designed peptide fiber provides insight into protein fibrillogenesis. Biochemistry 39(30):8728–8734
58. Pang AH, Obiero JM, Kulczyk AW, Sviripa VM, Tsodikov OV (2018) A crystal structure of coil 1b of vimentin in the filamentous form provides a model of a high-order assembly of a vimentin filament. FEBS J 285(15):2888–2899
59. Papapostolou D, Smith AM, Atkins ED, Oliver SJ, Ryadnov MG, Serpell LC, Woolfson DN (2007) Engineering nanoscale order into a designed protein fiber. Proc Natl Acad Sci USA 104(26):10853–10858
60. Pauling L, Corey RB, Branson HR (1951) The structure of proteins; two hydrogen-bonded helical configurations of the polypeptide chain. Proc Natl Acad Sci USA 37(4):205–211
61. Perutz MF (1951) New x-ray evidence on the configuration of polypeptide chains. Nature 167(4261):1053–1054
62. Ponstingl H, Henrick K, Thornton JM (2000) Discriminating between homodimeric and monomeric proteins in the crystalline state. Proteins 41(1):47–57
63. Potekhin SA, Melnik TN, Popov V, Lanina NF, Vazina AA, Rigler P, Verdini AS, Corradin G, Kajava AV (2001) De novo design of fibrils made of short alpha-helical coiled coil peptides. Chem Biol 8(11):1025–1032
64. Ragonis-Bachar P, Axel G, Blau S, Ben-Tal N, Kolodny R, Landau M (2024) What can alphafold do for antimicrobial amyloids? Proteins 92(2):265–281
65. Rhys GG, Wood CW, Lang EJM, Mulholland AJ, Brady RL, Thomson AR, Woolfson DN (2018) Maintaining and breaking symmetry in homomeric coiled-coil assemblies. Nat Commun 9(1):4132
66. Russ WP, Engelman DM (2000) The gxxxg motif: a framework for transmembrane helix-helix association. J Mol Biol 296(3):911–919
67. Salinas N, Tayeb-Fligelman E, Sammito MD, Bloch D, Jelinek R, Noy D, Uson I, Landau M (2021) The amphibian antimicrobial peptide uperin 3.5 is a cross-alpha/cross-beta chameleon functional amyloid. Proc Natl Acad Sci USA 118(3):e2014442118
68. Sawaya MR, Hughes MP, Rodriguez JA, Riek R, Eisenberg DS (2021) The expanding amyloid family: structure, stability, function, and pathogenesis. Cell 184(19):4857–4873
69. Scholey JE, Nithianantham S, Scholey JM, Al-Bassam J (2014) Structural basis for the assembly of the mitotic motor kinesin-5 into bipolar tetramers. elife 3:e02217
70. Scott AJ, Niitsu A, Kratochvil HT, Lang EJM, Sengel JT, Dawson WM, Mahendran KR, Mravic M, Thomson AR, Brady RL et al (2021) Constructing ion channels from water-soluble alpha-helical barrels. Nat Chem 13(7):643–650
71. Senes A, Gerstein M, Engelman DM (2000) Statistical analysis of amino acid patterns in transmembrane helices: the gxxxg motif occurs frequently and in association with beta-branched residues at neighboring positions. J Mol Biol 296(3):921–936

72. Sharp TH, Bruning M, Mantell J, Sessions RB, Thomson AR, Zaccai NR, Brady RL, Verkade P, Woolfson DN (2012) Cryo-transmission electron microscopy structure of a gigadalton peptide fiber of de novo design. Proc Natl Acad Sci USA 109(33):13266–13271
73. Shen H, Fallas JA, Lynch E, Sheffler W, Parry B, Jannetty N, Decarreau J, Wagenbach M, Vicente JJ, Chen J et al (2018) De novo design of self-assembling helical protein filaments. Science 362(6415):705–709
74. Shen H, Lynch EM, Akkineni S, Watson JL, Decarreau J, Bethel NP, Benna I, Sheffler W, Farrell D, DiMaio F et al (2024) De novo design of pH-responsive self-assembling helical protein filaments. Nat Nanotechnol 19(7):1016–1021
75. Smith AM, Acquah SF, Bone N, Kroto HW, Ryadnov MG, Stevens MS, Walton DR, Woolfson DN (2004) Polar assembly in a designed protein fiber. Angew Chem Int Ed Engl 44(2):325–328
76. Smith KM, Tsimbalyuk S, Edwards MR, Cross EM, Batra J, Soares da Costa TP, Aragao D, Basler CF, Forwood JK (2018) Structural basis for importin alpha 3 specificity of w proteins in hendra and nipah viruses. Nat Commun 9(1):3703
77. Sodek J, Hodges RS, Smillie LB, Jurasek L (1972) Amino-acid sequence of rabbit skeletal tropomyosin and its coiled-coil structure. Proc Natl Acad Sci USA 69(12):3800–3804
78. Squire JM, Parry DA (2017) Fibrous protein structures: hierarchy, history and heroes. Subcell Biochem 82:1–33
79. Stedall TP, Butler MF, Woolfson DN, Hanna S (2011) Computer simulations of the growth of synthetic peptide fibres. Eur Phys J E Soft Matter 34(1):5
80. St-Pierre SA, Hodges RS (1976) A sequential polyheptapeptide as a model for the double-stranded alpha-helical coiled-coil structure of tropomyosin. Biochem Biophys Res Commun 72(2):581–588
81. Tarafder AK, von Kügelgen A, Mellul AJ, Schulze U, Aarts DGAL, Bharat TAM (2020) Phage liquid crystalline droplets form occlusive sheaths that encapsulate and protect infectious rod-shaped bacteria. Proc Natl Acad Sci USA 117(9):4724–4731
82. Tayeb-Fligelman E, Tabachnikov O, Moshe A, Goldshmidt-Tran O, Sawaya MR, Coquelle N, Colletier JP, Landau M (2017) The cytotoxic Staphylococcus aureus PSMalpha3 reveals a cross-alpha amyloid-like fibril. Science 355(6327):831–833
83. Tayeb-Fligelman E, Salinas N, Tabachnikov O, Landau M (2020) Staphylococcus aureus psmalpha3 cross-alpha fibril polymorphism and determinants of cytotoxicity. Structure 28(3): 301–313 e306
84. Thomson AR, Wood CW, Burton AJ, Bartlett GJ, Sessions RB, Brady RL, Woolfson DN (2014) Computational design of water-soluble alpha-helical barrels. Science 346(6208): 485–488
85. Tian Y, Polzer FB, Zhang HV, Kiick KL, Saven JG, Pochan DJ (2018) Nanotubes, plates, and needles: pathway-dependent self-assembly of computationally designed peptides. Biomacromolecules 19(11):4286–4298
86. Walshaw J, Woolfson DN (2001a) Open-and-shut cases in coiled-coil assembly: alpha-sheets and alpha-cylinders. Protein Sci 10(3):668–673
87. Walshaw J, Woolfson DN (2001b) Socket: a program for identifying and analysing coiled-coil motifs within protein structures. J Mol Biol 307(5):1427–1450
88. Walshaw J, Woolfson DN (2003) Extended knobs-into-holes packing in classical and complex coiled-coil assemblies. J Struct Biol 144(3):349–361
89. Walters RF, DeGrado WF (2006) Helix-packing motifs in membrane proteins. Proc Natl Acad Sci USA 103(37):13658–13663
90. Walther D, Eisenhaber F, Argos P (1996) Principles of helix-helix packing in proteins: the helical lattice superposition model. J Mol Biol 255(3):536–553
91. Wang F, Gnewou O, Modlin C, Beltran LC, Xu C, Su Z, Juneja P, Grigoryan G, Egelman EH, Conticello VP (2021) Structural analysis of cross alpha-helical nanotubes provides insight into the designability of filamentous peptide nanomaterials. Nat Commun 12(1):407

92. Wang F, Cvirkaite-Krupovic V, Vos M, Beltran LC, Kreutzberger MAB, Winter JM, Su Z, Liu J, Schouten S, Krupovic M et al (2022a) Spindle-shaped archaeal viruses evolved from rod-shaped ancestors to package a larger genome. Cell 185(8):1297–1307 e1211
93. Wang F, Gnewou O, Solemanifar A, Conticello VP, Egelman EH (2022b) Cryo-em of helical polymers. Chem Rev 122:14055
94. Watson JL, Juergens D, Bennett NR, Trippe BL, Yim J, Eisenach HE, Ahern W, Borst AJ, Ragotte RJ, Milles LF et al (2023) De novo design of protein structure and function with rfdiffusion. Nature 620(7976):1089–1100
95. Whitby FG, Phillips GN Jr (2000) Crystal structure of tropomyosin at 7 angstroms resolution. Proteins 38(1):49–59
96. Wood CW, Woolfson DN (2018) Ccbuilder 2.0: powerful and accessible coiled-coil modeling. Protein Sci 27(1):103–111
97. Woolfson DN (2017) Coiled-coil design: updated and upgraded. Subcell Biochem 82:35–61
98. Wu Y, Norberg PK, Reap EA, Congdon KL, Fries CN, Kelly SH, Sampson JH, Conticello VP, Collier JH (2017) A supramolecular vaccine platform based on alpha-helical peptide nanofibers. ACS Biomater Sci Eng 3(12):3128–3132
99. Xu C, Liu R, Mehta AK, Guerrero-Ferreira RC, Wright ER, Dunin-Horkawicz S, Morris K, Serpell LC, Zuo X, Wall JS et al (2013) Rational design of helical nanotubes from self-assembly of coiled-coil lock washers. J Am Chem Soc 135(41):15565–15578
100. Xu J, Dayan N, Goldbourt A, Xiang Y (2019) Cryo-electron microscopy structure of the filamentous bacteriophage ike. Proc Natl Acad Sci USA 116(12):5493–5498
101. Yin H, Slusky JS, Berger BW, Walters RS, Vilaire G, Litvinov RI, Lear JD, Caputo GA, Bennett JS, DeGrado WF (2007) Computational design of peptides that target transmembrane helices. Science 315(5820):1817–1822
102. Zaccai NR, Chi B, Thomson AR, Boyle AL, Bartlett GJ, Bruning M, Linden N, Sessions RB, Booth PJ, Brady RL et al (2011) A de novo peptide hexamer with a mutable channel. Nat Chem Biol 7(12):935–941
103. Zhang SQ, Kulp DW, Schramm CA, Mravic M, Samish I, DeGrado WF (2015) The membrane- and soluble-protein helix-helix interactome: similar geometry via different interactions. Structure 23(3):527–541
104. Zhang HV, Polzer F, Haider MJ, Tian Y, Villegas JA, Kiick KL, Pochan DJ, Saven JG (2016) Computationally designed peptides for self-assembly of nanostructured lattices. Sci Adv 2(9): e1600307
105. Zhang SQ, Huang H, Yang J, Kratochvil HT, Lolicato M, Liu Y, Shu X, Liu L, DeGrado WF (2018) Designed peptides that assemble into cross-alpha amyloid-like structures. Nat Chem Biol 14(9):870–875
106. Zheng F, Grigoryan G (2017) Sequence statistics of tertiary structural motifs reflect protein stability. PLoS One 12(5):e0178272
107. Zheng W, Pena A, Low WW, Wong JLC, Frankel G, Egelman EH (2020) Cryoelectron-microscopic structure of the pKpQIL conjugative pili from carbapenem-resistant Klebsiella pneumoniae. Structure 28(12):1321–1328 e1322
108. Zhou J, Grigoryan G (2015) Rapid search for tertiary fragments reveals protein sequence–structure relationships. Protein Sci 24(4):508–524
109. Zhou J, Panaitiu AE, Grigoryan G (2020) A general-purpose protein design framework based on mining sequence-structure relationships in known protein structures. Proc Natl Acad Sci USA 117(2):1059–1068
110. Zhu J, Avakyan N, Kakkis A, Hoffnagle AM, Han K, Li Y, Zhang Z, Choi TS, Na Y, Yu CJ et al (2021) Protein assembly by design. Chem Rev 121(22):13701–13796
111. Zimenkov Y, Dublin SN, Ni R, Tu RS, Breedveld V, Apkarian RP, Conticello VP (2006) Rational design of a reversible pH-responsive switch for peptide self-assembly. J Am Chem Soc 128(21):6770–6771

Chapter 16
The Many Lives of a Single Sequence: Functional Plasticity Through Amyloid Polymorphism

Mariana Pigozzi Cali, Jim Monistrol, Fabio Strati, Janina Schiller, Rinat Indig, Ronja Markworth, and Meytal Landau

Abstract Amyloids play critical functional roles in biology, including microbial virulence, innate immunity, and cellular organization, broadening their traditional association with neurodegenerative and systemic diseases. This chapter explores the structural and functional plasticity of amyloids, emphasizing how a single protein sequence can adopt multiple fibrillar conformations, termed polymorphs, each with distinct biological outcomes. We synthesise recent high-resolution structural insights from cryo-EM, NMR, and microcrystallography that elucidate the polymorphic behaviour of amyloids in both pathogenic and functional contexts. Particular focus is placed on bacterial functional amyloids that stabilise biofilms and modulate host-pathogen interactions and on antimicrobial peptides that form reversible fibrils with cytotoxic or immune-stimulatory functions. We also highlight the emerging paradigm of amyloid–nucleic acid co-assemblies and their role in immune recognition, autoimmunity, and possibly the origin of life. By examining structure-function relationships across a broad evolutionary spectrum, we argue that amyloid

Authors Mariana Pigozzi Cali, Jim Monistrol and Fabio Strati have equally contributed to this chapter.

M. P. Cali · J. Monistrol · F. Strati · J. Schiller · R. Markworth
CSSB Centre for Structural Systems Biology, Deutsches Elektronen-Synchrotron DESY, Hamburg, Germany

R. Indig
Department of Biology, Technion-Israel Institute of Technology, Haifa, Israel

M. Landau (✉)
CSSB Centre for Structural Systems Biology, Deutsches Elektronen-Synchrotron DESY, Hamburg, Germany

Department of Biology, Technion-Israel Institute of Technology, Haifa, Israel

The Center for Experimental Medicine, Universitätsklinikum Hamburg-Eppendorf (UKE), Hamburg, Germany

European Molecular Biology Laboratory (EMBL), Hamburg, Germany
e-mail: meytal.landau@cssb-hamburg.de

D. A. D. Parry, P. K. Luther (eds.), *Fibrous Proteins: Structures and Mechanisms*, Subcellular Biochemistry 113, https://doi.org/10.1007/978-3-032-05273-5_16

polymorphism constitutes a general mechanism of biological regulation. Understanding how these fibrils shift between states, including cross-β, cross-α, nanotubular, or phase-separated condensates, offers insight into their dual roles in health and disease. This perspective repositions amyloids not merely as pathological end-products but as versatile, ancient scaffolds for structural adaptation and functional innovation.

Keywords Amyloids · Antimicrobial peptides · Virulence factors · Fibrils · Polymorphism · Biofilm · Cryo-EM · X-ray microcrystallography

Abbreviations

Aap	Accumulation-associated protein
AMPs	Antimicrobial peptides
Aβ	Amyloid beta
cryo-EM	Cryo-electron microscopy
eDNA	Extracellular DNA
Fap	Functional Amyloid in Pseudomonas
LLPS	Liquid–liquid phase separation
MRSA	Methicillin-resistant *Staphylococcus aureus*
PSM	Phenol-soluble modulin
Sbp	Small Basic Protein

Introduction

Amyloids in Pathology and Physiology

Amyloids are protein aggregates characterized by a fibrillar ordered morphology typically adopting a cross-β architecture composed of tightly mated β-sheets where β-strands align perpendicularly to the fibril axis (Fig. 16.1). In the human body, amyloid fibrils are widely associated with several pathological conditions. Bundles of amyloid fibrils, known as amyloid plaques, are commonly found in the brains of patients with various neurodegenerative diseases such as Parkinson's disease, Alzheimer's disease, and amyotrophic lateral sclerosis, among others [20, 38]. For a long time, the strong correlation between amyloid plaques and neuronal death was taken as evidence for their direct involvement in disease pathogenesis [121]. Similar deposits can also accumulate in other organs, including the kidneys, spleen, and heart, leading to systemic conditions collectively known as amyloidoses [206].

In Alzheimer's disease, the presence of amyloid fibrils led to the creation of the "amyloid cascade hypothesis", which proposed that amyloid deposition was the primary cause of cognitive decline. The current version of the Alzheimer cascade hypothesis differs in several ways from the original one proposed in the 1990s, and it describes the amyloid beta (Aβ) oligomers as the most likely pathogenic species, in

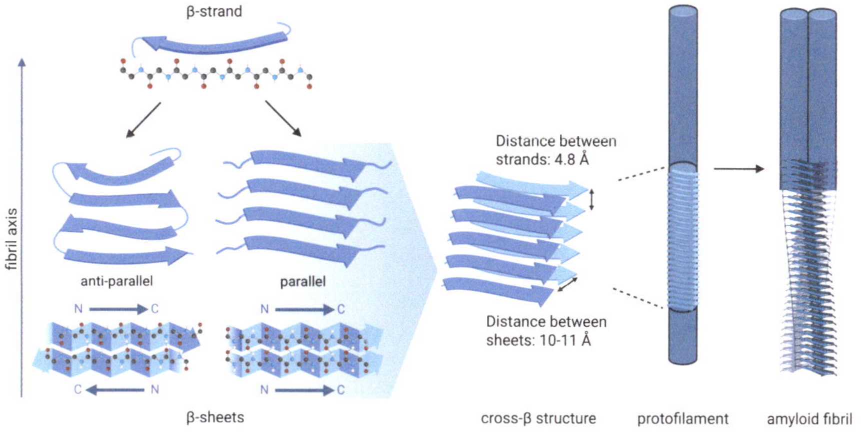

Fig. 16.1 Cartoon depiction of the assembly blocks of a classical amyloid fibril. β-strands that run perpendicular to the fibril axis stack to form either antiparallel or parallel β-sheets. These β-sheets then mate to a cross-β structure, often with a dry and tight interface, that grows into a protofilament. Several of these protofilaments make up a mature amyloid fibril

contrast to plaques, and considers the role of risk factors that modulate penetrance, for example, apolipoprotein E, hyperphosphorylated tau protein, and tau-Aβ interactions [76]. Along with many newly discovered genetic factors, a strong support for the involvement of Aβ in the disease progression are mutations in the Aβ precursor protein gene, linked to early-onset familial Alzheimer's disease, which increase the production of Aβ or promote the formation of longer, more aggregation-prone forms [150].

In the last decade, into the 2020s of the twenty-first century, 55–60 million people are estimated to suffer from dementia, and perhaps 30–40 million of them are clinically estimated to have Alzheimer's disease. Dementia kills in the order of a million people every year globally and is one of the most significant tolls on healthcare budgets in the developed world [55, 109]. Consequently, significant research efforts have been dedicated to developing effective therapies that would facilitate the clearance of Aβ aggregates in Alzheimer's disease patients. A swift analysis of the available trials involving anti-amyloid immune compounds prompts that the clinical benefits have fallen short of expectations, despite accomplishing the expected biological outcome, i.e., amyloid inhibition. Criticism ranges from identifying flaws in experimental design to questioning core neurobiological assumptions within the disease model. Ultimately, some authors argue that the "amyloid cascade hypothesis" may be insufficient to fully explain the pathogenesis of sporadic Alzheimer's disease. Others suggest that the limited success of anti-amyloid therapies may stem from their late administration, as the disease likely begins many years before clinical symptoms emerge.

Along with research regarding the pathogenesis and aetiology of amyloid fibrils in neurodegenerative diseases, many scientists have also revealed a functional aspect of amyloidogenic proteins. In fact, amyloids considered as pathological might also serve important physiological functions; yet for some, these functions are still unclear. In addition, there are proteins whose amyloid form has beneficial effects to the secreting organism, ranging from humans to microorganisms. Functional amyloids participate in diverse activities such as hormone storage, transcriptional and translational regulation, memory consolidation, antimicrobial activities, virulent functions acting against human and microbial cells, and microbial biofilm formation, revealing their broad biological relevance [26, 29, 47, 103, 114, 122, 123, 133, 167, 176, 185].

For example, in mammals, certain peptide hormones are stored in secretory granules in an amyloid-like state, ensuring their stability and controlled release [14, 103, 123, 133, 173]. Orb2, a prion-like RNA-binding protein, is potentially involved in long-lasting memories in *Drosophila melanogaster* [67]. Microorganisms also utilize functional amyloids to aid in surface adhesion and protection against environmental stresses, while some contribute to virulence by facilitating host cell invasion and immune system evasion [40, 123, 133]. These functions are physiological from the perspective of bacteria but pathological in terms of human infections. Notably, and perhaps in contrast to certain pathological amyloids, some functional amyloid systems exhibit tightly regulated secretion and assembly [41, 164] (see Section "Virulent bacterial amyloids in biofilm and as cytotoxins" for details). This tight regulation is critical to pursue functional specificity [10]. For instance, β-endorphin is stored as an amyloid in secretory granules, but upon release into the bloodstream, environmental shifts trigger disassembly into active monomers [119, 173]. Similar amyloid-based storage and functional release mechanisms have been described for oxytocin, prolactin, vasopressin, and other peptide hormones [103].

In some cases, aggregation of the same protein can be involved in both physiology and pathophysiology, induced by mutations or environmental triggers. For example, RNA-binding proteins such as FUS, TDP-43, hnRNPA1, and tau can induce phase separation as a part of their function, while mutations in these proteins lead to neurodegenerative syndromes [27, 54, 92, 100–102, 145]. Lastly, growing number of amyloids from various organisms, including ones associated with neurodegeneration and amyloidosis, display antimicrobial activity [15, 19, 32, 68, 79, 83, 86, 87, 127, 128, 179, 199, 209]. Conversely, antimicrobial peptides (AMPs) from different organisms can self-assemble into amyloid fibrils (see Section "The amyloid-antimicrobial link, a growing fundamental paradigm").

Overall, the mere occurrence of protein aggregation does not offer a reliable means to distinguish between physiological and pathological effects, as some proteins can serve both roles [27, 54, 92, 100–102, 145]. Even with the growing body of structural data on amyloids from diverse organisms and functions, a definitive classification separating pathological from functional amyloids remains elusive, if such a distinction exists at all. The variety of amyloids from different species, from bacteria to human, also highlight the potential for microbial mimicry of human structural motifs and emphasize the functional versatility and evolutionary significance of amyloids as ancient, information-encoding biomaterials [85]. Nevertheless, the wide range of amyloid functions highlights their relevance in both health and disease, underscoring the need to better understand the mechanisms and contexts that govern their biological roles.

Amyloids as Functional Materials

Amyloid fibrils can form remarkably stable structures that can resist harsh denaturing conditions such as high temperatures, extreme pH, high salinity, and pressure, making them attractive as functional materials. They have been explored for diverse applications, including gene delivery systems [220], conductive nanowires [78], carbon dioxide capture [96], aerogels for water purification [132], and catalytic agents [3, 8, 89, 113, 117, 157]. For example, a catalytic amyloid fibril formed by short peptides, shown by cryo-EM to self-assemble into polymorphic cross-β architectures, is capable of ester bond hydrolysis, uncovering a novel arrangement of catalytic sites and advancing the design of amyloid-based enzymatic materials [66].

Amyloids and Phase Separation

Amyloid formation typically proceeds through a nucleation-dependent polymerization process involving three distinct phases: an initial lag phase during which monomers form aggregation-competent nuclei; a rapid exponential growth phase characterized by the formation of oligomers, protofibrils, and fibrils; and a final stationary phase where a dynamic equilibrium is established between monomers and fibrils, often regulated by fragmentation and secondary nucleation mechanisms [33, 34]. However, emerging evidence highlights an alternative pathway involving liquid–liquid phase separation (LLPS), wherein proteins first condense into dynamic, membrane-less droplets before transitioning into fibrillar or amorphous aggregates [115, 134, 146, 207]. LLPS is a reversible, diffusion-driven process in which biomolecules spontaneously demix into dense liquid condensates with high internal mobility, spherical morphology, and the ability to fuse and recover after photobleaching [224]. This process underpins the formation of membrane-less organelles central to cellular processes such as metabolism, RNA regulation, and stress response [129, 224].

Proteins undergoing LLPS often contain intrinsically disordered regions, low-complexity domains, or prion-like motifs, which promote multivalent, weak interactions that drive phase separation. The LLPS behaviour is sensitive to protein concentration, pH, temperature, and cofactors such as nucleic acids, especially RNA [1, 25, 115, 146, 207]. In LLPS-mediated amyloidogenesis, proteins first condense into liquid droplets, which concentrate the local environment. These droplets can transition into more rigid gel-like states and, eventually, mature into amyloid fibrils or amorphous aggregates. This pathway has been observed in numerous disease-associated proteins such as α-synuclein, tau, FUS, and TDP-43, indicating that LLPS may serve as a universal intermediate in both functional and pathological amyloid formation [43, 115, 116, 146, 147, 154, 207].

Structures and Polymorphism of Amyloids

Amyloids are currently distinguished from other protein fibrils based primarily on structural criteria, most notably the presence of tightly packed β-sheets in which β-strands run perpendicular to the fibril axis, forming a dry interface between β-sheets, named by Eisenberg and co-workers "steric zippers" [118] (Fig. 16.1). While amyloids were historically identified by their affinity to dyes such as Thioflavin T and Congo Red, and by their association with extracellular pathological deposits, these criteria are now considered outdated. Functional amyloids involved in physiological processes and intracellular amyloids, such as tau, do not always conform to these classical markers. Moreover, dye-binding assays are not specific: non-amyloid structures may also bind these dyes, and some amyloids may not, due to differences in fibril cavity architecture. Therefore, high-resolution structural methods remain essential for definitive amyloid identification.

Historically, X-ray fibre and powder diffraction were the foundational techniques for identifying the characteristic cross-β pattern of amyloid fibrils with diffraction peaks at ~4.75 Å and 10–12 Å [71, 172]. However, the detailed arrangement of the two tightly mated β-sheets forming the dry interface was only elucidated years later through high-resolution structures [118]. Since the early twenty-first century, the advent of structural methods, such as X-ray microcrystallography, solid-state NMR, micro-electron diffraction, and, more recently, cryogenic electron microscopy (cryo-EM), has revolutionized our ability to resolve amyloid structures at near-atomic detail [4]. Moreover, although the cross-β architecture is predominant among amyloids, some functional and virulence-associated peptides form cross-α fibrils, where α-helices align perpendicularly to the fibril axis [82, 160, 188, 189, 222].

X-ray microcrystallography and microcrystal electron diffraction (micro-ED) can achieve resolutions near 1.5 Å, and even below 1 Å, but require peptide crystallization, limiting their use to short fragments rather than full-length amyloid-forming proteins. In contrast, advances in detector technology and image processing have driven a "resolution revolution" in cryo-EM, enabling near-atomic resolution without crystallization. Although historically lower in resolution, cryo-EM has now reached comparable levels, as shown by the 1.86 Å structure of tau fibrils [98]. Over the past decade, cryo-EM has become the leading technique for reconstructing amyloid fibrils in their native, non-crystalline states, including those extracted from brains, hearts, liver, and fat tissues [2, 5, 46, 49–51, 62, 70, 75, 80, 81, 93, 107, 111, 120, 136–138, 165, 169–171, 175, 181, 182, 186, 190, 213, 215, 216, 223, 231].

Polymorphism Compared in Pathological and Functional Amyloids

Unlike globular proteins, which adopt a single low-energy three-dimensional fold, amyloid-forming proteins often exhibit polymorphism: the same amino acid sequence can give rise to structurally distinct fibrils (Fig. 16.2). Cryo-EM ability to analyse full-length fibrils from heterogeneous and in situ sources has revealed novel polymorphs and provided critical insights into amyloid quaternary architecture, making it indispensable in structural amyloid research. Variations between amyloid polymorphs often involve differences in the two-dimensional fold, i.e., the specific conformation adopted by the peptide backbone within a single layer of the fibril stack. In addition, since amyloids commonly assemble as protofilaments that twist around each other, polymorphism can also arise from variations in protofilament number, their relative arrangement, and the degree of helical twist. Beyond these, some polymorphs form complex supramolecular architectures, including hollow nanotubes and multilamellar assemblies (Fig. 16.2) [84, 95, 98, 184, 194, 214]. This complexity challenges predictions of amyloid formation and structure based solely on sequence and underscores the necessity of detailed experimental characterization [95, 141].

Polymorphism is influenced by environmental factors such as pH, ionic strength, temperature, and cofactors (e.g., lipids, heparin, metal ions, RNA, and membrane receptors) and results in fibrils with varied morphologies, seeding efficiencies, stabilities, and biological activities, including cytotoxicity, antimicrobial function, and signalling. Beyond the structural variations of the same sequence, mutations and post-translational modifications further modulate structure and function [163, 166, 183]. For example, phosphorylation at specific residues of α-synuclein and tau has been shown to yield fibrils with distinct neurotoxic properties [106, 125, 143, 227]. It is important to note, however, that postmortem phosphorylation and dephosphorylation events may bias the assessment of proteins in the fibrillar state, as these aggregates are often inaccessible to kinases and phosphatases. For example, reduced

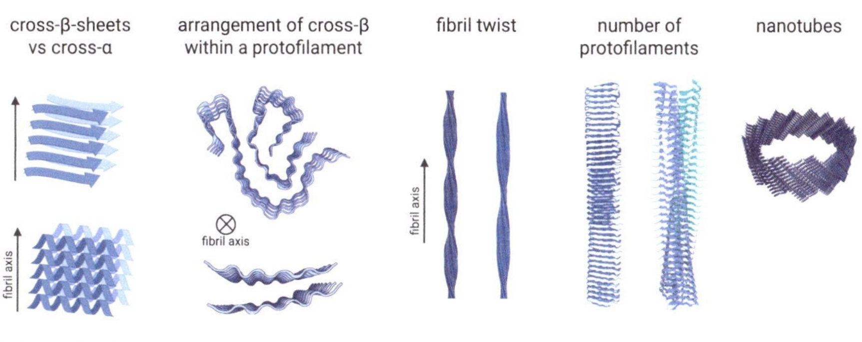

Fig. 16.2 Examples of several types of polymorphism in amyloid fibrils. Amyloid fibrils can structurally deviate from one another on several levels, examples of which are schematically depicted (PDB IDs of structures used as examples: 6HUD, 2M5N, 8ENR, 9ATW, 7T0X [22, 48, 64, 82, 186])

phosphorylation levels of tau observed in postmortem brain tissue may reflect processing artefacts rather than physiological states [200].

The structural and functional differences between polymorphs may contribute to the diversity of pathological outcomes observed in amyloid-associated diseases, beyond differences in overall plaque burden [104, 112, 131, 183]. Notably, similar fibril polymorphs have been isolated from different patients with the same disease, further supporting a correlation between fibril structure and disease phenotype [49]. Fibrils extracted from diseased tissues often display less polymorphism than those formed in vitro, although this may reflect selective pressures during extraction or in vivo constraints.

Functional amyloids were regarded to form less polymorphs compared to pathogenic amyloids, as they are more tightly controlled in their assembly [41, 164]. This regulation is often mediated by co-encoded chaperones and nucleators that spatially restrict fibril formation and may also constrain polymorphism. Nevertheless, the growing number of structures of functional amyloids displays extensive polymorphism similar to that observed for in vitro structures of pathological amyloids. This, like in pathological amyloids, might be related to environmental context which plays a pivotal role in determining amyloid behaviour. Notably, β-endorphin can form multiple polymorphs in the absence of cofactors, but a single, stable polymorph in the presence of heparin [119], highlighting how environmental cues and cofactors can lock amyloids into specific functional states, similar to within the brains. These findings suggest that the extent of polymorphism in vitro is not a reliable criterion for distinguishing between pathological and functional amyloids and that more in situ structures are needed to determine what encodes a biological outcome.

Another important point of comparison between functional and pathological amyloid polymorphs relates to their stability and reversibility. Pathological amyloids are commonly thought to form highly stable, irreversible fibrils, often accumulating as persistent plaques. This stability is often associated with structural features such as the tightness and complementarity of the cross-β steric zipper interface, as well as hydrophobicity of the interacting surfaces. In contrast, functional amyloids are generally assumed to be more dynamic and reversible, allowing for regulation and responsiveness in biological contexts. Cryo-EM structures suggest that some functional amyloids indeed tend to have more polar and flexible interfaces compared to the hydrophobic, tightly packed interfaces typical of pathological amyloids [139]. These features support reversible fibril formation, often regulated by environmental factors such as pH and salt concentration. Examples include the pH-sensitive, hydrophilic core of Orb2 and the flexible [67] and kinked interface of hnRNPA2 fibrils [100]. However, even though some functional amyloids exhibit tightly regulated secretion and assembly [10], they form highly stable and robust fibrils, which is consistent with their structural role as scaffolds within bacterial biofilms [22, 58, 73, 177]. Thus, stability is not necessarily exclusive to pathological assemblies, and their functional relevance may depend on biological context rather than on absolute structural features.

Polymorphism as a Sophisticated Level of Functional Control

This chapter explores the concept that polymorphism in amyloids introduces a sophisticated level of functional control. While a single monomer may have limited activity on its own, its assembly into amyloid structures significantly expands its physiological potential. The capacity of these structures to adopt distinct conformations in response to environmental cues adds an additional layer of regulation. In this context, amyloid polymorphism is not only structurally fascinating but also biologically consequential. Understanding the determinants of polymorph formation and their functional implications is critical for deciphering the dual roles of amyloids in health and disease and for harnessing their unique properties in biomedical and biotechnological applications. The following sections will examine the diverse roles of functional amyloids, introducing them as a dynamic mechanism of biological control.

Virulent Bacterial Amyloids in Biofilm and as Cytotoxins

Bacterial biofilms are structured communities of bacterial cells encased in a self-produced extracellular matrix. The development of biofilms by bacteria facilitates their attachment to surfaces and to each other, while also helping them retain nutrients and moisture [202]. Moreover, biofilms serve as a protective shield against antimicrobial treatments and the host's immune responses, making them more resilient, infectious, and harder to eliminate, including in chronic infections [105]. Functional amyloids play a central role in the formation, stabilization, and maturation of biofilms, indicating that their inhibition could serve as an effective anti-virulence strategy [94, 195]. Targeting biofilms and their amyloid components presents a promising strategy to combat antimicrobial resistance, a global health crisis responsible for approximately one million deaths annually and projected to surpass cancer as a leading cause of death by 2050 [6]. Despite this, there is a significant lack of structural knowledge about biofilms and their fibrillar constituents, as well as their interactions with other proteins. Gaining insight into these structures and their activity relationships could pave the way for more effective therapeutic interventions. In the following section, we will look at biofilms from different pathogens, exploring proteins and peptides involved in functional amyloid formation within biofilm and what we know so far about their structures (Fig. 16.3).

Bacillus subtilis

Bacillus subtilis is a bacterium which has a propensity to form biofilm as a response to different environmental stimuli. Unlike many other bacteria such as *Escherichia*

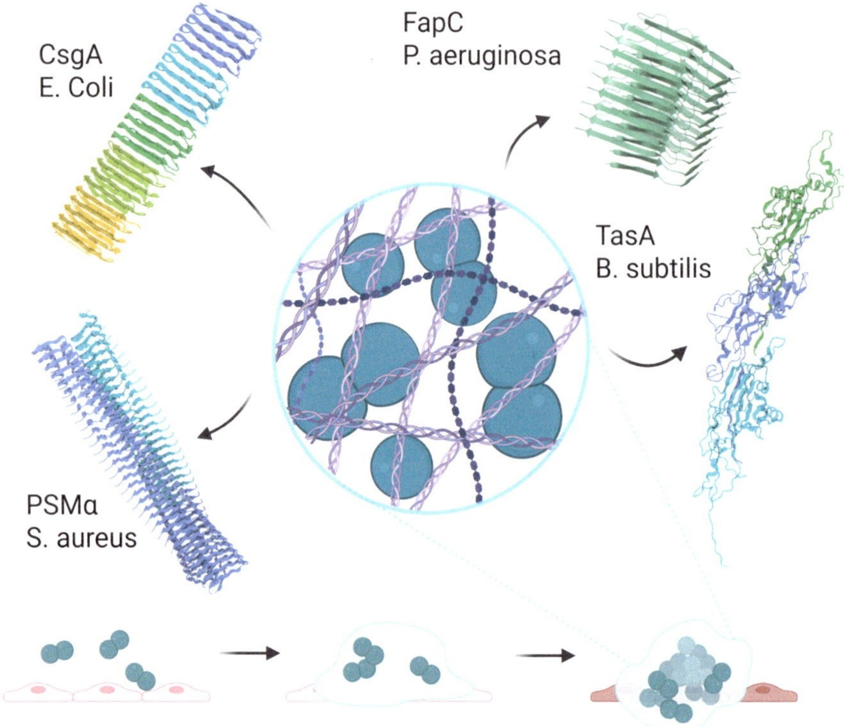

Fig. 16.3 Functional amyloids in biofilms. Schematic representation of biofilm maturation and the functional amyloids found within biofilms of different bacteria: CsgA: PDB id 8ENR [22], TasA: PDB id 8AUR [18], PSMα1: PDB id 9ATW [64], and FapC id 9U4U [73]

coli and *Pseudomona aeruginosa*, *B. subtilis* does not depend on polymeric fibres like pili or fimbriae to stabilize its biofilm, relying however on exopolysaccharides and the protein TasA, a 28.5 kDa protein, to stabilize its extracellular matrix [153]. While filaments were observed and isolated from *B. subtilis* biofilm, they were absent from *tasA* mutant strains [153]. Furthermore, the addition of purified TasA to the *tasA* mutant restored pellicle formation, indicating that TasA fibres are a key component to the extracellular matrix, providing a structural backbone to the biofilm. The fibrils showed binding to the amyloid dye Congo Red, while X-ray diffraction of TasA fibres indicated a weak β-sheet pattern both in vitro and in situ. Circular dichroism spectroscopy indicated a mixture of β-sheets, α-helix, and random coil characteristics, indicating polymorphism in the soluble and fibril forms [153].

Donor-Strand Exchange Drives TasA Fibre Assembly via β-Strand Complementation

Unlike most amyloids, which are intrinsically disordered in their monomeric state, TasA adopts a stable, globular conformation featuring a jellyroll fold. During biofilm formation, TasA assembles into β-sheet-rich fibrils, as demonstrated by NMR spectroscopy on live biofilms [39]. Cryo-EM analysis of biofilm-extracted TasA filaments revealed a distinctive assembly mechanism in which each monomer contributes an N-terminal β-strand, termed a "donor strand", that inserts into the neighbouring subunit, completing a β-sheet. This inter-subunit β-strand exchange recurs along the fibre axis [18]. Although TasA fibrils do not conform to the classical cross-β amyloid architecture, they display hallmark amyloid-like properties such as dye binding and high stability. This structure represents a milestone in resolving high-resolution architectures of functional fibrils directly isolated from native bacterial biofilms.

Escherichia coli

Escherichia coli is a common biofilm-forming bacterium also found in the human digestive system [16]. Curli amyloid fibres are found within the *E. coli* biofilm extracellular matrix and play an important role in the colonization of surfaces and host-tissue contact [29]. Curli fibres are composed of two proteins, CsgA and CsgB, with CsgA making up the majority [208].

Although purified CsgA can independently and efficiently form fibrils in vitro, the presence of the nucleator protein CsgB is crucial for proper aggregation of CsgA within living cells [63]. Curli assembly is a highly regulated process involving five additional proteins beyond CsgA and CsgB. These proteins are encoded by two divergently arranged operons, *csgBAC* and *csgDEFG*, which work in coordinated fashion to ensure the precise production of curli fibres that extend from the outer membrane of *E. coli* [13]. The transcriptional regulator CsgD promotes the expression of curli subunits CsgA and CsgB from the *csgBAC* operon, as well as the periplasmic chaperone CsgC. CsgB shares structural similarity with CsgA and provides a binding interface for CsgA, which is secreted from the cell as an intrinsically disordered protein. Upon encountering CsgB, however, CsgA rapidly forms fibrils [63, 196]. The periplasmic chaperone CsgC acts as a potent inhibitor of amyloid formation by interacting with both CsgA and CsgB during their transport, keeping them in a monomeric state and preventing premature fibrillation [45].

Structure of the CsgA Fibril Displaying a β-Solenoid Fold and a Cross-β Architecture

The investigations of the structural organization of curli amyloid fibres produced by *E. coli* proved to be quite challenging. Solid-state NMR and electron microscopy studies predicted that CsgA adopts a structure in which each molecule spans several β-sheet spacings, resulting in a repeat distance along the fibril axis that is a multiple of 4.7 Å. These findings suggested a β-helix-like architecture [174].

Notably, in addition to forming individual protofilaments, curli fibrils also frequently assemble into irregular higher-order structures, which complicate structural analyses [22, 177]. Initial imaging of native curli within the extracellular matrix revealed dense, tangled networks that impeded cryo-EM high-resolution structural analysis [177]. To address issues with aggregation and heterogeneity, the authors expressed and analysed a CsgA homologue, R15.5, hypothesizing that its high acidic residue content would reduce fibril entanglement. R15.5 formed well-dispersed, curli-like fibres under various pH and ionic conditions, and cryo-EM imaging yielded high-quality 2D class averages and a 7.6 Å resolution 3D reconstruction [177]. The overall curli protofibril adopts a highly ordered supramolecular β-solenoid structure with little to no detectable helical twist. The reconstructed map, assisted by AlphaFold modelling, confirmed the predicted β-solenoid fold of each CsgA monomer with the β-strands corresponding to five imperfect sequence repeats, forming non-centrosymmetric cross-β architecture. The model suggested head-to-tail monomer stacking along the fibril.

Building on the initial structure, a disulfide-engineered recombinant CsgA, aimed to reduce heterogeneity in fibril entanglement, was reconstructed using cryo-EM [22]. The resulting map confirmed the β-solenoid fold. While multiple CsgA–CsgA interaction modes, namely head-to-head, tail-to-tail, and head-to-tail, were observed, the moderate map resolution (~4–6 Å) prevented definitive assignment of monomer connectivity, highlighting the need for higher-resolution structures to resolve atomic details. Fibrils composed of approximately 7–8 CsgA monomers displayed notable stability and uniformity, suggesting this oligomer size may represent a characteristic intermediate or structural unit in CsgA aggregation. The study also unveiled two different spatial arrangements of CsgA fibrils, one with 2-CsgA-fibril pairs and another with 3-CsgA-fibril bundles, in which individual fibrils interacted at a specific angle of 34° [22]. Ultimately, solving the high-resolution structure of CsgA, and probably of other biofilm-associated amyloids, remains a challenge.

Pseudomonas aeruginosa

Like *E. coli*, *Pseudomonas aeruginosa* bacteria also produce functional amyloid fibrils that contribute to biofilm development and structural integrity, acting as extracellular scaffolding components. Amyloid expression in *Pseudomonas* is governed by the Functional Amyloid in Pseudomonas (Fap) operon system [40]. The Fap operon contains six genes encoding for the six Fap proteins, FapA through FapF. Among these, FapC constitutes the principal subunit of amyloid fibrils [40] while FapB was suggested as a supporting and regulatory amyloid component [40]. FapE is also secreted outside the bacterial cells, and while initially considered as a fibrillar component interacting with FapC via a conserved cysteine residue in its C-terminus [40, 42], later findings suggest that FapE is required for FapC secretion on the bacterial surface [155]. However, further investigations are required to elucidate its function and mechanism.

FapA acts as a chaperone preventing the formation of premature FapC amyloid fibrils in the bacterial periplasm and its deletion results in the presence of extracellular fibrils uniquely composed of FapB [42, 144]. FapD has a hypothetical cysteine protease function, which is necessary for the secretion of FapC [42, 155]. Lastly, the structure of FapF, resolved by X-ray crystallography, confirmed its function as a trimeric β-barrel channel localized in the outer membrane [155]. Each subunit of the trimeric pore is able to operate independently to mediate the secretion of FapB, FapC and FapE. Similar to the curli system, the initial translocation of Fap proteins across the inner membrane involves the Sec transport pathway [155].

In contrast to the nucleator and the main amyloid component roles of CsgB and CsgA, respectively, the roles and interactions of the homologous FapB and FapC are more complex. While FapB functions as a nucleator that accelerates FapC fibril elongation in some strains [42], in *P. aeruginosa* strain PAO1, it instead inhibits FapC fibrillation, despite colocalizing with FapC and possibly being nucleated by the kinetically faster FapC itself [58]. Moreover, the fibrillation behaviour in specific conditions emphasized the adaptability of Fap proteins across strains and environments. FapB and FapC both exhibit extraordinary resilience to extreme thermal and chemical conditions, raising concerns about biofilm persistence after standard sterilization while also highlighting their potential for biotechnological applications [58]. Understanding the biofilm amyloid properties in the different systems is crucial for both combating bacterial infections and exploiting amyloids in material science.

The Structure of FapC Fibrils Display a Triple-Layer Cross-β Solenoid

Due to the improvements in structure determination of amyloid fibrils by cryo-EM, two research laboratories have solved the structure of FapC amyloid fibrils, displaying a cross-β solenoid composed of a single protofibril, with the rigid core adopting a Greek key motif [65, 73, 124]. The rigid core of the β-solenoid structure is composed of the three imperfect repeating sequence motifs of FapC, each motif being a layer of β-strands. This arrangement resembles the CsgA β-solenoid fold composed of five imperfect sequence repeats forming five layers of β-strands. The rigid core of the FapC β-solenoid is mainly hydrophobic, and the intermolecular interactions between adjacent subunits rely on parallel hydrogen bonds [65, 73, 124]. The β-solenoid arrangement of CsgA and FapC resembles that of another functional amyloid, the HET-s(218–289) prion, whose structure was determined by solid-state NMR. In HET-s, each molecule adopts a two-turn intramolecular β-solenoid fold, which assembles into a repeating intermolecular architecture [205].

Staphylococcus aureus

Staphylococcus aureus is a pathogenic bacterium known for its ability to form biofilms. The persistence of *S. aureus* biofilms is associated with chronic and recurrent infections, including endocarditis, osteomyelitis, and device-related infections [77, 99, 228]. Bacterial amyloids play a critical role in the structural integrity and stability of *S. aureus* biofilms. In *S. aureus*, one of the key amyloid-forming components is the phenol-soluble modulin (PSM) family of peptides. PSMs are α-helical amphipathic peptides which grant them surfactant-like properties. These peptides are genome-encoded and are therefore found in most staphylococcal strains, oftentimes with minor differences in sequence, with PSM expression being regulated by the bacteria quorum-sensing system Agr [108, 130, 197, 217].

S. aureus produces four PSMα peptides, which are encoded in the psmα operon, and two PSMβ peptides encoded in the psmβ operon. PSMγ, otherwise known as δ-hemolysin or δ-toxin, is RNAIII-encoded [197]. The *S. aureus* PSMα peptides and the δ-toxin belong to the group of short α-type PSMs (~20–25 aa), whereas the PSMβ peptides belong to the β-type PSMs, which are about double in size compared with α-type PSMs. PSMαs are notable for their strong cytolytic activity against leukocytes, with high levels of PSMα expression being associated with increased virulence in methicillin-resistant *S. aureus* (MRSA) [135].

PSMs from *S. aureus* can self-assemble into amyloid-like fibrils that play a central role in biofilm formation, promoting cohesion, surface adhesion, and resistance to mechanical and chemical stress. These amyloid structures not only support the integrity of the biofilm matrix but also protect bacterial communities from host immune responses and antimicrobial agents. For instance, PSMs stabilize biofilms by preventing disassembly caused by matrix-degrading enzymes and are highly cytotoxic to human immune cells [167, 203].

All four PSMα peptides form insoluble aggregates in a yeast model, with PSMα4 exhibiting the highest toxicity, and microscopy revealed vesicle-like assemblies resembling *S. aureus* extracellular vesicles. Engineered variants of the disaggregase Hsp104 were shown to dissolve these aggregates and disrupt preformed biofilms, highlighting a potential enzymatic approach for biofilm clearance [69]. Another support for the critical role of the amyloid state of PSMα1 in biofilm formation was obtained using a phage display screen which revealed a peptide, KG7, that binds the fibrillation hotspot and blocks assembly. When conjugated to photothermal CuS nanoparticles, KG7 enabled targeted disruption of mature biofilms, prevented regrowth, and promoted wound healing [31]. These studies overall show how PSMα peptides contribute to amyloid-driven biofilm architecture and toxicity in *S. aureus*.

Structural and Functional Polymorphism in PSM Amyloids

X-ray microcrystallography studies have shown that segments from *S. aureus* PSMα1 and PSMα4 form canonical cross-β amyloid fibrils [159]. Cryo-EM studies resolved the structure of full-length PSMα1, revealing a cross-β architecture comprising two U-shaped protofilaments twisted around one another [64]. In contrast, PSMα3 adopts a unique cross-α amyloid structure, as shown by the crystal structure of the full-length peptide and its single-point mutants, where α-helices stack laterally to form tightly mated sheets [188, 189]. This structural motif has been linked to the potent cytotoxicity of PSMα3 towards human cells. Cryo-EM studies of both PSMα3 and PSMβ2 confirmed this cross-α architecture, showing spiral nanotubes of differing diameters formed by the stacked α-helices of mated sheets [82].

PSMα1 and PSMα3 exhibit cytotoxic activity in their soluble forms, and fibrillation generally reduces their toxicity. Similar to human amyloids, mature PSM fibrils are presumably largely non-toxic [218, 229]. Yet, fibrillation appears essential for activity: mutants unable to fibrillate or samples treated with fibrillation inhibitors lose toxicity [148, 188, 189]. This suggests that cytotoxicity is not driven by a static species, such as monomers or mature fibrils, but rather by the dynamic process of fibril formation itself. In particular, membrane-associated fibrillation may disrupt membranes by recruiting and extracting lipids, leading to structural damage. A similar mechanism has been proposed for human amyloids such as islet amyloid polypeptide, where lipid molecules are incorporated into growing amyloid deposits, indicating a potentially conserved cytotoxic pathway across diverse amyloid systems [180].

PSMα3 undergoes reversible disaggregation upon heat shock, regaining its cytotoxic potential as it returns to soluble forms [148]. In contrast, heat shock promotes irreversible cross-β fibril formation in PSMα1, further decreasing its toxicity and reinforcing its structural role in biofilm scaffolds. These observations imply that different fibril polymorphs enable distinct biological functions: PSMα1 contributes to both biofilm stability and (in specific contexts) cytotoxicity, while PSMα3 primarily acts as a cytotoxic agent [148]. Remarkably, PSMα3 fibrils have also been shown to catalyse the hydrolysis of β-lactam antibiotics, expanding their functional repertoire beyond structural and cytotoxic roles [8]. Overall, the full extent of how amyloid fibrillation governs the multifaceted activities of PSM peptides remains an open and exciting area for future investigation.

Staphylococcus epidermidis

Like *S. aureus*, *Staphylococcus epidermidis* is another staphylococcal bacterium involved in hospital-acquired infections, in particular in surgical operations involving the implantation of artificial devices such as joint prosthesis, heart valves, or venous catheters [35]. Twenty years ago, using scanning electron microscopy,

S. epidermidis was found to form biofilms on ex vivo prosthetic joints [232]. Later, a novel 18-kDa Small Basic Protein (Sbp) was found as a crucial component of the extracellular matrix of *S. epidermidis* biofilms [37]. Sbp acts as a scaffolding protein required for lab-grown *S. epidermidis* biofilm formation, with its deletion preventing the biofilm growth. In the early stages of the biofilm formation, Sbp localises at the bacterial-substrate interface. In the later stages, such as biofilm accumulation and maturation, Sbp co-localises with polysaccharide intercellular-adhesin and accumulation-associated protein (Aap), two other key biofilm extracellular components [37].

Sbp is partially unfolded and monomeric in vitro [201]. However, Sbp was suggested by amyloid indicator dye binding and atomic force microscopy to form amyloid fibrils in vitro. Furthermore, *S. epidermidis* biofilms are positive to Thioflavin S fluorescence whereas the bacterial strain inhibiting Sbp expression is Thioflavin S-negative, suggesting that Sbp forms amyloid fibrils within biofilm. However, more evidence is required, as, for instance, Aap, the potential binding partner of Sbp, can form zinc-dependent amyloid fibrils and could be the cause of the Thioflavin S fluorescence in the biofilms [219]. Therefore, additional studies are required to clarify the structure/function relationship of Sbp within biofilms.

S. epidermidis also expresses PSMs like *S. aureus*, including PSMαs, PSMβ1, PSMβ2, PSMδ, PSMε, and PSMγ/δ-toxin [110, 217]. PSMβs are highly expressed during biofilm formation and are involved in both biofilm structuring and dispersal [198]. However, unlike PSMs expressed in *S. aureus* whose ability to form amyloids in vitro has been established, PSMs expressed by *S. epidermidis* do not form amyloid fibrils in vitro as assessed by Thioflavin T assay [88]. Moreover, their deletion leads to the formation of larger but weaker lab-grown *S. epidermidis* biofilms.

High-Resolution Structures Suggest Differences Between Human and Bacterial Amyloids

Structures of bacterial amyloids reveal notable differences from their human counterparts. While amyloid fibrils universally consist of twisted protofilaments formed by stacked β-sheet layers characterized by diffraction peaks at ~4.75 Å and 10–12 Å [71, 172], their molecular architectures vary significantly across species. In bacteria, for example, *Enterobacterial* CsgA and *Pseudomonas* FapC adopt β-solenoid folds [22, 65, 124, 177], PSMα3 forms a unique cross-α architecture [188], and TasA assembles through donor-strand exchange between folded monomers [18]. In contrast, PSMα1 exhibits the canonical cross-β conformation [64].

To fully elucidate the roles, morphologies, and molecular interactions of these fibrils in biofilm structuring, it is essential to study them in their native cellular environment. As observed in neurodegeneration research, discrepancies between in vitro and ex vivo amyloid structures underscore the need for context-specific analysis. A similar approach applied to biofilms, whether laboratory-grown or ex vivo, may help bridge the gap between isolated fibril structures and their functional states within microbial communities. Advances in super-resolution fluorescence microscopy, cryo-electron tomography, and computational image analysis now make it possible to resolve the spatiotemporal organization of amyloids in situ. Such structural–functional insights are critical for deciphering biofilm biology and for the development of targeted antimicrobial strategies.

The Amyloid-Antimicrobial Link, a Growing Fundamental Paradigm

A striking convergence has emerged between amyloids and AMPs, two molecular classes traditionally viewed as functionally distinct. AMPs are potent broad-spectrum antibiotics and immunomodulators capable of targeting bacteria, mycobacteria, enveloped viruses, fungi, protozoans, and, in some cases, transformed or cancerous host cells. Self-assembly of AMPs into amyloid-like fibrils is conserved across major evolutionary groups, suggesting a synergistic relationship between antimicrobial activity and amyloid formation [11].

Both amyloids and AMPs disrupt membranes using similar mechanisms, either by directly forming pores (barrel-stave, toroidal) or by carpet-like insertion [11, 74, 87]. These interactions rely on shared physicochemical features: amphipathicity, hydrophobic amino acid content, and, in the case of AMPs, high positive charge density that promotes electrostatic interactions with negatively charged bacterial membranes [7, 210]. Importantly, membranes can also accelerate fibrillation [225]. In 2024, a cryo-EM study demonstrated phospholipid molecules embedded within Aβ1–40 fibrils, providing the first structural visualization of a heterotypic peptide-lipid amyloid complex and reinforcing the relevance of lipid-peptide interplay in fibril nucleation and stabilization in both functional and pathological contexts [51].

In accordance with the shared membrane activity and some physicochemical features, several disease-associated amyloids exhibit direct antimicrobial activity. Specifically, Aβ can inhibit the growth of diverse pathogens in vitro [179], and its expression enhances survival in infection models ranging from cell culture to nematodes and mice [83]. These effects are mediated through classic AMP mechanisms such as microbial entrapment and reduced adhesion, with Aβ fibrils acting as physical barriers. Similar antimicrobial activity has since been reported for α-synuclein, prion proteins, hIAPP, serum amyloid A, and β2-microglobulin, supporting a potential evolutionary origin of these proteins as immune effectors [15, 19, 32, 68, 74, 79, 83, 86, 87, 126–128, 146, 179, 187, 199, 209, 221, 230].

Conversely, many classical AMPs have been shown to form amyloid-like fibrils. Protegrin-1 and protegrin-4, for instance, were demonstrated to self-assemble into Thioflavin-T-positive fibrils with morphologies similar to Aβ [60, 72]. Dermaseptins and other AMPs have also been shown to exhibit amyloid-like aggregation under physiological conditions, suggesting that fibrillation may enhance their stability, potency, or immunomodulatory properties [9, 11, 59]. Moreover, the antimicrobial pheromone Plantaricin A was shown to form supramolecular protein–lipid amyloid-like fibres upon membrane contact [226], supporting the hypothesis that cell toxicity and antibacterial activity are not mediated by a single static species, such as monomers or mature fibrils, but rather by the dynamic process of fibril formation itself, during which lipid molecules become incorporated into the growing amyloid assemblies [160, 180, 189].

Amyloid-Forming AMPs Displaying a β/α Structural Switch

Uperin 3.5, secreted by the amphibian *Uperoleia mjobergii* (Mjoberg's toadlet), is an AMP shown to form amyloid fibrils [24]. Its crystal structure revealed a cross-α architecture with antiparallel α-helices stacked along the fibril axis (Fig. 16.4). Biophysical analyses indicated that uperin 3.5 also adopts a β-rich conformation, with a switch to the α-helical form triggered by lipid membranes [160]. Cryo-EM analysis further confirmed that in aqueous solution, uperin 3.5 forms cross-β fibrils with a propeller-like morphology [23]. These findings support the notion that uperin 3.5 operates via a lipid-induced structural switch from β-sheets to α-helices, classifying it as a "chameleon" peptide capable of environmentally driven secondary structure transitions.

To further explore the link between amyloid formation and antimicrobial function, a bioinformatic pipeline was employed to identify short AMPs predicted to form amphipathic α-helices, which also show propensities to form β-strands [140]. Fourteen peptides were experimentally confirmed to form fibrils, with a strong correlation observed between α-helical fibrillar content and cytotoxic activity. These results highlight the functional relevance of structural plasticity, particularly α-helicity, in antimicrobial efficacy. Using this approach, the cryo-EM structure of the amphibian aurein 3.3 was determined, revealing a cross-β amyloid architecture with kinked β-sheets typical of functional reversible amyloids [23] (Fig. 16.4).

The amphibian citropin 1.3 was another fibril-forming AMP identified through the computational platform [140]. Cryo-EM structures obtained under aqueous conditions and in the presence of lipid vesicles revealed striking quaternary polymorphism, with at least ten distinct fibril morphologies [184]. These included canonical amyloids and nanotubular assemblies. Five polymorphs were resolved at high resolution, all displaying cross-β structures with diverse lateral packing arrangements (Fig. 16.4). A crystal structure of citropin 1.3 further revealed a supramolecular assembly of α-helices, underscoring its secondary structure polymorphism. Notably, one low-resolution cryo-EM class exhibited a unique nanotubular assembly composed of three concentric layers, suggesting a hybrid fibrillar arrangement of β-sheets and α-helices, an architecture not previously reported for AMPs. Nanotubular assemblies composed of α-helices have been observed in *S. aureus* PSMα3 and PSMβ2 [82], as well as other AMPs [84], pointing

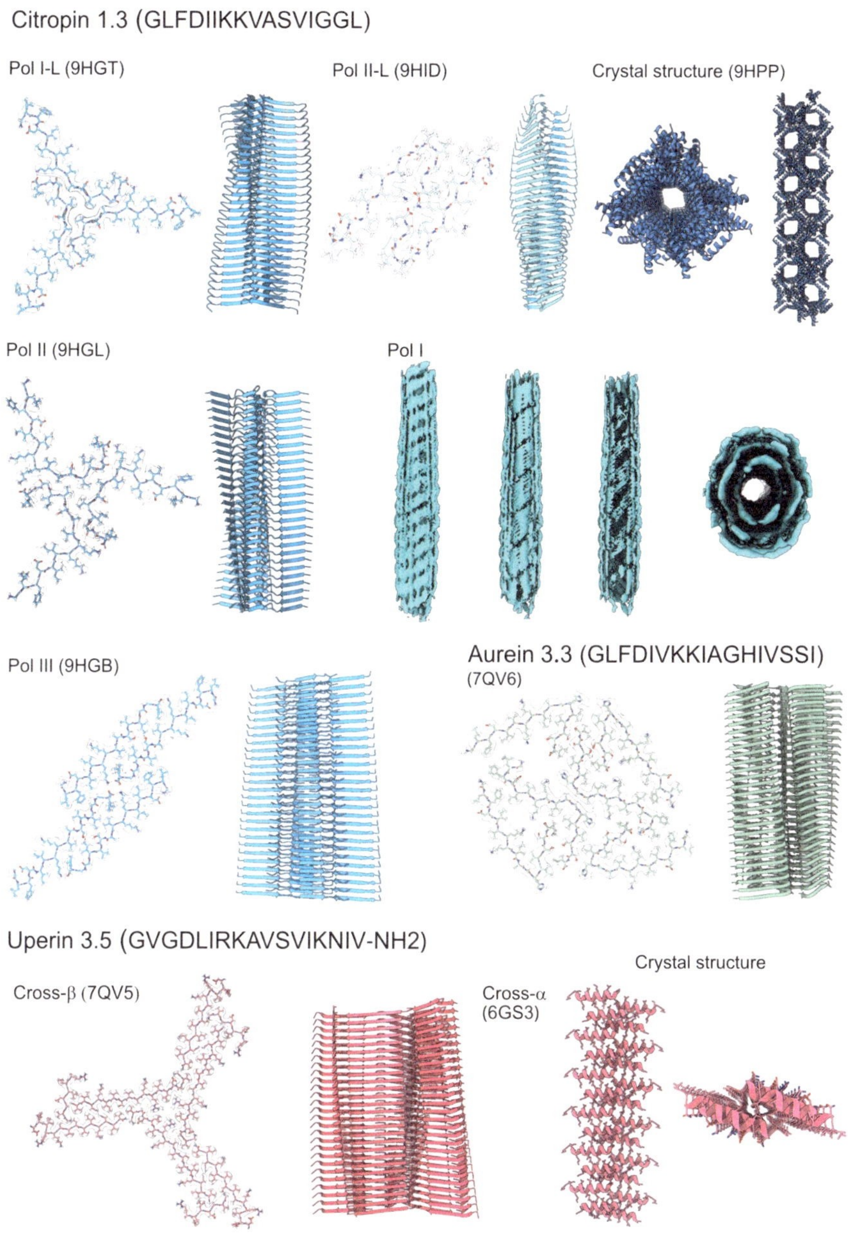

Fig. 16.4 High-resolution structures of fibril forming AMPs [23, 160, 184]. For each AMP a top view of the model fitting into the Coulomb density and a side view of the entire filament are reported. Within brackets, the PDB entry for each found polymorph is shown. If not specified, the reconstruction has been realized via single particle cryo-EM reconstruction

to a potential structural motif relevant to membrane interaction. Nanotubes are also known to serve as advanced nanomaterials and delivery systems [97].

While the functional implications of the fibril polymorphs and nanotubes remain unclear, citropin 1.3 was shown to induce vesicle fusion, supporting a membrane-disruptive mechanism. Additionally, the peptide accumulates in the nucleolus of mammalian cells, a site enriched in nucleic acids and ribosomal proteins, and undergoes LLPS, forming condensates. LLPS is known to accelerate amyloid formation by concentrating peptides locally, which might assist antibacterial activity, further linking structural polymorphism to biological functions.

Biological Relevance of AMP Amyloid Formation

A central question in understanding AMP amyloids is whether their aggregation and fibril morphologies are biologically relevant in vivo. Amphibians, for instance, secrete AMPs via specialized skin glands (also known as granular, serous, or poison glands), both constitutively and in response to infection or stress stimuli [17, 30]. Within these glands, AMPs are stored as inactive pro-peptides [152]. Upon stimulation, the precursors are enzymatically cleaved into their active forms, and the glands contract to release their contents, which include AMPs along with other bioactive molecules such as alkaloids and amines, into the mucus layer on the skin surface.

Remarkably, the concentrations of secreted AMPs can be extremely high: *Xenopus laevis*, for example, secretes AMPs at baseline levels of ~3.3 mg/mL, which can increase up to 20 mg/mL under stress, such as fungal infection [142]. These concentrations are orders of magnitude higher than the critical concentrations required for amyloid fibrillation of many AMPs. For instance, uperin 3.5 forms fibrils at concentrations far below 100μM (~0.19 mg/mL), and even at 4μM (~0.0076 mg/mL) in the presence of lipids [160], indicating that amyloid formation is highly feasible under physiological conditions. Of note, amphibians do not secrete individual AMPs in isolation, but rather a complex cocktail of AMPs, along with macromolecules such as mucins and polysaccharides that may influence peptide aggregation [211].

Despite the strong biophysical evidence for amyloid formation, the biological function of fibrillar AMPs remains poorly defined. Similar to ongoing questions surrounding virulent amyloids like PSMs, it is unclear whether the fibrils themselves are directly antimicrobial or serve as reservoirs of inactive peptides that disassemble upon interaction with lipid membranes to release active monomers or oligomers. Alternatively, the antimicrobial activity may be primarily mediated by soluble species, with membrane disruption driven by dynamic processes during self-assembly, such as lipid recruitment or phase transitions. In this framework, mature fibrils may be biologically inert, while the process of fibrillation, and its coupling to membrane components, could be central to AMP function and cytotoxicity.

The observation that many amyloids, across diverse organisms and sequences, exhibit antimicrobial activity raises important questions about their evolutionary function. It remains unclear whether the antimicrobial properties of human amyloids represent a dedicated physiological role or are merely by-products of their inherent propensity to self-assemble. Similarly, it is yet to be determined whether amyloid formation by AMPs reflects a mechanistically regulated, evolutionarily selected feature, or an incidental consequence of their structural plasticity. Further research is needed to disentangle these possibilities and clarify the functional relevance of amyloid formation in innate immunity.

Amyloid–Nucleic Acid Interactions and Immunomodulatory Roles

Both microbial and human amyloids, like many AMPs, can self-assemble into supramolecular fibrils that modulate the innate immune system. These assemblies contribute not only to host defence but also to inflammation and autoimmunity. A key mechanism involves their ability to form structured complexes with immune ligands, particularly nucleic acids such as DNA and dsRNA, which enhance recognition by pattern recognition receptors, including Toll-like receptors TLR2, TLR3, and TLR9 [91]. Such interactions suggest that amyloidogenesis may represent an evolutionarily conserved strategy for immune amplification, one that extends beyond direct antimicrobial activity.

However, dysregulation of these assemblies can shift the balance from protective immunity to pathology. For example, cytosolic DNA sensors, originally evolved to detect microbial DNA, are increasingly implicated in neurodegenerative disorders due to aberrant activation by self-DNA [212]. Pathological amyloid–nucleic acid complexes have been implicated in systemic lupus erythematosus, Alzheimer's disease, and Parkinson's disease [91, 204].

Several bacterial amyloids have been shown to engage directly with nucleic acids. The RNA chaperone Hfq, for instance, forms amyloid-like fibrils via its intrinsically disordered C-terminal region, which also binds DNA. DNA binding promotes Hfq amyloid assembly, though the mature fibrils themselves do not enhance nucleic acid affinity, pointing to monomeric species as key drivers of interaction [192]. Similarly, DNA induces amyloidogenesis of the bacterial replication protein RepA through an α-to-β conformational switch, without becoming part of the fibril, highlighting a nucleic acid–triggered but structurally independent mechanism [56].

Biofilm-associated amyloids show particularly strong interactions with extracellular DNA (eDNA). In *E. coli* and *Salmonella*, curli fibres form co-aggregates with eDNA, enhancing fibrillation and reinforcing biofilm integrity [52]. These curli–DNA complexes activate TLR2 and TLR9, leading to the production of anti-dsDNA autoantibodies and have been implicated in triggering autoimmunity [52, 53, 61, 193]. *S. aureus* PSMs also form amyloid-eDNA complexes. eDNA promotes the formation of amyloid fibres in *S. aureus* biofilms by facilitating the aggregation of the PSMs [168]. More specifically, CpG-rich DNA promotes PSMα3 fibrillation, and the resulting co-aggregates colocalize with DNA in biofilms. These complexes can trigger anti-dsDNA autoantibody responses in mice through TLR2/TLR9 signalling, mimicking the autoimmune effects of curli–DNA complexes. Notably, chronic exposure to biofilm-associated PSM-DNA aggregates, but not acute planktonic infection, drives this autoimmunity. The response is abrogated in mice infected with *S. aureus* Δpsm mutants or lacking TLR2/TLR9, establishing a direct mechanistic link between bacterial amyloids, biofilms, and autoimmune pathogenesis [61].

AMPs were also shown to target nucleic acids and to form LLPS condensates in vitro and inside bacteria, which might have implications for inhibiting transcription and translation, and for the mechanism of antimicrobial activity [178]. Human AMPs such as LL-37 and β-defensins 2 and 3 also form higher-order assemblies that organize DNA or RNA into nanostructures with defined periodicity. These complexes potently activate TLR9 and TLR3, underscoring that innate immune signalling depends not only on ligand identity but also on nanoscale geometry and spatial arrangement [90, 162, 191]. LL-37 also exhibits intriguing crosstalk with Aβ: it can bind Aβ across its aggregation states, inhibiting fibril formation [36]. This suggests that reduced LL-37 expression, such as during ageing or immune suppression, might exacerbate Aβ aggregation and neurodegeneration. Conversely, LL-37 upregulation during infection could preserve Aβ solubility and modulate its immune functions.

Strikingly, LL-37 and *S. aureus* PSMα3 share sequence similarity, a propensity to form α-helical fibrils [44, 161], and overlapping immunomodulatory activities. Both peptides activate SIRL-1 (Signal Inhibitory Receptor on Leukocytes-1), a negative regulator of myeloid cell function [158]. This functional convergence raises the possibility of molecular mimicry, whereby bacterial peptides hijack host regulatory pathways through structural and functional resemblance to host AMPs [57].

Studies show that *S. aureus* PSMα3 undergoes RNA-mediated LLPS, which stabilizes its dual antimicrobial and cytotoxic functions. In contrast, LL-37 undergoes heat-shock-induced RNA-dependent LLPS, while the RNA reduces its cytotoxicity towards human cells while preserving its antimicrobial activity. These effects are driven by RNA-induced modulation of aggregation kinetics and fibril morphology [149]. Thus, RNA acts as a context-specific regulator of peptide bioactivity, supporting toxicity and virulence in microbial peptides like PSMα3, while preserving host protection in human AMPs like LL-37. These findings suggest that LLPS and nucleic acid interactions serve as conserved mechanisms to fine-tune amyloid functionality across species, with implications for both innate immunity and neurodegenerative disease.

Overall, amyloid and AMP assemblies not only mediate antimicrobial activity but also shape immune responses through nucleic acid interactions, supramolecular organization, and receptor engagement. These dual-edged roles may underlie both protective immunity and pathological autoimmunity, with implications for infection biology, inflammation, and neurodegenerative disease.

Amyloids–Nucleic Acid Interactions and the Origin of Life

Functional interactions between proteins and nucleic acids are hypothesized to have played a central role since the earliest stages of life. Basic polypeptides, for example, can catalyse RNA hydrolysis, suggesting primitive catalytic activity [12]. Remarkably, even very short peptides such as diphenylalanine can self-assemble into amyloids that bind and stabilise nucleic acids, indicating a potential mechanism for early peptide–RNA co-evolution and compartmentalization [28].

Amyloid structures may have served as primitive scaffolds in the prebiotic world, providing protective environments, stabilizing nucleic acids, and promoting selective interactions. Short, basic peptides form amyloid fibrils in the presence of nucleic acids, yielding complexes that not only enhance peptide aggregation but also facilitate nucleic acid hybridization. These co-assemblies exhibit unique biophysical properties that could have enabled primitive cycles of replication, growth, and division, supporting Darwinian selection prior to the evolution of complex cellular systems. This hypothesis parallels the role of mineral surfaces such as clays in prebiotic chemistry and broadens the conceptual relevance of amyloids beyond pathology [21].

Co-assembly of amyloidogenic peptides and nucleic acids is thought to be driven by electrostatic complementarity between cationic peptide surfaces and the anionic phosphate backbone of nucleic acids, leading to the formation of ordered structures such as fibrils and multilamellar nanotubes [21, 151]. These interactions are also relevant in disease contexts, where nucleic acids can accelerate pathological amyloid formation [151].

Strikingly, RNA trinucleotides, namely codon-length sequences, bind peptide amyloid fibrils with high affinity, requiring at least three phosphate groups and three nucleobases for stable interactions. Solid-state NMR studies reveal that RNA interacts along the hydrophobic edges of amyloid fibrils, engaging both the phosphate backbone and nucleobases [156]. These repetitive, multivalent peptide surfaces likely stabilized early RNA molecules and promoted molecular co-assembly. The binding is mutually stabilizing: amyloids shield RNA from hydrolysis, while RNA facilitates peptide fibril formation under specific conditions.

One Sequence Many Functions

In this chapter we have accumulated polymorphic behaviours of amyloids, particularly on bacterial functional amyloids that target human cells, stabilise biofilms, and modulate host-pathogen interactions, as well as antimicrobial peptides that form reversible fibrils with cytotoxic or immune-stimulatory functions. This demonstrates several possible roles for amyloids that deviate far from their standard view as pathogenic components. The emerging paradigm of amyloid–nucleic acid co-assemblies and their role in immune recognition, autoimmunity, and possibly

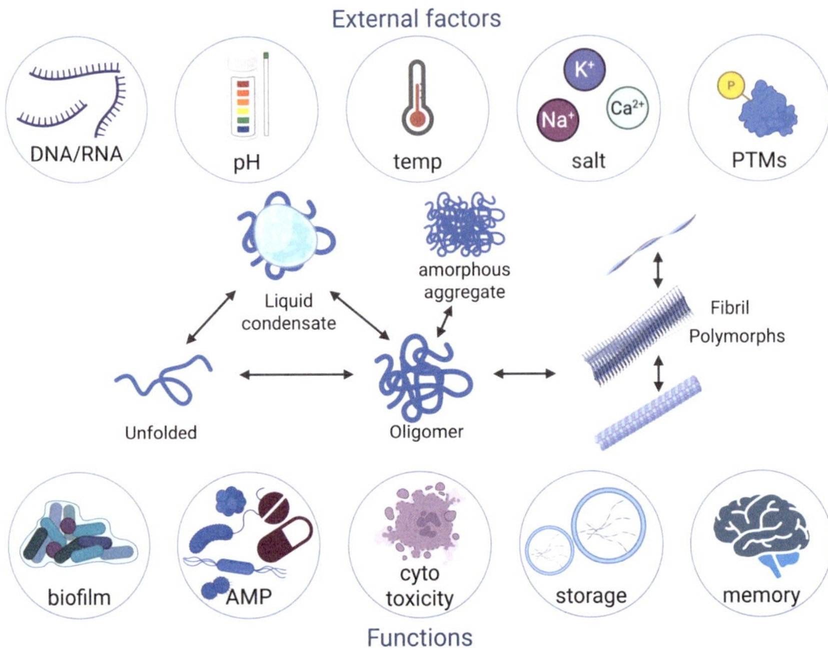

Fig. 16.5 Overview of structure function link among the fibril landscape. PTMs: Post-translational modifications

the origin of life are only the start of an exciting new research focus. By examining structure-function relationships across a broad evolutionary spectrum, we argue that amyloid polymorphism constitutes a general mechanism of biological regulation (summarized in Fig. 16.5). Understanding how these fibrils shift between states, including cross-β, cross-α, nanotubular, or phase-separated condensates, would offer profound insight into their dual roles in health and disease, and their potential as functional materials. This perspective repositions amyloids not merely as pathological end-products but as versatile, ancient scaffolds for structural adaptation and functional innovation.

Acknowledgements M.L. acknowledges research support from the Israel Science Foundation, Grant No. 2111/20, the Cure Alzheimer's Fund, the Forschungskooperation Niedersachsen–Israel, Volkswagenstiftung, No: 76251-4659/2022 (ZN 4042). Funded/co-funded by the European Union (ERC, FuncAmyloid, 101087140). Views and opinions expressed are however those of the author (s) only and do not necessarily reflect those of the European Union or the European Research Council. Neither the European Union nor the granting authority can be held responsible for them. J.M. acknowledges the funding of PIER (PIF-2024-04). We acknowledge the use of OpenAI's GPT-4 model to improve the writing quality of parts of this manuscript. Figures 16.1, 16.2, 16.3, and 16.5 were created in BioRender; Landau, M. (2025) https://BioRender.com/wbcx4tt, https://BioRender.com/3o3y73l, https://BioRender.com/6wm8azn, https://BioRender.com/vcg118i.

Copyright

References

1. Abskharon R, Sawaya MR, Boyer DR, Cao Q, Nguyen BA, Cascio D, Eisenberg DS (2022) Cryo-EM structure of RNA-induced tau fibrils reveals a small C-terminal core that may nucleate fibril formation. Proc Natl Acad Sci USA 119:e2119952119
2. Alam P, Hoyt F, Artikis E, Soukup J, Hughson AG, Schwartz CL, Barbian K, Miller MW, Race B, Caughey B (2024) Cryo-EM structure of a natural prion: chronic wasting disease fibrils from deer. Acta Neuropathol 148:56
3. Al-Garawi ZS, McIntosh BA, Neill-Hall D, Hatimy AA, Sweet SM, Bagley MC, Serpell LC (2017) The amyloid architecture provides a scaffold for enzyme-like catalysts. Nanoscale 9: 10773–10783
4. Amyloid Atlas (2025) https://people.mbi.ucla.edu/sawaya/amyloidatlas/. Accessed 19 May 2025
5. Andreotti G, Baur J, Ugrina M, Pfeiffer PB, Hartmann M, Wiese S, Miyahara H, Higuchi K, Schwierz N, Schmidt M, Fändrich M (2024) Insights into the structural basis of amyloid resistance provided by cryo-EM structures of AApoAII amyloid fibrils. J Mol Biol 436: 168441
6. Antimicrobial Resistance Collaborators (2022) Global burden of bacterial antimicrobial resistance in 2019: a systematic analysis. Lancet 399:629–655
7. Apostolidou M, Jayasinghe SA, Langen R (2008) Structure of alpha-helical membrane-bound human islet amyloid polypeptide and its implications for membrane-mediated misfolding. J Biol Chem 283:17205–17210
8. Arad E, Pedersen KB, Malka O, Mambram Kunnath S, Golan N, Aibinder P, Schiøtt B, Rapaport H, Landau M, Jelinek R (2023) Staphylococcus aureus functional amyloids catalyze degradation of β-lactam antibiotics. Nat Commun 14:8198
9. Auvynet C, El Amri C, Lacombe C, Bruston F, Bourdais J, Nicolas P, Rosenstein Y (2008) Structural requirements for antimicrobial versus chemoattractant activities for dermaseptin S9. FEBS J 275:4134–4151
10. Balistreri A, Kolli D, Jayaweera SW, Lundahl D, Han Y, Kalcec L, Goetzler E, Alessio R, Ruotolo B, Olofsson A, Chapman MR (2025) The bacterial chaperone CsgC inhibits functional amyloid CsgA formation by promoting the intrinsically disordered pre-nuclear state. J Biol Chem 301:110217
11. Baltutis V, O'Leary PD, Martin LL (2022) Self-assembly of linear, natural antimicrobial peptides: an evolutionary perspective. ChemPlusChem 87:e202200240
12. Barbier B, Brack A (1988) Basic polypeptides accelerate the hydrolysis of ribonucleic acids. J Am Chem Soc 110:6880–6882
13. Barnhart MM, Chapman MR (2006) Curli biogenesis and function. Ann Rev Microbiol 60: 131–147
14. Bavdek A, Kostanjsek R, Antonini V, Lakey JH, Dalla Serra M, Gilbert RJ, Anderluh G (2012) pH dependence of listeriolysin O aggregation and pore-forming ability. FEBS J 279: 126–141
15. Beatman EL, Massey A, Shives KD, Burrack KS, Chamanian M, Morrison TE, Beckham JD (2015) Alpha-Synuclein expression restricts RNA viral infections in the brain. J Virol 90: 2767–2782
16. Beloin C, Roux A, Ghigo JM (2008) Escherichia coli biofilms. Curr Top Microbiol Immunol 322:249–289
17. Bevins C (1990) Peptides From Frog Skin. Annu Rev Biochem 59:395–414

18. Böhning J, Ghrayeb M, Pedebos C, Abbas DK, Khalid S, Chai L, Bharat TAM (2022) Donor-strand exchange drives assembly of the TasA scaffold in Bacillus subtilis biofilms. Nat Commun 13:7082
19. Bourgade K, Garneau H, Giroux G, Le Page AY, Bocti C, Dupuis G, Frost EH, Fülöp T Jr (2015) β-Amyloid peptides display protective activity against the human Alzheimer's disease-associated herpes simplex virus-1. Biogerontology 16:85–98
20. Braak H, Braak E (1990) Cognitive impairment in Parkinson's disease: amyloid plaques, neurofibrillary tangles, and neuropil threads in the cerebral cortex. J Neural Transm Park Dis Dement Sect 2:45–57
21. Braun S, Humphreys C, Fraser E, Brancale A, Bochtler M, Dale TC (2011) Amyloid-associated nucleic acid hybridisation. PLoS One 6:e19125
22. Bu F, Dee DR, Liu B (2024) Structural insight into Escherichia coli CsgA amyloid fibril assembly. MBio 15:e0041924
23. Bücker R, Seuring C, Cazey C, Veith K, García-Alai M, Grünewald K, Landau M (2022) The Cryo-EM structures of two amphibian antimicrobial cross-β amyloid fibrils. Nat Commun 13: 4356
24. Calabrese AN, Liu Y, Wang T, Musgrave IF, Pukala TL, Tabor RF, Martin LL, Carver JA, Bowie JH (2016) The amyloid fibril-forming properties of the amphibian antimicrobial peptide Uperin 3.5. Chembiochem 17:239–246
25. Calamai M, Kumita JR, Mifsud J, Parrini C, Ramazzotti M, Ramponi G, Taddei N, Chiti F, Dobson CM (2006) Nature and significance of the interactions between amyloid fibrils and biological polyelectrolytes. Biochemistry 45:12806–12815
26. Cámara-Almirón J, Navarro Y, Díaz-Martínez L, Magno-Pérez-Bryan MC, Molina-Santiago-C, Pearson JR, de Vicente A, Pérez-García A, Romero D (2020) Dual functionality of the amyloid protein TasA in Bacillus physiology and fitness on the phylloplane. Nat Commun 11: 1859
27. Cao Q, Boyer DR, Sawaya MR, Ge P, Eisenberg DS (2019) Cryo-EM structures of four polymorphic TDP-43 amyloid cores. Nat Struct Mol Biol 26:619–627
28. Carny O, Gazit E (2011) Creating prebiotic sanctuary: self-assembling supramolecular Peptide structures bind and stabilize RNA. Orig Life Evol Biosph 41:121–132
29. Chapman MR, Robinson LS, Pinkner JS, Roth R, Heuser J, Hammar M, Normark S, Hultgren SJ (2002) Role of Escherichia coli curli operons in directing amyloid fiber formation. Science 295:851–855
30. Chen X, Liu S, Fang J, Zheng S, Wang Z, Jiao Y, Xia P, Wu H, Ma Z, Hao L (2022) Peptides isolated from amphibian skin secretions with emphasis on antimicrobial peptides. Toxins (Basel) 14:722
31. Chen C, Gao Y, Qiao X, Feng Y, Yu X, Cai J, Hu Q, Lin X, Xuan Q, Li H (2025) Functional amyloid phenol-soluble modulin α1-targeting photothermal nanoplatform for effective elimination of biofilm-associated infections. ACS Nano. https://doi.org/10.1021/acsnano.5c01761
32. Chiou S-J, Ko H-J, Hwang C-C, Hong Y-R (2021) The double-edged sword of Beta2-microglobulin in antibacterial properties and amyloid fibril-mediated cytotoxicity. Int J Mol Sci 22. https://doi.org/10.3390/ijms22126330
33. Chiti F, Dobson CM (2006) Protein misfolding, functional amyloid, and human disease. Annu Rev Biochem 75:333–366
34. Cohen SIA, Vendruscolo M, Dobson CM, Knowles TPJ (2012) From macroscopic measurements to microscopic mechanisms of protein aggregation. J Mol Biol 421:160–171
35. Darouiche RO (2004) Treatment of infections associated with surgical implants. N Engl J Med 350:1422–1429
36. De Lorenzi E, Chiari M, Colombo R, Cretich M, Sola L, Vanna R, Gagni P, Bisceglia F, Morasso C, Lin JS, Lee M, McGeer PL, Barron AE (2017) Evidence that the human innate immune peptide LL-37 may be a binding partner of amyloid-β and inhibitor of fibril assembly. J Alzheimer's Dis 59:1213–1226

37. Decker R, Burdelski C, Zobiak M, Büttner H, Franke G, Christner M, Saß K, Zobiak B, Henke HA, Horswill AR, Bischoff M, Bur S, Hartmann T, Schaeffer CR, Fey PD, Rohde H (2015) An 18 kDa scaffold protein is critical for Staphylococcus epidermidis biofilm formation. PLoS Pathog 11:e1004735
38. Dickson DW (1997) The pathogenesis of senile plaques. J Neuropathol Exp Neurol 56:321–339
39. Diehl A, Roske Y, Ball L, Chowdhury A, Hiller M, Molière N, Kramer R, Stöppler D, Worth CL, Schlegel B, Leidert M, Cremer N, Erdmann N, Lopez D, Stephanowitz H, Krause E, van Rossum B-J, Schmieder P, Heinemann U, Turgay K, Akbey Ü, Oschkinat H (2018) Structural changes of TasA in biofilm formation of Bacillus subtilis. Proc Natl Acad Sci USA 115:3237–3242
40. Dueholm MS, Petersen SV, Sønderkær M, Larsen P, Christiansen G, Hein KL, Enghild JJ, Nielsen JL, Nielsen KL, Nielsen PH, Otzen DE (2010) Functional amyloid in pseudomonas. Mol Microbiol 77:1009–1020
41. Dueholm MS, Nielsen SB, Hein KL, Nissen P, Chapman M, Christiansen G, Nielsen PH, Otzen DE (2011) Fibrillation of the major curli subunit CsgA under a wide range of conditions implies a robust design of aggregation. Biochemistry 50:8281–8290
42. Dueholm MS, Søndergaard MT, Nilsson M, Christiansen G, Stensballe A, Overgaard MT, Givskov M, Tolker-Nielsen T, Otzen DE, Nielsen PH (2013) Expression of Fap amyloids in Pseudomonas aeruginosa, P. fluorescens, and P. putida results in aggregation and increased biofilm formation. MicrobiologyOpen 2:365–382
43. Elbaum-Garfinkle S (2019) Matter over mind: liquid phase separation and neurodegeneration. J Biol Chem 294:7160–7168
44. Engelberg Y, Landau M (2020) The Human LL-37(17-29) antimicrobial peptide reveals a functional supramolecular structure. Nat Commun 11:3894
45. Evans ML, Chorell E, Taylor JD, Åden J, Götheson A, Li F, Koch M, Sefer L, Matthews SJ, Wittung-Stafshede P, Almqvist F, Chapman MR (2015) The bacterial curli system possesses a potent and selective inhibitor of amyloid formation. Mol Cell 57:445–455
46. Falcon B, Zhang W, Schweighauser M, Murzin AG, Vidal R, Garringer HJ, Ghetti B, Scheres SHW, Goedert M (2018) Tau filaments from multiple cases of sporadic and inherited Alzheimer's disease adopt a common fold. Acta Neuropathol 136:699–708
47. Fioriti L, Myers C, Huang Y-Y, Li X, Stephan JS, Trifilieff P, Colnaghi L, Kosmidis S, Drisaldi B, Pavlopoulos E, Kandel ER (2015) The persistence of hippocampal-based memory requires protein synthesis mediated by the prion-like protein CPEB3. Neuron 86:1433–1448
48. Fitzpatrick AWP, Debelouchina GT, Bayro MJ, Clare DK, Caporini MA, Bajaj VS, Jaroniec CP, Wang L, Ladizhansky V, Müller SA, MacPhee CE, Waudby CA, Mott HR, De Simone A, Knowles TPJ, Saibil HR, Vendruscolo M, Orlova EV, Griffin RG, Dobson CM (2013) Atomic structure and hierarchical assembly of a cross-β amyloid fibril. Proc Natl Acad Sci USA 110:5468–5473
49. Fitzpatrick AWP, Falcon B, He S, Murzin AG, Murshudov G, Garringer HJ, Crowther RA, Ghetti B, Goedert M, Scheres SHW (2017) Cryo-EM structures of tau filaments from Alzheimer's disease. Nature 547:185–190
50. Fowler SL, Behr TS, Turkes E, O'Brien DP, Cauhy PM, Rawlinson I, Edmonds M, Foiani MS, Schaler A, Crowley G, Bez S, Ficulle E, Tsefou E, Fischer R, Geary B, Gaur P, Miller C, D'Acunzo P, Levy E, Duff KE, Ryskeldi-Falcon B (2025) Tau filaments are tethered within brain extracellular vesicles in Alzheimer's disease. Nat Neurosci 28:40–48
51. Frieg B, Han M, Giller K, Dienemann C, Riedel D, Becker S, Andreas LB, Griesinger C, Schröder GF (2024) Cryo-EM structures of lipidic fibrils of amyloid-β (1-40). Nat Commun 15:1297
52. Gallo PM, Rapsinski GJ, Wilson RP, Oppong GO, Sriram U, Goulian M, Buttaro B, Caricchio R, Gallucci S, Tukel C (2015) Amyloid-DNA composites of bacterial biofilms stimulate autoimmunity. Immunity 42:1171–1184

53. Ganguly D, Chamilos G, Lande R, Gregorio J, Meller S, Facchinetti V, Homey B, Barrat FJ, Zal T, Gilliet M (2009) Self-RNA-antimicrobial peptide complexes activate human dendritic cells through TLR7 and TLR8. J Exp Med 206:1983–1994
54. Garcia-Pardo J, Bartolomé-Nafría A, Chaves-Sanjuan A, Gil-Garcia M, Visentin C, Bolognesi M, Ricagno S, Ventura S (2023) Cryo-EM structure of hnRNPDL-2 fibrils, a functional amyloid associated with limb-girdle muscular dystrophy D3. Nat Commun 14:239
55. GBD 2016 Dementia Collaborators (2019) Global, regional, and national burden of Alzheimer's disease and other dementias, 1990–2016: a systematic analysis for the Global Burden of Disease Study 2016. Lancet Neurol 18:88–106
56. Giraldo R (2007) Defined DNA sequences promote the assembly of a bacterial protein into distinct amyloid nanostructures. Proc Natl Acad Sci USA 104:17388–17393
57. Golan N, Engelberg Y, Landau M (2022) Structural mimicry in microbial and antimicrobial amyloids. Annu Rev Biochem 91:403–422
58. Golan N, Parizat A, Tabachnikov O, Barnea E, Olsen WP, Otzen DE, Landau M (2025) Resilience and charge-dependent fibrillation of functional amyloid: interactions of Pseudomonas biofilm-associated FapB and FapC amyloids. J Biol Chem 301:108096
59. Gössler-Schöfberger R, Hesser G, Muik M, Wechselberger C, Jilek A (2009) An orphan dermaseptin from frog skin reversibly assembles to amyloid-like aggregates in a pH-dependent fashion: self-assembly of anionic dermaseptin. FEBS J 276:5849–5859
60. Gour S, Kumar V, Singh A, Gadhave K, Goyal P, Pandey J, Giri R, Yadav JK (2019) Mammalian antimicrobial peptide protegrin-4 self assembles and forms amyloid-like aggregates: assessment of its functional relevance. J Pept Sci 25:e3151
61. Grando K, Nicastro LK, Tursi SA, De Anda J, Lee EY, Wong GCL, Tükel Ç (2022) Phenol-soluble modulins from Staphylococcus aureus biofilms form complexes with DNA to drive autoimmunity. Front Cell Infect Microbiol 12:884065
62. Hallinan GI, Hoq MR, Ghosh M, Vago FS, Fernandez A, Garringer HJ, Vidal R, Jiang W, Ghetti B (2021) Structure of tau filaments in Prion protein amyloidoses. Acta Neuropathol 142:227–241
63. Hammer ND, Schmidt JC, Chapman MR (2007) The curli nucleator protein, CsgB, contains an amyloidogenic domain that directs CsgA polymerization. Proc Natl Acad Sci USA 104: 12494–12499
64. Hansen KH, Byeon CH, Liu Q, Drace T, Boesen T, Conway JF, Andreasen M, Akbey Ü (2024) Structure of biofilm-forming functional amyloid PSMα1 from Staphylococcus aureus. Proc Natl Acad Sci USA 121:e2406775121
65. Hansen KH, Golcuk M, Byeon CH, Tunc A, Plechinger EB, Conway JF, Andreasen M, Gur M, Akbey U (2025) Structural basis of pseudomonas FapC biofilm-forming functional amyloid formation. bioRxiv
66. Heerde T, Bansal A, Schmidt M, Fändrich M (2023) Cryo-EM structure of a catalytic amyloid fibril. Sci Rep 13:4070
67. Hervas R, Rau MJ, Park Y, Zhang W, Murzin AG, Fitzpatrick JAJ, Scheres SHW, Si K (2020) Cryo-EM structure of a neuronal functional amyloid implicated in memory persistence in Drosophila. Science 367:1230–1234
68. Hirakura Y, Carreras I, Sipe JD, Kagan BL (2002) Channel formation by serum amyloid a: a potential mechanism for amyloid pathogenesis and host defense. Amyloid 9:13–23
69. Howard MK, Miller KR, Sohn BS, Ryan JJ, Xu A, Jackrel ME (2023) Probing the drivers of Staphylococcus aureus biofilm protein amyloidogenesis and disrupting biofilms with engineered protein disaggregases. MBio 14:e0058723
70. Hoyt F, Standke HG, Artikis E, Schwartz CL, Hansen B, Li K, Hughson AG, Manca M, Thomas OR, Raymond GJ, Race B, Baron GS, Caughey B, Kraus A (2022) Cryo-EM structure of anchorless RML prion reveals variations in shared motifs between distinct strains. Nat Commun 13:4005
71. Jahn TR, Makin OS, Morris KL, Marshall KE, Tian P, Sikorski P, Serpell LC (2010) The common architecture of cross-beta amyloid. J Mol Biol 395:717–727

72. Jang H, Arce FT, Mustata M, Ramachandran S, Capone R, Nussinov R, Lal R (2011) Antimicrobial protegrin-1 forms amyloid-like fibrils with rapid kinetics suggesting a functional link. Biophys J 100:1775–1783
73. Jiang Y, Peña-Díaz S, Zhang Z, Daugberg AOH, López Hernández M, Nielsen J, Huang Q, Qin S, Dueholm MKD, Dong M, Pedersen JS, Cao Q, Otzen DE, Wang H (2025) Natural design of a stabilized cross-β fold: structure of the FuA FapC from Pseudomonas Sp. UK4 reveals a critical role for stacking of imperfect repeats. Adv Mater 37:e2505503
74. Kagan BL, Jang H, Capone R, Arce FT, Ramachandran S, Lal R, Nussinov R (2011) Antimicrobial properties of amyloid. Peptides. https://doi.org/10.1021/mp200419b
75. Karimi-Farsijani S, Sharma K, Ugrina M, Kuhn L, Pfeiffer PB, Haupt C, Wiese S, Hegenbart U, Schönland SO, Schwierz N, Schmidt M, Fändrich M (2024) Cryo-EM structure of a lysozyme-derived amyloid fibril from hereditary amyloidosis. Nat Commun 15:9648
76. Kepp KP, Robakis NK, Høilund-Carlsen PF, Sensi SL, Vissel B (2023) The amyloid cascade hypothesis: an updated critical review. Brain 146:3969–3990
77. Khatoon Z, McTiernan CD, Suuronen EJ, Mah T-F, Alarcon EI (2018) Bacterial biofilm formation on implantable devices and approaches to its treatment and prevention. Heliyon 4: e01067
78. Knowles TPJ, Buehler MJ (2011) Nanomechanics of functional and pathological amyloid materials. Nat Nanotechnol 6:469–479
79. Kobayashi N, Masuda J, Kudoh J, Shimizu N, Yoshida T (2008) Binding sites on tau proteins as components for antimicrobial peptides. Biocontrol Sci 13:49–56
80. Kollmer M, Close W, Funk L, Rasmussen J, Bsoul A, Schierhorn A, Schmidt M, Sigurdson CJ, Jucker M, Fändrich M (2019) Cryo-EM structure and polymorphism of Aβ amyloid fibrils purified from Alzheimer's brain tissue. Nat Commun 10:4760
81. Kraus A, Hoyt F, Schwartz CL, Hansen B, Artikis E, Hughson AG, Raymond GJ, Race B, Baron GS, Caughey B (2021) High-resolution structure and strain comparison of infectious mammalian prions. Mol Cell 81:4540–4551.e6
82. Kreutzberger MAB, Wang S, Beltran LC, Tuachi A, Zuo X, Egelman EH, Conticello VP (2022) Phenol-soluble modulins PSMα3 and PSMβ2 form nanotubes that are cross-α amyloids. Proc Natl Acad Sci USA 119:e2121586119
83. Kumar DKV, Choi SH, Washicosky KJ, Eimer WA, Tucker S, Ghofrani J, Lefkowitz A, McColl G, Goldstein LE, Tanzi RE, Moir RD (2016) Amyloid-β peptide protects against microbial infection in mouse and worm models of Alzheimer's disease. Sci Transl Med 8: 340ra72
84. Laguera B, Golden MM, Wang F, Gnewou O, Tuachi A, Egelman EH, Wuest WM, Conticello VP (2025) Amphipathic antimicrobial peptides illuminate a reciprocal relationship between self-assembly and cytolytic activity. Angew Chem Int Ed Engl 64:e202500040
85. Landau M (2018) Mimicking cross-α amyloids. Nat Chem Biol 14:833–834
86. Landreh M, Johansson J, Jörnvall H (2014) Separate molecular determinants in amyloidogenic and antimicrobial peptides. J Mol Biol 426:2159–2166
87. Last NB, Miranker AD (2013) Common mechanism unites membrane poration by amyloid and antimicrobial peptides. Proc Natl Acad Sci USA 110:6382–6387
88. Le KY, Villaruz AE, Zheng Y, He L, Fisher EL, Nguyen TH, Ho TV, Yeh AJ, Joo H-S, Cheung GYC, Otto M (2019) Role of phenol-soluble modulins in Staphylococcus epidermidis biofilm formation and infection of indwelling medical devices. J Mol Biol 431:3015–3027
89. Lee M, Wang T, Makhlynets OV, Wu Y, Polizzi NF, Wu H, Gosavi PM, Stöhr J, Korendovych IV, DeGrado WF, Hong M (2017) Zinc-binding structure of a catalytic amyloid from solid-state NMR. Proc Natl Acad Sci USA 114:6191–6196
90. Lee EY, Zhang C, Di Domizio J, Jin F, Connell W, Hung M, Malkoff N, Veksler V, Gilliet M, Ren P, Wong GCL (2019) Helical antimicrobial peptides assemble into protofibril scaffolds that present ordered dsDNA to TLR9. Nat Commun 10:1012

91. Lee EY, Srinivasan Y, de Anda J, Nicastro LK, Tükel Ç, Wong GCL (2020a) Functional reciprocity of amyloids and antimicrobial peptides: rethinking the role of supramolecular assembly in host defense, immune activation, and inflammation. Front Immunol 11:1629
92. Lee M, Ghosh U, Thurber KR, Kato M, Tycko R (2020b) Molecular structure and interactions within amyloid-like fibrils formed by a low-complexity protein sequence from FUS. Nat Commun 11:5735
93. Leistner C, Wilkinson M, Burgess A, Lovatt M, Goodbody S, Xu Y, Deuchars S, Radford SE, Ranson NA, Frank RAW (2023) The in-tissue molecular architecture of β-amyloid pathology in the mammalian brain. Nat Commun 14:2833
94. Levkovich SA, Gazit E, Laor Bar-Yosef D (2021) Two decades of studying functional amyloids in microorganisms. Trends Microbiol 29:251–265
95. Li D, Liu C (2023) Molecular rules governing the structural polymorphism of amyloid fibrils in neurodegenerative diseases. Structure 31:1335–1347
96. Li D, Furukawa H, Deng H, Liu C, Yaghi OM, Eisenberg DS (2014) Designed amyloid fibers as materials for selective carbon dioxide capture. Proc Natl Acad Sci USA 111:191–196
97. Liu B, Li X, Zhang JP, Li X, Yuan Y, Hou GH, Zhang HJ, Zhang H, Li Y, Mezzenga R (2024) Protein nanotubes as advanced material platforms and delivery systems. Adv Mater 36: e2307627
98. Lövestam S, Koh FA, van Knippenberg B, Kotecha A, Murzin AG, Goedert M, Scheres SHW (2022) Assembly of recombinant tau into filaments identical to those of Alzheimer's disease and chronic traumatic encephalopathy. elife 11. https://doi.org/10.7554/eLife.76494
99. Lowy FD (1998) Staphylococcus aureus infections. N Engl J Med 339:520–532
100. Lu J, Cao Q, Hughes MP, Sawaya MR, Boyer DR, Cascio D, Eisenberg DS (2020) CryoEM structure of the low-complexity domain of hnRNPA2 and its conversion to pathogenic amyloid. Nat Commun 11:4090
101. Luo F, Gui X, Zhou H, Gu J, Li Y, Liu X, Zhao M, Li D, Li X, Liu C (2018) Atomic structures of FUS LC domain segments reveal bases for reversible amyloid fibril formation. Nat Struct Mol Biol. https://doi.org/10.1038/s41594-018-0050-8
102. Lye YS, Chen Y-R (2022) TAR DNA-binding protein 43 oligomers in physiology and pathology. IUBMB Life 74:794–811
103. Maji SK, Perrin MH, Sawaya MR, Jessberger S, Vadodaria K, Rissman RA, Singru PS, Nilsson KP, Simon R, Schubert D, Eisenberg D, Rivier J, Sawchenko P, Vale W, Riek R (2009) Functional amyloids as natural storage of peptide hormones in pituitary secretory granules. Science 325:328–332
104. Malinchik SB, Inouye H, Szumowski KE, Kirschner DA (1998) Structural analysis of Alzheimer's beta(1-40) amyloid: protofilament assembly of tubular fibrils. Biophys J 74: 537–545
105. Malone M, Bjarnsholt T, McBain AJ, James GA, Stoodley P, Leaper D, Tachi M, Schultz G, Swanson T, Wolcott RD (2017) The prevalence of biofilms in chronic wounds: a systematic review and meta-analysis of published data. J Wound Care 26:20–25
106. Mammeri NE, Dregni AJ, Duan P, Hong M (2024) Structures of AT8 and PHF1 phosphomimetic tau: insights into the posttranslational modification code of tau aggregation. Proc Natl Acad Sci USA 121:e2316175121
107. Manka SW, Zhang W, Wenborn A, Betts J, Joiner S, Saibil HR, Collinge J, Wadsworth JDF (2022) 2.7 Å cryo-EM structure of ex vivo RML prion fibrils. Nat Commun 13:4004
108. Marinelli P, Pallares I, Navarro S, Ventura S (2016) Dissecting the contribution of Staphylococcus aureus α-phenol-soluble modulins to biofilm amyloid structure. Sci Rep 6:34552
109. Masters CL, Bateman R, Blennow K, Rowe CC, Sperling RA, Cummings JL (2015) Alzheimer's disease. Nat Rev Dis Primers 1:1–18
110. Mehlin C, Headley CM, Klebanoff SJ (1999) An inflammatory polypeptide complex from Staphylococcus epidermidis: isolation and characterization. J Exp Med 189:907–918

111. Merz GE, Chalkley MJ, Tan SK, Tse E, Lee J, Prusiner SB, Paras NA, DeGrado WF, Southworth DR (2023) Stacked binding of a PET ligand to Alzheimer's tau paired helical filaments. Nat Commun 14:3048
112. Meyer-Luehmann M, Coomaraswamy J, Bolmont T, Kaeser S, Schaefer C, Kilger E, Neuenschwander A, Abramowski D, Frey P, Jaton AL, Vigouret J-M, Paganetti P, Walsh DM, Mathews PM, Ghiso J, Staufenbiel M, Walker LC, Jucker M (2006) Exogenous induction of cerebral ß-amyloidogenesis is governed by agent and host. Science 313:1781–1784
113. Monasterio O, Nova E, Diaz-Espinoza R (2017) Development of a novel catalytic amyloid displaying a metal-dependent ATPase-like activity. Biochem Biophys Res Commun 482: 1194–1200
114. Morris VK, Linser R, Wilde KL, Duff AP, Sunde M, Kwan AH (2012) Solid-state NMR spectroscopy of functional amyloid from a fungal hydrophobin: a well-ordered beta-sheet core amidst structural heterogeneity. Angew Chem Int Ed Engl 51:12621–12625
115. Mukherjee S, Poudyal M, Dave K, Kadu P, Maji SK (2024) Protein misfolding and amyloid nucleation through liquid-liquid phase separation. Chem Soc Rev 53:4976–5013
116. Murakami T, Qamar S, Lin JQ, Schierle GSK, Rees E, Miyashita A, Costa AR, Dodd RB, Chan FTS, Michel CH, Kronenberg-Versteeg D, Li Y, Yang S-P, Wakutani Y, Meadows W, Ferry RR, Dong L, Tartaglia GG, Favrin G, Lin W-L, Dickson DW, Zhen M, Ron D, Schmitt-Ulms G, Fraser PE, Shneider NA, Holt C, Vendruscolo M, Kaminski CF, St George-Hyslop P (2015) ALS/FTD mutation-induced phase transition of FUS liquid droplets and reversible hydrogels into irreversible hydrogels impairs RNP granule function. Neuron 88:678–690
117. Nandi A, Zhang A, Arad E, Jelinek R, Warshel A (2024) Assessing the catalytic role of native glucagon amyloid fibrils. ACS Catal 14:4656–4664
118. Nelson R, Sawaya MR, Balbirnie M, Madsen AØ, Riekel C, Grothe R, Eisenberg D (2005) Structure of the cross-β spine of amyloid-like fibrils. Nature 435:773–778
119. Nespovitaya N, Gath J, Barylyuk K, Seuring C, Meier BH, Riek R (2016) Dynamic assembly and disassembly of functional β-endorphin amyloid fibrils. J Am Chem Soc 138:846–856
120. Nguyen BA, Singh V, Afrin S, Singh P, Pekala M, Ahmed Y, Pedretti R, Canepa J, Lemoff A, Kluve-Beckerman B, Wydorski PM, Chhapra F, Saelices L (2024) Cryo-EM confirms a common fibril fold in the heart of four patients with ATTRwt amyloidosis. Commun Biol 7: 905
121. Niikura T, Tajima H, Kita Y (2006) Neuronal cell death in Alzheimer's disease and a neuroprotective factor, humanin. Curr Neuropharmacol 4:139–147
122. Oh J, Kim J-G, Jeon E, Yoo C-H, Moon JS, Rhee S, Hwang I (2007) Amyloidogenesis of type III-dependent harpins from plant pathogenic bacteria. J Biol Chem 282:13601–13609
123. Otzen D, Riek R (2019) Functional amyloids. Cold Spring Harb Perspect Biol 11. https://doi.org/10.1101/cshperspect.a033860
124. Otzen DE, Peña-Díaz S, Widmann J, Daugberg AOH, Zhang Z, Jiang Y, Mittal C, Dueholm MKD, Louros N, Wang H, Javed I (2025) Interactions between pathological and functional amyloid: a match made in heaven or hell? Mol Asp Med 103:101351
125. Paleologou KE, Oueslati A, Shakked G, Rospigliosi CC, Kim H-Y, Lamberto GR, Fernandez CO, Schmid A, Chegini F, Gai WP, Chiappe D, Moniatte M, Schneider BL, Aebischer P, Eliezer D, Zweckstetter M, Masliah E, Lashuel HA (2010) Phosphorylation at S87 is enhanced in synucleinopathies, inhibits alpha-synuclein oligomerization, and influences synuclein-membrane interactions. J Neurosci 30:3184–3198
126. Park S-C, Moon JC, Shin SY, Son H, Jung YJ, Kim N-H, Kim Y-M, Jang M-K, Lee JR (2016) Functional characterization of alpha-synuclein protein with antimicrobial activity. Biochem Biophys Res Commun 478:924–928
127. Pasupuleti M, Davoudi M, Malmsten M, Schmidtchen A (2009a) Antimicrobial activity of a C-terminal peptide from human extracellular superoxide dismutase. BMC Res Notes 2:136
128. Pasupuleti M, Roupe M, Rydengård V, Surewicz K, Surewicz WK, Chalupka A, Malmsten M, Sörensen OE, Schmidtchen A (2009b) Antimicrobial activity of human prion protein is mediated by its N-terminal region. PLoS One 4:e7358

129. Patel A, Lee HO, Jawerth L, Maharana S, Jahnel M, Hein MY, Stoynov S, Mahamid J, Saha S, Franzmann TM, Pozniakovski A, Poser I, Maghelli N, Royer LA, Weigert M, Myers EW, Grill S, Drechsel D, Hyman AA, Alberti S (2015) A liquid-to-solid phase transition of the ALS protein FUS accelerated by disease mutation. Cell 162:1066–1077
130. Periasamy S, Joo H-S, Duong AC, Bach T-HL, Tan VY, Chatterjee SS, Cheung GYC, Otto M (2012) How Staphylococcus aureus biofilms develop their characteristic structure. Proc Natl Acad Sci USA 109:1281–1286
131. Petkova AT, Leapman RD, Guo Z, Yau W-M, Mattson MP, Tycko R (2005) Self-propagating, molecular-level polymorphism in Alzheimer's ß-amyloid fibrils. Science 307:262–265
132. Peydayesh M, Vogt J, Chen X, Zhou J, Donat F, Bagnani M, Müller CR, Mezzenga R (2022) Amyloid-based carbon aerogels for water purification. Chem Eng J 449:137703
133. Pham CLL, Kwan AH, Sunde M (2014) Functional amyloid: widespread in Nature, diverse in purpose. Essays Biochem 56:207–219
134. Prince PR, Hochmair J, Brognaro H, Gevorgyan S, Franck M, Schubert R, Lorenzen K, Yazici S, Mandelkow E, Wegmann S, Betzel C (2023) Initiation and modulation of tau protein phase separation by the drug suramin. Sci Rep 13:3963
135. Qi R, Joo H-S, Sharma-Kuinkel B, Berlon NR, Park L, Fu C-L, Messina JA, Thaden JT, Yan Q, Ruffin F, Maskarinec S, Warren B, Chu VH, Fortes CQ, Giannitsioti E, Durante-Mangoni E, Kanafani ZA, Otto M, Fowler VG Jr (2016) Increased in vitro phenol-soluble modulin production is associated with soft tissue infection source in clinical isolates of methicillin-susceptible Staphylococcus aureus. J Infect 72:302–308
136. Qi C, Hasegawa M, Takao M, Sakai M, Sasaki M, Mizutani M, Akagi A, Iwasaki Y, Miyahara H, Yoshida M, Scheres SHW, Goedert M (2023) Identical tau filaments in subacute sclerosing panencephalitis and chronic traumatic encephalopathy. Acta Neuropathol Commun 11:74
137. Qi C, Kobayashi R, Kawakatsu S, Kametani F, Scheres SHW, Goedert M, Hasegawa M (2024) Tau filaments with the chronic traumatic encephalopathy fold in a case of vacuolar tauopathy with VCP mutation D395G. Acta Neuropathol 147:86
138. Radamaker L, Karimi-Farsijani S, Andreotti G, Baur J, Neumann M, Schreiner S, Berghaus N, Motika R, Haupt C, Walther P, Schmidt V, Huhn S, Hegenbart U, Schönland SO, Wiese S, Read C, Schmidt M, Fändrich M (2021) Role of mutations and post-translational modifications in systemic AL amyloidosis studied by cryo-EM. Nat Commun 12:6434
139. Ragonis-Bachar P, Landau M (2021) Functional and pathological amyloid structures in the eyes of 2020 cryo-EM. Curr Opin Struct Biol 68:184–193
140. Ragonis-Bachar P, Rayan B, Barnea E, Engelberg Y, Upcher A, Landau M (2022) Natural antimicrobial peptides self-assemble as α/β chameleon amyloids. Biomacromolecules. https://doi.org/10.1021/acs.biomac.2c00582
141. Ragonis-Bachar P, Axel G, Blau S, Ben-Tal N, Kolodny R, Landau M (2024) What can AlphaFold do for antimicrobial amyloids? Proteins 92:265–281
142. Ramsey JP, Reinert LK, Harper LK, Woodhams DC, Rollins-Smith LA (2010) Immune defenses against Batrachochytrium dendrobatidis, a fungus linked to global amphibian declines, in the South African clawed frog, Xenopus laevis. Infect Immun 78:3981–3992
143. Rani L, Mallajosyula SS (2021) Phosphorylation-induced structural reorganization in tau-paired helical filaments. ACS Chem Neurosci 12:1621–1631
144. Rasmussen HØ, Kumar A, Shin B, Stylianou F, Sewell L, Xu Y, Otzen DE, Pedersen JS, Matthews SJ (2023) FapA is an intrinsically disordered chaperone for Pseudomonas functional amyloid FapC. J Mol Biol 435:167878
145. Ratti A, Buratti E (2016) Physiological functions and pathobiology of TDP-43 and FUS/TLS proteins. J Neurochem 138(Suppl 1):95–111
146. Ray S, Singh N, Kumar R, Patel K, Pandey S, Datta D, Mahato J, Panigrahi R, Navalkar A, Mehra S, Gadhe L, Chatterjee D, Sawner AS, Maiti S, Bhatia S, Gerez JA, Chowdhury A, Kumar A, Padinhateeri R, Riek R, Krishnamoorthy G, Maji SK (2020) α-Synuclein aggregation nucleates through liquid-liquid phase separation. Nat Chem 12:705–716

147. Ray S, Mason TO, Boyens-Thiele L, Farzadfard A, Larsen JA, Norrild RK, Jahnke N, Buell AK (2023) Mass photometric detection and quantification of nanoscale α-synuclein phase separation. Nat Chem 15:1306–1316
148. Rayan B, Barnea E, Khokhlov A, Upcher A, Landau M (2023) Differential fibril morphologies and thermostability determine functional roles of Staphylococcus aureus PSMα1 and PSMα3. Front Mol Biosci 10:1184785
149. Rayan B, Barnea E, Indig R, Pantoja CF, Gayk J, Haber YL, Upcher A, Argoetti A, Larsen JA, Buell AK, Zweckstetter M, Landau M (2025) RNA selectively modulates activity of virulent amyloid PSMα3 and host defense LL-37 via phase separation and aggregation dynamics. Biophysics. https://doi.org/10.1101/2025.06.17.660072
150. Reitz C, Pericak-Vance MA, Foroud T, Mayeux R (2023) A global view of the genetic basis of Alzheimer disease. Nat Rev Neurol 19:261–277
151. Rha AK, Das D, Taran O, Ke Y, Mehta AK, Lynn DG (2020) Electrostatic complementarity drives amyloid/nucleic acid co-assembly. Angew Chem Int Ed Engl 59:358–363
152. Rollins-Smith LA (2023) The importance of antimicrobial peptides (AMPs) in amphibian skin defense. Dev Comp Immunol 142:104657
153. Romero D, Aguilar C, Losick R, Kolter R (2010) Amyloid fibers provide structural integrity to Bacillus subtilis biofilms. Proc Natl Acad Sci USA 107:2230–2234
154. Röntgen A, Toprakcioglu Z, Dada ST, Morris OM, Knowles TPJ, Vendruscolo M (2025) Aggregation of α-synuclein splice isoforms through a phase separation pathway. Sci Adv 11: eadq5396
155. Rouse SL, Hawthorne WJ, Berry J-L, Chorev DS, Ionescu SA, Lambert S, Stylianou F, Ewert W, Mackie U, Morgan RML, Otzen D, Herbst F-A, Nielsen PH, Dueholm M, Bayley H, Robinson CV, Hare S, Matthews S (2017) A new class of hybrid secretion system is employed in Pseudomonas amyloid biogenesis. Nat Commun 8:263
156. Rout SK, Cadalbert R, Schröder N, Wang J, Zehnder J, Gampp O, Wiegand T, Güntert P, Klingler D, Kreutz C, Knörlein A, Hall J, Greenwald J, Riek R (2023) An analysis of nucleotide-amyloid interactions reveals selective binding to codon-sized RNA. J Am Chem Soc 145:21915–21924
157. Rufo CM, Moroz YS, Moroz OV, Stöhr J, Smith TA, Hu X, DeGrado WF, Korendovych IV (2014) Short peptides self-assemble to produce catalytic amyloids. Nat Chem 6:303–309
158. Rumpret M, von Richthofen HJ, van der Linden M, Westerlaken GHA, Talavera Ormeño C, van Strijp JAG, Landau M, Ovaa H, van Sorge NM, Meyaard L (2021) Signal inhibitory receptor on leukocytes-1 recognizes bacterial and endogenous amphipathic α-helical peptides. FASEB J 35:e21875
159. Salinas N, Colletier J-P, Moshe A, Landau M (2018) Extreme amyloid polymorphism in Staphylococcus aureus virulent PSMα peptides. Nat Commun 9:3512
160. Salinas N, Tayeb-Fligelman E, Sammito MD, Bloch D, Jelinek R, Noy D, Usón I, Landau M (2021) The amphibian antimicrobial peptide uperin 3.5 is a cross-α/cross-β chameleon functional amyloid. Proc Natl Acad Sci USA 118. https://doi.org/10.1073/pnas.2014442118
161. Sancho-Vaello E, Gil-Carton D, François P, Bonetti E-J, Kreir M, Pothula KR, Kleinekathöfer U, Zeth K (2020) The structure of the antimicrobial human cathelicidin LL-37 shows oligomerization and channel formation in the presence of membrane mimics. Sci Rep 10:17356
162. Schmidt NW, Jin F, Lande R, Curk T, Xian W, Lee C, Frasca L, Frenkel D, Dobnikar J, Gilliet M, Wong GCL (2015) Liquid-crystalline ordering of antimicrobial peptide-DNA complexes controls TLR9 activation. Nat Mater 14:696–700
163. Schneider A, Biernat J, von Bergen M, Mandelkow E, Mandelkow EM (1999) Phosphorylation that detaches tau protein from microtubules (Ser262, Ser214) also protects it against aggregation into Alzheimer paired helical filaments. Biochemistry 38:3549–3558
164. Schubeis T, Yuan P, Ahmed M, Nagaraj M, van Rossum B-J, Ritter C (2015) Untangling a repetitive amyloid sequence: correlating biofilm-derived and segmentally labeled curli fimbriae by solid-state NMR spectroscopy. Angew Chem Int Ed Engl 54:14669–14672

165. Schulte T, Chaves-Sanjuan A, Speranzini V, Sicking K, Milazzo M, Mazzini G, Rognoni P, Caminito S, Milani P, Marabelli C, Corbelli A, Diomede L, Fiordaliso F, Anastasia L, Pappone C, Merlini G, Bolognesi M, Nuvolone M, Fernández-Busnadiego R, Palladini G, Ricagno S (2024) Helical superstructures between amyloid and collagen in cardiac fibrils from a patient with AL amyloidosis. Nat Commun 15:6359
166. Schwalbe M, Biernat J, Bibow S, Ozenne V, Jensen MR, Kadavath H, Blackledge M, Mandelkow E, Zweckstetter M (2013) Phosphorylation of human Tau protein by microtubule affinity-regulating kinase 2. Biochemistry 52:9068–9079
167. Schwartz K, Syed AK, Stephenson RE, Rickard AH, Boles BR (2012) Functional amyloids composed of phenol soluble modulins stabilize Staphylococcus aureus biofilms. PLoS Pathog 8:e1002744
168. Schwartz K, Ganesan M, Payne DE, Solomon MJ, Boles BR (2016) Extracellular DNA facilitates the formation of functional amyloids in Staphylococcus aureus biofilms. Mol Microbiol 99:123–134
169. Schweighauser M, Shi Y, Tarutani A, Kametani F, Murzin AG, Ghetti B, Matsubara T, Tomita T, Ando T, Hasegawa K, Murayama S, Yoshida M, Hasegawa M, Scheres SHW, Goedert M (2020) Structures of α-synuclein filaments from multiple system atrophy. Nature 585:464–469
170. Schweighauser M, Shi Y, Murzin AG, Garringer HJ, Vidal R, Murrell JR, Erro ME, Seelaar H, Ferrer I, van Swieten JC, Ghetti B, Scheres SHW, Goedert M (2024) Novel tau filament folds in individuals with MAPT mutations P301L and P301T. bioRxivorg
171. Seidler PM, Murray KA, Boyer DR, Ge P, Sawaya MR, Hu CJ, Cheng X, Abskharon R, Pan H, DeTure MA, Williams CK, Dickson DW, Vinters HV, Eisenberg DS (2022) Structure-based discovery of small molecules that disaggregate Alzheimer's disease tissue derived tau fibrils in vitro. Nat Commun 13:5451
172. Serpell LC, Sunde M, Benson MD, Tennent GA, Pepys MB, Fraser PE (2000) The protofilament substructure of amyloid fibrils. J Mol Biol 300:1033–1039
173. Seuring C, Verasdonck J, Gath J, Ghosh D, Nespovitaya N, Wälti MA, Maji SK, Cadalbert R, Güntert P, Meier BH, Riek R (2020) The three-dimensional structure of human β-endorphin amyloid fibrils. Nat Struct Mol Biol 27:1178–1184
174. Shewmaker F, McGlinchey RP, Thurber KR, McPhie P, Dyda F, Tycko R, Wickner RB (2009) The functional curli amyloid is not based on in-register parallel beta-sheet structure. J Biol Chem 284:25065–25076
175. Shi Y, Murzin AG, Falcon B, Epstein A, Machin J, Tempest P, Newell KL, Vidal R, Garringer HJ, Sahara N, Higuchi M, Ghetti B, Jang M-K, Scheres SHW, Goedert M (2021) Cryo-EM structures of tau filaments from Alzheimer's disease with PET ligand APN-1607. Acta Neuropathol 141:697–708
176. Si K, Lindquist S, Kandel ER (2003) A neuronal isoform of the aplysia CPEB has prion-like properties. Cell 115:879–891
177. Sleutel M, Pradhan B, Volkov AN, Remaut H (2023) Structural analysis and architectural principles of the bacterial amyloid curli. Nat Commun 14:2822
178. Sneideris T, Erkamp NA, Ausserwöger H, Saar KL, Welsh TJ, Qian D, Katsuya-Gaviria K, Johncock MLLY, Krainer G, Borodavka A, Knowles TPJ (2023) Targeting nucleic acid phase transitions as a mechanism of action for antimicrobial peptides. Nat Commun 14:7170
179. Soscia SJ, Kirby JE, Washicosky KJ, Tucker SM, Ingelsson M, Hyman B, Burton MA, Goldstein LE, Duong S, Tanzi RE, Moir RD (2010) The Alzheimer's disease-associated amyloid β-protein is an antimicrobial peptide. PLoS One 5:e9505
180. Sparr E, Engel MFM, Sakharov DV, Sprong M, Jacobs J, de Kruijff B, Höppener JWM, Killian JA (2004) Islet amyloid polypeptide-induced membrane leakage involves uptake of lipids by forming amyloid fibers. FEBS Lett 577:117–120
181. Steinebrei M, Gottwald J, Baur J, Röcken C, Hegenbart U, Schönland S, Schmidt M (2022) Cryo-EM structure of an ATTRwt amyloid fibril from systemic non-hereditary transthyretin amyloidosis. Nat Commun 13:6398

182. Stern AM, Yang Y, Jin S, Yamashita K, Meunier AL, Liu W, Cai Y, Ericsson M, Liu L, Goedert M, Scheres SHW, Selkoe DJ (2023) Abundant Aβ fibrils in ultracentrifugal supernatants of aqueous extracts from Alzheimer's disease brains. Neuron 111:2012–2020.e4
183. Stewart KL, Radford SE (2017) Amyloid plaques beyond Aβ: a survey of the diverse modulators of amyloid aggregation. Biophys Rev 9:405–419
184. Strati F, Pigozzi Cali M, Bloch Y, Mostafavi S, Monistrol J, Golubev A, Rayan B, Gustavsson E, Landau M (2025) Structural and functional versatility of the amyloidogenic antimicrobial peptide citropin 1.3. bioRxiv
185. Sunde M, Kwan AH, Templeton MD, Beever RE, Mackay JP (2008) Structural analysis of hydrophobins. Micron 39:773–784
186. Swuec P, Lavatelli F, Tasaki M, Paissoni C, Rognoni P, Maritan M, Brambilla F, Milani P, Mauri P, Camilloni C, Palladini G, Merlini G, Ricagno S, Bolognesi M (2019) Cryo-EM structure of cardiac amyloid fibrils from an immunoglobulin light chain AL amyloidosis patient. Nat Commun 10:1269
187. Tang Y, Zhang Y, Zhang D, Liu Y, Nussinov R, Zheng J (2024) Exploring pathological link between antimicrobial and amyloid peptides. Chem Soc Rev 53:8713–8763
188. Tayeb-Fligelman E, Tabachnikov O, Moshe A, Goldshmidt-Tran O, Sawaya MR, Coquelle N, Colletier J-P, Landau M (2017) The cytotoxic Staphylococcus aureus PSMα3 reveals a cross-α amyloid-like fibril. Science 355:831–833
189. Tayeb-Fligelman E, Salinas N, Tabachnikov O, Landau M (2020) Staphylococcus aureus PSMα3 cross-α fibril polymorphism and determinants of cytotoxicity. Structure 28:301–313.e6
190. Tetter S, Arseni D, Murzin AG, Buhidma Y, Peak-Chew SY, Garringer HJ, Newell KL, Vidal R, Apostolova LG, Lashley T, Ghetti B, Ryskeldi-Falcon B (2024) TAF15 amyloid filaments in frontotemporal lobar degeneration. Nature 625:345–351
191. Tewary P, de la Rosa G, Sharma N, Rodriguez LG, Tarasov SG, Howard OMZ, Shirota H, Steinhagen F, Klinman DM, Yang D, Oppenheim JJ (2013) β-Defensin 2 and 3 promote the uptake of self or CpG DNA, enhance IFN-α production by human plasmacytoid dendritic cells, and promote inflammation. J Immunol 191:865–874
192. Turbant F, Hamoui OE, Partouche D, Sandt C, Busi F, Wien F, Arluison V (2021) Identification and characterization of the Hfq bacterial amyloid region DNA interactions. BBA Adv 1: 100029
193. Tursi SA, Puligedda RD, Szabo P, Nicastro LK, Miller AL, Qiu C, Gallucci S, Relkin NR, Buttaro BA, Dessain SK, Tükel Ç (2020) Salmonella Typhimurium biofilm disruption by a human antibody that binds a pan-amyloid epitope on curli. Nat Commun 11:1007
194. Tycko R (2015) Amyloid polymorphism: structural basis and neurobiological relevance. Neuron 86:632–645
195. Van Gerven N, Van der Verren SE, Reiter DM, Remaut H (2018) The role of functional amyloids in bacterial virulence. J Mol Biol 430:3657–3684
196. Wang X, Chapman MR (2008) Sequence determinants of bacterial amyloid formation. J Mol Biol 380:570–580
197. Wang R, Braughton KR, Kretschmer D, Bach T-HL, Queck SY, Li M, Kennedy AD, Dorward DW, Klebanoff SJ, Peschel A, DeLeo FR, Otto M (2007) Identification of novel cytolytic peptides as key virulence determinants for community-associated MRSA. Nat Med 13:1510–1514
198. Wang R, Khan BA, Cheung GYC, Bach T-HL, Jameson-Lee M, Kong K-F, Queck SY, Otto M (2011) Staphylococcus epidermidis surfactant peptides promote biofilm maturation and dissemination of biofilm-associated infection in mice. J Clin Invest 121:238–248
199. Wang L, Liu Q, Chen J-C, Cui Y-X, Zhou B, Chen Y-X, Zhao Y-F, Li Y-M (2012) Antimicrobial activity of human islet amyloid polypeptides: an insight into amyloid peptides' connection with antimicrobial peptides. Biol Chem 393:641–646

200. Wang Y, Zhang Y, Hu W, Xie S, Gong C-X, Iqbal K, Liu F (2015) Rapid alteration of protein phosphorylation during postmortem: implication in the study of protein phosphorylation. Sci Rep 5:15709
201. Wang Y, Jiang J, Gao Y, Sun Y, Dai J, Wu Y, Qu D, Ma G, Fang X (2018) Staphylococcus epidermidis small basic protein (Sbp) forms amyloid fibrils, consistent with its function as a scaffolding protein in biofilms. J Biol Chem 293:14296–14311
202. Wang X, Liu M, Yu C, Li J, Zhou X (2023a) Biofilm formation: mechanistic insights and therapeutic targets. Mol Biomed 4:49
203. Wang X, Uppu DS, Dickey SW, Burgin DJ, Otto M, Lee JC (2023b) Staphylococcus aureus delta toxin modulates both extracellular membrane vesicle biogenesis and amyloid formation. MBio 14:e0174823
204. Wani NA, Gazit E, Ramamoorthy A (2024) Interplay between antimicrobial peptides and amyloid proteins in host defense and disease modulation. Langmuir 40:25355–25366
205. Wasmer C, Lange A, Van Melckebeke H, Siemer AB, Riek R, Meier BH (2008) Amyloid fibrils of the HET-s(218-289) prion form a beta solenoid with a triangular hydrophobic core. Science 319:1523–1526
206. Wechalekar AD, Gillmore JD, Hawkins PN (2016) Systemic amyloidosis. Lancet 387:2641–2654
207. Wegmann S, Eftekharzadeh B, Tepper K, Zoltowska KM, Bennett RE, Dujardin S, Laskowski PR, MacKenzie D, Kamath T, Commins C, Vanderburg C, Roe AD, Fan Z, Molliex AM, Hernandez-Vega A, Muller D, Hyman AA, Mandelkow E, Taylor JP, Hyman BT (2018) Tau protein liquid-liquid phase separation can initiate tau aggregation. EMBO J 37. https://doi.org/10.15252/embj.201798049
208. White AP, Collinson SK, Banser PA, Gibson DL, Paetzel M, Strynadka NC, Kay WW (2001) Structure and characterization of AgfB from salmonella enteritidis thin aggregative fimbriae. J Mol Biol 311:735–749
209. White MR, Kandel R, Tripathi S, Condon D, Qi L, Taubenberger J, Hartshorn KL (2014) Alzheimer's associated β-amyloid protein inhibits influenza a virus and modulates viral interactions with phagocytes. PLoS One 9:e101364
210. Wieprecht T, Dathe M, Schümann M, Krause E, Beyermann M, Bienert M (1996) Conformational and functional study of magainin 2 in model membrane environments using the new approach of systematic double-D-amino acid replacement. Biochemistry 35:10844–10853
211. Woodhams DC, McCartney J, Walke JB, Whetstone R (2023) The adaptive microbiome hypothesis and immune interactions in amphibian mucus. Dev Comp Immunol 145:104690
212. Xie J, Cheng J, Ko H, Tang Y (2024) Cytosolic DNA sensors in neurodegenerative diseases: from physiological defenders to pathological culprits. EMBO Mol Med 16:678–699
213. Yan NL, Candido F, Tse E, Melo AA, Prusiner SB, Mordes DA, Southworth DR, Paras NA, Merz GE (2025) Cryo-EM structure of a novel α-synuclein filament subtype from multiple system atrophy. FEBS Lett 599:33–40
214. Yang Y, Arseni D, Zhang W, Huang M, Lövestam S, Schweighauser M, Kotecha A, Murzin AG, Peak-Chew SY, Macdonald J, Lavenir I, Garringer HJ, Gelpi E, Newell KL, Kovacs GG, Vidal R, Ghetti B, Ryskeldi-Falcon B, Scheres SHW, Goedert M (2022a) Cryo-EM structures of amyloid-β 42 filaments from human brains. Science 375:167–172
215. Yang Y, Shi Y, Schweighauser M, Zhang X, Kotecha A, Murzin AG, Garringer HJ, Cullinane PW, Saito Y, Foroud T, Warner TT, Hasegawa K, Vidal R, Murayama S, Revesz T, Ghetti B, Hasegawa M, Lashley T, Scheres SHW, Goedert M (2022b) Structures of α-synuclein filaments from human brains with Lewy pathology. Nature 610:791–795
216. Yang Y, Zhang W, Murzin AG, Schweighauser M, Huang M, Lövestam S, Peak-Chew SY, Saito T, Saido TC, Macdonald J, Lavenir I, Ghetti B, Graff C, Kumar A, Nordberg A, Goedert M, Scheres SHW (2023) Cryo-EM structures of amyloid-β filaments with the Arctic mutation (E22G) from human and mouse brains. Acta Neuropathol 145:325–333

217. Yao Y, Sturdevant DE, Otto M (2005) Genomewide analysis of gene expression in Staphylococcus epidermidis biofilms: insights into the pathophysiology of S. epidermidis biofilms and the role of phenol-soluble modulins in formation of biofilms. J Infect Dis 191:289–298
218. Yao Z, Cary BP, Bingman CA, Wang C, Kreitler DF, Satyshur KA, Forest KT, Gellman SH (2019) Use of a stereochemical strategy to probe the mechanism of phenol-soluble Modulin alpha3 toxicity. J Am Chem Soc 141:7660–7664
219. Yarawsky AE, Johns SL, Schuck P, Herr AB (2020) The biofilm adhesion protein Aap from Staphylococcus epidermidis forms zinc-dependent amyloid fibers. J Biol Chem 295:4411–4427
220. Yolamanova M, Meier C, Shaytan AK, Vas V, Bertoncini CW, Arnold F, Zirafi O, Usmani SM, Müller JA, Sauter D, Goffinet C, Palesch D, Walther P, Roan NR, Geiger H, Lunov O, Simmet T, Bohne J, Schrezenmeier H, Schwarz K, Ständker L, Forssmann W-G, Salvatella X, Khalatur PG, Khokhlov AR, Knowles TPJ, Weil T, Kirchhoff F, Münch J (2013) Peptide nanofibrils boost retroviral gene transfer and provide a rapid means for concentrating viruses. Nat Nanotechnol 8:130–136
221. Zhang M, Zhao J, Zheng J (2014) Molecular understanding of a potential functional link between antimicrobial and amyloid peptides. Soft Matter 10:7425–7451
222. Zhang S-Q, Huang H, Yang J, Kratochvil HT, Lolicato M, Liu Y, Shu X, Liu L, DeGrado WF (2018) Designed peptides that assemble into cross-α amyloid-like structures. Nat Chem Biol 14:870–875
223. Zhang W, Tarutani A, Newell KL, Murzin AG, Matsubara T, Falcon B, Vidal R, Garringer HJ, Shi Y, Ikeuchi T, Murayama S, Ghetti B, Hasegawa M, Goedert M, Scheres SHW (2020) Novel tau filament fold in corticobasal degeneration. Nature 580:283–287
224. Zhang Z, Huang G, Song Z, Gatch AJ, Ding F (2023) Amyloid aggregation and liquid-liquid phase separation from the perspective of phase transitions. J Phys Chem B 127:6241–6250
225. Zhao H, Tuominen EKJ, Kinnunen PKJ (2004) Formation of amyloid fibers triggered by phosphatidylserine-containing membranes. Biochemistry 43:10302–10307
226. Zhao H, Sood R, Jutila A, Bose S, Fimland G, Nissen-Meyer J, Kinnunen PKJ (2006) Interaction of the antimicrobial peptide pheromone Plantaricin a with model membranes: implications for a novel mechanism of action. Biochim Biophys Acta 1758:1461–1474
227. Zhao K, Lim Y-J, Liu Z, Long H, Sun Y, Hu J-J, Zhao C, Tao Y, Zhang X, Li D, Li Y-M, Liu C (2020) Parkinson's disease-related phosphorylation at Tyr39 rearranges α-synuclein amyloid fibril structure revealed by cryo-EM. Proc Natl Acad Sci USA 117:20305–20315
228. Zheng Y, He L, Asiamah TK, Otto M (2018a) Colonization of medical devices by staphylococci. Environ Microbiol 20:3141–3153
229. Zheng Y, Joo HS, Nair V, Le KY, Otto M (2018b) Do amyloid structures formed by Staphylococcus aureus phenol-soluble modulins have a biological function? Int J Med Microbiol 308:675–682
230. Zheng H, Li H, Zhang J, Fan H, Jia L, Ma W, Ma S, Wang S, You H, Yin Z, Li X (2020) Serum amyloid a exhibits pH dependent antibacterial action and contributes to host defense against Staphylococcus aureus cutaneous infection. J Biol Chem 295:2570–2581
231. Zielinski M, Peralta Reyes FS, Gremer L, Schemmert S, Frieg B, Schäfer LU, Willuweit A, Donner L, Elvers M, Nilsson LNG, Syvänen S, Sehlin D, Ingelsson M, Willbold D, Schröder GF (2023) Cryo-EM of aβ fibrils from mouse models find tg-APPArcSwe fibrils resemble those found in patients with sporadic Alzheimer's disease. Nat Neurosci 26:2073–2080
232. Zimmerli W, Trampuz A, Ochsner PE (2004) Prosthetic-joint infections. N Engl J Med 351:1645–1654

MIX
Papier aus verantwortungsvollen Quellen
Paper from responsible sources
FSC® C105338

FSC
www.fsc.org

If you have any concerns about our products,
you can contact us on
ProductSafety@springernature.com

In case Publisher is established outside the EU,
the EU authorized representative is:
Springer Nature Customer Service Center GmbH
Europaplatz 3, 69115 Heidelberg, Germany

Printed by Libri Plureos GmbH
in Hamburg, Germany